Advances in Intelligent Systems and Computing

Volume 549

Series editor

Janusz Kacprzyk, Polish Academy of Sciences, Warsaw, Poland
e-mail: kacprzyk@ibspan.waw.pl

About this Series

The series "Advances in Intelligent Systems and Computing" contains publications on theory, applications, and design methods of Intelligent Systems and Intelligent Computing. Virtually all disciplines such as engineering, natural sciences, computer and information science, ICT, economics, business, e-commerce, environment, healthcare, life science are covered. The list of topics spans all the areas of modern intelligent systems and computing.

The publications within "Advances in Intelligent Systems and Computing" are primarily textbooks and proceedings of important conferences, symposia and congresses. They cover significant recent developments in the field, both of a foundational and applicable character. An important characteristic feature of the series is the short publication time and world-wide distribution. This permits a rapid and broad dissemination of research results.

More information about this series at http://www.springer.com/series/11156

Tutut Herawan · Rozaida Ghazali
Nazri Mohd Nawi · Mustafa Mat Deris
Editors

Recent Advances on Soft Computing and Data Mining

The Second International Conference
on Soft Computing and Data Mining (SCDM-2016),
Bandung, Indonesia, August 18–20, 2016,
Proceedings

Editors
Tutut Herawan
Department of Information System
University of Malaya
Kuala Lumpur
Malaysia

Rozaida Ghazali
Universiti Tun Hussein Onn Malaysia
Batu Pahat
Malaysia

Nazri Mohd Nawi
Universiti Tun Hussein Onn Malaysia
Batu Pahat
Malaysia

Mustafa Mat Deris
Universiti Tun Hussein Onn Malaysia
Batu Pahat
Malaysia

ISSN 2194-5357 ISSN 2194-5365 (electronic)
Advances in Intelligent Systems and Computing
ISBN 978-3-319-51279-2 ISBN 978-3-319-51281-5 (eBook)
DOI 10.1007/978-3-319-51281-5

Library of Congress Control Number: 2016960777

Printed on acid-free paper

This Springer imprint is published by Springer Nature
The registered company is Springer International Publishing AG
The registered company address is: Gewerbestrasse 11, 6330 Cham, Switzerland

Preface

Soft computing became a formal computer science area of study in the early 1990s. Earlier computational approaches could model and precisely analyze only relatively simple systems. More complex systems arising often remained intractable to conventional mathematical and analytical methods. Soft computing refers to a consortium of computational techniques in computer science (like fuzzy systems, neural networks, etc.) that deals with imprecision, uncertainty, partial truth, and approximation to achieve tractability, robustness and low solution cost. Soft computing tools, individually or in integrated manner, are turning out to be strong candidates for performing tasks in the area of data mining, decision support systems, supply chain management, medicine, business, financial systems, automotive systems and manufacturing, image processing and data compression, etc.

Data mining is a process that involves using a database along with any required selection, preprocessing, subsampling and transformations of it; applying data mining methods to enumerate patterns from it; and evaluating the products of data mining to identify the subset of the enumerated patterns deemed knowledge.

The SCDM 2016 is very significant in a sense that it starts a host of activity in which Faculty of Computer Science & Information Technology, Universiti Tun Hussein Onn Malaysia, Soft Computing & Data Mining research group, and Faculty of Industrial Engineering at Telkom University Indonesia collaborated. After the success of our four previous workshop events in 2010 until 2013 and our first SCDM edition in 2014, we hope to continuously move in this journey of success through this Second International Conference. Indeed, we are honored to host this event and the fact that we are getting more papers, commitments, contributions and partnerships indicate a continuous support from researchers throughout the globe. We are much honored to be an integral part of this conference with the theme “Navigating the Future with Precise Knowledge”.

The SCDM-2016 was held in Bandung, Indonesia during August 18–20, 2016. We received 122 regular papers submissions from 11 countries around the world. The conference also approved two special sessions—Ensemble Methods and Their Applications (EMTA) which accepted 12 manuscripts and Web Mining, Services and Security (WMSS) which accepted 10 manuscripts. Each paper in regular submission was screened by the proceedings' chairs and carefully peer reviewed by two experts from the Program Committee. Meanwhile for workshop submissions, the paper has been peer reviewed by two experts from the Program Committee. Finally, only 42 papers (for regular-giving an acceptance rate of 34%) and 22 papers (for workshops) with the highest quality and merit were accepted for oral presentation and publication in this proceedings volume.

The papers in these proceedings are grouped into two sections and two in conjunction workshops:

- Soft Computing
- Data Mining
- Workshop on Ensemble Methods and Their Applications
- Workshop on Web Mining, Services and Security

On behalf of SCDM-2016, we would like to express our highest gratitude to Soft Computing & Data Mining research group, Universiti Tun Hussein Onn Malaysia, Telkom University, Steering Committee, General Chairs, Program Committee Chairs, Organizing Chairs, Workshop Chairs, all Program and Reviewer Committee members for their valuable efforts in the review process that helped us to guarantee the highest quality of the selected papers for the conference.

We also would like to express our thanks to the three keynote speakers, Prof. Dr Jemal H. Abawajy from Deakin University, Australia; Prof. Mochamad Ashari, from Telkom University, Bandung, Indonesia; and Dr. Aida Mustapha from Universiti Tun Hussein Onn Malaysia.

Our special thanks are due also to Mr. Suresh Rettagunta and Dr. Thomas Ditzinger for publishing the proceedings in Advances in Intelligent Systems and Computing of Springer. We wish to thank the members of the Organizing and Student Committees for their substantial work, especially those who played essential roles.

We cordially thank all the authors for their valuable contributions and other participants of this conference. The conference would not have been possible without them.

Editors
Tutut Herawan
Rozaida Ghazali
Nazri Mohd Nawi
Mustafa Mat Deris

Conference Organization

Patron

Mohd Noh Dalimin	Vice Chancellor of Universiti Tun Hussein Onn Malaysia
Mochamad Ashari	The Rector of Telkom University

Honorary Chairs

Witold Pedrycz	University of Alberta, Canada
Junzo Watada	Waseda University, Japan
Ajith Abraham	Machine Intelligence Research Labs, USA
Fazel Famili	National Research Council of Canada
Hamido Fujita	Iwate Prefectural University, Japan
Nikola Kasabov	KEDRI, Auckland University of Technology, New Zealand
Hojjat Adeli	University of Ohio, USA
Mustafa Mat Deris	SMC, Universiti Tun Hussein Onn Malaysia

Steering Committee

Nazri Mohd Nawi	SMC, Universiti Tun Hussein Onn Malaysia
Dida Diah Damayanti	Telkom University, Indonesia
Jemal H. Abawajy	Deakin University, Australia
Prabhat K. Mahanti	University of New Brunswick, Canada

General Chairs

Tutut Herawan	University of Malaya
Rozaida Ghazali	Universiti Tun Hussein Onn Malaysia
Irfan Darmawan	Telkom University, Indonesia

Organizing Committee

Tatang Mulyana	Telkom University, Indonesia
Luciana Andrawina	Telkom University, Indonesia
Sampurno Wibowo	Telkom University, Indonesia
Nurdinintya	Telkom University, Indonesia
Albi Fitransyah	Telkom University, Indonesia
Adityas Widjajarto	Telkom University, Indonesia
Yuli Adam Prasetyo	Telkom University, Indonesia
Faishal Mufied Al-Anshary	Telkom University, Indonesia
M. Azani Hasibuan	Telkom University, Indonesia
Seno Adi Putra	Telkom University, Indonesia
R. Wahjoe Witjaksono	Telkom University, Indonesia
Warih Puspitasari	Telkom University, Indonesia
Tien Fabrianti K.	Telkom University, Indonesia
Soni Fajar Surya Gumilang	Telkom University, Indonesia
Riza Agustiansyah	Telkom University, Indonesia
Taufik Nur Adi	Telkom University, Indonesia
Ali Rosidi	Telkom University, Indonesia
Silvi Nurhayati	Telkom University, Indonesia

Program Committee Chairs

Shahreen Kassim	SMC, Universiti Tun Hussein Onn Malaysia
Nureize Arbaiy	SMC, Universiti Tun Hussein Onn Malaysia
Mohd Farhan Md Fudzee	SMC, Universiti Tun Hussein Onn Malaysia

Proceeding Chairs

Dedeh Witarsyah Jacob	Telkom University, Indonesia
M. Teguh Kurniawan	Telkom University, Indonesia
Nia Ambarsari	Telkom University, Indonesia

Workshop Chairs

Hairul Nizam Mahdin	SMC, Universiti Tun Hussein Onn Malaysia
Raden Rohmat Saedudin	Telkom University, Indonesia
Murahartawaty	Telkom University, Indonesia

Sponsorship

Umar Yunan	Telkom University, Indonesia
Luthfi Ramadani	Telkom University, Indonesia
Ilham Perdana	Telkom University, Indonesia
Nur Ichsan Utama	Telkom University, Indonesia

Program Committee

Soft Computing

Abdullah Khan	Abdul Wali Khan University, Pakistan
Abir Jaafar Hussain	Liverpool John Moores University, UK
Adamu I. Abubakar	Bayero University Kano, Nigeria
Adel Al-Jumaily	University of Technology Sydney, Australia
Ahmad Nazari Mohd Rose	Universiti Sultan Zainal Abidin, Malaysia
Aida Mustapha	SMC, Universiti Tun Hussein Onn Malaysia
Alexander Mendiburu	University of the Basque Country, Spain
Amelia Zafra Gomez	Universidad de Córdoba, Spain
Anan Banharnsakun	Kasetsart University, Thailand
Anca Ralescu	University of Cincinnati, USA
Andrei Petrovski	Robert Gordon University, UK
Andrew L. Nelson	Androtics LLC, USA
Anhar Risnumawan	Politeknik Elektronika Negeri Surabaya, Indonesia
Anthony Brabazon	University College Dublin, UK
António Dourado	University of Coimbra, Spain
Antonio LaTorre	Universidad Politécnica de Madrid, Spain
Asoke Nath	St. Xavier's College, India
Azizul Azhar Ramli	SMC, Universiti Tun Hussein Onn Malaysia
Carlos Pereira	Instituto Superior de Engenharia de Coimbra, Portugal
Cesar Torres	CINVESTAV, Mexico
Chuan Shi	Beijing University of Posts and Telecommunications, China
Dariusz Krol	Wroclaw University, Poland
Dhiya Al-Jumeily	Liverpool John Moores University, UK

Duoqian Miao	Tongji University, PR China
Eiji Uchino	Yamaguchi University, Japan
El-Sayed M. El-Alfy	KFUPM, Saudi Arabia
Elena N. Benderskaya	Saint-Petersburg State Polytechnical University, Russia
Elpida Tzafestas	University of Athens, Greece
Ender Ozcan	University of Nottingham, UK
Eulanda Dos Santos	Universidade Federal do Amazonas, Brazil
Fabrício Olivetti De França	Universidade Estadual de Campinas, Brazil
Fouad Slimane	EPFL, Switzerland
George Papakostas	Democritus University of Thrace, Greece
Gloria Cerasela Crisan	University of Bacau, Romania
Habib Shah	Islamic University of Medinah, Saudi Arabia
Haruna Chiroma	Federal College of Education (Technical), Nigeria
Heder S. Bernardino	Universidade Federal de Juiz de Fora, Brazil
Ian M. Thornton	University of Swansea, UK
Iwan Tri Riyadi Yanto	Universitas Ahmad Dahlan, Indonesia
Jan Platos	VSB-Technical University of Ostrava, Czech Republic
János Botzheim	Tokyo Metropolitan University, Japan
Jian Xiong	National University of Defense Technology, PR China
Jiye Liang	Shanxi University, PR China
Jon Timmis	University of York Heslington, UK
Jose-Raul Romero	University of Córdoba, Spain
Jose Francisco Chicano	Universidad de Málaga, Spain
José Ramón Villar	University of Oviedo, Spain
José Santos Reyes	Universidad de A Coruña, Spain
Kazumi Nakamatsu	University of Hyogo, Japan
Katsuhiro Honda	Osaka Prefecture University, Japan
Lim Chee Peng	Deakin University, Australia
M.Z. Rehman	SMC, Universiti Tun Hussein Onn Malaysia
Ma Xiuqin	Northwest Normal University, PR China
Mamta Rani	Krishna Engineering College, India
Maurice Clerc	Independent Consultant on Optimisation, France
Md Nasir Sulaiman	Universiti Putra Malaysia
Meghana R. Ransing	University of Swansea, UK
Mohamed Habib Kammoun	University of Sfax, National Engineering School of Sfax (ENIS), Tunisia
Muh Fadel Jamil Klaib	Jadara University, Jordan
Natthakan Iam-On	Mae Fah Luang University, Thailand
Noorhaniza Wahid	SMC, Universiti Tun Hussein Onn Malaysia
Qi Wang	Northwestern Polytechnical University, PR China
Qin Hongwu	Northwest Normal University, PR China
R.B. Fajriya Hakim	Islamic University of Indonesia

Rajesh S. Ransing	University of Swansea, UK
Richard Jensen	Aberystwyth University, UK
Russel Pears	Auckland University of Technology, New Zealand
Theresa Beaubouef	Southeastern Louisiana University, USA
Yusuke Nojima	Osaka Prefecture University, Japan

Data Mining

Alberto Cano	University of Córdoba, Spain
Anand Kulkarni	Maharashtra Institute of Technology, India
Alicia Troncos Lora	Pablo de Olavide University, Spain
Bac Le	University of Science, Ho Chi Minh City, Vietnam
Bay Vo	Ho Chi Minh City University of Technology, Vietnam
Beniamino Murgante	University of Basilicata, Italy
Binod Kumar	JSPM's Jayawant Technical Campus, India
Biju Issac	Teesside University, UK
Cheqing Jin	East China Normal University, PR China
Daisuke Kawahara	Kyoto University, Japan
David Taniar	Monash University, Australia
Eric Pardede	LaTrobe University, Australia
Gai-Ge Wang	Jiangsu Normal University, PR China
Gang Li	Deakin University, Australia
George Coghill	University of Auckland, New Zealand
Guokang Zhu	Chinese Academy of Sciences, PR China
Guowei Shen	Harbin Engineering University, PR China
Hongzhi Wang	Harbin Institute of Technology, China
Ildar Batyrshin	Mexican Petroleum Institute, Mexico
Ines Bouzouita	Faculty of Science of Tunis, Tunisia
Jeng-Shyang Pan	National Kaohsiung University of Applied Sciences, Taiwan
Jianwu Fang	Chinese Academy of Sciences, PR China
Jianyuan Li	Tongji University, PR China
José María Luna	University of Córdoba, Spain
Jose M. Merigo Lindahl	University of Barcelona, Spain
Kang Tai	Nanyang Technological University, Singapore
Kitsana Waiyamai	Kasetsart University, Thailand
La Mei Yan	ZhuZhou Institute of Technology, PR China
Md Anisur Rahman	Charles Sturt University, Australia
Mustafa Mat Deris	SCDM UTHM, Malaysia
Nadjet Kamel	University of Setif 1, Algeria

Nazri Mohd Nawi	SCDM UTHM, Malaysia
Naoki Fukuta	Shizuoka University, Japan
Noraziah Ahmad	Universiti Malaysia Pahang
Qi Wang	University of Science and Technology of China
Qinbao Song	Xi'an Jiaotong University, China
P. Shivakumara	Universiti Malaya, Malaysia
Patrice Boursier	University of La Rochelle, France
Patricia Anthony	Lincoln University, New Zealand
Rozaida Ghazali	SCDM UTHM, Malaysia
Sanah Abdullahi Muaz	Bayero University Kano, Nigeria
Sebastián Ventura	University of Córdoba, Spain
Sofian Maabout	Universitat Bordeaux, France
Tadashi Nomoto	National Institute of Japanese Literature, Japan
Tao Li	Florida International University, USA
Tetsuya Yoshida	Hokkaido University, Japan
Toshiro Minami	Kyushu University, Japan
Tutut Herawan	University of Malaya
Vera Yuk Ying Chung	University of Sydney, Australia
Wenny Rahayu	La Trobe University, Australia
Xiaoqiang Lu	Chinese Academy of Sciences, PR China
Xin Jin	University of Illinois at Urbana-Champaign, USA
Xue Li	The University of Queensland, Australia
Xuelong Li	Chinese Academy of Sciences, PR China
Yang-Sae Moon	Kangwon National University, Korea
Yingjie Hu	Auckland University of Technology, New Zealand
Yongsheng Dong	Chinese Academy of Sciences, PR China
You Wei Yuan	ZhuZhou Institute of Technology, PR China
Zailani Abdullah	Universiti Malaysia Kelantan
Zakaria Maamar	Zayed University, United Arab Emirates
Zhiang Wu	Nanjing University of Finance and Economics, China

Workshop on Ensemble Methods and Their Applications

Mokhairi Makhtar (Chair)	UniSZA, Malaysia
Mumtazimah Mohamad	UniSZA, Malaysia
Mohamad Afendee Mohamed	UniSZA, Malaysia
Engku Fadzli Hasan Syed Abdullah	UniSZA, Malaysia
Norita Md Norwawi	USIM, Malaysia
Madihah Mohd Saudi	USIM, Malaysia
Mohd Faizal Abdollah	UTEM, Malaysia

Workshop on Web Mining, Services and Security

Hairulnizam Mahdin (Chair)	UTHM, Malaysia
Isredza Rahmi Ab Hamid	UTHM, Malaysia
Masitah Ahmad	UiTM, Malaysia
Izuan Hafez Ninggal	UPM, Malaysia
Kamaruddin Malik Mohamad	UTHM, Malaysia
Harinda Fernando	APIIT, Sri Lanka

Contents

Soft Computing

Web Mining, Services and Security

Soft Computing

Cluster Validation Analysis on Attribute Relative of Soft-Set Theory

Rabiei Mamat[1(✉)], Ahmad Shukri Mohd Noor[1], Tutut Herawan[2], and Mustafa Mat Deris[3]

[1] School of Informatics and Applied Mathematic, University Malaysia Terengganu, 21300 Kuala Terengganu, Terengganu, Malaysia
{rab,ashukri}@umt.edu.my

[2] Faculty of Computer Science and Information Technology, University Malaya, 50000 Kuala Lumpur, Malaysia
tutut@um.edu.my

[3] Faculty of Computer Science and Information Technology, University Tun Hussein Onn Malaysia, 86400 Batu Pahat, Johore, Malaysia
mmustafa@uthm.edu.my

Abstract. Data clustering on categorical data pose a difficult challenge since there are no-inherent distance measures between data values. One of the approaches that can be used is by introducing a series of clustering attributes in the categorical data. By this approach, Maximum Total Attribute Relative (MTAR) technique that is based on the attribute relative of soft-set theory has been proposed and proved has better execution time as compared to other equivalent techniques that used the same approach. In this paper, the cluster validity analysis on the technique is explained and discussed. In this analysis, the validity of the clusters produced by MTAR technique is evaluated by the entropy measure using two standards dataset: Soybean (Small) and Zoo from University California at Irvine (UCI) repository. Results show that the clusters produce by MTAR technique have better entropy and improved the clusters validity up to 33%.

Keywords: Soft set · Data clustering · Attribute Relative · Cluster validity

1 Introduction

Organizing data into sensible groups is one of the foundations of learning and understanding. Through the similarity and dissimilarity of the created groups, the knowledge can be discovered. These activities are known by various name such as cluster analysis, segmentation analysis and taxonomy analysis is an important component in the decision making process. According to Ru Xui et al., cluster analysis may include the activities of feature selection, clustering algorithm execution, results validations and interpretation of the results [1].

T. Herawan et al. (eds.), *Recent Advances on Soft Computing and Data Mining*, Advances in Intelligent Systems and Computing 549, DOI 10.1007/978-3-319-51281-5_1

But the most important step in clustering process is clustering algorithm selection since it refers to the activity of choosing a suitable algorithm that later will reflects the definition of a good clustering scheme. In addition, the correctness of the algorithm is then verified by the results validation step where the output is analyzed using an appropriate criteria and technique. In [2], the technique which is called Maximum Total Attribute Relative (MTAR) that is based on the attribute relative of soft-set theory has been proposed. The technique which is used to determine the clustering attribute show a significant improvement in terms of executions time. However, the validity of the cluster produced by the proposed technique has never been evaluated. In this paper, the evaluation of the technique in term of cluster validation is carried out. In Sect. 2, the basic about the topic in discussion is elaborated; it is including the introduction to soft-set theory, the concept of clustering attribute selections and the cluster validation method. After that, in Sect. 3, the attribute relative of soft-set theory is described followed by Sect. 4 where the result and discussion is presented. Finally, some conclusions is elaborated in Sect. 5.

2 Preliminaries

2.1 Soft-Set Theory

Soft-set theory is a new mathematical tools for dealing with vagueness and uncertainties proposed by Molodtsov in 1999 [3]. Following is some basic definition that related to the softset theory. In this section, let U refers to an initial universe, where each objects in U described by a set of parameters E, $P(U)$ is the power set of U and $A \subseteq E$.

Definition 1 *(see [3]). A pair (F, A) is called a soft-set over U where F is a mapping given by*

$$F : A \rightarrow P(U) \tag{1}$$

Therefore, it is clear that a soft-set over the universe U is a parameterized family of the universe U.

Definition 2 *(see [4]). An information system U is a quadruple (U, A, V, f), where S is a non-empty finite set of interested objects, A is a non-empty finite set of attributes,V_i is the value set of the attribute a_i and f is an information function where $f_i : U \times a_i \rightarrow V_i$.*

Definition 3 *(see [5]). Let S is an information system as define in Definition 2, if V_a =\{0, 1\} for every $a \subseteq A$ then $S = (U, A, V, f)$ is called Boolean-valued information system.*

Proposition 1. *Every soft-set may be considered as a Boolean-valued information system.*

Proof. Let (F,E) is a soft-set over the universe U and $S = (U,A,V,f)$ is an information system. Obviously, if the universe U in S is equal to universe U in (F,E), then the parameter E can be considered as an attribute A. Thus, the information function f can be defined as

$$f = \begin{cases} 1\ , h \in F(e), \\ 0\ , h \in F(e). \end{cases} \tag{2}$$

Thus, we have $V = \{0,1\}$. Therefore, a soft-set (F,E) may be considered as Boolean-valued information system $S = (U,A,V_{\{0,1\}},f)$.

Definition 4 *(see [5]). Multi-soft set is the decomposition of multi-value information system.*

If $S = (U,A,V,f)$ is a multi-valued information system, then for each $a_i \in A$ exists $S^i = (U,a_i,v_{\{0,1\}},f)$ such that $\left(S^1 = (U,a_1,v_{\{0,1\}},f) = (F,e_1),\cdots,S^{|A|} = (U,a_{|A|},v_{\{0,1\}},f) = (F,e_{|A|})\right)$ and $\left((F,e_1),\cdots,(F,e_{|A|}) \in (F,E)\right)$. Thus, $(F,E) = \left\{(F,e_1),\cdots,(F,e_{|A|})\right\}$ is defined as a multi-soft set over the universe U representing a multi-valued information system $S = (U,A,V,f)$.

2.2 Clustering Attribute Selection

One of the approach in the categorical data clustering is by introducing a series of clustering attribute which had exploited the data uncertainties in the multi-valued information system. In other word, this approach requires the process of selecting clustering attribute and then using the different within categorical variables of the selected attribute as a basis for the partitioning process.

By this approach, Mazlack et al. proposed a technique called Total Roughness (TR) to handle multi-valued attributes and Bi-Clustering to handle binary value attribute [6]. Unfortunately, this separation requires user to determine the type of attributes manually which become the main drawback of this technique. To overcome this issue, Parmar et al. have proposed a new technique called Min-Mean Roughness (MMR) [7]. As compare to TR, this technique has producing a higher quality clusters. But, in different view, this technique consumes more execution time which is proportional to the square of the number of attributes. Herawan proposed a new technique called Maximum Attribute Dependency (MDA) [4,8] that used the attribute dependency to determine the partition attribute of the given information system. All those three techniques above are based on the theory of rough set in their solutions.

Based on the theory of soft-set, Qin et al. proposed a technique called Novel Soft-Set approach (NSS) [9] that redefined the attribute dependency of rough set theory in the view of soft-set theory. Consequently, results obtain from this technique does not have much different as compare to MDA technique.

2.3 Cluster Validation Method

According to Sripada and Rao [10], cluster validation measures can be categorized into two: internal validation and external validation. In external validation, the measures evaluate the extent to which the clustering structure discovered by a clustering algorithm matches some external structure, e.g., the one specified by the given class labels. For internal validation, however, the cluster evaluation is merely based on the clusters themselves, without any additional information out of the data. For most applications, the external validations are much more appropriate. This fact is confirmed by [11,12], where they also nominated entropy technique and purity technique are frequently used external validation measures in data mining community. In this paper a methods of measurement used is entropy index(entropy) (which is referred to the Shannon entropy [13] index). Entropy is a measure of the uncertainty associated with a random variable. In the cluster analysis, entropy measures the quality of the cluster with respect to the given class labels or in other word, entropy measure the distribution of various clusters within each class. Entropy method has been used in measuring the validity of HICAP [14], comparing K-Means and fuzzy C-Means and measuring hierarchical clustering document algorithm for the document dataset in [15]. In the meantime, [16] used this measures to evaluate the performance of their alternate least square NMF algorithm.

Let i is the number of class and j is the number of clusters. The entropy of a set of clusters j can be calculated by computing the class distribution of the objects in each cluster by computing the probability that a member of cluster j belongs to class i denoted by p_{ij} . The entropy of cluster j denoted by E_j is calculated using the standard entropy formula;

$$E_j = -\sum_i p_{ij} log(p_{ij}) \tag{3}$$

The total entropy for a set of clusters is computed as the weighted sum of the entropies of each cluster, denoted by E as the following equation:

$$E = \sum_{j=1}^{m} \frac{n_j}{n} \times E \tag{4}$$

where n is the size of cluster j, m is the number of clusters and n is the total number of data points. Based on the entropy value, the clustering quality is interpreted as a better clustering when the value is near to zero, and otherwise when the value is nearing to one, then the quality of clustering is in-doubt. By the other means, the lower entropy is achieved when the method used in the clustering process have successful reduce/managed the uncertainties among data, otherwise, the approach failed to dealing with uncertainties that occurs in data.

3 Attribute Relative of Soft-Set Theory

The definition of the attribute relative of soft-set is given as the following. In the definition, a pair (F, A) is refer to multi-soft sets over the universe U.

Definition 5. *Let* $((F, a_0), \cdots, (F, a_{|A|}) \subseteq (F, A))$ *and* $((F, a_{0_0}), \cdots, (F, a_{0_{|A_0|}}) \subseteq (F, A_0), \cdots, (F, a_{|A|_0}), \cdots, (F, a_{|A|_{|A|}}) \subseteq (F, A_{|A|}))$. *A soft-set* (F, a_{j_k}) *is said to be relative to* (F, a_{p_q}) *and vice-versa if* $((F, a_{j_k}) \bigcap (F, a_{p_q}) \neq 0)$

Definition 6. *If a soft-set* (F, a_{j_k}) *is relative to* (F, a_{p_q}), *then exists the relative support value of* F, a_{p_q} *by* (F, a_{j_k}) *denoted by* $RSup_{(F,a_{j_k})}(F, a_{p_q})$ *which is defined as*

$$RSup_{(F,a_{j_k})}(F, a_{p_q}) = \frac{|((F, a_{j_k}) \bigcap (F, a_{p_q})|}{|(F, a_{j_k})|}. \tag{5}$$

It is clear that from above definition, relative support value can be categorized into three types: full relative, partly relative and zero relative (no-relative).

Definition 7. *Total relative support is a summation of all full relative support for each soft-set* $\left(((F, a_{i_j}) \subseteq (F, a_i)) \subseteq (F, A)\right)$. *Hence, the total relative support of soft-set* (F, a_{i_m}) *which is denoted by* $TRS_{(F,a_{i_m})}$ *is computed as the following*

$$\sum_{i=0,j=0}^{i=|A|,j=|a_i|} (FullRelative_{(F,a_{i_j})})(F, a_{i_m})) \tag{6}$$

Definition 8. *Total attribute relative is a summation of all total relative support for each soft-set* $((F, a_i) \subseteq (F, A))$. *The total attribute relative for* (F, a_0) *which is denoted by* $TAR_{(F,a_0)}$ *is defined as:*

$$TAR_{(F,a_0)} = \sum_{i=0,j=0}^{i=|A|,j=|a_i|} TRS_{(F,a_{i_j})}. \tag{7}$$

Definition 9. *Maximum total attribute relative is the maximum value of* TAR *in the probability distribution which is denoted by* $MTAR$ *and defined as:*

$$MTAR = Max(TAR_{(F,a_0)}, \cdots, TAR_{(F,a_{|A|})}) \tag{8}$$

where Max is refer to maximum function.

Proposition 2. *If* $Max(MTAR = (F, a_i)) = 1$ *then* (F, a_i) *is a partition attribute.*

Proof. If (F, a_i) is the maximum of total attribute relative, then it is obvious that (F, a_i) have more full relative as compare to others. Thus, from Definitions 1, 2, 3, 4, 5, 6, 7, 8 and 9, (F, a_i) is selected as a clustering attribute.

Corollary 1. *If* $Mode\big(MTAR((F,a_i),\cdots,(F,a_{|A|}))\big) > 1$, *then the clustering attribute is* $Max(TRS_{(MTAR_0)},\cdots,TRS_{(MTAR_{|A|})})$.

Proof. Let (F,a_i) and (F,a_j) be two soft-set over the universe U and let (F,a_i) and (F,a_j) is a member of $MTAR$, then $TAR_{(F,a_i)} = TAR_{(F,a_j)}$ are maximum. Both attributes cannot be used as a clustering attribute unless it is proven that both attribute have the same full relative support value which can only be proved by the TRS value at the categorical level. Hence, if $TRS_{(F,a_i)}$ of $MTAR$ is maximum then it is clear that $TRS_{(F,a_i)}$ is most relative to all other categorical soft-set and is selected as the clustering attribute.

Thus, to form the cluster, each object in the information system will be partitioned accordingly to the selected attribute by a method known as "divide and conquer". The method will be applied recursively to get the next cluster where at each iteration the leaf node having more objects with mixed category is selected for further partitioning.

4 Result and Discussion

As discussed earlier, although there are lot of clustering techniques that are based on the attribute selection such as TR [6], MMR [7], MDA [4,8] and NSS [9]. However, TR and MMR have been proved produced lower validity clusters as compare to MDA, while NSS result is equivalent to MDA. Thus, in this paper, the comparisons only be made between MTAR and MDA using two benchmark datasets from UCIMLR i.e. Soybean (small) [17] and Zoo [18]. In summary, the soybean (small) dataset contains 47 objects of soybean on diseases. The dataset is in the completed state without any missing values. Each object is classified into one of the four diseases either Diaporthe Stem Canker (D1), Charcoal Rot (D2), Rhizoctonia Root Rot (D3) or Phyrophthora Rot (D4). The dataset is comprised of ten objects of D1, ten objects of D2, 10 objects of D3 and 17 objects of D4. Meanwhile, the Zoo dataset is comprised of 101 objects of animals. Each animal is then described by the terms of 18 categorical-valued attributes. The dataset is in the completed state without any missing values. Each animals is classified into one of the seven animal class ranged from one to seven.

Figures 1 and 2 respectively shows the entropy index of MTAR and MDA using Soybean (small) and Zoo dataset.

Generally both figures clearly shows that the validity of the clusters produced by MTAR is better when compared to the clusters produced by MDA technique. In both cases, a reduction of entropy value shows that both techniques are able to manage the uncertainty that exist in the data sets. Unfortunately, in a case of Soybean (Small), the changes of entropy value which is avoiding zero (0) is severe and thus shows that the uncertainty back to be a problem to the MDA technique and implicated that MDA failed to manage the uncertainty properly. On the other hand, it does not happen to the entropy value of MTAR which is directly approaching zero (0) and thus demonstrate the ability of MTAR that have better uncertainty management. But, since the number of clusters is

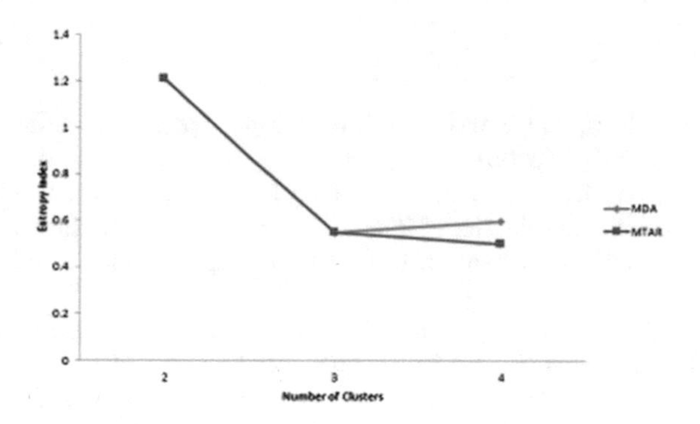

Fig. 1. Entropy index of MTAR and MDA on Soybean (Small) dataset

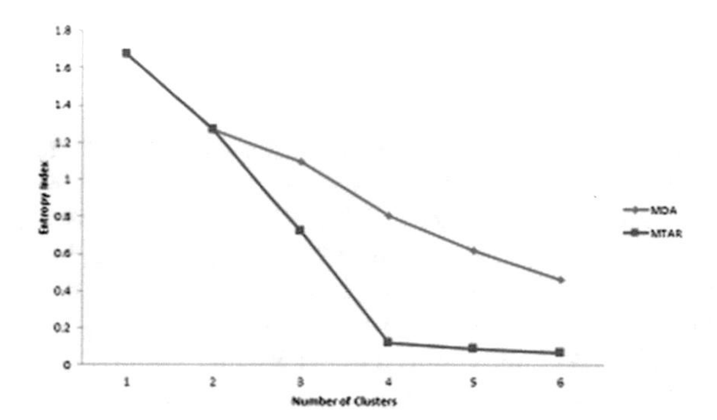

Fig. 2. Entropy index of MTAR and MDA on Zoo dataset

small then the differences in the entropy index are not very significant. However, MTAR technique still shows an improvement up to approximately 4 % of validity. Meanwhile, in a Zoo dataset that have more objects as compare to Soybean (Small) dataset, MDA does not repeat the above behavior. But, MDA still shows some lacking in the management of uncertainty through the small entropy index.

5 Conclusions

Providing a higher validity cluster is very important in clustering process. Producing higher validity cluster basically related to the readiness of the technique to dealing with uncertainty in data. Soft-set theory is a mathematical tool developed to work with uncertainties. In this paper, a clustering attribute selection technique that is based on the attribute relative of soft-set theory called MTAR is evaluated in term of the validity of the cluster. Comparison using the entropy index value is made with the rough set-based technique called MDA. The results shows that the attribute relative of softset based technique had better data uncertainty handling and deliver higher cluster validity.

References

1. Xui, R., Wunsch II, D.: Survey of Clustering Algorithms. IEEE Trans. Neural Netw. **16**(3), 645–678 (2005)
2. Mamat, R., Deris, M.M., Herawan, T.: MAR maximum attribute relative of soft-set for partition attribute selection. Knowl. Based Syst. **52**, 11–20 (2013)
3. Molodtsov, D.: Soft set theory - first results. Comput. Math. Appl. **37**(4/5), 19–31 (1999)
4. Herawan, T., Deris, M.M., Abawajy, J.H.: A rough set approach for selecting clustering attribute. Knowl. Based Syst. **23**(3), 220–231 (2010)
5. Herawan, T., Deris, M.M.: On multi-soft sets construction in information systems. In: Huang, D.-S., Jo, K.-H., Lee, H.-H., Kang, H.-J., Bevilacqua, V. (eds.) ICIC 2009. LNCS (LNAI), vol. 5755, pp. 101–110. Springer, Berlin, Heidelberg (2009). doi:10.1007/978-3-642-04020-7_12
6. Mazlack, L.J., He, A., Zhu, Y., Coppock, A.S.: Rough sets approach in choosing partitioning attributes. In: Proceeding of ICSA 13th International Conference, pp. 1–6 (2000)
7. Parmar, D., Wu, T., Blackhurst, J.: MMR: an algorithm for clustering categorical data using rough set theory. Data Knowl. Discov. **63**(3), 879–893 (2007)
8. Herawan, T., Ghazali, R., Yanto, I.T.R., Deris, M.M.: Rough set approach for categorical data clustering. Int. J. Database Theor. Appl. **3**(1), 33–52 (2010)
9. Qin, H., Ma, X., Zain, J.M., Herawan, T.: A novel soft set approach in selecting clustering attribute. Knowl. Based Syst. **36**, 139–145 (2012)
10. Sripada, S.C., Rao, S.M.: Comparison of purity and entropy of KMeans clustering and fuzzy C-Means clustering. Indian J. Comput. Sci. Eng. **2**(3), 343–346 (2011)
11. Steinbach, M., Karypis, G., Kumar, V.: A comparison of document clustering techniques. In: Workshop on Text Mining, the 6th ACM SIGKDD International Conference on Knowledge Discovery and Data Mining (2000)
12. Zhao, Y., Karypis, G.: Hierarchical clustering algorithms for document datasets. Data Min. Knowl. Discov. **10**(2), 141168 (2005)
13. Shannon, C.E.: A mathematical theory of communication. Bell Syst. Tech. J. **27**(3), 379–423 (1948)
14. Steinbach, M., Tan, P.-N., Kumpar, V., Xiong, H.: Hicap : hierarchial clustering with pattern preservation. In: Proceeding of 2004 SIAM International Conference on Data Mining (SDM), pp. 279–290 (2004)
15. Zhao, Y., Karypis, G., Fayyad, U.: Hierarchical clustering algorithms for document datasets. Data Min. Knowl. Discov. **10**(2), 141–168 (2005)
16. Kim, H., Park, H.: Sparse non-negative matrix factorizations via alternating non-negativity-constrained least squares for microarray data analysis. Bioinformatics **23**(12), 1495–1502 (2007)
17. http://archive.ics.uci.edu/ml/datasets/soybean+small
18. http://archive.ics.uci.edu/ml/datasets/zoo

Optimizing Weights in Elman Recurrent Neural Networks with Wolf Search Algorithm

Nazri Mohd. Nawi(✉), M.Z. Rehman, Norhamreeza Abdul Hamid, Abdullah Khan, Rashid Naseem, and Jamal Uddin

Faculty of Computer Science and Information Technology, Soft Computing and Data Mining Centre (SMC), Universiti Tun Hussein Onn Malaysia (UTHM), P.O. Box 101 Parit Raja, 86400 Batu Pahat, Johor, Malaysia
nazri@uthm.edu.my, zrehman862060@gmail.com, norhamreeza@gmail.com, abdullahdirvi@gmail.com, rnsqau@gmail.com, jamalmaths2014@gmail.com

Abstract. This paper presents a Metahybrid algorithm that consists of the dual combination of Wolf Search (WS) and Elman Recurrent Neural Network (ERNN). ERNN is one of the most efficient feed forward neural network learning algorithm. Since ERNN uses gradient descent technique during the training process; therefore, it is not devoid of local minima and slow convergence problem. This paper used a new metaheuristic search algorithm, called wolf search (WS) based on wolf's predatory behavior to train the weights in ERNN to achieve faster convergence and to avoid the local minima. The performance of the proposed Metahybrid Wolf Search Elman Recurrent Neural Network (WRNN) is compared with Bat with back propagation (Bat-BP) algorithm and other hybrid variants on benchmark classification datasets. The simulation results show that the proposed Metahybrid WRNN algorithm has better performance in terms of CPU time, accuracy and MSE than the other algorithms.

Keywords: Elman recurrent network · Wolf search algorithm · Local minima · Metaheuristic optimization · Nature inspired algorithms

1 Introduction

Artificial Neural Networks (ANN) consists of interconnected nodes that simulates that calculation process inside human brain [1–3]. This interconnected calculation process makes ANN flexible enough to perform large complex problems such as pattern recognition, engineering, biological modelling, health & medicine, decision control, manufacturing & management, marketing, and sea & space modelling etc. [4–10].

Usually ANN structures are fully connected feed-forward networks that starts from one layer through the connected layer and finally generates results on the output layer [4, 11]. This ability of learning makes ANN quite useful but an addition of short-term memory makes them more powerful and dynamic. Recurrent Neural Networks (RNN) provides short-term memory or states and show dynamic temporal behavior [11].

T. Herawan et al. (eds.), *Recent Advances on Soft Computing and Data Mining*,
Advances in Intelligent Systems and Computing 549, DOI 10.1007/978-3-319-51281-5_2

RNN have the ability to store earlier patterns thus paving way for the modelling of active systems [12, 13].

Until today, full and partial RNN have been extensively used to conduct associative memories, spatio-temporal pattern classification, optimization and prediction [14–19]. When the ease of use is considered, partial recurrent neural network (RNN) especially Elman recurrent neural network (ERNN) memorize previous states, executes, and learn very efficiently. ERNN performs efficiently on normal networks but when network size becomes larger, the calculation becomes more complex thus making ERNN more prone to converging to local minima. It also consumes more memory during experimentation due to extra connection weights [11].

Therefore, to overcome the problem of inefficient weights that cause the ERNN to converge to local minima due to flat surfaces and to avoid CPU overheads. This paper proposed a hybrid Wolf Search Elman Recurrent Neural Network (WRNN). The proposed WRNN algorithm is tested on classification datasets and compared with Artificial Bee Colony Back Propagation (ABCBP) [20], Artificial Bee Colony-Levenberg-Marquardt algorithm (ABC-LM) [21], and Bat with Back-Propagation [22] algorithms. The selected performance metrics are mean squared error (MSE), CPU time, accuracy, and convergence rate.

The remaining paper is organized as follows: Sect. 2 explains the Wolf Search Recurrent Neural Network's (WRNN) implementation. Meanwhile, the Sect. 3 discuss the results and finally, the paper is concluded in the Sect. 4.

2 Wolf Search (WS) Algorithm

Wolf Search (WS) algorithm is a metaheuristic algorithm inspired by the silent predatory behavior of the wolves and the intelligent avoidance of the enemy. Proposed by Rui Tang [23], WS follows three idealized rules for finding the best solution or prey;

(a) Each wolf has a fixed visual area with a radius. In 2D, the coverage would simply be the area of a circle by the radius v. In hyper-plane, where multiple attributes dominate, the distance would be estimated by Minkowski distance.
(b) The result or the fitness of the objective function represents the quality of the wolf's current position. The wolf always tries to move to better terrain but rather than choose the best terrain it opts to move to better terrain that already houses a companion. If the new position is better, the incentive is stronger provided that it is already inhabited by a companion wolf.
(c) At some point, it is possible that the wolf will sense an enemy. The wolf will then escape to a random position far from the threat and beyond its visual range.

2.1 The Proposed WRNN Algorithm

In the proposed Wolf Search Elman Recurrent Neural Network (WRNN) algorithm, each best position in WS algorithm represents a possible solution (i.e., the initial weight space and the corresponding biases for Elman Recurrent Neural Network (ERNN)).

The weight optimization problem and the size of the population represents the quality of the solution. In the first epoch, the best weights and biases are initialized with WS and then those weights are passed on to the ERNN. The weights in the ERNN are calculated. In the next cycle, WS updates the weights with the best possible solution and WS will continue searching the best weights until the last cycle/epoch of the network is reached or either the MSE is achieved. Figure 1 shows the proposed flowchart for the WRNN algorithm.

The WS is a population based optimization algorithm. It starts with a random initial population. In the proposed WRNN algorithm, the weight value of a matrix is calculated as following;

$$W_n = U_n = \sum\nolimits_{n=1}^{N} a.(rand - \frac{1}{2}) \tag{1}$$

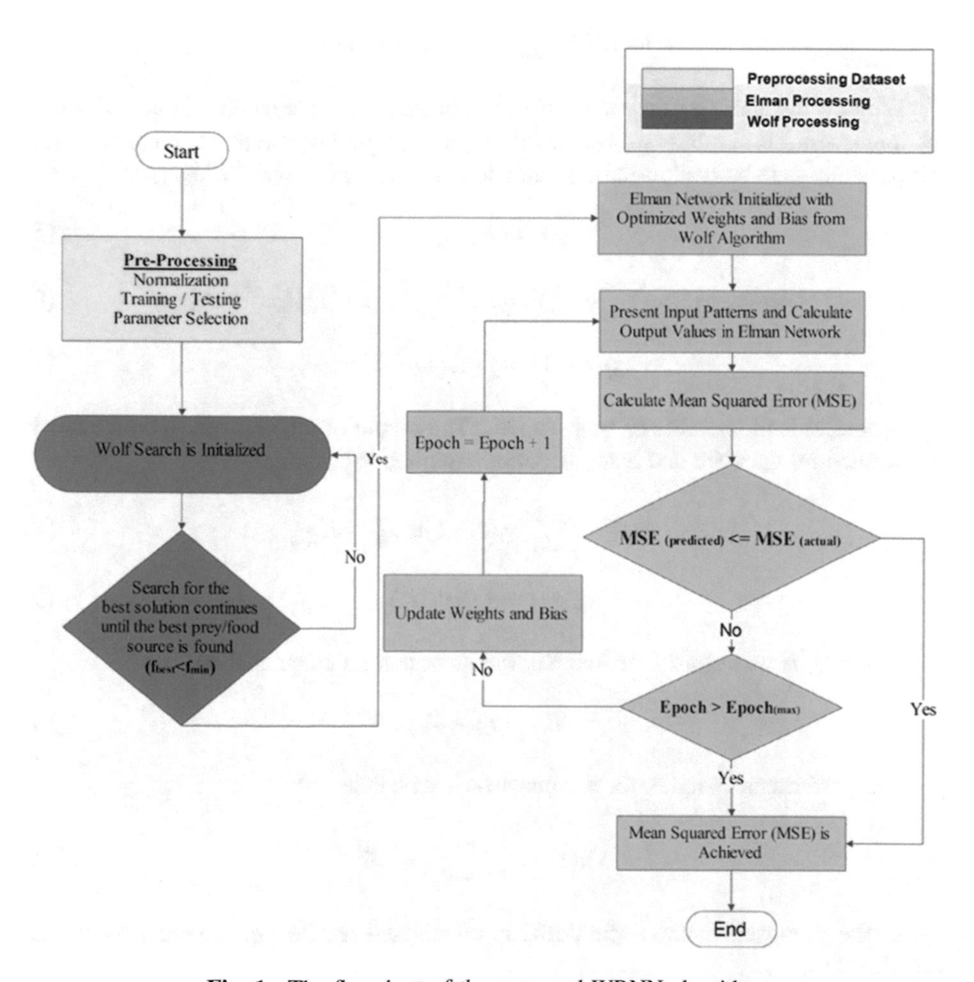

Fig. 1. The flowchart of the proposed WRNN algorithm

$$B_n = \sum_{n=1}^{N} a.(rand - \frac{1}{2}) \tag{2}$$

Where, $W_n = N^{th}$ is the weight value in a weight matrix. The *rand* in the Eq. (1) is the random number between [0, 1], *a* is any constant parameter for the proposed method it being less than one, and B_n is bias value. So the list of weight matrix is as follows;

$$W^c = [W_n^1, W_n^2, W_n^3, \ldots\ldots W_n^{N-1},] \tag{3}$$

Now from neural network process, MSE is easily calculated from every weight matrix, in W^c. For the ERN structure three layered network consisting of one input layer, one hidden or 'state' layer, and one 'output' layer is used. Each layer have its own index variable: *k* for output nodes, *j* and *l* for hidden, and i for input nodes. In a feed forward network, the input vector, *x* is propagated through a weight layer.

$$net_j(t) = \sum_i^n x_i(t) w_{n(ji)} + B_{n(j)} \tag{4}$$

Where *n* the number of inputs is, $B_{n(j)}$ is a bias. In a simple recurrent network (RN), the input vector is similarly propagated through a weight layer, but also combined with the previous state activation through an additional recurrent weight layer, *U*;

$$y_j(t) = f(net_j(t)) \tag{5}$$

$$net_j(t-1) = \sum_i^n x_i(t) W_{n(ji)} + \sum_l^m y_l(t-1) U_{n(jl)} + B_{n(j)} \tag{6}$$

$$y_j(t-1) = f(net_j(t-1)) \tag{7}$$

Where, *m* is the number of 'state' nodes. The output of the network in both cases is determined by the state and a set of output weights, *W*;

$$net_k(t) = \sum_j^M y_j(t-1) w_{n(kj)} + B_{n(k)} \tag{8}$$

$$Y_k(t) = g(net_k(t)) \tag{9}$$

Where, *g* is an output function. So, the error can be calculated as;

$$E = (T_k - Y_k) \tag{10}$$

The performances index for the network is calculated as;

$$V_F(x) = \frac{1}{2} \sum_{k=1}^{K} E^T.E \tag{11}$$

In the proposed method, the MSE is considered as the performance index and calculated as;

$$V_{\mu}(x) = \frac{\sum_{j=1}^{N} V_F(x)}{P_i} \tag{12}$$

Where, y_r is the output of the network when the kth input net_i is presented. And $E = (T_k - Y_k)$ is the error for the kth input, $V_{\mu}(x)$ is the average performance, $V_F(x)$ is the performance index, and P_i is the number of Wolf population in i^{th} iteration. At the end of each epoch, the list of average sum of Mean Square Error of ith iteration MSE can be calculated as;

Begin

Step 1: Initialize Wolf population size, dimensions and ERNN structure

Step 2: Load the training and testing data

Step 3: **While** MSE<stopping criteria

Step 4: Pass the wolf preys as weights to network

Step 5: Feed forward network runs using the weights initialized with Wolf Search

Step 6: Calculate the error using the Equation (10)

Step 7: Minimize the error by adjusting network parameter using Wolf Search.

Step 8: Generate Wolf prey (x_j) by using Brownian motion from a certain position.

Step 9: If enemy detected, abandon the current position and jump to a safe distance.

Step 10: Evaluate the fitness of the prey, Chose a random wolf i

If

a. $X_j > X_i$ *Then*

b. $x_i \leftarrow x_j$

c. $X_i \leftarrow X_j$

End if

Step 11: Wolf keeps on calculating the best possible weight at each epoch until the network is converged.

End While

Step 12: Post Process Results and Visualization.

End

Fig. 2. The pseudo code of the proposed WRNN algorithm

$$MSE_i = \{V_u^1(x), V_\mu^2(x), V_\mu^3(x)\ldots..V_\mu^n(x)\} \tag{13}$$

The Wolf search duplicates the MSE and is found when all the inputs are processed for each Wolf in the population. The Wolf search position x_j is calculated as;

$$x_j = Min\{V_\mu^1(x), V_\mu^2(x), V_\mu^3(x)\ldots..V_\mu^n(x)\} \tag{14}$$

And the rest of the Average MSE is considered as other wolves. A new solution x_i^{t+1} for Wolf i is generated using the following Equation;

$$x_i^{t+1} = x_{min}^t + rand * (x_{max}^t - x_{min}^t) \tag{15}$$

So, the movement of the other Wolf x_i^{t+1} toward x_j can be drawn from Eq. (16);

$$X_i = \begin{cases} x_i^{t+1} + rand \cdot *stepb * \left(x_j - x_i^{t+1}\right) \\ x_i^{t+1}\ else \end{cases} \tag{16}$$

Where, ∇V_i is a small movement of x_i towards x_j. The weights and biases for each layer is then adjusted as;

$$W_n^{s+1} = U_n^{s+1} = W_n^s - X_i \tag{17}$$

$$B_n^{s+1} = B_n^s - X_i \tag{18}$$

The pseudo code for the WRNN is given in the Fig. 2:

3 Results and Discussion

Basically, the main focus of this paper is to compare different algorithms based on mean squared error (MSE), accuracy during network convergence and the CPU time. Before discussing the simulation results, there are certain things that needs to be explained such as tools and technologies, network topologies, testing methodology and the classification problems used for the entire experimentation. The discussion is as follows:

3.1 Preliminary Study

In order to demonstrate the performance of the proposed BARNN algorithm for training RNN. The proposed algorithms are tested on breast cancer, thyroid, Iris and Australian credit card datasets taken from University of California, Irvine Machine Learning Repository (UCIMLR). The simulation experiments are carried out on Intel Core i5 Processor with 8 GB main memory. Simulation software is MATLAB 2012 running on Windows 7 home edition. The performance measure for each algorithm is based on the Mean Squared Error (MSE), standard deviation and accuracy. The three layers feed forward neural network architecture (i.e. input layer, one hidden layer, and output layers.) is used for each problem. The number of hidden nodes is kept fixed to 5 node. In the network structure, the bias nodes and the log-sigmoid activation function are used. For each problem, trial is limited to 1000 epochs. A total of 20 trials are run for each dataset with target error set to 0.0001. The network results are stored in the result file for each trial.

3.2 Classification Datasets

The first datasets selected is breast cancer that consists of a total 699 instances [24]. The selected network architecture used for this dataset has 9 inputs nodes, 5 hidden nodes and 2 output nodes. Table 1, shows the simulation results of all the algorithms used in this study. From the Table 1, it can be easily seen that the proposed WRNN algorithm achieves better convergence than the ABCBP, ABC-LM, and Bat-BP in terms of MSE, CPU time, epochs and accuracy. The proposed WRNN converged to global minima with an MSE of 0.0001, and 99.99% accuracy.

The second dataset for classification is Thyroid which consists of a total of 7200 instances [25]. The selected network architecture for Thyroid has 21 input nodes, 5 hidden nodes and 3 output nodes. The simulation result of Thyroid classification problem are shown in the Table 1. From Table 1, it can be seen that the proposed

Table 1. Summary of algorithms performance on classification problems

	Algorithms	Epochs	CPU Time	Accuracy	MSE
	ABC-BP	1000	1482.9	92.02	0.184
Breast Cancer	**ABC-LM**	1000	1880.64	93.83	0.0139
	BAT-BP	1000	345.42	97.81	0.0219
	WRNN	37	79	99.99	0.0001
	ABC-BP	1000	1747.23	93.28	0.046
Thyroid	**ABC-LM**	1000	1382.91	91.66	0.0409
	BAT-BP	1000	4610.79	99.35	0.01
	WRNN	114	1563	99.99	5.33E-05
	ABC-BP	1000	156.43	86.88	0.155
IRIS	**ABC-LM**	1000	171.52	79.56	0.058
	BAT-BP	1000	475.38	98.06	0.021
	WRNN	105	161	99.99	6.62E-05
Australian	**ABC-BP**	1000	6894	89.99	0.173
Credit Card	**ABC-LM**	1000	4213	77.78	0.05
Approval	**BAT-BP**	967.03	892.31	99.67	0.0033
	WRNN	44	177	99.99	4.90E-05

Note: The highlighted area represents the proposed WRNN algorithm

WRNN algorithm outperforms the other comparison algorithms in terms of epochs, CPU time, MSE, and accuracy. The proposed algorithm converged to the global minima within a mere 114 epochs with an MSE of 5.33E-05.

Collected by Fisher, IRIS classification datasets is one of the best pattern recognition dataset in use since 1936 [26]. It consists of 150 samples collected from three Iris species. The selected network architecture for Iris has 4 inputs, 5 hidden, and 3 outputs nodes. Table 1 shows the MSE, SD, Epochs, CPU time and accuracy of the proposed WRNN algorithm when compared with other hybrid variants. From the Table 1, it is clear that the proposed WRNN algorithm has better performance in terms of MSE, CPU time, and accuracy. WRNN converged with an MSE of 6.62E-05, an accuracy of 99.99%, and within 105 epochs when compared with the other algorithms.

The last dataset is Australian Credit Card Approval which consists of 690 instances [27]. The network architecture selected has 51 inputs, 5 hidden nodes, and 2 outputs. This dataset contains the information of the bank clients. All the attributes are changed to hide the client's data. Table 1 shows the proposed WRNN algorithm's performance comparison algorithms with other algorithms. From Table 1, it can be clearly realized that WRNN achieved higher accuracy of 99.99% with an MSE of 4.90E-05 within 44 epochs. Whereas, the other algorithms fell behind the proposed WRNN algorithm in terms of MSE, accuracy and epochs.

4 Conclusion

In this paper, Nature Inspired Metaheuristic Wolf search algorithm is used to train the weights in Elman Recurrent Neural Network (ERNN). The proposed Metahybrid Wolf Search with Elman Recurrent Neural Network (WRNN) algorithm has been able to overcome the inherent slow convergence problem of the ERNN. The performance of the proposed WRNN is compared with Bat with back propagation (Bat-BP) and other hybrid variant algorithms on benchmark classification datasets taken from UCIMLR. The simulation results show that the proposed WRNN algorithm has better convergence performance with high accuracy, small MSE and less CPU time than other algorithms.

Acknowledgments. The Authors would like to thank Office of Research, Innovation, Commercialization and Consultancy (ORICC), Universiti Tun Hussein Onn Malaysia (UTHM) and Ministry of Education (MOE) Malaysia for financially supporting this Research under Fundamental Research Grant Scheme (FRGS) vote no. 1236. This research is also supported by Gates IT Solution Sdn. Bhd under its publication scheme.

References

1. Nawi, N.M., Khan, A., Rehman, M.Z.: A new optimized Cuckoo Search Recurrent Neural Network (CSRNN) algorithm. In: The 8th International Conference on Robotic, Vision, Signal Processing and Power Applications, pp. 335–341. Springer, Singapore (2013)

2. Radhika, Y., Shashi, M.: Atmospheric temperature prediction using support vector machines. Int. J. Comput. Theory Eng. **1**, 55–58 (2009)
3. Akcayol, M.A., Cinar, C.: Artificial neural network based modeling of heated catalytic converter performance. Appl. Therm. Eng. **25**, 2341–2350 (2005)
4. Rehman, M.Z., Nawi, N.M.: Improving the accuracy of Gradient Descent Back Propagation algorithm (GDAM) on classification problems. Int. J. New Comput. Archit. Appl. **1**, 861–870 (2011)
5. Kosko, B.: Neural Network and Fuzzy System. Prentice Hall, Upper Saddle River (1994)
6. Krasnopolsky, V.M., Chevallier, F.: Some neural network applications in environmental sciences. part II: advancing computational efficiency of environmental numerical models. Neural Netw. **16**, 335–348 (2003)
7. Coppin, B.: Artificial Intelligence Illuminated. Jones and Bartlett Publishers Inc., Sudbury (2004)
8. Basheer, I.A., Hajmeer, M.: Artificial neural networks: fundamentals, computing, design, and application. J. Microbiol. Methods **43**, 3–31 (2000)
9. He, Z., Wu, M., Gong, B.: Neural network and its application on machinery fault diagnosis. In: IEEE International Conference on Systems Engineering. pp. 576–579 (1992)
10. Li, B., Chow, M.Y., Tipsuwan, Y., Hung, J.C.: Neural-network-based motor rolling bearing fault diagnosis. IEEE Trans. Ind. Electron. **47**, 1060–1069 (2000)
11. Nawi, N.M., Khan, A., Rehman, M.Z.: CSBPRNN: a new hybridization technique using cuckoo search to train back propagation recurrent neural network. In: Herawan, T., Deris, M. M., Abawajy, J. (eds.) Proceedings of the First International Conference on Advanced Data and Information Engineering (DaEng-2013). LNEE, vol. 285, pp. 111–118. Springer, Singapore (2014). doi:10.1007/978-981-4585-18-7_13
12. Zhang, J., Lok, T., Lyu, M.R.: A hybrid particle swarm optimization-back-propagation algorithm for feedforward neural network training. Appl. Math. Comput. **185**, 1026–1037 (2007)
13. Ab Aziz, M.F., Hj Shamsuddin, S.M., Alwee, R.: Enhancement of particle swarm optimization in elman recurrent network with bounded Vmax function. In: Proceedings 2009 3rd Asia International Conference on Modelling and Simulation, AMS 2009, pp. 125–130 (2009)
14. Sutskever, I., Hinton, G., Taylor, G.: The Recurrent temporal restricted Boltzmann machine. Neural Inf. Process. Syst. **21**, 1601–1608 (2008)
15. Gupta, L., McAvoy, M., Phegley, J.: Classification of temporal sequences via prediction using the simple recurrent neural network. Pattern Recognit. **33**, 1759–1770 (2000)
16. Saad, E.W., Prokhorov II, D., Donald, C.W.: Comparative study of stock trend prediction using time delay, recurrent and probabilistic neural networks. IEEE Trans. Neural Netw. **9**, 1456–1470 (1998)
17. Guo, L., Rivero, D., Pazos, A.: Epileptic seizure detection using multiwavelet transform based approximate entropy and artificial neural networks. J. Neurosci. Methods **193**, 156–163 (2010)
18. Güler, N.F., Übeyli, E.D., Güler, I.: Recurrent neural networks employing Lyapunov exponents for EEG signals classification. Expert Syst. Appl. **29**, 506–514 (2005)
19. Übeyli, E.D.: Recurrent neural networks employing Lyapunov exponents for analysis of doppler ultrasound signals. Expert Syst. Appl. **34**, 2538–2544 (2008)
20. Karaboga, D., Akay, B., Ozturk, C.: Artificial Bee Colony (ABC) optimization algorithm for training feed-forward neural networks. In: Torra, V., Narukawa, Y., Yoshida, Y. (eds.) MDAI 2007. LNCS (LNAI), vol. 4617. Springer, Heidelberg (2007). doi:10.1007/978-3-540-73729-2

21. Karaboga, D., Akay, B.: A comparative study of Artificial Bee Colony algorithm. Appl. Math. Comput. **214**, 108–132 (2009)
22. Nawi, N.M., Rehman, M.Z., Khan, A.: A new Bat Based Back-Propagation (BAT-BP) algorithm. In: Swiątek, J., Grzech, A., Swiątek, P., Tomczak, J.M. (eds.) Advances in Systems Science, pp. 395–404. Springer, Cham (2014)
23. Tang, R., Fong, S., Yang, X.-S., Deb, S.: Wolf search algorithm with ephemeral memory. In: Seventh International Conference on Digital Information Management (ICDIM 2012), pp. 165–172 (2012)
24. Wolberg, W.H., Mangasarian, O.L.: Multisurface method of pattern separation for medical diagnosis applied to breast cytology. Proc. Natl. Acad. Sci. U.S.A. **87**, 9193–9196 (1990)
25. Quinlan, J.R.: Induction of decision trees. Mach. Learn. **1**, 81–106 (1986)
26. Fisher, R.: The use of multiple measurements in taxonomic problems. Ann. Eugen. **7**, 179–188 (1936)
27. Quinlan, J.R.: Simplifying decision trees. Int. J. Man-Mach. Stud. Spec. Issue: Knowl. Acquisition Knowl.-Based Syst. Part 5 **27**(3), 221–234 (1987)

Optimization of ANFIS Using Artificial Bee Colony Algorithm for Classification of Malaysian SMEs

Mohd. Najib Mohd. Salleh(✉), Kashif Hussain, Rashid Naseem, and Jamal Uddin

Faculty of Computer Science and Information Technology, Universiti Tun Hussein Onn Malaysia, Parit Raja, 86400 Batu Pahat, Johor, Malaysia
najib@uthm.edu.my

Abstract. Adaptive Neuro-Fuzzy Inference System (ANFIS) has been widely applied in industry as well as scientific problems. This is due to its ability to approximate every plant with proper number of rules. However, surge in auto-generated rules, as the inputs increase, adds up to complexity and computational cost of the network. Therefore, optimization is required by pruning the weak rules while, at the same time, achieving maximum accuracy. Moreover, it is important to note that over-reducing rules may result in loss of accuracy. Artificial Bee Colony (ABC) is widely applied swarm-based technique for searching optimum solutions as it uses few setting parameters. This research explores the applicability of ABC algorithm to ANFIS optimization. For the practical implementation, classification of Malaysian SMEs is performed. For validation, the performance of ABC is compared with one of the popular optimization techniques Particle Swarm Optimization (PSO) and recently developed Mine Blast Algorithm (MBA). The evaluation metrics include number of rules in the optimized rule-base, accuracy, and number of iterations to converge. Results indicate that ABC needs improvement in exploration strategy in order to avoid trap in local minima. However, the application of any efficient metaheuristic with the modified two-pass ANFIS learning algorithm will provide researchers with an approach to effectively optimize ANFIS when the number of inputs increase significantly.

Keywords: ANFIS · Neuro-fuzzy · Rule optimization · Artificial Bee Colony · Optimization

1 Introduction

ANFIS has been widely applied to solve non-linear problems as compared to other fuzzy systems. The applications range from modeling control systems, expert systems to other complex decision system in a wide variety of areas [1,2].

T. Herawan et al. (eds.), *Recent Advances on Soft Computing and Data Mining*,
Advances in Intelligent Systems and Computing 549, DOI 10.1007/978-3-319-51281-5_3

Besides popularity, ANFIS suffers from exponential growth in the number of rules when inputs increase significantly. This makes ANFIS based models complex and negatively effects the performance [3]. The number of inputs n and membership function (MF) m determine the size of ANFIS rule-base by m^n. This rule-base consists of strong as well as weak rules. Unlike strong rules, the weak ones do not significantly contribute to output. Hence, mere increase in the number of rules will only result in network's complexity. On the other hand, over-reducing rules may boost ANFIS performance in terms of execution time, however, it will diminish accuracy [4–7]. Optimizing ANFIS by maintaining balance between rule-base minimization and accuracy maximization is crucial, as achieving both the objectives simultaneously is a trade-off problem [8].

Many researchers have proposed and applied techniques on dataspace and ANFIS rules to obtain the trade-off balance. Those employed on dataspace are clustering techniques that group input or output data, or conjunct input-output data. Whereas, others put threshold on firing strength of a rule. The later approach eliminates the less contributing or weak rules from rule-base and selects the strong or potential rules [3,9]. An efficient technique for optimizing ANFIS parameters faster along with reducing rule-base to foster maximum accuracy is still to come. For this purpose, this paper modifies the standard two-pass ANFIS learning algorithm. Instead of gradient methods; gradient descent and least square estimation, the proposed approach explores the applicability of Artificial Bee Colony (ABC) algorithm to ANFIS optimization.

The remainder of this paper is organized as follows. In Sect. 2, we briefly describe ANFIS network and its learning process. Section 3 provides an overview of Artificial Bee Colony (ABC) algorithm. Section 4 describes the modified two-pass ANFIS learning algorithm. The application of ABC on ANFIS parameter training and optimizing rule-base is also presented in this section. The experimental results are discussed in Sect. 5. Finally, Sect. 6 duly concludes this research.

2 The ANFIS Concept

ANFIS, developed by Jang [11], is a universal approximator based on adaptive technique. ANFIS architecture comprises of m^n rules, and jth rule can be expressed as:

$$Rule_j : \text{If } x^1 \text{ is } M^1_{j1} \text{ and } x^2 \text{ is } M^2_{j2}, ... \text{ and } x^n \text{ is } M^n_{jn} \text{ then } f \text{ is } C_j$$

where x^n are n input variables, M^n_{jn} are j membership functions (MFs), f is the output of ANFIS network, and C_j is the consequence of the jth rule. The aggregated output of all fuzzy rules can be calculated by:

$$f = \frac{\sum_{j=1}^{m^n} w_j c_j}{\sum_{j=1}^{m^n} w_j} \quad (1)$$

$$\bar{w}_j = \frac{w_j}{w_1 + ... + w_{m^n}} \quad (2)$$

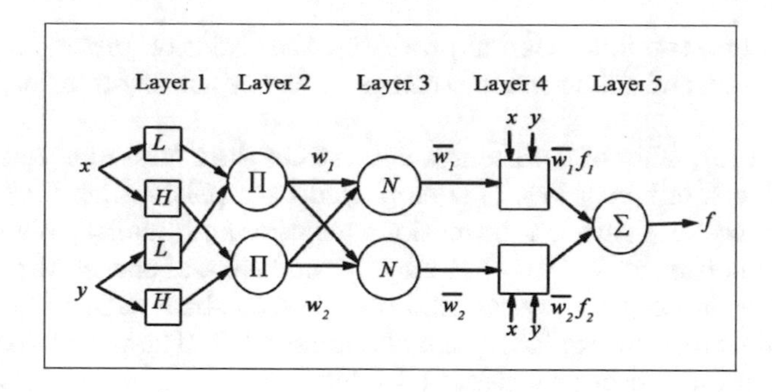

Fig. 1. ANFIS architecture [11]

Table 1. Two-pass ANFIS learning algorithm [11]

	Forward pass	Backward pass
Antecedent parameters	Fixed	GD
Consequent parameters	LSE	Fixed
Signals	Node outputs	Error signals

where w_j and $\bar{w}_j$ are firing strength and normalized firing strength of jth rule, respectively. As shown in Fig. 1, it is a five layer network: Layer 1 computes the MF $M_{ji}^{i}(x^i)$. Layer 2 computes the firing strength w_j of each rule in fuzzy rule-base. Layer 3 normalizes the firing strength of rules using Eq. (2). Layer 4 determines the consequent part of each rule by $\bar{w}_j c_j$. Lastly, Layer 5 aggregates consequents of rules by $\sum \bar{w}_j c_j$. The proposed modification adds two additional layers in the standard five layer ANFIS architecture, see Sect. 4.

ANFIS is optimized by adjusting MF parameters (c, σ) and consequent parameter (p, q, r) for approximating the desired output. The standard two-pass learning algorithm (Table 1) uses a hybrid of gradient descent (GD) and least squares estimator (LSE) for updating MF and consequent parameters, respectively. In forward pass, functional signals go forward till the 4th layer and update the consequent parameters by LSE, keeping premise parameters unchanged. In backward pass, MF parameters are updated using GD, keeping consequent parameters fixed by tracking the error backwards until 1st layer. This paper modifies the standard hybrid ANFIS learning algorithm in order to obtain optimized rule-base with maximum accuracy. Section 4 explains the proposed modification in greater detail.

3 Artificial Bee Colony Algorithm

Artificial Bee Colony (ABC) optimization algorithm, introduced by Karaboga in 2005 [10], is one of the swarm based probabilistic approaches for solving optimization problems. Inspired by the foraging phenomenon of honey bees,

ABC is based on two fundamental processes: the variation process for exploring the search-space, and the selection strategy to ensure the exploitation of previous process [12].

In ABC, the quality of food source determines the fitness of a solution. There are three groups of honey bees: (a) employed bees (EB) which look all around the search space and gather information (location and quality) about the food source, (b) onlooker bees (OB) that stay in the hive and choose the food source based on information given by the EB, and (c) scout bees (SB) randomly search for new food sources for replacing the abandoned EB. The pseudo-code of ABC algorithm is presented in Algorithm 1.1 [10].

As shown in Algorithm 1.1, ABC starts with generating randomly distributed initial population of SN food sources (solutions), where each solution x_i is a D-dimensional vector. Here, D is the number of variables in an optimization problem. After initialization, the positions of food sources of the EBs, OBs, and SBs are repeatedly updated until maximum iterations. In each iteration, EBs utilize Eq. (3) to search new food sources and compute their nectar. The OBs choose food sources based on probability using Eq. (6), and produce new food sources by Eq. (3) in the selected neighborhood. EBs, then, exploit the new neighborhood for better solutions. In the end of each iteration, OBs are sent to check the quality of food source by computing nectar amount. A food source is considered to be exhausted if the number of maximum cycles is reached, or the quality of it cannot be further improved. In this case, the SBs are sent to search for new food sources randomly using Eq. (3).

The following section explains the methodology adopted in this work. In this section, ABC is employed in the modified two-pass ANFIS learning algorithm to train ANFIS parameters for achieving optimized rule-base with maximization accuracy.

4 Methodology

The performance of ANFIS depends on how well its structure learning and parameter tuning is performed. The prior refers to the process of obtaining fuzzy rule-base with which ANFIS can approximate the desired output. Whereas, parameter tuning helps in maximizing accuracy of the model. This paper targets both the objectives of ANFIS optimization. Rule-base minimization is achieved in forwards pass, whereas accuracy maximization is pursued in backward pass of the modified two-pass ANFIS learning algorithm. The Subsect. 4.1 explains the proposed technique in greater detail.

4.1 ANFIS Rule-Base Minimization and Accuracy Maximization Using ABC

The main idea behind minimizing ANFIS rule-base is to consider the potential rules only. Thus, the rules that contribute significantly to the overall decision of ANFIS are potential rules, and are retained in the rule-base. Whereas, the

Algorithm 1.1. Pseudo-code of ABC Algorithm

1: Initialize control parameters and initial population.

$$x_{ij} = x_j^{min} + rand(0,1)(x_j^{max} - x_j^{min}), \text{where} \begin{cases} i = 1, ..., SN \\ j = 1, ..., D \\ SN = \text{number of food sources} \\ D = \text{number of paramters} \\ x_j^{min} = \text{lower limit of search space} \\ x_j^{max} = \text{upper limit of search space} \end{cases} \quad (3)$$

2: Calculate the fitness of population.

$$fitness_i = \begin{cases} \frac{1}{1+f_i}, f \geq 0 \\ \frac{1}{abs(f_i)}, f < 0 \end{cases} \quad (4)$$

3: **while** iteration number is less than maximum iterations **do**
4:     **for** $i = 1 \text{to} SN$ **do**
5:         Explore the neighborhood of food sources by EBs using Eq. (5), and calculate fitness by Eq. (4).

$$v_{ij} = x_{ij} +_{ij} (x_{ij} - x_{kj}), \text{where} \begin{cases} i, k = 1, ..., SN \\ j = 1, ..., D \\ SN = \text{number of food sources} \\ D = \text{number of paramters} \\ _{ij} = [-1, 1] \\ v_{ij} = \begin{cases} x_j^{min}, v_{ij} < x_j^{min} \\ v_{ij}, \\ x_j^{max}, x_j^{max} < v_{ij} \end{cases} \end{cases} \quad (5)$$

6:         Apply greedy selection between v_i and x_i.
7:     **end for**
8:     Select food source for EBs

$$p_i = \frac{fitness_i}{\sum_{j=1}^{SN} fitness_i} \quad (6)$$

9:     **for** $t = 1 \text{to} SN$ **do**
10:         **if** $random < p_i$ **then**
11:             Repeat steps 5 and 6.
12:         **end if**
13:     **end for**
14:     Memorize the best food source found so far
15:     **if** an abandoned source exists **then**
16:         Assign new food source to SB by Eq.(3).
17:     **end if**
18: **end while**

less or non-contributing rules are pruned. Each of the potential rules has output closest (with minimum squared error) to the desired output. On the other hand, the weak rules have error measure greater than the predefined error tolerance (θ). The value for θ is decided carefully, as selecting very small value for error tolerance may cause ignorance of the potential rules; resulting in the loss of accuracy. On the other hand, being generous in tolerance will increase the size of ANFIS rule-base. Therefore, the value of θ should be chosen based on multiple trials. In the modified two-pass ANFIS learning algorithm (Table 2), the reduced rule-base is obtained in forward pass, while the improvement in accuracy is

Table 2. Modified two-pass ANFIS learning algorithm

	Forward pass	Backward pass
Antecedent parameters	Fixed	ABC
Consequent parameters	ABC	Fixed
Rule-base	Reduced number of rules	Fixed
Signals	Rule-base minimization	Accuracy maximization

Algorithm 1.2. Pseudo-code of ANFIS optimization algorithm using ABC

Forward Pass:

1: Initialize ABC parameters and initial population.
2: Update consequent parameters of each rule by ABC.
3: Select rules that meet error tolerance θ specified for rules; Eq. (7).
4: Prune rest of the rules.
5: Calculate average output of selected rules; Eq. (8).
6: Calculate the error measure between actual output derived in step 5 and the desired output using Eq. (8). If the error measure is within the overall error tolerance, then select the rule-set along with rules consequent parameters and goto step 7. Else, goto step 2.

Backward Pass:

7: Tune membership functions parameters via ABC.
8: Calculate average output of the rules selected in step 5.
9: Calculate the error measure between actual output derived in step 8 and the desired output using Eq. (9). If the error measure is within the overall error tolerance, then stop. Else, goto step 7.

made in backward pass. Both the passes use metaheuristic algorithm ABC as an optimizer.

As per modified two-pass ANFIS learning algorithm (see Algorithm 1.2), ABC is first employed in forward pass to update the consequent parameters. Subsequently, the error measure between each rule's output and the desired output is calculated using Squared Error (SE) as Eq. (7).

$$SE = (O^t - O^r)^2 \tag{7}$$

where O^t and O^r are, target output, the output of a rule, respectively. The rules that have $SE > \theta$ are weak rules; hence, are pruned. The remaining rules are considered as strong or potential rules, and are retained as they have $SE \leq \theta$. Once the potentially contributing rules are selected, then the overall output of ANFIS is calculated by taking average O_{avg} of outputs of all the selected rules by Eq. (8).

$$O_{avg} = \frac{\sum_{i=1}^{n} O_i^r}{n} \tag{8}$$

where O_i^r and n are output of ith rule and the number of selected rules, respectively. The process of updating consequent parameters and selecting the

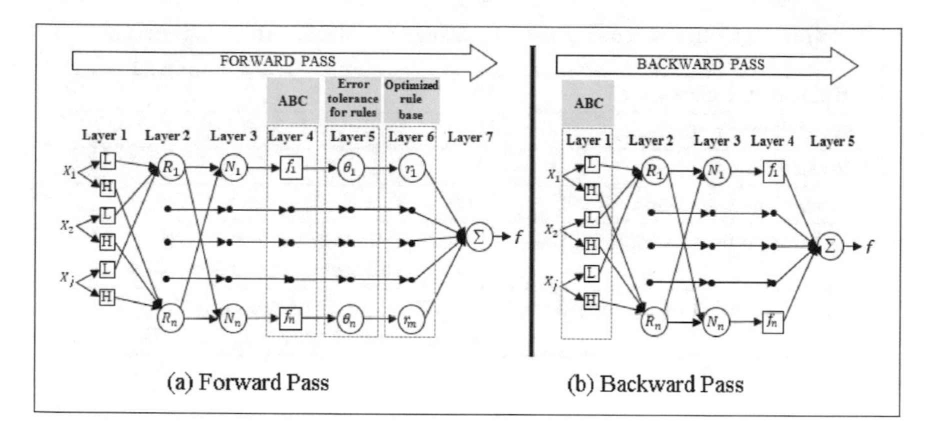

Fig. 2. Modified two-pass ANFIS learning algorithm

potential rules continues till the last iteration of ABC. The rule-set with the best Mean Square Error (MSE), as Eq. (9), is considered as the minimized rule-set, and it is selected for backward pass.

$$MSE = \frac{\sum_{i=1}^{m}(O_{avg} - O_{m}^{t})^2}{m} \tag{9}$$

where O_{avg}, O_m^t, and m are the average of selected rules outputs, target output of mth training pair, and the number of training pairs, respectively. As shown in Fig. 2(a), the forward pass of the modified ANFIS learning algorithm adds two extra layers to the standard ANFIS architecture. Layer 5 represents threshold nodes that filter the potential rules from weak ones by applying Eq. (7). This way, Layer 6 contains only the selected (strong) rules. After updating consequent parameters and obtaining the minimized rule-set, the backward pass (Fig. 2(b)) tunes MFs parameters to pursue accuracy maximization.

5 Simulation Results and Analysis

For practical implementation and computational verification of the proposed methodology, the ANFIS-based classification model was designed to classify Malaysian SMEs based in strength. For performance comparison, one of the popular optimization algorithms Particle Swarm Optimization (PSO) [13], and recently developed Mine Blast Algorithm (MBA) [14] were compared with ABC. The performance validation criteria were number of rules in optimized rule-base, accuracy, and iterations to converge.

SME dataset contained 7 inputs: Business Performance (BP), Financial Capability (FC), Technical Capability (TC), Production Capability (PC), Innovation (IN), Quality System (QS), and Management Capability (MC). The single output, denoted the class of an SME, ranged from 1–5. 1 represented the weakest and 5 was the strongest SME [15]. The designed ANFIS-based classification

Table 3. Common configuration settings for optimization algorithms

Rule error tolerance Θ	0.01
Overall error tolerance	0.005
Swarm size	15
Maximum iterations	100
Total variables in forward pass	Consequent parameters 8, Rules 128, Total parameters $128 \times 8 = 1024$
Total variables in backward pass	MFs 7×2, MF parameters (c,o) 2, Total parameters $7 \times 2 \times 2 = 28$

Table 4. Experimental results

Optimization algorithm	Optimized rule-base	Training MSE	Accuracy %	Iterations	Testing MSE
ABC	36	0.006499	99.67505	100	0.010335
PSO	51	0.006620	99.66900	100	0.007601
MBA	87	0.004910	99.75100	32	0.006515

model used Guassian type MFs. The universe of discourse for each input variable was 1–5, whereas each input was divided into 2 MFs. Initially, auto-generated rules-base contained 2^7=128 rules which were reduced by optimizers. Table 3 contains the common configuration settings for each optimization algorithm.

The simulation results are presented in Table 4, and figuratively illustrated by Fig. 3. The optimization techniques ABC and PSO performed equally deficient

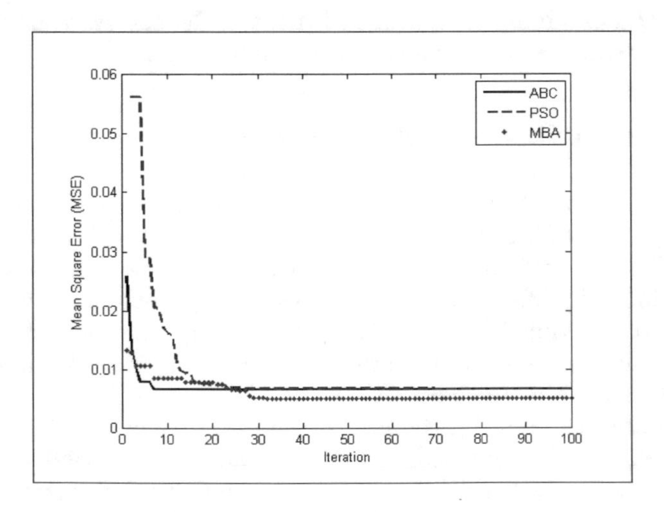

Fig. 3. Learning curve of ABC, PSO, and MBA while optimizing ANFIS

as compared to MBA in terms of selecting potential rules and pursuing accuracy while employed in the modified two-pass ANFIS learning algorithm. In relation to the identification of consequent parameters, ABC could not generate optimum weights. Thus, many of the potential rules were ignored and the ANFIS rule-base was over-reduced. Therefore, the designed ANFIS-based SME classification model suffered from the loss of accuracy. The case of PSO was no exception. ABC and PSO selected only 36 and 51 rules, respectively, out of total 128 rules in the ANFIS rule-base. Additionally, ABC and PSO were also trapped in local minima in their earlier iterations, and could not find other optimal locations or new food sources in the search area until last iteration. This proves that ABC and PSO need to improve their exploration capability. The efficient exploration strategy adopted by MBA helped it produce optimum weights for rules and achieve smallest overall error - MSE.

6 Conclusion

Even though Artificial Neuro-Fuzzy Inference System (ANFIS) has been applied widely in non-linear problems, it suffers from the exponential surge in rules when its inputs increase significantly. Moreover, the standard two-pass ANFIS learning algorithm has the drawback of gradient based learning. The integration of any efficient metaheuristic with the modified two-pass ANFIS learning algorithm can help in maintaining the trade-off balance between rule-base minimization and accuracy maximization. This research explored the applicability of Artificial Bee Colony (ABC) in the modified two-pass ANFIS learning algorithm for selecting optimum rule-set with maximum accuracy. When comparing the performance of ABC with Particle Swarm Optimization (PSO) and Mine Blast Algorithm (MBA), it was noticed that for finding optimum weights for ANFIS parameters, ABC needs efficient exploration strategy for EBs and OBs. To improve ABC in this regards, both the EBs and OBs should be introduced to some variation in exploration strategy instead of following just the same approach. Many have improved and developed variants of ABC; such as [12,16] which can be explored for their efficiency in optimizing ANFIS parameters.

In our future research, we intend to investigate the effectiveness of ABC variants in optimizing ANFIS network for solving classification problems.

Acknowledgments. The authors would like to thank CGS of Universiti Tun Hussein Onn Malaysia (UTHM) for supporting this research under.

References

1. Taylan, O., Karagözoğlu, B.: An adaptive neuro-fuzzy model for prediction of students academic performance. Comput. Ind. Eng. **57**, 732–741 (2009)
2. Kar, S., Das, S., Ghosh, P.K.: Applications of neuro fuzzy systems: a brief review and future outline. Appl. Soft Comput. **57**, 243–259 (2014)
3. Hussain, K., Salleh, M., Najib, M.: Analysis of techniques for anfis rule-base minimization and accuracy maximization. ARPN J. Eng. Appl. Sci. (2015)

4. Liu, P., Leng, W., Fang, W.: Training anfis model with an improved quantum-behaved particle swarm optimization algorithm. Math. Prob. Eng. **2013** (2013)
5. Neshat, M., Adeli, A., Sepidnam, G., Sargolzaei, M.: Predication of concrete mix design using adaptive neural fuzzy inference systems and fuzzy inference systems. Int. J. Adv. Manuf. Technol. **63**, 373–390 (2012)
6. Rini, D.P., Shamsuddin, S.M., Yuhaniz, S.S.: Balanced the trade-offs problem of anfis using particle swarm optimisation. TELKOMNIKA (Telecommun. Comput. Electron. Control) **11**, 611–616 (2013)
7. Gorzalczany, M.B.: Computational intelligence systems and applications: neuro-fuzzy and fuzzy neural synergisms. TELKOMNIKA (Telecommun. Comput. Electron. Control) **86** (2012)
8. Ishibuchi, H., Nojima, Y.: Multiobjective genetic fuzzy systems. In: Mumford, C.L., Jain, L.C. (eds.) Computational Intelligence. ISRL, vol. 1, pp. 131–173. Springer, Heidelberg (2009)
9. Teshnehlab, M., Shoorehdeli, M.A., Sedigh, A.K.: Novel hybrid learning algorithms for tuning ANFIS parameters as an identifier using fuzzy PSO. In: IEEE International Conference on Networking, Sensing and Control, ICNSC 2008, pp. 111–116 (2008)
10. Karaboga, D.: An idea based on honey bee swarm for numerical optimization. Technical report-tr06, Erciyes university, engineering faculty, computer engineering department (2005)
11. Jang, J.-S.R.: ANFIS: adaptive-network-based fuzzy inference system. IEEE Trans. Syst. Man Cybern. **23**, 665–685 (1993)
12. Sharma, H., Bansal, J.C., Arya, K.V., Yang, X.-S.: Lévy flight artificial bee colony algorithm. Int. J. Syst. Sci., 1–19 (2015). ahead-of-print
13. Eberhart, R.C., Kennedy, J.: A new optimizer using particle swarm theory. In: Proceedings of the Sixth International Symposium on Micro Machine and Human Science, pp. 39–43 (1995)
14. Sadollah, A., Bahreininejad, A., Eskandar, H., Hamdi, M.: Mine blast algorithm for optimization of truss structures with discrete variables. Comput. Struct. **102–103**, 49–63 (2012)
15. Malaysia, S.C. SME Competitiveness Rating For Enhancement (SCORE) (2014). http://www.smecorp.gov.my/vn2/node/48
16. Shah, H., Herawan, T., Ghazali, R., Naseem, R., Aziz, M.A., Abawajy, J.H.: An Improved Gbest Guided Artificial Bee Colony (IGGABC) algorithm for classification and prediction tasks. In: Neural Information Processing, pp. 559–569 (2014)

Forecasting of Malaysian Oil Production and Oil Consumption Using Fuzzy Time Series

Riswan Efendi[1,2(✉)] and Mustafa Mat Deris[1]

[1] Faculty of Computer Science, Universiti Tun Hussein Onn Malaysia, 86400 Batu Pahat, Johor, Malaysia
riswan.efendi@uin-suska.ac.id, mmustafa@uthm.edu.my
[2] Mathematics Department, Faculty of Science and Technology, State Islamic University of Sultan Syarif Kasim Riau, Panam, 28294 Pekanbaru, Indonesia

Abstract. Many statistical models have been implemented in the energy sectors, especially in the oil production and oil consumption. However, these models required some assumptions regarding the data size and the normality of data set. These assumptions give impact to the forecasting accuracy. In this paper, the fuzzy time series (FTS) model is suggested to solve both problems, with no assumption be considered. The forecasting accuracy is improved through modification of the interval numbers of data set. The yearly oil production and oil consumption of Malaysia from 1965 to 2012 are examined in evaluating the performance of FTS and regression time series (RTS) models, respectively. The result indicates that FTS model is better than RTS model in terms of the forecasting accuracy.

Keywords: Fuzzy time series · Regression time series · Oil production · Oil consumption · Interval number

1 Introduction

The oil, gas and energy are very essential sectors to force the modern economy in the world, especially in Malaysia [1]. The decision makers and researchers should pay attention seriously to enhance the studies in organizing and managing the oil production and oil consumption, respectively. This is due to the fact that, both components are very determinative sectors for the future of this country. Additionally, the forecasting models and the good planning have to be considered in maintaining both components continuously.

Some of the previous studies have been focused on the forecasting of the oil price and the oil consumption by using the specific models. For example, the forecasting of crude oil prices by using econometric model, neural network (ANN), fuzzy regression (FR), support vector machine (SVM), fuzzy neural network (FNN), Markov Chain, and wavelet analysis [2], energy price and consumption for ASEAN countries [3], statewide fuel consumption forecast models [4], short-term energy outlook [5], peak models oil forecast China's supply, demand by Hubbert, generalized Weng, and HCZ models [6]. Recently, the oil production consumption from the middle east countries already

T. Herawan et al. (eds.), *Recent Advances on Soft Computing and Data Mining*,
Advances in Intelligent Systems and Computing 549, DOI 10.1007/978-3-319-51281-5_4

modelled by using Artificial Bee Colony algorithm (ABC-LH) and Levenberg-Marquardt Neural Network (LMNN) [7]. From these studies, there are some perspectives can be identified such as, the model types, the time intervals, the forecasting accuracy, the non-statistical approaches, and the gap between theoretical models and their implementations in real situation.

In the forecasting area, the accuracy of models are still in issue and very important, both components are very challenging task to achieve by the researchers and academicians in the world, especially to forecast the sensitive data and the dynamic variables. Because, not easy to get the historical data accurately and many factors may influence these data. The conditions above occurred in the oil production and oil consumption data set. On the others hand, the prediction for the future of both data should be provided by government seriously. The studies are still going on to find out the systematic procedure and models in the oil, gas and energy sectors as mentioned the second paragraph.

In this paper, we present the univariate-FTS and multiple-RTS models. For the first model, there are some merits such as, no assumptions should be provided regarding data set, no explanatory variables needed, and it can be used to forecast the linguistic values [8]. This model has been frequently implemented to forecast some real data sectors such as, education [9–14], economics [15–18], energy [19–22], others. In this model, the forecasting accuracy is improved by using modification of interval numbers of data set and also out-sample model for linguistic time series. On the other hands, multiple-RTS model is used to evaluate and verify the performance of FTS model by using the historical data of yearly oil production and consumption of Malaysia from 1965 to 2012. From the second model, the causal relationship between oil production, oil consumption and time (year) may be investigated. Through this paper, we want to show how the mathematical and statistical approaches may be implemented in modelling the real time series data.

The rest of paper is organized as follows: In Sect. 2, the theories of FTS and RTS are described. The proposed method in modification of interval number of data set is presented in Sect. 3. In Sect. 4, the empirical analysis of oil production and consumption are discussed. In the end of this paper, the conclusion is mentioned briefly.

2 The Basic Theories of FTS and RTS

2.1 FTS Theory and Its Procedure

Fuzzy time series (FTS) is an implementation of the fuzzy theory to the time series data which the historical data are the linguistic values. From literature, no the conventional time series methods can be used to forecast this data type [8]. There are some definitions related to FTS as follows:

Definition 1. Fuzzy time series [8]
Let $Y(t)$ ($t = 0, 1, 2,\ldots$), a subset of real numbers, be the universe of discourse on which fuzzy sets $f_i(t)$ ($i = 1, 2,\ldots$) are defined in the universe of discourse. $Y(t)$ and $F(t)$ is a collection of $f_i(t)$ ($i = 1, 2,\ldots$). Then $F(t)$ is called a fuzzy time series defined on $Y(t)$ ($t = 0, 1, 2,\ldots$). Therefore, $F(t)$ can be understood as a linguistics time series variable, where $f_i(t)$ ($i = 1, 2,\ldots$), are possible linguistics values of $F(t)$.

Definition 2. Fuzzy relations [8]
If there exists a fuzzy relationship $R\ (t-1, t)$, such that $F(t) = F(t-1) \circ R\ (t-1, t)$, then $F(t)$ is said to be caused by $F(t-1)$ as denoted as

$$F(t-1) \rightarrow F(t). \quad (1)$$

Definition 3. Fuzzy logical relationship [15]
Let $F(t-1) = A_i$ and $F(t) = A_j$. The relationship between two consecutive data (called a FLR), i.e., $F(t)$ and $F(t-1)$, can be denoted as $A_i \rightarrow A_j$, $i, j = 1, 2, \ldots, p$ is called the LHS, and A_j is the RHS of the FLR.

Definition 4. Fuzzy logical group [15]
Let $A_i \rightarrow A_{j1}$, $A_i \rightarrow A_{j2}, \ldots$, $A_i \rightarrow A_{jn}$ are FLRs with the same LHS which can be grouped into an ordered FLG (called a fuzzy logical group) by putting all their RHS together as on the RHS of the FLG. It can be written as:

$$A_i \rightarrow A_{j1}, A_i \rightarrow A_{j2}, \ldots, A_{i,} \rightarrow A_{jn}\ ; i, j, \ldots, p = 1, 2, \ \ldots, \ n. \quad (2)$$

FTS Procedure

The basic of FTS forecasting algorithm can be derived by using steps as follows [9, 10, 15]:

Step 1: Define the universe of discourse (U) and divide it into several equal length interval.

Step 2: Fuzzify each interval into linguistic time series values (A_i, $i = 1, 2, \ldots, p$, p is the partition number).

Step 3: Establish fuzzy logical relationships (FLRs) among linguistic time series values ($A_i \rightarrow A_j$, $i, j = 1, 2, \ldots, p$).

Step 4: Establish forecasting rule. Actually, there are two rule for forecasting as follows:

Rule 1: IF no relationship occurred among linguistic THEN the final forecast is equal with the midpoint interval value of $A_{i.}$
Rule 2: otherwise, the final forecast is determined by Step 5.

Step 5: Determine the forecast value. Basically, there are three models in determining the final forecast by Song and Chissom [9], Chen [10] and Yu [15] as follows:
Song and Chissom's model:

$$F(t) = F(t-1)\,.\,R(t,\, t-1), \quad (3)$$

where $F(t)$ is the forecasted data of year t represented by fuzzy sets, $F(t-1)$ is the fuzzified of data year t-1, "." is max-min composition operator, R is union of fuzzy relations.
Chen's model:

$$F(t+1) = \text{Average}\left(m_1, m_2,\ \ldots, m_p\right), \quad (4)$$

where m_1, m_2,…, m_p are the midpoint interval values from fuzzy relationships.
Yu's model:

$$F(t) = \mathbf{M}(t) \times \mathbf{W}(t)^T, \tag{5}$$

where $\mathbf{M}(t)$ is the midpoint matrix ($1 \times n$) and $\mathbf{W}(t)$ is the weight matrix ($n \times 1$).

2.2 RTS Theory and Its Procedure

Let we have time series data available on two variables, say y and x, where y_t and x_t are dated contemporaneously. A static model relating y and x is

$$y_t = a + bx_t + u_t,\ t = 1, 2, \ldots, n. \tag{6}$$

Equation (6) is known as a "static model" which comes from the fact that we are modelling a contemporaneous relationship between y and x. Actually, a static model is postulated when a change in x at time t is believed to have an immediate effect on y. Static regression models are also used when we are interested in knowing the tradeoff between y and x. If we want to know the effect of the series of time to y, then it can be written as:

$$y_t = a + bt + u_t,\ t = 1, 2, \ldots, n. \tag{7}$$

Equations (6) and (7) may be combined, if any a time trend in a regression model as written as:

$$y_t = a + bx_t + ct + u_t,\ t = 1, 2, \ldots, n. \tag{8}$$

Equation (8) is called as a regression time series model with a linear trend (time trend). Three equations above may be applied to forecast the real data [23]. The forecasting algorithm can be calculated by following steps as:

Step 1: Identify the visual trend and relationship of x, y to t by scatter plot.
Step 2: Check the correlations among x, y and t.
Step 3: Estimate parameters a, b and c by using ordinary least square method (OLS).
Step 4: Check the validity of parameters a, b and c by t-test and f-test.

3 Proposed Interval Adjustment and Algorithm in FTS

In FTS forecasting, the interval length and partition number of data are very important to be considered, because their contribution in reducing the forecasting error is very significant. In previous studies, many rules and approaches have been discussed

[9, 10, 15], but they are still not standard to be followed. In 2013, Ismail *et al.* [18] proposed a new approach by using inter-quartile range approach, this approach more compatible if we compared with the existing approaches, however this approach has a problem when the ranges between quartiles are too long. In this section, we suggest to modify it as follows:

Let X_t ($t = 1, 2, 3, \ldots, n$) be a time series data and ($X_{min,}$ X_{max}) element of X_t. The effective interval length and partition number can be determined as:

Step 1: Determine the quartiles Q_1 Q_2 and Q_3 from data set, X_t.
Step 2: Divide data sct into four-intervals and determine their frequency respectively.
Step 3: Re-divide each interval from Table 1 by using each frequency respectively.

Thus, the total of interval number is equal with total of frequency also. Mathematically, the comparison of our proposed interval (b) with Ismail's *et al.* (a) can be written as:

$$\text{Total of interval number}_{(a)} < \text{Total of interval number}_{(b)}. \quad (9)$$

Thus,

$$l_a > l_b, \quad (10)$$

$$R_a/k_a > R_b/k_b. \quad (11)$$

By taking $a = b$, thus

$$R_a/(1+3.3\ \log(f_a)) > R_a/f_a, \quad (12)$$

$$1/(1+3.3\ \log(f_a)) > 1/f_a, \quad (13)$$

which l_a, k_a, R_a, F_a are interval length, number of interval, range, and frequency data from Ismail's *et al.* approach, otherwise from our proposed approach. Equation (13) shows that the interval length from proposed approach is smaller than Ismail's approach. The final forecast can be derived by using an algorithm as follows:

Step 1: Follow Steps 1–3 as mentioned in the first paragraph of Sect. 3.
Step 2: Transform the actual data to be linguistic time series values.
Step 3: Forecast the series of linguistic by using statistical time series methods.

Table 1. Quartile, range, and frequency of data set.

Interval	Range between quartiles	Frequency
$[X_{min}, Q_1]$	R_1	f_1
$[Q_2, Q_2]$	R_2	f_2
$[Q_2, Q_3]$	R_3	f_3
$[Q_3, X_{max}]$	R_4	f_4

Step 4: Forecast the numerical time series by using midpoint interval values.
Step 5: Verify the forecasting accuracy by using mean square error (MSE) and compare with regression time series model.

$$MSE = \sum (\text{actual}_t - \text{forecast}_t)^2 / n, \tag{14}$$

where actual_t is actual data at time t, forecast_t is a forecast value at time t, and n is a number of observation/testing data.

4 Empirical Analysis

In this section, the FTS and RTS models are implemented to forecast the yearly oil production and oil consumption of Malaysia, the period 1965 to 2012 which are used as model building. By using the proposed algorithm given in Sect. 3, the forecast values can be calculated as follows:

Step 1: Calculate the quartiles of oil production and oil consumption data sets (Table 2).
Step 2: Divide each data set into four intervals, and determine the ranges and frequencies (Table 3).
Step 3: Re-divide into sub-interval and determine the midpoint interval values (Table 4).
Step 4: Transform the actual data to be linguistic time series values (Table 5).
Step 5: Forecast the series of each linguistic by using statistical models. In this paper, regression time series model is considered, because there is strong linear correlation between oil production, oil consumption and year, respectively.

Table 6 shows the forecasted indexes of oil production and oil consumption are obtained using regression time series model as follows:

$$\text{Index OP} = 0.002 + 0.994(\text{Year}). \tag{15}$$

$$\text{Index OC} = -0.602 + 1.059(\text{Year}). \tag{16}$$

Step 6: Forecast the numerical time series value for each linguistic time series by using the midpoint interval values (Table 7).
Step 7: Verify and compare the forecasting results from FTS and RTS models.

Table 2. The quartiles of oil production and consumption

Data	Q_1	Q_2	Q_3
Oil production (barrels)	171	564.5	703.75
Oil consumption (barrels)	99.25	214.50	519.50

Table 3. The intervals, ranges, and frequencies of oil production and consumption

Oil Production		
Interval	Range between quartiles	Frequency
[1.00, 171.5]	170	12
[171.5, 564.5]	393.5	12
[564.5, 703.7]	139.5	12
[703.7, 776.0]	72.25	12
Oil Consumption		
[46, 99.25]	52.25	12
[99.25, 214.50]	115.25	12
[214.50, 519.50]	61	12
[519.50, 718.00]	39.7	12

Table 4. The sub-interval and the linguistic time series values

Oil Production		
Sub-interval	Midpoint interval value	Linguistic value
[1, 14.16]	7.58	A_1
[14.16, 28.32]	21.24	A_2
[28.32, 42.68]	35.40	A_3
…	…	…
[763.90, 769.90]	766.90	A_{47}
[769.90, 776.00]	772.90	A_{48}
Oil Consumption		
[46, 50.35]	48.175	A_1
[50.35, 54.70]	52.52	A_2
[54.70, 59.05]	56.875	A_3
…	…	…
[701.44, 718.00]	709.72	A_{48}

Table 5. The transformation actual data to be linguistic time series

Year	Actual oil production	Linguistic time series	Year	Actual oil consumption	Linguistic time series
1965	1	A_1	1965	46	A_1
1966	1	A_1	1966	54	A_2
1967	1	A_1	1967	54	A_2
…	…	…	…	…	…
2012	670	A_{34}	2012	283	A_{48}

Table 6. The actual and forecasted indexes of linguistic time series value

Year	Oil production		Oil consumption	
	Actual index	Forecasted index	Actual index	Forecasted index
1965	1	1	1	1
1966	1	2	2	2
1967	1	3	2	3
…	…	…	…	…
2012	34	48	48	49

Table 7. The forecasted real time series for each linguistic

Year	Oil production		Oil consumption	
	Forecasted index	Forecasted real time series	Forecasted index	Forecasted real time series
1965	A_1	8	A_1	48
1966	A_2	21	A_2	53
1967	A_3	35	A_3	57
…	…	…	…	…
2012	A_{48}	773	A_{49}	79

Table 8 shows three different models are used to forecast the yearly oil production and consumption which derived by regression time series model. For the actual data, the oil consumption and oil production models can be written as:

$$(\text{Oil Consumption})_t = 21.962(\text{Year}) - 0.30(\text{Oil Production})_t - 43213.97. \quad (17)$$

$$(\text{Oil Production})_t = 33.299(\text{Year}) - 0.891(\text{Oil Consumption})_t - 65478.573. \quad (18)$$

Table 8. The forecasting results using RTS and FTS models

Year	Oil production			
	Actual	RTS	FTS-1	FTS-2
1965	1	9	0	8
1966	1	28	18	21
…	…	…	…	…
2012	670	897	754	773
Year	Oil consumption			
	Actual	RTS	FTS-1	FTS-2
1965	46	-73	0	48
1966	54	-53	51	53
…	…	…	….	….
2012	712	755	738	721

Table 9. The comparison of forecasting accuracy from three different models

MSE (mean square error)					
	RTS-(OC, T)	RTS-(OP, T)	RTS-(T)	FTS-1	FTS-2
Oil Consumption		34014.199	6855.874	521.5045	232.6588
Oil Production	6295.419		8666.020	2328.861	1881.947
Rank	3	6	4, 5	2	1

In Eqs. (16) and (17), the interval numbers is 48. Additionally, the comparison between these models in term of forecasting accuracy can be shown in Table 9.

Table 9 indicates that MSE from FTS-1 and FTS-2 are smaller than RTS models, but FTS-2 has smaller MSE if compared with FTS-1.

5 Conclusion

In achieving the higher forecasting accuracy, we adjusted the interval length based on Ismail's *et al.* approach. The performance of adjusting interval length can be shown through Eq. (13) and the comparison errors with regression time series and existing FTS models. From MSE of data training and testing, the contribution of proposed interval length is very significant in reducing of the forecasting error. Therefore, the optimization of the partition number is still in issue to be considered in FTS forecasting procedure.

Acknowledgment. The authors are grateful to Research and Innovation Fund, UTHM for their financial support.

References

1. Economic Transformation Programme Chap. 1. http://www.etp.pemandu.gov.my
2. Gabralla, L.A., Abraham, A.: Computational modeling of crude oil price forecasting: a review of two decades of research. Int. J. Comput. Inf. Syst. Ind. Manag. Appl. **5**, 729–740 (2013)
3. Kimura, S.: The 2nd ASEAN Energy Outlook. The Energy Data and Modeling Centre, Japan (2009)
4. Washington State Department of Transportation – Economic Analysis: Statewide Fuel Consumption Forecast Models (2010)
5. US Energy Information Administration: Short-Term Energy Outlook. Independent Statistics & Analysis (2015). Czajkowski, K., Fitzgerald, S., Foster, I., Kesselman, C.: Grid information services for distributed resource sharing. In: 10th IEEE International Symposium on High Performance Distributed Computing, pp. 181–184. IEEE Press, New York (2001)
6. Liangyong, F., Junchen, L., Xiongqi, P., Xu, T., Lin, Z.: Peak oil models forecast China's oil supply, demand. Oil Gas J. 43–47 (2008)

7. Chiroma, H., et al.: An intelligent modeling of oil consumption. In: El-Alfy, El-Sayed M., Thampi, S.M., Takagi, H., Piramuthu, S., Hanne, T. (eds.). AISC, vol. 320, pp. 557–568 Springer, Cham (2015). doi:10.1007/978-3-319-11218-3_50
8. Song, Q., Chissom, B.S.: Fuzzy time series and its models. Fuzzy Sets Syst. **54**, 269–277 (1993)
9. Song, Q., Chissom, B.S.: Forecasting enrollments with fuzzy time series – Part 1. Fuzzy Sets Syst. **54**, 1–9 (1993)
10. Chen, S.M.: Forecasting enrollments based on fuzzy time series. Fuzzy Sets Syst. **81**, 311–319 (1996)
11. Singh, S.R.: A robust method for forecasting based on fuzzy time series. Int. J. Comput. Math. **188**, 472–484 (2007)
12. Kuo, I.: An improved method for forecasting enrollments based on fuzzy time series and particle swarm optimization. Expert Syst. Appl. **36**, 6108–6117 (2009)
13. Ismail, Z., Efendi, R.: Enrollment forecasting based on modified weight fuzzy time series. J. Artif. Intell. **4**, 110–118 (2011)
14. Ismal, Z., Efendi, R., Deris, M.M.: Inter-quartile range approach to length – interval adjustment of enrollment data. Int. J. Comput. Intell. Appl. **3**, 10 p. (2013)
15. Yu, H.K.: Weighted fuzzy time series models for TAIEX forecasting. Phys. A **349**, 609–624 (2005)
16. Yu, H.K., Huarng, K.H.: A bivariate fuzzy time series model to forecast the TAIEX. Expert Syst. Appl. **34**, 2945–2952 (2008)
17. Lee, H.L., Liu, A., Chen, W.S.: Pattern discovery of fuzzy time series for financial prediction. IEEE Trans. Syst. Man Cybern. Part B **18**, 613–625 (2006)
18. Efendi, R., Ismail, Z., Deris, M.M.: Improved weight fuzzy time series used in the exchange rates forecasting US Dollar to Ringgit Malaysia. Int. J. Comput. Intell. Appl. **12**, 19 p. (2013)
19. Bolturuk, E., Oztayzi, B, Sari, I.U.: Electricity consumption forecasting using fuzzy time series. In: IEEE 13th International Symposium on Computer Intelligence and Informatics, 20–22 November 2012, Istanbul, Turkey, pp. 245–249
20. Alpaslan, F., Cagcag, O.: Seasonal fuzzy time series forecasting method based on Gustafson-Kessel fuzzy clustering. J. Soc. Econ. Stat. **2**, 1–13 (2012)
21. Azadeh, A., Saberi, M., Gitiforouz, A.: An integrated simulation-based fuzzy regression-time series algorithm for electricity consumption estimation with non-stationary data. J. Chin. Inst. Eng. **34**, 1047–1066 (2012)
22. Efendi, R., Ismail, Z., Deris, M.M.: New linguistic out-sample approach of fuzzy time series for daily forecasting of Malaysian electricity load demand. Appl. Soft Comput. **28**, 422–430 (2015)
23. Wooldridge, J.M.: Introductory Econometrics A Modern Approach, 3rd edn. Thomson South Western, Mason (2006)

A Fuzzy TOPSIS with Z-Numbers Approach for Evaluation on Accident at the Construction Site

Nurnadiah Zamri(✉), Fadhilah Ahmad, Ahmad Nazari Mohd Rose, and Mokhairi Makhtar

Faculty of Infomatics and Computing, University Sultan Zainal Abidin, Tembila Campus, 22200 Besut, Terengganu, Malaysia
{nadiahzamri, fad, anm, mokhairi}@unisza.edu.my

Abstract. The construction industry has been identified as one of the most risky industries where involves fatalities accidents. Identifying the causes that lead to the accidents implicates a lot of uncertain and imprecise cases. Z-numbers involve more uncertainties than Fuzzy Sets (FSs). They provide us with additional degree of freedom to represent the uncertainty and fuzziness of the real situations. In this paper, we introduce a Fuzzy TOPSIS (FTOPSIS) with Z-numbers to handle uncertainty in the construction problems. Five criteria and six alternatives are used to evaluate the causes of workers' accident at the construction sites. Data in form of linguistic variables were collected from three authorised personnel of three agencies. From the analysis, it shows that the FTOPSIS with Z-numbers provides us with an another useful way to handle Fuzzy Multi-Criteria Decision Making (FMCDM) problems in a more intelligent and flexible manner due to the fact that it uses Z-numbers with FTOPSIS.

Keywords: Z-numbers · FTOPSIS · Accidents at the construction sites

1 Introduction

All while, construction industry is deemed to be an industry with high accident risk [1]. Construction is a hazardous occupation due to the unique nature of activities involved and the repetitiveness of several field behaviors [2]. Within the construction industry, the risk of a fatality is 5 times higher than in manufacturing, whilst the risk of a major injury is 2.5 times higher [3]. Malaysia is one of the develop counties that also faces with the high risk accidents at the construction industry. Abdullah and Wern [4] revealed that Malaysia faces with a seriousness of accidents in construction sector. Construction workers who work within the construction industry face a greater risk of fatality than workers in other industries. Therefore, it is a need to a method that can identify the causes of accidents in the Malaysian construction industry.

Identifying the causes that lead to the accidents implicates a lot of uncertain and imprecise cases. In the real world, uncertainty is a permeating phenomenon that exists in every current state of knowledge that the order or nature of things is unknown or vague. Safety at the construction sites is one of the complex phenomenon and a

T. Herawan et al. (eds.), *Recent Advances on Soft Computing and Data Mining*,
Advances in Intelligent Systems and Computing 549, DOI 10.1007/978-3-319-51281-5_5

subjective area of study that involves uncertainties. It is this challenge that motivates the concepts and ideas throughout this paper.

Decision making could be viewed to include Multi-Criteria Decision Making (MCDM) [5]. MCDM is one of the widely used approaches to deal with a number of uncertainties problems. This approach often requires the experts to provide qualitative and quantitative measurements for determining the performance of each alternative with respect to the criteria and the relative importance of evaluation criteria with respect to overall judgments [6]. One of the related research area of MCDM is Technique for Order Preference by Similarity to Ideal Solution (TOPSIS) [7–9]. The main advantage of the TOPSIS is its stability and ease of use with cardinal information [7–10]. Then, Fuzzy TOPSIS (FTOPSIS) was introduced to handle uncertainty in linguistic judgment. Initial research on FTOPSIS was conducted by Chen [11], who discussed the use of linguistic variables in Fuzzy Set (FS). Previous studies by [12–14] considered MCDM to solve the construction accidents' problems. However, the previous methods lacks with the consideration on reliability of the decision relevant information [15].

Z-numbers proposed by Zadeh [16] relates to the issue of reliability of information. The Z-numbers are an ordered pair of fuzzy numbers $\left(\tilde{A}, \tilde{Z}\right)$, which includes the fuzzy restriction and reliability. The form of Z-numbers are in line with the everyday decisions in the real world. Therefore, various authors used Z-numbers with FTOPSIS. For example, Kang et al., [17] proposed a new MCDM method based on Z-number to deal with linguistic decision making problems. Aliev and Zeinalova [15] developed a Choquet Integral based decision making using Z-information. Besides, Zeinalova [18] contributed Choquet aggregation based decision making under Z-information. If we can use Z-numbers FTOPSIS for handling fuzzy group decision making of construction problems, then there is a room for more flexibility due to the fact that Z-numbers combines with FSs are more suitable to represent uncertainties.

In this paper, we evaluate the Z-numbers with FTOPSIS method to handle accidents' problems at the construction sites. Five criteria and six alternatives are used to evaluate the causes of workers' accident at the construction sites. Data in form of linguistic variables were collected from three authorised personnel of three agencies. Therefore, the Z-numbers with FTOPSIS approach has seen creating a new dimension for construction area studies.

The rest of this paper is organized as follows; Sect. 2 discusses on the basic definition of Z-numbers. In Sect. 3, we starts with the construction of Z-numbers with FTOPSIS. In Sect. 4, an application of accidents at the construction sites is presented to demonstrate the feasibility and applicability of the proposed method. Section 5 concludes.

2 The Steps of Z-Numbers with Fuzzy TOPSIS

A Z-number by Zadeh [15] is an ordered pair of fuzzy numbers denoted as $Z = \left(\tilde{A}, \tilde{R}\right)$. The first component $\tilde{A}$, a fuzzy restriction on the values, is a real-valued uncertain variable X. The second component $\tilde{R}$ is a measure of fuzzy reliability for the first component.

Motivated from the definition on Z-numbers [15], this section presents the six steps of Z-numbers with FTOPSIS to determine the ranking of the outputs. The proposed steps are used to guide experts step-by-step. The overview of the proposed Z-numbers with FTOPSIS steps is shown below:

Step 1: Construct a hierarchical diagram of FMCDM problem.
The hierarchical diagram of FMCDM problem is constructed. Data for criteria and alternatives must be identified as part of a FMCDM problem.

Step 2: Scaling the relative of data.
In FMCDM problems, responses from experts are mainly focused on the opinion of the experts regarding rating of the attributes of the problems based on the identified criteria. The experts were asked to specify rating using seven of the new linguistic scales varying from 'Very Low' to 'Very High' and from 'Not Sure' to 'Very Confident' over the factors associated with FMCDM problems. The new linguistic scale of Z-numbers with FTOPSIS is used to define the experts measurement of each criteria and alternatives of the FMCDM problems. The new linguistic scale is shown in Tables 1 and 2.

Table 1 shows the linguistic terms of fuzzy restrictions and their corresponding Z-numbers with FSs.

Table 1. Linguistic terms of Fuzzy restriction and their corresponding Z-numbers with FSs

Linguistic terms	Fuzzy set with Z-numbers
Very Low (VL)	(0, 0, 1)
Low (L)	(0, 1, 3)
Medium Low (ML)	(1, 3, 5)
Medium (M)	(3, 5, 7)
Medium High (MH)	(5, 7, 9)
High (H)	(7, 9, 10)
Very High (VH)	(9, 10, 10)

Table 2 shows the linguistic terms of fuzzy reliability and their corresponding Z-numbers with FSs.

Table 2. Linguistic terms of Fuzzy reliability and their corresponding Z-numbers with IT2FSs

Linguistic terms	Fuzzy set with Z-numbers
Not Sure (NS)	(0, 0, 1)
Quite Sure (QS)	(1, 3, 5)
Sure (S)	(5, 7, 9)
Very Confident (VC)	(9, 10, 10)

Both Tables 1 and 2 help to construct the decision matrices as follows:

$$Y_p = \left(\tilde{f}_{ij}^p, r\tilde{f}_{ij}^p\right)_{m\times n} = \begin{array}{c} \\ f_1 \\ f_2 \\ \vdots \\ f_m \end{array} \begin{array}{c} \begin{array}{cccc} x_1 & x_2 & \cdots & x_n \end{array} \\ \begin{bmatrix} \left(\tilde{f}_{11}^p, r\tilde{f}_{11}^p\right) & \left(\tilde{f}_{12}^p, r\tilde{f}_{12}^p\right) & \cdots & \left(\tilde{f}_{1n}^p, r\tilde{f}_{1n}^p\right) \\ \left(\tilde{f}_{21}^p, r\tilde{f}_{21}^p\right) & \left(\tilde{f}_{22}^p, r\tilde{f}_{22}^p\right) & \cdots & \left(\tilde{f}_{2n}^p, r\tilde{f}_{2n}^p\right) \\ \vdots & \vdots & \vdots & \vdots \\ \left(\tilde{f}_{m1}^p, r\tilde{f}_{m1}^p\right) & \left(\tilde{f}_{m2}^p, r\tilde{f}_{m2}^p\right) & \cdots & \left(\tilde{f}_{mn}^p, r\tilde{f}_{mn}^p\right) \end{bmatrix} \end{array} \quad (1)$$

$$Y = \left(\tilde{f}_{ij}, r\tilde{f}_{ij}\right)_{m\times n}, \quad (2)$$

where $\left(\tilde{f}_{ij}, r\tilde{f}_{ij}\right) = \left(\frac{\left(\tilde{f}_{ij}^1, r\tilde{f}_{ij}^1\right)\oplus\left(\tilde{f}_{ij}^2, r\tilde{f}_{ij}^2\right)\oplus\ldots\oplus\left(\tilde{f}_{ij}^k, r\tilde{f}_{ij}^k\right)}{k}\right)$, $\left(\tilde{f}_{ij}, r\tilde{f}_{ij}\right)$ is a Z-numbers with FS, $1 \le i \le m,\ 1 \le j \le n,\ 1 \le p \le k$, and k denotes the number of experts.

Step 3: Construct the weighting and weighted decision matrix.
Construct the weighting matrix $\bar{W}_p$ of the criteria of the experts and construct the pth average weighting matrix $\bar{W}$, respectively, shown as follows:

$$\overline{W}_p = \left(\tilde{w}_i^p, r\tilde{w}_i^p\right)_{1\times m} = \begin{array}{c} \begin{array}{cccc} f_1 & f_2 & \cdots & f_n \end{array} \\ \left[\left(\tilde{w}_1^p, r\tilde{w}_1^p\right) \ \left(\tilde{w}_2^p, r\tilde{w}_2^p\right) \ \cdots \ \left(\tilde{w}_m^p, r\tilde{w}_m^p\right)\right] \end{array} \quad (3)$$

$$\bar{W} = (\tilde{w}, r\tilde{w})_{1\times m} \quad (4)$$

where $\left(\tilde{w}_i^p, r\tilde{w}_i^p\right) = \frac{\left(\tilde{w}_i^1, r\tilde{w}_i^1\right)\oplus\left(\tilde{w}_i^2, r\tilde{w}_i^2\right)\oplus\ldots\oplus\left(\tilde{w}_i^k, r\tilde{w}_i^k\right)}{k}$, $\left(\tilde{w}_i^p, r\tilde{w}_i^p\right)$ is a Z-numbers with IT2FS, $1 \le i \le m,\ 1 \le p \le k$ and denotes the number of experts.

Next, the weighted decision matrix with respect to aggregated matrix comparison of each criterion and alternatives is constructed and the importance of the experts is considered as linguistic variable. Thus,

$$\bar{Y}_w = \left(\tilde{v}_{ij}, r\tilde{v}_{ij}\right)_{m\times n} = \begin{array}{c} \\ f_1 \\ f_2 \\ \vdots \\ f_m \end{array} \begin{array}{c} \begin{array}{cccc} x_1 & x_2 & \cdots & x_n \end{array} \\ \begin{bmatrix} (\tilde{v}_{11}, r\tilde{v}_{11}) & (\tilde{v}_{12}, r\tilde{v}_{12}) & \cdots & (\tilde{v}_{1n}, r\tilde{v}_{1n}) \\ (\tilde{v}_{21}, r\tilde{v}_{21}) & (\tilde{v}_{22}, r\tilde{v}_{22}) & \cdots & (\tilde{v}_{2n}, r\tilde{v}_{2n}) \\ \vdots & \vdots & \vdots & \vdots \\ (\tilde{v}_{m1}, r\tilde{v}_{m1}) & (\tilde{v}_{m2}, r\tilde{v}_{m2}) & \cdots & (\tilde{v}_{mn}, r\tilde{v}_{mn}) \end{bmatrix} \end{array}, \quad (5)$$

where $\left(\tilde{v}_{ij}, r\tilde{v}_{ij}\right) = (\tilde{w}_i, r\tilde{w}_i) \otimes \left(\tilde{f}_{ij}, r\tilde{f}_{ij}\right)$, $1 \le i \le m$, and $1 \le j \le n$.

Step 4: Determine the Positive Ideal Solutions (PIS) and Negative Ideal Solutions (NIS).

Determine the PIS and NIS respectively;

$$f^* = \{v_1^*, \ldots, v_n^*\} = \{(\max_j v_{ij} | \, i \in I'), (\min_j v_{ij} | \, i \in I)\}, \quad (6)$$

$$f^{-} = \{v_1^{-}, \ldots, v_n^{-}\} = \{(\max_j v_{ij} \mid i \in I'), (\min_j v_{ij} \mid i \in I'')\}, \quad (7)$$

where I' is associated with the positive attribute, and I'' is associated with the negative attribute.

Step 5: Calculate the distance of PIS and NIS.
The separation of each alternative Z-numbers with IT2FS from the PIS is given as

$$\left(D_j^{*}, rD_j^{*}\right) = \left(\sqrt{\sum_{i=1}^{n} (\tilde{v}_{ij} - \tilde{v}_i^{*})^2}, \sqrt{\sum_{i=1}^{n} (r\tilde{v}_{ij} - \tilde{v}_i^{*})^2}\right), j = 1, \ldots, J. \quad (8)$$

Similarly, the separation of each alternative Z-numbers with IT2FS from the NIS is given as

$$\left(D_j^{-}, rD_j^{-}\right) = \left(\sqrt{\sum_{i=1}^{n} (\tilde{v}_{ij} - \tilde{v}_i^{-})^2}, \sqrt{\sum_{i=1}^{n} (r\tilde{v}_{ij} - r\tilde{v}_i^{-})^2}\right), j = 1, \ldots, J. \quad (9)$$

Step 6: Calculate the relative closeness.
The relative closeness of the alternative x_i with respect to f^* is defined as

$$\left(C_j^{*}, rC_j^{*}\right) = \frac{\left(\frac{D_j^{-}}{D_j^{*} + D_j^{-}}, \frac{rD_j^{-}}{rD_j^{*} + rD_j^{-}}\right)}{2}, j = 1, \ldots, J. \quad (10)$$

A large value of closeness coefficient N_j indicates a good performance of the alternative f_i. The best alternative is the one with the greatest relative closeness to the ideal solution.

3 An Application

In reviewing the literature, data for criteria and alternatives must be identified. This study set out six causes that lead to workers' accidents at the construction sites as the alternatives i.e. attitude (A_1) [19–22], age (A_2) [23, 24], drug abuse (A_3) [25, 26], competency and ability (A_4) [19, 27, 28], psychological (A_5) [29, 30] and experience (A_6) [22, 20, 29]. After considered all the journals and articles above, five alternatives have been decided.

Then, the criteria were selected based on the alternatives. The five criteria that considered in this study were striking or struck by object (C_1), falls (C_2), machine and Equipment error (C_3), loading and unloading error (C_4) and other types of accidents (C_5). This five criteria were selected based on the number of workers involved in accidents. Furthermore, a committee of three experts, D_1, D_2 and D_3 has been identified to seek reliable data over the accidents. Data in form of linguistics variables were collected through interviewing of three authorised personnel from Malaysian Government agencies, contractor and worker. The interview was conducted in three

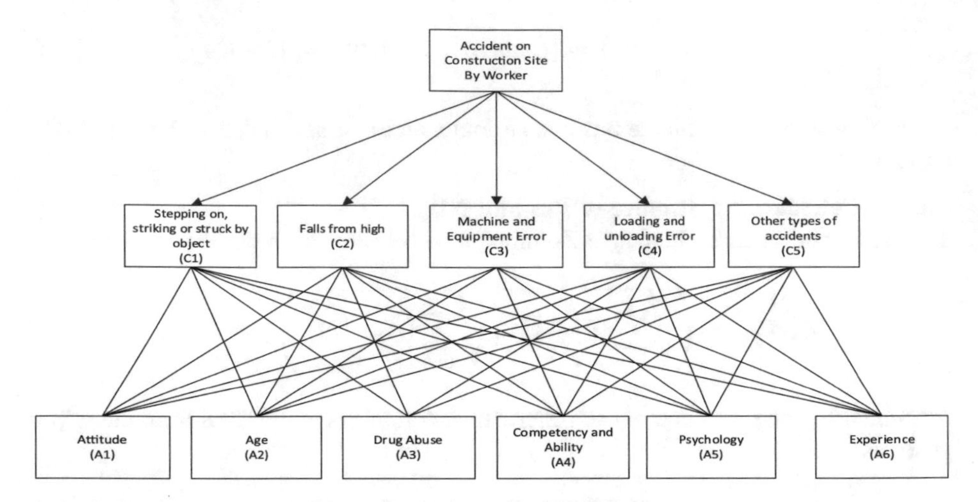

Fig. 1. Hierarchical structure of accident on construction site by workers

separated sessions to elicit the information about causes that regularly involve in accident. The three authorized personnel are chosen based on their experienced in the construction site project cases, for many years. The hierarchical structure of this experiment can be seen in Fig. 1.

Step 1: Construct the hierarchy structure for evaluating the most causes that lead to accident at the construction sites.
The hierarchical structure of evaluating the highest causes lead to construction site accidents by workers is given in Fig. 1, where all the criteria and alternatives are drew horizontally

Step 2: Scaling the relative of data.
The data that considers Z-numbers with FSs and its conversion (see Tables 1 and 2) are referred in order to construct matrix of attributes. Therefore, let's take the example on calculating the $\tilde{\tilde{f}}_{11}$.

$$\begin{aligned}\tilde{f}_{11} = (\mathrm{VH}, \mathrm{VC}) &= ((0.9, 1, 1), (9, 10, 10)) \\ (\mathrm{VH}, \mathrm{S}) &= ((0.9, 1, 1), (5, 7, 9)) \\ (\mathrm{H}, \mathrm{S}) &= ((0.7, 0.9, 1), (5, 7, 9))\end{aligned}$$

Then, the average for $\tilde{f}_{11}$ is ((0.8333, 0.9667, 1), (0.6333, 0.8, 0.9333)).

Use the other similar calculation to the other attributes. Then, the fuzzy decision matrix is shown as Table 3.

Table 3. Fuzzy decision matrix

	A1	A2	A3	A4	A5	A6
C1	((0.8333, 0.9667,1), (0.6333, 0.8, 0.9333))	((0.2667, 0.4333, 0.6333), (0.7667, 0.9, 0.9667))	((0.3333, 0.5, 0.6667), (0.5, 0.6667, 0.8))	((0.2, 0.3333, 0.5), (0.9, 1, 1))	((0.3333, 0.5, 0.6667), (0.6333, 0.8, 0.9333))	((0.1333, 0.2667, 0.4333), (0.6333, 0.8, 0.9333))
C2	((0.5667, 0.7667, 0.9), (0.3667, 0.5667, 0.7667))	((0.3333, 0.5, 0.7), (0.5, 0.7, 0.9))	((0.4, 0.5333, 0.6667), (0.3667, 0.5667, 0.7667))	((0.2, 0.3667, 0.5667), (0.9, 1, 1))	((0.2, 0.3333, 0.5), (0.7667, 0.9, 0.9667))	((0.2667, 0.4333, 0.6333), (0.9, 1, 1))
C3	((0.5667, 0.7667, 0.9333), (0.9, 1, 1))	((0.3667, 0.5667, 0.7667), (0.9, 1, 1))	((0.2333, 0.3333, 0.4667), (0.9, 1, 1))	((0.1333, 0.3, 0.5), (0.6333, 0.8, 0.9333))	((0.2667, 0.4, 0.5667), (0.6333, 0.8, 0.9333))	((0.2667, 0.4, 0.5333), (0.9, 1, 1))
C4	((0.1, 0.3, 0.5), (0.7667, 0.9, 0.9667))	((0.2667, 0.4333, 0.6333), (0.7667, 0.9, 0.9667))	((0.2, 0.3333, 0.5), (0.7667, 0.9, 0.9667))	((0.4, 0.5333, 0.6667), (0.6333, 0.8, 0.9333))	((0.1333, 0.2667, 0.4333), (0.6333, 0.8, 0.9333))	((0.2, 0.3667, 0.5667), (0.5, 0.6667, 0.8))
C5	((0.2667, 0.4333, 0.6333), (0.6333, 0.8, 0.9333))	((0.2667, 0.4, 0.5667), (0.6333, 0.8, 0.9333))	((0.2, 0.3667, 0.5667), (0.6333, 0.8, 0.9333))	((0.1333, 0.2667, 0.4333), (0.7667, 0.6, 0.6667))	((0.2667, 0.4, 0.5667), (0.9, 1, 1))	((0.4, 0.5667, 0.7333), (0.6333, 0.8, 0.9333))

Step 3: Construct the weighting and weighted decision matrix.
Next the weighted of decision matrix is constructed with respect to aggregated matrix comparison of each criterion and alternatives (Table 4).

Table 4. Weighted decision matrix

	Weight
C1	((0.4, 0.5667, 0.7333), (0.9, 1.0, 1.0))
C2	((0.3333, 0.5, 0.6667), (0.6333, 0.8, 0.9333))
C3	((0.0333, 0.1667, 0.3667), (0.7667, 0.9, 0.9667))
C4	((0, 0.1, 0.3) (0.2333, 0.4333, 0.6333))
C5	((0.1, 0.2, 0.3667), (0.7667, 0.9, 0.9667))

Step 4: Determine the PIS and NIS.
Then, the PIS A^+ and the NIS A^- are successfully determined, as follows:

$$f^+ = (1,\ 1,\ 1) \quad f^- = (0,\ 0,\ 0)$$

Step 5: Calculate the distance of PIS and NIS.
The separation measures are calculated using the n-dimensional Euclidean distance, stated as follows (Table 5);

Table 5. The Distance of PIS and NIS

	D^*	D^-
A1	(3.2928, 0.4557)	(0.8167, 2.6293)
A2	(3.4612, 0.3716)	(1.0130, 2.1915)
A3	(3.3962, 0.3658)	(0.7230, 2.8348)
A4	(3.8375, 0.1538)	(0.6744, 2.8500)
A5	(3.6228, 0.2287)	(0.9991, 2.1984)
A6	(3.5863, 0.2381)	(1.034, 2.2066)

Step 6: Calculate the relative closeness.

The results for the relative closeness coefficients is successfully stated in Table 6 as follows,

Table 6. Closeness coefficients of the proposed method

	Closeness coefficients	Final results
A1	(0.1216, 0.7630)	0.4423
A2	(0.0970, 0.6839)	0.3904
A3	(0.0972, 0.7968)	0.4470
A4	(0.0385, 0.8087)	0.4236
A5	(0.0594, 0.6875)	0.3735
A6	(0.0623, 0.6809)	0.3716

As a conclusion, the best alternative selection is A3 and the ranking order of the alternative of selecting the causes lead to accidents at the construction sites is given by $A3 \succ A1 \succ A4 \succ A2 \succ A5 \succ A6$.

4 Comparative Studies

The system output (Sect. 3) can be compared with other existed FMCDM methods. The TOPSIS [11] is used to employ this comparison. Since the proposed model introduced a new equilibrium standardized approach in the evaluation process, it is important to compare it with the existing approach, particularly fuzzy sets.

Thus, the results of the comparative analysis for numerical example of Chen [11] with other different existed FMCDM methods were obtained as given in Table 7.

Table 7. Comparative analysis

Methods	Ranking order according to correlation coefficient
FTOPSIS Chen [11]	$A1 \succ A2 \succ A3 \succ A5 \succ A4 \succ A6$
Proposed method	$A3 \succ A1 \succ A4 \succ A2 \succ A5 \succ A6$

In light of the result from Table 7, it can be concluded that the proposed method is comparable with other methods. It seems that the output from proposed method is a way close to consistency with the output proposed by FTOPSIS Chen [11]. The minor inconsistency occurs, perhaps due to the effect of the Z-numbers in the proposed method. After all, this shows that the FTOPSIS and Z-numbers can plays an important role in the production to enhance the FTOPSIS method. Besides, this proposed ranking values for FTOPSIS and Z-numbers method provides a new perspective in the decision making area especially in Z-numbers area. This decision system also appropriate in solving the complicated decision process appearing in the majority real world applications.

5 Conclusion

In this paper, we have presented a Z-numbers and FTOPSIS method to handle accidents at the construction sites problems. Five criteria with six alternatives were evaluated thoroughly by three experts from Malaysian Agencies. In light of the result, it can be concluded that the proposed method is able to reach the optimum decision based on a group of experts' agreement. Besides, it allows the experts to easily embed their rational decisions of knowledge into the language used to represent solutions. The proposed Z-numbers with FTOPSIS creates a new dimensions in the constructions area. It helps solving certain cases and would certainly make an interesting study in relation to knowledge capture and representation, as a guidance to the experts' own interests. This is just the beginning of Z-numbers with FSs in construction industry, thus for our future work, we aim to apply other case studies such as health studies, infrastructure planning and environmental studies.

References

1. Zhenghui, C., Yaoxing, W.: Explaining the Causes of Construction Accidents and Recommended Solutions. IEEE (2010)
2. Khosravi, Y., Asilian-Mahabadi, H., Hajizadeh, E.: Factors influencing unsafe behaviors and accidents on construction sites: a review. Int. J. Occup. Saf. Ergon. (JOSE) **20**(1), 111–125 (2014)
3. Sawacha, E., Naoum, S., Fong, D.: Factors affecting safety performance on construction sites. Int. J. Proj. Manage. **17**(5), 309–315 (1999)
4. Abdullah, D.N.M.A., Wern, G.C.M.: An analysis of accidents statistics in Malaysian construction sector. In: International Conference on E-business, Management and Economics, vol. 3. IACSIT Press, Hong Kong (2010)
5. Syibrah, N., Hani, H.: A general type-2 fuzzy logic based approach for multi-criteria group decision making. In: IEEE International Conference on Fuzzy Systems, pp. 353–358 (2005)
6. Abdullah, L., Sunadia, J., Imran, T.: A new analytic hierarchy process in multi-attribute group decision making. Int. J. Soft Comput. **4**(5), 208–2014 (2009)
7. Wang, Z.-X., Wang, Y.-Y.: Evaluation of the provincial competitiveness of the Chinese high-tech industry using an improved TOPSIS method. Expert Syst. Appl. **41**(6), 2824–2831 (2014)
8. Lourenzutti, R., Krohling, R.A.: The Hellinger distance in multicriteria decision making: an illustration to the TOPSIS and TODIM methods. Expert Syst. Appl. **41**(9), 4414–4421 (2014)
9. Schneider, E.R.F.A., Krohling, R.A.: A hybrid approach using TOPSIS, Differential Evolution, and Tabu Search to find multiple solutions of constrained non-linear integer optimization problems. Expert Syst. Appl. **62**, 47–56 (2014)
10. Hwang, C.L., Yoon, K.S.: Multiple Attribute Decision Making: Methods and Applications. Springer, Berlin (1981)
11. Chen, C.T.: Extension of the TOPSIS for group decision-making under fuzzy environment. J. Fuzzy Sets Syst. **114**(1), 1–9 (2000)
12. Rezakhani, P.: Fuzzy MCDM model for risk factor selection in construction projects. Eng. J. **16**, 5 (2012)

13. Lu, S.-T., Yu, S.-H., Chang, D.-S.: Using fuzzy multiple criteria decision-making approach for assessing the risk of railway reconstruction project in Taiwan. Sci. World J. **2014**, 1–14 (2014)
14. Zhou, J.-L., Bai, Z.-H., Sun, Z.-Y.: A hybrid approach for safety assessment in high-risk hydropower-construction-project work systems. Saf. Sci. **64**, 163–172 (2013)
15. Aliev, R.A., Zeinalova, L.M.: Decision making under Z-information. Hum.-Centric Decis.-Making Models **502**, 233–252 (2013)
16. Zadeh, L.A.: A note on Z-numbers. Inf. Sci. **181**, 2923–2932 (2011)
17. Kang, B., Wei, D., Li, Y., Deng, Y.: Decision making using Z-numbers under uncertain environment. J. Comput. Inf. Syst. **8**(7), 2807–2814 (2012)
18. Zeinalova, L.M.: Chouquet aggregation based decision making under Z-information. ICTAT J. Soft Comput. Spec. Issue Soft Comput. Syst. Anal. Decis. Control **4**, 819–824 (2014)
19. Aksorn, T., Hadikusumo, B.H.W.: Critical success factors influencing safety program performance in Thai construction projects. Saf. Sci. **46**(4), 709–727 (2008)
20. Kaskutas, V., Dale, A.M., Lipscomb, H., Gaal, J., Fuchs, M., Evanoff, B.A.: Fall prevention among apprentice carpenters. Scand. J. Work Environ. Health **36**(3), 258–265 (2010)
21. Cheng, C.W., Leu, S.S., Lin, C.C., Fan, C.: Characteristic analysis of occupational accidents at small construction enterprises. Saf. Sci. **48**(6), 698–707 (2010)
22. Lai, D.N.C., Liu, M., Ling, F.Y.Y.: A comparative study on adopting human resource practices for safety management on construction projects in the United States and Singapore. Int. J. Proj. Manage. **29**(8), 1018–1032 (2011)
23. Siu, O.L., Phillips, D.R., Leung, T.W.: Age differences in safety attitudes and safety performance in Hong Kong construction workers. J. Saf. Res. **34**(2), 199–205 (2003)
24. López, M.A.C., Ritzel, D.O., Fontaneda, I., Alcantara, O.J.G.: Construction industry accidents in Spain. J. Saf. Res. **39**(5), 497–507 (2008)
25. Hinze, J., Devenport, J.N., Giang, G.: Analysis of construction worker injuries that do not result in lost time. J. Constr. Eng. Manage. **132**(3), 321–326 (2006)
26. Zheng, L., Xiang, H., Song, X., Wang, Z.: Nonfatal unintentional injuries and related factors among male construction workers in central China. Am. J. Ind. Med. **53**(6), 588–595 (2010)
27. Törner, M., Pousette, A.: Safety in construction—a comprehensive description of the characteristics of high safety standards in construction work, from the combined perspective of supervisors and experienced workers. J. Saf. Res. **40**(6), 399–409 (2009)
28. Mohamed, S., Ali, T.H., Tam, W.Y.V.: National culture and safe work behaviour of construction workers in Pakistan. Saf. Sci. **47**(1), 29–35 (2009)
29. Choudhry, R.M., Fang, D.: Why operatives engage in unsafe work behavior: investigating factors on construction sites. Saf. Sci. **46**(4), 566–584 (2008)
30. Abbe, O.O., Harvey, C.M., Ikuma, L.H., Aghazadeh, F.: Modeling the relationship between occupational stressors, psychosocial/physical symptoms and injuries in the construction industry. Int. J. Ind. Ergon. **41**(2), 106–117 (2011)

Formation Control Optimization for Odor Localization

Bambang Tutuko(✉), Siti Nurmaini, Rendyansyah, P.P. Aditya, and Saparudin

Robotic and Control Lab, Computer Engineering Department, Universitas Sriwijaya, Jln. Raya Palembang-Prabumulih KM 32 Indralaya, Ogan Ilir, Indonesia
tutukocn235@gmail.com, sitinurmaini@gmail.com, rendyansyah@gmail.com, aditrecca@gmail.com, saparudinmasyarif@gmail.com

Abstract. This paper presents a swarm robots formation control by using a new hybrid algorithm Fuzzy-Kohonen Networks and Particle Swarm Optimization (FKN-PSO). The FKN-PSO approach is proposed, to overcome the formation control problem due to the loss of a source of odor, caused by the failure in sensor detection, fail in robot motion control and environmental uncertainty in odor localization. The experiments are conducted by using simple swarm robots in the real environment with on-board sensor and processor. The results are compared between FKN-PSO and Fuzzy-PSO to look at the performance of the swarm robots in the process of odor localization. As the results found that the propose algorithm produce fast response and efficiently process than Fuzzy-PSO, they are able to locate the source of odor in a short time and capable for keeping in formation to find the target.

Keywords: Pattern recognition · Swarm robots · Odor localization · Optimization

1 Introduction

Formation control of swarm robots systems performing a coordinated task has become a challenging research field. It is considered as one of the steps of more complex distributed tasks in the robotics application [1–3]. The formation control problem is defined as finding a control algorithm ensuring that swarm robots can uphold a specific formations while traversing a trajectory, avoiding collisions simultaneously and achieving the target. without a designated leader with limited communication and simple algorithm. Rather than equipping an individual robot with a control mechanism that enables it to solve a complex task on its own. It has been studied extensively in the literature with the hope that through efficient coordination many inexpensive and simple, can achieve better performance than a single powerful monolithic robot.

The swarm robots formation control approach can be roughly categorized as leader-follower, behavioral, and virtual structures [4–6]. In the leader-follower approach, one of the robot is designated as the leader, and the other as followers.

T. Herawan et al. (eds.), *Recent Advances on Soft Computing and Data Mining*,
Advances in Intelligent Systems and Computing 549, DOI 10.1007/978-3-319-51281-5_6

Some of techniques are discussed in the previous article [4], including leader tracking, nearest neighbor tracking, barycenter tracking, and other tree topologies. In the behavioral approach where control strategies are derived by averaging several competing behaviors including goal seeking, collision avoidance, and formation keeping. Since competing behaviors are averaged, occasionally strange and unpredicted behaviors may occur [5]. In the virtual structure approach, the entire formation is treated as a single structure and the control is derived in three steps: (i) the desired dynamics of the virtual structure are defined, (ii) the motion of the virtual structure is translated into the desired motion for each agent, and (iii) tracking controls for each robot are derived [6].

Swarm robots usually controlled interactions among the robots by a simple algorithm, as well as between the robots and the environment. In unknown environment, it needs an approach that can deal with uncertain situation [7, 8]. It is very complicated and highly nonlinear, due to different types of sensors have different measurement characteristics [9]. Especially in odor localization, it has received much attention due to its practical significance for human security for searching for the sources of toxic gas leaks [10]. This system entities that possess the ability of robots to move within their environment, to interact with other robots, to perceive the information of the environment and to process this information for localizing the odor source.

The environmental mapping for odor localization is a representation of the physical environments through the swarm robots sensory data into spatial models. It means for finding the absolute or rational location of robot in the spatial models generated which must be clear of obstacles and attend all the optimizations criteria [11]. The problems that addresses mapping and localization has been referred to as conventional approach with simultaneous localization and mapping (SLAM) or concurrent mapping and localization (CML) [12–14]. However, the robot lacks a global positioning sensor and the algorithm produce the hard computational resources. Due to the onboard sensing and processing, the control strategy in swarm robots formation must be simple algorithm with less computational ability. Therefore, when designing swarm robotics system, complex control algorithms is avoided, and instead principles such as locality of sensing and communication, homogeneity and decentralized, are followed [15]. Thus simple control strategy with limited processing speed and memory space is desirable.

Moreover, odor localization studies, only priority odor detection behavior but does not consider the control of the swarm robots movement. Not many researchers conduct the experiments about the loss of a source of odor, caused by the failure in sensor detection, worst in robot motion control and environmental uncertainty. Therefore, the time of travelled for finding the source of odor, will be slow. Due to, the parameters used only affecting change in the direction of swarm robots movement as a result of changes in the distribution of odor [16]. Thus, the loss of odor detection by the robot can be affect the overall performance.

In this research, such problems will be discussed as an important parameter in the control design. Improvisation performed by combining formation control with heuristic approach and route optimization by using Fuzzy-Kohonen Networks (FKN) and Particle Swarm Optimization (PSO) approach. Meanwhile, to optimize the swarm robots route, FKN approach will be combine with PSO algorithm, that includes the possibility of sensor uncertainty in it.

2 Odor Localization Design

Generally, odor localization tasks using a group of agents can be divided into 3 subtasks [17, 18]: (1) *Plume Finding,* the robots do not know or have no contact with the plume and try to contact it; (2) *Plume Transversal,* the robots are already know the plume and try to maintain the connection; (3) *Source Declaration,* the robots declare the location of the source that has been found. Some researchers have been successful in making the experiments using multiple robots [19, 20] and using swarm robots [21–24]. However, the successful of swarm robots in localizing the odor has not arrived to the best one. Moreover, the experiments mostly is conducted in simulation.

From Fig. 1, there are five phases separate in the process localization, such as searching the odor source, follow the flume along concentration gradient, tracking the odor plume, searching for flume across the wind and finally, declaration the odor source. In this paper only four step is employed, because the influence of the wind in terms of turbulence and instability wind in swarm robots environment not discuss in this paper. Moreover the focus of this research is plume finding. It was done in a very simplest way, the robot did not move to search the target, until the sensor detect the gas concentration. If one robot senses the gas density which is beyond a certain threshold value, it means that the gas source location is specified; and hence, the searching behavior is terminated. In the other word, the target search is terminated if with the maximum iteration time step is reached and the process localization is fail.

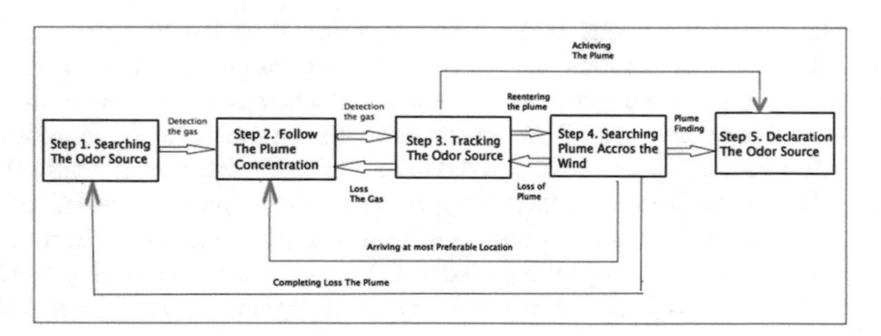

Fig. 1. Odor localization process

2.1 Odor Localization Process

Solving odor localization problems in dynamic environments requires hardware and software platforms. The essential problem is that the available odor sensors lack the combination of speed and sensitivity necessary to perceive the complex and dynamic structure of odor plume. The approach of moving slowly and continually sampling odor and flow data to reduce environmental noise is used in nature (starfish) and has been applied to robotic systems [23] but there are environmental constraints, significant plume sparseness or meander with time critical performance. It can be rendered these systems ineffective [24, 25]. Therefore to enhance the odor sensor performance, the researchers of odor localization systems must be able to focus on algorithm design to achieve better result.

If a single robot wants to find an odor source, it should wander around to sample a series of odor concentration to find the densest place. This sample process is a temporal process. If a group of robots dispersed within a broad range sample the odor simultaneously and they can share their information of concentration rapidly, the sample process becomes a spatial one. The key point of odor localization process is information exchange among robots and the control system design as shown in Fig. 2.

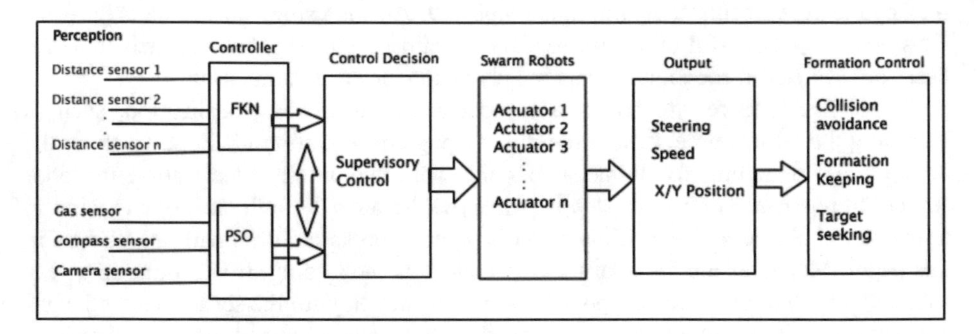

Fig. 2. Control system design process

2.2 The Localization Process-Based FKN-PSO Algorithm

Many intelligent autonomous systems for exploring their environment require to estimate the positions of surrounding objects as precisely as possible. Due to real-world optimization problems are dynamic, and change stochastically over time. In recent year, several researchers used a PSO that can be applied in managing the cooperation of multi robots and it was produced successful in localizing several odor sources [21–25]. However, it was analyzed that there were some problems occurred when trying to localize multiple odors: (1) Having no cooperation strategy would give a chance for the robots to search the same region repeatedly; (2) The efficiency of tracing would be decreased due to the collision of robots in the same region; (3) Traditional methods have difficulties in searching for more than one odor source in each run.

In this paper, heuristic method of reasoning is adopted instead of analytical solution. The strategy is using fuzzy rules has major benefits compared to the mathematical model. Basic principles of our algorithm are described in [25] in detail, however the FKN algorithm was developed in our previous research in single robot. In the study presented here, we investigated the resulting emergent properties of the collective behaviors of our proposed control algorithm. FKN approach is used for environmental recognition and to optimize the performance in odor localization process, the FKN approach is combined with particle swarm optimization (PSO). The approach has been proposed in simple environment [26]. The PSO concept will be applied to follow the cues determined by the sensed gas distribution. The proposed algorithm is expected to make swarm robots moving safer, easier, and efficient.

This algorithm can switch each other in some environmental conditions, such as (1) there is an obstacle, though the one of odor sensor is active; (2). the odor sensor is loss the concentration or the odor concentration become low; (3) no obstacle detect in

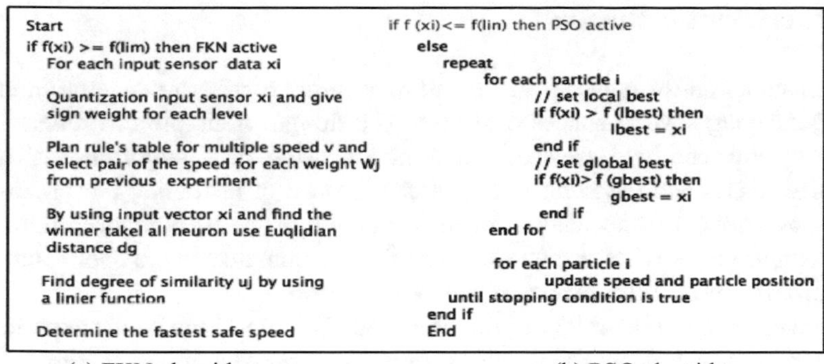

(a) FKN algorithm (b) PSO algorithm

Fig. 3. FKN-PSO algorithm

the environment. In general the FKN-PSO strategy for odor localization can be shown in Fig. 3.

From the Fig. 3, initially all the robots will navigate on the environment with seven obstacles configuration as a initially pattern as shown in Fig. 4. When one robot detects the presence of a source of odor, then the PSO algorithm will be activated. The robot will make the communication among the robots, to share the information and coordinate the other robot to move towards the odor sources. The robot will tracking the concentration of odor value to the maximum threshold and then will stop moving when in the coverage area that has been determined as the source of the odor. The speed control changes only is associate with seven conditions, and PSO algorithm is activated especially in pattern 7. The strategy produce simple algorithm, because the patterns are recognized by the robot using the rule base tables with supervised learning strategies.

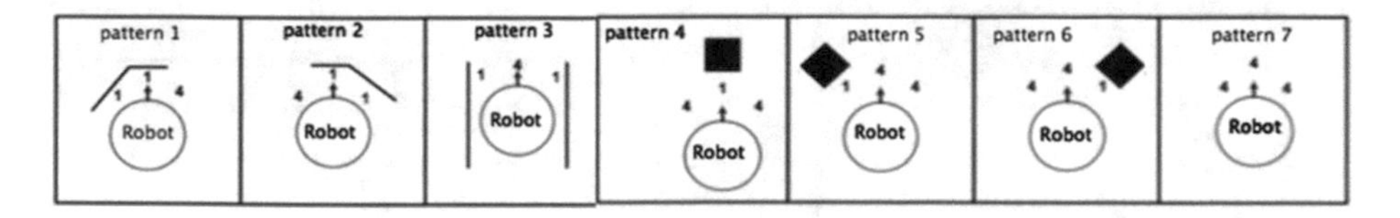

Fig. 4. Environmental patterns

From Fig. 4, it will be created the rule base table, to connect between the pattern and speed of the motor. In our previous research, the FKN-based supervised learning, recognize the current environmental pattern and compare with a rule base table, but only in one robot [25]. It will produce the patterns formed and the amount of input to the motor control.

3 Experimental Results

The formation control is designed that allow swarms robots to converge in environments containing several shapes obstacles. The design of simple controllers for formation control and decentralized coordination with FKN algorithm is used to synthesize shapes and environmental patterns, instead of restricting our environment. Three simple robots competencies for the execution of the FKN algorithm including, (i) to recognize the environmental situation; (ii) to communicate the odor source to the whole swarm; and (iii) to localize it self or the group.

To investigate the feasibility of the proposed FKN algorithm, some experiments is conducted by using three small robots with three gas sensors in it. From the real experiments produce swarm robots have the ability to synthesize several obstacle pattern shapes in these environments as show in Fig. 5, by using robot 1 (R1), robot 2 (R2) and robot 3 (R3) respectivelly. When they explore the environment to do the localization, they explicit coordination allowed robots to successfully navigate and move to the odor source. From experimental result in Fig. 5, it can be seen that by using FKN algorithm, 3 robots (named R1, R2 and R3) with sample of environment have the ability to recognize each environmental patterns during the localization process. Particularly on the pattern 7, these conditions make PSO algorithm becomes active, because without obstacles all robots try to search the odor concentration. Therefore the robot move to find the target quickly. The patterns output from R1, R2 and R3 are different each other relation to the control action when they move in the group formation. The results show that R1 has produce 3 patterns, R2 has produce 6 patterns and R3 has produce 4 patterns. It mean's, the swarm robots localization process successfully performed in sample environment and the odor source can be find in a short time without collision anything.

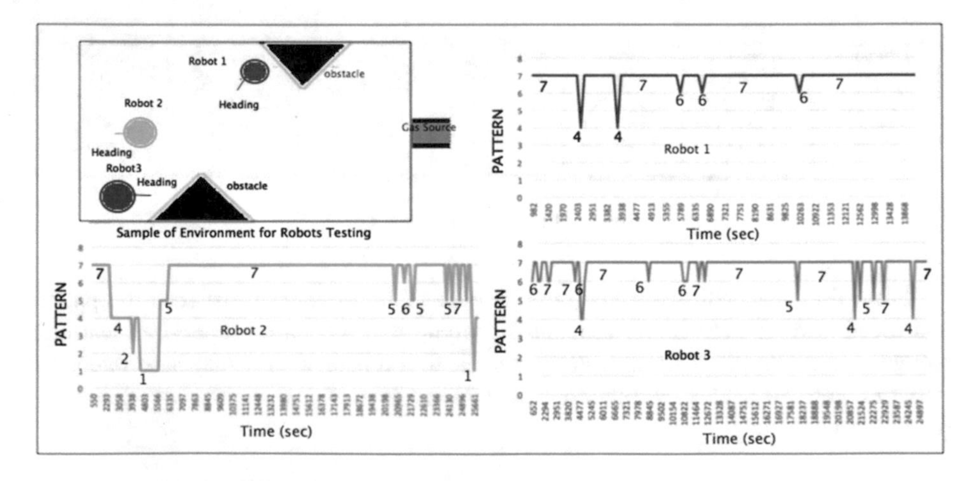

Fig. 5. Environmental recognition-based FKN algorithm

To know the response of the gas sensor, they tested by using chemical formula butane and octane, as shown in Fig. 6. While, Fig. 7 describes the sensor response when it move on mobile robot. The sensor setup minimum value at 40 decimal and maximum value at 80 decimal and the response of odor concentration dependent on the distance between the sensor and the source of odor. The odor sensor is tested by using single robot to see the sensor response when the it moves in circle area. It produce stable data and good response.

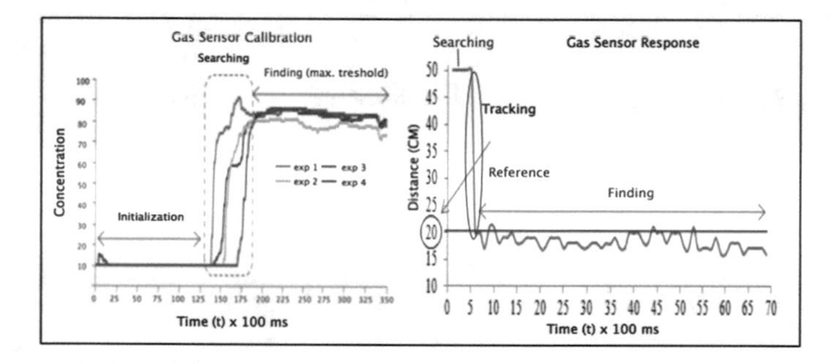

Fig. 6. Gas sensor and single robot response with octane gas source

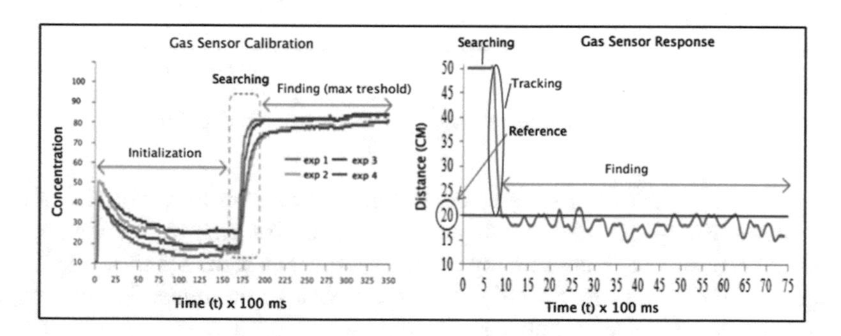

Fig. 7. Gas sensor and single robot response with butane gas source

From Figs. 8(a)–(b) and 9(a)–(b), the odor localization process through a several stages, the initial position of the robots that spread out and random direction from the source of artificial odor. Than, robots moving in divergent for searching the source of the odor and successfully tracking the source of the gas leak. But to avoid obstacles that exist in the environment still priority task. By using FKN-PSO algorithm, the swarm robots more faster to find the odor source, due to the environmental situation have been recognized by the robots. Thus the swarm robots directly execute the target. The odor detection by sensor at robot 1 about 3 s, and time travelled to find the target about 14 s. The other robots 2 and robot s follow the target 24 s and 26 s respectivelly, the data can be show in Fig. 8(a). However by using Fuzzy-PSO, the odor is detected by Robot 1 about 14 s (Fig. 8(b)), it's very long time in process searching and but faster when to track the odor source about 20 s. Because the robots must be access every rules in the

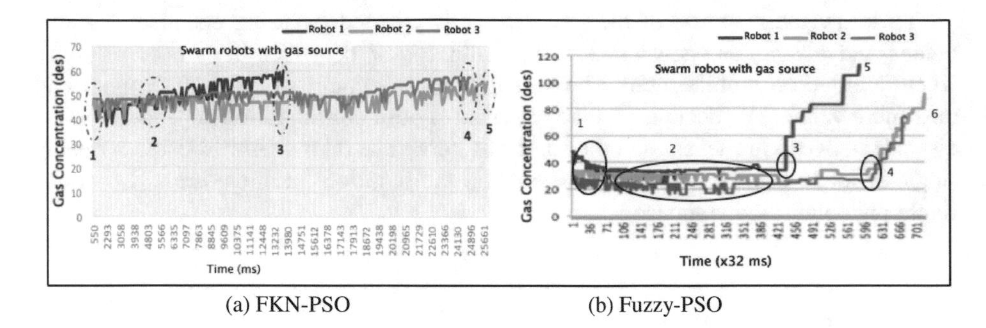

(a) FKN-PSO (b) Fuzzy-PSO

Fig. 8. Comparison of odor localization process in experiment 1

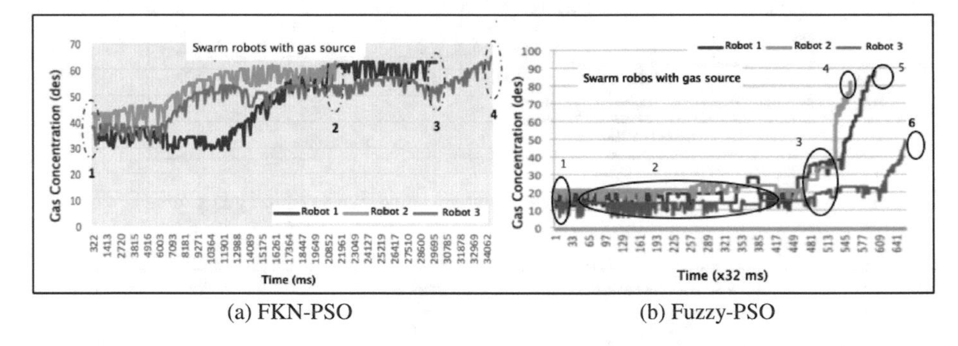

(a) FKN-PSO (b) Fuzzy-PSO

Fig. 9. Comparison of odor localization process in experiment 2

fuzzy process. The number of rules utilize in Fuzzy-PSO algorithm about 27 rules, but only 7 rules in FKN-PSO. Hence, this process increase the time to find the target and computational resources. The summary of the experimental results is shown in Table 1.

Figure 9(a), describes that FKN-PSO can't recognize the dynamic obstacle, thus when they move the other robots can become obstacle. Therefore in complex environment the FKN-PSO produce good performance, however the time travelled to find the odor source about 20 s, unfortunately Fuzzy-PSO only 19 s (Fig. 9(b)). Due to, in complex environment, the swarm robots hard to recognize the environmental pattern with 'local optima' condition. In the future this problem will be include in the rules table and selected as pattern prototype. In the searching condition FKN-PSO produce faster time compare to Fuzzy-PSO.

From Table 1, total time for localization almost the same between FKN-PSO and Fuzzy-PSO approach. However, by using Fuzzy-PSO, the swarm robots very slow to search from the first detection the concentration of odor in the environment. If the concentration odor is detected by one robot, then they tracking very fast utilize FKN-PSO approach compare to FKN-PSO approach for finding the source of odor. In the future, the wind instability and turbulence become two important parameters to determined the performance. The FKN-PSO with 7 environmental pattern generate a satisfactory performance to recognize the environment, to avoid the obstacle and

Table 1. Swarm robots performance in odor localization process

Description	Fuzzy-PSO	FKN-PSO
Time to search the odor/first detection	1.0–14.0 s (env. 1)	2.0–3.0 s (env. 1)
	2.0–17.0 s (env. 2)	0.2 s – 0.5 s (env. 2)
Time to tracking the odor source	14.0–20.0 s (env. 1)	3.0–14.0 s (env. 1)
	17.0–18.5 s (env. 2)	3.0–16.0 s (env. 2)
Time to find the odor source	19.0 s (env 1.)	14.0 s (env. 1)
	19.0 s (env 2.)	20.0 s (env. 2)
Rule base of algorithm	27.0 rules	7.0 rules
Resources of algorithm	60.0 kbytes	43.0 kbytes
Time to process localization for three robots	65.0 s	63.0 s

always move in the group, and to find the odor source, with simple algorithm. The formation control of swarm robots based pattern recognition approach become a promising technique to develop further.

4 Conclusion and Future Work

Formation control for swarm robots with optimization approach have been presented in this paper, for keeping the swarm robots move in a group when they localize and find the odor source. It process based on real-time sensory information and successfully implemented on three simple mobile robots. Hence, optimization of the prototype-environmental mapping become important role in the propose algorithm. By employing only 7 rules base for making the control decision, satisfactory performance has been achieved. In this research Fuzzy-PSO become a benchmark for comparing the swarm robots results. By using FKN-PSO approach the computational resource is reduced and this enhances process localization performance. Many aspects of this method are worth further investigation in the future. Although the present design can cope with moving swarm robots as well as stationary obstacles, more accurate perception sensors are required for more complex environmental configurations.

Acknowledgments. The Authors thank to The Ministry of Technology Research and Higher Education (Kemenristek-Dikti), Indonesia and Sriwijaya University (UNSRI) for their financial support in Competitive Grants Project.

References

1. Brambilla, M., Ferrante, E., Birattari, M., Dorigo, M.: Swarm robotics: a review from the swarm engineering perspective. Swarm Intell. **7**(1), 1–41 (2013)
2. Xu, D., Zhang, X., Zhu, Z., Chen, C., Yang, P.: Behavior-based formation control of swarm robots. Math. Probl. Eng. (2014)
3. Meng, Y., Guo, H., Jin, Y.: A morphogenetic approach to flexible and robust shape formation for swarm robotic systems. Robot. Autonom. Syst. **61**(1), 25–38 (2013)

4. Yamaguchi, H., Burdick, J.W.: Asymptotic stabilization of multiple nonholonomic mobile robots forming group formations. In: Proceedings of IEEE International Conference on Robotics Automation, Leuven, Belgium, pp. 3573–3580, May 1998
5. Balch, T., Arkin, R.C.: Behavior-based formation control for multi- robot teams. IEEE Trans. Robot. Automat. **14**, 926–939 (1998)
6. Lewis, M.A., Tan, K.H.: High precision formation control of mobile robots using virtual structures. Auton. Robot. **4**, 387–403 (1997)
7. Senanayake, M., Senthooran, I., Barca, J.C., Chung, H., Kamruzzaman, J., Murshed, M.: Search and tracking algorithms for swarms of robots: A survey. Robot. Auton. Syst. **75**, 422–434 (2016)
8. Hayes, A.T., Martinoli, A., Goodman, R.M.: Swarm robotic odor localization. In: Proceedings of IEEE/RSJ International Conference on Intelligent Robots and Systems, vol. 2, pp. 1073–1078 (2001)
9. Martinez, S., Cortes, J., Bullo, F.: Motion coordination with distributed information. IEEE Control Syst. Mag. **27**(4), 75–88 (2007)
10. Lu, Q., Luo, P.: A learning particle swarm optimization algorithm for odor source localization. Int. J. Autom. Comput. **8**(3), 371–380 (2011)
11. Mohan, Y., Ponnambalam, S.G.: An extensive review of research in swarm robotics. In: IEEE World Congress on Nature & Biologically Inspired Computing, NaBIC, pp. 140–145 (2009)
12. Bailey, T., Durrant-Whyte, H.: Simultaneous localization and mapping (SLAM): Part II. IEEE Robot. Autom. Mag. **13**(3), 108–117 (2006)
13. Mondada, F., Gambardella, L.M., Floreano, D., Nolfi, S., Deneuborg, J.L., Dorigo, M.: The cooperation of swarm-bots: physical interactions in collective robotics. IEEE Robot. Autom. Mag. **12**(2), 21–28 (2005)
14. Martinez, S., Cortes, J., Bullo, F.: Motion coordination with distributed information. IEEE Control Syst. Mag. **27**(4), 75–88 (2007)
15. Iyer, A., Rayas, L., Bennett, A.: Formation control for cooperative localization of MAV swarms. In: Proceedings of the International Conference on Autonomous Agents and Multi-Agent Systems, pp. 1371–1372 (2013)
16. Marques, L., Nunes, U., de Almeida, A.T.: Particle swarm-based olfactory guided search. Auton. Robots **20**(3), 277–287 (2006)
17. Lochmatter, T., Martinoli, A.: Understanding the potential impact of multiple robots in odor source localization. In: Asama, H., Kurokawa, H., Ota, J., Sekiyama, K. (eds.) Distributed Autonomous Robotic Systems 8, pp. 239–250. Springer, Heidelberg (2009)
18. Lu, Q., He, Y., Wang, J.: Localization of unknown odor source based on Shannon's entropy using multiple mobile robots. IEEE (2014)
19. Jiu, H.F., Li, J.L., Pang, S., Han, B.: Odor plume localization with a pioneer 3 mobile robot in an indoor airflow environment. IEEE (2014)
20. Nurmaini, S., Zaiton, S., Zarkasih, A., Tutuko, B., Triadi, A.: Intelligent mobile olfaction of swarm robots. Int. J. Robot. Autom. (IJRA) **2**(4), 189–198 (2013)
21. Marjovi, A., Nunes, J., Souse, P., Faria, R., Marques, L.: An olfactory-based robot swarm navigation method. In: IEEE Interntional Conference on Robotics and Automation, pp. 4958–4963 (2010)
22. Dadgar, M., Jafari, S., Hamzeh, A.: A PSO-based multi-robot cooperation method for target searching in unknown environments. Neurocomputing **177**, 62–74 (2015)
23. Men, M.C., Chen, L.W.: An approach for active odor source localization based on particle swarm optimization. Appl. Mech. Mater. **738**, 493–498 (2015)
24. Hayes, A.T., Martinoli, A., Goodman, R.M.: Swarm robotic odor localization: off-line optimization and validation with real robots. Robotica **21**(04), 427–441 (2003)

25. Nurmaini, S.: Memory-based reasoning algorithm based on Fuzzy-Kohonen self organizing map for embedded mobile robot navigation. Int. J. Control Autom. **5**(3), 47–63 (2012)
26. Nurmaini, S., Saparudin, Tutuko, B., Aditya, P.P.: Pattern recognition approach for swarm robots reactive control with fuzzy-kohonen networks and particle swarm optimization algorithm. In: 3rd International Conference on Communication and Computer Engineering. LNEE. Springer (2016)

A New Search Direction for Broyden's Family Method in Solving Unconstrained Optimization Problems

Mohd Asrul Hery Ibrahim[1(✉)], Zailani Abdullah[1],
Mohd Ashlyzan Razik[1], and Tutut Herawan[2]

[1] Faculty of Entrepreneurships and Business, Universiti Malaysia Kelantan, Kota Bharu, Malaysia
{hery.i,ashlyzan}@umk.edu.my, zailania@umt.edu.my
[2] Faculty of Science Computer and Information Technology, Universiti Malaya, Kuala Lumpur, Malaysia
tutut@um.edu.my

Abstract. The conjugate gradient method plays an important role in solving large scale problems and the quasi-Newton method is known as the most efficient method in solving unconstrained optimization problems. Hence, in this paper, we proposed a new hybrid method between conjugate gradient method and quasi-Newton method known as the CG-Broyden method. Then, the new hybrid method is compared with the quasi-Newton methods in terms of the number of iterations and CPU-time using Matlabin Windows 10 which has 4 GB RAM and running using an Intel ® Core ™ i5. Furthermore, the performance profile graphic is used to show the effectiveness of the new hybrid method.. Our numerical analysis provides strong evidence that our CG-Broyden method is more efficient than the ordinary Broyden method Besides, we also prove that the new algorithm is globally convergent.

Keywords: Broydenmethod · CG-Broyden method · CPU time · Conjugate gradient method

1 Introduction

Quasi-Newton methods are well-known methods in solving the unconstrained optimization method which uses the updating formulas for approximation of the Hessian. These methods were introduced by Davidon in1959, and later popularised by Fletcher and Powell in 1963, but the Davidon-Fletcher-Davidon (DFP) method is rarely used nowadays. However, in 1970 Broyden, Fletcher, Goldfarb and Shanno developed the idea of a new updating formula, known as BFGS, which has become widely used and recently the subject of many modifications. Then, Broyden [1] proposed a family of quasi-Newton methods in 1970.

T. Herawan et al. (eds.), *Recent Advances on Soft Computing and Data Mining*,
Advances in Intelligent Systems and Computing 549, DOI 10.1007/978-3-319-51281-5_7

In general, the unconstrained optimization problems are described as follows:

$$\min_{x \in R^n} f(x) \tag{1}$$

where R^n is an n-dimensional Euclidean space and $f : R^n \rightarrow R$ is continuously differentiable. The gradient and Hessian for (1) are denoted as g and G, respectively. In order to display the updated formula of Broyden's family, the step-vectors s_k and y_k are defined as

$$\begin{aligned} s_k &\stackrel{def}{=} x_{k+1} - x_k \\ y_k &\stackrel{def}{=} g(x_{k+1}) - g(x_k) \\ &= g_{k+1} - g_k \end{aligned} \tag{2}$$

The Broyden's algorithm for unconstrained optimization problem uses the matrices B_i which is updated by the formula

$$B_{k+1} = B_k - \left(\frac{B_k s_k s_k^T B_k}{s_k^T B_k y_k} \right) + \frac{y_k y_k^T}{s_k^T y_k} + \phi_k \left(s_k^T B_k s_k \right) v_k v_k^T, \tag{3}$$

where ϕ is a scalar and

$$v_k = \left[\frac{y_k}{s_k^T y_k} - \frac{B_k s_k}{s_k^T B_k s_k} \right].$$

The choice of the parameter ϕ is important, since it can greatly affect the performance of the method [2]. When $\phi_i = 1$ in Eq. (3), we obtain the DFP algorithm and $\phi_i = 0$, we get the BFGS algorithm. But, Byrd et al. [3] extended his result to $\phi \in (0, 1]$. Based on [4], the Broyden's algorithm is one of the most efficient algorithm for solving the unconstrained optimization problem. It's also well known that the matrix B_{k+1} is generated by (3) to satisfy the Secant Equation

$$B_{k+1} s_k = y_k, \tag{4}$$

which may be regarded as an approximate version of the relation. Note that it is only possible to fulfil the secant equation if

$$s_k^T y_k > 0, \tag{5}$$

which is known as the curvature condition.

Realising the possible non-convergence for general objective functions, some authors have considered modifying quasi-Newton methods to enhance the convergence. For example, Li and Fukushima [5] modify the BFGS method by skipping the update when certain conditions are not satisfied and prove the global convergence of the resulted BFGS method with a "cautious update" (which is called the CBFGS method).

However, their numerical tests show that the CBFGS method does not perform better than the ordinary BFGS method. Then, Mamat et al. [6], Ibrahim et al. [7] and Sofi et al. [8] proposed a new search direction for quasi-Newton methods in solving unconstrained optimization problems. Generally, the search direction focused on the hybridization of quasi-Newton methods with the steepest descent method. The search direction proposed by Mustafa et al. [6] is $d_k = -\eta B_k^{-1} g_k - \delta g_k$, where $\eta > 0$ and $\delta > 0$. They realised that the hybrid method is more effective compared with the ordinary BFGS in terms of computational cost. Hence, the delicate relationships between the conjugate gradient and the BFGS method have been explored in the past.

In this paper, motivated by the idea of conjugate gradient methods, we propose a line search algorithm for solving (1), where the search direction of the quasi-Newton methods will be modified using the search direction of the conjugate gradient method approach. We prove that our algorithm with the Wolfe line search is globally convergent for general objective function. Then, we test the new approach on standard test problems, comparing the numerical results with the results of applying the quasi-Newton methods to the same set of test problems.

The organization of this paper is organized as follows. Section 2 describes the basic algorithm involved. Section 3 discusses the proposed method. This is followed by experimental results Sect. 4. Finally, conclusion and future direction are reported in Sect. 5.

2 Preliminaries

The iterative method is used to solve unconstrained optimization problems in order to get the minimal value of the function where the gradient is 0. Hence, the iterative formula for the quasi-Newton methods will be defined as

$$x_{k+1} = x_k + \alpha_k d_k, \tag{6}$$

where the a_k and d_k denote the step size and the search direction, respectively. The step size must always have a positive value such that $f(x)$ is sufficiently reduced. The success of a line search depends on the effective choices of both the search direction d_k and the step size a_k. There are a lot of formulas in calculating the step size, which are divided into an exact line search and an inexact line search.

The ideal choice would be the exact line search formula, which is defined as $\alpha_k = \arg\min(f(x_k + \alpha_k d_k)\, \alpha > 0$, but in general it is too expensive to identify this value. Generally, it requires too many evaluations of the objective function f and also its gradient g. The inexact line search has a few formulas which have been presented by previous researchers, such as the Armijo line search [9], Wolfe condition [10, 11], and Goldstein condition [12]. Shi [13] claims that among several well-known inexact line search procedures, the Armijo line search is the most useful and the easiest to implement in the computational calculation. It is also easy to implement it in programming like Matlab and Fotran. The Armijo line search is described as follows:

Given $s > 0, \lambda \in (0,1), \sigma \in (0,1)$ and $\alpha_i = \max\{s, s\lambda, s\lambda^2, \ldots\}$ such that

$$f(x_k) - f(x_k + \alpha_k d_k) \geq -\sigma \alpha_k g_k^T d_k, \tag{7}$$

$k = 0, 1, 2, 3, \ldots$. The reduction in f should be proportional to both the step size and directional derivative $g_k^T d_k$.

The search directions are also important in order to determine the value of f, which decreases along the direction. Moreover, the search direction of the quasi-Newton methods often has the form

$$d_k = -B_k^{-1} g_k \tag{8}$$

where B_k is a symmetric and non-singular matrix of approximation of the Hessian (3). Initial matrix B_0 is chosen by an identity matrix which subsequently is updated by an update formula. When d_k is defined by (8) and B_k is a positive definite, we have $d_k^T g_k = -g_k^T B_k^{-1} g_k < 0$, and therefore d_k is a descent direction. Hence, the algorithm for an iteration method of ordinary Broyden is described in Fig. 1.

Step 0. Given a starting point x_0 and $B_0 = I_n$. Choose values for s, β and σ.

Step 1. Terminate if $\|g(x_{k+1})\| < 10^{-6}$.

Step 2. Calculate the search direction by (8).

Step 3. Calculate the step size α_k by the Armijo Line Search (7).

Step 4. Compute the difference $s_k = x_{k+1} - x_k$ and $y_k = g_{k+1} - g_k$.

Step 5. Update B_k by (3) to obtain B_{k+1}.

Step 6. Set $k = k + 1$ and go to Step 1.

Fig. 1. Broyden algorithm

3 Proposed Model

In this section, we will discuss the new search direction for the quasi-Newton methods, which will be proposed by using the concept of the conjugate gradient method. The search direction of conjugate gradient method is

$$d_k = \begin{cases} -g_k & k = 0 \\ -g_k + \beta_k d_{k-1} & k \geq 1, \end{cases} \tag{9}$$

where β_k is a coefficient of the conjugate gradient method. So, the concept of the conjugate gradient method's search direction will be implemented into the new search direction as introduced in [14]. Therefore, the new search direction for the quasi-Newton method, known as the CG-Broyden method, is

$$d_k = \begin{cases} -B_k^{-1} g_k & k = 0 \\ -B_k^{-1} g_k + \lambda_k d_{k-1} & k \geq 1, \end{cases} \tag{10}$$

where $\lambda_k = \eta g_k^T g_k / g_k^T d_{k-1}$ and $\eta \in (0, 1]$. With these considerations in mind, we shall now propose the algorithm for the CG-Broyden method as shown in Fig. 2.

Step 0. Given a starting point x_0 and $B_0 = I_n$. Choose values for s, β and σ.
Step 1. Terminate if $\|g(x_{k+1})\| < 10^{-6}$.
Step 2.Calculate the search direction by (10).
Step 3. Calculate the step size α_k by (7).
Step 4. Compute the difference $s_k = x_{k+1} - x_k$ and $y_k = g_{k+1} - g_k$.
Step 5. Update B_k by (3) to obtain B_{k+1}.
Step 6. Set $k = k + 1$ and go to Step 1.

Fig. 2. CG-Broyden Algorithm

Based on Algorithm 2, we assume that every search direction d_k satisfied the descent condition

$$g_k^T d_k < 0, \tag{11}$$

for all $k \geq 0$. If there exists a constant $c_1 > 0$ such that

$$g_k^T d_k \leq c_1 \|g_k\|^2 \tag{12}$$

for all $k \geq 0$, then the search directions satisfy the sufficient descent condition which can be proof in Theorem 3.1.

Assumption 3.1
H1: The objective function f is twice continuously differentiable.
H2: The level set L is convex. Moreover, positive constants c_1 and c_2 exist, satisfying

$$c_1 \|z\|^2 \leq z^T F(x) z \leq c_2 \|z\|^2,$$

for all $z \in R^n$ and $x \in L$, where $F(x)$ is the Hessian matrix for f.

H3: The Hessian matrix is Lipschitz continuous at the point x^*, that is, the positive constant c_3 exists, satisfying

$$\|G(x) - G(x^*)\| \leq c_3 \|x - x^*\|$$

for all x in a neighborhood of x^*.
If the iterates $\{x_k\}$ are converging to a point x^*, it is to be expected that y_k is approximately equal to $G(x^*)s_k$.

Theorem 3.1. Suppose that Assumption 3 hold. Then, condition (12) hold for all $k \geq 0$.

Proof. From (9), we see that

$$\begin{aligned} g_k^T d_k &= -g_k^T B_k^{-1} g_k - \eta g_k^T d_{k-1} \\ &= -g_k^T B_k^{-1} g_k - \eta \frac{g_k^T g_{k-1}}{g_k^T d_{k-1}} g_k^T d_{k-1}. \end{aligned}$$

and using the Cauchy inequality, we get

$$\begin{aligned} g_k^T d_k &\leq -g_k^T \delta_k g_k - \eta g_k^T g_k \\ &\leq -\delta_k \|g_k\|^2 - \eta \|g_k\|^2 \\ &\leq c_1 \|g_k\|^2, \end{aligned}$$

where $c_1 = -(\delta_k + \eta)$ which is bounded away from zero. Hence, (12) holds, and the proof is completed.

4 Experimental Result

In this section, we use a large number of test problem considered in Andrei [15], Zbigniew [16] and More et al. [17] in Table 1 to analyse the improvement of the CG-Broyden method with the Broyden method. The dimensions of the tests range between 2 and 1,000 only.

The comparison between Algorithm 1 (Broyden) and Algorithm 2 (CG-Broyden) uses the cost of computation based on the number of iterations and CPU-time. As suggested by More et al. [17], for each of the test problems, the initial point x_0 will take further away from the minimum point and we analyse three of initial points of each of test problems. In doing so, it leads us to test the global convergence properties and the robustness of our method. For the Armijo line search, we use $s = 1$, $\beta = 0.5$ and $\sigma = 0.1$. In our implementation, the programs are all written in Matlab. The stopping criteria that we used in both algorithms are $\|g(x_{i+1})\| \leq 10^{-6}$. The Euclidean norm is used in the convergence test to make these results comparable.

The performance results will be shown in Figs. 3 and 4, respectively, using the performance profile introduced by Dolan and More [18]. The performance profile seeks

Table 1. A list of problem functions

Test Probem	N-Dimensional	Sources
Powell Badly Scaled	2	More et al. [17]
Beale	2	More et al. [17]
Biggs Exp6	6	More et al. [17]
Chebyquad	4,6	More et al. [17]
Colville Polynomial	4	Michalewicz [16]
Variably Dimensioned	4,8	More et al. [17]
Freudenstein And Roth	2	More et al. [17]
Goldstein Pricepolynomial	2	Michalewicz [16]
Himmelblau	2	Andrei [15]
Penalty 1	2,4	More et al. [17]
Extended Powell Singular	4,8	More et al. [17]
Extended Rosenbrock	2,10,100,200,500,1000	Andrei [15]
Trigonometric	6	Andrei [15]
Watson	4,8	More et al. [17]
Six-Hump Camel Back Polynomial	2	Michalewicz [16]
Extended Shallow	2,4,10,100,200,500,1000	Andrei [15]
Extended Strait	2,4,10,100,200,500,1000	Andrei [15]
Scale	2	Michalewicz [16]
Raydan 1	2,4	Andrei [15]
Raydan 2	2,4	Andrei [15]
Diagonal 3	2	Andrei [15]
Cube	2,10,100,200	More et al. [17]

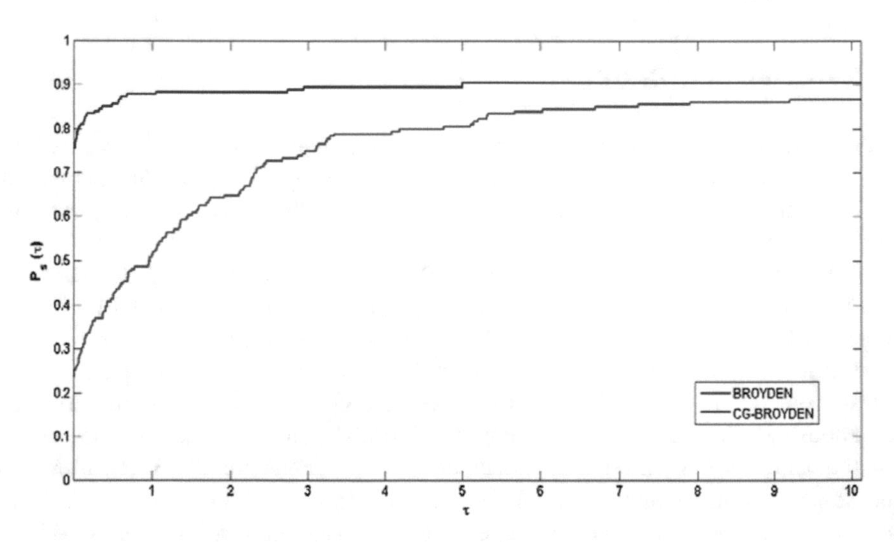

Fig. 3. Performance profile in a $\log_{10}$ scaled based on iteration.

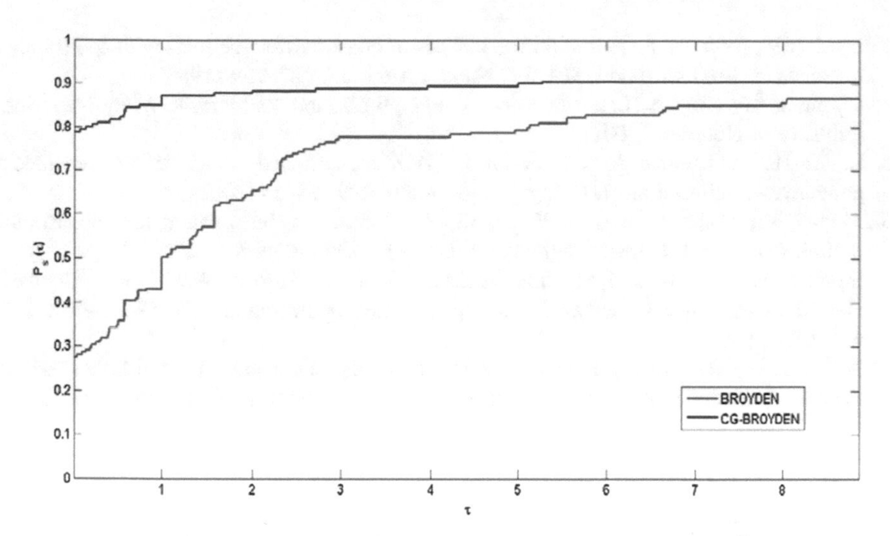

Fig. 4. Performance profile in a $\log_{10}$ scaled based on CPU time.

to find how well the solvers perform relative to the other solvers on a set of problems. In general, $P(\tau)$ is the fraction of problems with performance ratio τ, thus, a solver with high values of $P(\tau)$ or one that is located at the top right of the figure is preferable.

Figures 3 and 4 show that the CG-Broyden method has the best performance since it can solve 91% of the test problems while the Broyden method only solve 86%. Moreover, we can also say that the CG-Broyden method is the fastest solver on approximately 76% of the test problems for iteration and 79% of CPU-time. Therefore, the CG-Broyden method is better in solving the unconstrained optimization problems compare to the original Broyden method.

5 Conclusion

We have presented a new hybrid method for solving unconstrained optimization problems. The numerical results for a small dimension of test problems show that the CG-Broyden method is efficient and robust in solving unconstrained optimization problems. The numerical results and figures from the programming are reported and analysed to show the characters of the proposed method. Our further interest is to try the CG-Broyden method with the coefficient of the conjugate gradient methods Fletcher-Reeves [19], Hestenes-Steifel [20] and the Liu-Storey [21] coefficient for β_k.

References

1. Broyden, C.G.: The convergence of a class of double-rank minimization algorithms 2. The new algorithm. IMA J. Appl. Math. **6**, 222–231 (1970)
2. Xu, D.-C.: Global convergence of the Broyden's class of quasi-newton methods with nonmonotone linesearch. Acta Math. Appl. Sinica Engl. Ser. **19**, 19–24 (2003)

3. Byrd, R.H., Nocedal, F.: A tool for the analysis of quasi-newton methods with application to unconstrained minimization. SIAM J. Numer. Anal. **26**, 727–739 (1989)
4. Chong, E.K.P., Zak, S.H.: An Introduction to Optimization, 2nd edn. A Wiley-Interscience Publication, Hoboken (2001)
5. Li, D.-H., Fukushima, M.: A modified BFGS method and its global convergence in nonconvex minimization. J. Comput. Appl. Math. **129**, 15–24 (2001)
6. Mamat, M., Mohd, I., June, L.W., Dasril, Y.: Hybrid broyden method for unconstrained optimization. Int. J. Numer. Methods Appl. **1**, 121–130 (2009)
7. Ibrahim, M.A.H., Mamat, M., Sofi, A.Z.M., Mohd, I., Ahmad, W.M.A.W.: Alternative algorithms of broyden familyami: for unconstrained optimization. AIP Conf. Proc. **1309**, 670–680 (2010)
8. Sofi, A.Z.M., Mamat, M., Ibrahim, M.A.H.: Reducing computation time in DFP (Davidon, Fletcher & Powell) update method for solving unconstrained optimization problems. AIP Conf. Proc. **1522**, 1337–1345 (2013)
9. Armijo, L.: Minimization of functions having Lipschitz continuous partial derivatives. Pac. J. Math. **16**, 1–3 (1966)
10. Wolfe, P.: Convergence conditions for ASCENT methods. SIAM Rev. **11**, 226–235 (1969)
11. Wolfe, P.: Convergence conditions for ASCENT methods. II: some corrections. SIAM Review **13**, 185–188 (1971)
12. Goldstein, A.: On steepest descent. J. Soc. Ind. Appl. Math. Ser. A Control **3**, 147–151 (1965)
13. Shi, Z.-J.: Convergence of quasi-Newton method with new inexact line search. J. Math. Anal. Appl. **315**, 120–131 (2006)
14. Ibrahim, M.A.H., Mamat, M., Leong, W.J.: The hybrid BFGS-CG method in solving unconstrained optimization problems. Abstr. Appl. Anal. 6 (2014)
15. Andrei, N.: An unconstrained optimization test functions, collection. Adv. Modell. Optimization **10**, 147–161 (2008)
16. Zbigniew, M.: Genetic Algorithms + Data Structures = Evolution Programs. Springer Verlag, Heidelberg (1996)
17. More, J.J., Garbow, B.S., Hillstrom, K.E.: Testing unconstrained optimization software. ACM Trans. Math. Softw. **7**, 17–41 (1981)
18. Dolan, E.D., Moré, J.J.: Benchmarking optimization software with performance profiles. Math. Program. **91**, 201–213 (2002)
19. Fletcher, R., Reeves, C.M.: Function minimization by conjugate gradients. Comput. J. **7**, 149–154 (1964)
20. Hestenes, M.R., Stiefel, E.: Method of conjugate gradient for solving linear equations. J. Res. Natl. Bur. Stan. **49**, 409–436 (1952)
21. Liu, Y., Storey, C.: Efficient generalized conjugate gradient algorithms, part 1: Theory. J. Optim. Theory Appl. **69**, 129–137 (1991)

Improved Functional Link Neural Network Learning Using Modified Bee-Firefly Algorithm for Classification Task

Yana Mazwin Mohmad Hassim(✉), Rozaida Ghazali, and Noorhaniza Wahid

Faculty of Computer Science and Information Technology, Universiti Tun Hussein Onn Malaysia (UTHM), 86400 Batu Pahat, Johor, Malaysia
{yana, rozaida, nhaniza}@uthm.edu.my

Abstract. Functional Link Neural Network (FLNN) has been becoming as an important tool used in many applications task particularly in solving a non-linear separable problems. This is due to its modest architecture which required less tunable weights for training as compared to the standard multilayer feed forward network. The most common learning scheme for training the FLNN is a Backpropagation (BP-learning) algorithm. However, learning method by BP-learning algorithm tend to easily get trapped in local minima especially when dealing with non-linearly separable classification problems which affect the performance of FLNN. This paper discussed the implementation of modified Artificial Bee Colony with Firefly algorithm for training the FLNN network to overcome the drawback of BP-learning scheme. The aim is to introduce an alternative learning scheme that can provide a better solution for training the FLNN network for classification task.

Keywords: Functional link neural network · Modified artificial bee colony · Firefly algorithm · Classification

1 Introduction

Functional Link Neural Network (FLNN) is a type of Higher Order Neural Networks (HONNs) which was first introduced by Giles and Maxwell [1]. It was make known as an alternative approach to standard multilayer feed forward network in Artificial Neural Networks (ANNs) by Pao [2]. One of the best known types of ANNs is the Multilayer Perceptron (MLP). The MLP structure consists of multiple layers of nodes which give the network the ability to solve problems that are not linearly separable. However, MLP usually requires a fairly large amount of available measures in order to achieve good classification ability. Difficulties in fixing appropriate number of neurons in layers and a challenging work on determining number of hidden layers has make the MLP architecture becomes not that easy to train. The increase number of hidden layers and neurons also make the MLP architecture become complex and resulting in slower operation. Hence these have prompted FLNN as an alternative network structure to overcome the drawback of MLP [3].

T. Herawan et al. (eds.), *Recent Advances on Soft Computing and Data Mining*,
Advances in Intelligent Systems and Computing 549, DOI 10.1007/978-3-319-51281-5_8

The FLNN is a flat network (without hidden layers) where it reduced the neural architectural complexity while at the same time possesses the ability to solve non-linear separable problems. The flat architecture of FLNN has also make the learning algorithm in the network less complicated [4]. In order to capture non-linear input-output mapping, the input vector of FLNN is extended with a suitable enhanced representation of the input nodes which artificially increase the dimension of input space [5].

The FLNN architecture discussed in this paper is based on generic basis architecture which uses a tensor representation. Figure 1 illustrates the FLNN structure of 2 inputs up to second order input enhancement. The first order consists of the original inputs which are x_1 and x_2. To provide the network with non-linear mapping capability the FLNN inputs are enhanced with additional higher order unit by extending them based on the product unit. The product unit of this network is x_1x_2 and is known as the second order input enhancement for the network.

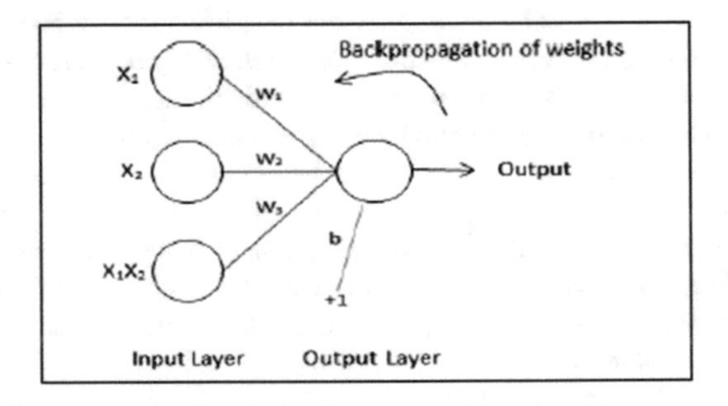

Fig. 1. FLNN structure up to 2nd order of 2 input nodes

Figure 2 show the network structure of MLP also with 2 input nodes. The network structure of FLNN presented in Fig. 1 was enhanced up to 2nd order (the highest order) make it employed only 4 trainable parameters (3 weights + 1 bias) in it structure. As compared to Fig. 2, the MLP with the same number of input nodes (2 inputs) and with a single hidden layer of 2 nodes (the least numbers of hidden nodes and layers) formed 9 trainable parameters (6 weights + 3 biases) in it structure. Of both this network,

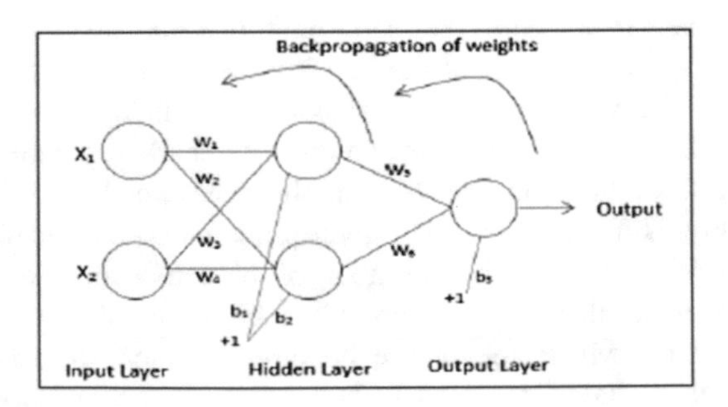

Fig. 2. Single layer MLP structure with 2 input nodes

FLNN need less trainable weights as compared to MLP for training and this make the learning scheme for FLNN less complicated.

FLNN network is usually trained by adjusting the weight of connection between neurons. The most common method for tuning the weight in FLNN is using a Back-propagation (BP) learning algorithm. However one of the crucial problems with the standard BP-learning algorithm is that it can easily get trapped in local minima especially for the non-linear problems [6], thus effect the performance of FLNN network. To overcome this, the modified Artificial Bee Colony (MABC) optimization algorithm is used to optimize the weights of FLNN instead of the BP-learning algorithm [7] for solving classification tasks. In this study, we proposed the integration of MABC algorithm with Firefly algorithm (FFA) to provide a thorough global and local search in order to get the optimal weights set for the FLNN training for enhancing classification accuracy.

2 Functional Link Neural Network Learning Scheme

Most previous learning algorithm used in the training of FLNN, is the BP-learning algorithm [4, 6, 8–10]. As shown as in Fig. 1. The weight values between enhanced input nodes and output node are randomly initialized. Let the enhanced input nodes x be represented as $x_t = \langle x_1, x_2, \ldots x_n, x_1x_2, x_1x_3, \ldots x_{n-1}x_n \rangle$. The output value of the FLNN is obtained by:

$$\hat{y} = f(wx_t + b) \tag{1}$$

where $\hat{y}$ is the output while f denote the output node activation function and b is the bias. In Eq. (1), wx_t is the aggregate value which is the inner product of weight, w and x_t. The squared error E, between the target output and the actual output will be minimized as:

$$E = \frac{1}{N}\sum_{i=1}^{N}(y_i - \hat{y}_i)^2 \tag{2}$$

where y_i is the target output and $\hat{y}_i$ is the actual output of the i th input training pattern, while N is the number of training pattern. During the training phase, the BP-learning algorithm will continue to update w and b until the maximum epoch or the convergent condition is reached.

Although BP-learning is the mostly used algorithm in the training of FLNN, the algorithm however has several limitations; It is tends to easily gets trapped in local minima especially for those non-linearly separable classification problems. The convergence speed of the BP learning also can gets is too slow even if the learning goal, a given termination error, can be achieved. Besides, the convergence behavior of the BP-learning algorithm depends on the choices of initial values of the network connection weights as well as the parameters in the algorithm such as the learning rate and momentum [6].

3 Modified Artificial Bee Colony

A modified Artificial Bee Colony (MABC) was designed to be used as a learning scheme to train the FLNN network for solving classification problems [7, 11]. The MABC is based on the original standard Artificial Bee colony Optimization algorithm (ABC) by Karaboga [12]. The original standard ABC simulates the intelligent foraging behavior of bees (employed, onlooker and scout bees) for solving multidimensional and multimodal optimization problem. In mABC, the modification was done on the employed and onlooker bees' search strategy in which the random selection of single neighbor, *j* as is (3) from the original standard ABC is removed from the equation.

$$v_{ij} = x_{ij} + \emptyset_{ij}(x_{ij} - x_{kj}) \tag{3}$$

The modified version of employed and onlooker bees' search strategy for MABC is presented as in (4). The modified search strategy prompted the bees' to exploit all weights in the FLNN weights vector instead of single random weight.

$$v_i = x_i + \emptyset_i(x_i - x_k) \tag{4}$$

In the FLNN-MABC, the position of a food source (FS) represents a possible solution to the optimization problem, and the nectar amount of a food source corresponds to the profitability (fitness) of the associated solution. The weight, *w* and bias, *b* of the network are treated as optimization parameters with the aim to finding minimum Error, *E* as in (2). This optimization parameters are represented as *D*-dimensional

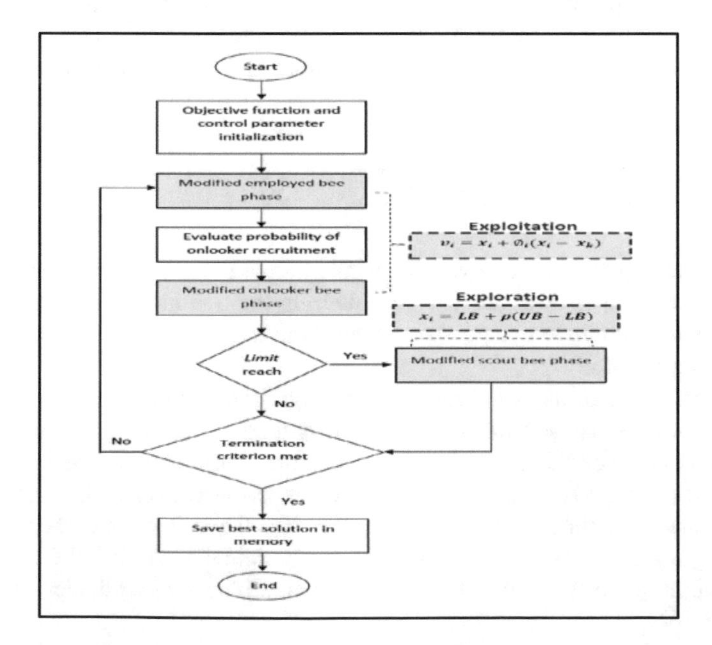

Fig. 3. The flow chart for FLNN-MABC learning scheme

solution vector, $x_{i,j}$ where $(i = 1, 2, \ldots, FS)$ and $(j = 1, 2, \ldots, D)$ and each vector i is exploited by only one employed and onlooker bee. Since we have removed the single random weight selection of j, both employed and onlooker are forced to update all parameters in j where, $j = 1, 2, \ldots, D$. In order to produce a candidate food source v_i from the old one x_i in memory, the MABC uses Eq. (4) where $\emptyset_i$ is a random number in the range [−1,1]. Figure 3 shows the flow chart for FLNN-MABC learning scheme.

4 Firefly Algorithm

Firefly algorithm (FA) was developed by Yang [13] which is inspired by the flashing behavior of fireflies. The purpose of flashing behavior is to act as signal system to attract other fireflies. Yang [13] formulated the algorithm base on three rules:

1. Fireflies are unisex so that one firefly will be attracted to other fireflies regardless of their sex.
2. The attractiveness is proportional to the brightness, and they both decrease as their distance increases. Thus for any two flashing fireflies, the less bright one will move towards the brighter one. If there is no brighter one than a particular firefly, it will move randomly.
3. The brightness of a firefly is determined by the landscape of the objective function.

In FA, the attractiveness of firefly is determined by its brightness or light intensity. The attractiveness, β of a firefly is defined in (5):

$$\beta(r) = \beta_0 e^{-\gamma r^2} \tag{5}$$

where β_0 is the attractiveness at distance, $r = 0$. The movement of a firefly i attracted to brighter firefly j at x_i and x_j is determined by Eq. (6).

$$x_i = x_i + \beta_0 e^{-\gamma r^2} \left(x_j - x_i\right) + \alpha_1 (rand - 0.5) \tag{6}$$

In (6), the second term is due to the attraction while the third term is randomization with α_1 being the randomization parameter and *rand* is random number generator uniformly distributed in [0,1].

5 Modified Bee-Firefly Algorithm

The MABC was introduced as an alternative learning scheme for training the FLNN network to overcome drawbacks on classifying a multiclass classification data. Although it gives promising results, sometimes it may be leads to slow convergence as the proposed search strategy in (4) tend to encourage the bees more on exploration initiatives in the search space. In population-based optimization algorithms, both exploration and exploitation are necessary and need to be well balanced to avoid local minima trapping and thus achieved good optimization performance [14].

According to the solution search equation of MABC algorithm described by Eq. (4), the new candidate solution is generated by moving the old solution towards another solution selected randomly from the neighboring population. However, the probability whether that the randomly selected neighbor solution k, is a good solution is the same as the probability of the randomly selected neighbor solution k, is a bad one. Therefore, the new candidate solution is not promising to be a solution better than the previous one. On the other hand, since Eq. (4) is used by both employed and onlooker bee, with x_k is a random selected neighboring bees in the population, therefore, the solution search dominated by Eq. (4) is random enough for exploration but poor at exploitation.

To overcome this disadvantage, we substitute the onlooker bee search strategy with firefly algorithm. The proposed modified bee-firefly algorithm (MBF) consists of two phases. The first phase employs the exploratory search by employed bee while the other phase employs the onlooker firefly search strategy to improve the solution quality. Figure 4 shows the flow chart for modified Bee-Firefly (MBF) algorithm for training the FLNN network.

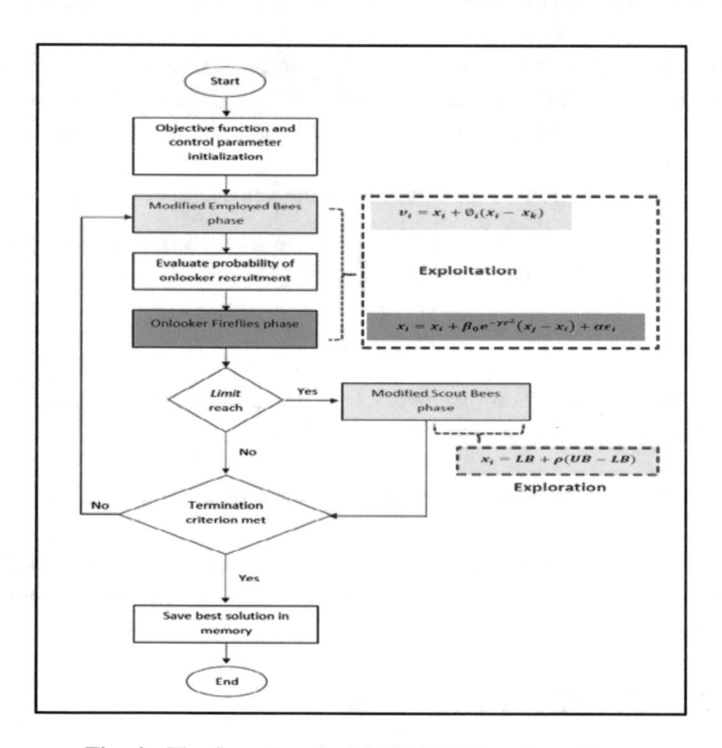

Fig. 4. The flow chart for FLNN-MBF learning scheme

6 Experimentation and Result

A set of experiments have been conducted, in order to show MBF algorithm effectiveness for the purpose of training the FLNN networks on classification tasks. Simulation experiments were carried out on a 2.30 GHz Core i5-2410 M Intel CPU with 8.0 GB RAM in a 64-bit Operating System. The Simulation experiments were performed on the training of FLNN with standard BP-learning (FLNN-BP), FLNN with original standard ABC (FLNN-ABC), FLNN with modified ABC (FLNN-MABC) and FLNN with our proposed modified bee-firefly algorithm (FLNN-MBF). The best training accuracy for every benchmark problems were noted from these simulations. The learning rate and momentum used for the FLNN-BP were between 0.1–0.9 with the maximum of 1000 epoch and the minimum error = 0.001 as for the stopping criteria. Meanwhile, parameters setup for the FLNN-ABC, FLNN-MABC and FLNN-MBF, are as follows; the colony size = 50, stopping criteria of maximum 1000 cycles and minimum error = 0.001. The activation function used for the FLNN network output is Logistic sigmoid function. Table 1 summarized parameters setting for the experiment.

Table 1. The parameter setting for the experiment

Parameters	FLNN-BP	FLNN-ABC	FLNN-MABC	FLNN-MBF
Learning rate	0.1 − 0.5	–	–	–
Momentum	0.1 − 0.9	–	–	–
Epoch	1000	1000	1000	1000
Minimum error	0.001	0.001	0.001	0.001

We evaluated our proposed learning scheme using five public classification datasets; BUPA Liver Disorders (BUPA), Balance Scale Dataset (BALANCE), Iris (IRIS), Pima Indians Diabetes (PIMA) and Thyroid Disease (THYROID) obtained from the University of California at Irvine (UCI) Machine Learning Repository [15]. The datasets were divided into 10 folds where, 9-folds were used as the training set and the remaining fold was used as the test set. For the sake of convenience we set our FLNN input enhancement up to second order.

Ten trials were performed on each simulation where 10 independent run were performed for every single fold. The average value of 10 fold cross validation of each data set are used for comparison with other learning scheme. The simulation result is presented as in Table 2 below.

The classification accuracy results presented in Table 2 shows that training the FLNN network with MBF gives better accuracy result both on training set and test set for BUPA, BALANCE, IRIS, PIMA and THYROID dataset. The FLNN-MBF outperformed the rest of FLNN-MABC, FLNN-ABC, FLNN-BP and MLP-BP models with significant result based on accuracy rate on unseen data. Hence it is clearly indicated that the proposed training scheme could facilitated better learning for FLNN network as compared to the standard ABC and standard BP-learning algorithm.

To evaluate the performance of our proposed training scheme we also carried out a statistical performance evaluation based on Freidman test [16]. We began with the null hypothesis stated all the training schemes are equivalent so that their rank, R_j should be equal. In this test we assign rank r_i^j to the jth of k training scheme on the ith of N datasets based on their classification accuracy on unseen data. We used the Friedman statistic as in (7).

Table 2. Simulation result on classification accuracy

Classification accuracy (%)				
Datasets	FLNN-BP	FLNN-ABC	FLNN-MABC	FLNN-MBF
BUPA	70.38	67.58	69.38	72.26
BALANCE	85.43	91.26	90.82	92.09
IRIS	96.80	94.20	96.73	97.27
PIMA	73.93	75.43	76.03	76.16
THYROID	93.46	94.77	94.81	95.45

Table 3. Rank of each training scheme on different dataset based on average classification accuracy on test set

Datasets	FLNN-BP	FLNN-ABC	FLNN-MABC	FLNN-MBF
BUPA	70.38 (2)	67.58 (4)	69.38 (3)	72.26 (1)
BALANCE	85.43 (4)	91.26 (2)	90.89 (3)	92.09 (1)
IRIS	96.80 (2)	94.20 (4)	96.73 (3)	97.27 (1)
PIMA	73.93 (4)	75.43 (3)	76.03 (2)	76.16 (1)
THYROID	93.46 (4)	94.77 (3)	94.81 (2)	95.45 (1)

$$F = \frac{12}{Nk(k+1)} \cdot \sum R^2 - 3N(k+1) \quad (7)$$

The ranks of each training scheme are presents as Table 3.

Based on formula in (7), we found the F value as 9.72. Meanwhile the critical value of $F(4, 5)$ for $\alpha = 0.05$ is 7.80 which is less than the obtained F value, thus the null hypothesis is rejected. As the null hypothesis is rejected, we proceed with a post-hoc procedure based on pairwise comparisons method [17]. The null hypothesis for this test stated the pair of training schemes compared are equivalent. The test statistic used in this procedure is based on equation as in (8), where R_i, R_j are the average rankings by Friedman of the algorithm that being compared.

$$z = \frac{(R_i - R_j)}{\sqrt{\frac{k(k+1)}{6N}}} \quad (8)$$

Based on Holm test, the probability p is calculated based on z value and compared with $\alpha/(k-i)$. The result of this test is presented as in Table 4.

Table 4. Post-hoc procedure based on pairwise comparisons method

i	Classifiers	z-values	p-values	$\alpha/(k-i)$
1	ABC-MBF	2.694	0.00706	0.017
2	BP-MBF	2.694	0.00706	0.025
3	MABC-MBF	1.225	0.049996	0.05

From Table 4, based on pairwise comparison with FLNN-ABC, FLNN-BP and FLNN-MABC it is clear that FLNN-MBF has significantly better performance. The classification result obtained with MBF training scheme can be considered to be better than the result obtained by the ABC, BP and MABC.

7 Conclusion

In this work, we evaluated the FLNN-MBF model for the task of pattern classification problems. The experiment has demonstrated that FLNN-MBF performs the classification task quite well. For the case of BUPA, BALANCE, IRIS, PIMA and THYROID, the simulation result shows that the proposed modified Bee-Firefly algorithm (MBF) can successfully train the FLNN for solving classification problems with better accuracy percentage on unseen data. This research work is carried out to introduce MBF as an alternative learning scheme for training the Functional link Neural Network that can give a promising result.

Acknowledgments. The authors would like to thank Universiti Tun Hussein Onn Malaysia (UTHM) and Ministry of Higher Education (MOHE) Malaysia for financially supporting this research under the Fundamental Research Grant Scheme (FRGS), Vote No. 1235.

References

1. Giles, C.L., Maxwell, T.: Learning, invariance, and generalization in high-order neural networks. Appl. Optics. **26**(23), 4972–4978 (1987)
2. Pao, Y.H., Takefuji, Y.: Functional-link net computing: theory, system architecture, and functionalities. Computer **25**(5), 76–79 (1992)
3. Klassen, M., Pao, Y.H., Chen, V.: Characteristics of the functional link net: a higher order delta rule net. In: 1988 IEEE International Conference on Neural Networks (1988)
4. Misra, B.B., Dehuri, S.: Functional link artificial neural network for classification task in data mining. J. Comput. Sci. **3**(12), 948–955 (2007)
5. Pao, Y.H.: Adaptive Pattern Recognition and Neural Networks. Addison-Wesley Longman Publishing Co. Inc., Reading (1989). Medium: X; Size (327 pages)
6. Dehuri, S., Cho, S.-B.: A comprehensive survey on functional link neural networks and an adaptive PSO–BP learning for CFLNN. Neural Comput. Appl. **19**(2), 187–205 (2010)
7. Hassim, Y.M.M., Ghazali, R.: A modified artificial bee colony optimization for functional link neural network training. In: Herawan, T., Deris, M.M., Abawajy, J. (eds.). LNEE, vol. 285, pp. 69–78. Springer, Singapore (2014). doi:10.1007/978-981-4585-18-7_8
8. Haring, S., Kok, J.: Finding functional links for neural networks by evolutionary computation. In: Van de Merckt, T., et al. (eds.) Proceedings of the fifth Belgian–Dutch conference on machine learning, BENELEARN 1995, pp. 71–78. Brussels, Belgium (1995)
9. Pengyi, G., Chuanbo, C., Sheng, Q., Yingsong, H.: An optimization method for neural network based on GA and TS algorithm. In: 2010 The 2nd International Conference on Computer and Automation Engineering (ICCAE) (2010)
10. Dehuri, S., Cho, S.-B.: Evolutionarily optimized features in functional link neural network for classification. Expert Syst. Appl. **37**(6), 4379–4391 (2010)

11. Hassim, Y.M.M., Ghazali, R.: Optimizing functional link neural network learning using modified bee colony on multi-class classifications. In: Jeong, H.Y., Obaidat, M.S., Yen, N. Y., Park, J.J. (eds.) Advances in Computer Science and Its Applications. LNEE, vol. 279, pp. 153–159. Springer, Heidelberg (2014). doi:10.1007/978-3-642-41674-3_23
12. Karaboga, D., Basturk, B.: A powerful and efficient algorithm for numerical function optimization: artificial bee colony (ABC) algorithm. J. Glob. Optim. **39**(3), 459–471 (2007)
13. Yang, X.-S., He, X.: Firefly algorithm: recent advances and applications. Int. J. Swarm Intell. **1**(1), 36–50 (2013)
14. Trelea, I.C.: The particle swarm optimization algorithm: convergence analysis and parameter selection. Inf. Process. Lett. **85**(6), 317–325 (2003)
15. Lichman, M.: UCI Machine Learning Repository, School of Information and Computer Science, University of California, Irvine, CA (2013). http://archive.ics.uci.edu/ml
16. Demsar, J.: Statistical comparisons of classifiers over multiple data sets. J. Mach. Learn. Res. **7**, 1–30 (2006)
17. García, S., Fernández, A., Luengo, J., Herrera, F.: Advanced nonparametric tests for multiple comparisons in the design of experiments in computational intelligence and data mining: experimental analysis of power. Inf. Sci. **180**(10), 2044–2064 (2010)

Artificial Neural Network with Hyperbolic Tangent Activation Function to Improve the Accuracy of COCOMO II Model

Sarah Abdulkarem Alshalif[1(✉)], Noraini Ibrahim[1], and Tutut Herawan[2]

[1] Universiti Tun Hussein Onn Malaysia, 86400 BatuPahat, Parit Raja, Johor, Malaysia
gi140034@siswa.uthm.edu.my, noraini@uthm.edu.my
[2] University of Malaya, 50603 Pantai Valley, Kuala Lumpur, Malaysia
tutut@um.edu.my

Abstract. In software engineering, Constructive Cost Model II (COCOMO II) is one of the most cited, famous and widely used model to estimate and predict some important features of the software project such as effort, cost, time and manpower estimations. Lately, researchers incorporate it with soft computing techniques to solve and reduce the ambiguity and uncertainty of its software attributes. In this paper, Artificial Neural Network (ANN) with Hyperbolic Tangent Activation Function is used to improve the accuracy of the COCOMO II model and the backpropagation learning algorithm used in the training process. In the experiment, COCOMO II SDR dataset is used for training and testing the model. The result shows that eight out of twelve projects have a closer effort value of actual effort. It shows that the proposed model produces better performance comparing to sigmodal function.

Keywords: Software cost estimation · COCOMO II · Artificial neural network · Hyperbolic tangent activation function · Backpropagation algorithm

1 Introduction

Software cost estimation is the estimating process of the effort required to develop a software engineering project [1]. This process becomes one of the most challenging tasks and complex activity in the area of software engineering and project management. Whereas software cost estimation may be easy in concept, it is in fact difficult and complicated [2]. Software project managers and developers interested to accurately estimate the cost, time and manpower requirement of software development. However, the accuracy of estimation models is still unsatisfactory in software project management. The source and the main reason of this inaccuracy comes from the ambiguity and uncertainty of software attributes at the early phase of software development process [3, 4]. In the process, the first step is to calculate the effort that's required for developing software system, then based on this effort, cost, time, and manpower will be calculated and estimating. Several software development cost estimation models have

T. Herawan et al. (eds.), *Recent Advances on Soft Computing and Data Mining*,
Advances in Intelligent Systems and Computing 549, DOI 10.1007/978-3-319-51281-5_9

been proposed and developed, which can be classified into two categories algorithmic and non-algorithmic models [3].

Algorithmic models were established based on statistical analysis of past project data such as cost factors and scale factors, also known as the conventional method. It provides the mathematics and experimental equations to compute software cost [5, 6]. The most popular algorithmic cost estimation models are Boehm's Constructive Cost Model (COCOMO I and II), Albrecht's Function Point, and Putnam's Software Life Cycle Management (SLIM) [7]. These models need many specific requirements such as inputs and attributes for example source line of code (SLOC), number of user screen, interfaces, complexities which are difficult to gain at the early stage of software development. Furthermore, they provide a fast estimation and their formula and calculation more easily comparing to non-algorithm models [3].

Non-algorithmic models which published on 1990's are based on new approaches, for example expert judgment, Price-to-win and machine learning approaches [8]. The machine learning approach is used to group together a set of techniques that represent some of the facts of the human mind such as evolutionary computation, Fuzzy Systems, genetic algorithms, Bayesian networks and artificial neural networks which classified as soft computing group [9]. As well as, they provide and show powerful linguistic representation that help to represent more accurate software attribute and can overcome the algorithm models defects when corporate together. Therefore, this research aims to estimate the software effort which furnishes the motivation to develop a modified model, COCOMO II model based on Artificial Neural Network with Hyperbolic Tangent Activation Function uses to estimate the software effort, cost, time and manpower which inclusive of the backpropagation learning algorithm.

This paper is organized as follows: Sect. 2 describes the related works. Section 3 describes the theoretical background for this research. The proposed model and the training algorithm are elaborated in Sect. 4. The results are discussed in Sect. 5. Finally, Sect. 6 concludes the paper.

2 Related Works

Accurate software estimates at the early phase of software development is one of the crucial objectives and great challenges in software project management because of the ambiguity and uncertainty of software attributes at the early phase of development [3]. Several models of software development cost estimation were proposed and developed to improve the estimation accuracy. As well as, several researcher attempts to enhance the existing models to produce more estimation accuracy by incorporating software cost estimation models with other techniques such as soft computing techniques, Maturity level, and others [9]. Therefore, an accurate software cost estimation model is highly required in software project management [10]. The COCOMO II has been updated and improved to have more accurate estimation results. There are many researchers interested to improve the estimation accuracy of COCOMO II.

Reddy and Raju [11] improved COCOMO based on Artificial Neural Networks (ANN) to improve the performance of the network that suits to the COCOMO model. They used multilayer feedforward neural network and the backpropagation learning algorithm for training the network. Identity function was used for all layers of the network. Kaushik, *et al.* proposed two models [12, 13]. In [12], they used intermediate COCOMO that has 15 Effort Multipliers (EM) and there is no scale factors attribute. They integrated the base of COCOMO with ANN using perceptron learning algorithm for training. They used identity function for activation function in the input layer and threshold activation function for hidden and output layers [12]. However, in [13] follows the idea of Reddy and Raju [11] with some changes in the activation functions. They used sigmoidal function in the hidden and output layers [13]. Attarzadeh and Ow [14] used Adaptive Artificial Neural Network with COCOMO II. The proposed model using 23 software inputs and 1 system output to estimate the effort in person-months. Moreover, three bias factors are added to each of the input data categories to reduce the error that affects those input data. Feed Forward ANN architecture and sigmoid activation function were used in [14].

3 Rudimentary

3.1 COCOMO II Model

COCOMO II is a model that helps to think about the cost and time of software development. It provides an accurate cost and schedule estimates for both current and likely future software projects. It involves three sub levels, which are Application-Composition model, Early Design model and Post-Architecture model. The Application-Composition model supports the earliest phases or spiral cycles that involve the prototyping. The Early Design model is a high level model that supports the next phase or spiral cycles that involving alternatives for exploring architecture or strategies for incremental development. Post-Architecture model is for a project that is ready to develop and it is more detailed and widely used model. Post-Architecture and Early Design models use the same approach for estimating time and effort. Boehm has introduced 7 to 17 multiplicative factors (Effort Multiplier) that are used on COCOMO II to adjust the nominal effort in person months (PM) to reflect the software product under development. For the post-architecture model, there are 17 EMs, 5 Scale Factor (SF) and 1 Software Size (Size). To calculate the amount of effort in person-months (PM), Eq. (1) is used.

$$\left.\begin{array}{c} Effort_{PM} = A \times Size^{E} \times \prod_{i=1}^{17} EM_i \\ E = B + 0.01 \times \sum_{j=1}^{5} SF, A = 2.94\, and\, B = 0.91. \end{array}\right\} \quad (1)$$

where

While, to calculate of time to develop with an estimated effort, Eq. (2) is used.

$$\left.\begin{array}{l} Schedule_{Months} = C \times Effort^{F} \\ \text{where} \\ F = D + 0.2 \times (E - B), C = 3.67 \text{ and } D = 0.28. \end{array}\right\} \quad (2)$$

In calculating manpower needed, Eq. (3) will be used.

$$Average\ Staffing_{People} = \frac{Effort}{Schedule} \tag{3}$$

Moreover, SIZE in Eq. (1) can be calculated by several methods such as Source Line of Code (SLOC), Function Point (FP) and Adaptation Adjustment Factors (AAF) [15].

3.2 Artificial Neural Network

Artificial Neural Networks (ANNs) is a model for data processing. The idea is taken from the human biological neural network such as brain and information processing. It is a collection of interconnected processing elements called neurons. Each simple neuron has many inputs, but one output [16]. In fact, the neuron is a mathematical function that consists of three elements which are inputs, formulas, and output [17]. There are two types of ANNs which are feed-forward networks and feedback networks. Feed-forward networks, where signals move in one direction from input to output, but for the feedback networks the signals move on both directions so they have looped in the movement [16].

The ANN can be a single layer or multi-layer perceptron. In the single-layer perceptron, input nodes are connected directly to the output layer. In multi-layer perceptron, there are at least three layers, which are input, hidden and output layers. ANN is composed of nodes organized into layers and connected through weight elements. At each node, the weighted inputs are aggregated, threshold and inputted to an activation function to generate an output of that node [18].

The activation function is used to transform the activation level of a neuron into an output signal. Some of the most commonly used activation functions are: Identity function, Uni-polar sigmoid, Bi-polar sigmoid and Hyperbolic Tangent Function (Tanh) [18].

3.2.1 Identity Function

This is the simplest activation function, one that is commonly used for the output layer activation function in regression problems. The identity/linear activation function is shown in Eq. (4).

$$f(x) = x. \tag{4}$$

This activation function simply maps the pre-activation to itself and the output values range between $(-\infty,\infty)$. After all, a multi-layered network with linear activations at each layer can be equally-formulated as a single-layered linear network as shown in Fig. 1.

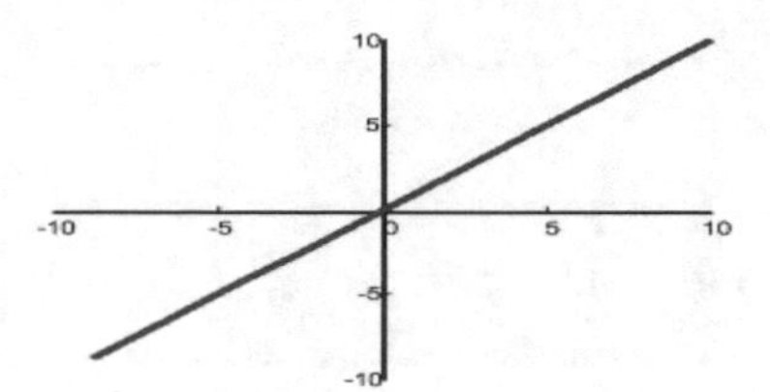

Fig. 1. Identity activation function

3.2.2 Uni-Polar Sigmoid Function

The Uni-polar sigmoid function is given as Eq. (5):

$$f(x) = \frac{1}{1 + e^{-x}}. \tag{5}$$

This function is especially advantageous to use in neural networks trained by back-propagation algorithms. Because it is easy to distinguish, and this can interestingly minimize the computation capacity for training. The term sigmoid means ‘S-shaped’, and logistic form of the sigmoid maps [18] (Fig. 2).

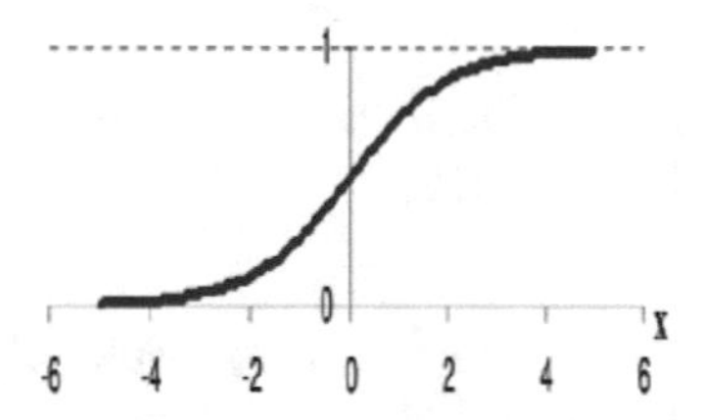

Fig. 2. Uni-polar sigmoid activation function

3.2.3 Bi-polar Sigmoid Function

The Bi-polar sigmoid activation function is given by Eq. (6).

$$f(x) = \frac{1 - e^{-x}}{1 + e^{-x}} \tag{6}$$

This function is similar to the Uni-polar sigmoid function. For this type of activation function described in Fig. 3, it goes well for applications that produce output values in the range of [− 1, 1] [18].

3.2.4 Hyperbolic Tangent Activation Function

Hyperbolic Tangent Function This function is easily defined as the ratio between the hyperbolic sine and the cosine functions or expanded as the ratio of the half-difference and half-sum of two exponential functions in the points x and $-x$ as Eq. (7):

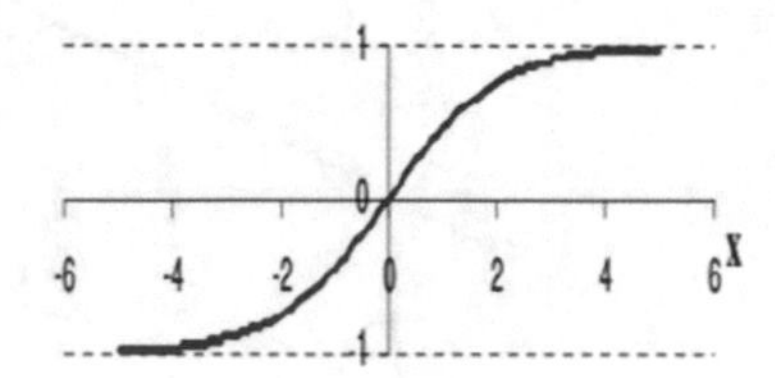

Fig. 3. Bi-polar sigmoid activation function

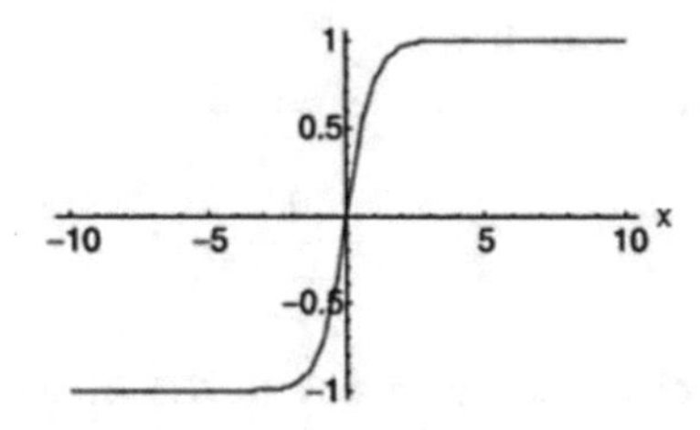

Fig. 4. Hyperbolic Tangent activation function

$$f(x) = \frac{1 - x^{-2x}}{1 + x^{-2x}} \tag{7}$$

Hyperbolic Tangent Function is similar to sigmoid function. Its range outputs between −1 and 1 as seen in Fig. 4 [18].

4 The Proposed Artificial Neural Network with Hyperbolic Tangent Activation Function

4.1 The Proposed COCOMO II Model

The model presented in this paper is inspired by the architecture referred by Reddy and Raju [11]. It explores the impact of variations of activation functions on software cost estimates. It accommodates the COCOMO II post architecture model, Unlike the work in [12] that used the sigmoidal activation function in the hidden and output layers [13]. The proposed model uses the Hyperbolic Tangent activation function at the hidden and the output layer. The use of neural network to estimate PM (person months) requires twenty four input nodes in the input layer. The proposed structure of the neural network accommodates to COCOMO II in Eq. (1). The COCOMO II model which is non-linear is transformed into a linear model using natural logarithms as Eq. (8).

$$\begin{aligned} \ln(PM) = {} & \ln(A) + \ln(EM_1) + \ln(EM_2) + \ldots + \ln(EM_{17}) \\ & + [1.01 + SF_1 + \ldots + SF_5] * \ln(size) \end{aligned} \tag{8}$$

After applying Eq. (8) to the neural network structure becomes as Eq. (9):

$$O_{PM} = [b_1 + wei_1 * x_1 + wei_2 * x_2 + \ldots + wei_{17} * x_{17}] + [b_2 + y_1 + \ldots + y_5] * [wsi_i + \ln(size)] \quad (9)$$

Where

$$O_{PM} = \ln(PM),$$
$$x_1 = \ln(EM_1); x_2 = \ln(EM_2); \ldots; x_{17} = \ln(EM_{17}),$$
$$y_1 = SF_1; \ldots; y_5 = SF_5.$$

The b_1 and b_2 are the biases and the coefficients wei_i and wsi_i are the additional terms used in the model which act as the weights from the input layer to the hidden layer. The COCOMO II model as given by Eq. (1) is used in the proposed work. The neural network consists of two hidden layer nodes H_{EM} and H_{SF} that take into account the contribution of effort multipliers and scale factors. The O_{PM} is the node of the output layer where we get the value of $\ln(PM)$ which is the desired output of the model. In this network all the original EM_i values of COCOMO II are preprocessed to $\ln(EM_i)$ and used as input nodes. The two bias values are denoted by b_1 and b_2, which are $\ln(A)$ and 1.01 respectively. The size of the product is not considered as one of the inputs to the network, but as a co-factor for the initial weights for scale factors (SF). The weights associated to the input nodes connected to the hidden layer are denoted by wei_i for $1 \leq i \leq 17$ for each input $\ln(EM_i)$ and b_1. On the other hand, the weights associated with the hidden layer for each $\ln(SF_i)$ input nodes and b_2 are $wsi_i + \ln(size)$ for $18 \leq i \leq 22$. These weights are initialized as $wei_i = 1$ and $wsi_i = 0$. The weights of the hidden layer to the output layer are denoted by weh and wsh initialized as $weh = wsh = 1$. As mentioned before, the activation function in the hidden layer and the output layer is the Hyperbolic Tangent activation function.

4.2 Training Algorithm

The feedforward backpropagation algorithm is used to train the network by iteratively processing a set of training samples and comparing the network's prediction with the actual value. For each training sample, the weights and bias are modified to minimize the error between the predicted value and the actual value. The following algorithm is used for training the proposed network and for calculating the new set of weights:

- Step 1: Initialize the weights and learning rate.
- Step 2: Perform Step 3–10 when stopping condition is false.
- Step 3: Perform Step 4–9 for each training pair.
- Step 4: Each input unit receives an input signal and sends it to the hidden unit.
- Step 5: Each hidden unit sums its weighted input signals to calculate net input given by Eqs. (10) and (11):

$$H_{EM} = b_1 + \sum_{i=1}^{17} x_i * wei_i \tag{10}$$

$$H_{SF} = b_2 + \sum_{j=18}^{22} y_j * \left(wsi_j + \ln(size)\right) \tag{11}$$

Then, hyperbolic tangent activation function will be applied and send the output signal from the hidden unit to the input of output layer units.

- Step 6: The output unit O_{PM} calculates the net input given by Eq. (12).

$$O_{PM} = H_{EM} * weh + H_{SF} * wsh \tag{12}$$

Apply hyperbolic tangent activation function over O_{PM} to compute the output signal E_{est}.

- Step 7: Calculate the error correction term *Error* as Eq. (13).

$$Error = E_{act} - E_{est} \tag{13}$$

Where E_{act} is the actual effort from the dataset and E_{est} is the estimated effort from step 6.

- Step 8: Update the weights between hidden and the output layers.
- Step 9: Update the weights and bias between input and hidden layers.
- Step 10: Check for the stopping condition. The stopping condition may be a certain number of epochs reached or if the error is smaller than a specific tolerance.

5 Results and Discussion

This section shows the experimental method and discusses the result that obtain when applying tangent activation function to the COCOMOII_SDR dataset for only one iteration. COCOMO_SDR dataset is from Turkish Software Industry it is available in the public domain from [19]. This dataset consists of 12 projects. It has 24 attributes: 22 attributes from COCOMO II model, one being the software size expressed in thousands of source line of code (SLOC) and the last being actual effort in man months. All the data is normalized to fit in the natural logarithmic space. The calculation was done using MATLAB. The evaluation consists in comparing the estimated effort with the actual effort. Estimated effort produced after applying Tangent Activation Function and Sigmoid Function were shown in Table 1. It shows that there are eight projects produced closer estimated effort to actual effort after implementing tangent activation function.

However, the result can be improved when they are applied to iteration more than one. Consequently, we expect that when we got the final values of weights by applying the backpropagation learning algorithm then apply final values to the proposed model it will produce more accurate result compared to other techniques.

Table 1. Comparison of actual and estimated effort using sigmoid and hyperbolic tangent functions

Project ID	Actual effort in log	Estimated effort using	
		Tangent activation function	Sigmoid activation function
1	0.1823	0.7742	0.8187
2	0.6931	0.5651	0.8282
3	1.5041	0.9280	0.8396
4	1.0986	0.9500	0.8541
5	1.3863	0.9489	0.8534
6	3.0910	0.5763	0.8040
7	0.6931	0.4625	0.7968
8	1.6094	0.9489	0.8533
9	2.8904	0.7980	0.8210
10	1.3863	0.9170	0.8398
11	0	0.5433	0.8014
12	0.7419	0.8296	0.8245

6 Conclusion

Software cost estimation is a complicated activity in software engineering due to the ambiguity and uncertainty of software attributes at the early stage of software development. Many models were proposed and developed in order to overcome ambiguous and uncertain software attributes. A COCOMO II is incorporated with soft computing techniques can reduce and overcome this ambiguity and uncertainty of the software attributes. This paper aims to incorporate the artificial neural network that uses hyperbolic tangent activation function in its hidden and output layers with the COCOMO II to handle the ambiguity of software attributes. Based on the obtained result, the proposed model produces more accurate results comparing to sigmoidal function. However, the result can be improved by applying to more iteration. In the future, several activation functions will be taken to be compared with the proposed model.

Acknowledgments. This paper was funded by Office for Research, Innovation, Commercialization and Consultancy Management (ORICC), UTHM.

References

1. Lindstrom, B.: A software measurement case study using GQM. J. Lund Univ., USA (2004)
2. Jones, C.: Software cost estimation in 2002. J. Def. Softw. Eng. **15**, 4–8 (2002)
3. Boehm, B.W.: Software Engineering Economics, vol. 197. Prentice-hall, Englewood Cliffs (1981)

4. Leung, H., Fan, Z.: Software cost estimation. In: Handbook of Software Engineering. Hong Kong Polytechnic University, pp. 1–14 (2002)
5. Strike, K., El Emam, K., Madhavji, N.: Software cost estimation with incomplete data. J. IEEE Trans. Softw. Eng. **27**, 890–908 (2001)
6. Garratt, P.W., Hodgkinson, A.C.: A Neurofuzzy Cost Estimator. IASTED/Acta Press, Anaheim (1999)
7. Putnam, L.H.: A general empirical solution to the macro software sizing and estimating problem. J. IEEE Trans. Softw. Eng. **4**, 345 (1978)
8. Boehm, B., Clark, B., Horowitz, E., Westland, C., Madachy, R., Selby, R.: Cost models for future software life cycle processes: COCOMO 2.0. J. Ann. Softw. Eng. **1**, 57–94 (1995)
9. Srinivasan, K., Fisher, D.: Machine learning approaches to estimating software development effort. J. IEEE Trans. Softw. Eng. **21**, 126–137 (1995)
10. Attarzadeh, I., Ow, S.H.: Improving estimation accuracy of the COCOMO II using an adaptive fuzzy logic model. In: 2011 IEEE International Conference on Fuzzy Systems (FUZZ), pp. 2458–2464. IEEE Press (2011)
11. Reddy, C.S., Raju, K.: A concise neural network model for estimating software effort. Int. J. Recent Trends Eng. **1**, 188–193 (2009)
12. Kaushik, A., Chauhan, A., Mittal, D., Gupta, S.: COCOMO estimates using neural networks. Int. J. Intell. Syst. Appl. **4**, 22 (2012)
13. Kaushik, A., Soni, A.K., Soni, R.: A simple neural network approach to software cost estimation. Global J. Comput. Sci. Technol. **13**, 23–30 (2013)
14. Attarzadeh, I., Ow, S.H.: Proposing an effective artificial neural network architecture to improve the precision of software cost estimation model. Int. J. Softw. Eng. Knowl. Eng. **24**, 935–953 (2014)
15. Boehm, B.W., Madachy, R., Steece, B.: Software cost estimation with Cocomo II with Cdrom. Prentice Hall PTR, Upper Saddle River (2000)
16. Stergiou, C., Siganos, D.: Neural networks 1996 (2010)
17. Moløkken, K., Jørgensen, M.: A review of software surveys on software effort estimation. In: 2003 International Symposium on Empirical Software Engineering, pp. 223–230. IEEE Press (2003)
18. Karlik, B., Olgac, A.V.: Performance analysis of various activation functions in generalized MLP architectures of neural networks. Int. J. Artif. Intell. Exp. Syst. **1**, 111–122 (2011)
19. PROMISE. http://openscience.us/repo/effort/cocomo/

A Study of Data Imputation Using Fuzzy C-Means with Particle Swarm Optimization

Nurul Ashikin Samat(✉) and Mohd Najib Mohd Salleh

Faculty of Computer Science and Information Technology,
Universiti Tun Hussein Onn Malaysia (UTHM),
Parit Raja, 86400 Batu Pahat, Johor, Malaysia
gi140022@siswa.uthm.edu.my, najib@uthm.edu.my

Abstract. An imputation method involving Fuzzy C-Means (FCM) with Particle Swarm Optimization (PSO) is implemented. The FCM is applied to identify the similar records in the complete dataset. Then, the records are optimized using PSO based on information from incomplete dataset. To evaluate the proposed method, experimental test are conducted using three datasets which is Cleveland Heart Disease, Iris and Breast Cancer dataset to verify the proposed method. Root Mean Square Error (RMSE) results of three different datasets are compared with seven different ratios of missing data. The results show the proposed approach can be used to the existing ones for imputation.

Keywords: Fuzzy C-Means · Particle swarm optimization · Imputation · Preprocessing

1 Introduction

Data mining works to extract the knowledge from a pile of data collection [1]. Various data mining methods such as classification, clustering and regression have been implemented into various real world field such as medical [2], banking [3], planting [4], education [5] and etc. Thus, data quality is very important concern for the researcher due to the bias towards the knowledge that be extracted can be happened. However, the real world data collection is not always complete and accurate [6]. The collection process might be involved and tangled with uncertain environments and consequently incomplete dataset have been appeared. This situation might arise from the several factors such as parallax error, human error, equipment error [7], noise data, irrelevant input data, and missing value recorded in the dataset [8]. Therefore, preprocessing process is considered important before training the data into the classifier models to maximize the accuracy and efficiency of machine learning techniques [9]. According to Rubin [10] missing data can be categorized to three which is (1) Missing completely at random (MCAR), (2) Missing at random (MAR), and (3) Missing not at random (MNAR). MCAR category is the missing value has no relationship or dependency towards other data set or variable. While MAR is the missing value that depended on other variables but the missing value can be obtained by estimated other complete

T. Herawan et al. (eds.), *Recent Advances on Soft Computing and Data Mining*,
Advances in Intelligent Systems and Computing 549, DOI 10.1007/978-3-319-51281-5_10

variables. The NMAR is the missing value that depended on other missing value, therefore the missing value cannot be estimated from existing data.

There are several ways to treat the incomplete dataset such as (1) Delete and ignore the missing data, (2) Parameter expectation, and (3) Imputation [10]. For delete and ignore method, it is not practical in several area of study as the data might contribute to the knowledge practical only when the data contain relatively small number of examples with missing values. While, for the second method are not efficient in computational time consuming and expensive.

In this paper, incomplete dataset due to missing value recorded in the dataset is focused. If the missing dataset is not treated, apart from less accuracy problems, it might create bias result, and loss efficiency of computational process due to the holes in the dataset. Thus, preprocessing process focused on handling the missing value in the dataset by using imputation method. This method has ability to manipulate and maximize the available information to find and fill up the missing attribute value with the most plausible value. Imputation methods have various types, from mean imputation until study the relationship among attributes. Choosing the right imputation is important as different methods will gives different performance.

This paper is organized as follows: Sect. 2 reviews the related works regarding FCM and PSO in imputation works, Sect. 3 overviews the Fuzzy C-Means Clustering algorithm and Particle Swarm Optimization algorithm. It also describes about the proposed methodology, FCMPSO used in this paper. Section 4 explained about experiments and result analysis. Finally, Sect. 5 provides the conclusion and recommendations.

2 Related Works

This section presents a brief summary related to imputation based on clustering methods and focused on Fuzzy C-Means and Particle Swarm Optimization that have been done by other researchers. From last decade, Hathaway and Bezdek [11] proposed that incomplete missing data can be obtained by calculate the distance between the missing and complete dataset. This estimation are then be proposed by other researcher in imputation method and we also applied this imputation idea in this study. Other than that, Li and Gu [12] used FCM with Nearest Neighbor (NN) intervals as the representation of missing data apart from numerical representation. They find the range of NN instead of using basic NN. It shows that minimum the range can increase the rate of accurate of imputation. In addition, Di Nuovo [13] applied FCM in psychological scenario for missing dataset because author mention that, analyzing and find relation between the data is important as they can contain valuable information. While Nishanth et al. [14] used K-Mean algorithm and multilayer perceptron (MLP) in solving missing data using imputation method. This method then is applied in predicting the severity of phishing alerts. Authors claimed that the proposed method yield better results in accuracy of data mining algorithm due to the imputed dataset. In addition, Aydilek and Arslan [15] used FCM together with support vector regression (SVM) and genetic algorithm (GA) to find the impute value. Author optimizes the FCM parameters with

GA which is the cluster size and weighting factor. They come out with conclusion that the ability of FCM is better when the parameters are optimized. But, FCM clustering imputation is just one iterative step to impute the missing data. It does not check how well the impute value to the dataset. Thus, we want to optimizes the imputed value within the range in the missing dataset, so that, the imputed value is more reliable.

Krishna *et al.* [16] use Particle swarm optimization with covariance matrix for imputation. Author want to preserve the covariance matrix of missing data and used PSO to minimize the mean squared error (MSE) and absolute difference. With help of PSO, the method gives better results for covariance matrix for complete data and imputes data. Gautam *et al.* [17] proposed imputation method using PSO with evolving clustering method (ECM) and auto associative extreme machine learning (AAELM). Authors also want to preserve the covariance structure and using PSO to choose the optimal value of *Dthr* needed by ECM. PSO can find better optimized value and give better results for imputation. Furthermore, Tang *et al.* [18] applied FCM with GA for missing traffic volume data. Author used GA to optimize the membership function and the centroids value in FCM model. With the help of GA, optimize FCM can give good imputation results for the problem. Rahman and Islam [19] proposed that missing value been imputed using fuzzy clustering based on Expectation Maximation (EM) approach. These approach been applied towards numerical and categorical missing values and they optimizes the FCM imputation based on the EM concept and yield a good results especially in time series tailoring problems.

From all the past researchers have done, it shows that FCM have an ability to find the imputation value with great results based on the centroid and membership function values. But, FCM just impute the missing data once in one time. Thus, we are proposed to used PSO to optimizes the impute value by recalculate the imputation value between the dataset until get the optimum value.

3 Fuzzy C-Means and Particle Swarm Optimization

The proposed method involves FCM and PSO are used to find the optimum value for finding the best value to replace the missing value in the dataset. The following section will be discussed about FCM and PSO approach before to the proposed method.

3.1 Fuzzy C-Means Clustering (FCM)

FCM was developed by Dunn in 1972 and improved by Bezdek in 1981. Fuzzy clustering algorithm allows one data to belong to two or more clusters. Traditional clustering algorithm are always partitioned the data into crisp division which membership function is either 0 or 1. Thus FCM come to overcome the problem with extend the membership function value in range 0 until 1 which eventually improve the result. But, FCM cannot handle missing value directly, thus, many researcher have done improvement towards this shortness. With the ability of FCM to separate the data towards different group with certain value, thus the missing value can be obtain by calculate the distance from complete dataset and used it.

FCM partitions set of n dataset $x = \{x_1, x_2, \ldots, x_n\}$ in R^d dimensional space into fuzzy cluster c, $1 < c < n$ with $r = \{r_1, r_2, \ldots, r_c\}$ cluster centers or centroids. Fuzzy clustering dataset is described in by fuzzy matrix μ with n rows and c columns which n is number of dataset and c is the number of clusters. While, μ_{ij} is the element in i^{th} row and j^{th}column in μ, shows the membership function of the ith dataset with the jth cluster. While μ was defines as follows,

$$\mu_{ij} \in [0, 1] \quad \forall i = 1, 2, \ldots, n; \forall i = 1, 2, \ldots, c \tag{1}$$

$$\sum\nolimits_{j=1}^{c} \mu_{ij} = 1, \forall i = 1, 2, \ldots, n \tag{2}$$

The objective function of FCM algorithm to minimize iteratively,

$$J_m = \sum\nolimits_{j=1}^{c} \sum\nolimits_{i=1}^{n} \mu_{ij}^m \left\| x_i - r_j \right\| \tag{3}$$

In which $m(m > 1)$ is a scalar termed the weighting exponent and controls the fuzziness of the resulting clusters and $\left\| x_i - r_j \right\|_2^2$ stands for the Euclidean distance from dataset, x_i to the cluster center,r_j. The centroid r_j of the jth cluster is obtained using,

$$r_j = \frac{\sum\nolimits_{i=1}^{n} \mu_{ij}^m x_i}{\sum\nolimits_{i=1}^{n} \mu_{ij}^m} \tag{4}$$

3.2 Particle Swarm Optimization (PSO)

PSO is swarm metaheuristic algorithm that can optimize the solutions of problems. Particle Swarm Optimization developed and introduced by Kennedy and Eberhart in 1995 based on the natural behavior of bird flocking or fish schooling to find the food. The flock of bird fly in a group follows the member that has closest distance to destination. In traditional PSO, population is called the swarm and the candidate of solutions in swarm is called particles while the food is called objective function.

The D-dimensional position for the particle i at iteration t can be represented as flows $x_i^t = \{x_{i1}^t, x_{i2}^t, \cdots, x_{iD}^t\}$ Each elements of particle contains parameter; own position, own velocity, and own historical information. Each particle is given random position in search space and random velocity for the particles to fly within the search space.

Let $p_i^t = \{p_{i1}^t, p_{i2}^t, \cdots, p_{iD}^t\}$ represent the best solution that particle i has obtained until iteration t, and $p_g^t = \{p_{g1}^t, p_{g2}^t, \cdots, p_{gD}^t\}$ denote the best solution obtained from p_i^t. in the population at iteration t. To search for the optimal solution, at each time step, each changes its velocity according to the *pbest* and *gbest* parts according to Eqs. (5) and (6), respectively:

$$v_{id}^{t} = v_{id}^{t-1} + c_1 r_1 (P_{id}^{t} - x_{id}^{t}) + c_2 r_2 (P_{gd}^{t} - x_{id}^{t}), d = 1, 2, \ldots, D \tag{5}$$

$$X_{id}^{t+1} = X_{id}^{t} + V_{id}^{t}, d = 1, 2, \ldots, D \tag{6}$$

Where c_1 indicates the cognition learning rates for individual ability, c_2 indicates the social learning factor and r_1, r_2 are random numbers uniformly distributed in the interval 0 and 1.

This advantage of PSO is applicable towards FCM imputation due to FCM used the closest distance knowledge between the complete data to find the imputable value for missing data.

4 The Proposed FCMPSO Method

In this section, a missing data imputation approach based on FCM and PSO is discussed. Figure 1 displays the proposed method in this study. The main operation is to group the data in the similar features with FCM to get the centroid values of each attribute and to calculate the distance between each missing data with each cluster. After that, the impute value will be optimized with PSO according to the information from missing dataset.

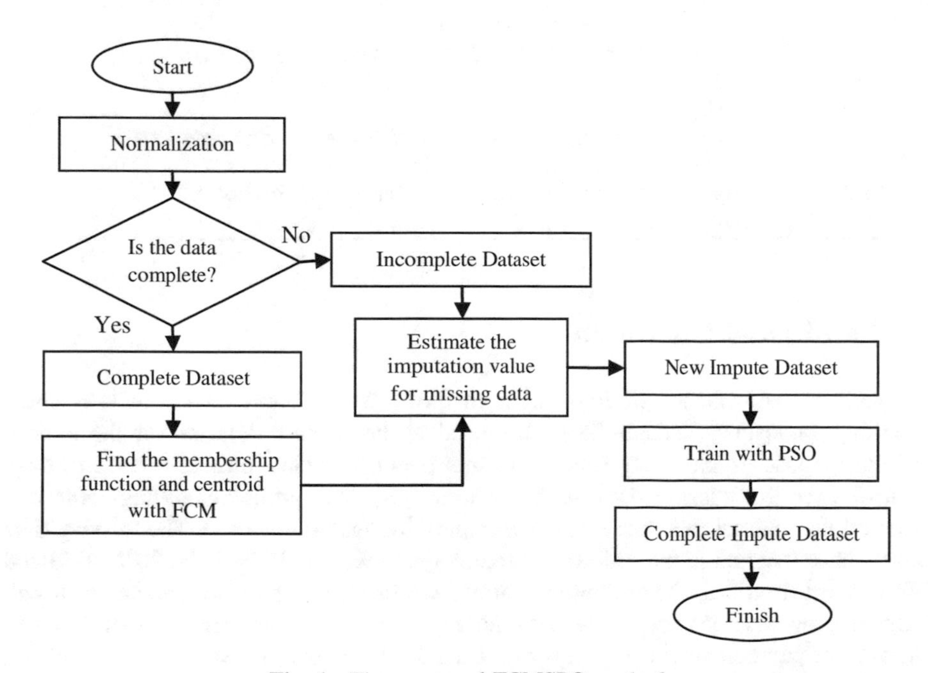

Fig. 1. The proposed FCMSPO method

The method of the proposed method is as follows:

```
Step 1:  Normalize the dataset using min-max
         normalization. Separate the Complete and
         Incomplete data.
Step 2:  For all Complete dataset,
           Calculate the cluster center using
           Equation 4.
           Compute the Euclidean distance ||x_i - r_j||_2^2
           Update the membership function using
           Equation 1, 2, and 3.
           If condition is not met, repeat Step 2.
Step 3:  For all Incomplete dataset
           Calculate the imputation value.
         End for;
Step 4:  For each Imputed dataset,
           Initialize the population size, N=50;
           Maximum Iterations, Iters=100;
           C1 and C2 = 2;
           And create a swarm with P particles
           Calculate the fitness each particle using
           equation (3);
           Calculate the pbest for the particles and
           best for the swarm;
           Update velocity for using equation (5) and
           u
           (6);
           Terminating if condition met otherwise
           repeats step 4.
Step 5:  The optimum value for Imputed dataset is
         obtained.
```

5 Results and Discussions

To test the performance of imputation method, three datasets from UCI Machine Learning Repository Dataset [20] were used as benchmark datasets for the experimental purpose as shown in Table 1. In this paper, complete dataset is used to have control over the missing data in the dataset [18]. The artificial missing ratio was inserted into dataset to analyse how the imputation method worked. The missing data value ratios inserted in the following percentages; 1%, 5%, 10%, 15%, 20%, 25% and 30%. Apart from that, the attribute is normalized using min-max normalization. It will helps to prevent larger range value attribute to overcome the smaller range attributes in distance calculation apart from helps to accelerate the training phase [15]. In this paper, FCMPSO results are compared with two other ways of imputation which is Nearest-Neighbor Imputation and Fuzzy C-Means imputation. The performance of proposed method will be evaluated with RMSE using Eq. 9 where n is the number of missing values, y_1 is the real value and y_2 is estimated value. A lower error value indicates better performance.

$$RMSE = \sqrt{\frac{1}{n}\sum_{i=1}^{n}(y_1 - y_2)^2} \tag{9}$$

5.1 Cleveland Heart Disease Dataset

Cleveland Heart Disease dataset was based on heart disease diagnosis in Cleveland Clinic Foundation by Dr. Robert Detrano. This dataset contains 297 records and 13 attributes, while the class contains two classes. Table 1 presents that proposed method FCMPSO displays improvement of RMSE values whereas, when the missing data ratios increased, the imputation gets better. Afterwards, Fig. 2 illustrates that FCMPSO outperforms compared to other two imputation methods by having lower RMSE values. A lower error value indicates better performance.

Table 1. RMSE results from cleveland heart disease.

Missing ratios	1	5	10	15	20	25	30
NN	0.05432	0.0577	0.0631	0.0695	0.0765	0.0832	0.0873
FCM	0.05207	0.05500	0.06130	0.06620	0.07110	0.07510	0.07780
FCMPSO	0.05231	**0.05360**	**0.06070**	**0.06540**	**0.06950**	**0.07350**	**0.07650**

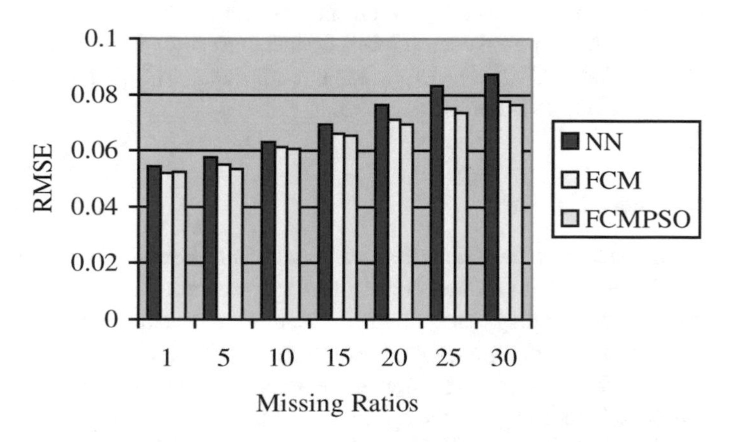

Fig. 2. Performance of RMSE comparison for cleveland heart disease

5.2 Iris Dataset

Iris is a flower data set from R.A. Fisher, which consists of 150 samples, four attributes and three classes which is Iris setosa, Iris virginica and Iris versicolor. Table 2 demonstrates that proposed method, FCMPSO shows a good improvement of RMSE

Table 2. RMSE results from Iris.

Missing ratios	1	5	10	15	20	25	30
NN	0.01305	0.0131	0.0487	0.0487	0.0629	0.0744	0.0784
FCM	0.00994	0.01370	0.01020	0.01970	0.02370	0.03170	0.11800
FCMPSO	0.00969	0.01390	0.02040	**0.01530**	**0.01860**	**0.02380**	**0.07600**

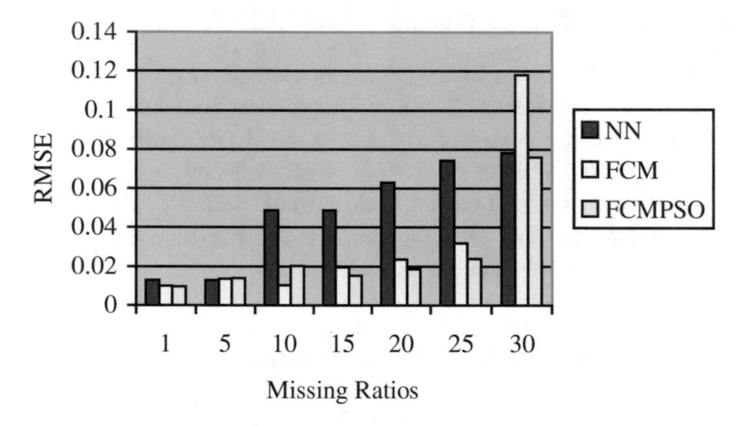

Fig. 3. Performance of RMSE comparison for Iris

whereas when the missing data ratios increased, the imputation gets better. Even though the proposed method not give a good RMSE values when the missing ratios is less than 10%, but when the missing ratios increase, FCMPSO performed better. It clearly point out that, proposed method can be used to high ratios missing dataset. Figure 3 exhibits that FCMPSO outperforms compared to other two imputation methods for 15%, 20%, 25% and 30%.

5.3 Breast Cancer Dataset

Breast cancer dataset was created based on Wisconsin Breast Cancer problem by Dr. William H. Wolberg. This dataset contain 683 records and 9 attributes. This dataset contains 2 classes. Table 3 displays the RMSE value of proposed method with the other 2 imputation methods. The proposed method not gives the best results for 1% and 5% missing ratios, but when the missing ratios are increasing, the RMSE value is smaller. It shows that, proposed method gives the optimum value for imputation (Refer to Fig. 4).

Consequently, from overall findings, it reveals that FCMPSO gives a lower RMSE compared to other imputation methods over the increasing of missing ratios. It highlights that, the missing dataset is important and can give information for better imputations. Thus, further studies can be done towards the sensitivity of FCMPSO towards the dataset properties and condition.

Table 3. RMSE results from breast cancer.

Missing ratios	1	5	10	15	20	25	30
NN	0.44794	0.44790	0.44780	0.44790	0.44790	0.44790	0.44800
FCM	0.44795	0.44770	0.44730	0.44660	0.44660	0.44660	0.44620
FCMPSO	0.44793	0.44780	**0.44720**	**0.44660**	**0.44650**	**0.44640**	**0.44580**

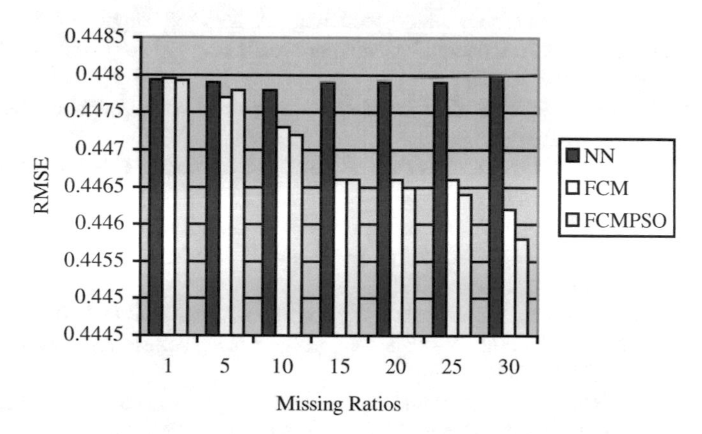

Fig. 4. Performance of RMSE comparison for breast cancer

6 Conclusion and Future Works

In this study, an imputation method using FCM and PSO are proposed which is aimed to optimize the imputation values for missing values in dataset. The effectiveness of proposed method is demonstrated on three different datasets which contains seven different missing ratios. PSO is used to find the best value for imputation by searching the best position of distance value and the result gives promising results as traditional clustering imputation.

Acknowledgment. This research was funded by Universiti Tun Hussein Onn Malaysia Research Contract Grant Scheme (VOT No U560).

References

1. Qiu, J., Wu, Q., Ding, G., Xu, Y., Feng, S.: A survey of machine learning for big data processing. EURASIP J. Adv. Sig. Process. **2016**, 1–16 (2016)
2. Teimouri, M., Farzadfar, F., Soudi Alamdari, M., Hashemi-Meshkini, A., Adibi Alamdari, P., Rezaei-Darzi, E., Varmaghani, M., Zeynalabedini, A.: Detecting diseases in medical prescriptions using data mining tools and combining techniques. Iranian J. Pharm. Res. **15**, 113–123 (2016)

3. Raju, P.S., Bai, D.V.R., Chaitanya, G.K.: Data mining: techniques for enhancing customer relationship management in banking and retail industries. Int. J. Innov. Res. Comput. Commun. Eng. **2**, 2650–2657 (2014)
4. Heung, B., Ho, H.C., Zhang, J., Knudby, A., Bulmer, C.E., Schmidt, M.G.: An overview and comparison of machine-learning techniques for classification purposes in digital soil mapping. Geoderma **265**, 62–77 (2016)
5. Romero, C., Ventura, S.: Data mining in education. Wiley Interdisc. Rev. Data Min. Knowl. Discov. **3**, 12–27 (2013)
6. Salleh, M.N.M.: Implementing fuzzy modeling of decision support for crop planting management. In: 2013 International Conference on Fuzzy Theory and Its Applications (iFUZZY), pp. 161–166 (2013)
7. Amiri, M., Jensen, R.: Missing data imputation using fuzzy-rough methods. Neurocomputing **205**, 152–164 (2016)
8. Kotsiantis, S.B., Zaharakis, I., Pintelas, P.: Supervised machine learning: a review of classification techniques (2007)
9. Saleem, A., Asif, K.H., Ali, A., Awan, S.M., Alghamdi, M.A.: Pre-processing methods of data mining. In: 2014 IEEE/ACM 7th International Conference on Utility and Cloud Computing (UCC), pp. 451–456 (2014)
10. Rubin, D.B.: Inference and missing data. Biometrika **63**, 581–592 (1976)
11. Bezdek, J.C., Ehrlich, R., Full, W.: FCM: the fuzzy c-means clustering algorithm. Comput. Geosci. **10**, 191–203 (1984)
12. Li, D., Gu, H., Zhang, L.: A fuzzy c-means clustering algorithm based on nearest-neighbor intervals for incomplete data. Expert Syst. Appl. **37**, 6942–6947 (2010)
13. Di Nuovo, A.G.: Missing data analysis with fuzzy C-Means: A study of its application in a psychological scenario. Expert Syst. Appl. **38**, 6793–6797 (2011)
14. Nishanth, K.J., Ravi, V., Ankaiah, N., Bose, I.: Soft computing based imputation and hybrid data and text mining: the case of predicting the severity of phishing alerts. Expert Syst. Appl. **39**, 10583–10589 (2012)
15. Aydilek, I.B., Arslan, A.: A hybrid method for imputation of missing values using optimized fuzzy c-means with support vector regression and a genetic algorithm. Inf. Sci. **233**, 25–35 (2013)
16. Krishna, M., Ravi, V.: Particle swarm optimization and covariance matrix based data imputation. In: 2013 IEEE International Conference on Computational Intelligence and Computing Research (ICCIC), pp. 1–6. IEEE (2013)
17. Gautam, C., Ravi, V.: Data imputation via evolutionary computation, clustering and a neural network. Neurocomputing **156**, 134–142 (2015)
18. Tang, J., Zhang, G., Wang, Y., Wang, H., Liu, F.: A hybrid approach to integrate fuzzy C-means based imputation method with genetic algorithm for missing traffic volume data estimation. Transp. Res. Part C: Emerg. Technol. **51**, 29–40 (2015)
19. Rahman, M.G., Islam, M.Z.: Missing value imputation using a fuzzy clustering-based EM approach. Knowl. Inf. Syst. **46**, 1–34 (2015)
20. Frank, A., Asuncion, A.: UCI machine learning repository (2010)

Utilizing Clonal Selection Theory Inspired Algorithms and *K*-Means Clustering for Predicting OPEC Carbon Dioxide Emissions from Petroleum Consumption

Ayodele Lasisi[1(✉)], Rozaida Ghazali[1], and Haruna Chiroma[2]

[1] Faculty of Computer Science and Information Technology, Universiti Tun Hussein Onn Malaysia, 86400 Parit Raja, Batu Pahat, Johor, Malaysia
lasisiayodele@yahoo.com, rozaida@uthm.edu.my

[2] Department of Computer Science, School of Science, Federal College of Education (Technical), Gombe, Gombe State, Nigeria
freedonchi@yahoo.com

Abstract. The prediction of carbon dioxide (CO_2) emissions from petroleum consumption inspired and motivated this research. Over the years, the rate of emissions of CO_2 continues to multiply, resulting in global warming. This paper thus proposes the use of clonal selection theory inspired algorithms; CLONALG and AIRS to forecast global CO_2 emissions. The *K*-means algorithm divides the data into groups of similar and meaningful patterns. Comparative simulations with multi-layer Perceptron, IBk, fuzzy-rough nearest neighbor, and vaguely quantified nearest neighbor reveal that the CLONALG and AIRS produced outstanding results, and are able to generate highest detection rates and lowest false alarm rates. As such, gathering useful information with the accurate prediction of CO_2 emissions can help to reduce the emission of CO_2 contributions to global warming which assist in policies on climate change.

Keywords: Clonal selection algorithm · Artificial immune recognition system · Clustering algorithm · Carbon dioxide emissions

1 Introduction

The emission of greenhouse gases is as a result of energy consumption [1], which in turn causes global warming. Components of the greenhouse emissions include carbon dioxide (CO_2) emissions from petroleum consumption. An increase in energy consumption by the Organization of the Petroleum Exporting Countries (OPEC) from 1970 to 2010 was recorded at 685%, whereas a 440% increment in the emissions of CO_2 due to burning fossils fuels surfaced in the concurrent years. This affirms that the CO_2 emissions has projected upward, and could be hazardous to human health.

Various attempts have been made to reduce the emission of CO_2 using computational intelligent algorithms. One of such propositions is the application of fuzzy regression and backpropagation neural network termed FRBPNN towards estimating

T. Herawan et al. (eds.), *Recent Advances on Soft Computing and Data Mining*,
Advances in Intelligent Systems and Computing 549, DOI 10.1007/978-3-319-51281-5_11

the global concentration of CO_2 [2], and shows to boost the accuracy of CO_2 estimation. A collaborative fuzzy neural network was used to improve FRBPNN in the prediction accuracy of CO_2 [3]. Experimental results inform that collaborative fuzzy neural network performed better than FRBPNN and other statistical methods. The relation of CO_2 emissions to energy in developing countries is predicted using genetic algorithm (GA) and grey model [4]. GA was incorporated to improve the fitting ability of the grey model. This hybridization was then combined with BPNN to improve nonlinear approximation ability, and able to produce an enhanced performance through energy load prediction.

To further upgrade the prediction accuracy performances of CO_2 emissions; this paper proposes the injection of clonal selection theory inspired algorithms namely; CLONal selection ALGorithm (CLONALG) [5] and Artificial Immune Recognition System (AIRS) [6], both of which has exhibited their proficiencies by training and testing in [7]. CLONALG and AIRS fall into the family of Artificial Immune System (AIS), with the work done on Immune Network [8] and Negative Selection Algorithm [9–11] bridging the gap between immunology and computing. The dataset gathered is unlabelled. Therefore, *K*-means clustering algorithm is applied to partition the dataset into clusters which makes it appropriate as input for training and testing the algorithms.

The structure of the paper is as follow: Sect. 2 discusses the clonal selection theory, clonal selection algorithm, and artificial immune recognition system. The *K*-means algorithm for clustering occupies Sect. 3. Experimentations are reported and analysed in Sect. 4. Conclusion sums up Sect. 5.

2 Clonal Selection Theory

The adaptive immunity plays a major role when the clonal selection theory is concerned. Postulated by Burnet [12], it demonstrate how the B-cells proliferate following recognizing an antigen. Not all the B-cells are initiated for proliferation, as only those capable of recognition are granted permission. The process of antigen recognition is by binding the antibodies secreted by the B-cells to the antigen, and this activates with differentiating into memory cells. Thus, these memory cells execute a rapid response if such an antigen is encounter again.

2.1 Clonal Selection Algorithm (CLONALG)

The proposition of CLONALG by De Castro and Von Zuben [5] is inspired from the inherent attributes of clonal selection principle. The objective of CLONALG is tied to the antibodies generated by the B-cells, which tend to multiply thereby producing a pool of antibodies for its recognition task. To further enhance its performance, a local search strategy and a global search strategy are applied. The local search is through affinity maturation (hypermutation) of cloned antibodies, and global search via the generation of random antibodies added to the population to amplify the search and avoid local optima. It should be noted that the number of cloned antibodies, created from each antibody, are proportional to their affinity by utilizing a rank-based measure and calculated using Eq. 1.

$$numClones = \left\lfloor \frac{\beta \cdot N}{i} + 0.5 \right\rfloor \tag{1}$$

where β is the clonal factor, N is the size of the antibody pool, and i is the antibody current rank for $i \in [1, n]$. For each antigen exposure to the system, the total number of clones produced is calculated in Eq. 2.

$$Nc = \sum_{i=1}^{n} \left\lfloor \frac{\beta \cdot N}{i} + 0.5 \right\rfloor \tag{2}$$

where Nc is the total number of clones, with n representing the selected antibodies.

2.2 Artificial Immune Recognition System (AIRS)

The Artificial Immune Recognition System (AIRS), proposed in [6] got inspiration from the clonal expansion and affinity maturation of the clonal selection. The clonal expansion stage entails the rapid proliferation of the B-cells, proportional to binding affinity of antigens. The B-cells undergo somatic hypermutation, and a selection process affirming a given cell to attain the memory cell status. In AIRS, the generation and population of memory cells is now used for classification. The Artificial Recognition Ball (ARB), used in the AIRS algorithm, is a representation of similar B-cells for the reduction of B-cells duplication and survival within the population.

Furthermore, by combing through the AIRS algorithm, the following stages are revealed. The first is data normalization and initialization. The goal is to normalize the data within the range of [0, 1]. The primary measure used for this task is inverted Euclidean distance. This is achieved by including Eq. 3 in the normalization process. Also, a different method could be used without data normalization for acquiring a distance value in the range of [0, 1]. This is accomplished via dividing the calculated Euclidean distance by the maximum distance between two vectors depicted in Eqs. 4 and 5 respectively.

$$normalisedValue = normalisedValue \cdot \sqrt{\frac{1}{n}} \tag{3}$$

where *normalisedValue* is data attribute in the range of [0, 1], and n is the number of attributes.

$$Eucdist = \sqrt{\sum_{i=1}^{n} (x1_i - x2_i)^2} \tag{4}$$

where *Eucdist* stands for Euclidean distance, $x1_i$ and $x2_i$ are two elements for affinity measurement between them, and n is the number of attributes.

$$\max Dist = \sqrt{\sum_{i=1}^{n} r_i^2} \tag{5}$$

where $\max Dist$ is the maximum distance, and r is the data range for attribute i.

The division of Euclidean distance by maximum distance ultimately translates to the affinity depicted in Eq. 6. Affinity value is inversely proportional to affinity which means the smaller the affinity value, the higher the affinity becomes.

$$affinity = \left(\frac{Eucdist}{\max Dist}\right) \tag{6}$$

The calculation of affinity threshold, defined as the average affinity value between antigens in the training dataset, follows immediately after normalization. Seeding of the memory cells and initial population of ARB culminates the first stage of AIRS. The second stage is training of antigen. The antigens are exposed to the memory pool one at a time and the memory cell with the greatest stimulation is selected for use in the affinity maturation process as the best match memory cell calculated in Eq. 7. Selected from the memory cells are a number of mutated clones which are added to the ARB pool. The number of best matched mutated clones generated is calculated in Eq. 8.

$$stimulation = 1 - affinity \tag{7}$$

$$numClones = stimulation \cdot clonalRate \cdot hypermutationRate \tag{8}$$

where *stimulation* is simulation between the best matched memory cell and antigen. *clonalRate* and *hypermutationRate* are parametric input by user.

The competition for resources in the development of a candidate memory cell occupies the third stage of AIRS algorithm. The final stage of AIRS is the introduction of potential candidate memory cell into the set of established memory cells. The pool of memory cell now becomes the core of the AIRS classifier. Classification is performed with k-Nearest Neighbor where the matching of the k best to data instances are located, and class determined using majority vote.

3 *K*-Means Clustering Algorithm

The algorithms applied for clustering aim to partition data into a specified number of clusters i.e. groups, subsets, or categories. The definition of clustering varies as there is no acclaimed one [13]. Basically, most researchers describe a cluster as exhibiting similar patterns within the same cluster, which differs from patterns in another cluster [14]. Two types of clustering exist namely; partitioning clustering and hierarchical clustering [14], but focus will be on partitioning clustering as K-means falls under this category of clustering. Mathematical description is as follows:

The notion is: Given a set of input patterns $X = \{x_1, \ldots, x_j, \ldots, x_N\}$, where $x_j = (x_{j1}, x_{j2}, \ldots, x_{jd})^T \in \Re^d$ and each measure x_{ji} is said to be a feature i.e. attribute, dimension, or variable.

(a) Partitioning clustering strives to acquire K-partition of X, $C = \{C_1,\ldots,C_K\}$ $(K \leq N)$, such that
  i. $C_i \neq \phi, i = 1, \ldots, K$;
  ii. $\cup_{i=1}^{K} C_i = X$;
  iii. $C_i \cap C_j = \phi, i,j = 1, \ldots, K$ and $i \neq j$.

The process of K-means clustering describes the best partitioning of a data containing k number of clusters. It is defined by its objective function with the goal of minimizing the sum of all squared distances within a cluster, for all clusters. The objective function is defined as:

$$\arg\min_{S} \sum_{i=1}^{k} \left(\sum_{x_J \in S_i} \left\| x_j - \mu_i \right\|^2 \right) \tag{9}$$

where x_j is a data point in the dataset, S_i is a cluster and μ_i is the cluster means (corresponding to the center point of cluster S_i). It should be noted that the K-means algorithm is also known as Lloyd's algorithm. In 1957, Stuart Lloyd proposed the first heuristic algorithm to perform K-means clustering [15]. It iteratively improves the position of the cluster centers in the following steps:

$$S_i = \{x_j : \left\| x_j - \mu_i \right\| \leq \left\| x_j - \mu_c \right\| \forall 1 \leq c \leq k\} \tag{10}$$

$$\mu_i = \frac{1}{|S_i|} \left(\sum_{x_j \in S_i} x_j \right) \tag{11}$$

In Eq. 10, all the data points $x_1 \ldots x_n$, are assigned to one of its closet center μ_i. With respect to Eq. 11, all centers $\mu_1 \ldots \mu_k$, are updated by calculating the mean of all data points in the cluster. As a summary, the pseudocode of K-means clustering is shown in Algorithm 1.

Algorithm 1. K-Means Clustering (Lloyd's Algorithm)

1: $\mu_1, \ldots, \mu_k \leftarrow$ randomly chosen centers
2: **while** Objective function still improves **do**
3: $S_1, \ldots, S_k \leftarrow \phi$
4: **for** $j \in 1, \ldots, n$ **do**
5: $i \leftarrow \arg\min_{i'} \|x_j - \mu_{i'}\|^2\}$
6: add i to S_j
7: **end for**
8: **for** $i \in 1, \ldots, k$ **do**
9: $\mu_i = \frac{1}{|S_i|}\left(\sum_{x_j \in S_i} x_j\right)$
10: **end for**
11: **end while**

4 Experimental Procedure and Analysis

The experiments focus on the application of clonal selection theory inspired algorithms for predicting OPEC CO_2 emission from petroleum consumption which are CLONALG and AIRS algorithms. The *K*-means algorithm's purpose is for clustering the unlabelled data with similar patterns. All analyses are carried out with Waikato Environment for Knowledge Analysis (WEKA). For comparison, the Multi-Layer Perceptron (MLP) [16], IBk (Instance-based method on *k*-NN neighbor) [17], Fuzzy-Rough Nearest Neighbor (FRNN) [18], and Vaguely Quantified Nearest Neighbor (VQNN) [19] are incorporated. A 10-fold cross validation is used for testing and evaluating. The process entails dividing the dataset into ten subsets of equal size, with nine subsets for the training data, and one subset as the test data. Performance statistics are calculated across all 10 trails. Parameters values for CLONALG are: antibody pool size (N) = 50, clonal factor (β) = 0.9, number of generations (G) = 10, selection pool size (n) = 20, and total replacement = 2. For AIRS, the parameters are: affinity threshold scalar (ATS) = 0.2, clonal rate = 10, hypermutation rate = 2, total resources = 150, stimulation threshold = 0.9, and *k*-NN = 3. The parameter settings for *K*-means clustering are number of cluster (k) = 2, and seed = 10.

The performance evaluation metrics are represented in Eq. 12 for Detection Rate (DR) and Eq. 13 for False Alarm Rate (FAR).

$$DR = \frac{TP}{TP + FN} \tag{12}$$

$$FAR = \frac{FP}{FP + TN} \quad (13)$$

where *TP* and *FN* depicts true positives and false negatives, while *FP* and *TN* depicts false positives and true negatives.

4.1 Dataset

The OPEC CO_2 emission dataset from 1980 to 2011 was retrieved through [20] which is a reliable reference of energy data [21]. Measurements in the dataset are measured in million metric tons (mmt). This is a yearly data resting on the fact that the availability of the data is on yearly basis. The data contains OPEC CO_2 emissions from 12 countries and the summation of all the countries OPEC CO_2 emissions. There exist 13 columns and 32 rows within the dataset.

In order to pass the data as input for algorithm's execution, the total OPEC CO_2 emissions is the dependent variable and the OPEC CO_2 emissions from the 12 countries are the independent variable serving as the input. Thus, the OPEC CO_2 emissions acquired by the 12 countries are used as input to predict the OPEC CO_2 emissions from petroleum consumption.

4.2 Simulation Results and Discussions

The simulations are implemented on 3.40 GHz Intel Pentium® Core i7 Processor with 4 GB of RAM. Results of experimentations are tabulated, graphed, and discussed.

Firstly, the result of the *K*-means clustering algorithm as applied on the unlabelled data of the OPEC CO_2 emission dataset with $k = 2$, is shown in Table 1. The data of 32 instances is partitioned into two groups of clusters containing 22 and 10 instances respectively.

Table 1. Data partition of OPEC CO_2 emission data using *K*-means

Algorithm	Size of data	Cluster instances	% of data size in each cluster	Cluster class
K-Means	32	22	69%	cluster0
		10	31%	cluster1

After the data has been categorized into groups and also labeled, it now becomes suitable to be uploaded as input for algorithms' execution. Table 2 reveals the simulation results of all algorithms on OPEC CO_2 emission data.

From the graph illustrations in Figs. 1 and 2, algorithms have been labeled 1 to 6 with CLONALG standing for label 1, AIRS for label 2, up until the last algorithm in Table 2 indicating VQNN for label 6 in ascending order. On the basis of detection rate

Table 2. Experimental Results for OPEC CO_2 Emission Prediction

Algorithms	Detection rate (%)	False alarm rate (%)
CLONALG	96.90	6.90
AIRS	100	0.00
MLP	96.90	1.40
IBk	96.90	1.40
FRNN	93.80	8.30
VQNN	93.80	13.80

for OPEC CO_2 emission, it is shown that all the algorithms performed above the 93.00% range. Two of the algorithms namely; FRNN and VQNN produced detection rates of 93.80% respectively. Also, the CLONALG, MLP, and IBk algorithms equally generated detection rates of 96.90%. The highest detection rate is attributed to AIRS at 100%. AIRS has proven to surpass all the other algorithms when trained and tested on the OPEC CO_2 emission data for detection rate. The results in Table 2 are diagrammatically represented in Fig. 1.

Taking an insight at the results with respect to false alarm rate, in ascending order of performance, VQNN gave the lowest rate at 13.80%. Next is FRNN algorithm at 8.30% false alarm rate. Following closely is CLONALG producing rate of 6.90%, and both the MLP and IBk algorithms account for false rate at 1.40% respectively. Finally, AIRS generated false rate of 0.00%, and this surmounts the other algorithms when false alarm rate is concerned because the lower the false rate, the better the performace. The Fig. 2 below shows the plot analysis for false alarm rate.

Overall, the clonal theory inspired algorithm of AIRS is capable of providing an enhanced prediction of OPEC CO_2 emissions from the consumption of petroleum in comparison to CLONALG, MLP, IBk, FRNN, and VQNN that also contributed good performance rates.

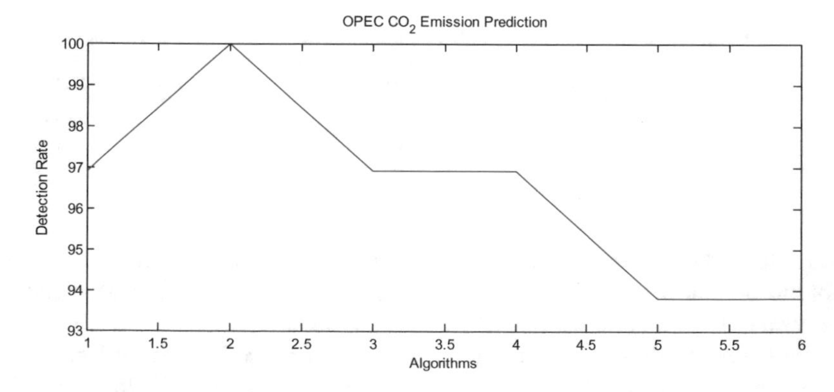

Fig. 1. Graph plots for detection rate

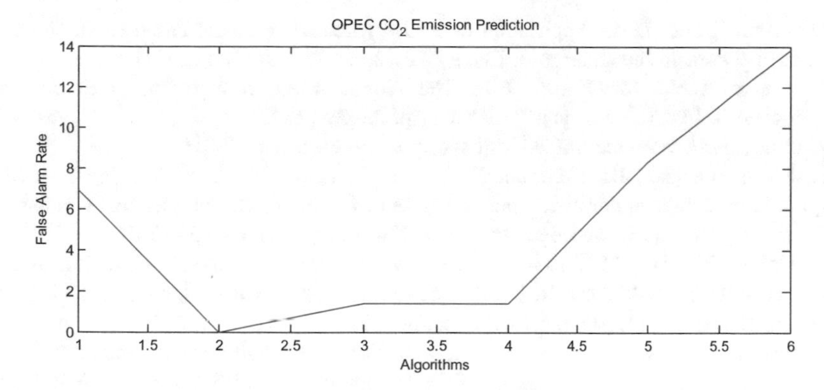

Fig. 2. Graph plots for false alarm rate

5 Conclusion

The proliferation and adverse effect of greenhouse gas emissions among OPEC countries channeled the proposition of CLONal selection ALGorithm (CLONALG) and Artificial Immune Recognition System (AIRS), with the adoption of *K*-means clustering algorithm, to increase the detection and lower false rates leading to better prediction of CO_2 emissions from the consumption of petroleum. The Energy Information Administration provided access to the dataset used. A total of four algorithms are employed for comparison purpose which are MLP, IBk, FRNN, and VQNN. Experimental performances indicated that while CLONALG performed on the same level as MLP and IBk, it outclassed FRNN and VQNN, in terms of detection rate. Best detection and false rates were generated by AIRS. These results ultimately propel better accurate predictions thereby causing a reduction in CO_2 emissions. Future research will be to combine metaheuristic algorithms with the clonal selection methods in this study to forecast the emission of CO_2.

Acknowledgements. This work is supported by the Office for Research, Innovation, Commercialization, and Consultancy Management (ORICC), Universiti Tun Hussein Onn Malaysia (UTHM), and Ministry of Higher Education (MOHE) Malaysia under the Fundamental Research Grant Scheme (FRGS) Vote No. 1235.

References

1. Sari, R., Soytas, U.: Are global warming and economic growth compatible? Evidence from five OPEC countries? Appl. Energy **86**, 1887–1893 (2009)
2. Chen, T., Wang, Y.C.: A fuzzy-neural approach for global CO2 concentration forecasting. Intell. Data Anal. **15**, 763–777 (2011)
3. Chen, T.: A collaborative fuzzy-neural system for global CO2 concentration forecasting. Int. J. Innov. Comput. Inf. Control **8**, 7679–7696 (2012)

4. Tang, N., Zhang, D.-J.: Application of a load forecasting model based on improved grey neural network in the smart grid. Energy Procedia. **12**, 180–184 (2011)
5. de Castro, L.N., von Zuben, F.J.: The clonal selection algorithm with engineering applications. In: Proceedings of GECCO, pp. 36–39 (2000)
6. Watkins, A.: A resource limited artificial immune classifier (2001)
7. Lasisi, A., Ghazali, R., Herawan, T., Lasisi, F., Deris, M.M.: Knowledge extraction of agricultural data using artificial immune system. In: 2015 12th International Conference on Fuzzy Systems and Knowledge Discovery (FSKD), pp. 1653–1658 (2015)
8. Bersini, H., Varela, F.J.: Hints for adaptive problem solving gleaned from immune networks. In: Schwefel, H.P., Männer, R. (eds.) PPSN 1990. LNCS, vol. 496, pp. 343–354. Springer, Berlin, Heidelberg (1991). doi:10.1007/BFb0029775
9. Forrest, S., Perelson, A.S., Allen, L., Cherukuri, R.: Self-nonself discrimination in a computer. In: 1994 Proceedings of IEEE Computer Society Symposium on Research in Security and Privacy, 1994, pp. 202–212 (1994)
10. Lasisi, A., Ghazali, R., Herawan, T.: Negative selection algorithm: a survey on the epistemology of generating detectors. In: Herawan, T., Deris, M.M., Abawajy, J. (eds.). LNEE, vol. 285, pp. 167–176. Springer, Singapore (2014). doi:10.1007/978-981-4585-18-7_20
11. Lasisi, A., Ghazali, R., Herawan, T.: Comparative performance analysis of negative selection algorithm with immune and classification algorithms. In: Herawan, T., Ghazali, R., Deris, M.M. (eds.). AISC, vol. 287, pp. 441–452. Springer, Cham (2014). doi:10.1007/978-3-319-07692-8_42
12. Burnet, S.F.M.: The Clonal Selection Theory of Acquired Immunity. University Press Cambridge, Cambridge (1959)
13. Everitt, B.S., Landau, S., Leese, M.: Cluster analysis. Arnold, London (2001)
14. Xu, R., Wunsch, D.: Survey of clustering algorithms. IEEE Trans. Neural Netw. **16**, 645–678 (2005)
15. Jain, A.K.: Data clustering: 50 years beyond K-means. Pattern Recogn. Lett. **31**, 651–666 (2010)
16. Fine, T.L.: Feedforward Neural Network Methodology. Springer Science & Business Media, New York (1999)
17. Aha, D.W., Kibler, D., Albert, M.K.: Instance-based learning algorithms. Mach. Learn. **6**, 37–66 (1991)
18. Jensen, R., Cornelis, C.: A new approach to fuzzy-rough nearest neighbour classification. In: Chan, C.C., Grzymala-Busse, J., Ziarko, W. (eds.) Rough Sets and Current Trends in Computing, pp. 310–319. Springer, Heidelberg (2008)
19. Jensen, R., Cornelis, C.: Fuzzy-rough nearest neighbour classification and prediction. Theor. Comput. Sci. **412**, 5871–5884 (2011)
20. Energy Information Administration of the US Department of Energy (2014)
21. Chiroma, H., Abdulkareem, S., Abubakar, A., Usman, M.J.: Computational intelligence techniques with application to crude oil price projection: a literature survey from 2001–2012. Neural Netw. World **23**, 523 (2013)

One-Way ANOVA Model with Fuzzy Data for Consumer Demand

Pei Chun Lin[1(✉)], Nureize Arbaiy[2], and Isredza Rahmi Abd. Hamid[2]

[1] Graduate School of Information, Production and System, Waseda University, 2-7 Hibikino, 808-0135 Kitakyushu, Fukuoka, Wakamatsu-ku, Japan
peichunpclin@gmail.com

[2] Faculty of Computer Science and Information Technology, University Tun Hussein Onn Malaysia, 86400 Batu Pahat, Johor, Malaysia
{nureize, rahmi}@uthm.edu.my

Abstract. This paper presents a statistical method which could distinguish the customer's demand into different type whereby fuzzy data is in consideration. A one-way analysis of variance (ANOVA) model for fuzzy data is introduced with hypothesis test, F-test, which is the pivot statistic in ANOVA model. In the experiment, several different factors in testing with one-way ANOVA model are considered. The results of this study indicate that the solution method introduced in this paper could give decision maker a result with favorable degree of each factor. This kind of result is beneficial to the decision maker and retailer to distinguish which factor is the most critical for the customer and with how much amount of products would be allocated for customers.

Keywords: ANOVA model · Fuzzy numbers · Multinomial analysis · Marketing survey · T-test · F-test

1 Introduction

In social science research, many decisions, evaluations and psychological tests are conducted using surveys and/or questionnaires to seek people's opinions. It is routine to ask people about their opinions according to binary, multiple-choice questions. In fact people have complex and/or vague thoughts [1, 18]. However there is no consistent statistical test which could be able to deal with vague data from market surveys. In view of this, we would like to introduce a statistical method which could deal with fuzzy data from marketing survey in this paper.

The analysis of variance (hereafter, denoted as ANOVA) model was originally used in the analysis of agricultural research data. Due to the strength and versatility of the technique, the ANOVA model is commonly used in almost all research areas especially in social science research and managerial decision making. Usually, ANOVA model is used to observe the difference between groups on certain variable and available for both parametric (score data) and non-parametric (ranking/ordering) data [2]. It is more extensively used for analyzing each kind of data. Hence, we consider building up an ANOVA model with fuzzy data in this paper.

T. Herawan et al. (eds.), *Recent Advances on Soft Computing and Data Mining*,
Advances in Intelligent Systems and Computing 549, DOI 10.1007/978-3-319-51281-5_12

There are few researcher works on variance of fuzzy data because it is hard to calculate the multiplication and division of fuzzy data. Considering various uncertainties into statistical data makes common mathematical analysis become difficulty, Arbaiy et al. [3] presents the fuzzy random regression approach to determine the coefficient whereby statistical data used contain fuzziness and randomness. Arbaiy and Rahman [4] proposed a fuzzy random regression method to estimate the coefficient values for which statistical data contains simultaneous fuzzy random information. Moreover, Arbaiy and Lin [5] defined a weight value to apply for linear fractional programming and solve possibilistic programming of the multi-objective decision-making problem. Those mathematical concepts gave us a anticipation in defining the ANOVA model.

Recently, Konishi et al. [6] proposed the method of ANOVA which can treat the fuzzy interval data by using the concept of fuzzy sets. Wu [1] used α-level set to define the hypothesis test of one factor ANOVA model for fuzzy data. Also that, Kalpanapriya and P. Pandian [7] improved the method proposed by Wu [1] and ignored the notions of pessimistic degree and optimistic degree to solve their problems. Moreover, in testing for fuzzy data, Montenegro et al. [8] have developed asymptotic methods for the two-sample case, and Gil et al. [7] have discussed the asymptotic ANOVA test. The outcome from Gil et al.'s [7] research work provides more criteria to build up the ANOVA for fuzzy data. Gil et al. [10] also proposed an approach of one-way ANOVA test based on fuzzy scale of measurement. They also illustrated the approaches by means of a real-life case study. Furthermore, Jiryaei etc. [11] deal with one-way ANOVA in fuzzy environment by using least squares method to estimate the fuzzy parameters and also proposed a method to check the adequacy of the proposed model. In all of above researches, they all consider the problem by using the fuzzy set theory concepts introduced by Zadeh [12] and tried to get simulative results of fuzzy data in ANOVA model. Different from their method, we would like to use the central point and radius to define the statistical hypothesis test in this paper. In Lin et al.,'s paper, they has developed a series of hypothesis test for fuzzy data by using central point and radius [13–16]. We would like to continue this idea and define one-way ANOVA model with fuzzy data in this paper.

This paper intends to investigate the factors which would affect customer decision for buying product from online shop. A statistical method is introduced to distinguish the customers' demand into different type. Firstly a hypothesis test, F-test is established, which is the pivot statistic in ANOVA model. Then, one-way ANOVA model with fuzzy data is initiated. In this experiment, several different factors for testing in one-way ANOVA model are provided. The results show that the proposed solution method could provide decision maker not only a result but also a favorable degree of each factor. This kind of result is beneficial to the decision maker and retailer to distinguish which factor is the most critical for the customer and with how much amount of products would be allocated for customers.

Considering the vague environment, it is apparent that customers also are difficult to remember their history of buying from certain shop. Based on this situation, a fuzzy based of questionnaire is developed and distributed to the customer. Such fuzzy questionnaire is useful to acquire fuzzy information. Then, a hypothesis test, F-test, for fuzzy data is employed to deal with fuzzy information and build an ANOVA model. A real-life case study is presented in the experiment analysis section. The results show that it is more flexible for decision maker to know which factor is important and helps them to plan the marketing strategy based on customer's demand.

The structure of this paper is organized as follows. Section 1 gives introduction of the study. Section 2 describes notation and preliminary definitions of one-way ANOVA model with fuzzy data. Section 3 explains the solution of numerical experiments. Section 5 gives a conclusion.

2 Notations and Preliminary Definitions

This section is spent to explain some theoretical explanations useful for the model development in this paper.

2.1 Fuzzy Statistical Algorism

Let us introduce some Definitions we will use in the following sections as follows:

Definition 1. Let $X = [a, b, c]$ be a triangular fuzzy number. Its membership function is written in the following form: [5]

$$\mu_X(t) = \begin{cases} \frac{t-a}{b-a}; a \leq t \leq b \\ \frac{c-t}{c-b}; b \leq t \leq c \, . \\ 0; \ otherwise \end{cases}$$

Definition 2. Let $X = [a, b, c]$ be a triangular fuzzy number, then the central point and radius are written as $o = \frac{a+b+c}{3}$ and $l = \frac{c-a}{4}$ respectively [17].

Definition 3. Expected Value for Triangular Fuzzy Numbers.

Let $X_i = [a_i, b_i, c_i]$, $\forall\, i = 1, 2, \ldots, n$, be a triangular fuzzy number. We denoted $\mathbf{x}_i \equiv (o_i, l_i)$ be a vector on the probability space $(\Omega,\ \mathcal{F},\ P)$, where o_i and l_i are random variables as central point and radius, respectively.

Suppose that we have $\mathbf{x} = (x_1, x_2, \ldots, x_n)'$ observations. Then, the fuzzy expected for triangular fuzzy number could be calculated as follows:

$$E(\mathbf{x}) = \sum_{i=1}^{n} E(\mathbf{x}_i) \equiv \left(\sum_{i=1}^{n} E(o_i), \sum_{i=1}^{n} E(l_i) \right).$$

Definition 4. Variance for Triangular Fuzzy Numbers.

The same assumption as Definition 3, we have the variance for triangular fuzzy number as follows:

$$var(\mathbf{x}) = \sum_{i=1}^{n} Var(\mathbf{x}_i) \equiv \left(var(\sum_{i=1}^{n} o_i), var(\sum_{i=1}^{n} l_i) \right),$$

where $var\left(\sum_{i=1}^{n} o_i x_i\right) = \sum_{i=1}^{n} \sigma_{o_{ii}}^2 x_i^2 + \sum_{j=1}^{n} \sum_{i=1,i\neq j}^{n} \sigma_{o_{ij}} x_i x_j, \quad \sigma_{o_{ij}} = cov(o_i, o_j),$ and $var\left(\sum_{i=1}^{n} l_i x_i\right) = \sum_{i=1}^{n} \sigma_{l_{ii}}^2 x_i^2 + \sum_{j=1}^{n} \sum_{i=1,i\neq j}^{n} \sigma_{l_{ij}} x_i x_j, \; \sigma_{l_{ij}} = cov(l_i, l_j), \forall\, i = 1, 2, \ldots, n.$

2.2 One-Way ANOVA Model

ANOVA test is the method used to compare continuous measurements to determine if the measurements are sampled from the same or different distributions. It is an analytical tool used to determine the significance of factors on measurements by looking at the relationship between a quantitative "response variable" and a proposed explanatory "factor". We first introduce the one-way ANOVA model in this section.

Definition 5. One-Way Analysis of Variance Model [2].

Consider the model in which we observe $Y_{ij} \sim N(\mu_i, \sigma^2)$, $j = 1, \ldots, n_i, i = i, \ldots, k$.

Note that, for each i, $Y_{i1}, \ldots, Y_{in_i}$ are a sample from a normal distribution with mean μ_i and that the k samples are independent. Since the observations have been classified on one subscript (i), we call this model the one-way analysis of variance model (one-way ANOVA model).

We give a form of organizing experimental observations Y_{ij} with groups in columns in Table 1. Moreover, we also give a summary statistics in the Table 1.

Definition 6. F-test [2].

Let N be the total number of observations, let $\overline{Y}_i$ be the average of the observations in the ith sample, and let $\overline{Y}$ be the average of all the observations. That is,

$$N = \sum n_i,\; \overline{Y}_i = \frac{\sum_i Y_{ij}}{n_i}, \overline{Y} = \frac{\sum_i \sum_j Y_{ij}}{N} = \frac{\sum_i n_i \overline{Y}_i}{N}.$$

(Note that $\overline{Y}$ is not the average of the $\overline{Y}_i$, unless the n_i are equal.)

Considering the null hypothesis $H_0 : \mu_1 = \mu_2 = \cdots = \mu_k$ against $\mu_i \neq \mu_j$ for some i and j.

Table 1. ANOVA data organization

Observations	1	2	$\cdots$	j	Sum
1	Y_{11}	Y_{12}	$\cdots$	Y_{1j}	$\overline{Y}_1$
2	Y_{21}	Y_{22}	$\cdots$	Y_{2j}	$\overline{Y}_2$
3	Y_{31}	Y_{32}	$\cdots$	Y_{3j}	$\overline{Y}_3$
$\vdots$	$\vdots$	$\vdots$	$\vdots$	$\vdots$	$\vdots$
i	Y_{i1}	Y_{i2}	$\cdots$	Y_{ij}	$\overline{Y}_i$
Mean	$\overline{Y}_1$	$\overline{Y}_2$	$\cdots$	$\overline{Y}_j$	$\overline{Y}$

Table 2. ANOVA table for fixed model, single factor, fully randomized experiment.

Source of variation	Sums of Squares (SS)	Degree of Freedom (DF)	Mean Square (MS)	F
Treatments	$SST = \sum_i n_i \overline{Y}_i^2 - N\overline{Y}^2$	$K-1$	$MST = \frac{SST}{K-1}$	$\frac{MST}{MSE}$
Error	$SSE = \sum_i \sum_j Y_{ij}^2 - \sum_i n_i \overline{Y}_i^2.$	$N-K$	$MSE = \frac{SSE}{N-K}$	
Total	$SS = \sum_i \sum_j Y_{ij}^2 - N\overline{Y}^2$	$N-1$		

*Note that SS = SST + SSE

The F-statistic is $F = \frac{SST/K-1}{SSE/N-K}$, which follows the F-distribution with $K-1$, $N-K$ degrees of freedom under the null hypothesis.

Here,

$$SST = \sum_i n_i (\overline{Y}_i - \overline{Y})^2 = \sum_i n_i \overline{Y}_i^2 - N\overline{Y}^2,$$

and

$$SSE = \sum_i \sum_j (\overline{Y}_{ij} - \overline{Y}_i)^2 = \sum_i \sum_j Y_{ij}^2 - \sum_i n_i \overline{Y}_i^2.$$

We reject hypothesis when

$$F > F(K-1, N-K).$$

Table 2 tabulates a summary statistics of ANOVA test.

3 One-Way ANOVA Model with Fuzzy Data

ANOVA test is a statistical model that will enable us to test the null hypothesis. The ANOVA model produces an F-statistic, the ratio of the variance calculated among the means to the variance within the samples. We first introduce the F-test for fuzzy data in the following.

Definition 7. F-Test for Fuzzy Data.
Let N be the total number of observations, $Y_{ij} = [a_{ij}, b_{ij}, c_{ij}]$ be triangular fuzzy numbers. By Definition 2, we have $o_{ij} = \frac{a_{ij}+b_{ij}+c_{ij}}{3}$ is a central point for Y_{ij} and $l_{ij} = \frac{c_{ij}-a_{ij}}{4}$ is average width for Y_{ij}. We denote the $\mathbf{Y}_{ij} \equiv (o_{ij},\ l_{ij})$ as an observed vector and $\boldsymbol{\mu} = (\mu_o,\ \mu_l)$ is a mean vector for $\mathbf{Y}_{ij}$. We have a hypothesis test for F-statistic as follows.

Considering the null hypothesis

$$H_0 : \boldsymbol{\mu} = \boldsymbol{\mu}_0 \text{ for statistic } \boldsymbol{F} \equiv (F_o, F_l).$$

The $\boldsymbol{F}$-statistic is

$$F_o = \frac{SST_o/K-1}{SSE_o/N-K} \text{ and } F_l = \frac{SST_l/K-1}{SSE_l/N-K}$$

Which follows the F-distribution with $K-1,\ N-K$ degrees of freedom under the null hypothesis.

Here,

$$SST_o = \sum_i n_i(\overline{o_i} - \overline{o})^2 = \sum_i n_i \overline{o}_i^2 - N\overline{o}^2,$$

$$SST_l = \sum_i n_i(\overline{l_l} - \overline{l})^2 = \sum_i n_i \overline{l}_i^2 - N\overline{l}^2,$$

$$SSE_o = \sum_i \sum_j (\overline{o}_{ij} - \overline{o}_i)^2 = \sum_i \sum_j o_{ij}^2 - \sum_i n_i \overline{o}_i^2,$$

and

$$SSE_l = \sum_i \sum_j (\overline{l}_{ij} - \overline{l}_i)^2 = \sum_i \sum_j l_{ij}^2 - \sum_i n_i \overline{l}_i^2.$$

We reject hypothesis when

$$\boldsymbol{F} > \frac{F_o(K-1, N-K) + F_l(K-1, N-K)}{2}.$$

On the other hand, we also could conclude the result by using P-value.

If $\mathbf{P} = \frac{P_o + P_l}{2} < P$-value, then we reject hypothesis. Here, P_o is P-value for central point o and P_l is P-value for radius l.

4 Numerical Experiments

In this section, we would like to show an example to present that how to use one-way ANOVA test for triangular fuzzy numbers. The process of experiment is as follows:

Step1 Sampling Survey

A manager wants to find out the factors which affect the customers for shopping online. We investigate the customers that "`how many times they are shopping online during about one month`". We also asked them by using questionnaire about some basic problems, such as educational level, age, occupation, blood type, gender, etc

Table 3. Triangular fuzzy numbers and their central point and radius.

	100 Observations				
$[a, b, c]$	[0.00,0.30,0.60]	[0.35,0.40,0.45]	[0.59,0.80,1.01]	$\cdots$	[0.47,0.69,0.91]
o	0.30	0.40	0.80	$\cdots$	0.69
l	0.15	0.025	0.105	$\cdots$	0.11

Step2 Data Gathering
We investigate 100 customers and get the following observed values of shopping during one month. The data in triangular fuzzy number is denoted as mean value for one month. Based on our problems in questionnaire, we got a regular triangular fuzzy number in this experiment (Table 3)

Step3 One-Way ANOVA Test

I. Suppose that the shopping behavior is relative to ages. The hypothesis H_0 is that customers' shopping frequencies during one month is relative to ages. We use Minitab 15 for analyzing the data. The results are shown in Table 4.

For the central point of observations, we got the F-statistic $F_o(4, 95) = 18.66$, and the P-value is $P_o = 0.000$. For the radius of observations, we got the F-statistic $F_l(4, 95) = 1.69$, and the P-value is $P_l = 0.158$. We get $\mathbf{P} = \frac{P_o + P_l}{2} = \frac{0.000 + 0.158}{2} = 0.079 > 0.05$ under 95% significant level. Hence, we do not reject H_0. It means that the age is relative to shopping behavior by considering the triangular fuzzy numbers with central point o and radius l.

We also consider other factors to check that whether it will affect the shopping behavior or not. We give the assumption and results in the following.

II. Suppose that the shopping behavior is relative to gender. The hypothesis H_0 is that customers' shopping frequencies during one month is relative to gender.

We use Minitab 15 for analyzing the data. The results are shown in Table 5. For the central point of observations, we got the F-statistic $F_o(1, 98) = 2.76$, and the P-value is

Table 4. One-Way ANOVA test for $\mathbf{H}_0 \equiv (H_{o_0}, H_{l_0})$.

One-Way ANOVA test for H_{o_0}					
	SS_o	DF_o	MS_o	F_o	P_o
Treatment(Age)	1.6669	4	0.4167	18.66	0.000
Error	2.1214	95	0.0223		
Total	3.7884	99			
S = 0.1494 R-Sq = 44.00% R-Sq(adj) = 41.64%					
One-Way ANOVA test for H_{l_0}					
	SS_l	DF_l	MS_l	F_l	P_l
Treatment(Age)	0.02968	4	0.00742	1.69	0.158
Error	0.41587	95	0.00438		
Total	0.44555	99			
S = 0.06616 R-Sq = 6.66% R-Sq(adj) = 2.73%					

$P_o = 0.100$. For the radius of observations, we got the F-statistic $F_l(1,98) = 0.29$, and the P-value is $P_l = 0.591$. We get $\mathbf{P} = \frac{P_o + P_l}{2} = \frac{0.100 + 0.591}{2} = 0.3455 > 0.05$ under 95% significant level. Hence, we do not reject H_0. It means that the gender is relative to shopping behavior by considering the triangular fuzzy numbers with central point o and radius l.

III. Suppose that the shopping behavior is relative to education. The hypothesis H_0 is that customers' shopping frequencies during one month is relative to education.

The results are shown in Table 6. For the central point of observations, we got the F-statistic $F_o(3,96) = 23.98$, and the P-value is $P_o = 0.000$. For the radius of observations, we got the F-statistic $F_l(3,96) = 3.40$, and the P-value is $P_l = 0.021$. We get $\mathbf{P} = \frac{P_o + P_l}{2} = \frac{0.000 + 0.021}{2} = 0.0105 < 0.05$ under 95% significant level. Hence, we reject H_0. It means that the education is not relative to shopping behavior by considering the triangular fuzzy numbers with central point o and radius l.

Table 5. One-Way ANOVA test for $\mathbf{H}_0 \equiv (H_{o_0}, H_{l_0})$.

One-Way ANOVA test for H_{o_0}					
	SS_o	DF_o	MS_o	F_o	P_o
Treatment (Gender)	0.1038	1	0.1038	2.76	0.100
Error	3.6846	98	0.0376		
Total	3.7884	99			
S = 0.1939 R-Sq = 2.74% R-Sq(adj) = 1.75%					
One-Way ANOVA test for H_{l_0}					
	SS_l	DF_l	MS_l	F_l	P_l
Treatment (Gender)	0.00132	1	0.00132	0.29	0.591
Error	0.44423	98	0.00453		
Total	0.44555	99			
S = 0.06733 R-Sq = 0.30% R-Sq(adj) = 0.00%					

Table 6. One-Way ANOVA test for $\mathbf{H}_0 \equiv (H_{o_0}, H_{l_0})$.

One-Way ANOVA test for H_{o_0}					
	SS_o	DF_o	MS_o	F_o	P_o
Treatment (Education)	1.6227	3	0.5409	23.98	0.000
Error	2.1657	96	0.0226		
Total	3.7884	99			
S = 0.1502 R-Sq = 42.83% R-Sq(adj) = 41.05%					
One-Way ANOVA test for H_{l_0}					
	SS_l	DF_l	MS_l	F_l	P_l
Treatment (Education)	0.04280	3	0.01427	3.40	0.021
Error	0.40275	96	0.00420		
Total	0.44555	99			
S = 0.06477 R-Sq = 9.61% R-Sq(adj) = 6.78%					

Table 7. One-Way ANOVA test for $\mathbf{H}_0 \equiv (H_{o_0}, H_{l_0})$.

One-Way ANOVA test for H_{o_0}					
	SS_o	DF_o	MS_o	F_o	P_o
Treatment (Blood type)	1.1364	3	0.3788	13.71	0.000
Error	2.6519	96	0.0276		
Total	3.7884	99			
S = 0.1662 R-Sq = 30.00% R-Sq(adj) = 27.81%					
One-Way ANOVA test for H_{l_0}					
	SS_l	DF_l	MS_l	F_l	P_l
Treatment(Blood type)	0.05832	3	0.01944	4.82	0.004
Error	0.38723	96	0.00403		
Total	0.44555	99			
S = 0.06351 R-Sq = 13.09% R-Sq(adj) = 10.37%					

Table 8. Results of One-Way ANOVA test for $\mathbf{H}_0$ based on different decision rule

Decision rule								
Significant level	$\mathbf{H}_0 \equiv (H_{o_0}, H_{l_0})$				$\mathbf{H}_0 = H_{o_0}$			
	I	II	III	IV	I	II	III	IV
	P-value				P_o-value			
	0.079	0.3455	0.0105	0.002	0.000	0.100	0.000	0.000
90%	Reject	Accept	Reject	Reject	Reject	Accept	Reject	Reject
95%	Accept	Accept	Reject	Reject	Reject	Accept	Reject	Reject
97.5%	Accept	Accept	Reject	Reject	Reject	Accept	Reject	Reject
99%	Accept	Accept	Accept	Reject	Reject	Accept	Reject	Reject
99.9%	Accept	Accept	Accept	Accept	Reject	Accept	Reject	Reject

IV. Suppose that the shopping behavior is relative to blood type. The hypothesis H_0 is that customers' shopping frequencies during one month is relative to blood type.

The results are shown in Table 7. For the central point of observations, we got the F-statistic $F_o(3, 96) = 13.71$, and the P-value is $P_o = 0.000$. For the radius of observations, we got the F-statistic $F_l(3, 96) = 4.82$ and the P-value is $P_l = 0.004$. We get $\mathbf{P} = \frac{P_o + P_l}{2} = \frac{0.000 + 0.004}{2} = 0.002 < 0.05$ under 95% significant level. Hence, we reject H_0. It means that the blood type is not relative to shopping behavior by considering the triangular fuzzy numbers with central point o and radius l.

5 Conclusions

We concluded the results in different significant level in Table 8. We also compared our method with other scholars' method which considered the defuzzification method by using only one point (here is the same data as our central point). The results tell us that, when we consider the radius in defuzzification formula, the results will be a little difference from only considering the central point. It also give us other information that decision maker should consider more factors which based on human complicated thought. In this paper, we consider one-way ANOVA model for distinguish the factors in customers' behavior in vague environment. We gave a series of experiment studies to show that how to use the one-way ANOVA model with fuzzy data. We concluded that the age and gender are the factors which affect customers for shopping online. Considering the complex of variance with fuzzy data, we will show the multinomial ANOVA mode with fuzzy data in the future work.

Acknowledgements. The authors express her appreciation to the University Tun Hussein Onn Malaysia (UTHM) and Research Acculturation Grant Scheme (RAGS) Vot R044. This research also supported by GATES IT Solution Sdn. Bhd. Under its publication scheme.

References

1. Wu, H.C.: Analysis of variance for fuzzy data. Int. J. Syst. Sci. **38**, 235–246 (2007)
2. Arnold, S.F.: Mathematical Statistics. Prentice-Hall, Englewood Cliffs (1990)
3. Arbaiy, N., Watada, J., Lin, P.-C.: Fuzzy random regression-based modeling in uncertain environment. Sustaining Power Resources through Energy Optimization and Engineering, **127** (2016)
4. Arbaiy, N., Rahman, H.M.: Fuzzy random regression to improve coefficient determination in fuzzy random environment. In: Herawan, T., Ghazali, R., Deris, M.M. (eds.) Recent Advances on Soft Computing and Data Mining. Advances in intelligent systems and computing, vol. 287, pp. 205–214. (2014)
5. Arbaiy, N., Lin, P.-C.: Weighted value assessment of linear fractional programming for possibilistic multi-objective problem. Int. J. Adv. Intell. Paradigms **8**(1), 42–58 (2016)
6. Konishi, M., Okuda, T., Asai, K.: Analysis of variance based on fuzzy interval data using moment correction method. Int. J. Innovative Comput. Inf. Control (IJICIC) **2**(1), 83–99 (2006)
7. Kalpanapriya, D., Pandian, P.: Fuzzy hypothesis testing of ANOVA model with fuzzy data. Int. J. Mod. Eng. Res. (IJMER) **2**(4), 2951–2956 (2012)
8. Montenegro, M., Casals, M.R., Lubiano, M.A., Gil, M.A.: Two-sample hypothesis tests of means of a fuzzy random variable. Inf. Sci. **113**, 89–100 (2001)
9. Gil, M.A., Montenegro, M., Gonzalez-Rodrlguez, G., Colubi, A., Casals, M.R.: Bootstrap approach to the classic oneway multi-sample test with imprecise data, Comput. Stat. Data Anal. (2006)
10. González-Rodrígueza, G., Colubib, A., Gilb, M.Á.: Fuzzy data treated as functional data: a one-way ANOVA test approach. Comput. Stat. Data Anal. **56**(4), 943–955 (2012)
11. Jiryaei, A., Parchami, A., Mashinchi, M.: One-way Anova and least squares method based on fuzzy random variables. Official J. Turk. Fuzzy Syst. Assoc. **4**(1), 18–33 (2013)
12. Zadeh, L.A.: Fuzzy sets. Inf. Control **8**(3), 338–353 (1965)
13. Lin, P.-C., Watada, J., Wu, B.: A parametric assessment approach to solving facility location problems with fuzzy demands. IEEJ Trans. Electron. Inf. Syst. **9**(5), 484–493 (2014)
14. Lin, P.-C., Watada, J., Wu, B.: Risk assessment of a portfolio selection model based on a fuzzy statistical test. IEICE Trans. Inf. Syst. **96**(3), 579–588 (2013)

15. Lin, P.-C., Watada, J., Wu, B.: Identifying the distribution difference between two populations of fuzzy data based on a nonparametric statistical method. IEEJ Trans. Electron. Inf. Syst. **8**(6), 591–598 (2013)
16. Lin, P.-C., Wu, B., Watada, J.: Kolmogorov-Smirnov two sample test with continuous fuzzy data. Integr. Uncertainty Manag. Appl. AISC **68**, 175–186 (2010)
17. Esogbue, A.O., Song, Q.: On optimal defuzzification and learning algorithms: theory and applications. Fuzzy Optim. Decis. Making **2**(4), 283 (2003)
18. Nguyen, H., Wu, B.: Fundamentals of Statistics with Fuzzy Data. Springer, Netherlands (2006)

Chicken S-BP: An Efficient Chicken Swarm Based Back-Propagation Algorithm

Abdullah Khan[1(✉)], Nazri Mohd Nawi[2], Rahmat Shah[1], Nasreen Akhter[1], Atta Ullah[1], M.Z. Rehman[2], Norhamreeza AbdulHamid[2], and Haruna Chiroma[3]

[1] Institute of Business and Management Science, Agricultural University Peshawar (AUP), KPK, Peshawar, Pakistan
abdullahdirvi@gmail.com, mr.rahmat.shah@gmail.com, researcher.aup@gmail.com, Atta12319@yahoo.com

[2] Soft Computing and Data Mining Centre (SCDM), Faculty of Computer Science and Information Technology, Universiti Tun Hussein Onn Malaysia (UTHM), 86400 Parit Raja, Batu Pahat, Johor, Malaysia
nazri@uthm.edu.my, zrehman862060@gmail.com, norhamreeza@gmail.com

[3] Faculty of Computer Science and IT, University of Malaya (UM), Kuala Lumpur, Malaysia
hchiroma@acm.org

Abstract. An innovative metaheuristic based algorithm Chicken Swarm Optimization (CSO) is inspired by characteristics of chicken flock. CSO is particularly suitable for the investigation in candidate solutions for large spaces. This paper hybridize the CSO algorithm with the Back Propagation (BP) algorithm to solve the local minimum problem and to enhance convergence to global minimum in BP algorithm. The proposed Chicken Swarm Back Propagation (Chicken S-BP) is compared with the Artificial Bee Colony Back-Propagation (ABCBP), Genetic Algorithm Neural Network (GANN) and traditional BPNN algorithms. In particular Iris, Australian Credit Card, and 7-Bit Party classification datasets are used in training and testing the performance of the Chicken S-BP hybrid network. Results of simulation illustrates that Chicken S-BP algorithm efficiently prevents local minima and provides optimal solution.

Keywords: Global minima · Gradient descent · Back-propagation · Chicken swarm optimization · Artificial bee colony · Genetic algorithm

1 Introduction

An optimization technique Back Propagation (BP) is applied to Multi-layered Artificial Neural-Networks (ANN) to accelerate the convergence of the network to global optima during the network training [1, 2]. Back Propagation Neural Network (BPNN) consists of single input layer, single or multiple hidden layers and a final single layer known as output layer. The architecture of BPNN, in which each node of a layer is associated to all other nodes of adjacent layers [3]. Unlike ANN, BPNN understands by evaluating

T. Herawan et al. (eds.), *Recent Advances on Soft Computing and Data Mining*,
Advances in Intelligent Systems and Computing 549, DOI 10.1007/978-3-319-51281-5_13

errors in output layer by finding errors in hidden layers [4], subsequently making BPNN very suitable for problems that have no association between inputs and outputs. Apart from providing useful results, BPNN still contains some convergence problems due to gradient descent learning that requires cautious selection of parameters i.e. topology of network, biases and initial weights, rate of learning, activation function and gained value in activation function etc. [5]. Extensive use of the aforementioned parameters may result into a slow network convergence, or a stagnant network [6, 7]. Numerous changes have been suggested by many researchers in order to improve network training time. Some of the solutions are; use of momentum and learning rate to stop the stagnation of the network and also to accelerate the optimal global network convergence. Both the mentioned parameters are frequently used for controlling the weight variations together with absolute descent and also to control oscillations [8, 9].

More recently, in-order to efficiently prevent local minima problem in BPNN, several evolutionary computational techniques are applied for training weights. Some of these techniques include PSO-BP [10], artificial bee colony back-propagation (ABC-BP) [11, 12], evolutionary artificial neural networks algorithm (EANN) [13], firefly [14], Bat [15], Bat-BP [16], CSLM [17], and genetic algorithms neural network (GANN) [18] etc. Therefore, this paper propose a new hybrid Chicken Swarm based Back Propagation (Chicken S-BP) which is a hybrid of Chicken Swarm optimization algorithm [19] and BPNN to find the optimal weights for the network. The proposed Metahybrid Chicken S-BP algorithm is used to train on Iris, 7-Bit Parity and Australian Credit Card classification datasets to avoid network stagnancy and to improve the network convergence. The proposed Chicken S-BP is compared with other similar algorithms such ABCBP, GANN, and BPNN.

Rest of the paper is arranged as follows: Sect. 2 of the paper illustrates Chicken Swarm Optimization (CSO) Algorithm. Section 3 discusses the weight updating process in the proposed Chicken S-BP algorithm and the simulation results are discussed in Sect. 4. Section 5 concludes this paper.

2 Chicken Swarm Optimization Algorithm

Chicken Swarm optimization (CSO) is a new bio-inspired algorithm developed by Meng [19]. For ease, the Chicken Swarm optimization algorithm selected the behaviors of chickens by the following rules.

(1) Chicken Swarm algorithm includes several clusters. Each cluster has, a couple of hens, chicks and a dominant rooster.
(2) The personality of chickens (hens, chicks and roosters) depends on their values of fitness. Chickens having best fitness will be graded as roosters. The chickens with numerous weakest fitness values would be chosen as chicks.
(3) Dominance relationship and mother-child relationship and hierarchal order in a cluster will continue unchanged.
(4) Chickens trail their group-mate rooster to hunt for food, while they can stop the ones from eating their own food. Assume chickens will randomly be stolen good food already searched by others. The chicks hunt for food around their mother (hen). The dominant individuals have an advantage in struggle for food.

3 Proposed Chicken S-BP Algorithm

Chicken Swarm Optimization algorithm is based on the natural behavior of chicken swarm that involves local skills for searching and self-determining flocking movement. Inside Chicken Swarm, there are different groups. Each group includes a dominant rooster, some hens and chickens. Chickens having greatest fitness value acted as roosters, which will lead the chicken to rank as head rooster of group. Worst chickens with different fitness values will be nominated as the chicks. Rest will be treated as hens. Chicks looking for food and stay near their mother (chicken). In the Suggested Chicken S-BP algorithm, every chicken characterizes a probable solution (e.g. weight space and associated basis for optimizing BPNN in this study). The problem of weight optimizing and the size of the pack that represents application of solutions.

Initially, best weights and biases are set with chicken swarm, then those weights are delivered to BPNN. Weights of BPNN are evaluated and compared with best solution

```
Chicken Swarm Initializes and pass the weights to BPNN
Load the Training data to the Network
While MSE< Stopping Criteria
   If (t%G=0)
         Rank the Chicken's fitness values and establish a hierarchical order in the
         swarm;
         Divide the swarm into different groups, and determine the relationship
         between the
         Chicks and mother hens in a group;
   End If
   For i=1:N
         If i==rooster
         Update its solution/location using Equation (1);
         End If
         If i==hen
         Update its solution/location using Equation (3);
         End If
         If i==hen
         Update its solution/location using Equation (6);
         End If
         Evaluate the New Solution
         If the new solution is better than the previous one, replace previous with
         the new solution
   End For
   Chicken Swam finds the best weights and pass to the network
   Chicken keeps on calculating the best possible weight at each epoch until the
   network is converged
End While
```

Fig. 1. Pseudo code of the proposed Chicken S-BP algorithm

in backward direction. In the next epoch, chicken swarm will update those weights with the possible best solution and will continue until the desired MSE is attained. The pseudo code of the proposed Chicken S-BP algorithm is given in Fig. 1.

4 Results and Discussions

Main theme of this article is to increase network convergence accuracy. Before debating the simulated results some of the terminologies need to be clarified like technologies and tools, testing methodology, network topologies and classification problems that is used in experiments. Further discussion are as follows,

4.1 Preliminary Study

This section discusses the experimentation, workstation used for experiments had an AMD APU E-450 with Radeon (1.65 GHz processor), having 2 GB of RAM, while Operating System is Microsoft Windows 7. Software used for simulations is MATLAB 2012 version. Performance Chicken Swarm based Back-Propagation (Chicken S-BP) is examined and compared with Artificial Bee Colony (ABC-BP), BPNN algorithms, Genetic Neural Network Algorithm (GANN) and also conventional BPNN algorithms. For verification of accuracy of the proposed algorithm is tested on Iris & 7-Bit Parity and Credit Card Classification problem. In the comparative analysis, three-layer back propagation neural network model with three layer is used to train model. Number of hidden is fixed to five while input and output layer nodes may vary conferring to dataset. In the entire set, 0.3 has been selected as a learning rate for global. Whereas activation function used from input to hidden and output layer is log sigmoid. One trial is restricted to 1000 epochs for every problem. A total of 5 trials were performed to verify the algorithm. For each trial, results are recorded in result file. Mean Square Error (MSE) and Accuracy is selected as parameter for comparison.

4.2 IRIS-Classification Problem

Iris classification dataset is designed by Fisher [20]. That is used to demonstrate the values of differentiate analysis. This can be the best famous database to be found in the pattern recognition literature. This contains 150 instances, 3 outputs and 4 inputs, in dataset. Iris dataset involving the data of petal length, the sepal length, petal width, and sepal width keen on three modules of species, that consists of Iris Santos, Iris Vermicular, and Iris Virginia. Chosen network structure for classification of Iris dataset is 4-5-3 that consists of five hidden nodes, three outputs nodes four inputs nodes, number of instances are set to 75 in training dataset and others are used in testing dataset.

Table 1 shows performance of proposed Chicken S-BP algorithm, compare with the ABCBP, GANN and BPNN algorithms. From simulation results, this is evident that proposed model performs better for both training and testing data set. In Table 1 the proposed Chicken S-BP algorithm achieved 99.91% accuracy with 0.0029 MSE for 70% of training data set. Similarly for 30% of testing data the proposed technique

Table 1. Accuracy and MSE of iris classification problem for 4-5-3 ANN structure

Algorithms	Training (70%)		Testing (30%)	
	MSE	Accuracy	MSE	Accuracy
Chicken S-BP	0.0029	99.91	0.0023	99.94
ABCBP	0.0448	82.91	0.0345	85.74
GANN	0.125	89.45	0.2134	88.45
BPNN	0.3336	94.47	0.3077	94.63

convergence with 0.0023 of MSE and get 99.94% of accuracy. While comparing algorithms such ABCP, GANN, and BPNN fall behind of the proposed model with respect to MSE and accuracy. For the training data the existing model ABCBP achieve 82.91% of accuracy, GANN has 89.45 and BPNN gain 94.47% of accuracy with convergence of (0.048, 0.125, 0.336) MSE. Furthermore, for the 30% of testing data the proposed chicken S-BP algorithm also perform better than the compare model w.r.t. MSE and accuracy. For the testing data the proposed model achieves 99.94% of accuracy with 0.0023 MSE, which is better than the used algorithm given in the Table 1.

4.3 Seven Bit Parity Problem

In parity problem if a given input vectors contain an odd number of one, the corresponding target value is 1, and otherwise the target value is 0. The N-bit parity training set consist of 2N training pairs, with each training pairs including an N-length input vector and a single binary target value. The input vector of 2N represents all possible groupings of N binary numbers uses the chosen architecture 7-5-1 for NN.

Tables 2 illustrate accuracy and MSE of the 7 Bit parity dataset together with five hidden neurons. This can be see clearly in Table 2 that proposed algorithm achieves a smaller MSE about 0.0027 with 99.95% of accuracy of 70% of training data. While ABCBP, comes second with an MSE of 0.2331 MSE and 88.50% accuracy. Similarly GANN and BPNN convergence with 0.2350 and 02470 MSE and achieve 88.50 and 95.06% of accuracy.

Similarly for 30% of testing data the proposed Chicken S-BP also outperform the compare algorithm Such as ABCBP, GANN and BPNN in term of MSE and accuracy. In Table 2 it's clearly show that the proposed model achieve 99.39% with 0.0025 MSE while the ABCBP, GANN and BPNN convergence with (0.2364, 0.2451, 0.2767) and gets (95.63, 88.69, 95.03) percent of accuracy which falls behind the proposed algorithm.

Table 2. Accuracy and MSE of 7 bit parity classification problem for 7-5-1 ANN structure

Algorithms	Training (70%)		Testing (30%)	
	MSE	Accuracy	MSE	Accuracy
Chicken S-BP	0.0027	99.95	0.0025	99.39
ABCBP	0.2331	95.11	0.2364	95.63
GANN	0.2350	88.50	0.2451	88.69
BPNN	0.2470	95.06	0.2767	95.03

4.4 Australian Credit Card Approval (ACCA)

This dataset has been selected from UCI Machine Learning Repository that contains all the detail on the subject of card and application. Dataset of Australian Credit Card include 690 instances, 2 outputs and 51 inputs. Every example of dataset represents a real detail about the application of credit card, if the bank or alike institute produced the credit card or not. Values and all the names of attributes that have been transformed to worthless signs to defend privacy of data. Selected architecture of NN is kept to 51-5-2.

Tables 3, confirms that MSE, and accuracy for the Credit Card Approval classification problem with five hidden neurons. The proposed Chicken S-BP converged to an MSE of 0.0035 with 99.93% of accuracy for the 70% of the training data set. While the comparison algorithms such as ABCBP achieve 0.2471 of MSE with 95% of accuracy. Similarly GANN and BPNN converge with (0.2534, 0.279) of MSE and (87.50, 94.93) percent of accuracy.

Table 3. Accuracy, and MSE of credit card approval classification problem for 51-5-2 ANN structure

Algorithms	Training (70%)		Testing (30%)	
	MSE	Accuracy	MSE	Accuracy
Chicken S-BP	0.0035	99.93	0.0024	98.73
ABCBP	0.2471	95	0.2473	95.28
GANN	0.2534	87.50	0.2643	87.84
BPNN	0.279	94.93	0.2440	95.23

Similarly for 30% of testing data the Table 3 illustrates performance of proposed Chicken S-BP algorithm compared with ABCPP, GANN and BPNN. From the result, this can be seen that proposed model executes better with respect to MSE and accuracy than comparison algorithm. For 30% of test data the proposed model converges with 0.0024 MSE and achieve 98.73% of accuracy.

5 Conclusion

Chicken Swarm Optimization (CSO) is a new bio inspired metaheuristic based optimization algorithm. It's inspired by the aggressive and enduring capabilities of the chicken candidate solutions to search large spaces. This paper examines the hybridization of a CSO with back-propagation neural network (BPNN) algorithm to overcome the problem of local minima and to optimally converge during gradient descent. The performance of proposed Chicken Swarm based Back-Propagation (Chicken S-BP) algorithm is compared with Artificial Bee Colony Back-Propagation (ABCBP), Genetic Algorithm Neural Network (GANN), and conventional BPNN algorithms. Specifically, Iris, Seven Bit Parity and Credit card classification datasets are used for training and testing the network. Simulation results shows that the proposed

Chicken S-BP algorithm performs better than ABCBP, GANN and BPNN algorithm in term accuracy and MSE. From the experiment, it is also clear that with CSO the BPNN effectively achieves global minima.

Acknowledgments. The Authors would like to thank Office of Research, Innovation, Commercialization and Consultancy (ORICC), Universiti Tun Hussein Onn Malaysia (UTHM) and Ministry of Education (MOE) Malaysia for financially supporting this Research under Fundamental Research Grant Scheme (FRGS) vote no. 1236. This research is also supported by Gates IT Solution Sdn. Bhd under its publication scheme.

References

1. Rumelhart, D.E., McClelland, J.L.: Parallel Distributed Processing: Explorations in the Microstructure of Cognition. Foundations. MIT Press, New York (1986)
2. Nawi, N.M., Khan, A., Rehman, M.Z.: A new back-propagation neural network optimized with cuckoo search algorithm. In: Murgante, B., Misra, S., Carlini, M., Torre, C.M., Nguyen, H.Q., Taniar, D., Apduhan, B.O., Gervasi, O. (eds.) ICCSA 2013. LNCS, vol. 7971, pp. 413–426. Springer, Berlin (2013). doi:10.1007/978-3-642-39637-3_33
3. Lahmiri, S.: A comparative study of back propagation algorithms in financial prediction. Int. J. Comput. Sci. Eng. Appl. **1**(4), 15–21 (2011)
4. Rehman, M.Z., Nawi, N.M.: The effect of adaptive momentum in improving the accuracy of gradient descent back propagation algorithm on classification problems. In: Mohamad Zain, J., Wan Mohd, W.M., El-Qawasmeh, E. (eds.) ICSECS 2011. CCIS, vol. 179, pp. 380–390. Springer, Berlin (2011). doi:10.1007/978-3-642-22170-5_33
5. Nandy, S., Sarkar, P.P., Das, A.: Analysis of a nature inspired firefly algorithm based back-propagation neural network training. Int. J. Comput. Appl. **43**(22), 0975–8887 (2012)
6. Xingbo, S., Pingxian, Y.: BP neural networks with improved activation function and its application in the micrographs classification. In: 2008 Fourth International Conference on Natural Computation, ICNC 2008 (2008)
7. Wang, X.G., Tang, Z., Tamura, H., Ishii, M., Sun, W.D.: An improved backpropagation algorithm to avoid the local minima problem. Neurocomputing **56**, 455–460 (2004)
8. Yu, C.C., Liu, B.D.: A Backpropagation algorithm with adaptive learning rate and momentum coefficient. In: IJCNN 2002, Honolulu, pp. 1218–1223 (2002)
9. Zhang, J.R., et al.: A hybrid particle swarm optimization–back-propagation algorithm for feedforward neural network training. Appl. Math. Comput. **185**(2), 1026–1037 (2007)
10. Mendes, R., et al.: Particle swarms for feedforward neural network training. In: 2002 Proceedings of the 2002 International Joint Conference on Neural Networks, IJCNN (2002)
11. Karaboga, D., Akay, B., Ozturk, C.: Artificial bee colony (ABC) optimization algorithm for training feed-forward neural networks. In: Torra, V., Narukawa, Y., Yoshida, Y. (eds.) MDAI 2007. LNCS (LNAI), vol. 4617, pp. 318–329. Springer, Berlin (2007). doi:10.1007/978-3-540-73729-2_30
12. Karaboga, D., Ozturk, C.: Neural networks training by artificial bee colony algorithm on pattern classification. Neural Netw. World **19**(3), 279–292 (2009)
13. Yao, X.: Evolutionary artificial neural networks. Int. J. Neural Syst. **4**(03), 203–222 (1993)
14. Yang, X.S.: Firefly algorithms for multimodal optimization. In: Watanabe, O., Zeugmann, T. (eds.) SAGA 2009. LNCS, vol. 5792, pp. 169–178. Springer, Heidelberg (2009). doi:10.1007/978-3-642-04944-6_14

15. Yang, X.S.: A new metaheuristic bat-inspired algorithm. In: González, J.R., Pelta, D.A., Cruz, C., Terrazas, G., Krasnogor, N. (eds.) Nature Inspired Cooperative Strategies for Optimization (NICSO 2010), pp. 65–74. Springer, Heidelberg (2010)
16. Nawi, N.M., Rehman, M.Z., Khan, A.: A new bat based back-propagation (BAT-BP) algorithm. In: Swiątek, J., Grzech, A., Swiątek, P., Tomczak, J.M. (eds.). AISC, vol. 240, pp. 395–404 Springer, Cham (2014). doi:10.1007/978-3-319-01857-7_38
17. Nawi, N.M., Khan, A., Rehman, M.Z.: CSLM: levenberg marquardt based back propagation algorithm optimized with cuckoo search. J. ICT Res. Appl. **7**(2), 103–116 (2013)
18. Montana, D.J., Davis, L.: Training feedforward neural networks using genetic algorithms. In: IJCAI (1989)
19. Meng, X., Liu, Y., Gao, X., Zhang, H.: A new bio-inspired algorithm: chicken swarm optimization. In: Tan, Y., Shi, Y., Coello, C.A. (eds.) ICSI 2014. LNCS, vol. 8794, pp. 86–94. Springer, Cham (2014). doi:10.1007/978-3-319-11857-4_10
20. Fisher, R.A.: The use of multiple measurements in taxonomic problems. Ann. Eugenics **7**, 179–188 (1936)

A Review on Violence Video Classification Using Convolutional Neural Networks

Ashikin Ali(✉) and Norhalina Senan(✉)

Faculty of Computer Science and Information Technology,
Universiti Tun Hussein Onn Malaysia (UTHM),
Parit Raja, 86400 Batu Pahat, Johor, Malaysia
gi150038@siswa.uthm.edu.my, halina@uthm.edu.my

Abstract. The volatile growth of social media content on the Internet is revolutionizing content distribution and social interaction. Social media exploded as a category of online discourse where people create content, share it, bookmark it and network it at prodigious rate. Examples comprise Facebook, MySpace, Youtube, Instagram, Digg, Twitter, Snapchat and others. Since it is easy to reach, use and the high velocity of spreading information among users. The internet as it is at present is made up of a vast array of protocols and networks where traffickers can anonymously share large volumes of illegal material amongst each other from locations with relaxed or non-existent laws that prohibit the possession or trafficking of illegal material. In this paper, a review of applications of deep networks techniques has been presented. Hence, the existing literature suggests that we do not lose sight of the current and future potential of applications of deep network techniques. Thus, there is a high potential for the use of Convolutional Neural Networks (CNN) for violence video classification, which has not been fully investigated and would be one of the interesting directions for future research in video classification.

Keywords: Social media · Convolutional Neural Network · Classification · Video classification

1 Introduction

In 2011, 70% of teens use social media sites on daily basis and nearly one in four teens hit their favorite social media sites 10 or more times a day. Whilst adolescent benefit from their use of social media by interacting with and learning from others, but in some way the risk of being exposed to large amounts of offensive online contents. While violence is not new to the human race, it is an increasing problem in modern society. The most recent research indicates that young people between the ages of 8 and 18 spend an average of approximately 50 min per day online pursuing activities other than games, including 22 min on social networking sites like Facebook, 15 min on video sites like YouTube, and 13 min visiting other types of websites [17].

There has been almost no research attempting to document the amount of violent content that young people encounter online, whether through social media or by way of popular video or other websites. One in four teen social media users say they often

T. Herawan et al. (eds.), *Recent Advances on Soft Computing and Data Mining*,
Advances in Intelligent Systems and Computing 549, DOI 10.1007/978-3-319-51281-5_14

encounter some type of hate speech online, such as racist, sexist, or homophobic remarks, but it is not clear what portion of that includes threats or discussions of physical violence [16]. A 2008 survey of 1500 adolescents aged between 10 to 15 years old found that 38% had been exposed to violent scenes on the Internet [21]. However, it is very little awareness about young people's exposure to violent content in social media such as multiplayer online games or other online content.

To address trepidations on children's access to violence content over Internet, the Malaysian Communication and Multimedia Commission (MCMC) or generally the administrators of popular social media pages often manually review online contents to detect and delete violence materials. However, the conventional review tasks of identifying offensive contents are labour intensive, time consuming, and thus not sustainable and scalable in reality. So there is an increasing demand for automated rating and tagging systems that can detect the violent videos and remove them. Some automated content filtering software packages, such as Internet Security Suite, have been developed and applied to detect and filter online offensive contents. Most of them simply blocks webpage and paragraphs that contain violence words and images. These approaches not only affect the readability and usability of web sites, but also fail to identify subtle offensive contents, such as words, images and it can be more demanding to detect or combat videos [2–4, 9]. To deal with this limitation, this study intended to propose a solution to improve the demand on violence video detection issues using Convolutional Neural Network (CNN), despite image, text, voice recognition CNN works well with video processing in term of classifications and clustering.

2 Theoretical Background

2.1 Convolutional Neural Network (CNN)

Neural networks, a beautiful biologically-inspired programming paradigm which enables a computer to learn from observational data. Deep learning is a powerful set of technique for learning in neural networks. Convolutional neural networks (CNNs), a variant of deep learning, were motivated by neurobiological research on locally sensitive and orientation-selective nerve cells in the visual cortex. Convolutional Neural Networks are a special kind of multi-layer neural networks, with the following characteristic: a CNN is a feed forward network that can extract topological properties from an image. Like almost every other neural network, it is trained with a version of the back-propagation algorithm. CNNs are designed to recognize visual patterns directly from pixel images with little-to-none preprocessing. They can recognize patterns with extreme variability, such as handwritten text and natural images. A CNN typically consists of a convolution layers, subsampling layers, and a fully connected layer. In CNN [10], successive layers of convolution and subsampling are typically alternated. Figure 1 shows a Schematic diagram of CNN is shown below.

Convolutional Neural Networks have only recently become mainstream in computer vision applications. Over the past 3 years CNNs have achieved state-of-the-art performance in a broad array of high-level computer vision problems, including image classification, object detection, fine-grained categorization, image segmentation, pose

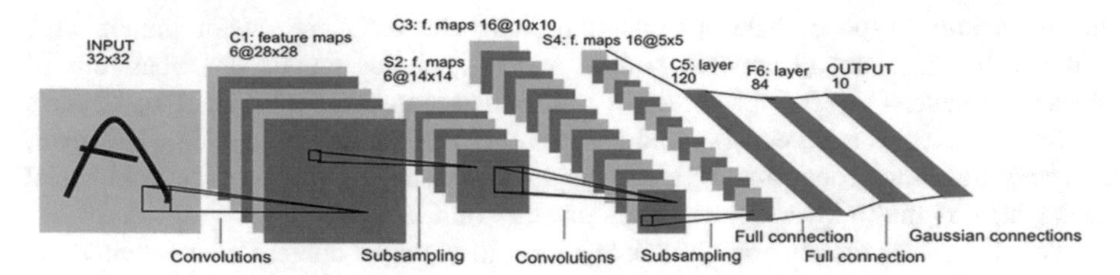

An early (Le-Net5) Convolutional Neural Network design, LeNet-5, used for recognition of digits

Fig. 1. One a Schematic diagram of CNN is shown below (Le Cun *et al.*, 1990).

estimation and OCR in natural images among others. CNNs are trained in an end-to-end manner and deliver strikingly better results than systems relying on carefully engineered representations. This success can be partially attributed to the built-in invariance of CNNs to local image transformations, which underpins their ability to learn hierarchical abstractions of the data. After the convolutional layers there may be any number of fully connected layers. The densely connected layers are identical to the layers in a standard multilayer neural network.

2.2 Backpropagation

After the convolutional layers there may be any number of fully connected layers. The densely connected layers are identical to the layers in a standard multilayer neural network or multilayer perceptron (MLP). The backpropagation algorithm [18] is used in layered feed-forward MLP. The backpropagation algorithm uses supervised learning, where the algorithm is provided with the inputs and outputs which the network has to compute and then the error is calculated [12]. The idea of the backpropagation algorithm is to reduce this error, until the MLP learns the training data. The training begins with random weights, and the goal is to adjust them so that the error will be minimal.

The weighted sum of a neuron is written as:

$$A_j(x,w) = \sum\nolimits_{i-0}^{n} X_i W_{ji}, \tag{2.1}$$

where the sum of input X_i is multiplied by their respective weights, W_{ji}. The activation depends only on the inputs and the weights. If the output function would be the identity, then the neuron would be called linear. The most used output function is sigmoid function [19]:

$$O_j(x,w) = \frac{1}{1+e^{-A}(x,w)} \tag{2.2}$$

The sigmoid function is very close to one for large positive numbers and very close to zero for large negative numbers. This allows a smooth transition between the low and high output of the neuron. The output depends only in the activation, which in turn depends on the values of the inputs and their respective weights. The goal of the training process is to obtain a desired output when certain inputs are given. Since the error is the difference between the actual and desired output, the error depends on the weights and preferred to be adjusted in order to minimize the error. The error function for the output of each neuron can be defined as:

$$E_j(x, w, d) = \left(O_j(x, w) - d_j\right)^2 \tag{2.3}$$

The output will be positive and the desired target will be greater if the difference is big and lesser if the difference is small. The error of the network will simply be the sum of the errors of all the neurons in the output layer:

$$E(x, w, d) = \sum_j \left(O_j(x, w) - d_j\right)^2 \tag{2.4}$$

where O_j is the target output and d_j is the target or desired output. After finding this, the weights can be adjusted using the method of gradient descent:

$$\Delta w_{ji} = -\eta \frac{\partial E}{\partial w_{ji}} \tag{2.5}$$

This equation can be inferred in the following way: the adjustment of each weight (Δw_{ji}) will be the negative of a constant eta (η), where η is the learning rate. Multiplied by the dependence of the previous weight on the error of the network, which is derivative of E in respect to w_{ji}. The size of the adjustment will depend on η, and on the contribution of the weight to the error of the function. This is, if the weight contributes a lot to the error, the adjustment will be greater than if it contributes in a smaller amount. Equation (2.5) is used until appropriate weights with minimal error founded.

Henceforth, derivative of E in respect to w_{ji} discovered. This is the goal of the backpropagation algorithm, since the backwards need to be achieved. First, calculate the error depends on the output, which is the derivate of E in respect to O_j from Eq. (2.3).

$$\frac{\partial E}{\partial O_j} = 2\left(O_j - d_j\right) \tag{2.6}$$

The reliance of the output on the activation depends on the weights from Eq. (2.1) and Eq. (2.2). Can be seen that from Eq. (2.6) and Eq. (2.7):

$$\frac{\partial O_j}{\partial w_{ji}} = \frac{\partial O_j}{\partial A_j}\frac{\partial A_j}{\partial w_{ji}} = O_j\left(1 - O_j\right)x_i \tag{2.7}$$

$$\frac{\partial E}{\partial w_{ji}} = \frac{\partial E}{\partial O_j}\frac{\partial O_j}{\partial w_{ji}} = 2\left(O_j - d_j\right)O_j\left(1 - O_j\right)x_i \tag{2.8}$$

The adjustment to each weight will begin from Eqs. (2.5) and (2.8):

$$\Delta w_{ji} = -2\eta(O_j - d_j)O_j(1 - O_j)x_i \tag{2.9}$$

Equation (2.9) can be used as it is for training ANN with two layers. For training the network with one more layer, some considerations are needed particularly on training time which can be affected by the architecture of the network. For practical reasons, ANNs implementing the backpropagation algorithm do not have too many layers, since the time for training the networks grows exponentially [10].

3 A Review of Violence Video Classification Using CNN

3.1 Violence Content Social Media Network

Violence clearly is an important part of the television and internet diet, but a rigorous and comprehensive account of the violent content publicly available in the Web video world is lacking. Researchers have focused on some specific forms of violence online, such as cyber bullying [12], self-inflicted pain, and verbal, however, there is little known about the general amount or context of violent content in Web videos. The study reported here aims to fill this gap with a systematic analysis of content available on YouTube. Although there are several sites where viewers can access Web-based videos, YouTube is the most successful and widely used site. Created in 2005, YouTube has grown rapidly; by some estimates YouTube garners about 137 million users per month, which ranks third in the United States behind only Google and Facebook. Moreover, YouTube accounts for approximately 60% of all videos watched online.

YouTube also has an international reach as one of the most viewed sites in countries spanning Europe, North and South America, and Southeast Asia. YouTube has become ubiquitous, accessible via mobile devices and routinely embedded and linked to other Websites and social networking sites. For instance, just on Facebook, a popular social networking site, there are approximately 17,155 of YouTube videos watched each day.

Meanwhile, mobile devices and the internet are capable of being influential tools, but there are also potential weapons. The United Nations has issued an alert about cyber violence against adolescents and women. It can be as damaging as physical abuse; in particular as technology can now reach remote comers all over the world. According to research from Pew Hispanic Center, 65% of young internet users have been suffered from online harassment. Online violence has subverted the original positive promise of the internet's freedoms and in too many circumstances has made it a comfort zone that permits anonymous cruelty and facilitates harmful acts towards adolescents and women. There are forms of violence that have been identified from the experts, namely online harassment from abusive SMS messages to tracking movement through geolocation, intimate partner violence threats of disclosure of intimate communication or revenge porn, technology is used to lure adolescents and women into situations that result in forms of violence [1]. Thus, it is essential to enquire about the amount of violence in the top-rated videos as a way to identify positively evaluated content.

3.2 Existing Works on Video Classifications

Essentially, video content is becoming a predominant part of user's daily lives on the Web. By allowing users to generate and distribute their own content to large audiences, the Web has been transformed into a major channel for the delivery of multimedia, leading society to a new multimedia age. Video pervades the Internet and supports new types of interaction among users, including video forums, video chats, video mail, and video blogs. Video processing is a particular case of signal processing, which often employs video filters and where the input and output signals are video files or video streams [18]. Table 1 illustrates the summary of existing works done by previous researchers.

Based on summary, [13] investigated a novel perspective which combines frame features to create a global descriptor. The study contributed fast algorithm to extract global frame features. The traditional *k*-means replaced random forest approach, also Vector of Locally Aggregated Descriptor (VLAD) combined with Fisher kernel in order to replace the Bag-of-Words technique. Meanwhile, [5] explored the deep learning methods to tackle this challenging problem. The system consists of several deep learning features. First, train a Convolutional Neural Network (CNN) model with a subset of ImageNet classes selected particularly for violence detection. Secondly, the research adopted a specially designed two-stream CNN framework to extract features on both static frames and motion optical flows. Third, Long Short Term Memory (LSTM) models are applied on top of the two-stream CNN features, which can capture the longer-term temporal dynamics. In addition, several conventional motion and audio features are also extracted as complementary information to the deep learning features. By fusing all the advanced features, this study achieved a mean average precision of 0.296 in the violence detection subtask, and an accuracy of 0.418 and 0.488 for arousal and valence respectively in the induced affect detection subtask. However, there is still room of improvements in model re-training and additional computation needs is required.

In Ding [6], the study claimed deep models can act directly on the raw inputs and automatically extract features. A 3D ConvNets model developed for violence detection in video without using any prior knowledge. Karpathy [8] have studied on Convolutional Neural Networks (CNNs) and realized that CNN is established as a powerful class of models for image recognition problems. Encouraged by the results, the study provided an extensive empirical evaluation of CNNs on large scale video classification using a new dataset of 1 million YouTube videos belonging to 487 classes.

From a less technical point of view, in the feature analysis it is found that features correlated to the arousal and valence dimensions are beneficial to violence labelling. Eyben [7] proposed affective video retrieval: video detection in Hollywood movies by large-scale segmental feature extraction. Violence detection in video using computer vision techniques was proposed to show that this method can construct a versatile accurate fight detector using a local descriptor approach [15]. Experiments on this database show that violence can be detected with 90% accuracy. This type of processing is critical in systems that have live video or where the video data is so large that loading the entire set into the workspace is inefficient. Giannakopoulus proposed a multimodal approach with k-NN classifier, using audio, text and visual features in order

Table 1. A summary of existing works

Authors/Years	Classification Features	Advantages/Feature Extraction	Disadvantages
CONVOLUTIONAL NEURAL NETWORK			
Dai *et al.*, 2015	Convolutional Neural Networks SVM Classifier	-CNN-violence -Two-stream CNN -Long Short Term Memory	-Requires additional computation -Needs improvements m model re-training
Ding *et a.l.*, 2014	3D ConvNets Classifier	-Automatically learn the video features -No any prior knowledge/pre-processing	-Needs more accuracy improvements - Possibility to detect mid-level concepts
Karpathy *et al.*, 2014	Convolutional Neural Networks classifier	-Large scale sports video classification -No any prior knowledge/pre-processing	-Computational complexity -In need of more desired level of accuracy
Authors/Years	Classification Features	Advantages/Feature Extraction	Disadvantages
OTHER CLASSIFIERS			
Mironica *et al.*, 2015	-Modified Vector of Locally Aggregated Descriptors (VLAD). -SVM Classifier	Frame based feature extraction	Computational complexity
Eyben *et al.*, 2013	Large-Scale Segmental Feature -SVM Classifier	-Audio-visual classifications -Peak-related audio feature extraction. -Low-level histrogram-based video analysis	Join feature and error analysis in cross domain
Nievas *et al.*, 2011	Computer vision techniques SVM Classifier	-Local descriptors approach -Discriminative classifier -New dataset of hockey video containing fights	Critical choice of descriptor, insensitive
Giannakopoulos *et al.* 2010	-Multimodal Approach -*k*-NN Classifier	-Audio features, text-based features, visual features	Emphasis to give on assisting modalities such as more sophisticated features

to improve classification. However, the approach needs to emphasize on assisting modalities such as more sophisticated features. Thus, it is identified that to handle and combat this type of data a classification of Convolutional Neural Network is proposed in order to overcome the efficiency and accuracy in video processing.

4 Proposed Framework

The process of the proposed research is illustrated according to the experimental need initiating from the data acquisition, theoretical study, and design of new approach, implementation and observation up to performance evaluation. The figure shows the summary of the research framework for this study. Each phase contains different steps and approach that will project a useful results which applicable for subsequent stages. Theoretical study performed at the beginning to obtain extensive coverage of the chosen field. The detailed experimental methodology have been shown in the figure, consisting of data collection, data pre-processing, data partitioning, classification and performance evaluation (Fig. 2).

The data augmentation is an effective way to reduce overfitting when training a large Convolutional Neural Network (CNN), which generates more training samples by cropping small-sized patches and horizontally flipping these patches from original

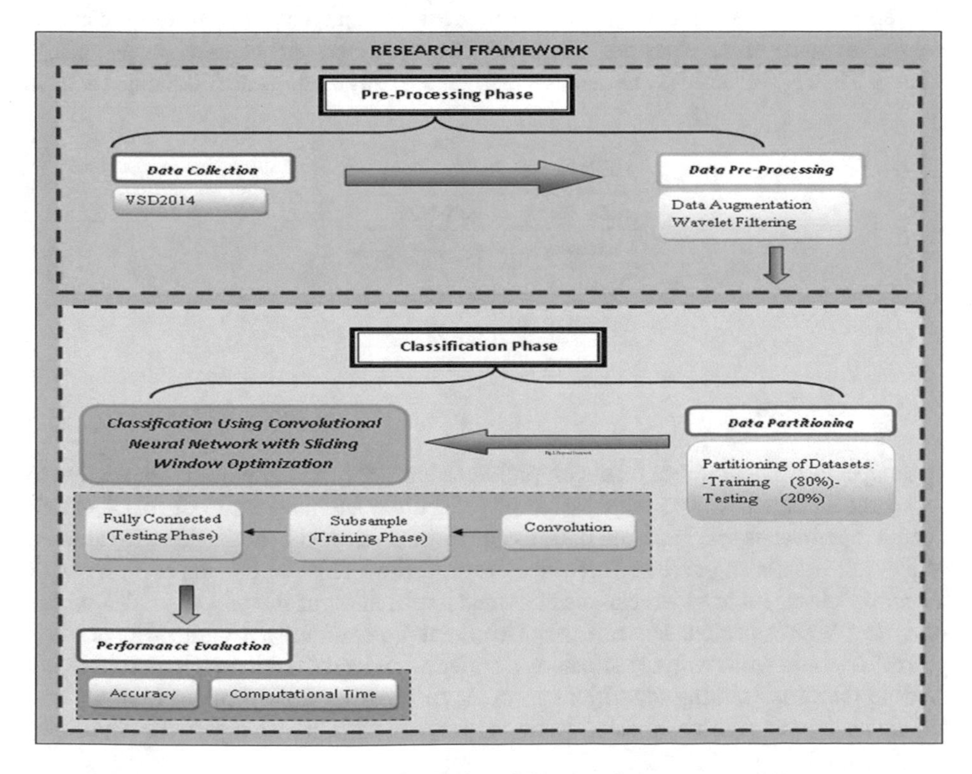

Fig. 2. Proposed framework

samples. Before presenting to a network, every dataset will be pre-processed to needed requirements in order to have a noise-less data, to accomplish that this study will be using wavelet filtering approach. However, video sampling involves taking samples along a new and different dimension such as time. As such, it involves some different concepts and techniques. Following a standard pipeline, several types of features will be extracted at all frames of the videos, discretize the videos using *k*-means vector quantization. Feature extraction is an important step in the construction of any pattern classification and aims at the extraction of the relevant information that characterizes each class. In this process relevant features are extracted from video to form feature vectors. These feature vectors are then used by classifiers to recognize the input unit with target output unit. Therefore, in this study the feature set will be based on audio and visual. As preliminary, dataset will be segregated by assigning 80% of the videos to the training set and 20% to a test set. At the same time as chances to restrain duplicate videos are visible, it is possible that the same video could appear in both the training and test set. The convolutional layers is implemented by convolving the input with a set of filters, followed by an element wise non-linear function, and generates a group of feature maps, each of which corresponds to the response of convolution between the input and a filter. The pooling layer, if max-pooling used, outputs the maximal values of spatially successive local religions on the feature maps. Its main function is to reduce the spatial size of feature maps and hence to reduce computation for upper layers. Thus, in order to quantify the performance of classification method, the performance metrics: classifier accuracy, precision, and recall will be used to access the performance of the classifier. There are four terms that will be used for this study, namely TP, TN, FP and FN. Evaluation measures of this study is as following Table 2.

Table 2. Evaluation measures

Measure	Formula
Accuracy (A)	$\frac{TP+TN}{TP+TN+FP+FN}$
Recall (R)	$\frac{TP}{TP+FN}$
Precision (P)	$\frac{TP}{TP+FP}$

5 Conclusion

This paper examines current practices, problems and prospects of video classification in social media network. The emphasis is placed on the summarization of major classification approaches that has been done by previous researchers. A considerable amount of literature has been published on video classification. Thus, in this paper a review of recent development in video classification and explanation of the proposed idea of this study has been reflected. Hence it is proven that Convolutional Neural Network is a powerful tool to work with classifications. However, we need more efficient, robust and flexible machine learning algorithms in order to improve classification accuracy and reduce uncertainties in video classification in term of prior knowledge in the algorithm.

Acknowledgments. This research funded by Ministry of Higher Education Malaysia MyBrain 15. This paper also was partly sponsored by the Center for Graduate Studies UTHM.

References

1. BBC NEWS. http://www.bbc.com/news/world-34911605
2. Benevenuto, F., Rodrigues, T., Almeida, V., Almeida, J., Zhang, C., Ross, K.: Identifying video spammers in online social networks. In: Proceedings of the 4th International Workshop on Adversarial Information Retrieval on the Web, pp. 45–52. ACM, April 2008
3. Benevenuto, F., Rodrigues, T., Almeida, V., Almeida, J., Gonçalves, M.: Detecting spammers and content promoters in online video social networks. In: Proceedings of the 32nd International ACM SIGIR Conference on Research and Development in Information Retrieval, pp. 620–627. ACM, July 2009
4. Chen, Y., Zhou, Y., Zhu, S., Xu, H.: Detecting offensive language in social media to protect adolescent online safety. In: Privacy, Security, Risk and Trust (PASSAT), 2012 International Conference on and Social Computing (SocialCom), pp. 71–80. IEEE, September 2012
5. Dai, Q., Zhao, R.W., Wu, Z., Wang, X., Gu, Z., Wu, W., Jiang, Y.G.: Fudan-Huawei at MediaEval 2015: Detecting Violent Scenes and Affective Impact in Movies with Deep Learning (2015)
6. Ding, C., Fan, S., Zhu, M., Feng, W., Jia, B.: Violence detection in video by using 3D convolutional neural networks. In: Bebis, G., et al. (eds.) ISVC 2014. LNCS, vol. 8888, pp. 551–558. Springer, Cham (2014). doi:10.1007/978-3-319-14364-4_53
7. Eyben, F., Weninger, F., Lehment, N., Schuller, B., Rigoll, G.: Affective video retrieval: violence detection in Hollywood movies by large-scale segmental feature extraction. PLoS ONE **8**(12), e78506 (2013)
8. Karpathy, A., Toderici, G., Shetty, S., Leung, T., Sukthankar, R., Fei-Fei, L.: Large-scale video classification with convolutional neural networks. In: 2014 IEEE Conference on Computer Vision and Pattern Recognition (CVPR), pp. 1725–1732. IEEE, June 2014
9. Law-To, J., Chen, L., Joly, A., Laptev, I., Buisson, O., Gouet-Brunet, V., Stentiford, F.: Video copy detection: a comparative study. In: Proceedings of the 6th ACM International Conference on Image and Video Retrieval, pp. 371–378. ACM, July 2007
10. Le Cun, B.B., Denker, J.S., Henderson, D., Howard, R.E., Hubbard, W., Jackel, L.D.: Handwritten digit recognition with a back-propagation network. In: Advances in Neural Information Processing Systems (1990)
11. Gershenson, C.: Artificial neural networks for beginners. Retrived from arXiv preprint cs/0308031 (2003)
12. Giannakopoulos, T., Pikrakis, A., Theodoridis, S.: A multimodal approach to violence detection in video sharing sites. In: 2010 20th International Conference on Pattern Recognition (ICPR), pp. 3244–3247. IEEE, August 2010
13. Mishna, F., Khoury-Kassabri, M., Gadalla, T., Daciuk, J.: Risk factors for involvement in cyber bullying: Victims, bullies and bully–victims. Child Youth Serv. Rev. **34**(1), 63–70 (2012)
14. Mironică, I., Duţă, I.C., Ionescu, B., Sebe, N.: A modified vector of locally aggregated descriptors approach for fast video classification. Multimedia Tools Appl., 1–28 (2015)
15. Bermejo Nievas, E., Deniz Suarez, O., Bueno García, G., Sukthankar, R.: Violence detection in video using computer vision techniques. In: Real, P., Diaz-Pernil, D., Molina-Abril, H., Berciano, A., Kropatsch, W. (eds.) CAIP 2011. LNCS, vol. 6855, pp. 332–339. Springer, Berlin, Heidelberg (2011). doi:10.1007/978-3-642-23678-5_39

16. Rideout, V.J.: Social media, social life: How teens view their digital lives (2012)
17. Rideout, V.J., Foehr, U.G., Roberts, D.F.: Generation M^2: Media in the Lives of 8-to 18-Year-Olds. Henry J. Kaiser Family Foundation (2010)
18. Rumelhart, D.E.: Back Propagation: Theory, Architectures, and Applications. Psychology Press, Cambridge (1995)
19. Sullivan, G.J., Ohm, J.R., Han, W.J., Wiegand, T.: Overview of the high efficiency video coding (HEVC) standard. IEEE Trans. Circuits Syst. Video Technol. **22**(12), 1649–1668 (2012)
20. Tommiska, M.T.: Efficient digital implementation of the sigmoid function for reprogrammable logic. In: IEE Proceedings of Computers and Digital Techniques, vol. 150, No. 6, pp. 403–411. IET, November 2003
21. Ybarra, M., Suman, M.: Reasons, assessments and actions taken: sex and age differences in uses of Internet health information. Health Educ. Res. **23**(3), 512–521 (2008)

Modified Backpropagation Algorithm for Polycystic Ovary Syndrome Detection Based on Ultrasound Images

Untari N. Wisesty(✉), Jondri Nasri, and Adiwijaya

School of Computing, Telkom University, Bandung, Indonesia
{untarinw, jondri, adiwijaya}@telkomuniversity.ac.id

Abstract. Polycystic Ovary Syndrome (PCOS) is an endocrine abnormality that occurred in the female reproductive cycle. In general, the approaches to detect PCO follicles are (1) stereology and (2) feature extraction and classification. In Stereology, two-dimensional images are viewed as projections of three-dimensional objects. In this paper, we use the second approach, namely Gabor Wavelet as a feature extractor and a modified backpropagation as a classifier. The modification of backpropagation algorithm which is proposed, namely Levenberg - Marquardt optimization and Conjugate Gradient - Fletcher Reeves to improve the convergence rate. Levenberg - Marquardt optimization produce the higher accuracy than Conjugate Gradient - Fletcher Reeves, but it has a drawback of running time. The best accuracy of Levenberg - Marquardt is 93.925% which is gained from 33 neurons and 16 vector feature and Conjugate Gradient - Fletcher Reeves is 87.85% from 13 neurons and 16 vector feature.

Keywords: PCOS · Neural network · Backpropagation · Levenberg – Marquardt · Conjugate Gradient – Fletcher Reeves

1 Introduction and Related Works

Infertility is a failed ovulation, i.e. a process when an egg is released from the ovary. There are a lot of factors causing infertility, one of which is the abnormal number and size of follicle growth in ovulation phase. This abnormality is the initial symptom of Polycystic Ovary Syndrome (PCOS). PCOS is an endocrine abnormality that occurred in female reproductive cycle [1]. According to the criteria of the National Institutes of Health (NIH), 6%–10% female has PCOS while Rotterdam can widely define as 15% prevalence of PCOS [2]. According to Rotterdam Consensus Criteria [2] in 2003, PCOS may be diagnosed if 2 of the 3 following criteria are identified: (1) oligo/amenorrhea, (2) clinical or biochemical symptom of excess androgen activity, and (3) PCO follicles according to Ultrasonography (USG) result. However, until this moment, the examination of USG results is still done manually based on human vision [3, 4]. This examination takes a long time because there is a diagnosis process for USG results and produce inconsistent results between USG experts, which cause the limits of human vision.

T. Herawan et al. (eds.), *Recent Advances on Soft Computing and Data Mining*,
Advances in Intelligent Systems and Computing 549, DOI 10.1007/978-3-319-51281-5_15

Based on those problems, the research on PCO detection has been done with its drawbacks and advantages. Furthermore, the PCO follicles have varied forms and sizes, so that to develop ideal computation and performance for PCO follicle detection is a difficult thing. According to a research conducted by Akkasaligar and Malagavi [5], PCOS ovary contains more than 10 PCO follicles. On the other hand, medical experts assisting in this research validate that PCOS ovary contains 10 or more PCO follicles. In PCO, the numerous growth of cyst-structured antral follicles cause the ovary to possess polycystic characteristic (multiple small cysts) [5]. The main difference from normal ovary is that antral follicles in polycystic ovary grow in number, but not in size (immature), resulting in failed ovulation.

In general, the approaches to detect PCO follicles are (1) stereology and (2) feature extraction and classification. In Stereology, two-dimensional images are viewed as projections of three-dimensional objects. Stereology relates three-dimensional parameters of structures to two-dimensional measurements that are obtained from 2D slices through the structures. A variety of geometric attributes of follicles can be calculated using stereology, such as the follicle count, distribution of follicles within the ovary, and follicle size [6, 7]. In [4], stereology is used to measure the follicle diameter and Euclidean distance is used for quantification of follicles. Another study conducted by Ashika Raj [8], implemented Fuzzy C-Means Clustering for ovarian follicle detection on PCOS. In [9], used an image clustering approach for follicles segmentation using Particle Swarm Optimization with a new modified no-parametric fitness function, i.e. Mean Structural Similarity Index (MSSIM) and Normalized Mean Square Error (NMSE) to produce more compact and convergent cluster. Also, Gabor Wavelet as a feature extraction method and three classification methods, which is Learning Vector Quantization, KNN - Euclidean distance, and Support Vector Machine used in paper [1].

This paper designs a detection system to classify PCOS based on follicle detection using USG images. In this paper, we propose modification of Backpropagation algorithm, namely Levenberg - Marquardt optimization [10] and Conjugate Gradient – Fletcher Reeves [11, 12], to improve the convergence rate of standard Backpropagation. Based on the two optimization methods, the performance of system will be analyzed based on the accuracy and running time criteria.

The rest of paper is organized as follows. Section 2 explains the proposed method of PCO detection. Section 3 explains the analysis and experimental results, and conclusion are provided in Sect. 4.

2 Proposed Scheme

This study uses ultrasound image of the ovary which is considered as input. The data were obtained from Permata Bunda Syariah clinic, Cirebon, and validate by gynecologists. These USG images will be preprocessed using gray scaling, histogram equalization, image binarization, morphology filtering, data cleaning, edge detection, and region props. The next step, the important features as output of preprocessing will be taken its important features using Gabor Wavelet. Finally, in the classification step used a modification of Backpropagation algorithm, namely Levenberg - Marquardt optimization and Conjugate Gradient - Fletcher Reeves. And the last process is output system.

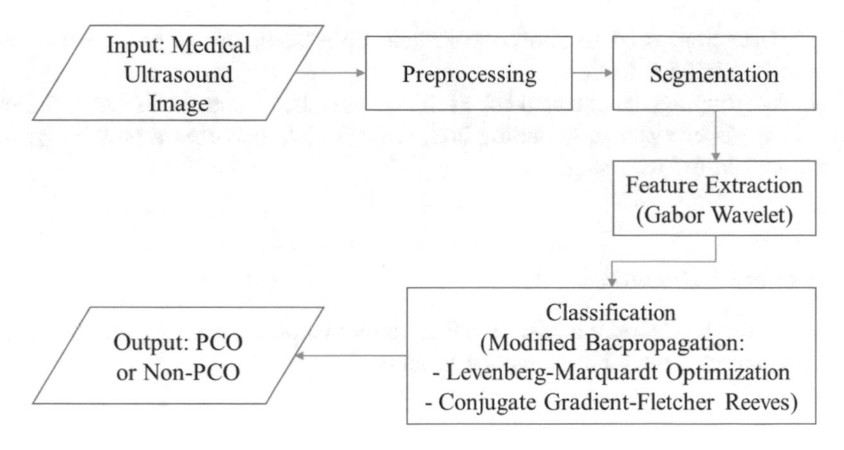

Fig. 1. General system design of PCO detection system

The output is represented as "0" and "1", where "0" represents non PCO follicle and "1" represents PCO follicle. Figure 1 shows the general design system to detect PCO follicles.

From Fig. 1, the steps of general system design of PCO detection system are described as follow:

2.1 Preprocessing

Preprocessing is essential to improve the input image quality in order to easily acquire important information to be processed in feature extraction. Here are the steps that used in preprocessing:

a. Grayscaling, used to start an enhancement process to be a two-dimensional matrix, in order to do the next step in preprocessing.
b. Histogram Equalization, used to produce the output image which has the same histogram values. This technique is intended to get the difference of some pixels that have low frequency, because that pixel will look like they have the same color.
c. Image Binarization, in this process, the image is changed to black and white image, therefore the image matrix contains only "1" or "0".
d. Image Morphology, the technique that used in this paper is dilation and erosion.
e. Invert image, invert the black to white and otherwise, because the object is easier to be detected in white with black background color.
f. Data Cleaning, used to eliminate unimportant object.

2.2 Segmentation

Segmentation used to separate the follicles object from the background image. Here is the steps used in this paper:

a. Edge Detection, used to show the follicle objects. In this paper, we use canny method for edge detection.
b. Follicle Cropping, the detected edge follicles are labeled and each labeled follicle in the image is cropped in bounding box. Each follicle becomes a new image to be processed in the next stage.

2.3 Feature Extraction

In this paper used Gabor Wavelet as a feature extraction method. Figure 2 shows the block diagram of the feature extraction process.

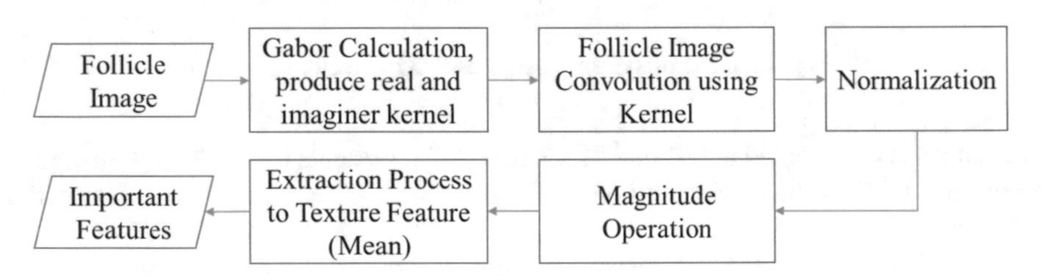

Fig. 2. Block diagram of Gabor wavelet

2.4 Classification

This paper proposed modification of Backpropagation algorithm, namely Levenberg - Marquardt optimization and Conjugate Gradient – Fletcher Reeves. Here is the explanation of each algorithm.

A. Networks Architecture

The network architecture that used in this paper is a multilayer neural network. This network has at least one hidden layer. In this system uses N input neurons, where the value N depends on the number of features from feature extraction process, 1 hidden layer consists of 1 until 50 neurons, and 1 output neuron.

B. Standard Backpropagation

Backpropagation is one of learning algorithm for Feed Forward Neural Network. The standard backpropagation has three phases [9]:

a. Phase I: Forward propagation
   Calculate the output for each neuron from input layer until the output layer using specified activation functions.
b. Phase II: Backward propagation
   The error between output network and data target denote the error of the network. The error was backward propagated, start from the weights in output layer to the input layer.

c. Phase III: Weight Modification
Update the weights to decrease the network error.

The standard backpropagation has a weak convergence rate. It needs many iterations to have minimum error. Hence, backpropagation is usually modified with other optimization method, one of them is Levenberg-Marquardt optimization and Conjugate Gradient optimization.

C. Levenberg-Marquardt Optimization
Here are pseudo-codes of Levenberg-Marquardt optimization [11]:
1. Initialize the weights and parameter μ (learning rate, μ = 0.01 is appropriate)
2. Compute the sum of the squared errors over all input F(w).
3. Solve (2) to obtain the increment of weights Δw.

$$\Delta \mathrm{w} = \left[J^T J + \mu \mathrm{I}\right]^{-1} J^T e \quad (1)$$

Where:
J: the Jacobian matrix
μ: the learning rate
4. Recomputed the sum of squared errors F(w)
Using w + Δw as trial w, and judge:
IF trial F(w) < F(w) in step 2 THEN

$$w = w + \Delta \mathrm{w} \quad (2)$$

$$\mu = \mu * \beta \quad (3)$$

Where β is the decay rate (β = 0.1 (0 < β < 1))
Go back to step 2.
ELSE

$$\mu = \mu / \beta \quad (4)$$

Go back to step 4
END IF

D. Conjugate Gradient – Fletcher Reeves
Conjugate Gradient algorithm for training process of Backpropagation algorithm are as follows [11, 12]:
1. Initialize all weights to small random number.
2. If the termination conditions were not satisfied, do steps 2–13.

Forward propagation phase:

3. Each input unit receives the signal and passes it to the hidden units on it.
4. Calculate all the outputs in the hidden units z_j $(j = 1, 2, \ldots, p)$.

$$z\_net_j = v_{j0} + \sum_{i=1}^{n} x_i v_{ji} \tag{5}$$

$$z_j = f\left(z\_net_j\right) = \frac{1}{1 + e^{-z\_net_j}} \tag{6}$$

5. Calculate all the output in units of output y_k (k = 1, 2, …, m).

$$y\_net_k = w_{k0} + \sum_{j=1}^{p} z_j w_{kj} \tag{7}$$

$$y_k = f(y\_net_k) = \frac{1}{1 + e^{-y\_net_k}} \tag{8}$$

Backpropagation Phase:

6. Calculate the error factor in the output unit based on the difference (error) value of actual and forecasted values (output from the input units).

$$\delta_k = (t_k - y_k) f'(y_{net_k}) = (t_k - y_k) y_k (1 - y_k) \tag{9}$$

7. Calculate the error factor in the hidden units based on the error factor in the unit above it.

$$\delta\_net_j = \sum_{k=1}^{m} \delta_k w_{kj} \tag{10}$$

$$\delta_j = \delta\_net_j f'\left(z_{net_j}\right) = \delta_{net_j} z_j (1 - z_j) \tag{11}$$

8. Calculate the gradient at the unit output of the objective function is established.

$$g_{k+1} = \frac{1}{N} \sum_{n=1}^{p} \delta_{nk} y_{nk} \tag{12}$$

9. Calculate the gradient at the hidden units.

$$g_{j+1} = \frac{1}{N} \sum_{n=1}^{p} \delta_{nj} y_{nj} \tag{13}$$

10. Calculate the parameter β for all neurons in the hidden units and output units. For Fletcher-Reeves optimization, the equation as follows:

$$\beta_{k+1} = \frac{g_{k+1}^T + 1(g_{k+1})}{g_k^T g_k} \tag{14}$$

Where

β_{k+1} = β in the recent iteration
g_{k+1} = gradient in the recent iteration
g_k = gradient in the prior iteration

11. Calculate the direction for all neurons in the hidden units and output units.

$$d_{t+1} = -g_{t+1} + \beta_t d_t \tag{15}$$

For the initial direction:

$$d_1 = -g_1 \tag{16}$$

12. Calculate the parameter α for all neurons in the hidden units and output units, which is how big the steps taken for each direction. This parameter can be searched by line search technique.

Changes in weight:

13. Weight updating is carried out by the following manner:

$$w_{t+1} = w_t + \alpha_{t+1} d_{t+1} \tag{17}$$

3 Experiment and Analysis

This experiment is conducted to analyze the performance of feature extraction and classification method. The performance will be measured with accuracy and running time. In this experiment, there are some variables that used to gain the best performance:

a. Number of features gained from Gabor Wavelet method: 16, 24, and 32 features, and also from each set of features will be sought the texture feature (used Mean) and will be measured the accuracy.

b. The comparison of classification method: Backpropagation learning algorithm with Levenberg - Marquardt optimization and Conjugate Gradient – Fletcher Reeves optimization.
c. Number of neurons used in the hidden layer for Neural Network Architecture: [1… 50] neurons.

3.1 The Number of Feature Experiment

Figure 3 shows the result of a number of feature experiment. In this experiment the feature gained from Gabor Wavelet method and we use 16, 24, and 32 features, and also from each set of features will be sought the texture feature (used Mean). Based on the experiment, the best accuracy gained from vector feature that have 16 features, that is 93.925%. However, if we use mean texture feature, the best accuracy decreases to 84.11% which is gained from 32 features texture features. This is happened because when we use mean texture feature instead of vector feature, the features are changed into one feature. So that, the classification method cannot classify the data optimally. And also, from the experiment can be concluded that if it's used vector feature, more features used than the accuracy will be decreased. On the other hand, if it's used Mean feature, the more features are used then the accuracy will be increased.

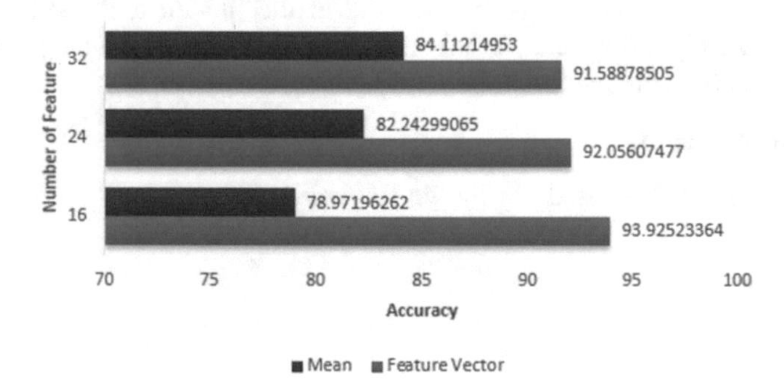

Fig. 3. The comparison result between feature vector and feature texture (Mean). The axis x denotes the accuracy and axis y denote the number of features.

3.2 The Classification Method Experiment

In this experiment, we used two optimization methods to optimize the convergence rate of Backpropagation. The two optimization methods are Levenberg - Marquardt optimization and Conjugate Gradient – Fletcher Reeves optimization.

Figure 4 shows the comparison accuracy of two optimization methods, and Fig. 5 shows the comparison of running time. Based on the experiment, it can be concluded that Levenberg - Marquardt can have the higher accuracy than Conjugate Gradient - Fletcher Reeves, but it has a drawback of running time. If we use more number of

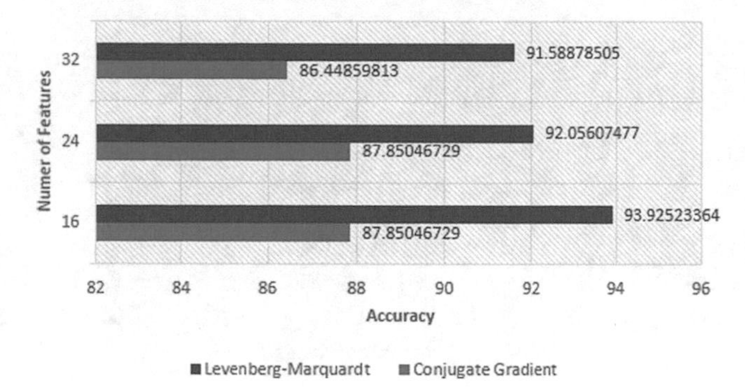

Fig. 4. The comparison accuracy of optimization method for backpropagation. The axis x denote the accuracy and axis y denote the number of feature.

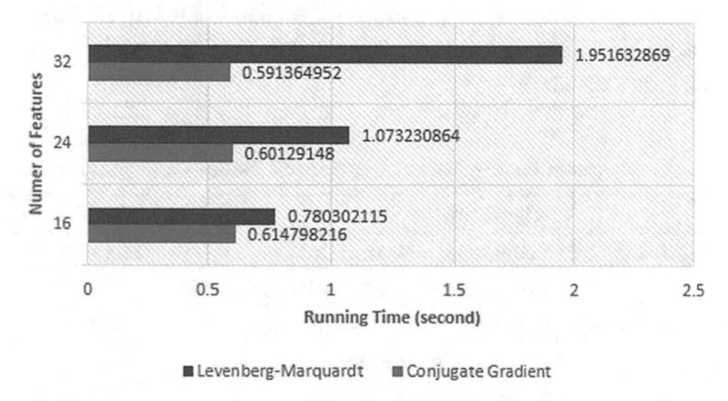

Fig. 5. The comparison running time of optimization method for backpropagation. The axis x denote the running time (second) and axis y denote the number of features.

features and number of neurons, the time which is used will increase significantly. The Conjugate Gradient - Fletcher Reeves shows contrarily. Conjugate Gradient - Fletcher Reeves have stable running time. The best accuracy of Levenberg - Marquardt is 93.925% and Conjugate Gradient - Fletcher Reeves is 87.85%.

3.3 Number of Neurons Experiment

Figure 6 shows the result of the number of neurons in hidden layer from the best accuracy gained from each optimization method. This experiment uses one hidden layer which contains [1…50] neurons. Based on the experiment, the two optimization methods need above 20 neurons to have stable accuracy.

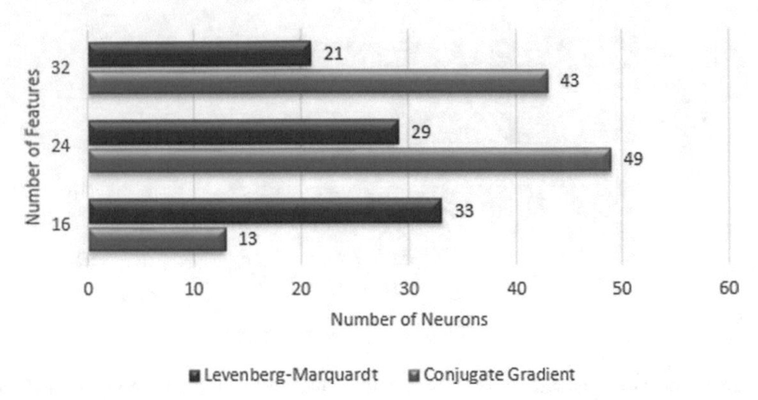

Fig. 6. The result of number of neurons experiment. The axis x denotes the number of neurons and axis y denotes the number of features.

Based on the experiments above, proposed method can have better accuracy than previous research [1]. Table 1 shows the comparison accuracy between proposed method and previous research.

Table 1. Comparison results between proposed method and previous research

	Algorithm	Accuracy (%)
Proposed method	Levenberg – Marquardt	93.925
	Conjugate Gradient – Fletcher Reeves	87.85
Previous research	Learning vector quantization	74.63
	K-nearest neighbors	80.73
	Support vector machine	82.55

4 Conclusions

Based on the analysis towards the experiment, it can be concluded that the proposed scheme can be implemented to detect the PCO follicles. From some the feature experiments can be concluded that if it's used vector feature, the number of features used than the accuracy will be decreased. On the other hand, if we use Mean feature, the number of features used then the accuracy will be increased. Levenberg - Marquardt optimization can have the higher accuracy than Conjugate Gradient - Fletcher Reeves, but it has a drawback of running time. If we use more data and number of neurons, the time which is used will increase significantly. In the other hand, the Conjugate Gradient - Fletcher Reeves shows contrarily. Conjugate Gradient - Fletcher Reeves have stable running time. The best accuracy of Levenberg - Marquardt is 93.925% it gained from 33 neurons and 16 vector feature and Conjugate Gradient - Fletcher Reeves is 87.85% from 13 neurons and 16 vector feature. For the future work, it is important to do some more experiment in parameter of Conjugate Gradient to have better accuracy.

References

1. Purnama, B., Wisesty, U.N., Adiwijaya, Nhita, F., Gayatri, A., Mutiah, T.: A classification of polycystic Ovary Syndrome based on follicle detection of ultrasound images. In: 3rd International Conference on Information and Communication Technology, pp. 396–401. IEEE Press, New York (2015)
2. Sheikhha, M.H., Seyed, M.K., Nasrin, G.: Genetics of polycystic ovary syndrome. Iran. J. Reprod. Med. **5**(1), 1–5 (2006)
3. Adiwijaya, Faoziyah, P.N., Permana, F.P., Wirayuda, T.A.B., Wisesty, U.N.: Tamper detection and recovery of medical image watermarking using modified LSB and Huffman compression. In: 2nd International Conference on Informatics and Applications (ICIA) 2013, pp. 129–132. IEEE Press, New York (2013)
4. Adiwijaya, Purnama, B., Wisesty, U.N., Hasyim, A., Septiani, M., Astuti, W.: Follicle detection on the USG images to support determination polycystic ovary syndrome. J. Phys: Conf. Ser. **622**, 012027 (2015). doi:10.1088/1742-6596/622/1/012027
5. Akkasaligar, P.T., Malagavi, G.V.: Detection of cysts in medical ultrasound images of ovary. In: Proceedings of SARC-IRF 5th International Conference Bangalore, India, ISBN 978-93-84209-13-1 (2014)
6. Lawrence, M.J., Pierson, R.A., Eramian, M.G., Neufeld, E.: Computer assisted detection of polycystic ovary morphology in ultrasound images. In: Fourth Canadian Conference on Computer and Roboot Vision (CRV). IEEE Press, New York (2007)
7. Prapas, N., Karkanaki, A.: Ultrasound and PCOS. In: Stadtmauer, L.A., Tur-Kaspa, I. (eds.) Ultrasound Imaging in Reproductive Medicine, pp. 75–91. Springer, New York (2014). doi:10.1007/978-1-4614-9182-8_7
8. Raj, A.: Ovarian follicle detection for polycystic ovary syndrome using fuzzy c-means clustering. Int. J. Comput. Trends Technol. (IJCTT), **4**(7), 2146–2149 (2013). ISSN 2231-2803. www.ijcttjournal.org. Published by Seventh Sense Research Group
9. Setiawati, E., Adiwijaya, Tjokorda, A.B.W.: Particle swarm optimization on follicles segmentation to support PCOS detection. In: 3rd International Conference on Information and Communication Technology, pp. 369–374. IEEE Press, New York (2015)
10. Suratgar, A.A., Tavakoli, B., Mohammad, B.A., Hoseinabadi, A.: Modified Levenberg-Marquardt method for neural network training. Int. J. Comput. Electr. Autom. Control Inf. Eng. **1**(6), 46–48 (2007)
11. Adiwijaya, Wirayuda, A.B., Wisesty, U.N., Baizal, Z.K.A., Haryoko, U.: An improvement of backpropagation performance by using conjugate gradient on forecasting of air temperature and humidity in Indonesia. Far East J. Math. Sci. **1**, 57–67 (2013). Pushpa Publishing House
12. Adiwijaya, Wisesty, U.N., Nhita, F.: Some line search techniques on the modified backpropagation for forecasting of weather data in Indonesia. Far East J. Math. Sci. **86**(2), 139–148 (2014). Pushpa Publishing House

An Implementation of Local Regression Smoothing on Evolving Fuzzy Algorithm for Planting Calendar Forecasting Based on Rainfall

Arizal Akbar Rahma Saputro(✉), Fhira Nhita, and Adiwijaya

Computational Science, School of Computing, Telkom University, Bandung 40257, Indonesia
arizal@bclaboratory.com,
{fhiranhita, adiwijaya}@telkomuniversity.ac.id

Abstract. The agricultural sector has an important role in the Indonesian economy. Agriculture provides a national food stocks, especially rice as a staple food of Indonesian people. Weather conditions, especially rainfall, severely affected when the right time to start planting. This is very important because this will affect the productivity of farmers. Therefore, rainfall forecasting system is required to create a calendar season, especially rice plant. In this paper, we propose a rainfall forecasting system based on Fuzzy which is optimized using Genetic Algorithms. Data preprocessing is handled by using Local Regression Smoothing for handling of fluctuating data. This paper implements the Local Regression Smoothing on Evolving Fuzzy algorithm with monthly rainfall data. Based on the accuracy of more than 80%, the result of next months rainfall forecasting could be used in the making of rice plant planting calendar in the Bandung regency with 3 periods of planting season, which are from November to February, from December to March, and from January to April given that a control of water needs in surplus of rainfall, and added water needs in rainfall deficiency.

Keywords: Evolving fuzzy · Forecasting · Local regression smoothing · Planting calendar · Rainfall

1 Introduction

Indonesia is an archipelago country traversed by the equator. As the result, Indonesia has a tropical climate with high rainfall. Rainfall is one important factor in agriculture that is to meet the plant's needs of water. A fluctuation of rainfall intensity becomes a problem for the farmers to determine the exact time to begin a planting process. Many farmers suffer from harvest failure because they do not consider rainfall probability in the planting season, problems occur, such as floods and drought, which is the impact of rainfall intensity fluctuation [1]. Therefore, a rainfall forecasting system that has a high degree of accuracy in order to help the farmers' problems is in need.

T. Herawan et al. (eds.), *Recent Advances on Soft Computing and Data Mining*,
Advances in Intelligent Systems and Computing 549, DOI 10.1007/978-3-319-51281-5_16

In forecasting, rainfalls can be determined by understanding the pattern of rainfalls in the past. Gomez in [2] presents an evolutionary approach for generating fuzzy rule based classifier. Many researchers have used Fuzzy Inference System for time series forecasting in many areas [3, 4]. But there is a weakness of Fuzzy Inference System such as determine the membership function [5]. Because of that, Genetic algorithm is used to optimize fuzzy parameters because it produces good results for optimization [6–8], so it will cover constraints in the use of Fuzzy Inference System. While using Evolving Fuzzy, it requires data input, namely rainfall. The fluctuating rainfall makes the data has a wide range value of data, given that, the results can not follow the predictions of actual data that has a that wide range of data [6]. Therefore, it requires a data filtering technique beforehand to build the system.

The main idea of our research is developing the planting calendar forecasting using Soft Computing algorithm with the several algorithm of preprocessing method. In this paper, the proposed method used Local Regression Smoothing (LRS) and Evolving Fuzzy. Local Regression Smoothing which capable of producing a better value with smoother curves than other data filtering techniques, such as Moving Average, Savitzky-Golay and Hamming Window [9, 10]. Ultimately, it is expected to provide a rainfall prediction system with high accuracy, which incorporate technical filtering data by Local Regression Smoothing and Fuzzy parameter optimization with Genetic Algorithms that can be used to create a planting calendar that could be beneficial for farmers to determine the exact time period of rice plant planting in the regency of Bandung regency.

2 Research Method

Rice plants grow excellent on the upper layer of soil thickness of 18–22 cm with a pH between 4–7, a temperature of 23°C, the height is somewhere between 0–1500 m above sea level and the average rainfall of 200 mm per month or more, with distribution for 4 months [11]. Rice plants have 3 phases of growth, namely the phase of vegetative, reproductive, and maturation. In the vegetative and reproductive phase is the stage of germination to the formation of flowers, which happens in 55 days with the needs of rainfall 240–264 mm per month. While of maturation phase is the phase of grain to mature until ready for harvest, which occurs in 65 days with the needs of rainfall 219–228 mm per month [12, 13].

In this paper, we propose a data preprocessing using Local Regression Smoothing (LRS). This method is a smoothing technique that produces values with less range with other data. The steps of this method are as follows [5]:

(1) Calculate regression weight of every data with the following formula:

$$w_i = \left(1 - \left|\frac{x - x_i}{d(x)}\right|^3\right)^3 \tag{1}$$

(2) Calculate sum of squares from weight with second degree polynomial regression method

As well-known, genetic algorithm (GA) is an algorithm from Evolutionary Algorithms (EAs) which is published by John Holland around 1975, which has a very simple form, thus called Simple GA (SGA), later developed David Goldberg, his apprentice [14].

In the proposed scheme, we would like to use the Fuzzy Inference system as a classifier. The stages of Fuzzy Inference System algorithm as follows [8, 14]:

(1) **Fuzzyfication**
Fuzzyfication is a technique to change the crisp inputs into fuzzy input. Crisp input is the form of inputs that have certain truth value, a number that indicates the size of a particular variable. While Fuzzy input is the degree of membership of each linguistic value which is calculated based on the specific membership functions in the interval [0,1].

(2) **Inference**
Inference is a fuzzy rule making with similar principles such as human reasoning on instinct. A Fuzzy rules are written as follows:

(IF *antecedent* THEN *consequent*)

In the Mamdani model, there are two kinds of inference which are clipping and scaling. Inference clipping methods have rules, while using the AND operator, the degree of membership used is the minimum value (min $\{\mu(x1), \mu(x2)\}$), and while using the OR operator, the degree of membership used is the maximum value (max $\{\mu(x1), \mu(x2)\}$).

(3) **Defuzzyfication**
Defuzzyfication is the process of transforming the Fuzzy output into crisp values. The first process performed is composition, which is an aggregation of clipping results of inference in order to obtain a single Fuzzy set. In Mamdani Model, Centroid method used to obtain the crisp value.

The data used is rainfall data monthly from the Indonesian Agency for Meteorology, Climatological and Geophysics (Badan Meteorologi, Klimatologi, dan Geofisika or simply BMKG) of Bandung Regency from January 2005 through December 2015. These data, then divided into two parts, January 2005 through December 2012 as training data, and January 2013 through December 2015 as testing data.

3 Proposed Scheme

We build an upcoming rainfall forecasting system to create a planting calendar by implementing Local Regression Smoothing on Evolving Fuzzy with monthly rainfall parameter data on Bandung Regency. Block diagram of proposed scheme as follows in Fig. 1:

The next step is the optimization process of fuzzy parameters by using Genetic Algorithm according to training data as seen in Fig. 2 below. This step produces optimal Fuzzy parameters so that it could be processed by Fuzzy and later could provide prediction results.

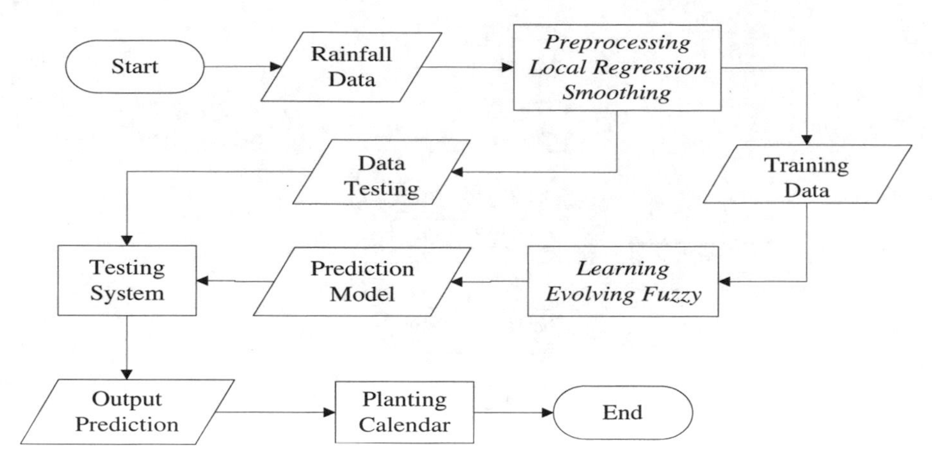

Fig. 1. Block diagram of proposed scheme

4 Testing and Analysis

This research is conducted four scenarios on fuzzy parameter optimization in the form of Membership Function (MF) and fuzzy rule, the scenarios are as follows:

(1) Membership Function Optimization for input and Membership Function to output with data preprocessing.
(2) Membership Function Optimization for input and same for output with data preprocessing.
(3) Membership Function Optimization for input and Membership Function to output with actual data.
(4) Membership Function Optimization for input and same for output with actual data.

Then all of the scenario has one similarity, which is their number of linguistic value is set to four rather than optimized. It is accustomed to BMKG's four classes of rainfall. GA's parameter combination used for training is Evolving Fuzzy Algorithm. Population sizes used are 50, 100, 200 and Maximum Generation at 400,200,100 so that the sum of individual utilized on every generation is equal to 20.000. Crossover Probability (Pc) used value combination of 0.7 and 0.9, while Mutation Probability (Pm) used value combination of 0.1 and 0.3, just as commonly used Pc and Pm value [15]. All combinations are tested 2 times for each and utilized the best result.

4.1 Results and Analysis

(1) Preprocessing Data

Data utilized in the testing phase were processed with Local Regression Smoothing (LRS) beforehand to obtain smoother results. Result as follows:

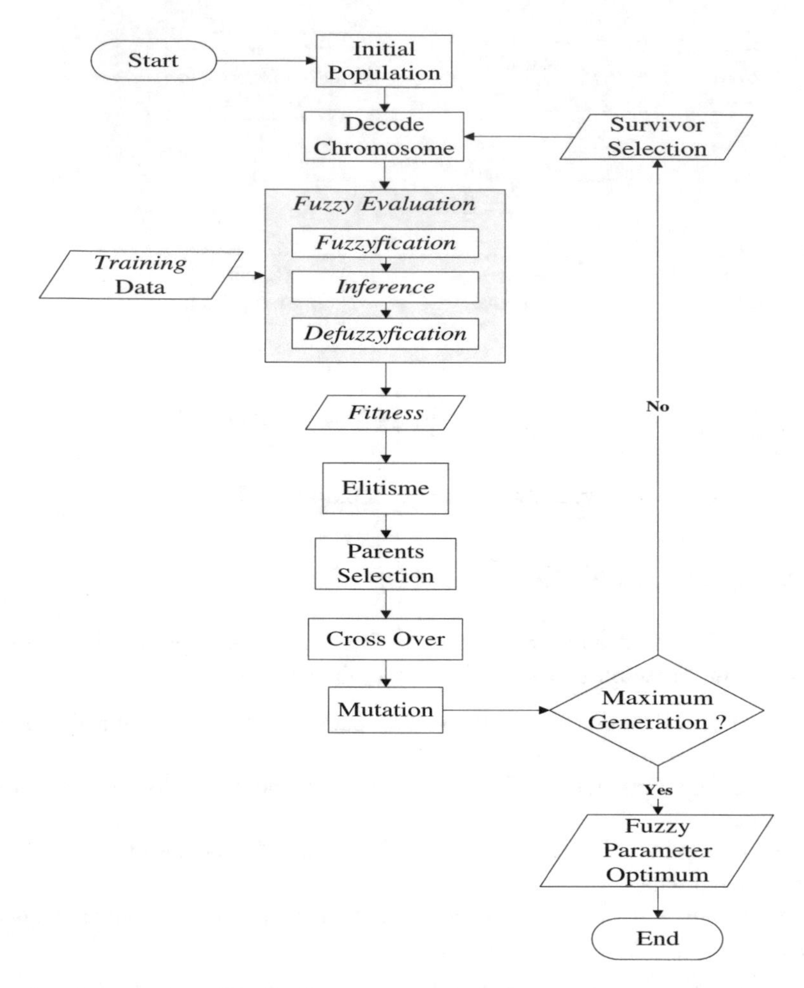

Fig. 2. Fuzzy parameters design steps

According to Fig. 3, it is stated that LRS produce smoother data, in other word its have a smaller range between data compared to the actual data. The result could ease the learning process so that the data produced by LRS are better to be implemented on Evolving Fuzzy Algorithm. Next normalization will be done to shorten the range to 0.1–0.9.

(2) **Learning Process**

Learning Process used Evolving Fuzzy algorithm which produce optimal WMAPE and fuzzy parameter. Result as follows:

Figure 4 below is an WMAPE comparison for each scenario:

According to Table 1 which is the best result from the input Membership Function optimization and different output Membership Function with preprocessing data, best

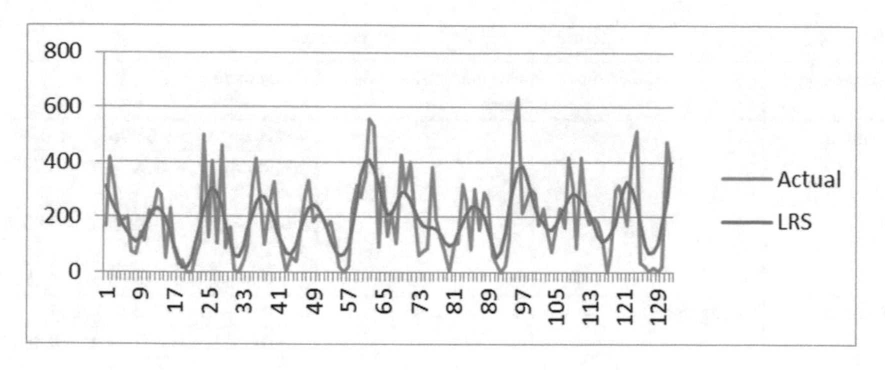

Fig. 3. Actual Data vs LRS results

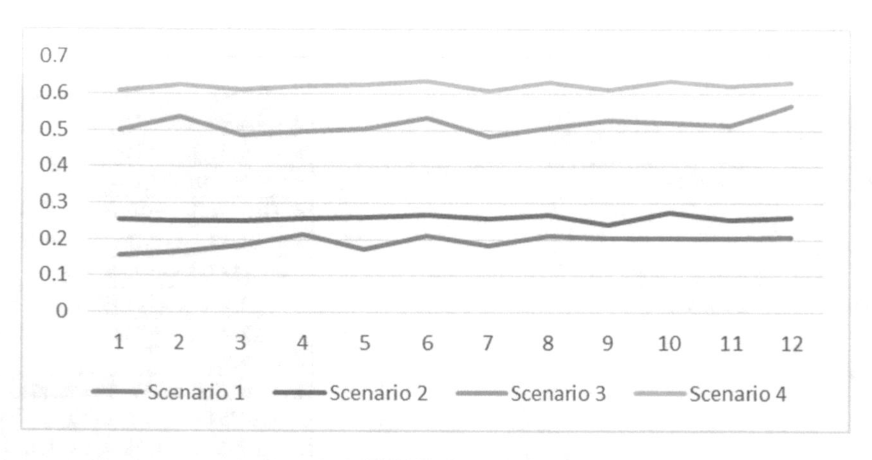

Fig. 4. WMAPE for each scenario

Table 1. Best WMAPE result

Max generation	Population size	Probability of crossover	Probability of mutation	WMAPE
400	50	0.9	0.1	0.1555

result with WMAPE 0.1555 produce optimal Fuzzy parameters. So the forecasting accuracy for training is 84.45%. It is shown in the table as follows:

Table 2 above contains optimally Fuzzy with smallest WMAPE produced on Evolving Fuzzy. The result could be utilized in the testing process with testing data to predict upcoming month's rainfall.

(3) **Testing Results**

Optimal fuzzy parameters are produced from the training process with best Genetic Algorithm parameter combination which is population size of 50, Crossover probability 0.9, and Mutation probability 0.1, then it is tested using test data from January

Table 2. Optimal fuzzy parameters

Attribute	Type of membership function	Number of linguistic value	Parameters
Rainfall M − 4	Phi	4	a = 0, b = 0, c = 0.9961, d = 0.9974
			a = 0.9961, b = 0.9974, c = 0.9982, d = 0.9991
			a = 0.9982, b = 0.9991, c = 0.9994, d = 0.9995
			a = 0.9994, b = 0.9995, c = 1, d = 1
Rainfall M − 3	Triangular	4	a = 0, b = 0.6133, c = 0.8044
			a = 0.6133, b = 0.8044, c = 0.8812
			a = 0.8044, b = 0.8812, c = 0.9371
			a = 0.8812, b = 0.9371, c = 1
Rainfall M − 2	Trapezoid	4	a = 0, b = 0, c = 0.0820, d = 0.3814
			a = 0.0820, b = 0.3814, c = 0.5349, d = 0.7456
			a = 0.5349, b = 0.7456, c = 0.8231, d = 0.8684
			a = 0.8231, b = 0.8684, c = 1, d = 1
Rainfall M − 1	Triangular	4	a = 0, b = 0.0781, c = 0.0853
			a = 0.0781, b = 0.0853, c = 0.1353
			a = 0.0853, b = 0.1353, c = 0.2096
			a = 0.1353, b = 0.2096, c = 1
Rainfall M	Trapezoid	4	a = 0, b = 0, c = 0.2422, d = 0.5841
			a = 0.2422, b = 0.5841, c = 0.6410, d = 0.6788
			a = 0.6410,b = 0.6788,c = 0.4026, d = 0.4854
			a = 0.4026, b = 0.4854, c = 1, d = 1
Rainfall M + 1	Trapezoid	4	a = 0, b = 0, c = 0.6523, d = 0.7888
			a = 0.6523, b = 0.7888, c = 0.8350, d = 0.8608
			a = 0.8350, b = 0.8608, c = 0.8774, d = 0.8920
			a = 0.8774, b = 0.8920, c = 1, d = 1

2013 through December 2015 so that upcoming month rainfall could be predicted. This testing produces testing accuracy of 81.94%. For upcoming month rainfall prediction example, in 2015 could be seen on the following table:

Table 3 shows an example of 2015 rainfall prediction. Overall, most prediction results could represent its actual value. That is due to actual data that have gone through preprocessing, it has a smaller range than other data. With smoother data, the prediction produced could reach the actual value. The comparison could be seen on following Fig. 5:

Table 3. Rainfall prediction example in 2015

Month	LRS results	Prediction
January	303.55	328.80
February	329.45	294.29
March	303.49	241.29
April	248.09	251.35
May	177.86	248.34
June	97.74	227.06
July	69.68	160.76
August	74.26	106.18
September	107.17	94.90
October	169.66	120.84
November	264.14	162.65
December	393.01	212.98

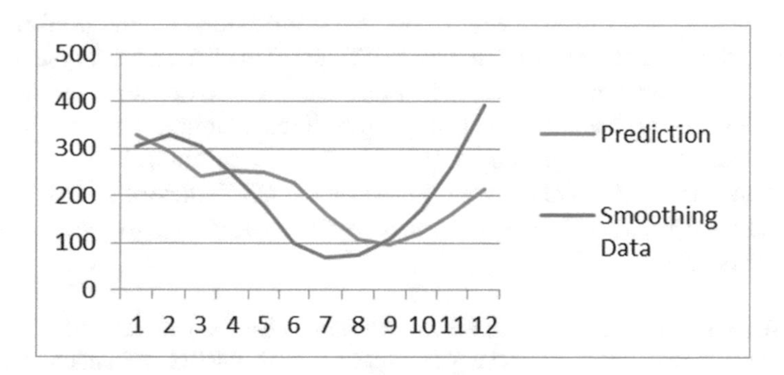

Fig. 5. Smoothing data (LRS results) vs prediction results

(4) Planting Calendar

Optimal Fuzzy Parameter produced from the training process with best Genetic Algorithm parameter combination which are Population Size 50, Crossover Probability 0.9, and Mutation Probability 0.1, which later tested with testing data from January 2013 through December 2015 so that upcoming month rainfall prediction is produced. The example of 2015 rainfall as follows: Table 4

Table 4. Rice plant Planting Calendar

Month / Data	Oct 2014	Nov 2014	Dec 2014	Jan 2015	Feb 2015	Mar 2015	Apr 2015	May 2015	Jun 2015	Jul 2015	Aug 2015	Sep 2015	Oct 2015
Prediction	138.8	193.9	279.5	328.8	294.2	241.2	251.3	248.3	227.1	160.7	106.1	94.9	120.8
Actual	65	297	316	233.5	170	425.5	514.6	35.2	22	5.5	20	5.5	21

The above table shows that the rainfall prediction result in over 200 mm is occurring in December 2014 until June 2015. However, May and June are not recommended, because the results are much different with the actual data. The actual data shows that rainfall of over 200 mm happening from November 2014. During that month, the prediction results show that it is not eligible for planting rice, because the rainfall intensity is still below 200 mm. But the differences are not significant enough to meet the minimum requirements rice plant crop water needs. Therefore, if the result of this prediction is used as a planting reference, then three periods of planting could be done. The first period is November to February, the second period is December to March, and the third period is January to April, given that water control to maintain water intake will be done.

5 Conclusion

According to the testing result, Local Regression Smoothing is good to be implemented on Evolving Fuzzy algorithm. Fuzzy parameter optimization is better performed on the input and output membership functions are different. This forecasting system produces over 80% training accuracy, so that the result could be used to develop the rice plant planting calendar for Bandung Regency with three planting period, November – February, December – March, and January – April with control of water intake in rainfall surplus and added water flow in case of rainfall defficiency. This information can be used by Agricultural Departement of Bandung Regency to create the planting target for next year.

Acknowledgments. Authors would like to thank Telkom University for support of this research. Also, the author would like to express a great appreciation to BMKG of Bandung Regency, for the data used in this study and kindly discussion.

References

1. Nhita, F., Adiwijaya, Annisa, S., Kinasih, S.: Comparative study of grammatical evolution and adaptive neuro-fuzzy inference system on rainfall forecasting in bandung, In: International Conference on Information and Communication Technology (ICoICT), pp. 6–10. IEEE Press, New York (2015)
2. Gomez, J.: Evolution of fuzzy rule based classifiers. In: Deb, K. (ed.) GECCO 2004. LNCS, vol. 3102, pp. 1150–1161. Springer, Berlin, Heidelberg (2004). doi:10.1007/978-3-540-24854-5_112
3. Ramli, A.A., Islam, M.R., Fudzee, M.F.M., Salamat, M.A., Kasim, S.: A practical weather forecasting for air traffic control system using fuzzy hierarchical technique. In: Herawan, T., Ghazali, R., Deris, M.M. (eds.) Recent Advances on Soft Computing and Data Mining. AISC, vol. 287, pp. 99–109. Springer, Heidelberg (2014)
4. Abdullah, L., Gan, C.L.: A fuzzy time series model in road accidents forecast. In: Herawan, T., Ghazali, R., Deris, M.M. (eds.) Recent Advances on Soft Computing and Data Mining. AISC, vol. 287, pp. 1–10. Springer, Heidelberg (2014)

5. Makridakis, S., Wheelwright, S.C., Hyndman, R.J.: Forecasting: Methods and Applications. Wiley, New York (1998)
6. Leung, H.: Artificial Intelligence and Soft Computing. Acta Press, Palma De Mallorca (2002)
7. Adiwijaya, Wisesty, U.N., Nhita, F.: Study of line search techniques on the modified backpropagation for forecasting of weather data in Indonesia. Far East J. Math. Sci. **86**(2) (2014). Pushpa Publishing House
8. Nhita, F., Adiwijaya: A rainfall forecasting using fuzzy system based on genetic algorithm. In: 2013 International Conference of Information and Communication Technology (ICoICT), pp. 111–115, 20–22 March 2013
9. Wettayaprasit, W., Laosen, N., Chevakidagarn. S.: Data filtering technique for neural networks forecasting. In: Proceedings of the 7th WSEAS International Conference on Simulation, Modelling and Optimization, Beijing (2007)
10. Nhita, F.; Saepudin, D.; Adiwijaya; Wisesty, U.N.: Comparative study of moving average on rainfall time series data for rainfall forecasting based on evolving neural network classifier. In: 2015 3rd International Symposium on Computational and Business Intelligence (ISCBI), pp. 112–116, 7–9 December 2015
11. Stigter, J., Winarto, Y.T.: What climate change means for farmers in Asia. Earthzine. IEEE Press (2012)
12. Uchida, S.: Monitoring of planting paddy rice with complex cropping pattern in the tropical humid climate region using landsat and modis data: a case of west java, Indonesia. In: International Archives of the Photogrammetry, Remote Sensing and Spatial Information Science, vol. XXXVIII, Part 8, Kyoto, Japan (2010)
13. Nhita, F.: Adiwijaya, Wisesty, U.N., Ummah I.: Planting calendar forecasting system using evolving neural network. Far East. J Electron. Commun. **14**(2), 81–92 (2015). Pushpa Publishing House, Allahabad
14. De Jong, K.A.: Evolutionary Computation. A Bradford Book. The MIT Press, Cambridge (2002)
15. Tettamanzi, A., Tomassini, M.: Soft Computing: Integrating Evolutionary, Neural, and Fuzzy Systems. Springer, Heidelberg (2010)

Chebyshev Multilayer Perceptron Neural Network with Levenberg Marquardt-Back Propagation Learning for Classification Tasks

Umer Iqbal(✉) and Rozaida Ghazali

Faculty of Computer Science and Information Technology, Universiti Tun Hussein Onn Malaysia, Batu Pahat, 86400 Parit Raja, Johor, Malaysia
umeriqbal15@gmail.com, rozaida@uthm.edu.my

Abstract. Artificial neural network has been proved among the best tools in data mining for classification tasks. Multilayer perceptron (MLP) neural network commonly used due to the fast convergence and easy implementation. Meanwhile, it fails to tackle higher dimensional problems. In this paper, Chebyshev multilayer perceptron neural network with Levenberg Marquardt back propagation learning is presented for classification task. Here, Chebyshev orthogonal polynomial is used as functional expansion for solution of higher dimension problems. Four benchmarked datasets for classification are collected from UCI repository. The computational results are compared with MLP trained by different training algorithms namely, Gradient Descent back propagation (MLP-GD), Levenberg Marquardt back propagation (MLP-LM), Gradient Descent back propagation with momentum (MLP-GDM), and Gradient Descent with momentum and adaptive learning rate (MLP-GDX). The findings show that, proposed model outperforms all compared methods in terms of accuracy, precision and sensitivity.

Keywords: Classification · Chebyshev multilayer perceptron · Levenberg Marquardt · Gradient descent

1 Introduction

Classification is appeared as revolutionary task in data mining. Classification problem occurs when an object needs to be assigned to a specific class or group on the base of their attributes that related to objects. The significance of classification can be seen in different real life phenomena's such as diagnosis of diseases in medical [1], marketing [2] and in stock exchange [3]. Classification task is based on two steps. First step is to construct the model, which represents the group of precogitated classes. Second step is the model usage, which is used for classifying the unknown objects.

Various techniques have been developed for classification, such as statistical and neural networks techniques, which are prominent. With the passage of time, artificial neural networks have gained much popularity as useful alternative of statistical techniques and due to having the variety of applications in real life [4]. Multilayer perceptron known as MLP, is a feed forward neural network that consists of input, hidden and output nodes. Every node is interconnected with other node in next layer which

T. Herawan et al. (eds.), *Recent Advances on Soft Computing and Data Mining*,
Advances in Intelligent Systems and Computing 549, DOI 10.1007/978-3-319-51281-5_17

makes a connection between them. Back propagation is one of the most used and old supervised learning algorithm, which is proposed by Rumelhart, Hinton and Williams [5]. MLP has been used to find the missing values from data [6]. Due to its ability of fast learning, MLP has been tested for stock trading problems for better prediction [7]. MLP is also successfully implemented for early fault detection in gearboxes [8].

Researchers used different types of learning algorithm with back propagation to train different types of algorithms such as adaptive momentum to improve accuracy of gradient descent [9]. Some trained the artificial bee colony (ABC) with LM for classification problems [10]. Besides the advantages of MLP, there are also some disadvantaged of multilayer perceptron such as firstly, this neural network can only be used for supervised learning, and secondly due to its multilayer structure, it cause computationally expensive training that stuck in local minima [11].

In this paper, we propose Chebyshev polynomials as functional expansion with multilayer perceptron. These polynomials are used to make standard MLP more accurate for classification tasks. We have compared proposed method with four types of learning algorithms that are used as training of multilayer perceptron. The rest of the paper is organized as follows: In Sect. 2, the proposed model is presented. Some experimental setups are presented in Sect. 3. Section 4 is devoted to results and discussions. Finally, Sect. 5 outlines the conclusion.

2 Proposed Model: Chebyshev Multilayer Perceptron Neural Network

According to approximation theory, the non-linear approximation capacity of Chebyshev orthogonal polynomial is very dominant [12]. The proposed method is a combination of the characteristics of Chebyshev orthogonal polynomial and multilayer perceptron, which is, named as CMLP. This method utilizes the MLP input-output pattern with non-linear capabilities of Chebyshev orthogonal polynomial for classification. The Chebyshev multilayer perceptron is multilayer neural network. The structure consists of two parts, first one is the transformation part and learning is the second part. Transformation means that from a lower feature space to higher feature space. The approximate transformable method is implemented on input feature vector to the hidden layer in transformable part. This transformation is also known as functional expansion, where Chebyshev polynomial basis can be seen as a new input layer. Levenberg Marquardt back propagation is used for as a learning part [1]. Table 1 describes the recurrence relation to find the Chebyshev polynomials of degree n.

Where $T_0(x), T_1(x)$ are the Chebyshev polynomials when $n = 0, 1$ and $T_{n+1}(x)$ is the equation for n^{th} polynomial. The reason for which we are using Chebyshev polynomials is that, when we are using polynomial of truncated power series then series represent the

Table 1. Recursive formula for Chebyshev polynomials

$T_0(x) = 1,$	(2.1)
$T_1(x) = x,$	(2.2)
$T_{n+1}(x) = 2xT_n(x) + T_{n-1}(x).$	(2.3)

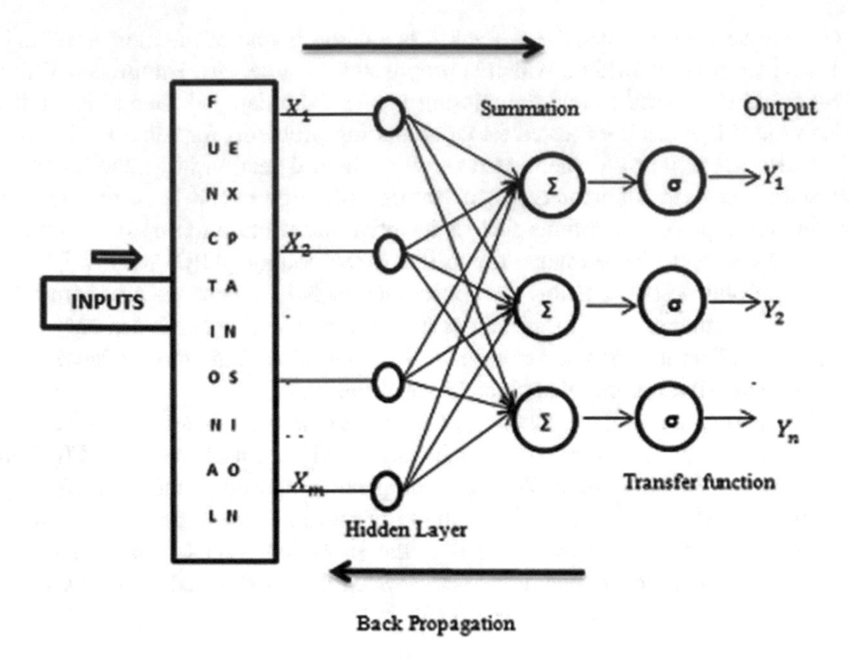

Fig. 1. Chebyshev multilayer perceptron neural network for classification

function with very small error near the point of expansion and when implemented on all points then it cause increase in error. On the other hand, computational economy gained by Chebyshev increases when power series is slowly convergent [12]. This reason makes Chebyshev polynomials more useful and effective for approximation to functions as compared to other polynomials. Chebyshev polynomials are more convergent than the other expansions of polynomials. These properties make Chebyshev polynomial more selective on other polynomials. The proposed method can be seen in Fig. 1.

Where, $X_1, X_2 \ldots X_m$ are the inputs and $Y_1, Y_2, \ldots Y_n$ indicates the output of the neural network.

3 Experimental Setup

This section will illustrate about the working of proposed method, comparison techniques, data sets and evaluation measures.

3.1 Data Collection

The data set for classification analysis, which is the requisite input to the models, is obtained from UCI Repository [13]. We have collected four data sets named as, Iris, Wine, Breast Cancer and Bank Authentication for classification tasks. The dataset is divided in two parts such as; training set and testing set. The data ratio of 70% and 30% was set for training and testing respectively. The Iris dataset consists of 4 features and

150 samples. This data is categorized into three classes named as, Iris Setosa, Iris Versicolour and Iris Virginica. Wine dataset consist of 13 features and 178 samples. This data is categorized into three classes. Similarly, breast cancer data consists of 10 features and 699 samples. This data is categorized into two classes such as, benign and malignant. Respectively Bank Authentication data is consisting of 1372 samples, 5 features and 2 classes. Detail of used datasets is described in Table 2.

Table 2. Description of datasets

Datasets	Number of samples	Number of attributes	Number of classes
Iris	150	4	3
Wine	178	13	3
Breast Cancer	699	10	2
Banknote Authentication	1372	5	2

3.2 Proposed Methodology

The Fig. 2 illustrates the working of proposed method for classification. Inputs are expanded from lower dimension to higher dimension by using Chebyshev functional expansion block [14]. To train the network, we supposed the random weights; choose number of hidden nodes, outputs nodes (according to desired classes) and activation function. Levenberg Marquardt back propagation is used as learning algorithm. If we get our desired error according to the classification task then training will be finished, otherwise, we have to start the training again until the desired results followed by the testing of data based on evaluation measures.

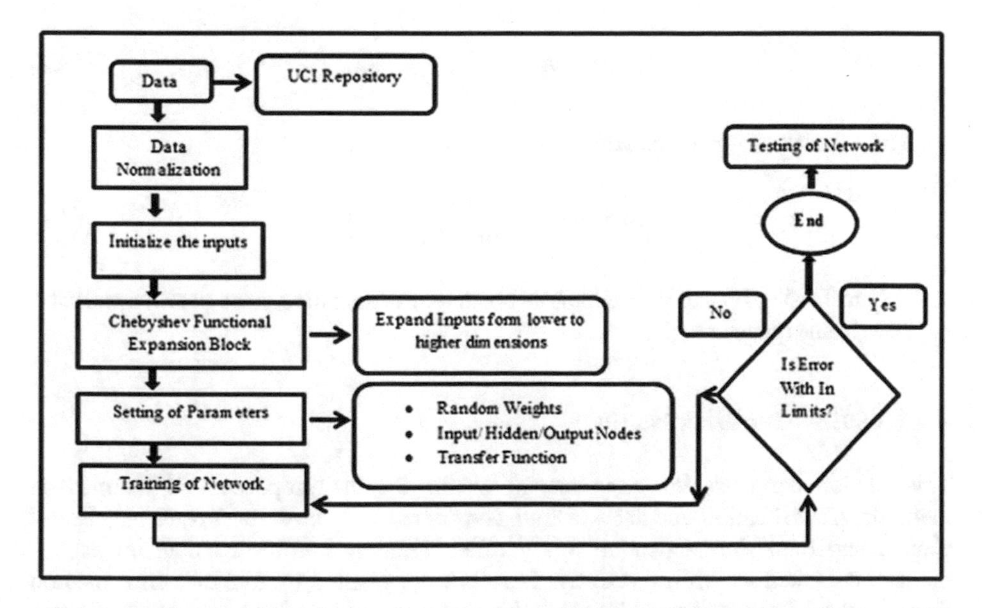

Fig. 2. Working of chebyshev multilayer perceptron neural network

3.3 Benchmarked Approaches

MLP appeared as a good competitor over the statistical methods due to its less computational cost and better accuracy [15]. Many researchers implemented multilayer perceptron neural network for different task of data mining by using different learning algorithms. These learning algorithms proved their strength to make this network stronger as compared to statistical methods. Researchers have used the Levenberg Marquardt back propagation (LM-BP) [1], gradient descent back propagation (GD-BP) [9], gradient descent with momentum back propagation (GDM-BP) [16], gradient descent with momentum and adaptive learning rate back propagation (GDX-BP) [17] to improve the efficiency of the neural network. We have compared our proposed method with multilayer perceptron neural network learned with these four mentioned learning algorithms.

3.4 Evaluation Measures

The performance of the proposed model against the MLP with four different learning algorithms will be claimed based on accuracy, precision and sensitivity for classification tasks. The equation to find out the accuracy is given as:

$$Accuracy = \frac{Tp + Tn}{Tp + Tn + Fp + Fn}. \quad (3.1)$$

For classification task, precision is defined as "The number of true positives is divided by the total number of true positives and false positives". The equation for precision is described as:

$$Precision = \frac{Tp}{Tp + Fp} \times 100. \quad (3.2)$$

The sensitivity for the classification is calculated as:

$$Senstivity = \frac{Tp}{Tp + Fn} \times 100. \quad (3.3)$$

Where Tp, Tn, Fp and Fn are the true positive, true negative, false positive and false negative values respectively.

4 Results and Discussion

This section describes the experimental results for the proposed and comparison methods. Classification accuracy of four considered data sets viz. Iris, Wine, Breast Cancer and Bank Authentication was verified. Training and testing data for experiments were taken as 7:3 respectively. Results were verified by trial and error method that was done 10 times for each experiment and then average for each experiment was considered for further processing. The number of epochs for each experiment was same that is 1000 epochs. Performance of proposed and comparison methods were based on accuracy, precision and sensitivity. We have represented our proposed model Chebyshev multilayer perceptron neural network as CMLP-LM. Here, LM is the learning algorithm that we used in our proposed method. On the other hand, four learning algorithms that we used with multilayer perceptron neural network as comparison techniques are written as MLP-LM, MLP-GD, MLP-GDM and MLP-GDX. It is evident from the results that CMLP-LM provides much better results with classification accuracy of 98%, 99%, 78.11% and 90% on Iris, Wine Breast Cancer and Bank Authentication data sets respectively. On considering the detail of classification accuracy, MLP-LM training proves to be 89% accurate on Iris classification results but it provides best results on wine data set, that is, 94.50%. MLP-GD classification accuracy was found to be 89.22% on wine data set. Similarly, MLP-GDM and MLP-GDX perform best on Iris dataset; those are 84% and 81.50% respectively. The Fig. 3 gives comparative accuracy results, which support the proposed model being more efficient on the comparison methods.

Figure 4 represents the precision results of proposed model compared with the existing methods. It can be inferred from the simulation results that MLP-LM and MLP-GD give their best precision values on Wine dataset, that are, 95% and 90.4% respectively while MLP-GDM and MLP-GDX precision values are 83.8% and 79% respectively on Iris data set. On the other hand, CMLP-LM precision results appear to

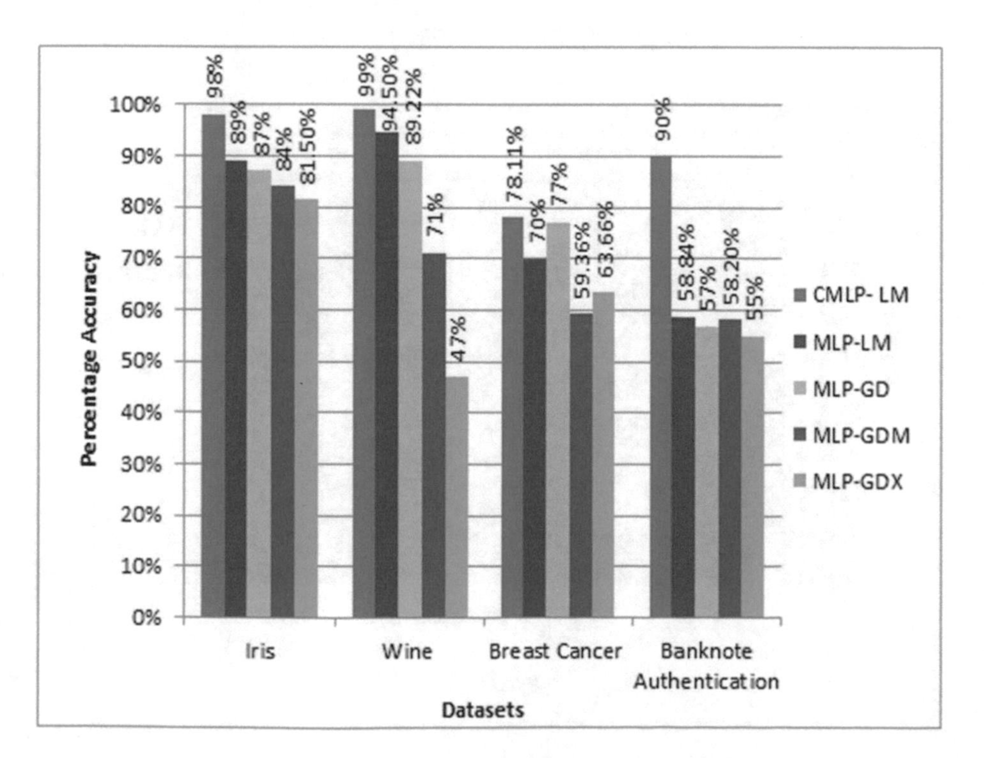

Fig. 3. Comparison in terms of classification accuracy

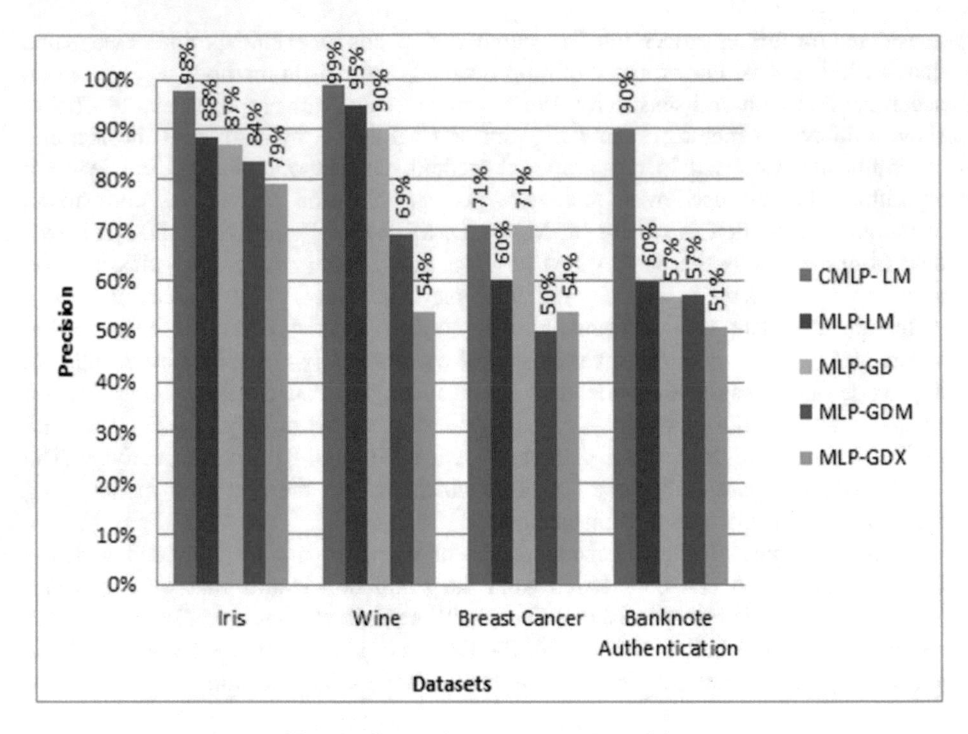

Fig. 4. Comparison in terms of Precision

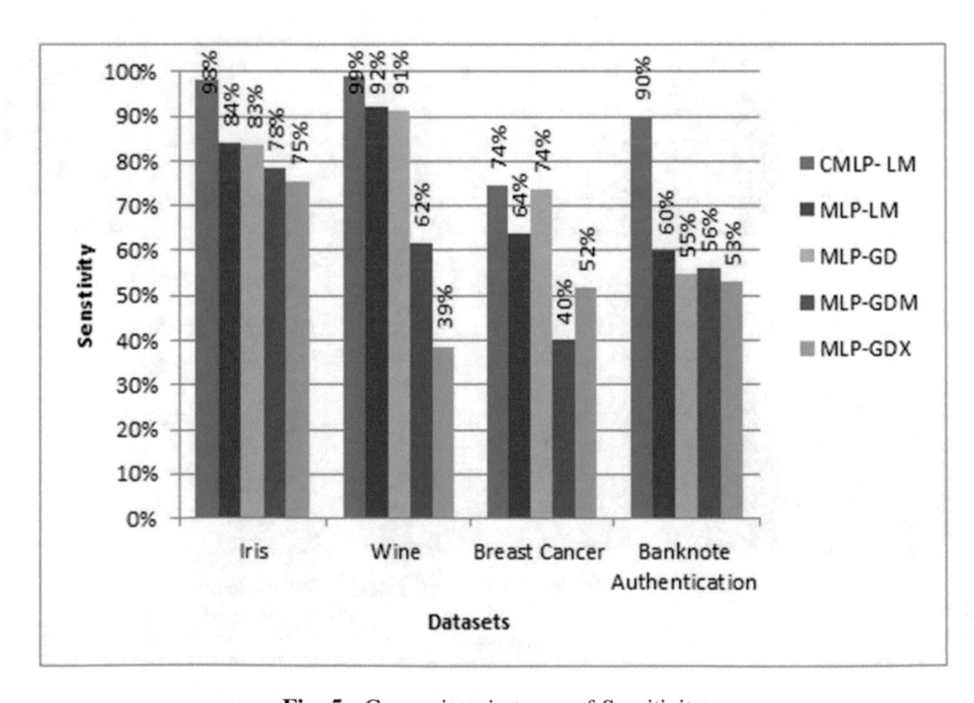

Fig. 5. Comparison in terms of Sensitivity

be more precise on all data sets, which are, 97.96% for Iris, 98.9% for Wine and 70.93% for Breast Cancer, 90% for Bank Authentication.

Comparative analysis of proposed and comparison methods based on sensitivity shows that MLP-LM and MLP-GD results are 92.29% and 91.36% on Wine dataset but MLP-GDM and MLP-GDX results for Iris are 78.44% and 75.42% respectively. Figure 5 shows the performance of CMLP-LM for Iris, that is, 98.06%; 98.99% for Wine, 74.44% for Breast Cancer and 90% for Bank Authentication.

5 Conclusion

In this paper, we have proposed the Chebyshev multilayer perceptron neural network with Levenberg Marquardt back propagation learning algorithm (CMLP-LM). This model was developed for classification task. Method was experimentally trained and tested on four benchmarked data sets which are taken from UCI repository. The performance of proposed method indicates the validity of classification task. The evaluation measures performance is showing that the CMLP-LM has better performance in terms of accuracy, sensitivity and precision over MLP-LM, MLP-GD, MLP-GDM and MLP-GDX.

Acknowledgments. The authors would like to thank Universiti Tun Hussein Onn Malaysia (UTHM) and Ministry of High Education (MOHE) for financially supporting this research under Fundamental Research Grant Scheme (FRGS), Vote No 1235.

References

1. Kumar, M., Singh, S., Rath, S.K.: Classification of microarray data using functional link neural network. Procedia Comput. Sci. **57**, 727–737 (2015)
2. Decker, R., Kroll, F.: Classification in marketing research by means of LEM2-generated rules. In: Advances in Data Analysis, pp. 425–432 (2007)
3. Bebarta, D.K., Biswal, B., Rout, A.K., Dash, P.K.: Forecasting and classification of Indian stocks using different polynomial functional link artificial neural networks. In: INDICON Annual IEEE, pp. 178–182 (2012)
4. Paliwal, M., Kumar, U.A.: Neural networks and statistical techniques: a review of applications. Expert Syst. Appl. **36**(1), 2–17 (2009)
5. Rumelhart, D.E., Hinton, G.E., Williams, R.J.: Learning representations by back-propagating errors. Cogn. Model. **5**(3), 533–536 (1988)
6. Silva-Ramírez, E.L., Pino-Mejías, R., López-Coello, M.: Single imputation with multilayer perceptron and multiple imputation combining multilayer perceptron and k-nearest neighbours for monotone patterns. Appl. Soft Comput. **29**, 65–74 (2015)
7. Mabu, S., Obayashi, M., Kuremoto, T.: Ensemble learning of rule-based evolutionary algorithm using multi-layer perceptron for supporting decisions in stock trading problems. Appl. Soft Comput. **36**, 357–367 (2015)
8. Jedliński, Ł., Jonak, J.: Early fault detection in gearboxes based on support vector machines and multilayer perceptron with a continuous wavelet transform. Appl. Soft Comput. **30**, 636–641 (2015)

9. Rehman, M.Z., Nawi, N.M.: The effect of adaptive momentum in improving the accuracy of gradient descent back propagation algorithm on classification problems. In: Software Engineering and Computer Systems, pp. 380–390 (2011)
10. Shah, H., Ghazali, R., Nawi, N.M., Deris, M.M., Herawan, T.: Global artificial bee colony-Levenberq-Marquardt (GABC-LM) algorithm for classification. Int. J. Appl. Evol. Comput. (IJAEC) **4**(3), 58–74 (2013)
11. Lee, T.T., Jeng, J.T.: The Chebyshev-polynomials-based unified model neural networks for function approximation. IEEE Trans. Syst. Man Cybern. Part B: Cybernet. **28**(6), 925–935 (1998)
12. Konstantinidis, S., Karampiperis, P., Sicilia, M.A.: Enhancing the Levenberg-Marquardt method in neural network training using the direct computation of the error cost function hessian. In: Proceedings of the 16th International Conference on Engineering Applications of Neural Networks (INNS) (2015)
13. Blake, C., Merz, C.J.: UCI repository of machine learning databases (1998)
14. Bui, D.T., Tuan, T.A., Klempe, H., Pradhan, B., Revhaug, I.: Spatial prediction models for shallow landslide hazards: a comparative assessment of the efficacy of support vector machines, artificial neural networks, kernel logistic regression, and logistic model tree. Landslides, pp. 1–18 (2015)
15. Liu, H., Tian, H.Q., Liang, X.F., Li, Y.F.: Wind speed forecasting approach using secondary decomposition algorithm and Elman neural networks. Appl. Energy **157**, 183–194 (2015)
16. Singh, B., De, S., Zhang, Y., Goldstein, T., Taylor, G.: Layer-specific adaptive learning rates for deep networks (2015)
17. Gates, G.W.: The reduced nearest neighbor rule. IEEE Trans. Inf. Theor. 431–435 (1972)

Computing the Metric Dimension of Hypercube Graphs by Particle Swarm Optimization Algorithms

Danang Triantoro Murdiansyah(✉) and Adiwijaya

School of Computing, Telkom University, Jl. Telekomunikasi, Bandung, Indonesia
{danangtri,adiwijaya}@telkomuniversity.ac.id

Abstract. In this paper, we present a PSO (Particle Swarm Optimization) algorithm for determining the metric dimension of graphs. We choose PSO because of its simplicity, robustness, and adaptability for various optimization problems [5]. Our PSO uses the binary valued vector for particles. The binary valued vector is used to represent which one of vertices of a graph is belong to resolving set. The feasibility is enforced by repairing particles. We tested our PSO by computing the metric dimension of hypercube graphs. The result is our PSO can achieve metric dimension known in literature [8] in reasonable amount of time.

Keywords: Graph theory · Hypercube graph · Metric dimension · Particle Swarm Optimization

1 Introduction

The metric dimension problem was introduced by Harary, and Melter [4]. The metric dimension problem arises in many diverse areas, such as robot navigation [7], telecommunication [2], and geographical routing protocol [9]. For an instance, in the area of robot navigation, a robot uses a signal to determine its position according to a set of fixed landmarks. Minimum number of landmarks is required to achieve efficiency. This problem is equivalent to the metric dimension problem. The minimum resolving set of a graph is used as the set of nodes where the landmarks are placed. The number of landmarks is equal to the metric dimension of the graph.

From computational complexity point of view, such as the determining of f-chromatic index [1], metric dimension is one of combinatorial optimization which are included in the NP-Complete problem [6]. In the determining of metric dimension of a graph, the size of its solution space grow exponentially with the problem dimension. To deal with this problem, approximation and heuristic approach can be applied. Although heuristic methods do not offer a guarantee of reaching the optimum solution, they give satisfactory solutions in a reasonable amount of time. The use of developed class of heuristic algorithms, that is metaheuristic algorithm, related to metric dimension problems have been investigated

T. Herawan et al. (eds.), *Recent Advances on Soft Computing and Data Mining*,
Advances in Intelligent Systems and Computing 549, DOI 10.1007/978-3-319-51281-5_18

by Jozef Kratica, Vera Kovačević-Vujčić, and Mirjana Čangalović [8], where the metric dimension was computed using genetic algorithm. In this paper, we present other meta-heuristic method, namely PSO algorithm for determining the metric dimension of graphs. Our PSO uses the binary valued vector for particles. The binary valued vector is used to represent which one of vertices of a graph is belong to resolving set. Infeasible particles are repaired by adding necessary vertices in order to become feasible.

2 Metric Dimension of Graphs

In this section we describe a definition of the metric dimension of graphs. Let G be a simple connected graph with $V(G)$ as its vertex set. Suppose that $W = \{w_1, w_2, \ldots, w_k\} \subseteq V(G)$ and $v \in V(G)$. A metric representation of v with respect to W is defined as $r(v|W) = (d(v, w_1), d(v, w_2), \ldots, d(v, w_k))$, where $d(v, w)$ is the distance between the vertices v and w. W is said to be a resolving set for G if and only if every $v \in V(G)$ has distinct representation $r(v|W)$. A resolving set containing a minimum number of vertices is called a minimum resolving set or a basis for G. The metric dimension of a graph G, denoted by $dim(G)$, is the number of vertices in a basis of G.

3 Hypercube Graphs

The k-cube or k-dimensional hypercube Q_k is defined as graph whose vertices set $V(Q_k)$ consists of the 2^k k-dimensional binary vectors, where 2 vertices are adjacent whenever they differ in exactly 1 coordinate. The k-cube or Q_k has $n = 2^k$ vertices and $m = k2^{k-1}$ edges.

For example, Q_3 has 8 vertices (0, 0, 0), (0, 0, 1), (0, 1, 0), (0, 1, 1), (1, 0, 0), (1, 0, 1), (1, 1, 0), and (1, 1, 1). Vertex (1, 1, 1) has adjacent vertices (1, 1, 0), (1, 0, 1), and (0, 1, 1).

Figure 1 are examples of graphical representation of hypercube graphs with k = 2 and k = 4.

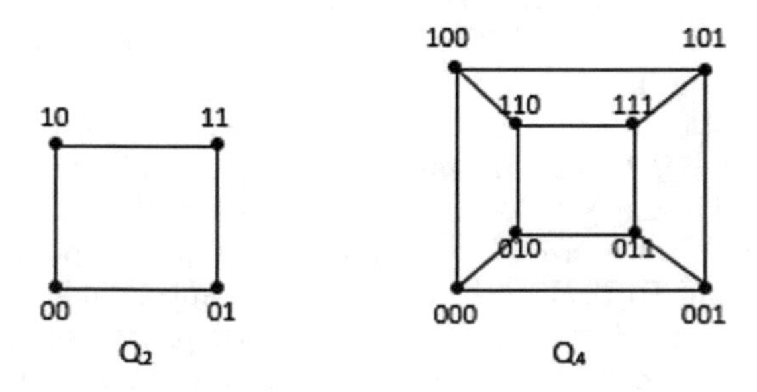

Fig. 1. Q_2 and Q_4

Table 1. The metric dimension of some hypercube graphs

Hypercube	Metric dimension
Q_2	2
Q_3	3
Q_4	4
Q_5	4
Q_6	5
Q_7	6
Q_8	6
Q_9	<8
Q_{10}	<8

According to the literature [3], the metric dimension of hypercube graphs is summarized in Table 1.

4 Particle Swarm Optimization for the Metric Dimension

4.1 The Discrete Particle Swarm Optimization Model

In standard PSO, a solution is represented by a particle in a swarm P moving through D-dimensional space with position vector $x_i = (x_{i1}^k, x_{i2}^k, ..., x_{iD}^k)$, where k is the iterative number. At each iteration, the particles adjust their velocity v_i^k along each dimension based on the previous best position of particle p_i^k and the best position of all swarms p_g. The position of each particle is modified according to the following function:

$$v_{ij}^{k+1} = w.v_{ij}^k + c_1.r_1.(p_{ij}^k - x_{ij}^k) + c_2.r_2.(p_{gj}^k - x_{ij}^k) \tag{1}$$

$$x_{ij}^{k+1} = x_{ij}^k + v_{ij}^k \tag{2}$$

In function (1), w is the inertia term, c_1 and c_2 are learning factors, r_1 and r_2 are random variables. The value of inertia w will determine the ability of a particle to explore the solution space [5]. At each iteration, a larger w will make a particle has tendency to explore a larger region of the solution space. It also makes a particle maximize global search ability. Whereas a smaller w will make a particle restrict its exploration ability to local search. The value of c_1 and c_2 will determine the convergence characteristic of the algorithm [5]. A larger value of c_1 will make exploration of a particle restricted to local regions of the best found solutions previously. Whereas a larger value of c_2 will make exploration of a particle closer to global best solution at each iteration. The value of r_1 and r_2 will determine the search direction [5].

4.2 The Binary Particle Swarm Optimization Model for the Metric Dimension

In this section, we present a binary PSO model for solving the metric dimension problem. We use linear function [10] as follows:

$$L(x_{ij}) = \frac{(x_{ij} - R_{min})}{(R_{max} - R_{min})} \tag{3}$$

$$xbinary_{ij} = \begin{cases} 1, random() \leq L(x_{ij}) \\ 0, otherwise \end{cases} \tag{4}$$

The value of linear function $L(x_{ij})$ is (0, 1). The random function random() has uniform distribution, its value is [0.0, 1.0]. Whilst R_{max} and R_{min} are predefined value with $R_{min} < R_{max}$. The vector x_{ij} is real valued position vector, and $xbinary_{ij}$ is binary valued position vector.

In binary valued position vector $xbinary_{ij} = (x_{i1}, x_{i2}, ..., x_{ij})$, if j-th element of the vector has value 1, it means that vertex j belongs to S. If every $v \in V(G)$ has distinct representation $r(v|S)$, then S is a resolving set. The binary valued position vector has correlation with both the real valued position vector and the real valued velocity vector. This correlation is occured because of the linear function. The value of binary valued position vector is produced by computing the value of linear function. The value of linear function is produced by involving the real valued position vector. The real valued position vector is produced by involving the real valued velocity vector. This correlation is very important to make the solution of this PSO convergent to the best solution.

In our PSO algorithm, when a particle is not feasible as a resolving set, that particle is repaired by adding a vertex from $V \backslash S$. This procedure is applied until that particle becomes a resolving set. Note that after a particle has been repaired, that particle is not updated according to the final resolving set. The reason is to maintain the correlation between binary valued vector and real valued vector. The outline of our PSO implementation is presented in Algorithm 1.

Algorithm 1. The basic scheme of the PSO Algorithm

for all particle i **do**
  Initialize real valued position vector $x_i = 1$, real valued velocity vector $v_i = 0$;
end for
while termination criteria not satisfied **do**
  for all particle i **do**
    update x_i, v_i, and compute the value of the linear function
    if value of random function < value of linear function **then**
      set the element of binary valued position vector $xbinary_i := 1$;
    else
      set the element of binary valued position vector $xbinary_i := 0$;
    end if
    while particle is not a resolving set **do**

adding a vertex from $V \backslash S$ to the particle;
end while
set objective value = cardinality of the resolving set;
if objective value < objective value of best local p_i **then**
objective value of best local := objective value;
binary valued position vector best local $pbinary_i$:= binary valued position vector $xbinary_i$;
real valued position vector best local p_i := real valued position vector x_i;
end if
end for
if objective value of best local < objective value of best global **then**
objective value of best global := objective value of best local;
binary valued position vector best global $pbinary_g$:= binary valued position vector best local $pbinary_i$;
real valued position vector best global p_g:= real valued position vector best local p_i;
end if
end while

5 Main Results

This section summarizes the result of our PSO applied to hypercube graphs. We set parameter $c_1 = c_2 = 5$, and $w = 0.8$. For R_{min} and R_{max}, we set $R_{min} = -50$, $R_{max} = 50$, these value are considered as the best value, because it can achieve the best optimization results [10]. The stopping criterion of our PSO is the cardinality of resolving set that reached known metric dimension of hypercube graph in literature [8].

The binary PSO has been run 20 times for each instance. The results are summarized in the Table 2. The Table 2 is organized as follows:

1. the first column contains the instance name of hypercube graph,
2. the second column contains n, the number of vertices,
3. the third column contains m, the number of edges,
4. the fourth column contains best solution, the best (minimum) cardinality of the resolving set,
5. the fifth column contains iteration, the average number of iterations for finishing the algorithms to achieve best solution,
6. the sixth column contains *agap*, the average solution quality, it is a gap percentage defined as $agap = \frac{1}{20} \sum_{i=1}^{20} gap_i$, where $gap_i = 100. (PSO_i - bestsolution)/bestsolution$, and PSO_i represents the PSO solution obtained in the i-th run,

7. the seventh column contains σ, standard deviation of gap_i, $i = 1, 2, \ldots, 20$, obtained by formula

$$\sigma = \sqrt{\frac{1}{20}\sum_{i=1}^{20}(gap_i - agap)^2},$$

8. the eighth column contains evaluation, the average number of objective function evaluations,

Regarding to our PSO results, Table 2 shows that for all hypercubes Q_r, $1 \leq r \leq 8$, our PSO has reached optimal solution known from the literature.

Table 2. PSO results

Instance	n	m	PSO best solution	Iteration	$agap$	σ	Evaluation
Q_1	2	1	1	1	0	0	24
Q_2	4	4	2	1	0	0	28
Q_3	8	12	3	2	0	0	29
Q_4	16	32	4	3	0	0	86
Q_5	32	80	4	3	0	0	122
Q_6	64	192	5	3	0	0	137
Q_7	128	448	6	4	0	0	146
Q_8	256	1024	6	10	0	0	486

Table 3 shows that for all hypercubes Q_r, $1 \leq r \leq 8$, the previous work that using GA [8] has also reached optimal solution known from the literature.

From Tables 2 and 3, we can see the differences between our PSO results and GA results [8]. The iteration and evaluation results between our PSO results

Table 3. GA results [8]

Instance	n	m	GA best solution	Iteration (generation)	$agap$	σ	Evaluation
Q_1	2	1	1	1	0	0	150
Q_2	4	4	2	346	0	0	66
Q_3	8	12	3	614	0	0	306
Q_4	16	32	4	2001	0	0	24486
Q_5	32	80	4	2001	0	0	48565
Q_6	64	192	5	2001	0	0	62323
Q_7	128	448	6	2001	0	0	66791
Q_8	256	1024	6	2002	0	0	71657

Table 4. PSO time results

Instance	Time (sec)
Q_1	0.19
Q_2	0.26
Q_3	0.52
Q_4	1.23
Q_5	3.45
Q_6	19.63
Q_7	189.89
Q_8	3429.52

and GA results [8] are very different. It is because our PSO stopping criterion is different from the previous work that using GA [8]. We use different stopping criterion to achieve the fastest time to reach the best solution.

The results in Table 4 shows the fastest (averaged) time needed to reach the best solution. The results in Table 4 is obtained by applying different stopping criterion from the previous work that using GA [8]. If using stopping criterion from the previous work [8], that is maximum generations or iterations equal to 5000 or at most 2000 generations or iterations without improvement of objective value, for example, for Q_4, the time needed for our PSO is 185.29 s. Different from our criterion, our stopping criterion is the cardinality of resolving set that reach known metric dimension of hypercube graph in literature [8]. For Q_4, the time needed for our PSO is 1.23 s. In general, the result is our PSO can achieve best (optimal) solution (known metric dimension) in reasonable amount of time, especially for Q_k, with $k \leq 6$. For $k > 6$, the time needed is significantly increased. We provide the complexity of computing the metric dimension increased as the number of vertices and edges increased.

6 Conclusions

The presented PSO algorithm which is obtained by modifying standard discrete PSO to match the metric dimension problem has solved the metric dimension problem. Binary valued vector, real valued vector, and linear function were used. Infeasible particles are repaired by adding some vertices until the particles become feasible. The performance of our PSO is tested by computing the metric dimension of hypercube graphs. Our PSO algorithm obtains solution that match theoretical results as well-known in literature. Moreover, our PSO can achieve optimal solution known from literature in reasonable amount of time.

References

1. Adiwijaya, Salman, A.N.M., Serra, O., Suprijanto, D., Baskoro, E.T.: Some graphs in c_f2 based on f-coloring. Int. J. Pure Appl. Math. **102**(2), 201–207 (2015)
2. Beerliova, Z., Eberhard, F., Erlebach, T., Hall, A., Hoffmann, M., Mihal'ak, M., Ram, L.S.: Network discovery and verification. IEEE J. Sel. Areas Commun. **24**(12), 2168–2181 (2006)
3. Cáceres, J., Hernando, C., Mora, M., Pelayo, I.M., Puertas, M.L., Seara, C., Wood, D.R.: On the metric dimension of Cartesian products of graphs. SIAM J. Discrete Math. **21**(2), 423–441 (2007)
4. Harary, F., Melter, R.: On the metric dimension of a graph. Ars Combin. **2**(191–195), 1 (1976)
5. Ilaya, O., Bil, C., Evans, M.: A particle swarm optimisation approach to graph permutations. In: Information, Decision and Control, IDC 2007, pp. 366–371. IEEE (2007)
6. Johnson, D.S.: The NP-completeness column: an ongoing guide. J. Algorithms **6**(3), 434–451 (1985)
7. Khuller, S., Raghavachari, B., Rosenfeld, A.: Landmarks in graphs. Discrete Appl. Math. **70**(3), 217–229 (1996)
8. Kratica, J., Kovačević-Vujčić, V., Čangalović, M.: Computing the metric dimension of graphs by genetic algorithms. Comput. Optim. Appl. **44**(2), 343–361 (2009)
9. Liu, K., Abu-Ghazaleh, N.: Virtual coordinates with backtracking for void traversal in geographic routing. In: Kunz, T., Ravi, S.S. (eds.) Ad-Hoc, Mobile, and Wireless Networks. LNCS, vol. 4104, pp. 46–59. Springer, Heidelberg (2006). doi:10.1007/11814764_6
10. Wang, L., Wang, X., Fu, J., Zhen, L.: A novel probability binary particle swarm optimization algorithm and its application. J. Softw. **3**(9), 28–35 (2008)

Non-linear Based Fuzzy Random Regression for Independent Variable Selection

Mohd Zaki Mohd Salikon(✉) and Nureize Arbaiy

Faculty of Computer Science and Information Technology,
Universiti Tun Hussein Onn, Batu Pahat, 86400 Parit Raja, Johor, Malaysia
{mdzaki,nureize}@uthm.edu.my

Abstract. This paper demonstrates a fuzzy random regression approach using genetic algorithm (FRR-GA) to select independent variable for regression model. The FRR-GA approach enables us to indicate the best coefficient values among regressor that indicate the best independent variable, which is important to build regression model. Additionally, the fuzzy random regression approach is employed to treat dual uncertainties due to the realization of such data in real application. This paper presents an algorithm reflecting the non-linear strategy in the fuzzy random regression model. A numerical example illustrates the proposed solution procedure whereby the result suggested several feasible solutions to the user.

Keywords: Fuzzy random variable · Fuzzy random regression · Regressor · Coefficient · Genetic algorithm

1 Introduction

Various real life applications including complex systems, such as finance and marketing, agriculture, industrial, and many more areas are exposed in an environment of uncertainty and imprecision [1, 2]. Such systems commonly require decisions in a mixed condition based on human thinking and judgment and involve human–machine interactions [3, 4]. In such environments, exact numerical data from a system is difficult to obtain [5]. Apart from that, existing solution model which is previously developed with rigid specification is unable to meet the uncertainties requirement. Mathematical model is typically built with numerical crisp values though it is not easy to attain such values in presence of uncertainties [6]. Additionally, uncertainties (i.e. fuzziness and randomness) are observed in various real life applications. It makes the existing evaluation model with precise values is incapable of handling such uncertainties and may generates misleading result. This is due to human expert's judgment which commonly includes uncertainty and probability of the event outcomes [5]. Thus fuzzy mathematical definition provides ability to describe the information with uncertainty.

The uncertainties should be properly treated before mathematical model is addressed. This is to ascertain that the mathematical model holds the uncertainty in finding the solution that is to carry the real meaning of problem nature. The theories of probability and possibility [1] are widely employed as it is competent to handle random

T. Herawan et al. (eds.), *Recent Advances on Soft Computing and Data Mining*,
Advances in Intelligent Systems and Computing 549, DOI 10.1007/978-3-319-51281-5_19

and fuzzy information respectively. A fuzzy set is ultimately able to treat uncertain and imprecise information in the mathematical programming that reflects the uncertainties.

The imprecision in the formulation of real-world problem is generally because of the inclusion of human judgments, and the arbitrary situation. Problem formulation usually involves a translation of real world problem into mathematical model to find solution. Parameter values are important in the development of mathematical models and must be determined earlier. However, determination of model parameter or coefficient becomes more complicated with fuzzy and random situation coexist. The statistical data and available information captured from this uncertain environment will contains fuzzy random data. Thus, the fuzzy random uncertainties contains in the data should be treated whilst formulating the mathematical model. The unawareness of such uncertainties makes the formulated model generates infeasible and improper solutions [5–7]. Thus, regression analysis can be used to obtain the coefficient using realistic data. Regression method provides a tool to estimate the quantitative effect of the independent variables upon the dependent variable [8]. However, conventional regression analysis is incapable of dealing data with dual uncertainties (fuzziness and randomness). Fuzzy Random Regression model [9] is developed based on fuzzy random data and the σ-confidence intervals. This solution model is beneficial to overcome the previous limitation of regression analysis. To simplify the explanation, we restrict ourselves to describe a concise introduction to Fuzzy Random Regression approach [9, 10] to estimate the model coefficient and develop the fuzzy random regression model.

Selecting independent variable and its coefficient in the regression model is sometimes complicated. Few approaches to regression model selection are previously introduced such as stepwise regression, stochastic search and others. The stepwise regression approach however has a drawback of local optima converges. This is due to the local search process. A meta-heuristic strategy such as genetic algorithm has successfully tackled the limitation of local search. Genetic algorithm strategy explores the search space simultaneously by a population of candidate solution. Here the solutions will compete and merge together [11, 12]. Motivated by the situation, this paper proposes a genetic algorithm method into fuzzy random regression analysis to select the independent variable. In this study, a fuzzy random regression [9] is used to treat statistical data which contains fuzzy random uncertainties. Subsequently, the genetic algorithm method has been employed to solve fuzzy random regression to select independent variable. We attempt to improve dual uncertainty treatment in data and independent variable selection in regression model. This emphasizes that the proposed method is facilitated by meta-heuristic strategy and is ably to signify independent variable selection under the presence of dual uncertainty situations.

The remainder of this paper is organized as follows. Section 2 describes the preliminary studies of regression model and genetic algorithm. Section 3 explains the solution model which is genetic algorithm based fuzzy random regression method to select independent variable for regression model. The methodology herein is applied to an illustrative example in presented in Sect. 4. Finally, Sect. 5 depicts conclusions.

2 Preliminary Studies

In this section, some theoretical explanations of Fuzzy Random Regression Analysis and Genetic Algorithm are provided.

2.1 Fuzzy Random Regression Analysis

Regression analysis is one of well established method which is able to find the relationship between regressors and its strength. A regression method examines statistical data to approximate the coefficients in developing effective models [8]. Regression analysis is able to predict a continuous dependent variable from a number of independent variable. However, it is more realistic to consider the estimated values of the coefficients as imprecise values rather than precise ones which is formerly used in conventional regression model. In practical systems, uncertain information exists in a probabilistic and vague situation such as predictions of future event and incomplete historical data. Therefore, the conventional crisp regression model is unable to deal with uncertain information. The fuzzy random regression model is introduced to solve the inability of previous model with the existence of the randomness and fuzziness in historical data used for the approximation [9, 13]. The characteristic of fuzzy random regression is utilized to allow the presence of fuzziness and randomness in the data. Thus, the approximated value which is deduced by the fuzzy random regression model is valuable to build a decision making model.

The triangular fuzzy number of a model coefficient is determined by fuzzy random regression approach. It is simultaneously able to handle the uncertainties in the statistical data used to determine the coefficients. Fuzzy random variables are characterized by the expected value and confidence interval. The detail explanation of fuzzy random variable and fuzzy random regression are given elsewhere [9, 14–17]. Regression based on fuzzy random data is widely used and many researcher researches this technique to solve problem with mixed type of data uncertainty [9, 18, 19]. Statistical inference with fuzzy random data transfers the fuzziness into parameter estimators. So it is necessary to defuzzify the vague parameter at level decision making [18]. Moreover, the fuzzy random regression method also has been proposed as an integral component of regression models in handling the existence of fuzzy random information [9, 15].

The fuzzy regression model can be mathematically written as the following linear programming problem:

$$
\begin{aligned}
\min_{\mathbf{a},\mathbf{h}} \quad & \sum_{j=1}^{n} w|\mathbf{x}_j|^t \\
\text{subject to:} \quad & y_j + L^{-1}(\alpha)d_j \leq \mathbf{a}\mathbf{x}_j^t + L^{-1}(a)w|\mathbf{x}_j|^t, j = 1, 2, \cdots, n, \\
& y_j - L^{-1}(\alpha)d_j \geq \mathbf{a}\mathbf{x}_j^t - L^{-1}(a)w|\mathbf{x}_j|^t, j = 1, 2, \cdots, n, \\
& w \geq 0.
\end{aligned}
\tag{1}
$$

Model (1) has been extended to include the expected value and variance of fuzzy random variable, that is, the one-sigma confidence $(1 \times \sigma)$ interval as shown in the following:

$$I[e_X, \sigma_X] \underline{\underline{\Delta}} \left[E(X) - \sqrt{\text{var}(X)}, E(X) + \sqrt{\text{var}(X)} \right] \tag{2}$$

Hence, the fuzzy random regression model with $\sigma-$ confidence intervals [9] is described as follows:

$$\begin{aligned}
&\min_{\mathrm{A}} \quad J(A) = \sum_{k=1}^{K} \left(A_k^r - A_k^l \right) \\
&A_k^r \geq A_k^l, \\
&\mathrm{Y}_\mathrm{j}^* = \mathrm{A}_\mathrm{j}^* I\left[e_{X_{j1}}, \sigma_{X_{j1}} \right] + \ldots + \mathrm{A}_\mathrm{K}^* I\left[e_{X_{jK}}, \sigma_{X_{jK}} \right] \underset{\tilde{h}}{\supseteq} I\left[e_{Y_j}, \sigma_{Y_j} \right] \\
&j = 1, \ldots, n; k = 1, \ldots, K,
\end{aligned} \tag{3}$$

The solution of the fuzzy random regression model (3) with confidence interval can be rewritten as a problem of n samples with one output and K input interval values [9]. Linear programming of simplex method can be used to solve Model (3). However to improve the selection of important independent variable in regression analysis, meta-heuristic search (i.e. Genetic Algorithm) will be used in the solution method proposed in this paper.

2.2 Genetic Algorithm

Genetic algorithm (hereafter, denoted as GA) have been used in the science and engineering fields by adapting algorithms to solve practical problems and as computational models of natural evolutionary systems [20, 21]. GA is classified as global search heuristics and a particular class of evolutionary algorithms. This strategy use techniques inspired by evolutionary biology. A genetic algorithm improves optimization and searching problem by implementing its strategy in a computer simulation which results a better solutions. GA has a capability to solve optimization problems (mutation and crossover operators applied to populations of chromosomes) either from the domain to specific aspects of a problem (the evaluation function for the chromosomes). In GAs method, the fittest is active based on the selection mechanisms and natural genetic in searching algorithms. General process of GA is illustrated in the procedure in Fig. 1.

Initialization:
Generate random population of n chromosomes (suitable solutions for the problem)
Fitness:
Evaluate the fitness $f(x)$ of each chromosome x in population
New population:
Create a new population by repeating the following steps until the new population is complete
Selection:
a. Select two parent chromosomes from the population according to their fitness (the better fitness, the bigger chance to be selected)
b. Crossover: With a crossover probability cross over the parents to form a new offspring. If no crossover was performed, offspring is the exact copy of parent
c. Mutation: With a mutation probability mutate new offspring at each locus (position in chromosome)
d. Accepting: Place new offspring in the new population
Replace:
Use new generated population for a further run of algorithm
Test:
If the end condition is satisfied, stop, and return the best solution in current population
Loop:
Go to step 2.

Fig. 1. Outline of genetic algorithm [22]

3 Regression Selection Model Using FRR-GA

The model selection problem in Regression is presented as follows:

Let $X \equiv \{X_1, \ldots, X_m\}$ be the set of independent m variables with n observations. Y is the dependent variable in multivariate regression model. The relationship between dependent and independent variables is explained as

$$Y = \beta_0 + \beta_1 \tilde{X}_1 + \ldots + \beta_p \tilde{X}_p + \varepsilon \quad (4)$$

where $\tilde{X} \subseteq X$ is the set of independent variables as regressors. β_p is the coefficient set.

In this paper, the coefficient set for Regression Model (4) is estimated by using Fuzzy Random Regression (3). Then, the selection to choose which independent variables should be included $\tilde{X} \subseteq X$ in is completed by genetic algorithm.

The following algorithm exemplifies the solution method in this paper.

Step 1 Problem Description and Data Preparation

Fuzzy input output data is gathered and prepared. Probability values are identified. Fuzzy random data Y_j (dependent) and X_{jk} (independent variable) for all $j = 1,\ldots,N$ and $k = 1,\ldots,K$ are defined as

$$Y_j = \bigcup_{t=1}^{M_{Y_j}} \left\{ \left(Y_j^t, Y_j^{t,l}, Y_j^{t,r} \right)_\Delta, p_j^t \right\}, \text{and } X_{jk} = \bigcup_{t=1}^{M_{X_{jk}}} \left\{ \left(X_j^t, X_j^{t,l}, X_j^{t,r} \right)_\Delta, q_{jk}^t \right\},$$

respectively. p_j^t and q_{jk}^t are probability for $j = 1,\ldots,n$, $k = 1,\ldots,K$ and $t = 1,\ldots,M$ or $t = 1,\ldots,M_{X_{jk}}$.

Fuzzy regression analysis was used to model an expert evaluation structure. Each respective criterion is obtained by using the Fuzzy Intensity of Compliance Scale [5].

Step 2 Genetic Algorithm based Fuzzy Random Regression Model for estimation

2.1 Compute the confidence interval

Compute the confidence interval of each fuzzy random variable to construct the one-sigma confidence interval, $I[e_X, \sigma_X]$.

2.2: Estimate coefficient and select independent variable using GA

[Initial setting]:

Initialize all the necessary data, including part of linear programming and GA: The trial count is set to be a necessary value and the terminate count is set to be and needed value in order to satisfy the precision. The input and output data are read from data files.

[Deduce] :

Determine the $\sigma -$ confidence intervals (Step 2.1) for each criterion of sample, and assign the signs of interval $\underline{a}_k^{(n)}$ and $\overline{a}_k^{(n)}$.

[Genetic algorithm]:

Use the evolution theory to improve the chromosome individual, memorize the best individual among a generation. Consider the signs of the intervals as the chromosomes. Each chromosome has 3 bits whose values are -1, 0, and 1. Here -1 stands for the interval of [-,-], 0 stands for the interval [-, +], and 1 stands for the interval of [+, +].

[Selection]:

Use the roulette-wheel selection method to select the candidate chromosome.

[Crossover]:

Do the crossover operation to the selected chromosomes according to some fixed possibility.

[Mutation]:

For only one chromosome, do the mutation operations when it satisfies some constrains. For example, we set the mutation possibility, if the random number is bigger than the possibility, we do this operation, otherwise do not.

[Termination]:

Terminate the algorithm.

Step 4 Analysis and Decision Making

Result yield from Step 3.2 are an estimated coefficients for β_p. Using this information, most important independent variable $\tilde{X}$ to be selected in the set $\tilde{X} \subseteq X$ can be decided.

Also, the estimated coefficient produced by fuzzy random regression can be used for several purposes such as to develop a production planning model and to calculate final score of evaluation to select the best samples among alternatives in multi-attribute decision making.

4 Numerical Experiments

We demonstrate the use of the proposed method on a milled rice quality evaluation. In Asian country, rice is an essential food commodity and continuously rice supply is required. To meet the demand and at the same time to maintain the rice quality are the main endeavor in rice production. However, the industry of rice production fails to meet the demand caused by the lower quantity and quality of rice production [23, 24].

Let us consider that the evaluation of rice grading is based on several criteria. Y is the total evaluation of fruit quality and j is the number of samples. The general form of the milled rice quality evaluation is written as following regression model:

$$\mathrm{Y} = [Y_j] = [c_1 x_{j1} + c_2 x_{j2} + \cdots + c_5 x_{j5}] = \mathbf{C}\mathbf{x}_j^t \tag{5}$$

C is the weight value that will be estimated using proposed method in this paper.

Uncertainties are observed by considering grader's knowledge and thoughts during the inspection process. Additionally, the evaluation given by a number of this expert grader may be different. One sample may be evaluated differently by a number of grader, in random. Such fuzzy and random situation does exist in the grading process.

Fuzzy intensity of compliance scale [5] is used to represent such fuzzy value from the grader. For example, grader may assign value 9 to criterion c_1. This information shows that the criterion sample is substantially exceed compliance with the quality standard. Additionally, different observation of inspection is discovered with presence of numerous graders. In some cases, graders are responsible to evaluate the same sample. For example, for one sample, one grader evaluates fuzzy intensity 9, and another grader evaluates fuzzy intensity 8. Such differences should account in the decision making process. In this cased probability is used to calculate the proportional occurrence. Capturing the aforementioned data, the observed statistical data at the present include stochastic and fuzzy information.

Let us consider four criteria were considered during the process of inspection for rice quality, which are (1) purity, (2) defectives, (3) moisture content, and (4) foreign matter. Table 1 shows fuzzy weight values of rice samples which were assigned based on an intensity of compliance scale. 200 samples were used for the weights against each criterion.

Table 2 tabulates the findings the weights obtained from regression model (4), where a_i and d_i denote a weight and its spread of attribute c_i. From the results exemplified in Table 2, we conclude that in the evaluation of rice quality, (1) purity, (2) defectives, and (3) moisture content, are evaluated important ones, followed by

Table 1. Data samples and total evaluation

Sample	(y_j, d_j)	c_1	c_2	c_3	c_4
A1	(8, 0.2)	7	5	7	5
A2	(6, 0.1)	5	7	7	5
⋮	⋮	⋮	⋮	⋮	⋮
A95	(6, 0.1)	5	4	8	6
A96	(6, 0.1)	5	8	4	6
⋮	⋮	⋮	⋮	⋮	⋮
A198	(9, 0.2)	9	5	9	5
A199	(5, 0.1)	5	5	7	6
A200	(9, 0.1)	9	5	8	6

$c_i = (a_i, d_i)$ where a_i is the central value and spread d_i. This value represents triangular fuzzy number.

Table 2. Weights of Criteria

Method	Fuzzy random regression		Fuzzy regression	
	Weight	Width	Weight	Width
Attributes	$a_1 = 0.53$	$d_1 = 0.05$	$a_1 = 0.53$	$d_1 = 0.00$
	$a_2 = 0.17$	$d_2 = 0.30$	$a_2 = 0.15$	$d_2 = 0.22$
	$a_3 = 0.28$	$d_3 = 0.10$	$a_3 = 0.30$	$d_3 = 0.06$
	$a_4 = 0.02$	$d_4 = 0.17$	$a_4 = 0.02$	$d_4 = 0.07$

where each weight $c_i = (a_i, d_i)$ for $i = 1, 2, \ldots, 4$.

(0.53, 0.050), (0.017, 0.300), and (0.28, 0.10), respectively. The other attribute i.e. foreign matter is not strongly considered though it contributes small portion of total quality evaluation. From the finding, it indicates that the attribute of foreign matter is however important and provide the flexibility covering from 0.02 to 0.17. Therefore, that evaluation should consider for decision of foreign matter judgment also. This weight value yielded from the fuzzy regression model is helpful to assist the grading process with minimal monitoring by human experts. Table 2 also tabulates the findings from comparable fuzzy regression method. Comparing the findings, results from proposed FRR-GA obtained larger width value as compared to the Fuzzy Regression method. This indicates that Fuzzy Random Regression model can captures more information under its judgment. This information is useful in decision making especially in fuzzy environment.

The fuzzy random regression model with confidence interval for the rice quality evaluation was then defined as follow:

$$\begin{aligned} Y_{rice\_quality} &= \left(A_i^{L,R}\right)_T I\left[e_{X_i}, \sigma_{X_i}\right] \\ &= (0.53, 0.05)_T I\left[e_{X_1}, \sigma_{X_1}\right] + (0.17, 0.30)_T I\left[e_{X_2}, \sigma_{X_2}\right] \\ &\quad + (0.28, 0.10)_T I\left[e_{X_3}, \sigma_{X_3}\right] + (0.02, 0.17)_T I\left[e_{X_4}, \sigma_{X_4}\right] \end{aligned} \tag{6}$$

This model (6) can be used to predict the rice quality in future for decision making.

5 Conclusion

The purpose of the experiment is to find the optimal solution that is to determine the coefficient and select the most important regressors. The evolution operation in GA made the result close to the best one. In addition, the experiment is carrying out to examine the effects to the optimal solution obtained in respect to related kinds of crossover, mutation, and population sizes. Experiments have been done with a default combination of crossover rate and mutation rate. It should be obtained with an optimal solution while the other parameters were fixed. In general, both fuzzy regression and fuzzy random regression is capable of dealing such data to estimate the coefficient values. The decision results in the form of coefficient and its width of decision factor, x_i. The fuzzy random regression model had an extensive coefficient width. This is as of the consideration of the confidence interval in its valuation. The width in this evaluation plays a significant role, as it reflects natural human judgment. An extensive width indicates that the evaluation can captures more information under fuzzy evaluations.

Acknowledgments. The authors express their appreciation to the Research, Innovation, Commercialization and Consultancy (RICC) Fund, University Tun Hussein Onn Malaysia (UTHM) and Research Acculturation Grant Scheme (RAGS) Vot R044. This research also supported by GATES IT Solution Sdn. Bhd. under its publication scheme.

References

1. Abiyev, R.H., Aliev, R., Kaynak, O., Turksen, I.B., Bonfig, K.W.: Fusion of computational intelligence techniques and their practical applications. Comput. Intell. Neurosci. **2015**, 463147:1–463147:3 (2015)
2. Stewart, T.J., Durbach, I.: Dealing with uncertainties in MCDA. In: Multiple Criteria Decision Analysis, pp. 467–496. Springer, New York (2016)
3. Dubois, D., Prade, H.: Possibility theory and its applications: where do we stand? In: Kacprzyk, J., Pedrycz, W. (eds.) Springer Handbook of Computational Intelligence, pp. 31–60. Springer, Heidelberg (2015)
4. Scholten, L., Schuwirth, N., Reichert, P., Lienert, J.: Tackling uncertainty in multi-criteria decision analysis–an application to water supply infrastructure planning. Eur. J. Oper. Res. **242**(1), 243–260 (2015)
5. Arbaiy, N.: Fuzzy regression for weight information extraction in fuzzy environment. In: Knowledge Management International Conference (KMICe) (2014)

6. Arbaiy, N., Watada, J., Lin, P.C.: Fuzzy random regression-based modeling in uncertain environment. In: Sustaining Power Resources through Energy Optimization and Engineering, p. 127 (2016)
7. Griffiths, T.L., Tenenbaum, J.B.: Predicting the future as Bayesian inference: people combine prior knowledge with observations when estimating duration and extent. J. Exp. Psychol. **140**(4), 725–743 (2011)
8. Sykes, A.O.: An introduction to regression analysis (1993)
9. Watada, J., Wang, S., Pedrycz, W.: Building confidence interval-based fuzzy random regression model. IEEE Trans. Fuzzy Syst. **11**(6), 1273–1283 (2009)
10. Nureize, A., Watada, J.: Building fuzzy random objective function for interval fuzzy goal programming. In: Proceedings of IEEE International Conference on Industrial Engineering and Engineering Management, pp. 980–984 (2010)
11. Oreski, S., Oreski, G.: Genetic algorithm-based heuristic for feature selection in credit risk assessment. Expert Syst. Appl. **41**(4), 2052–2064 (2014)
12. Aydilek, I.B., Arslan, A.: A hybrid method for imputation of missing values using optimized fuzzy c-means with support vector regression and a genetic algorithm. Inf. Sci. **233**, 25–35 (2013)
13. Nureize, A., Watada, J.: Multi-level multi-objective decision problem through fuzzy random regression based objective function. In: Fuzzy Systems (FUZZ) (2011)
14. Liu, B., Liu, Y.-K.: Expected value of fuzzy variable and fuzzy expected value models. IEEE Trans. Fuzzy Syst. **10**(4), 445–450 (2002)
15. Liu, Y.-K., Liu, B.: Fuzzy random variable: a scalar expected value operator. Fuzzy Optim. Decis. Making **2**(2), 143–160 (2003)
16. Kwakernaak, H.: Fuzzy random variables-I. Definitions and theorems. Inf. Sci. **15**(1), 1–29 (1978)
17. Guo, H., Wang, X.: Variance of uncertain random variables. J. Uncertainty Anal. Appl. **2**(1), 1 (2014)
18. Näther, W.: Regression with fuzzy random data. Comput. Stat. Data Anal. **51**(1), 235–252 (2006)
19. González-Rodríguez, G., Blanco, Á., Colubi, A., Lubiano, M.A.: Estimation of a simple linear regression model for fuzzy random variables. Fuzzy Sets Syst. **160**(3), 357–370 (2009)
20. Dasgupta, D., Michalewicz, Z. (eds.): Evolutionary Algorithms in Engineering Applications. Springer, Heidelberg (2013)
21. Hoque, M.S., Mukit, M., Bikas, M., Naser, A.: An implementation of intrusion detection system using genetic algorithm (2012). arXiv preprint arXiv:1204.1336
22. Melanie, M.: An Introduction to Genetic Algorithms. A Bradford Book. The MIT Press, Cambridge (1999). Fifth printing
23. Fahmi, Z., Samah, B.A., Abdullah, H.: Paddy industry and paddy farmers well-being: a success recipe for agriculture industry in Malaysia. Asian Soc. Sci. **9**(3), 177 (2013)
24. Daño, E.C., Samonte, E.D.: Public sector intervention in the rice industry in Malaysia. State intervention in the rice sector in selected countries: Implications for the Philippines (2005)

Time Series Forecasting Using Ridge Polynomial Neural Network with Error Feedback

Waddah Waheeb[1,2](✉), Rozaida Ghazali[1], and Tutut Herawan[3]

[1] Faculty of Computer Science and Information Technology, Universiti Tun Hussein Onn Malaysia, Batu Pahat, 86400 Parit Raja, Johor, Malaysia
waddah.waheeb@gmail.com, rozaida@uthm.edu.my

[2] Computer Science Department, Hodeidah University, Alduraihimi, 3114, Hodeidah, Yemen

[3] Department of Information Systems, University of Malaya, 50603 Pantai Valley, Kuala Lumpur, Malaysia
tutut@um.edu.my

Abstract. Time series forecasting gets much attention due to its impact on many practical applications. Higher-order neural network with recurrent feedback is a powerful technique which used successfully for forecasting. It maintains fast learning and the ability to learn the dynamics of the series over time. In general, the most used recurrent feedback is the network output. However, no much attention has been paid to use network error instead of the network output. For that, in this paper, we propose a novel model which is called Ridge Polynomial Neural Network with Error Feedback (RPNN-EF) that combines the properties of higher order and error feedback recurrent neural network. Three signals have been used in this paper, namely heat wave temperature, IBM common stock closing price and Mackey–Glass equation. Simulation results show that RPNN-EF is significantly faster than other RPNN-based models for one-step ahead forecasting and its forecasting performance is more significant than these models for multi-step ahead forecasting.

Keywords: Time series forecasting · Higher order neural networks · Recurrent neural networks · Ridge Polynomial Neural Network with Error Feedback · Error feedback

1 Introduction

Time series forecasting aims to build a model that use past observations to predict the future. It takes a series of data $x_{t-n}, \ldots, x_{t-2}, x_{t-1}, x_t$ to forecasts data values $x_{t+1}, \ldots, x_{t+m}$. It gets much attention due to its impact on many practical applications such as the forecasting of financial trends, stock market or exchange rate; and natural phenomena, sunspots number or heat waves. Based on the literature, time series forecasting approaches can be classified into statistical-based and intelligent-based approaches. Intelligent-based approaches have shown better performance than statistical approaches in time series forecasting due to the nonlinear nature of most of time series signals [1].

T. Herawan et al. (eds.), *Recent Advances on Soft Computing and Data Mining*,
Advances in Intelligent Systems and Computing 549, DOI 10.1007/978-3-319-51281-5_20

Artificial Neural Network (ANN) is an intelligent-based approach which is inspired by biological nervous systems; it can learn from historical data and adjust its weight matrices to build model that can predict the future. ANNs have been applied successfully for time series forecasting due to their useful properties and capabilities such as their nonlinear nature and the ability to produce complex nonlinear input-output mapping [2]. Based on network structure, ANNs can be grouped into two groups, feedforward and recurrent networks [2]. In feedforward networks, the information moves in one direction only from the input nodes to the output nodes through network connections (i.e. weights). Unlike feedforward networks, the connections between the units in recurrent networks form a cycle.

One of the most used feedforward ANNs in forecasting tasks is Multilayer perceptron (MLP) [3]. However, due to the multilayered structure of MLP, it needs a large number of units to solve complex nonlinear mapping problems, which results in low learning rate and poor generalization [4]. To overcome the drawbacks of multilayered networks, different types of single layer higher order feedforward neural networks with product neurons were introduced.

Ridge Polynomial Neural Networks (RPNNs) [5] are a higher-order feedforward neural networks that maintain fast learning and powerful mapping properties which make them suitable for solving complex problems [3]. A recurrent version of RPNNs is the Dynamic Ridge Polynomial Neural Networks (DRPNNs) [6]. DRPNNs use the output value from the output layer as a feedback connection to the input layer. Output feedback connection is the most used type of feedback connections in many studies [7–10]. The idea behind such networks is learning the network the dynamics of the series over time. As a result, the network should use this memory when forecasting [11]. RPNNs and DRPNNs have been successfully used for time series forecasting [1, 3, 6, 12].

A variation of feedback connection that uses forecasting error instead of network output as an additional input to the network was used with state space Neural Network (ssNN) and Adaptive Neuro-Fuzzy Inference System (ANFIS) models [13, 14]. The forecasting error can be calculated by subtracting the desired value from the predicted value. These two models were applied effectively for time series forecasting [13, 14].

Due to the success of RPNNs, DRPNNs and the models that use forecasting error for time series forecasting, in this paper we propose a model that combine the properties of RPNNs and error feedback recurrent neural networks. This model is called Ridge Polynomial Neural Network with Error Feedback (RPNN-EF). Three time series from different categories have been used in this paper, namely the daily heat wave temperatures for Oklahoma City (Heatwave) [15], IBM common stock closing price (IBM) [16] and the well-known chaotic Mackey–Glass differential delay equation (Mackey–Glass). We forecast the Heatwave and IBM series in one step ahead and multistep ahead forecasting horizon for Mackey–Glass series.

This paper consists of six sections. Section 2 introduces the basic concepts of RPNNs and DRPNNs. We describe the proposed model in Sect. 3. Section 4 covers the experimental settings. Section 5 is about results and discussion. And finally, Sect. 6 concludes the paper.

2 Ridge Polynomial Neural Network Based Models

This section describes Ridge Polynomial Neural Networks (RPNNs) and Dynamic Ridge Polynomial Neural Networks (DRPNNs) which constructed based on RPNNs.

2.1 Ridge Polynomial Neural Networks

RPNNs [5] are an example of higher order feedforward neural networks that use a single layer of adjustable weights. They maintain fast learning properties and powerful mapping capabilities of single layer high order feedforward neural networks [3]. RPNNs are constructed by adding different degrees of Pi-Sigma Neural Networks (PSNNs) blocks [17] until a specified goal is achieved. They can approximate any continuous function on a compact set in multidimensional input space with arbitrary degree of accuracy [5]. RPNNs utilize univariate polynomials which help to avoid an explosion of free parameters that found in some types of higher order feedforward neural networks [5]. In the context of training RPNNs, the mostly used algorithm is the BackPropagation learning algorithm (BP) with gradient descent method.

2.2 Dynamic Ridge Polynomial Neural Network

DRPNNs are a recurrent version of RPNNs [6] as shown in Fig. 1. DRPNNs use the network output value from the output layer as a feedback connection to the input layer. DRPNNs are provided with memories which give them the ability of retaining information to be used later [6]. The real time recurrent learning algorithm [18] is used to train DRPNNs.

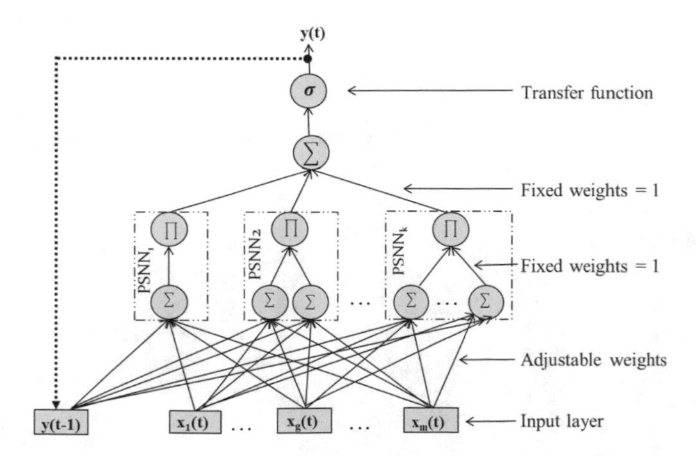

Fig. 1. Dynamic Ridge Polynomial Neural Networks. *PSNN* stands for *Pi-Sigma Neural Network.*

DRPNNs are more suitable than RPNNs due the fact that the behavior of some time series signals related to some past inputs on which the present inputs depends. Therefore, explicit treatment of dynamics is needed [6]. Interested readers may be referred to [1, 6, 12] for more details about DRPNNs and their application in physical and financial time series forecasting.

3 The Proposed Model: Ridge Polynomial Neural Network with Error Feedback

Learning from error is not a new concept to neural network. Backpropagation algorithm (BP) is based on the concept of learning from error. BP calculates the difference between the desired output and network output, which is called error, and uses this error to direct the training.

Another way to learn from the error is using this error as a feedback connection as an additional spatial dimension into input space. This concept was used successfully with state space Neural Network (ssNN) and Adaptive Neuro-Fuzzy Inference System (ANFIS) models for time series forecasting [13, 14].

Due to the success of RPNN, DRPNNs and the models that used error feedback for time series forecasting, we propose Ridge Polynomial Neural Networks with Error Feedback (RPNN-EF) that combine the properties of RPNNs, recurrence and error feedback concept.

Figure 2 shows generic network architecture of the RPNN-EF using Pi-Sigma neural networks as basic building blocks. RPNN-EF uses constructive learning method as other RPNN based models. That means the network structure grows from a small network and the network becomes larger as learning proceeds until the desired level of specified error is reached [5].

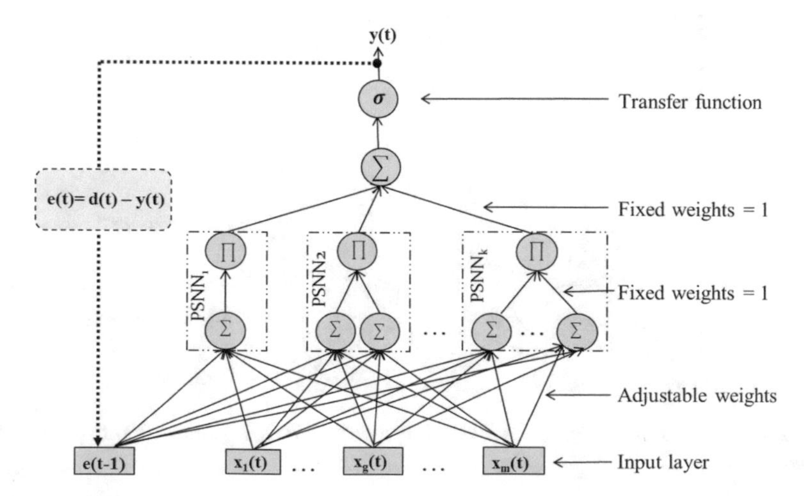

Fig. 2. Ridge Polynomial Neural Networks with Error Feedback. *PSNN, d(t)* stands for *Pi-Sigma Neural Network* and *the desired output at time t.*

Like DRPNNs, the real time recurrent learning algorithm [18] is used to train RPNN-EF. The output of RPNN-EF, which is denoted by *y(t)*, is calculated as follows:

$$\begin{aligned} & y(t) = \sigma\left(\sum_{i=1}^{k} P_i(t)\right), \\ \text{where} \quad & P_i(t) = \prod_{j=1}^{i} h_j(t), \\ \text{and} \quad & h_j(t) = \sum_{g=1}^{M+1} \left(w_{gj} * Z_g(t)\right) + w_{0j}. \end{aligned} \tag{1}$$

where $P_i(t)$ is the output of Pi-Sigma block, σ the transfer function, $h_j(t)$ is the net sum of the sigma unit *j*, w_{0j} is the bias, w_{gj} is the weights between input and sigma units, and *Z(t)* is the inputs which given as follow:

$$Z_g(t) = \begin{cases} x_g(t), 1 \le g \le M \\ e(t-1) = d(t-1) - y(t-1), g = M+1 \end{cases} \tag{2}$$

Network error is calculated using the sum squared error as follows:

$$E(t) = \frac{1}{2}\sum e(t)^2, \tag{3}$$

$$\text{where } e(t) = d(t) - y(t) \tag{4}$$

where *d* is the desired output and *y* is the predicted output. At every time *t*, the weights changes are calculated as follows:

$$\Delta w_{gl} = -\eta * \left(\frac{\partial E(t)}{\partial w_{gl}}\right) \tag{5}$$

where η is the learning rate. The value of $\left(\frac{\partial E(t)}{\partial w_{gl}}\right)$ is determined as:

$$\frac{\partial E(t)}{\partial w_{gl}} = e(t) * \frac{\partial e(t)}{\partial w_{gl}} \tag{6}$$

The value of $\left(\frac{\partial e(t)}{\partial w_{gl}}\right)$ is determined as:

$$\frac{\partial e(t)}{\partial w_{gl}} = \frac{\partial e(t)}{\partial y(t)} * \frac{\partial y(t)}{\partial P_i(t)} * \frac{\partial P_i(t)}{\partial w_{gl}} \tag{7}$$

From Eqs. (1) and (4), we have

$$\begin{aligned} \frac{\partial e(t)}{\partial y(t)} * \frac{\partial y(t)}{\partial P_i(t)} * \frac{\partial P_i(t)}{\partial w_{gl}} = -1 * (y(t))' * \left(\prod_{\substack{j=1 \\ j \neq l}}^{i} h_j(t)\right) \\ * \left(w_{(M+1)l} * \frac{\partial e(t-1)}{\partial w_{gl}} + Z_g(t)\right) \end{aligned} \tag{8}$$

Assume D as the dynamic system variable, where D is:

$$D_{gl}(t) = \frac{\partial e(t)}{\partial w_{gl}} \tag{9}$$

Substituting Eq. (8) into Eq. (7), so Eq. (9), we have:

$$D_{gl}(t) = \frac{\partial e(t)}{\partial w_{gl}} = -1 * (y(t))' * \left(\prod_{\substack{j=1 \\ j \neq l}}^{i} h_j(t) \right) * \left(w_{(M+1)l} * D_{gl}(t-1) + Z_g(t) \right) \tag{10}$$

Finally, the weights change is derived by substituting Eq. (10) into Eq. (6) then the resulted equation into Eq. (5), we have

$$\Delta w_{gl} = -\eta * e(t) * D_{gl}(t) \tag{11}$$

4 Experimental Design

4.1 Time Series Used in the Experiments

Three time series have been used in this paper, namely the daily heat wave temperatures for Oklahoma City [15], IBM common stock closing price [16] and the well-known chaotic Mackey–Glass differential delay equation. The details of these series are given in Table 1. We segregated the data into two partitions; training and out-of-sample with a distribution of 75% and 25%, respectively. The forecast horizon for the series was used based on previous studies [1, 19, 20].

Table 1. Time series used.

Time series data	Total points	Forecast horizon
Heat wave temperatures for Oklahoma City (Heatwave) 01/05/2012–30/09/2012	153	1
IBM common stock closing price (IBM) 17/05/1961–02/11/1962	360	1
Mackey–Glass delay equation (Mackey–Glass) for t = 118 to1117, $\alpha = 0.2$, $\beta = -0.1$, $c = 10$, $x(0) = 1.2$, $\tau = 17$	1000	6

4.2 Data Preprocessing

The data were scaled in the range [0.2, 0.8] in order to avoid getting network output too close to the two endpoints of sigmoid transfer function [1]. The calculation for the minimum and maximum normalization method is given by:

$$\hat{x} = (max_2 - min_2) * \left(\frac{x - min_1}{max_1 - min_1} \right) + min_2. \quad (12)$$

where x refers to the observed (original) value, $\hat{x}$ is the normalized version of x, min_1 and max_1 are the respective minimum and maximum values of all observations, and min_2 and max_2 refer to the desired minimum and maximum of the new scaled series.

4.3 Network Topology

The network topology that we used in this paper is shown in Table 2. Most of the settings are either based on the previous works with RPNNs and DRPNNs that found in the literature [1, 6, 12] or by trial and error.

Table 2. Network topology.

Setting	Value
Number of input units for Heatwave & IBM	Five units represents x(t–4), x(t–3), …, x(t)
Number of input units for Mackey–Glass	Four units represents x(t–18), x(t–12), x(t–6) and x(t)
Activation function	Sigmoid function
Number of Pi-Sigma block (PSNN)	Incrementally grown from 1 to 5
Stopping criteria	Maximum number of epochs = 3000 or after accomplishing the 5th order network learning
Initial weights	[–0.5, 0.5]
Momentum	[0.4–0.8]
Learning rate (n)	[0.01–1]
Decreasing factors for n	0.8
Threshold of successive PSNN addition (r)	0.0001
Decreasing factors for r	0.1

Table 3. Performance metrics and their calculations.

Root mean squared error (RMSE)	Normalized mean squared error (NMSE)	Single to noise ratio (SNR)
$RMSE = \sqrt{\left(\frac{\sum_{i=1}^{N}\left(Y_i - \widehat{Y}_i\right)^2}{N}\right)}$	$NMSE = \frac{1}{\sigma^2 N}\sum_{i=1}^{N}\left(Y_i - \widehat{Y}_i\right)^2$ $\sigma^2 = \frac{1}{N-1}\sum_{i=1}^{N}\left(Y_i - \overline{Y}\right)^2$ $\overline{Y} = \frac{1}{N}\sum_{i=1}^{N} Y_i$	$SNR = 10\log_{10} Sigma$ $Sigma = \frac{m^2 N}{\sum_{i=1}^{N}\left(Y_i - \widehat{Y}_i\right)^2}$ $m = \max(Y_i)$

4.4 Performance Metrics

In this paper, three widely used metrics were chosen namely, Root Mean Squared Error (RMSE), Normalized Mean Squared Error (NMSE) and Signal to Noise Ratio (SNR) as shown in Table 3. The number of epoch for training is also reported.

Table 4. The average simulation results. Bold and underlined values indicate that the performance of the RPNN-EF is better than RPNN and DRPNN with t test, respectively.

Series: Heatwave	RPNN	DRPNN	RPNN-EF
RMSE	0.0859	0.0864	0.0861
NMSE	0.4697	0.4755	0.4722
SNR	18.34	18.29	18.32
Order	Order 1	Order 1	Order 1
Epoch	286.4	247.4	**<u>151.9</u>**
Series: IBM	RPNN	DRPNN	RPNN-EF
RMSE	0.0175	0.017	0.0171
NMSE	0.2028	0.1907	0.1934
SNR	27.44	27.72	27.63
Order	Order 2	Order 2	Order 1
Epoch	903.8	828.7	**<u>601.1</u>**
Series: Mackey–Glass	RPNN	DRPNN	RPNN-EF
RMSE	0.0078	0.0088	**<u>0.0061</u>**
NMSE	0.0026	0.0033	**<u>0.0017</u>**
SNR	40.10	39.06	**<u>42.55</u>**
Order	Order 4	Order 4	Order 4
Epoch	2999	2245.6	2817.2

5 Results and Discussion

The average simulation results of 10 simulations are shown in Table 4. We carried out t-test with a significance level of 0.05 between RPNN-EF and the other two models.

As shown in Table 4, there is no significant in the forecasting performance between RPNN-EF and the other models for one step ahead forecasting in Heatwave and IBM

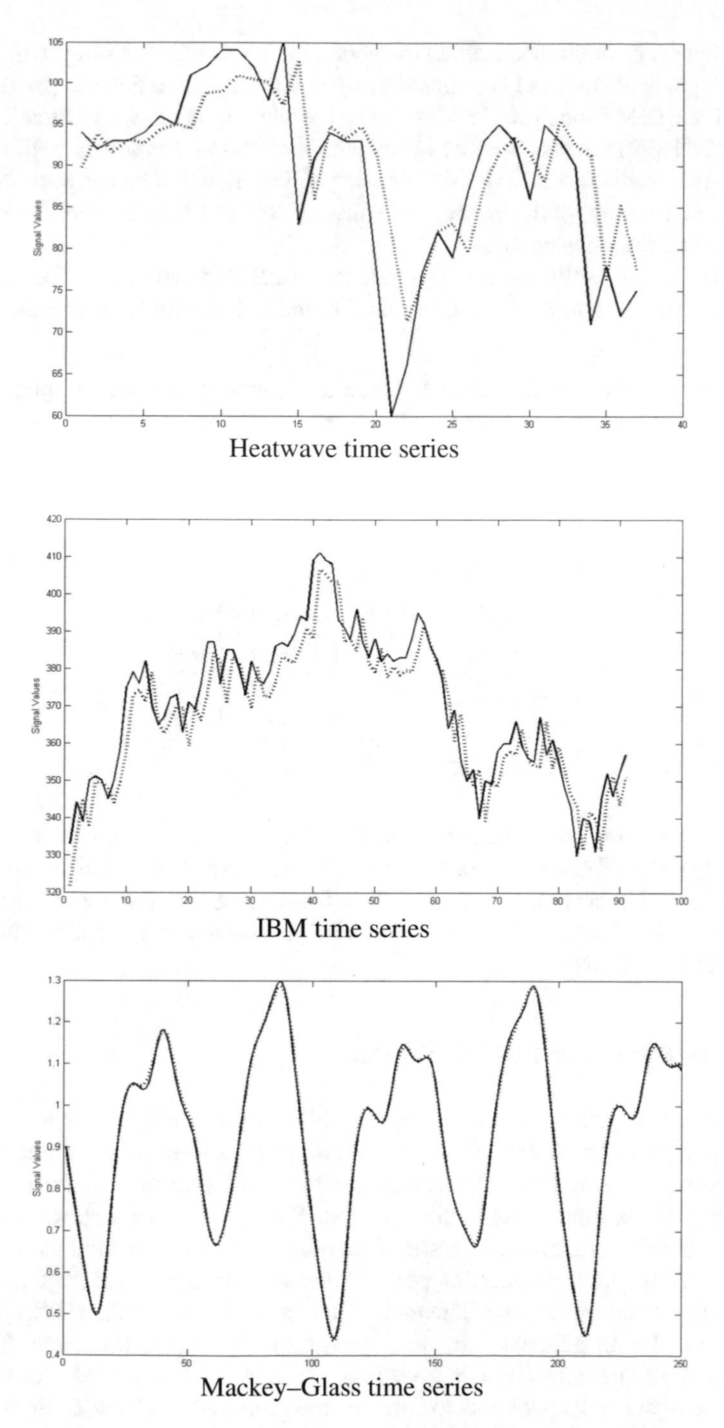

Fig. 3. The best forecasting results for the signals using RPNN-EF. *Solid line* is the original series while the *dotted line* is the forecast series.

signals. However, the number of needed epoch for RPNN-EF learning with Heatwave and IBM signals are reduced significantly. Also, we can notice that the performance of RPNN-EF for IBM time series is with Order 1 while the others with Order 2. And that means RPNN-EF uses less number of weights than the other models with this series. For multi-step ahead forecasting using Mackey–Glass signal, it can be seen that there is a significant filtering performance in terms of RMSE, NMSE and SNR between RPNN-EF and the other models.

Based on these results, we can conclude that the RPNN with Error Feedback can be an effective way to improve forecasting performance. The network offers the following advantages:

1. For one step ahead forecasting: It needs less number of training epochs in comparison to other neural networks. Also, it can reduce the number of needed weights for training in some cases.
2. For multi-step ahead forecasting: It provides better understanding for the signals and smaller error in comparison to other neural networks.

Table 5. Best forecasting results.

Time series data	RMSE	NMSE	SNR
Heatwave	0.0856	0.4660	18.37
IBM	0.0156	0.1609	28.43
Mackey–Glass	0.0045	0.00085	44.93

Table 5 presents performance values using the used metrics for the best out-of-sample data forecasting for the three signals using RPNN-EF model. We also plot in Fig. 3 the best out-of-sample data forecasting for the three signals using RPNN-EF model. These figures indicate that RPNN-EF model can follow the dynamic behavior of the signals.

6 Conclusions and Future Works

This paper investigated the forecasting capability of the Ridge Polynomial Neural Network with Error Feedback (RPNN-EF) for time series forecasting. Three time series have been used, namely the heat wave temperatures, IBM common stock closing price and Mackey–Glass differential delay equation. Simulation results show that the proposed RPNN-EF is significantly faster than other RPNN-based models for one step ahead forecasting and its forecasting performance is more significant than these models for multi-step ahead forecasting. Therefore, we can conclude that the RPNN with Error Feedback can be an effective way to improve forecasting performance. The future works will be expanding the network settings to search for best settings for all models. Using other higher order networks to compare the proposed model with them and using more time series signals to ensure the significant performance of the proposed model will be taken in consideration too.

Acknowledgments. The authors would like to thank Universiti Tun Hussein Onn Malaysia (UTHM) and Ministry of Higher Education (MOHE) Malaysia for financially supporting this research under the Fundamental Research Grant Scheme (FRGS), Vote No. 1235.

References

1. Al-Jumeily, D., Ghazali, R., Hussain, A.: Predicting physical time series using dynamic ridge polynomial neural networks. PLoS ONE **9**(8), e105766 (2014)
2. Haykin, S.S.: Neural Networks and Learning Machines. Prentice Hall, New Jersey (2009)
3. Ghazali, R., Hussain, A.J., Liatsis, P., Tawfik, H.: The application of ridge polynomial neural network to multi-step ahead financial time series prediction. Neural Comput. Appl. **17** (3), 311–323 (2008)
4. Yu, X., Tang, L., Chen, Q., Xu, C.: Monotonicity and convergence of asynchronous update gradient method for ridge polynomial neural network. Neurocomputing **129**, 437–444 (2014)
5. Shin, Y., Ghosh, J.: Ridge polynomial networks. IEEE Trans. Neural Netw. **6**(3), 610–622 (1995)
6. Ghazali, R., Hussain, A.J., Nawi, N.M., Mohamad, B.: Non-stationary and stationary prediction of financial time series using dynamic ridge polynomial neural network. Neurocomputing **72**(10), 2359–2367 (2009)
7. Hsieh, T.J., Hsiao, H.F., Yeh, W.C.: Forecasting stock markets using wavelet transforms and recurrent neural networks: an integrated system based on artificial bee colony algorithm. Appl. Soft. Comput. **11**(2), 2510–2525 (2011)
8. Cao, Q., Ewing, B.T., Thompson, M.A.: Forecasting wind speed with recurrent neural networks. Eur. J. Oper. Res. **221**(1), 148–154 (2012)
9. Chen, P.A., Chang, L.C., Chang, F.J.: Reinforced recurrent neural networks for multi-step-ahead flood forecasts. J. Hydrol. **8**(497), 71–79 (2013)
10. Anbazhagan, S., Kumarappan, N.: Day-ahead deregulated electricity market price forecasting using recurrent neural network. IEEE Syst. J. **7**(4), 866–872 (2013)
11. Samarasinghe, S.: Neural Networks for Applied Sciences and Engineering: from Fundamentals to Complex Pattern Recognition. CRC Press, New York (2006)
12. Ghazali, R., Hussain, A.J., Liatsis, P.: Dynamic ridge polynomial neural network: forecasting the univariate non-stationary and stationary trading signals. Expert Syst. Appl. **38**(4), 3765–3776 (2011)
13. Lanza, P.A.G., Cosme, J.M.Z.: A short-term temperature forecaster based on a state space neural network. Eng. Appl. Artif. Intel. **15**(5), 459–464 (2002)
14. Mahmud, M.S., Meesad, P.: An innovative recurrent error-based neuro-fuzzy system with momentum for stock price prediction. Soft. Comput. **20**(10), 1–19 (2015)
15. National Oceanic and Atmosphere Administration. http://www.srh.noaa.gov/oun/?n=climate-okc-heatwave
16. DataMarket. https://datamarket.com/data/set/2322/ibm-common-stock-closing-prices-daily-17th-may-1961-2nd-november-1962#!ds=2322&display=line
17. Shin, Y., Ghosh, J.: The Pi-Sigma network: an efficient higher-order neural network for pattern classification and function approximation. In: International Joint Conference Neural Networks, IJCNN 1991, Seattle, vol. 1, pp. 13–18. IEEE (1991)
18. Williams, R.J., Zipser, D.: A learning algorithm for continually running fully recurrent neural networks. Neural Comput. **1**(2), 270–280 (1989)

19. Samsudin, R., Shabri, A., Saad, P.: A comparison of time series forecasting using support vector machine and artificial neural network model. J. Appl. Sci. **10**, 950–958 (2010)
20. Aizenberg, I., Luchetta, A., Manetti, S.: A modified learning algorithm for the multilayer neural network with multi-valued neurons based on the complex QR decomposition. Soft. Comput. **16**(4), 563–575 (2012)

Training ANFIS Using Catfish-Particle Swarm Optimization for Classification

Norlida Hassan(✉), Rozaida Ghazali, and Kashif Hussain

Faculty of Computer Science and Information Technology, Universiti Tun Hussein Onn Malaysia (UTHM), Batu Pahat, 86400 Parit Raja, Johor, Malaysia
{norlida, rozaida}@uthm.edu.my,
gi150040@siswa.uthm.edu.my

Abstract. ANFIS performance depends on the parameters it is trained with. Therefore, the training mechanism needs to be faster and reliable. Many have trained ANFIS parameters using GD, LSE, and metaheuristic techniques but the efficient one are still to be developed. Catfish-PSO algorithm is one of the latest successful swarm intelligence based technique which is used in this research for training ANFIS. As opposed to standard PSO, Catfish-PSO has string exploitation and exploration capability. The experimental results of training ANFIS network for classification problems show that Catfish-PSO algorithm achieved much better accuracy and satisfactory results.

Keywords: Catfish · Particle Swarm Optimization · ANFIS training

1 Introduction

Recently, fuzzy systems have attracted researchers in both theory and practice [1]. These systems in integrations with neural networks have evolved in their characteristics like flexibility, speed, and adaptability. Adaptive Neuro-Fuzzy Inference System (ANFIS) being first order sugeno type Fuzzy Inference System (FIS) produces results as robust as those of statistical models [2]. After designing and testing the FIS and ANFIS systems, Neshat in [3] found that the results of ANFIS were better than other fuzzy expert systems. Moreover, ANFIS can be interpreted as local linearization model for model estimation, thus it has a good applicability in system modeling.

However, when designing ANFIS based model, the major concern of researchers is to train its parameters efficiently so that enhanced accuracy can be achieved. Researchers have developed various methods to train these parameters. These methods are usually categorized as deterministic and probabilistic techniques. Deterministic techniques, including gradient descend (GD) and lease squares estimation (LSE), are slow and sometimes will never converge. Whereas, metaheuristic algorithms with global search ability are population based algorithms. The individual in population represents a potential solution [4]. Particle Swarm Optimization (PSO) is one of the stochastic techniques which can find better solutions as compared to deterministic ones. Although PSO performs well on nonlinear function optimization problems but sometimes it exhibits poor local search ability as well premature convergence [5]. To overcome this problem, many researchers have improved the performance of PSO by

T. Herawan et al. (eds.), *Recent Advances on Soft Computing and Data Mining*,
Advances in Intelligent Systems and Computing 549, DOI 10.1007/978-3-319-51281-5_21

proposing several of its variants – Catfish-PSO is one of those which is proposed by [6]. In Catfish-PSO, the catfish effect is applied to improve the performance of PSO. Here, the solution individuals having worst solution candidate are replaced by new individuals. This way, better optimal solutions are found if global best solutions are repetitively appears to be the same for a specific number of iterations. The process of replacement opens up new opportunities to find better optimal solutions as well as increase convergence speed.

As discussed above, updating and training ANFIS parameters is crucial step when modeling fuzzy model [1]. While training ANFIS network, the total number of ANFIS modifiable parameters is an important factor because it influences computational effort required for the adaptation process. Therefore, the type of membership functions should be chosen carefully. Gaussian type membership function which takes only two parameters namely, width and center, is more preferable than other types of membership functions [3]. Based on this argument, this paper also used Gaussian membership functions. Since, it is time consuming to determine the correct set of parameters, this research employs Catfish-PSO to efficiently identify system parameters.

The rest of the paper is organized as follows: The next section gives an overview of ANFIS and Catfish-PSO algorithm. How to train ANFIS with Catfish-PSO is explained in the subsequent section. Results of ANFIS training on benchmark problems are presented later on. Finally, we summarize the contribution of this paper and offers concluding remarks.

2 ANFIS and Catfish-PSO Algorithm

In this paper, Catfish approach combined with the PSO algorithm is embedded to adjust the weights and parameters of ANFIS network because metaheuristic algorithms have powerful parallel searching capability. Our objective is to train the premise and consequent parameters of ANFIS so that the fitness function is based on mean squared error (MSE) of ANFIS in each iteration.

2.1 ANFIS Concept

The most commonly used fuzzy system in ANFIS architectures is the Sugeno model since it is less computationally exhaustive and more transparent than other models. In addition, the defuzzification process in Sugeno fuzzy models is a simple weighted average calculation [7].

Jang in [7] introduced ANFIS architecture which is a universal approximator based on adaptive technique to assist learning and adaptation [8].

The network architecture of ANFIS comprises of two types of nodes: fixed and adaptable. Nodes of membership functions (layer 1) and consequent part (layer 4) are modifiable, while the fixed nodes are of product (layer 2) and normalization (layer 3). The network applies least square method to train consequent parameters, in forward pass, and uses gradient descent for tuning membership function parameters in backward pass of network training process. Figure 1 illustrates ANFIS architecture. For explaining the ANFIS, the two fuzzy if-then rules are considered here.

$$\text{Rule 1 : If } x \text{ is } A_1 \text{ and } y \text{ is } B_1 \text{ then } f = p_1x + q_1y + r_1$$

$$\text{Rule 2 : If } x \text{ is } A_2 \text{ and } y \text{ is } B_2 \text{ then } f = p_2x + q_2y + r_2$$

where A_i and B_i are fuzzy sets in the antecedent, p_i, q_i and r_i are the design parameters that are identified during training process. To execute the above two rules, let's examine five layer architecture of ANFIS:

Layer 1: Every node i in this layer is adaptive membership function.

$$\begin{aligned} O_{1,i} &= \mu_{A_i}(x), \quad i = 1,2 \\ O_{1,i} &= \mu_{B_{i-2}}(y), \quad i = 3,4 \end{aligned} \tag{1}$$

Typically, membership function for a fuzzy set or linguistic label can be any parameterized membership function, i.e. Triangle, Trapezoidal, Gaussian, or generalized Bell function. For example Gaussian membership function which is specified by two parameters $\{c, \sigma\}$ as follows:

$$guassian(x; c, \sigma) = e^{-\frac{1}{2}\left(\frac{x-c}{\sigma}\right)^2} \tag{2}$$

where center c and width σ are controlling parameters of Gaussian membership function. These parameters are also referred to as premise or antecedent parameters.

Layer 2: These nodes are fixed and represent simple product $\prod$ to calculate firing strength of a rule.

$$O_{2,i} = w_i = \mu_{A_i}(x)\mu_{B_i}(y), \quad i = 1,2. \tag{3}$$

Rules are generated here using grid partitioning which divides the data space into rectangular subspaces using axis-paralleled partition based on predefined number of membership functions and their types in each dimension [10]. The number of rules is equal to m^n where m is the number of membership functions in each input and n is inputs to ANFIS.

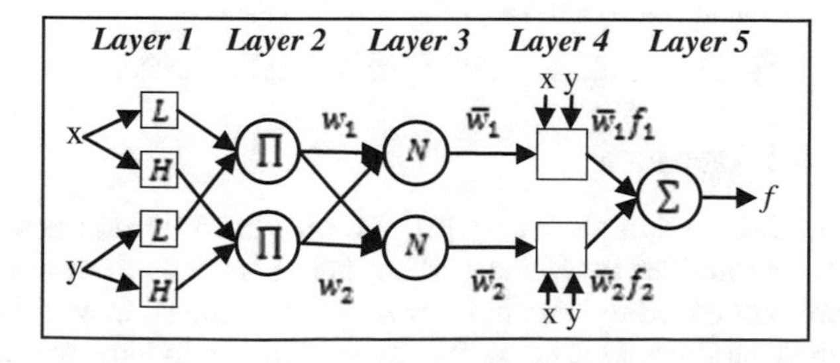

Fig. 1. ANFIS architecture

Layer 3: This layer comprises of fixied nodes which are represented as *N*. Each node of this layer normalizes firing strength of rule by calculating the ratio of the *i*th rule's firing strength to the sum of all rules' firing strength.

$$O_{3,i} = \bar{w}_i = \frac{w_i}{w_1 + w_2}, \quad i = 1, 2. \tag{4}$$

where $\bar{w}$ is normalized firing strength of rule.

Layer 4: The nodes in this layer are modifiable with node functions:

$$f_i = p_i x + q_i y + r_i \tag{5}$$
$$O_{4,i} = \bar{w}_i f_i = \bar{w}_i (p_i x + q_i y + r_i), \quad i = 1, 2.$$

where $\bar{w}$ is rule's normalized firing strength and $\{p_i, q_i, r_i\}$ is a first order polynomial. Parameters in this layer are consequent parameters and are determined during the training process.

Layer 5: The node in this layer is the output node which summarizes the outputs of all rules in layer 4.

$$O_{5,i} = \sum_{i=1}^{2} \overline{w_i f_i} = \frac{\sum_{i=1}^{2} \overline{w_i f_i}}{w_1 + w_2} \tag{6}$$

The training process of ANFIS determines all modifiable parameters such as c, σ and $\{p_i, q_i, r_i\}$. It uses a combination of gradient descent (GD) and least squares estimator (LSE). The parameter update process uses a two pass learning algorithm.

In forward pass of ANFIS training process presented in Table 1, functional signals go forward till Layer 4 and the consequent parameters are updated by LSE. In this pass premise parameters remain unchanged or fixed. Whereas, in the backward pass, membership function parameters are updated using GD. In this pass, the consequent parameters are kept fixed.

Table 1. Two pass hybrid learning algorithm of ANFIS.

	Forward pass	Backward pass
Antecedent parameters	Fixed	GD
Consequent parameters	LSE	Fixed
Signals	Node outputs	Error signals

2.2 Catfish-PSO Algorithm

In Catfish-PSO which is introduced by [6], the idea is to introduce new movement among particles or solutions individuals when there happens any constant stagnancy. When catfish are introduced to large holding tanks of sardines, they initiate a renewed search by the other sardines. This helps PSO avoid trapping in local minima by moving towards new search regions.

The basic idea reveals from sardines, transportation process, fish tank, and catfish of the Norwegian fishermen. In this algorithm, as standard PSO, the swarm is initialized at first with particles randomly distributed across the search space dimensions. Each particle has two characteristic, its velocity and its positions. Each particle has its best solution (fitness) called personal best *pBest*. Whereas, the particle which has the best solution value *gBest*, is supposed to be the leader of the swarm. The velocity and the position of the PSO particle is updated using following equations:

$$v_i(k+1) = w \times v_j(k) + c_1 \times r_1(p_i - x_i(k)) + c_2 \times r_2(p_g - x_i(k)) \tag{7}$$

where v_i is ith velocity element in the vector v in iteration k. w is inertia weight, r_i and $r_2 \sim U(0,1)$ are random variables, whereas $c_1 > 0$ and $c_2 > 0$ are acceleration coefficients influencing particle's personal best location p_i and global best location p_g, respectively. Then particles move to new location $x_i(k+1)$ as follows:

$$x_i(k+1) = x_i(k) + v_i(k+1) \tag{8}$$

where $x_i(k)$ is the current position of the ith particle.

In standard PSO, if the distance between *gBest* particle and the other particles is small then each particle will move only a very small distance in the next iterations. This will lead to premature convergence. To avoid this, Catfish particles are introduced which replace any specific number of particles which are having worst optimal solutions in the swarm. This process is illustrated by the Fig. 2.

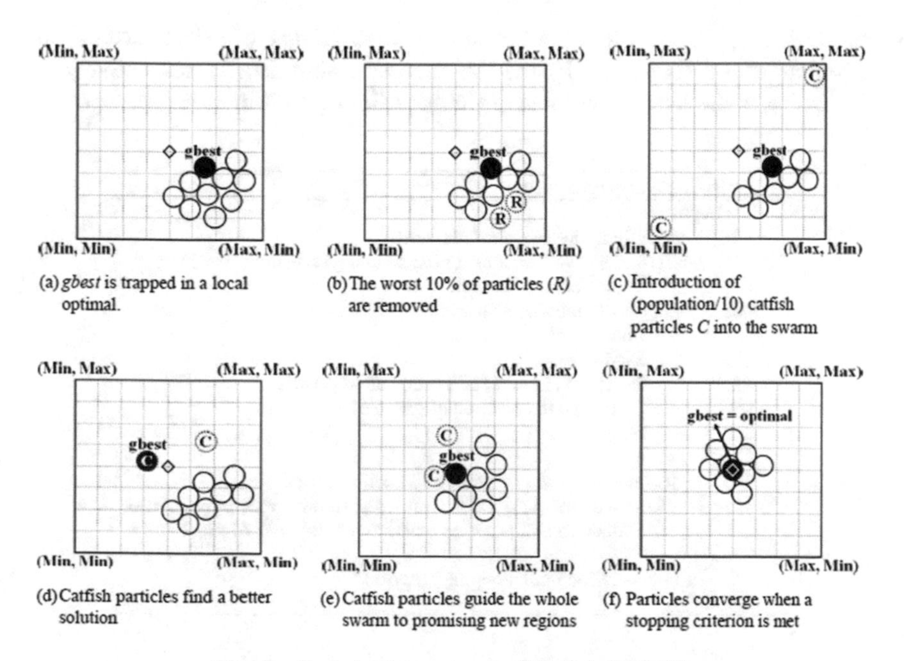

Fig. 2. Optimization process of Catfish-PSO [6]

The Fig. 2-a shows the particles are converged near *gBest* particle after a certain period. PSO is considered to be stuck in local minima when the value of *gBest* particle does not change for multiple iterations. In Catfish-PSO, 10% particles are chosen with worst optimal solutions R (Fig. 2-b) and these particles are replaced by those which are newly initialized catfish particles C (Fig. 2-c). The Catfish particles (Fig. 2-d) lead *gBest* particles to new search region where possible optimal solution may reside. Figure 2-e shows that the particles converge where the desired optimal solution is found.

3 ANFIS Training by Catfish-PSO

In this paper, Catfish-PSO is employed to tune both the premise and consequent parameters of ANFIS network. Since PSO techniques are computationally less expensive, thus we apply Catfish-PSO to update ANFIS parameters in an easier and faster way as compared to gradient-based methods. The position of each particle in Catfish-PSO represents a complete set of parameters for the ANFIS network. These parameters comprise of both the membership function parameters and the linear coefficients of the consequent part of fuzzy rule. The fitness is defined as mean squared error (MSE) between actual output and desired output, it can expressed as:

$$MSE = \frac{\sum_{i=1}^{m} (O_{avg} - O_m^t)^2}{m}, \tag{9}$$

where MSE, O_{avg}, O_m^t, and m are mean square error, average of selected rules' output, target output of *m*th training pair, and the size of training dataset, respectively.

Following is the listed step by step Catfish-PSO algorithm:

Catfish-PSO Algorithm
01: begin
02: Randomly initialize particles swarm
03: **while**(number of iterations, or the stopping criterion is not met)
04: Evaluate fitness of particle swarm
05: **for** *n*=1 to number of particles
06: Find *pBest*
07: Find *gBest*
08: **for** *d*=1 to number of dimension of particle
09: update the position of particles
10: **next** *d*
11: **next** n
12: **if** fitness of *gBest* is the same Max times **then**
13: Remove 10% particles R with worst fitness value in population
14: Introduce catfish particles which represent 10% of population
15: **end if**
16: **next** generation until stopping criterion
17: end

The ANFIS network trained by Catfish-PSO algorithm is illustrated in Fig. 3 and outlined as below:

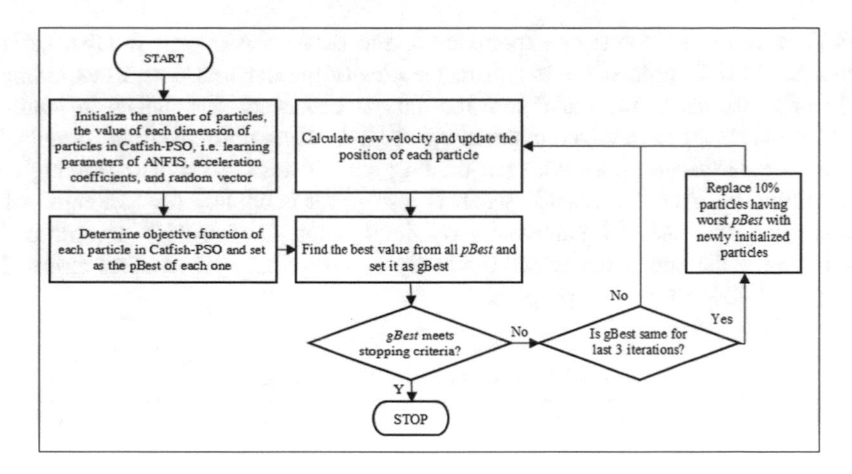

Fig. 3. An overview of ANFIS training process by Catfish-PSO

Step 1. Initialize the swarm of particles such that the position of each particle are uniformly distributed with the search scope, set acceleration coefficients, random vector, and the iteration number $n = 0$

Step 2. Set the position of each particle $X_{i,n}$ as the set of parameters comprising of premise and the consequent parameters. Then calculate fitness value of each particle, and set each particle's personal position as $P_{i,0} = X_{i,0}$

Step 3. For each particle in the population, execute from to Step 8

Step 4. Calculate each particle's fitness value, fitness $\left(X_{i,n}\right)$, and then compare it with the fitness of its personal best position fitness $\left(P_{i,n} - 1\right)$. If fitness $\left(X_{i,n}\right) <$ fitness $\left(P_{i,n} - 1\right)$, then $P_{i,0} < X_{i,n}$, otherwise $P_{i,n} < P_{i,n} - 1$

Step 5. Compare the fitness value of each particle's personal best position fitness $\left(P_{i,n}\right)$ with that of the global best position, fitness $\left(P_{i,n}\right)$ with that of the global best position, fitness $(G_n - 1)$. If fitness $\left(P_{i,n}\right) <$ fitness $(G_n - 1)$, then $G_n = P_{i,n}$, otherwise $G_n = G_n - 1$

Step 6. Calculate V_i, for each particle according to (7)

Step 7. Update the position of each particle X_i using (8)

Step 8. If the termination condition is met, go to exit, otherwise go to Step 2, and set $n = n + 1$

4 Experimental Results

To investigate the efficiency of Catfish-PSO for training ANFIS network, three benchmark datasets were tested. These datasets are Iris Flower, Post-Operative, Hayes-Roth taken from UCI machine learning repository [9]. ANFIS performance hinders due to curse of dimensions when using grid partitioning method for generating rules. Therefore, our focus of experiment is the datasets with less number of input variables. Table 2 summarizes the characteristics of the datasets used in experiments.

In order to ensure the make sure the data is consistent, the values in input and output variables are normalized in the range of [0, 1] using normalization formula. The datasets are partitioned into two sets: training set and testing set. The training set is used to train the ANFIS network, whereas testing set is applied for validation of the performance over ANFIS parameters obtained using Catfish-PSO algorithm. The partitioning of datasets is performed randomly in a way that 80% samples reserved for training and 20% for testing purpose.

Table 2. Characteristic of datasets.

Dataset	Examples	Input No.	Output No.	Class	No. of instance in each class
Iris	150	4	1	3	C1 = 50, C2 = 50, C3 = 50
Post-operative	87	8	1	3	C1 = 01, C2 = 62, C3 = 24
Hayes-roth	160	4	1	3	C1 = 65, C2 = 64, C3 = 31

For training ANFIS with Catfish-PSO, Table 3 mentions the initial values for Catfish-PSO algorithm as well as ANFIS network. Because Catfish-PSO is a stochastic algorithm thus the experiments of training of ANFIS with Catfish-PSO were performed ten times on each dataset with different random seeds. After these experiments we selected the best results to report. In this section, we report the results in the form of MSE and percentage of accuracy.

Table 4 shows the MSE of ANFIS trained by PSO and Catfish-PSO algorithm. It reveals that the best results are obtained in terms of highest accuracy in case of Post-Operative data. In this case Catfish-PSO best performed by achieving 99.96%

Table 3. Specification of ANFIS and Catfish-PSO Algorithm.

Catfish-PSO paramater	Value	ANFIS parameter	Value
Number of particles	15	Number of inputs	As per datasets mentioned in Table 1
Maximum iterations	50	Number of membership functions	3 for each input
Max. matching time for gBest value	3		
Catfish particles %	10		
Objective function	Mean Square Error (MSE)		
Acceleration coefficients	$C_1 =$, $C_2 =$		
Random vector R_1 and R_2	Random		

accuracy as compared to ANFIS-PSO (99.71%) and ANFIS (98.48%). For Iris datasets, the highest accuracy was shown by ANFIS-CatfishPSO (99.71%) and followed by ANFIS-PSO and ANFIS that obtained 97.77% and 91.40% respectively. The similar effect shown by using ANFIS-CatfishPSO learning, that exhibit the highest accuracy 98.48% on Hayes-roth dataset. These indicate that all the benchmarks datasets is performed better accuracy by using ANFIS-CatfishPSO.

Figure 4 illustrates learning curve of ANFIS when trained by Catfish-PSO on benchmark datasets. All benchmark datasets achieved least MSE which are 0.005758, 0.000892, and 0.03032 for Post-operative, Iris, and Hayes-roth respectively.

Table 4. Experimental results of ANFIS.

	Accuracy %			MSE		
	Post-operative	Iris	Hayes-roth	Post-operative	Iris	Hayes-roth
ANFIS	90.21	91.40	90.39	0.19251	0.172025	0.19570
ANFIS-PSO	96.15	97.77	96.39	0.079218	0.054828	0.06043
ANFIS-CatfishPSO	99.96	99.71	98.48	0.005758	0.000892	0.03032

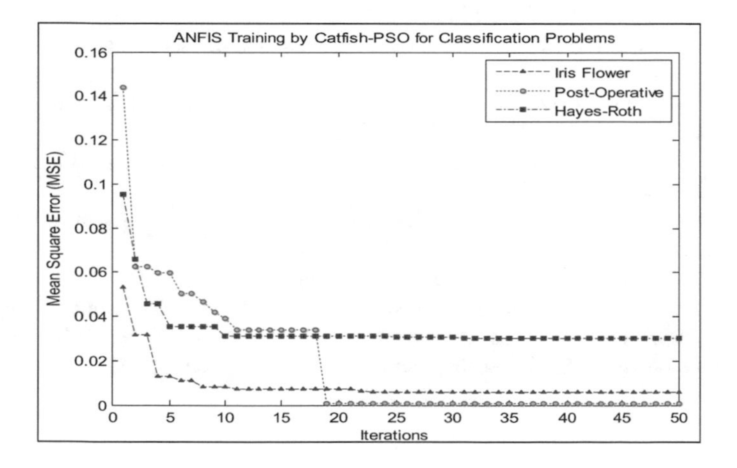

Fig. 4. Catfish-PSO performance when training ANFIS

5 Conclusion

In this paper, we used catfish-PSO for effectively training ANFIS network for solving benchmark classification problems. For efficient learning, both the ANFIS parameters, premise and consequent parameters were trained by Catfish-PSO.

The simulation results show that Catfish-PSO has better and stable convergence while training ANFIS network against benchmark datasets. Catfish-PSO showed high capability of achieving better accuracy in shorter few number of iterations. The approach of ANFIS training, proposed in this paper, is less complex than training both the parts of fuzzy rule by different learning algorithms. In future, more metaheuristic algorithms can be utilized more efficiently in this regard.

Acknowledgments. The authors would like to thank Universiti Tun Hussein Onn Malaysia (UTHM) and Ministry of Higher Education (MOHE) Malaysia for financially supporting this research under the Fundamental Research Grant Scheme (FRGS), Vote No. 1235.

References

1. Nhu, H.N., Nitsuwat, S., Sodanil, M.: Prediction of stock price using an adaptive neuro-fuzzy inference system trained by Firefly Algorithm. In: 2013 International Computer Science and Engineering Conference (ICSEC) (2013)
2. Taylan, O., Karagözoğlu, B.: An adaptive neuro-fuzzy model for prediction of student's academic performance. Comput. Ind. Eng. **57**(3), 732–741 (2009)
3. Neshat, M., Adeli, A., Masoumi, A., Sargolzae, M.: A Comparative Study on ANFIS and Fuzzy Expert System Models for Concrete Mix Design (2011)
4. Orouskhani, M., Mansouri, M., Orouskhani, Y., Teshnehlab, M.: A hybrid method of modified cat swarm optimization and gradient descent algorithm for training ANFIS. Int. J. Comput. Intell. Appl. **12**(2), 1350007 (2013)
5. Liu, Y., Qin, Z., Shi, Z., Lu, J.: Center particle swarm optimization. Neurocomputing **70**(4), 672–679 (2007)
6. Chuang, L.Y., Tsai, S.W., Yang, C.H.: Catfish particle swarm optimization. In: Swarm Intelligence Symposium, SIS 2008. IEEE (2008)
7. Jang, J.S.R.: ANFIS: adaptive-network-based fuzzy inference system. IEEE Trans. Syst. Man Cybern. **23**(3), 665–685 (1993)
8. Kothandaraman, R., Ponnusamy, L.: PSO tuned adaptive neuro-fuzzy controller for vehicle suspension systems. J. Adv. Inf. Technol. **3**(1), 57–63 (2012)
9. Asuncion, A., Newman, D.: UCI machine learning repository (2007)
10. Teshnehlab, M., Shoorehdeli, M.A., Sedigh, A.K.: Novel hybrid learning algorithms for tuning ANFIS parameters as an identifier using fuzzy PSO. In: IEEE International Conference on Networking, Sensing and Control, ICNSC 2008 (2008)

Data Mining

FCA-ARMM: A Model for Mining Association Rules from Formal Concept Analysis

Zailani Abdullah[1(✉)], Md Yazid Mohd Saman[2], Basyirah Karim[2], Tutut Herawan[3], Mustafa Mat Deris[4], and Abdul Razak Hamdan[5]

[1] Faculty of Entrepreneurship and Business / Centre of Computing and Informatics, Universiti Malaysia Kelantan, 16100 Kota Bharu, Kelantan, Malaysia
zailania@umk.edu.my

[2] School of Informatics and Applied Mathematics, Universiti Malaysia Terengganu, 21030 Kuala Terengganu, Terengganu, Malaysia
yazid@umt.edu.my, syira_karim@yahoo.com

[3] Faculty of Computer Science and Information Technology, Universiti Malaya, 50603 Kuala Lumpur, Malaysia
tutut@um.edu.my

[4] Faculty of Science Computer and Information Technology, Universiti Tun Hussein Onn Malaysia, 86400 Batu Pahat, Johor, Malaysia
mmustafa@uthm.edu.my

[5] Faculty of Information Science and Technology, Universiti Kebangsaan Malaysia, 43600 Bangi, Selangor, Malaysia
arh@ftsm.ukm.my

Abstract. The evolution of technology in this era has contributed to a growing of abundant data. Data mining is a well-known computational process in discovering meaningful and useful information from large data repositories. There are various techniques in data mining that can be deal with this situation and one of them is association rule mining. Formal Concept Analysis (FCA) is a method of conceptual knowledge representation and data analysis. It has been applied in various disciplines including data mining. Extracting association rule from constructed FCA is very promising study but it is quite challenging, not straight forward and nearly unfocused. Therefore, in this paper we proposed an Integrated Formal Concept Analysis–Association Rule Mining Model (FCA-ARMM) and an open source tool called FCA-Miner. The results show that FCA-ARMM with FCA-Miner successful in generating the association rule from the real dataset.

Keywords: Data mining · Association rule · Formal concept analysis

1 Introduction

Data mining is a process of analyze and extract meaningful knowledge from data repositories. It aims at identifying the trend or pattern in the data. Recently, data mining becomes more important study especially in an organization, because the usefulness knowledge extracted and can be used in assisting the decision making process.

T. Herawan et al. (eds.), *Recent Advances on Soft Computing and Data Mining*,
Advances in Intelligent Systems and Computing 549, DOI 10.1007/978-3-319-51281-5_22

For example, the marketing officers in hypermarket can increase the products sales by arranging them according to the information of association rules. In fact, the association rules can also provide some interesting, expected and meaningful information the organizational projections.

One of the fundamental elements in knowledge discovery is association rules mining. Association rule mining is a technique to extract the pattern from database. By definition, it is a process of finding frequent associations among itemset in transaction databases, relational databases, and other data repositories [1]. There are three terms that related to association rule which are itemset, support and confidence. Itemset is referred to a set of item that occurs together. The strength of association rule is measure by the support and confidence. In association rule, itemset that meet a minimum support threshold is called frequent itemsets. There are two steps in mining the association rule. First, generate all itemset whose support more than or equal to minimum support. The second step is to generate high confidence rules from each of the frequent itemset, where each rule is a binary partitioning of a frequent itemset. Each itemset in the lattice is called as candidate frequent itemset. In Apriori algorithm, pruning technique is always applied to reduce the number of candidate itemset.

Association rule is if/then statements that help to discover relationships between seemingly unrelated data. It has two parts, an antecedent (if) and a consequent (then). The antecedent and the consequent are disjoint whereby they have no items in common. After more than a decade, Apriori-based is still commonly used algorithm in association rule mining with certain enhancements [2–4]. Association rule can be mine from various sources of data such as dataset, relational database, XML and Formal Concept Analysis (FCA).

According to Poelmans *et al.* [5], FCA is concerned with formalization of concepts and conceptual thinking in which it has been applied in various disciplines. The founder of FCA, Wille [6] defined FCA as a mathematical method of analyzing data which visualized the relationship between a set of objects and a set of attributes. It basically represents in a form of a line diagram that shows the relationship between attribute and object sets, as well as representation of knowledge. The FCA line diagram that connects the nodes of object and attributes sets is called lattice concept. Lattice concept is partial order over the concepts. It is partially used because some concepts or itemsets have no meaningful knowledge.

Mining association rules from FCA is one of the promising approaches because it can eliminate the dependency against the original datasets. Moreover, it can also provide an alternative platform rather than using common algorithms in mining the association rules. Therefore, the contributions of this are as follows. First, we proposed a Formal Concept Analysis–Association Rule Mining Model (FCA-ARMM) as an alternative approach for mining pertinent rules. Second, we developed FCA-Miner tool based on the proposed model and evaluated its performance against real dataset. Third, we depicted representation of Textual FCA in complementing with the 2D visualization of FCA.

The reminder of this paper is organized as follows. Section 2 describes the related work. Section 3 describes the proposed model. This is followed by the results and discussion in Sect. 4. Finally, conclusions of this work are reported in Sect. 5.

2 Related Works

Formal Concept Analysis (FCA) was first developed by Wille [6]. The core data structure in FCA is concept lattice. Nowadays, it has been widely employed in various disciplines such as in machine learning, data mining and database (KDD), knowledge management, sciences etc. FCA received much attention from KDD researchers because it uses visualization to show the relationship and pattern of the data. In concept lattice, every node is a formal concept and is defined as a pair. It consists of extent a set of objects (extent) and a set of attributes (intent). Until this, there are quite a number of algorithms [7–11] of association rule using concept lattice.

Ourida et al. [12] proposed Semantic Based on Formal Concept Analysis Approach (SFC2A) to extract the interesting itemsets based on Formal Concept Analysis (FCA). A sematic aspect in term of confidence is employed during generating association rule. A new quality-criteria, the releance measure is also introduced as a value added for FCA by considering the width and length of Formal Concept in its formula.

Pasquier et al. [13] introduced an efficient algorithm called Close to prune the closed itemset lattice which is a sub-order of the subset lattice. This lattice is quite similar to Wille's concept lattice in FCA. From the experiments, the Close algorithm performs very well for mining various datasets. The key features of the Close algorithm are it can reduce the database passes and the CPU overhead during mining process.

Zaki and Hasio [14] proposed CHARM algorithm for mining all closed frequent itemsets. The algorithm intensively utilizes the itemset and transaction spaces. Basically, CHARM quickly identifies the closed frequent itemsets by avoiding many unnecessary levels.

Zaki and Hsiao [15] later proposed an extension of CHARM algorithm called CHARM-L to directly construct a frequent closed itemset lattice while generating candidate itemsets. CHARM-L exploits a vertical encoding of the database by taking the advantages of asymmetry between the number of transactions and its items.

Stumme et al. [16] introduced TITANIC algorithm for computing iceberg concept lattices with a level-wise approach. Iceberg concept lattice is refered to top-most part of the concept lattice. Titanic computes the closure system of frequent concept intents using the support function. TITANIC algorithm is based on Apriori algorithm. However, this algorithm does not address clearly address the incremental construction of concept lattice.

Pasquier et al. [17] later suggested A-Close algorithm to extract frequent closed itemsets based on closure mechanism. The algorithm pruns the closed itemset lattice instead of the general itemset lattice. Thus, by utilizing closure mechanism of Galois connection, the number of generated sub-order of the itemset lattice called closed itemset lattice becomes lasser.

Liang et al. [18] introduced Incremental Building algorithm and closed label lattice structure to find association. This algorithm integrates boths features which are the batch and incremental implementation. From the experiments, the number of non-redundent rules produced smaller as compared to Godin et al. [19] and Xie et al. [20] as well as its computational times.

Burdick et al. [21] suggested MAFIA algorithm intends to compute all maximal frequent itemsets. The algorithm uses vertical bitmaps to compress the transaction id list, thus improving the counting efficiency. MAFIA employs two pruning strategies called parent-equivalent pruning and superset checking to reduce the search space. It is quite excellent for mining a *superset* of all maximal frequent itemsets.

3 Proposed Model

3.1 FCA-Miner Model

There are four major components in FCA-Miner model, which are dataset, preprocessing, FCA and association rule. A complete overview of FCA-ARMM is shown in Fig. 1.

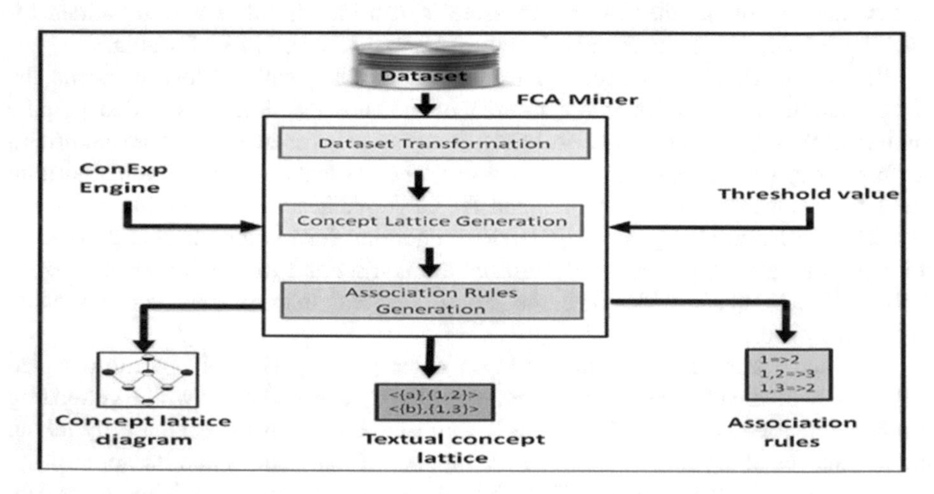

Fig. 1. The overview of FCA-ARMM

Dataset Transformation. This process in FCA Miner is to convert the standard dataset into a format that compatible with the ConExp engine. The standard dataset in .txt will converted into .cxt format. In transformation process from standard dataset into ConExp engine format dataset, the first two lines after the table name in ".cxt" format are referring to the number of attribute and object respectively. Then the next line list all the attribute name and object name. Symbol "X" represent the cross refer to the empty cell.

Concept Lattice Generation. Concept lattice diagram is produced by the ConExp engine by importing the transformed dataset. The Textual FCA is generated by FCA Miner from then Semi Textual FCA after removing the duplicates intent.

Association Rules Generation. Association rules are generated from the Semi Textual FCA and Textual FCA. There are five thresholds to measure the strength and importance of the association rules. Which are support, confidence, lift, Definite Factor

(DF) and Critical Relative Support (CRS). The minimum thresholds need to be specified to reduce the total of association rules. Only association rules that fulfill the minimum threshold will be display.

3.2 The Algorithms

In FCA-ARMM we divide the algorithms into three parts which are Dataset Transformation, Textual FCA Generation and Association Rules Generation.

3.2.1 Dataset Transformation Algorithm

Figure 2 shows the algorithm for transforming the standard dataset into the ConExp engine format. The output then will be saved in .cxt format.

```
1. Input: dataset.txt
2. Output: dataset.cxt
3. Determine attributes
4. for each (input)
5.     (attribute)
6.         if (attribute ∈ input) then
7.                 print "X"
8.         else
9.                 print "."
10.        end if
11. end of loop
```

Fig. 2. Dataset transformation algorithm

3.2.2 Textual FCA Generation Algorithm

Figure 3 shows the algorithm to generate Textual FCA. The Textual FCA is generated from Semi Textual FCA after removing all the duplicate intent and obtained the longest intent.

```
1. Input: semi textual FCA (object, attribute)
2. Output: textual FCA
3. for each (semi textual FCA)
4.         if (object>1) then
5.             textual FCA add(object, getLongest (attributes))
6.         else
7.             textual FCA add(object, attributes)
8.         end if
9. end for loop
```

Fig. 3. Association rule algorithm

3.2.3 Associations Rules Generation Algorithm

Figure 4 shows the algorithm to extract the association rules from the Textual FCA and Semi Textual FCA. The association rules that fulfill the minimum support will be extracted.

```
1.  Input: Textual FCA, Semi Textual FCA and minimum thresholds
2.  Output: Association rule
3.  for each (textual FCA)
4.      getSubset (textual FCA)
5.          for each (semi textual FCA)
6.              if ((textual FCA∈ semi)>minimum threshold)
7.               print Association rule
8.              end if
9.          end for loop
10. end of loop
```

Fig. 4. Association rules generation algorithm

4 Results and Discussions

4.1 Experimental Setup

The experiments were carried out on Intel® Core^{TM}i5-321 M CPU at 2.50 GHz speed with 4 GB RAM, running on Window 7 Home Premium. All algorithms have been developed using Java as a programming language and NetBeans IDE 8.0 with JDK 1.7.09 as platform. We evaluate the proposed algorithm to air pollution data taken in Kuala Lumpur on July 2002 as presented and used in [22]. The ARs of the presented results are based on a set of air pollution data items, i.e. $\{CO_2, O_3, PM_{10}, SO_2, NO_2\}$. The value of each item is with the unit of part per million (*ppm*) except PM_{10} is with the unit of micro-grams (μgm). The data were taken for every one-hour every day. The actual data is presented as the average amount of each data item per day. For brevity, each data item is mapped to parameters 1, 2, 3, 4 and 5 respectively. There are 30 transactions involved with respect to this parameters. Each transaction is defined as a set of data items corresponds to such numbers.

4.2 Implementation of FCA Miner

In this section the output produced by FCA Miner for KLAP dataset is presented. Figure 5 shows the Semi Textual FCA of KLAP dataset. This output is then used as an input for generation association rule.

```
1)<{ 1 2 3 4 5 6 7 8 9 10 11 12 14 15 16 17 18 19 20 21 22 23 24 25 27 28 29 30 } , { 1 }>
2)<{ 6 7 18 } , { 1 2 }>
3)<{ 6 7 18 } , { 1 2 3 }>
4)<{ 6 7 18 } , { 1 2 3 4 }>
5)<{ 6 7 18 } , { 1 2 3 4 5 }>
6)<{ 6 7 18 } , { 1 2 3 5 }>
7)<{ 6 7 18 } , { 1 2 4 }>
8)<{ 6 7 18 } , { 1 2 4 5 }>
9)<{ 6 7 18 } , { 1 2 5 }>
10)<{ 5 6 7 8 14 15 16 17 18 19 20 21 22 28 29 } , { 1 3 }>
11)<{ 5 6 7 14 16 17 18 19 20 21 22 28 29 } , { 1 3 4 }>
12)<{ 5 6 7 14 16 17 18 19 20 21 22 28 29 } , { 1 3 4 5 }>
13)<{ 5 6 7 8 14 15 16 17 18 19 20 21 22 28 29 } , { 1 3 5 }>
14)<{ 3 5 6 7 9 10 11 12 14 16 17 18 19 20 21 22 25 27 28 29 30 } , { 1 4 }>
15)<{ 5 6 7 10 11 14 16 17 18 19 20 21 22 25 27 28 29 } , { 1 4 5 }>
16)<{ 1 5 6 7 8 10 11 14 15 16 17 18 19 20 21 22 23 25 27 28 29 } , { 1 5 }>
17)<{ 6 7 18 } , { 2 }>
18)<{ 6 7 18 } , { 2 3 }>
19)<{ 6 7 18 } , { 2 3 4 }>
20)<{ 6 7 18 } , { 2 3 4 5 }>
21)<{ 6 7 18 } , { 2 3 5 }>
22)<{ 6 7 18 } , { 2 4 }>
23)<{ 6 7 18 } , { 2 4 5 }>
24)<{ 6 7 18 } , { 2 5 }>
25)<{ 5 6 7 8 13 14 15 16 17 18 19 20 21 22 28 29 } , { 3 }>
26)<{ 5 6 7 13 14 16 17 18 19 20 21 22 28 29 } , { 3 4 }>
27)<{ 5 6 7 14 16 17 18 19 20 21 22 28 29 } , { 3 4 5 }>
28)<{ 5 6 7 8 14 15 16 17 18 19 20 21 22 28 29 } , { 3 5 }>
29)<{ 3 5 6 7 9 10 11 12 13 14 16 17 18 19 20 21 22 25 26 27 28 29 30 } , { 4 }>
30)<{ 5 6 7 10 11 14 16 17 18 19 20 21 22 25 27 28 29 } , { 4 5 }>
31)<{ 1 5 6 7 8 10 11 14 15 16 17 18 19 20 21 22 23 25 27 28 29 } , { 5 }>
```

Fig. 5. Semi textual FCA for KL air pollutions dataset

Figure 6 shows the output of textual lattice concept generated by FCA Miner from the standard dataset. There are 10 numbers of concepts from KL Air Pollutions dataset.

```
1) <{ 5 6 7 8 14 15 16 17 18 19 20 21 22 28 29 }> , <{ 1 3 5 }>
2) <{ 1 5 6 7 8 10 11 14 15 16 17 18 19 20 21 22 23 25 27 28 29 }> , <{ 1 5 }>
3) <{ 3 5 6 7 9 10 11 12 14 16 17 18 19 20 21 22 25 27 28 29 30 }> , <{ 1 4 }>
4) <{ 5 6 7 10 11 14 16 17 18 19 20 21 22 25 27 28 29 }> , <{ 1 4 5 }>
5) <{ 6 7 18 }> , <{ 1 2 3 4 5 }>
6) <{ 5 6 7 14 16 17 18 19 20 21 22 28 29 }> , <{ 1 3 4 5 }>
7) <{ 5 6 7 13 14 16 17 18 19 20 21 22 28 29 }> , <{ 3 4 }>
8) <{ 3 5 6 7 9 10 11 12 13 14 16 17 18 19 20 21 22 25 26 27 28 29 30 }> , <{ 4 }>
9) <{ 5 6 7 8 13 14 15 16 17 18 19 20 21 22 28 29 }> , <{ 3 }>
10) <{ 1 2 3 4 5 6 7 8 9 10 11 12 14 15 16 17 18 19 20 21 22 23 24 25 27 28 29 30 }> , <{ 1 }>
```

Fig. 6. Textual lattice concept for KL air pollutions dataset

Figure 7 shows the concept lattice diagram that generated by ConExp engine from the transformed dataset. This concept lattice shows the relationship of type of air pollutions by day in a month.

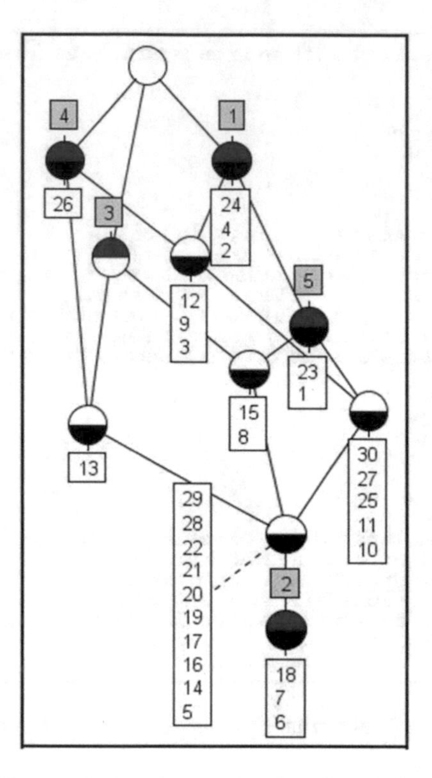

Fig. 7. Concept lattice diagram for KL air pollutions dataset

Figure 8 shows the association rules for the minimum support and minimum confidences are set to 50%. There are 14 association rules produced from this threshold. The most dominant antecedent and consequence are item *1* and *5*, respectively. Item *1* and *2* represent CO_2 and NO_2, respectively. Only two negative association rules occurred from the list which is based on the Lift values that less than 1.00.

Association Rules		Support	Confidence	Lift	DF	CRS
1) 1	==> 3	50.00	53.57	1.00	0.40	0.91
2) 3	==> 1	50.00	93.75	1.00	0.40	0.91
3) 1	==> 5	70.00	75.00	1.07	0.25	1.00
4) 5	==> 1	70.00	100.00	1.07	0.25	1.00
5) 3	==> 5	50.00	93.75	1.34	0.22	0.89
6) 5	==> 3	50.00	71.43	1.34	0.22	0.89
7) 1 3	==> 5	50.00	100.00	1.43	0.29	1.00
8) 5	==> 1 3	50.00	71.43	1.43	0.29	1.00
9) 1	==> 4	70.00	75.00	0.98	0.16	0.85
10) 4	==> 1	70.00	91.30	0.98	0.16	0.85
11) 4	==> 5	56.67	73.91	1.06	0.07	0.69
12) 5	==> 4	56.67	80.95	1.06	0.07	0.69
13) 1 4	==> 5	56.67	80.95	1.16	0.00	0.68
14) 5	==> 1 4	56.67	80.95	1.16	0.00	0.68

Fig. 8. Association rules for 50% of minimum support and minimum confidences

4.3 Analysis

There were 52 of association rules produced from the KL Air Pollutions dataset. From this numbers of association rules, some rules are insignificant, thus by increasing the minimum thresholds it can reduce the existing association rules. Figure 9 shows the association rules by minimum CRS of 0.8% and 50% support. These 10 association rules were sorted according to descending order of CRS measure. There are four rules with the highest CRS value. From this list, two of them are classified as negative association rules with CRS value equal to 0.85.

Association Rules		Support	Confidence	Lift	DF	CRS
1) 1 3	=> 5	50.00	100.00	1.43	0.29	1.00
2) 5	=> 1 3	50.00	71.43	1.43	0.29	1.00
3) 1	=> 5	70.00	75.00	1.07	0.25	1.00
4) 5	=> 1	70.00	100.00	1.07	0.25	1.00
5) 1	=> 3	50.00	53.57	1.00	0.40	0.91
6) 3	=> 1	50.00	93.75	1.00	0.40	0.91
7) 3	=> 5	50.00	93.75	1.34	0.22	0.89
8) 5	=> 3	50.00	71.43	1.34	0.22	0.89
9) 1	=> 4	70.00	75.00	0.98	0.16	0.85
10) 4	=> 1	70.00	91.30	0.98	0.16	0.85

Fig. 9. Top ten association rules with minimum CRS of 0.8% and 50% support.

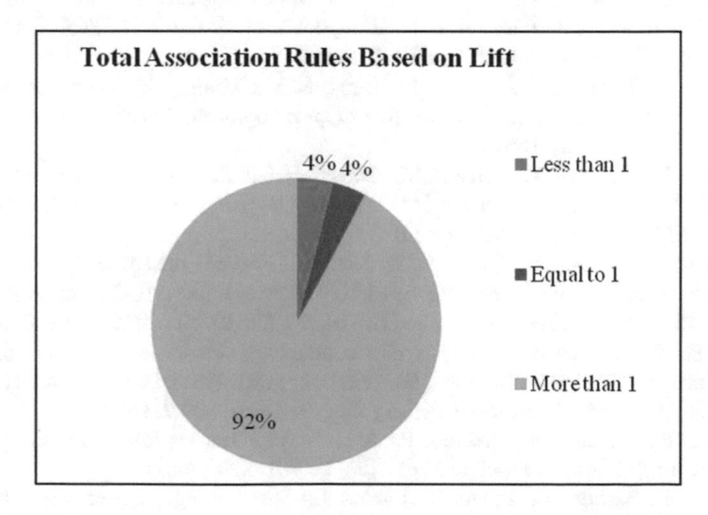

Fig. 10. Total association rules of KL air pollutions dataset based on lift measure.

Figure 10 illustrates the pie chart of total association rules based on lift measure. From the chart, lift more than 1 is the highest rank with total of 97%. This shows that most of the association rules are positively correlated. There are only 4% of the rules that are independent which lift value is equal to 1.

5 Conclusion

There are few studies have been carried out on mining association rules from Formal Concept Analysis (FCA). Most of them are attracted on exploring the potential of utilizing the FCA into the domain of Knowledge Discovery in Data Mining (KDD). However, mining the association rules from the components of FCA is not straight forward and required certain approaches. Therefore, in this paper we proposed a Formal Concept Analysis - Association Rule Mining Model (FCA-ARMM) and an open source tool called FCA-Miner. The results show that the proposed FCA-ARMM and FCA-Miner are able to produce the association rules from the real dataset. In the near future, we plan to apply them into several real datasets and benchmarked datasets.

Acknowledgement. This work is supported by the research grant from Research Acceleration Center Excellence (RACE) of Universiti Kebangsaan Malaysia.

References

1. Agrawal, R., Srikant, R.: Fast algorithms for mining association rules. In: Proceedings of 20th International Conference Very Large Data Bases (VLDB), vol. 1215, pp. 487–499. VLDB Endowment (1994)
2. Abdullah, Z., Herawan, T., Deris, M.M.: An alternative measure for mining weighted least association rule and its framework. In: Zain, J.M., et al. (eds.) ICSECS 2011, vol. 188, pp. 475–485. Springer, Heidelberg (2011). Part II
3. Abdullah, Z., Herawan, T., Ahmad, N., Deris, M.M.: Mining significant association rules from educational data using critical relative support approach. Procedia Soc. Behav. Sci. **28**, 97–101 (2011). Science Direct
4. Abdullah, Z., Herawan, T., Deris, M.M.: Detecting definite least association rule in medical database. In: Herawan, T., Deris, M.M., Abawajy, J. (eds.) DaEng-2013. LNEE, vol. 285, pp. 127–134. Springer, Heidelberg (2014)
5. Poelmans, J., Elzinga, P., Viaene, S., Dedene, G.: Formal concept analysis in knowledge discovery: a survey. In: Croitoru, M., Ferré, S., Lukose, D. (eds.) ICCS-ConceptStruct 2010. LNCS, vol. 6208, pp. 139–153. Springer, Heidelberg (2010). 10.1007/978-3-642-14197-3_15
6. Wille, R.: Formal concept analysis as mathematical theory of concepts and concept hierarchies. In: Ganter, B., Stumme, G., Wille, R. (eds.) Formal Concept Analysis. LNCS, vol. 3626, pp. 1–33. Springer, Heidelberg (2005). doi:10.1007/11528784_1
7. Valtchev, P., Missaoui, R., Lebrun, P.: A partition-based approach towards constructing Galois (concept) lattices. Discrete Math. **256**(3), 801–829 (2002)
8. Pasquier, N., Bastide, Y., Taouil, R., Lakhal, L.: Discovering frequent closed itemsets for association rules. In: Beeri, C., Buneman, P. (eds.) ICDT 1999. LNCS, vol. 1540, pp. 398–416. Springer, Heidelberg (1999). doi:10.1007/3-540-49257-7_25

9. Stumme, G.: Efficient data mining based on formal concept analysis. In: Hameurlain, A., Cicchetti, R., Traunmüller, R. (eds.) DEXA 2002. LNCS, vol. 2453, pp. 534–546. Springer, Heidelberg (2002). doi:10.1007/3-540-46146-9_53
10. Pasquier, N., Bastide, Y., Taouil, R., Lakhal, L.: Efficient mining of association rules using closed lattices. Inf. Syst. **24**(1), 25–46 (1999). Elsevier
11. Stumme, G., Taouil, R., Bastide, Y., Pasquier, N., Lakhal, L.: Fast computation of concept lattices using data mining techniques. In: Proceeding of 7th International Workshop on Knowledge Representation Meets Databases, pp. 129–139 (2000)
12. Ourida, B.B.S., Waf, T.: Formal concept analysis based association rules extraction. Int. J. Comput. Sci. Issues **8**(4), 490–497 (2011). No. 2
13. Pasquier, N., Bastide, Y., Taouil, R., Lakhal, L.: Pruning closed itemset lattices for association rules. In: N Actes Bases De Données Avancées, pp. 177–196 (1998)
14. Zaki, M.J., Hsiao, C.-J.: Chaarm: an efficient algorithm for closed association rule mining, Technical report, Computer Science Department, Rensselaer Polytechnic, pp. 1–20 (1999)
15. Zaki, M.J., Hsiao, C.-J.: Efficient algorithms for mining closed itemsets and their lattice structure. IEEE Trans. Knowl. Data Eng. **17**(4), 462–478 (2005)
16. Stumme, G., Taouil, R., Bastide, Y., Pasquier, N., Lakhal, L.: Computing Iceberg concept lattices with titanic. Data Knowl. Eng. **42**(2), 189–222 (2002)
17. Pasquier, N., Bastide, Y., Taouil, R., Lakhal, L.: Discovering frequent closed itemsets for association rules. In: Proceeding of the 7th International Conference on Database Theory (ICDT 1999), pp. 398–416 (1999)
18. Godin, R., Missaoui, R., Alaoui, H.: Incremental concept formation algorithms based on Galois (concept) lattices. Comput. Intell. **11**(2), 246–267 (1995)
19. Xie, Z.P., Liu, Z.T.: Concept lattice and association rule discovery. J. Comput. Res. Dev. **37** (12), 1415–1421 (2000). (in Chinese)
20. Liang, J., Wang, J.: A new lattice structure and method for extracting association rules based on concept lattice. Int. J. Comput. Sci. Netw. Secur. **6**(11), 107–114 (2006)
21. Burdick, D., Calimlim, M., Gehrke, J.: Mafia: a maximal frequent itemset algorithm for transactional databases. In: Proceeding of the 17th International Conference on Data Engineering, pp. 443–452. IEEE Computer Society (2001)
22. Abdullah, Z., Herawan, T., Deris, M.M.: Mining significant least association rules using fast SLP-growth algorithm. In: Kim, T.H., Adeli, H. (eds.) AST/UCMA/ISA/ACN 2010. LNCS, vol. 6059, pp. 324–336. Springer, Heidelberg (2010)

ELP-M2: An Efficient Model for Mining Least Patterns from Data Repository

Zailani Abdullah[1(✉)], Amir Ngah[2], Tutut Herawan[3], Noraziah Ahmad[4], Siti Zaharah Mohamad[5], and Abdul Razak Hamdan[5]

[1] Faculty of Entrepreneurship and Business / Centre of Computing and Informatics, Universiti Malaysia Kelantan, 16100 Kota Bharu, Kelantan, Malaysia
zailania@umk.edu.my

[2] School of Informatics and Applied Mathematics, Universiti Malaysia Terengganu, 21030 Kuala Terengganu, Malaysia
amirnma@umt.edu.my

[3] Faculty of Computer Science and Information Technology, Universiti Malaya, 50603 Kuala Lumpur, Malaysia
tutut@um.edu.my

[4] Faculty of Computer System and Software Engineering, Universiti Malaysia Pahang Malaysia, 26300 Kuantan, Pahang, Malaysia
noraziah@ump.edu.my

[5] Faculty of Information Science and Technology, Universiti Kebangsaan Malaysia, 43600 Bangi, Selangor, Malaysia
zarabusiness87@gmail.com, arh@ftsm.ukm.my

Abstract. Most of the algorithm and data structure facing a computational problem when they are required to deal with a highly sparse and dense dataset. Therefore, in this paper we proposed a complete model for mining least patterns known as Efficient Least Pattern Mining Model (ELP-M2) with LP-Tree data structure and LP-Growth algorithm. The comparative study is made with the well-know LP-Tree data structure and LP-Growth algorithm. Two benchmarked datasets from FIMI repository called Kosarak and T40I10D100K were employed. The experimental results with the first and second datasets show that the LP-Growth algorithm is more efficient and outperformed the FP-Growth algorithm at 14% and 57%, respectively.

Keywords: Model · Least patterns · Data mining · Efficient

1 Introduction

Data mining is a process of automatically discover hidden and useful information from large database repositories. It is a core step in Knowledge Discovery in Data Mining (KDD) process. For more broaden definition, data mining is referred to "the nontrivial process of identifying valid, novel, potentially useful, and ultimately comprehensible knowledge from database" to assist in the decision making process [1]. Various techniques in data mining are employed to search for novel, unknown and useful patterns.

T. Herawan et al. (eds.), *Recent Advances on Soft Computing and Data Mining*,
Advances in Intelligent Systems and Computing 549, DOI 10.1007/978-3-319-51281-5_23

In addition, it also provides the functionality to predict the future result. Hence, data mining techniques have been broadly deployed to enhance the ability of information retrieval systems. As mentioned earlier, data mining is part of the KDD process model which transforms the input data into meaning information. These input data can be stored in a variety of formats such as in flat files, spreadsheets, XML, relational tables, etc.

Association Rules (ARs) mining is one of the most important and well researched techniques of data mining. It was first invented in [2]. Until today, mining of ARs has been extensively studied in the literature [3]. It aims at discovering correlations, frequent patterns, associations or casual structures among sets of items in the transaction databases or other data repositories [4]. Association is a rule, which implies certain association relationships among a set of objects such as occur together or one implies the other [5]. Its main goal is to find associations among items from transactional database. Usually, ARs are considered to be interesting if they satisfy both a Minimum Support (MinSupp) threshold and a Minimum Confidence (MinConf) threshold. The most common approach to finding ARs is to break up the problem into two parts [6]. First, find all frequent itemsets: By definition, each of these itemsets will occur at least as frequently as a pre-determined MinSupp count [7]. Second, generate strong ARs from the frequent itemsets: By definition, these rules must satisfy MinSupp and MinConf [7].

Until this recent, several works have been carried out on this area [8–28]. In fact, various approaches have been proposed in the literature such as Cfarm Algorithm [29], Automatic Item Weight Generation [30], Weak Ratio Rules [31], FGP Algorithm [32], ConSP [33], Multiple Support-based Approach [34], Non-Coincidental Sporadic Rules [35], ODAM [36], Fixed Antecedent and Consequent [37], Exceptionality Measure [38], Apriori-inverse [39], Relative Support Apriori Algorithm [40], Multiple Minimum Support [41], Pushing Support Constraints [42], Transactional Co-Occurrence Matrix [43].

Even though there are quite a number of improvements that have been achieved, there are still three major drawbacks that have been encountered. The first two non-trivial costs are contributed by the implementation of Apriori-like algorithm. First, the cost of generating a complete set of candidate itemsets. For k-itemset, Apriori will produce up to 2k−2 total candidates. Second, cost of repeatedly scanning the database and check all candidates by pattern matching activities. The last drawback is nearly all of the proposed measures to discover the least patterns are embedded with standard Apriori-like algorithm. The point is this algorithm is undoubtedly may suffer from the "rare item problem". Frequent pattern tree (FP-Tree) [44] has become one of the great alternative data structure to represent the vast amount of transactional database in compressed manner. Afterwards, numerous enhancements of FP-Tree have been suggested according to the implementation of multiple and single database scans. For the first category and including FP-Tree [44], the related studies are Ascending Frequency Ordered Prefix-Tree (AFOPF) [45], Adjusting FP-Tree for Incremental Mining (AFPIM) [46] and Extending FP-tree for Incremental Mining (EPFIM) [47]. The related researches in the second category are Compressed and Arranged Transaction Sequence (CATS) tree [48], Fast Updated FP-Tree (FUFPT) [49], Branch Sorting Method (BSM) [50] and Batch Incremented Tree (BIT) [51].

However, there are still three major shortcomings encountered. First, FP-Tree may not fit into memory due to the characteristics (e.g. highly sparse and dense) of the transactions. Second, it is very expensive to build FP-Tree based on the multi-iterations of transactions. Third, FP-Tree is purposely developed for mining the frequent patterns rather than the least patterns. Therefore and in summary, highly computational cost in constructing FP-Tree is still an outstanding issue in pattern mining. Therefore, in this paper we proposed an Efficient Least Pattern Mining Model (ELP-M2) to alleviate the mentioned above problems. We employed Union-Intersect from set operations to build the trie data structure in ELP-M2 called Least Pattern Tree (LP-Tree) and LP-Growth algorithm to mine the desired patterns. The performance evaluation of the model is made based on two benchmarked datasets from Frequent Itemset Mining Dataset Repository (FIMI) [52].

The rest of the paper is organized as follows. Section 2 explains the details of the proposed model. This is followed by the comparison tests in Sect. 3. Finally, conclusion and future direction are reported in Sect. 4.

2 Proposed Model

There are three major components involved in producing the least ARs. All these components are interrelated and the process flow is moving in one-way direction. The datasets used and least ARs produced in this model are in a format of flat file. A complete overview model of mining the least pattern is shown in Fig. 1.

(i) Transaction Data (Dataset): Transaction data employed in this model are in a format of flat file. For every single transaction, it is written in a line (horizontal) and each item is separated by a single space.

(ii) LF-Itemsets: The first component of the model is to construct Least Frequent Itemsets (LF-Itemsets) and finally will be sorted in support descending order.

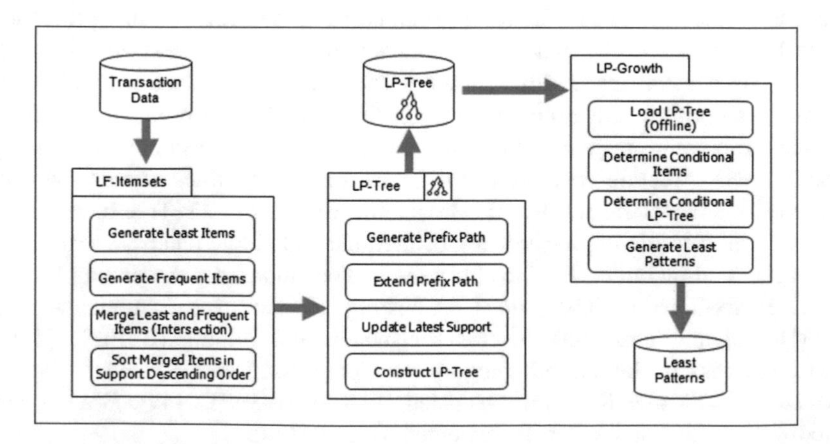

Fig. 1. An overview of ELP-M2

These itemsets are constructed based on the implementations of *ILSupp* and intersection operation. There are four sub-processes at this level.

(iii) LP-Tree: The second component is to construct LP-Tree based on the particular LF- itemsets provided from the previous component. The final structure of the LP-Tree is quite similar to FP-Tree but it might be more compact and less number of nodes (items).

(iv) LP-Growth: The third component is to mining the complete set of significant patterns from LP-Tree by pattern fragment-growth and without generating of candidate itemsets.

(v) Least Patterns: All least patterns are stored in a flat file for further processing in data mining such as to apply interesting measures, etc.

The pseudocode for constructing LP-Tree and executing LP-Growth are shown in Figs. 2 and 3, respectively.

```
Pseudocode LP-Tree
Input:  Transaction data
Output: LP-Tree
1. Begin
2.     Load line of transaction
3.     Generate least patterns
4.     Generate frequent patterns
5.     Sort frequent and least patterns
6.     Intersection operations
7.     Intersected Itemset
8.     Generate prefix paths
9.     Extend prefix paths
10.    Construct LP-Tree
11. End
```

Fig. 2. Steps in generating LP-Tree

```
Pseudocode LP-Growth
Input:  LP-Tree
Output: least patterns
1. Begin
2.     Load LP-Tree
3.     Determine Conditional Items
4.     Determine Conditional LP-Tree
5.     Generate frequent patterns
6. End
```

Fig. 3. Steps in executing LP-Growth

3 Comparison Tests

In this section, we do comparison tests between benchmarked FP-Growth and our LP-Growth. The performance analysis is made by comparing the computational time required to completely mine the patterns (including both least and frequent patterns). We conducted our experiment in two benchmarked datasets. The experiment has been performed on Intel® Core™ 2 Quad CPU at 2.33 GHz speed with 4 GB main memory, running on Microsoft Windows Vista. All algorithms have been developed using C# as a programming language.

Two benchmarked datasets from Frequent Itemset Mining Dataset Repository [15] were employed in the experiment. The first benchmarked dataset was T40I10D100K and was generated using IBM synthetic data generator (http://www.almaden.ibm.com/cs/projects/iis/hdb/Projects/data_mining/datasets/syndata.html). It is a sparse dataset and contains a high proportion of items with a density approximately around 4.2. Thus, it is very suitable to assess the performance of the algorithm when the itemsets can't be magnificently depicted to demonstrate any particular patterns. For the second experiment was Kosarak dataset. The dataset contains the click-stream data obtained from a Hungarian on-line news portal. Kosarak is a sparse dataset with density of less than

Table 1. Main characteristics of Kosarak and T40I10D100K datasets

Data sets	Size	#Trans	Avg length	#Items	Density
Kosarak	15,116 KB	990980	8.1	119	0.02
T40I10D100K	32,246 KB	100000	39.6	942	4.2

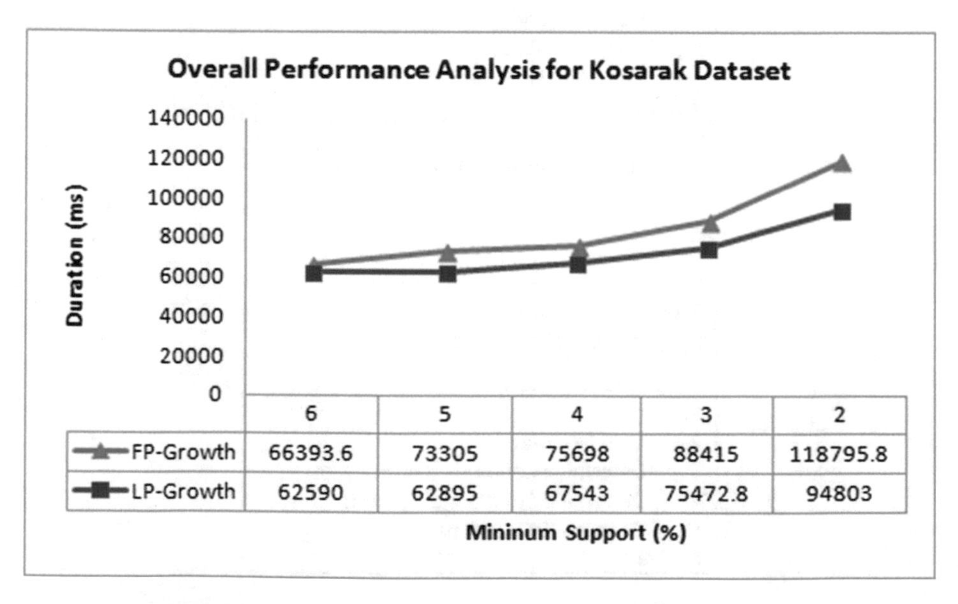

Fig. 4. Performance analysis of the algorithms against Kosarak dataset

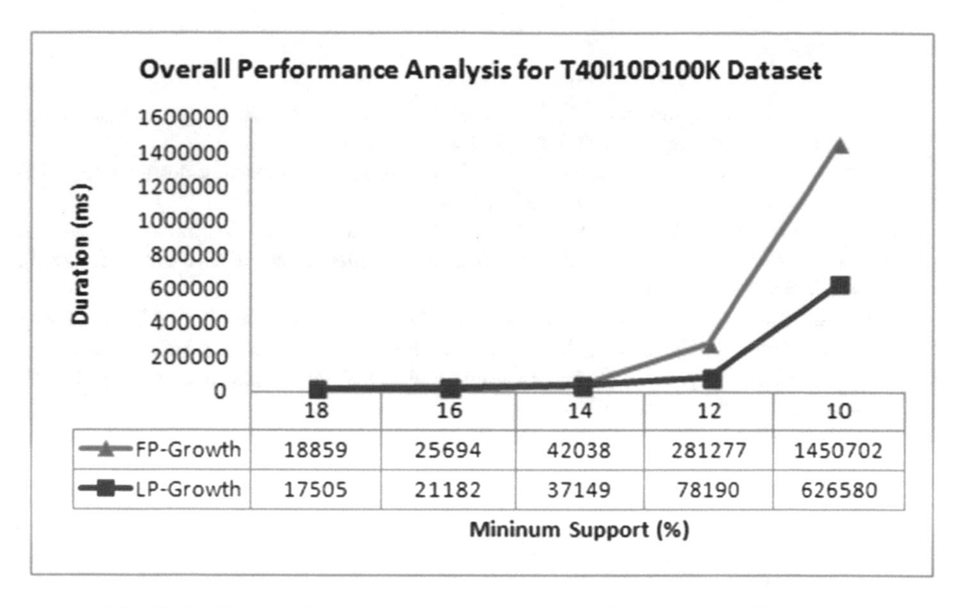

Fig. 5. Performance analysis of the algorithms against T40I10D100K dataset

0.02 as compared to T40I10D100K. This dataset contains enormous number of very long transactions (itemsets) and also rich of distinct items. Table 1 shows the fundamental characteristics of the both datasets.

In this experiment, α is set to equivalent to Minimum Support (MinSupp) and β is set to 100%. Generally, the processing time is decreasing once the MinSupp is increasing. This fact is applicable for both FP-Growth and LP-Growth. Figure 4 reveals computational performance between both algorithms against Kosarak dataset. In average, the time taken for implementing LP-Growth algorithm was 14% faster than FP-Growth algorithm. In term of T40I10D100K, the computational performance between FP-Growth and LP-Growth are presented in Fig. 5. In average, the duration for implementing LP-Growth algorithm was 57% faster than FP-Growth algorithm.

4 Conclusion

In this paper we proposed Efficient Least Pattern Mining Model (ELP-M2) with LP-Tree data structure and LP-Growth algorithm. The experimental results with the well known FP-Growth algorithm against Kosarak and T40I10D100K datasets show that the LP-Growth algorithm in ELP-M2 is outperformed the FP-Growth algorithm at 14% and 57%, respectively. These results indicate that LP-Growth algorithm in the ELP-M2 is more efficient against benchmarked algorithm. In the near future, we plan to apply the model against the several big data repository in order to verify its efficiency.

Acknowledgement. This work is supported by the research grant from Research Acceleration Center Excellence (RACE) of Universiti Kebangsaan Malaysia.

References

1. Fayyad, U., Patesesky-Shapiro, G., Smyth, P., Uthurusamy, R.: Advances in Knowledge Discovery and Data Mining. MIT Press, Cambridge (1996)
2. Agrawal, R., Imielinski, T., Swami, A.: Database mining: a performance perspective. IEEE Trans. Knowl. Data Eng. **5**(6), 914–925 (1993)
3. Hipp, J., Guntzer, U. Nakhaeizadeh, G.: Algorithms for association rule mining – a general survey and comparison. In: Proceedings of SIGKDD Explorations. ACM, New York (2000). ACM SIGKDD **2**(1), 58–64
4. Zhou, L., Yau, S.: Association rule and quantitative association rule mining among infrequent items. In: Koh, Y.S., Rountree, N. (eds.) Rare Association Rule Mining and Knowledge Discovery: Technologies for Infrequent and Critical Event Detection, pp. 15–32. IFI-Global, Pennsylvania (2010)
5. Tan, P.-N., Steinbach, M., Kumar, V.: Introduction in Data Mining. Addison Wesley, Boston (2006)
6. Dunham, M.H.: Data Mining: Introductory and Advanced Topics. Prentice-Hall, New Jersey (2003)
7. Han, J., Kamber, M.: Data Mining Concepts and Techniques. Morgan Kaufmann, San Francisco (2001)
8. Abdullah, Z., Herawan, T., Noraziah, A., Deris, M.M.: Extracting highly positive association rules from students' enrollment data. Procedia Soc. Behav. Sci. **28**, 107–111 (2011)
9. Abdullah, Z., Herawan, T., Noraziah, A., Deris, M.M.: Mining significant association rules from educational data using critical relative support approach. Procedia Soc. Behav. Sci. **28**, 97–101 (2011)
10. Yun, U., Ryu, K.H.: Approximate weighted frequent pattern mining with/without noisy environments. Knowl. Based Syst. **24**, 73–82 (2011)
11. Duraiswamy, K., Jayanthi, B.: A novel preprocessing algorithm for frequent pattern mining in multidatasets. Int. J. Data Eng. (IJDE) **2**(3), 111–118 (2011)
12. Leung, C.K.-S., Jiang, F.: Frequent pattern mining from time-fading streams of uncertain data. In: Cuzzocrea, A., Dayal, U. (eds.) DaWaK 2011. LNCS, vol. 6862, pp. 252–264. Springer, Heidelberg (2011). doi:10.1007/978-3-642-23544-3_19
13. Leung, C.K.-S., Jiang, F., Hayduk, Y.: A landmark-model based system for mining frequent patterns from uncertain data streams. In: Proceedings IDEAS 2011, pp. 249–250. ACM Press (2011)
14. Abdullah, Z., Herawan, T., Deris, M.M.: Scalable model for mining critical least association rules. In: Zhu, R., Zhang, Y., Liu, B., Liu, C. (eds.) ICICA 2010. LNCS, vol. 6377, pp. 509–516. Springer, Heidelberg (2010). doi:10.1007/978-3-642-16167-4_65
15. Abdullah, Z., Herawan, T., Deris, M.M.: Mining significant least association rules using fast SLP-growth algorithm. In: Kim, T.-h., Adeli, H. (eds.) ACN/AST/ISA/UCMA -2010. LNCS, vol. 6059, pp. 324–336. Springer, Heidelberg (2010). doi:10.1007/978-3-642-13577-4_28
16. Abdullah, Z., Herawan, T., Mat Deris, M.: An alternative measure for mining weighted least association rule and its framework. In: Zain, J.M., Wan Mohd, W., El-Qawasmeh, E. (eds.) ICSECS 2011. CCIS, vol. 180, pp. 480–494. Springer, Heidelberg (2011). doi:10.1007/978-3-642-22191-0_42
17. Abdullah, Z., Herawan, T., Deris, M.M.: Visualizing the construction of incremental disorder trie itemset data structure (DOSTrieIT) for frequent pattern tree (FP-Tree). In: Badioze Zaman, H., Robinson, P., Petrou, M., Olivier, P., Shih, Timothy, K., Velastin, S., Nyström, I. (eds.) IVIC 2011. LNCS, vol. 7066, pp. 183–195. Springer, Heidelberg (2011). doi:10.1007/978-3-642-25191-7_18

18. Herawan, T., Yanto, I.T.R., Mat Deris, M.: Soft set approach for maximal association rules mining. In: Ślęzak, D., Kim, T.-h., Zhang, Y., Ma, J., Chung, K.-i. (eds.) DTA 2009. CCIS, vol. 64, pp. 163–170. Springer, Heidelberg (2009). doi:10.1007/978-3-642-10583-8_19
19. Herawan, T., Yanto, I.T.R., Deris, M.M.: SMARViz: soft maximal association rules visualization. In: Badioze Zaman, H., Robinson, P., Petrou, M., Olivier, P., Schröder, H., Shih, T.K. (eds.) IVIC 2009. LNCS, vol. 5857, pp. 664–674. Springer, Heidelberg (2009). doi:10.1007/978-3-642-05036-7_63
20. Herawan, T., Deris, M.M.: A soft set approach for association rules mining. Knowl. Based Syst. **24**(1), 186–195 (2011)
21. Herawan, T., Vitasari, P., Abdullah, Z.: Mining interesting association rules of student suffering mathematics anxiety. In: Zain, J.M., Wan Mohd, W., El-Qawasmeh, E. (eds.) ICSECS 2011. CCIS, vol. 180, pp. 495–508. Springer, Heidelberg (2011). doi:10.1007/978-3-642-22191-0_43
22. Abdullah, Z., Herawan, T., Deris, M.M.: Efficient and scalable model for mining critical least association rules. J. Chin. Inst. Eng. **35**(4), 547–554 (2012). In a special issue from AST/UCMA/ISA/ACN 2010
23. Herawan, T., Noraziah, A., Abdullah, Z., Deris, M.M., Abawajy, J.H.: IPMA: indirect patterns mining algorithm. In: Nguyen, N.-T., Hoang, K., Jędrzejowicz, P. (eds.) ICCCI 2012. SCI, vol. 457, pp. 187–196. Springer, Heidelberg (2012). doi:10.1007/978-3-642-34300-1_18
24. Herawan, T., Noraziah, A., Abdullah, Z., Deris, M.M., Abawajy, J.H.: EFP-M2: efficient model for mining frequent patterns in transactional database. In: Nguyen, N.-T., Hoang, K., Jędrzejowicz, P. (eds.) ICCCI 2012. LNCS (LNAI), vol. 7654, pp. 29–38. Springer, Heidelberg (2012). doi:10.1007/978-3-642-34707-8_4
25. Noraziah, A., Abdullah, Z., Herawan, T., Deris, M.M.: Scalable technique to discover items support from trie data structure. In: Liu, B., Ma, M., Chang, J. (eds.) ICICA 2012. LNCS, vol. 7473, pp. 500–507. Springer, Heidelberg (2012). doi:10.1007/978-3-642-34062-8_65
26. Noraziah, A., Abdullah, Z., Herawan, T., Deris, M.M.: WLAR-Viz: weighted least association rules visualization. In: Liu, B., Ma, M., Chang, J. (eds.) ICICA 2012. LNCS, vol. 7473, pp. 592–599. Springer, Heidelberg (2012). doi:10.1007/978-3-642-34062-8_77
27. Herawan, T., Abdullah, Z.: CNAR-M: a model for mining critical negative association rules. In: Li, Z., Li, X., Liu, Y., Cai, Z. (eds.) ISICA 2012. CCIS, vol. 316, pp. 170–179. Springer, Heidelberg (2012). doi:10.1007/978-3-642-34289-9_20
28. Abdullah, Z., Herawan, T., Noraziah, A., Deris, M.M.: DFP-growth: an efficient algorithm for mining frequent patterns in dynamic database. In: Liu, B., Ma, M., Chang, J. (eds.) ICICA 2012. LNCS, vol. 7473, pp. 51–58. Springer, Heidelberg (2012). doi:10.1007/978-3-642-34062-8_7
29. Khan, M.S., Muyeba, M.K., Coenen, F., Reid, D., Tawfik, H.: Finding associations in composite data sets: the Cfarm algorithm. Int. J. Data Warehous. Min. **7**(3), 1–29 (2011)
30. Koh, Y.S., Pears, R., Dobbie, G.: Automatic item weight generation for pattern mining and its application. Int. J. Data Warehous. Min. **7**(3), 30–49 (2011)
31. Jiang, B., Hu, X., Wei, Q., Song, J., Han, C., Liang, M.: Weak ratio rules: a generalized Boolean association rules. Int. J. Data Warehous. Min. **7**(3), 50–87 (2011)
32. Giannikopoulos, P., Varlamis, I., Eirinaki, M.: Mining frequent generalized patterns for web personalization in the presence of taxonomies. Int. J. Data Warehous. Min. **6**(1), 58–76 (2010)
33. Lu, J., Chen, W., Keech, M.: Graph-based modelling of concurrent sequential patterns. Int. J. Data Warehous. Min. **6**(2), 41–58 (2010)

34. Kiran, R.U., Reddy, P.K.: An improved frequent pattern-growth approach to discover rare association rules. In: Proceedings of the International Conference on Knowledge Discovery and Information Retrieval (ICKDIR 2009), pp. 43–52. INSTICC Press, Funchal
35. Koh, Y.S., Rountree, N., O'keefe, R.A.: Finding non-coincidental sporadic rules using Apriori-inverse. Int. J. Data Warehous. Min. **2**(2), 38–54 (2006)
36. Ashrafi, M.Z., Taniar, D., Smith, K.A.: ODAM: an optimized distributed association rule mining algorithm. IEEE Distrib. Syst. Online **5**(3), 1–18 (2004)
37. Ashrafi, M.Z., Taniar, D., Smith, K.A.: Redundant association rules reduction techniques. Int. J. Bus. Intell. Data Min. **2**(1), 29–63 (2007)
38. Taniar, D., Rahayu, W., Lee, V.C.S., Daly, O.: Exception rules in association rule mining. Appl. Math. Comput. **205**(2), 735–750 (2008)
39. Koh, Y.S., Rountree, N.: Finding sporadic rules using Apriori-inverse. In: Ho, T.B., Cheung, D., Liu, H. (eds.) PAKDD 2005. LNCS (LNAI), vol. 3518, pp. 97–106. Springer, Heidelberg (2005). doi:10.1007/11430919_13
40. Yun, H., Ha, D., Hwang, B., Ryu, K.H.: Mining association rules on significant rare data using relative support. J. Syst. Softw. **67**(3), 181–191 (2003)
41. Liu, B., Hsu, W., Ma, Y.: Mining association rules with multiple minimum support. In: Proceedings of the 5th ACM SIGKDD International Conference on Knowledge Discovery and Data Mining (KDD 1999), pp. 337–341. ACM, San Diego (1999)
42. Wang, K., He, Y., Han, J.: Pushing support constraints into association rules mining. IEEE Trans. Knowl. Data Eng. **15**(3), 642–658 (2003)
43. Ding, J.: Efficient association rule mining among infrequent items. Ph.D. Thesis, University of Illinois at Chicago (2005)
44. Han, J., Pei, H., Yin, Y.: Mining frequent patterns without candidate generation. In: Proceedings of the 2000 ACM SIGMOD, pp. 1–12. ACM, Texas (2000)
45. Liu, G., Lu, H., Lou, W., Xu, Y., Yu, J.X.: Efficient mining of frequent patterns using ascending frequency ordered prefix-tree. Data Min. Knowl. Disc. **9**, 249–274 (2004)
46. Koh, J.-L., Shieh, S.-F.: An efficient approach for maintaining association rules based on adjusting FP-Tree structures. In: Lee, Y., Li, J., Whang, K.-Y., Lee, D. (eds.) DASFAA 2004. LNCS, vol. 2973, pp. 417–424. Springer, Heidelberg (2004). doi:10.1007/978-3-540-24571-1_38
47. Li, X., Deng, Z.-H., Tang, S.: A fast algorithm for maintenance of association rules in incremental databases. In: Li, X., Zaïane, O.R., Li, Z. (eds.) ADMA 2006. LNCS (LNAI), vol. 4093, pp. 56–63. Springer, Heidelberg (2006). doi:10.1007/11811305_5
48. Cheung, W., and Zaïane, O.R.: Incremental Mining of Frequent Patterns without Candidate Generation of Support Constraint. In Proc. of the 7th International Database Engineering and Applications Symposium (IDEAS'03), IEEE Computer Society, New York, 111–117 (2003)
49. Hong, T.-P., Lin, J.-W., We, Y.-L.: Incrementally fast updated frequent pattern trees. Int. J. Expert Syst. Appl. **34**(4), 2424–2435 (2008)
50. Tanbeer, S.K., Ahmed, C.F., Jeong, B.-S., Lee, Y.-K.: CP-tree: a tree structure for single-pass frequent pattern mining. In: Washio, T., Suzuki, E., Ting, K.M., Inokuchi, A. (eds.) PAKDD 2008. LNCS (LNAI), vol. 5012, pp. 1022–1027. Springer, Heidelberg (2008). doi:10.1007/978-3-540-68125-0_108
51. Totad, S.G., Geeta, R.B., Reddy, P.P.: Batch processing for incremental FP-Tree construction. Int. J. Comput. Appl. **5**(5), 28–32 (2010)
52. Frequent Itemset Mining Dataset Repository. http://fimi.cs.helsinki.fi/data/

Integration of Self-adaptation Approach on Requirements Modeling

Aradea[1(✉)], Iping Supriana[2], Kridanto Surendro[2], and Irfan Darmawan[3]

[1] Department of Informatics Engineering, Faculty of Engineering, Siliwangi University, Tasikmalaya, Indonesia
aradea.informatika@gmail.com
[2] School of Electrical Engineering and Informatics, Bandung Institute of Technology, Bandung, Indonesia
[3] Department of Information System, School of Industrial and System Engineering, Telkom University, Bandung, Indonesia

Abstract. Self-adaptation approaches appear to respond to environmental complexity and uncertainty of today's software systems. However, in order to prepare the system with the capability of self-adaptation requires a specific strategy, including when conducting stage requirements modeling. Activity of requirements modeling to be very decisive, when selecting and entering new elements to be added. Here we adopt a feedback loop as a strategy of self-adaptation, which is integrated into a goal-based approach as an approach to requirements. This paper discusses the integration of the two approaches, with the aim of obtaining a new model, which has the advantages of both.

Keywords: Self-adaptive systems · Requirements modeling · Goal-based · Feedback loop · Rule-based systems · ECA rules

1 Introduction

Standish Group International [1], in the report shows that the challenge of software development continues to increase, reaching 43%, while employment requirements in capturing, selecting, and implementing a custom development applications is the most difficult activity. This is partly due to the requirements modeling for adaptation needs very different from the requirements for conventional requirements, which only aims to understand the problem domain, and is done only at design time [2]. Besides the cost of system maintenance continue developing States from 60% to 80% [3]. Thus, the issue of maintenance-related system configuration and reconfiguration of the system repeated has been a challenge that requires settlement.

These facts indicate that the software system must have the ability to adapt to a dynamic environment. In order to meet these needs, it is important to establish a perspective that can guide us in understanding the domain and identify possible changes [4], which need to be predicted from the beginning. Therefore, modeling changes to the system must be controlled completely, including the evaluation of needs, especially related to the design of the control system [5], it can be met, one of

T. Herawan et al. (eds.), *Recent Advances on Soft Computing and Data Mining*,
Advances in Intelligent Systems and Computing 549, DOI 10.1007/978-3-319-51281-5_24

them during the activity of requirements modeling in order to realize an adaptability of the system independently, or known as self-adaptive systems (SAS). Activity of requirements modeling can be directed to arrest and formulate the needs of the system at design time, but is able to accommodate the needs at run-time. The second section of this paper discusses the related work, then on the third section we describe the proposed model, followed by a discussion of related work in section four, and we conclude this paper in section five.

2 Related Work

The focus of the research discussed in this section relate to the goal oriented requirements engineering (GORE) approach, based on a literature review and paper survey, this approach has had many successes as a basis for forming autonomous behavior adaptation needs. However, this approach to get its own challenges when it comes to meeting the needs of the nature of SAS. Table 1 shows the comparison of related research, in general model of GORE expanded by integrating elements that can bring the ability of self-adaptation. Among them, GORE collaborated with the principles of fuzzy logic as used in the model FLAGS and ADS-i*, this work focuses on addressing the uncertainty of the system, similar to LoREM, but the elements are adopted is model-driven approaches. Additionally, GORE approaches through an agent concept has been widely adopted, such as Tropos4AS, CARE, SOTA, GASD, STSs including Lorem, FLAGS, and ADS-i*, these works exploit the advantages of the concept of an agent to capture variability context and develop the behavior of self-configure. While the work GOCC and ZANSHIN, enter the control theory to bring a generic function feedback loops.

Table 1. Comparison of related work.

Model	Specification of design-time	Specification of run-time
LoREM [6] (Goal-based, Model-driven) Goldsby, 2008	i* model	Application-driven process
	Forth concept of requirements: goal, requirements, mechanism selection, infrastructure adaptation	Technology-driven process
GOCC [7] (Goal-based, Control theory) Naka, 2008, 2011	KAOS model	Control loop pattern (stimulus-respond)
	Modelling of configuration architecture generator (compiler): goal relation	Parser machine detector (new pattern and conflict)
	Three-layer (collect, analyze, act)	
FLAGS [8] (Goal-based, Fuzzy goal)	KAOS model and LTL	ECA rules
	Adaptive goal, run-time trigger	Supervision manager: mapping goal to BPEL

(*continued*)

Table 1. (*continued*)

Model	Specification of design-time	Specification of run-time
Baresi, 2010	Fuzzy goal, goal operator temporal	
ADS-i* [9] (Goal-based, Fuzzy goal) Serrano, 2011	i* model	Implementation BDI abstraction to JADEX
	Strategic dependency and rationale	
	Heuristic of i* abstraction into BDI	Reasoning machine: qualitative reasoning
	Fuzzy logic: task analysis and softgoal	
ARML [10] (Goal-based, Ontology) Qureshi, 2012	Techne model	ECA rules
	Integration of goal and shared ontology	Integration of PDDL3 and Hierarchical task network
	Context, resources, domain assumption	
GASD [11] (Goal-based, Model-driven) Wang, 2012	Prometheus model	OWL ontology
	Goal tree concept (selection algorithm): goal, plan, restriction model	Mechanism of knowledge based
	Designing of BDI architecture	JADE run-time
STS [12] (Goal-based, Ontology) Dalpiaz, 2013	Tropos/i* model	BDI agent
	Goal model, context model, plan specification, domain assumptions	Monitor-diagnose-reconcile-compensate
Zanshin [13] (Goal-based, Control theory) Souza, 2013	Techne model	PID controller
	Awarness requirements	Qualitative adaptation. Evolution requirements: ECA rules
	Variaton point (VP), control variable (CV), diferensial relation	
SOTA [14] (Goal-based, Natural language) Dhaminda, 2014	i* model and context-free grammar	Checking: labeled transition system analyzer.
	Goal pattern: achieve, maintain avoid	Mapping goal to event
	Precondition and postcondition goal, actor utility, entity and dependency	
Tropos4AS [15] (Goal-based, Model-driven) Morandini, 2015	Tropos/i* model	BDI agent
	Goal type, environment, failure	Transitions rules: goal state-intuition
	Environment and goal condition	
	Variability design	Inference rules

A variety of approaches are integrated self-adaptation to approach these goals, in addition to having the advantages of each still has some shortcomings, such as the principle of fuzzy logic and natural language constructions require approaches are quite complex and highly dependent on the specification requirements are very complete. In addition, the agency approach and model-driven substantially less can be directly mapped at run-time concepts related needs evolution cycle. While the application of control theory is very focused on the control feedback loop, in which the needs of the entity of the problem domain that is represented in the domain of models has not been included, as well as the need to determine variability and system independent, requires a special architectural design, so the need for further research to analyze the relationship. Inspired from the above description, arises an idea of how to formulate the mechanism of the process of adaptation of generic (providing a formal framework that can give rise generic function feedback loop for adaptation needs at run-time), but can accommodate variability context and behavior of real-world systems (through the ability of an agent in forming a knowledge of design time).

Here, we see loop feedback implementation has a promising chance if be equipped with data mining approach. A data mining algorithm can be functioned to face uncertain condition at run-time. these views become one of our future research concern to formulate a model that we are developing.

3 Integration of Self-adaptation Approach and Requirements

One approach GORE adopted in concept of requirements modeling are Tropos [16]. Stages requirements in Tropos consists of phases, (a) early requirements, a view to identifying the needs of stakeholders and how it relates to each other, (b) late requirements, capturing changes in a domain that is caused by the need for the system to-be and the true nature of the system. While abstraction of the model feedback loop which is used as the concept of self-adaptation, consists of monitors, analyzers, planners, executors, and knowledge (MAPE-K) [17]. Taking into account the context-aware scenario, MAPE-K implements architectural patterns ECA (event-condition-action), consists of three main components [18], which is a component context processor, the component controller, and component action performers, wherein (a) events, modeled and observed by one or more context processor, this component depends on the definition and modeling of context information, (b) condition, described the behavior of the application and observe the event, by empowering component of the controller is represented by the behavior description component, (c) action, triggered by action performers through a rule that has been specified.

3.1 The Concept of Integration

In order to realize the adaptability of the system, we steer stages in Tropos requirements may have the capability to monitor the variables of each goal decomposition of changes occurring, and can manage the changes at run-time (Fig. 1). This is done to supplement the capability requirements Tropos models, in capturing and representing variability of context and behavior of the system.

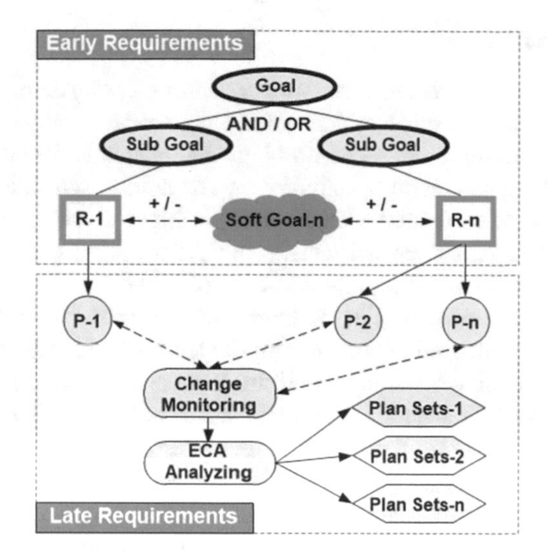

Fig. 1. Integration of self-adaptation approach on requirements modeling activity.

According to Zhuo-qun [19] in the modeling of i* model, variables can be derived from the actor's inner task set, the task is an entity that can be detected and the source of the parameter type, if a task has uncertainty, then all tasks that have a relationship dependencies with the task of the need to be monitored, and when determining what is monitored, the values of the parameters can be used to represent them. Based on these opinions, if the theory is applied to the model Tropos then this task equivalent to the plan, in which each plan may have a dependency relationship with the goals, resources, or other notation. While monitoring the variable and handling needs of adaptation, the model was developed as a criterion of feedback loop mechanism guarantees that the process will be established.

Early requirements other than the function to identify the needs of stakeholders, this phase also aimed to capture the monitoring requirements of any entity that allows the change. Thus modeling the goal at this stage is expected to represent the needs of early to determine the mechanism of adaptation. Begins with decomposition AND/OR towards goal into sub-goals, identify requirements (R−1, R−2, R−n) of each goal that may change and affect the parameters, and has contributed a positive or negative (+/−) to one or more soft goals (non-functional) requirements, topped off by defining variables and parameters of each of these goals (P−1, P−2, P−n). In the late phase requirements, delegates goal in addition to capturing the needs of system-to-be, directed also to analyze the needs of adaptation. Starting with the restructure dependency goals, monitor changes that affect the parameters of each goal (P−1, P−2, P−n), analysis by ECA methods to analyze behavioral changes, and to determine the variation of adaptation is based on the establishment of the rule which is defined as a plan (Plan Sets-1, Plan Sets-2, Plan Sets-n).

3.2 Case Illustration

Cases discussed is related to the adaptation needs of the system of lectures at universities. Where there is a class system actors who represent the needs of lecturer, students, and the program (prodi). The problem that occurs is the system must be able to ensure that teaching and learning activities can be held as scheduled. In addition, the number of face-to-face classroom sessions must achieve a minimum 14 and a maximum of 16 sessions, if it does not satisfy the range, it must be done through the establishment of additional tuition replacement schedule.

In the illustrative model of Fig. 2, hard goal or goal "to monitor the course" decomposed into two sub-goal of "organizing the lecture" and "cancellation of lecture" through OR-decomposition, meaning that if the lecture was held in accordance with the criteria, then the goal "cancellation lecture" does not need to be achieved but if otherwise, then the solutions developed is to determine the replacement schedule. Modeling for the replacement schedule need not be discussed in this paper, in the illustration Fig. 2 needs replacement schedule delegated to other actors through the goal "choose schedule". Goal "organizing the lecture" decomposed with AND-decomposition, the three sub-goals of the (start on schedule, completed on schedule, filling of attendance and the minutes) must all be achieved through the plan "monitoring" and an additional plan "check the number of lectures" for the goal "fill the attendance and the minutes", which contribute positively/full satisfaction (++) against soft goal "availability of classroom" and "lecture targets". While the plan for

Fig. 2. Modeling of lecture monitoring system.

achieving the goal "lecture cancellation" has contributed partial negative (−) against both soft goal, considering the cancellation of the lecture could adversely affect the achievement of the lecture and meeting room availability, but will not impact negatively if the determination of the replacement schedule can be achieved.

Based on the modeling, we can capture the issue of adaptation to be resolved, namely the third sub-goal of "organizing the lecture" is a goal that must be monitored, because the decomposition made an AND-decomposition, meaning that variable in every goal that is linked dependencies to each other. This can be represented by defining each parameter value in the plan that will be developed, namely by setting a rule to analyze aspects of dynamic behavior. For example the study program will be given a notification message when lectures in class was held or not held. Assumptions for the lecture was held, when the course of time according to the schedule. Based on the concept in the pattern of the ECA, we consider the situation "lecturer and students enter the classroom" as an event (e) that triggers an evaluation by the ECA rules:

```
If <lecturer and students enter the classroom (e1) AND
start as scheduled (c1) AND completed on schedule (c2)
AND filling attendance (c3)>, then <send notification
{prodi} (a1),"course achieved">
```

Elements of c1, c2 and c3 is a condition (c) to ensure that the lecture was held based on the criteria or not, as an example is late more than 10 min of schedule, was considered not fit the criteria. This condition represents a situation where the rules of action (a) will be activated, based on the condition (c) specified when the event occurs (when the "lecturer and students enter the classroom"), and a1 is "send notification (in the program)" is an action that is executed when a condition that occurs in accordance with the criteria or is true. Figure 3 illustrates the flow of information among the components in the pattern of ECA.

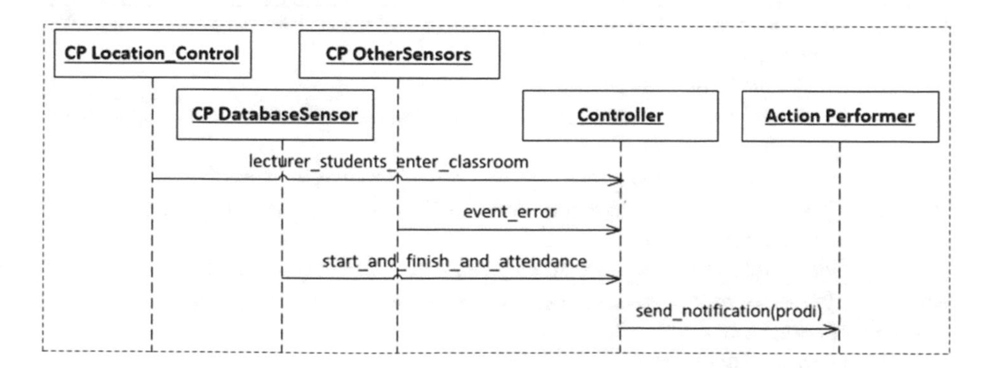

Fig. 3. The dynamic behaviour of ECA pattern.

The controller observed the occurrence of an event when lecturer and students enter the classroom, this event was captured by the location controller component, which is an instance of context processor (CP). With the use of sensors in the classroom, location controller can sense when lecturer and students entering the class. When this happens, the event "lecturer_students_enter_classroom" generated. Then, the controller evaluates "start_and_finish_and_attendance" and "event_error". Finally, if the third condition is true, the controller will trigger action "send_notification" which has been specified in the ECA, this action is the executed by the performers.

In the requirements activity, evaluation controller begins by defining variables that must be monitored, which is derived from the resource "course schedule". Based on the path dependencies, all the variables associated with "time" holding "course", i.e. start, finish, filling attendance and event news, as well as delay time tolerance. Tolerance delay time is context variables are added. In addition, to maximize the system's behavior, also added requirements to detect "errors" associated with "event" unexpected, e.g. sending error messages (msg_error) or lost connections (conn_error). Thus obtained variables are represented as: course_time and error_event = (course_time, event_error).

As for the needs analysis process, we should set the value of each parameter and the rules for the lecture can be considered established criteria or not. The action was carried out for both of these criteria is, if the lecture is held it will be sent a notification to the course that lectures held, and if the college is not established then the notifications are delivered is not the implementation of tuition based on a schedule, and the goal "choose the schedule" will be delegated to actors program a study to establish a replacement schedule. The criteria that the college is considered established, if the college began as scheduled, completed on schedule, charging absent and the minutes of a maximum of 15 min before the lecture is finished, and the delay tolerance for all these parameters is 10 min. Thus, the rule can be specified as in Table 2.

Table 2. Rule of course monitoring.

Rule	Statement
Rule-1:	*If (course_time = time.start and time.finish and time.attendance) and (event_error = null), then send notification {prodi}"course achieved"*
Rule-2:	*If (course_time = time.start and time.finish and time.attendance > 10 minutes) and (event_error = null), then send notification {lecturer} "check your teaching schedule"*
Rule-3:	*If (course_time = not time.start and time.finish and time.attendance) and (event_error = null), then send notification "course not conducted" and delegate goal to actor {prodi}*
Rule-4:	*If (course_time = time.start and time.finish and time.attendance) and (event_error = not null), then send notify user*
Rule-5:	*If not [event], then send notification "course not conducted" and delegate goal to actor {prodi}*
Rule-6:	*If not [criteria], then send notify user*

Based on these rules, it can be mapped into the ECA table (Table 3), so that there are four action as an alternative solution for the needs of adaptation. In order to define a comprehensive planning, determination of other parameters that can affect the system, as well as the formulation of reasoning mechanisms can be developed further.

Table 3. ECA of course monitoring.

Event	Condition	Action
"lecturer_students_enter_classroom"	*course_time = time. start and time.finish and time.attendance; event_error = null;*	*send notification {prodi} "course achieved"*
"lecturer_students_enter_classroom"	*course_time > 10 min; event_error = null;*	*send notification {lecturer} "check your teaching schedule"*
"lecturer_students_enter_classroom"	*course_time = not time.start and time. finish and time. attendance; event_error = null; not [event];*	*send notification "course not conducted" and delegate goal to actor {prodi}*
"lecturer_students_enter_classroom"	*event_error = not null; not [criteria] {msg_error};*	*send notify user*

4 Discussion and Future Work

The proposed framework when compared to the Tropos framework adopted, basically can be a complement to enhance the ability of adaptation requirements. When evaluated in general, the framework of this proposal can be developed as a generic framework that can be applied to various types of systems, in addition to the feedback loop automation through reconfiguration and evolution mechanism can improve the adaptability. So that this concept can reduce maintenance costs and increase flexibility in dealing with change factors for requirements engineering activities.

Comparison adaptation framework through two phases of requirements can be seen in Table 4. The work that has been done, is an initial foundation for the work ahead more specific, namely (a) identifies the need for an extension of the modeling language is adopted, taking into account the initial concept that has composed, (b) the alignment of the model which has been developed into a model that is adopted, to define the operational semantics that can accommodate the abilities of adaptation mechanisms, mainly dealing with strategy reconfiguration and evolution of software devices. Data mining approach become one of our future study, to determine appropriate time in adaptation, (c) evaluating the model in more detail to determine the level of success of approaches that have been proposed.

Table 4. Comparison of requirements modeling framework.

Phase	Tropos framework	Proposed framework
Early requirements	Stakeholder requirements • Description: actors, goals, plans, resources • Strategic dependency • Decompositions, means-ends, contributions	Adaptation requirements • Identify stakeholder requirements • Defining goals that could potentially change, as the requirements (R−n) • Determine the variable and parameters goals (P−n)
Late requirements	Delegating goal as a representation system-to-be • Introducing the actors • Specification of dependencies • Decompositions, means-ends, contributions to the actors	Delegating goal as adaptation requirements • Identify the cause of the goal parameters change (change monitoring) • Analyzing the behavior change • Determine the variation changes (ECA) • Determine the plan (plan sets-n)

5 Conclusion

Activity of requirements modeling at design time for the adaptation needs of the system, in essence proposes concepts and analytical techniques for designing monitoring needs and adaptation at run-time. Uncertainty at design time managed to combine design with specifications variability monitoring, through achievement criteria satisfaction goals. Here we propose an extension of an existing model. We see the feedback loop approach through ECA rules have the advantage to realize a generic function in reasoning at run-time. While Tropos models with the agent's concept has advantages in terms of capturing the variability of context and system behaviour, particularly related to the needs of the entity of the problem domain that is represented in the domain models.

We assume the integration of the two approaches can be complementary disadvantages and advantages of each. Therefore, we propose a framework as our preliminary study, and this approach still requires more in-depth study. Mainly deal with how to integrate a feedback loop with a centralized approach to the design needs of independent software-based agents. Currently, we are evaluating and formulating the system in more detail, including committing study toward alignment from both approach with formulating analysis technique using data mining at sensor data through data history, and context inference in determining context needs as run-time at the knowledge base.

References

1. The Standish Group: Chaos Manifesto: Think Big, Act Small. The Standish Group International (2013)
2. Perini, A.: Self-adaptive service based applications: challenges in requirements engineering. In: RCIS, CIT-Irst., FBK, Trento, Italy (2012)

3. Sherry, J., Hasan S., Scott, C., Krishanmurthy, A., Ratnasamay, S., Sekar, V.: Making middleboxes someone else's problem. In: Proceedings of the ACM SIGCOMM 2012, New York, USA, p. 13. ACM Press (2012)
4. Aradea, Supriana, I., Surendro, K.: Prinsip Paradigma Agen Dalam Menjamin Keberlangsungan Hidup Sistem. Prosiding KNSI. Universitas Klabat Sulawesi Utara (2015)
5. Aradea, Supriana, I., Surendro, K.: An overview of multi agent system approach in knowledge management model. In: Proceedings of the ICITSI, STEI, ITB (2014)
6. Goldsby, H.J., Cheng, B.H.C.: Automatically generating behavioral models of adaptive systems to address uncertainty. In: Czarnecki, K., Ober, I., Bruel, J.-M., Uhl, A., Völter, M. (eds.) MODELS 2008. LNCS, vol. 5301, pp. 568–583. Springer, Heidelberg (2008). doi:10.1007/978-3-540-87875-9_40
7. Nakagawa, H., Ohsuga, A., Honiden, S: GOCC: a configuration compiler for self-adaptive systems using goal-oriented requirements description. In: Proceedings of the 6th International Symposium on SEAMS, May 21–28, pp. 40–49. ACM, USA (2011)
8. Baresi, L., Pasquale, L., Spoletini, P.: Fuzzy goals for requirements-driven adaptation. In: Proceedings of RE, pp. 125–134. IEEE (2010)
9. Serrano, M., Sampaio, J.C.: Development of agent-driven systems: from i* architectural models to intentional agents' code. In: CEUR Proceedings of the International i* Workshop (2011)
10. Qureshi, N.A., Jureta, I.J., Perini, A.: Towards a requirements modeling language for self-adaptive systems. In: Regnell, B., Damian, D. (eds.) REFSQ 2012. LNCS, vol. 7195, pp. 263–279. Springer, Berlin (2012). doi:10.1007/978-3-642-28714-5_24
11. Wang, T., Li, B., Zhao, L., Zhang, X.: A goal-driven self-adaptive software system design framework based on agent. Phys. Procedia **24**, 2010–2016 (2012). ICAPIE Organization Commite, Elsevier B.V.
12. Dalpiaz, F., Giorgini, P., Mylopoulos, J.: Adaptive socio-technical systems: a requirements-based approach. J. Requirements Eng. **18**(1), 1–24 (2013)
13. Souza, V.E.S., Lapouchnian, A., Angelopoulos, K., Mylopoulos, J.: Requirements-driven software evolution. J. Comput. Sci. Res. Dev. **28**(4), 311–329 (2013). Springer
14. Dhaminda B., Hoch, N., Zambonelli, F.: An integrated eclipse plug-in for engineering and implementing self-adaptive systems. In: 23rd International WETICE. IEEE (2014)
15. Morandini, M., Penserini, L., Perini, A., Marchetto, A.: Engineering requirements for adaptive systems. J. Requirements Eng. **21**, 1–27 (2015). Springer
16. Bresciani, P., Perini, A., Giorgini, P., Giunchiglia, F., Mylopoulos, J.: TROPOS: an agent-oriented software development methodology. J. Auton. Agent. Multi-Agent Syst. **8**(3), 203–236 (2004)
17. Kephart, J.O., Chess, D.M.: The vision of autonomic computing. IEEE Comput. **36**(1), 41–50 (2003)
18. Daniele, L.M.: Towards a rule-based approach for context-aware applications. Master thesis. University of Cagliari, Italy (2006)
19. Zhuo-Qun, Y., Zhi, J.: Requirements modeling and system reconfiguration for self-adaptation of internet-ware. In: Fourth Asia-Pacific Symposium on Internet-ware. ACM (2012)

Detection of Redundancy in CFG-Based Test Cases Using Entropy

Noor Fardzilawati Md Nasir[1(✉)], Noraini Ibrahim[1], and Tutut Herawan[2]

[1] Universiti Tun Hussein Onn Malaysia, Batu Pahat, 86400 Parit Raja, Johor, Malaysia
fasha_nasir@yahoo.com.sg, noraini@uthm.edu.my
[2] University of Malaya, Pantai Valley, 50603 Kuala Lumpur, Malaysia
tutut@um.edu.my

Abstract. Testing is an activity conducted by the software tester to validate the behavior of the system, whether it is working correctly or not. The effectiveness of generating test cases becomes a crucial task where there are an increment of source code and the rapid change of the requirement. Therefore, to select the effective test cases become a problem when the test cases are redundant. It creates a new challenge on how to reduce the unnecessary test cases that will increase the cost and maintenance of the software testing process. Thus, this paper proposed the usage of entropy in detecting and removing the redundancy of test cases generated from Control Flow Graph (CFG). The result shows that the proposed approach reduced 61% of test cases compared to the original test suite. In conclusion, entropy can be an alternative approach in detecting and reducing the redundant test cases.

Keywords: Redundancy · Test cases · Entropy · Control Flow Graph (CFG)

1 Introduction

In software testing, test case generation is the most challenging and important step compared with other part [1]. In the other word, testing is an activity conducted by the software tester to validate the behavior of the system whether it is working correctly or not. Test cases are always known as one of the challenging tasks in software testing [2] which are important and need to be provided as a baseline in acceptance testing so that the entire system requirements are tested. According to [3], a test case is a set of test input, condition and expected result running of the software to verify the quality specification of the system. The results from the test will represent the view of the system, whether it meets the user and system specification or not. The combination of test cases is called a test suite which is developed to test the source code and also can be used to test every successive path of the system [4].

Nowadays, software testing provides a new challenge with the evolution of software systems, so the mechanism to determine the system has passed or failed becomes an interesting subject of research. Since there have been new changes in system requirement, new test case(s) should be developed and added to the test suite. However, it becomes a

T. Herawan et al. (eds.), *Recent Advances on Soft Computing and Data Mining*,
Advances in Intelligent Systems and Computing 549, DOI 10.1007/978-3-319-51281-5_25

problem when a large number of test cases generated and somehow redundant test cases are also generated too, even they are referring to the same requirements of the system. Therefore, there is a need to find a minimum size of test cases, but provides the same requirement coverage as the original test suites. This can be done by identifying all the redundant test cases and eliminating them from the generated test suites. Since the redundant test cases will increase the size of test suites, the cost will also increase. Hence, it is important that the unnecessary or useless test cases are removed before being executed by the tester, so that the maintenance cost will be reduced.

This paper is organized as follows: Sect. 2 described the related works. The theoretical background is described in Sect. 3. The proposed methodology is discussed in Sect. 4 and its result and discussion are discussed in Sect. 5. Section 6 contains the conclusion and future work.

2 Related Works

In this paper, the existing reduction techniques that related to this research will be reviewed. This research is focusing on detecting the redundancy and removing the test cases. Saif-ur-Rehman, Nadeem & Awais [5] detect the redundancy of test cases from a program called *Triangle* when there is same associated set of test case referring to the same test suites and the same requirements. They proposed a new reduction technique called *Test Filter* that uses weight criteria for each test case. Chaurasia & Thirunavukkarasu [6] improve the limitation of the previous technique by considering the input generated automatically by the compiler. They proposed a reduction technique called *Test Exude* where the test cases are generated according to the given requirement. Then, the frequency of the test cases is calculated by counting the associated test case occurred in test condition. They come out with the reduced test case where only the test cases that satisfied the corresponding condition are taken and will discard the one that not fulfilled the requirement needed. Both techniques in [5, 6] are only focused on finding redundant set of test cases but do not remove them. In Panday *et al.* [7], they proposed a new technique that reduced the set and remove them as well. The test cases are generated from Control Flow Graph (CFG) and the association between test cases and paths were calculated using coverage matrix. They define the matrix as $C(t_i;p_j)$; where t_i is a test case (TS) and p_j is a path. The matrix is all the digits '0' and '1' which are denoting the path coverage information. "1" in row i and column j means test case t_i executes path p_j and '0' means test case t_i did not execute path p_j. The subset of test cases is identified where the highest coverage is selected and if there are equal score, random choice are selected based on the requirement of the program. The redundant test case is extracted, where the useful test cases are saved and removed the useless test cases. Other redundancy test cases occur in user based testing, where there is a redundant user request that overload the test case generation [8]. They proposed the entropy gain theory in test cases reduction and satisfy all base requests as an original test suite.

Entropy theory has been used in various fields of research and also been introduced in software testing. Herbold [9] defined entropy as a measure of uncertainty of random variables. He proposed one of the novel approaches for the usage based test case generated by applying the concept of entropy rate for stochastic process and typical sequences. He also combined several concepts, which are High Order Markov model and prediction by Partial Match to claim that the technique offer better results in optimize test case generation in usage based testing. Another research done by Maung & Win [19] uses entropy gain theory in test case reduction process for user session based testing. His research has shown that the new reduction technique based on entropy value analysis can reduce the test cases and satisfies all base requests as original test suites. Thus, this research focused on identifying the redundancy of the test cases, where entropy will be embedded to reduce the original test suite.

3 Rudimentary

3.1 Control Flow Graph (CFG)

Control Flow Graph (CFG) or also known as "flow graph" or "program graph" represented the possible control flow of the program where each node has some properties and edges are based on property relations [10]. Figure 1 shows a sample of code and the respective constructed CFG where a node represents a program statement, an edge represents the ability for a program to flow from its current statement to the statement at the other end of the edge. If an edge is associated with a conditional, the edge will be labeled as conditional's value, either true or false. From the CFG, we can now extract all possible paths.

```
1    function P return INTEGER is
2    begin
3     X, Y: INTEGER;
4     READ(X); READ(Y);
5    while (X > 10) loop
6      X := X - 10;
7    exit when X = 10;
8    end loop;
9    if (Y < 20 and then X mod 2 = 0) then
10   Y := Y + 20;
11  else
12   Y := Y - 20;
13  end if;
14  return 2 □□X □□Y□
15          end P;
```

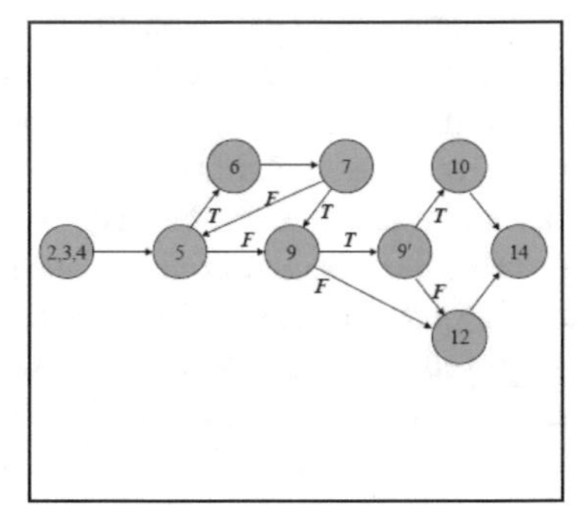

Fig. 1. Sample program with CFG

3.2 Test Case Redundancy

Understanding and evaluating the testing phase become very tough, especially in random testing. Test cases are the important part in software testing and a small modification in test case effect the test suite. A set of test suite contains a set of test cases can have a redundant test cases. A set of test cases in a test suite might contain redundant test cases. When different test cases in a test suite test the same requirement of the systems, it is considered as redundant [11]. With the evolution of the system, the source codes are growing larger and the requirements of the system is always changing may lead the test cases redundant. Moreover, the redundant test cases wills affects the quality of system testing in term of bug. A redundant test case is once, which if removed it will not affect the fault detection effectiveness of the system. Redundant test cases can give us serious consequences on test maintenance. For example, we have two test cases, test the same features of a unit of code, and if one of them is updated correctly and not the other one and one test may fail while the other may pass. So, it will confuse the test result. Detecting the redundant test cases give us some motivation on how to deal with redundant test case and remove it, so that it will reduce the cost maintenance without affecting the performance of the system.

Test case redundancy reduction is one of the test case generation techniques which mention by many researchers. There are various techniques for test redundancy detection. The researchers in [5, 6, 12, 13] used statement coverage in detecting redundant test case. Research in [7] tends to overcome the large number of false-positive error problem which is a test case being identified as redundant while it is not by using the data flow coverage or control flow criteria. Few researches applied a collaborative approach in which semi-automated systems were used to human interaction [14, 15].

3.3 Test Case Redundancy Reduction

One of the challenges faced by organizations is to optimize the test suites [16] in order develop a high quality of software. As we have mentioned before, due to many versions of development of the system, the possibility of redundant test cases occurred in the test suite is high [17]. It is really important to have a technique that will manage the test suite by removing the redundant test case without affecting the performance of the software. Test case reduction or can also be called as minimization is one of the technique to optimize the test case generation techniques [10]. Thus, this research focused on identifying the redundancy of test cases using entropy which are explained in the next section.

3.4 Entropy

Entropy is a concept derived from information theory that will describe the amount of uncertainty and disorder of the system [8]. Entropy can be expressed as follows Eq. 1 and assume that N objects and a variable containing K category (Bailey) [18]:

$$H = -\sum_{i=0}^{n} P_i \log_k P_i \tag{1}$$

Entropy theory has been applied in many areas such as physic and engineering, complex system, chemistry and biology, machine learning and system theory and much more. The application of entropy in information theory become a new heuristic for optimization and reduces redundancy [19]. Deris *et al.* [20], proposed a new approach of rough condition entropy for attribute selection and obtain a minimal reduction in incomplete information system. A conditional entropy equation will be used to identify similarity between all test cases in a test suite. A low entropy value indicates that the test cases have more similarity or redundant and vice-versa.

4 Proposed Approach

The proposed approach described two important concepts: redundancy detection and remove the redundancy of test cases. This proposed method starts with identifying the redundancy of test. The steps are (i) generate test cases using CFG and (ii) find possible path from CFG. Then, (iii) entropy value for each test case is calculated. Next, (iv) the association of entropy value will be compared with possible path and the subset table will be created for each test case. The entropy value for each test case is calculated by using the formula in Eq. (1). (v) First, we start to select from singleton set S_i to find the related path and put in primary test set $T_{p.}$ Next, (vi) test cases with two elements from unmarked subset are identified and test cases with higher entropy value are selected. If the values are equal, the test case is selected by random choice. (vii) Repeat the process for remaining unmarked subset. The redundant test cases now will be extracted from the subset table. Table 1 shows the algorithm to remove the redundant test cases using entropy to detect the redundant path from the CFG based test case.

Table 1. The proposed algorithm to remove the redundant test cases

Algorithm to remove the redundant test cases.
Input: Sample program/source code **Output**: Reduced Test suite T=(t_0, t_1,…, t_n) Step 1: Create a CFG from the source code Step 2: Generate possible test cases from CFG Step 3: Calculate possible path between test cases and path Step 4: Calculate the entropy value for each test case using the equation (1) Step 5: Create a subset table Step 6: Remove redundant test cases - start to select from singleton set S_i to find the related path and put in primary test set T_p - test cases with two elements from unmarked subset are identified - test cases with higher entropy value are selected - If the values are equal, the test case is selected by random choice - Repeat the process for remaining unmarked subset - Extract the redundant test cases from a subset table

There are four main phases as illustrated in Fig. 2. The first phase is, CFG is created from the source code. The successive paths are converted into test cases in a second phase. The third phase is, the redundant paths are analyzed using entropy value. Finally, some heuristic are applied in reducing test cases.

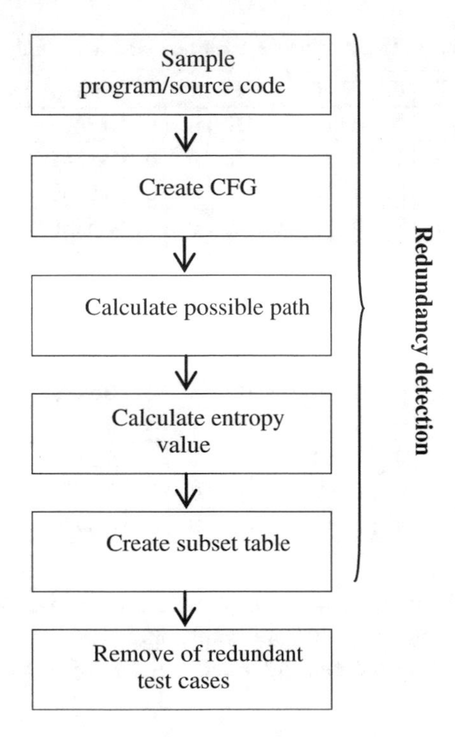

Fig. 2. Proposed approach model

5 Result and Discussion

We perform an experimental study from the example in Panday [7] to test the usage of entropy which detect and remove the redundancy of test cases. Table 2 shows all possible paths from a program. Test suite, $TS = t_1, t_2, t_3, \ldots, t_{13}$ are generated from CFG. Using the path and test cases, an entropy value for each test case is calculated. For example, the entropy value for each test case is calculated by using entropy in Eq. (1):

$$(t_1 - t_{13}) = -\left[\left(\frac{1}{8}\right)\log_2\left(\frac{1}{8}\right) + \left(\frac{7}{8}\right)\log_2\left(\frac{7}{8}\right)\right] = 0.5436 \tag{2}$$

$$t_2 = -\left[\left(\frac{3}{8}\right)\log_2\left(\frac{3}{8}\right) + \left(\frac{5}{8}\right)\log_2\left(\frac{5}{8}\right)\right] = 0.9544 \tag{3}$$

$$t_{11} = -\left[\left(\frac{2}{8}\right)\log_2\left(\frac{2}{8}\right) + \left(\frac{6}{8}\right)\log_2\left(\frac{6}{8}\right)\right] = 0.8113 \quad (4)$$

The entropy value of each test case is shown in Table 2.

Table 2. Entropy value for each test case according to path

	P1	P2	P3	P4	P5	P6	P7	P8	Entropy value
t_1	1	0	0	0	0	0	0	0	0.5436
t_2	0	1	1	1	0	0	0	0	0.9544
t_3	0	0	0	0	1	0	0	0	0.5436
t_4	0	0	0	0	0	0	0	1	0.5436
t_5	0	0	0	0	0	0	0	1	0.5436
t_6	0	0	0	0	0	0	0	1	0.5436
t_7	0	0	0	0	0	1	0	0	0.5436
t_8	0	0	0	0	0	1	0	0	0.5436
t_9	0	0	0	0	0	0	1	0	0.5436
t_{10}	0	0	0	0	0	0	1	0	0.5436
t_{11}	1	0	0	0	0	1	0	0	0.8113
t_{12}	0	0	1	0	0	0	0	0	0.5436
t_{13}	0	0	0	1	0	0	0	0	0.5436

As the example calculation, low entropy value indicated that test cases have more similarity or redundant and high entropy value indicated no similarity or redundant among test cases. By using the entropy value, subsets of test case are identified and shown in Table 3.

Table 3. Entropy value and subset of test cases

Path	Subset	Entropy value
P1	S1 = $\{t_1, t_{11}\}$	t_1 = 0.5436, t_{11} = 0.8113
P2	S2 = (t_2)	t_2 = 0.9544
P3	S3 = $\{t_2, t_{12}\}$	t_2 = 0.9544, t_{12} = 0.5436
P4	S4 = (t_2)	t_2 = 0.9544
P5	S5 = (t_3)	t_3 = 0.5436
P6	S6 = $\{t_7, t_8, t_{11}\}$	t_7 = 0.5436, t_8 = 0.5436, t_{11} = 0.8113
P7	S7 = $\{t_9, t_{10}\}$	t_9 = 0.5436, t_{10} = 0.5436
P8	S8 = $\{t_4, t_5, t_6\}$	t_4 = 0.5436, t_5 = 05.436, t_6 = 0.5436

Next, the singleton set S2, S4, S5 are identified and put t_2, t_3 in primary test set T_p. All related paths and subset that have these test cases are marked P1, P2, P3, P4, P5.

Next, test cases with two elements from subsets are identified from unmarked subsets which are t_1, t_{11} and t_9, t_{10} from S1 and S7. Test cases with the higher entropy value are selected. Therefore, from S1, t_{11} is selected. In S7, both have equal entropy value, so by random choice, t_9 is selected. Then, all related paths that have these test cases are marked P1, P6, P7.

The same process is repeated for the remaining unmarked subset S8 and all test cases in S8 have equal value and by random choice, t_4 is selected.

Finally, the reduce test suite $T_p = t_2, t_3, t_4, t_9, t_{11}$ is created. Compared to the original test suites, we observed that the average reduction is 61% of test cases.

6 Conclusion and Future Work

This paper presents an approach to detect the redundancy of CFG based test cases using entropy. We have proposed the use of entropy to analyze the entropy value of each test case. We implement entropy value to analyze the redundant test cases in CFG and the result shows that this approach also can reduce test cases because it produced the same result as an approach by Panday [7]. This approach is tested in simple size program and manually analyzed. These are preliminary results and need a lot of extensive research to prove the approach is a good technique to reduce test cases, while cover the path selection criteria such as statement coverage, branch coverage and path coverage. It is to show that, however we success to reduce redundancy of test cases, but we have considered to fulfill the coverage criteria. The proposed approach offers several avenues for future research. First, this paper is only considered for one example, and we plan to compare the result with other methodologies. Then, we plan to evaluate the performance of our approach in presence of coverage criteria. Finally, our current experiment or case study of test suite needs to be improved. Thus, future work will focus on how to prove that entropy can be an alternative approach in detecting and reducing the test cases in other more program code with the result of coverage criteria. Thus, all these works motivated us to do some new works in this area.

Acknowledgments. This paper was funded by Office for Research, Innovation, Commercialization and Consultancy Management (ORICC), UTHM.

References

1. Singh, R.: Test case generation for object-oriented systems: a review. In: Fourth International Conference on Communication Systems and Network Technologies (CSNT), Bhopal, pp. 981–989 (2014)
2. Zeng, B., Tan, L.: Test criteria for model-checking-assisted test case generation: a computational study. In: IEEE 13th International Conference on Information Reuse and Integration (IRI), Las Vegas, NV, pp. 600–607 (2012)
3. Liu, Y., Li, Y., Wang, P.: Design and implementation of automatic generation of test cases based on model driven architecture. In: Second International Conference on Information Technology and Computer Science (ITCS), Kiev, pp. 344–347 (2010)
4. Kumar, G., Bhatia, P.K.: Software testing optimization through test suite reduction using fuzzy clustering. CSI Trans. ICT **1**, 253–260 (2013)

5. Saif-ur-Rehman, K., Nadeem, A., Awais, A.: TestFilter: a statement-coverage based test case reduction technique. In: Multitopic Conference INMIC 2006, Islamabad, pp. 275–280 (2006)
6. Chaurasia, V., Thirunavukkarasu, K.: Test exude: approach for test case reduction. IOSR-JCE **16**, 16–23 (2014)
7. Panday, A., Gupta, M., Singh, M.K., Ali, N.: Test case redundancy detection and removal using code coverage analysis. MIT Int. J. Comput. Sci. Inf. Technol. **3**, 6–10 (2013)
8. Hirsh, J.B., Mar, R.A., Peterson, J.B.: Psychological entropy: a framework for understanding uncertainty-related anxiety. Psychol. Rev. **119**, 304–320 (2012)
9. Herbold, S.: Usage-based testing of event-driven software (Doctoral Dissertation), Niedersächsische Staats-und Universitätsbibliothek Göttingen (2012)
10. Filho, R.S.S., Budnik, C.J., Hasling, W.M., McKenna, M., Subramanyan, R..: Supporting concern-based regression testing and prioritization in a model-driven environment. In: IEEE 34th Annual on Computer Software and Applications Conference Workshops (COMPSACW), Seoul, pp. 323–328 (2010)
11. Singh, N.P., Mishra, R., Yadav, R.R.: Analytical review of test redundancy detection techniques. Int. J. Comput. Appl. **27**, 30–33 (2011)
12. Harrold, M.J., Gupta, R., Soffa, M.L.: A methodology for controlling the size of a test suite. ACM Trans. Softw. Eng. Methodol. (TOSEM) **2**, 270–285 (1993)
13. Jones, J., Harrold, M.J.: Test-suite reduction and prioritization for modified condition/decision coverage. IEEE Trans. Softw. Eng. **29**, 195–209 (2003)
14. Ruhe, G.: A systematic approach for solving the wicked problem of software release planning. Soft Comput. **12**, 95–108 (2008)
15. Koochakzadeh, N., Garousi, V.: A tester-assisted methodology for test redundancy detection. Adv. Softw. Eng. (2010)
16. Ilkhani, A., Abaee, G.: Extraction test cases by using data mining; reducing the cost of testing. In: International Conference on Computer Information Systems and Industrial Management Applications (CISIM), Krackow, pp. 620–625 (2010)
17. Mohapatra, S.K., Prasad, S.: Finding representative test case for test case reduction in regression testing. Int. J. Intell. Syst. Appl. **11**, 60–65 (2015)
18. Bailey, K.D.: Entropy system theory. systems science and cybernetics. In: Encyclopedia of Life Support Systems (EOLSS). Eolss Publishers, Oxford (2001). Developed under the Auspices of the UNESCO
19. Maung, H.M., Win, K.T.: Entropy based test cases reduction algorithm for user session based testing. In: Ninth International Conference on Genetic and Evolutionary Computing, Yangon, Myanmar, pp. 365–373 (2015)
20. Deris, M.M., Abdullah, Z., Mamat, R., Yuan, Y.: An attribute selection using similarity limited tolerance relation for incomplete information systems. In: Web Proceeding on ICT Innovation (2015)

Variety of Approaches in Self-adaptation Requirements: A Case Study

Aradea[1(✉)], Iping Supriana[2], Kridanto Surendro[2], and Irfan Darmawan[3]

[1] Department of Informatics Engineering, Faculty of Engineering, Siliwangi University, Tasikmalaya, Indonesia
aradea.informatika@gmail.com
[2] School of Electrical Engineering and Informatics, Bandung Institute of Technology, Bandung, Indonesia
[3] Department of Information System, School of Industrial and System Engineering, Telkom University, Bandung, Indonesia

Abstract. Self-adaptation requirements are requirements engineering studies to develop self-adaptive systems. This approach provides a way how activity at design-time requirements to meet stakeholder needs and system-to-be. Currently, there is a variety of approaches were proposed to the researchers through the development of goal-oriented requirements engineering. The ideas expressed through the expansion of this model into a way that is quite promising, however the various approaches proposed, does not mean no shortage. This paper describes in detail the variety of approaches available today through the implementation of a case study, and analysis of the results, we found 5 main features that can be used as consideration in formulating self-adaptation requirements, namely goal concept, environment model, behavior analysis, run-time dependencies, and adaptation strategy. Besides that, we saw of future research chance through deep study at goal-based modeling and loop feedback with utilizing data mining technique.

Keywords: Self-adaptive systems · Adaptation requirements · Goal oriented requirements engineering

1 Introduction

Requirements engineering approach used for the development of self-adaptive systems (SAS), has the form of a different approach to the traditional requirements engineering, which only represents the understanding of the problem domain requirements at design-time. In SAS, attention to changes in requirements that may occur at run-time, a problem that must be anticipated and determined handling solutions. In general, the model of goal-oriented requirements engineering (GORE) widely adopted as the basic concept to develop an alternative solution to the issue. This model was expanded through the establishment of requirement specifications are prepared to requirements at design-time and run-time. Design-time specification is realized through various proposals to create monitoring mechanisms and adaptation, while the run-time

T. Herawan et al. (eds.), *Recent Advances on Soft Computing and Data Mining*,
Advances in Intelligent Systems and Computing 549, DOI 10.1007/978-3-319-51281-5_26

specification is represented through the implementation of the mechanism of dynamic systems, for example through reconfiguration and evolution solutions. Survey and preliminary study of the research topic have been discussed in our previous paper [1], and we conclude importance of feature specification determining in more detail from both those design-time and run-time specification, to gain pattern more clearly in formulating self-adaptation requirements.

In the discussion of this paper, we develop a case study approach is applied to the four previous researchers then evaluated to identify opportunities perfected. The main objective of this paper is to focus on the discussion of the stages of modeling requirements, and the proposed extension of the findings of the case study discussion. Discussion begins with a case study (Sect. 2), four approaches adaptation requirements (Sect. 3), proposed the expansion and future work (Sect. 4), and conclusion (Sect. 5).

2 Case Study: Goal Model

The case studies developed as an illustration of the adaptation needs of the lecture at the college. Where there are two actors of the system (the classroom system and the majors system) representing the needs of lecturer, students, and majors. The system must be able to ensure that the activities of the lecture can be held as scheduled, the number of sessions to be in the range of 14 sessions – 16 sessions. If the range is not met, then the replacement schedule should be established, it relates to the willingness of lecturer time and space.

Figure 1 illustrates the modeling of the case through Tropos models [2], in which the actor classroom system has a major goal “lecture monitoring” which can be achieved through one of its sub-goal. Goal “organizing the lecture” can be achieved if the activity of lectures “start on scheduled”, “completed on schedule” as well as the lecturer and students “filling of attendance”, this may be a achieved through the plan “monitoring” and “check the numbers of lectures”, which can contribute to positive/full satisfaction (++) against soft goal “availability of classroom” and “lecture targets”. While the plan for achieving the goal “cancelation of lecture” has contributed partial negative (-) against both soft goal, considering the cancelation of the lecture could adversely affect the achievement of the lecture and meeting classroom availability, but will not influence negatively if the determination of the replacement schedule can be achieved. Meanwhile, if the goal of “organizing the lecture” is not achieved, then the scenario that was developed was a “determining a replacement schedule” by delegating goal “choose schedule” to the majors system actor. This goal can be achieved by doing “collect time-lecturer”, “find free classroom”, “choose the schedule”, and “confirmation message”.

3 Adaptation Requirements Modeling

In this part of the case study mapped the standpoint of adaptation requirements, based on the four works most closely with the research that is being done.

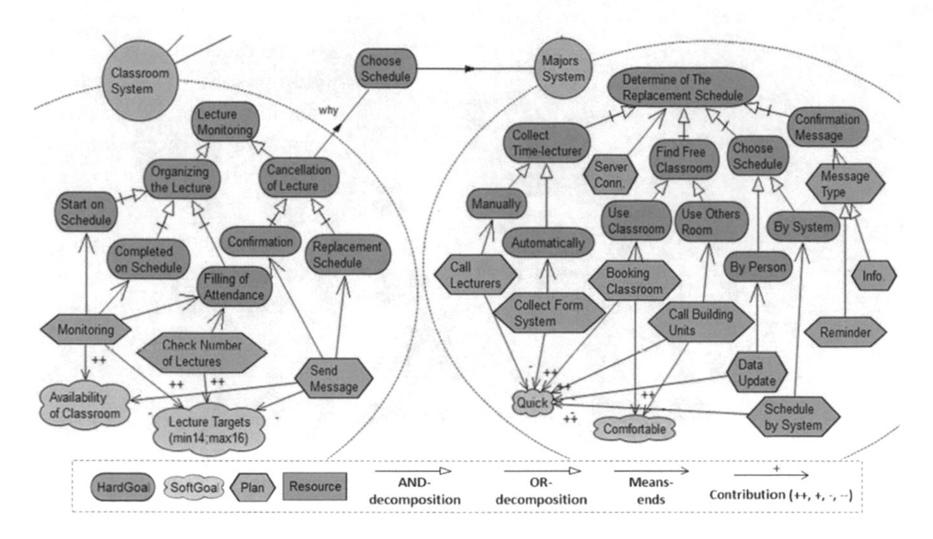

Fig. 1. Modeling of system monitoring and scheduling college tuition replacement.

3.1 Tropos for Adaptive Systems (TROPOS4AS)

Tropos4AS proposed by Morandini et al. [3, 4], the approach is to expand the model Tropos [2] by introducing (a) the model type of goal and conditions of satisfaction, (b) the model environment that will affect the satisfaction of a goal that must be monitored, (c) a model of failure through the selection of alternative behavior when the goal should be achieved not met. Figure 2 shows the modeling of lecture monitoring. Satisfaction goal adapted into three types of goal, namely (a) achievement (A), namely, when certain circumstances is reached or reached the state at a time, (b) maintain (M) is the need to maintain a specific state in a period of time or the state from time to time, (c) perform (P) is successfully implementing an activity or one of the model (plan or subgoal). Meanwhile, to represent the run-time dependencies between goals, realized through, (a) <<sequence>>, for example, the goal "began on schedule" is reached before activating the goal "is completed on schedule", (b) <<inhibits>>, for example goal "completed on schedule" cannot be active as long as the goal "fill out absentee news show" active.

Environmental models represented into a UML class, and can provide functionality to feel and act as an evaluation of a condition, for example, associated resources (database/schedule), or system device used (sensors in the classroom). Each goal and plan to connect into artifacts environment through conditions of state transition, for example (a) precondition, namely the goal or plan can only be activated if true, (b) Creation Condition, which was to determine the criteria to activate the goal or starting the process of satisfaction goals, please note that the sub-goal activated implicitly through decomposition, such as "creation-c: 09:00 AM", "creation-c: 11:00 AM", "creation-c: max 15 min before the end", (c) Achieve Condition, which determines when the goal is reached so that these goals will be dropped, like "achieve-c: course achieved", (d) Failure Condition, which shows the situation where it is impossible to achieve a goal, such as "failure-c: > 10 min of activation". In the

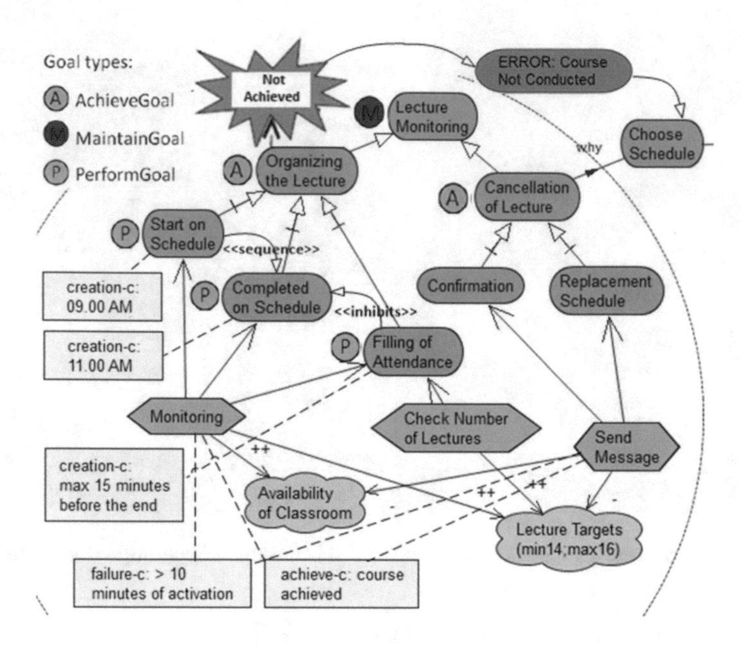

Fig. 2. Modeling of Tropos4AS in lecture monitoring.

scenario developed, if such failure occurs, the system will change the plan to give a warning to the lecturer, and if the goal remains "not achieved", it will set a replacement schedule, by delegating goal "choose the schedule" to the major actor.

3.2 Adaptive STSs (Socio-Technical Systems)

This work was developed by Dalpiaz et al. [5], which defines the need for adaptation of a socio-technical system (STS). This model proposes architecture based approaches to requirements so that the STS becomes a self-reconfigured. Architecture that was developed to expand the model Tropos [2], the concept utilizes UML 2.0 component diagrams to show the architectural components and connections, which is determined by the cycle MDRC (monitoring, diagnosis, reconcile, compensation). Model goals and requirements of stakeholders to express dependencies actor and multi-agent plan to meet the requirements, while the BDI paradigm to guide diagnosis and selection plan for each actor, and a compensation mechanism exploited to handle failure. Interaction among actors is supported by elements of context sensors, agent, and context actuators.

The behavior of the model in Fig. 3. using the goal models, context models, plan specifications, and domain assumptions (core ontology), which together capture system requirements. Model expressed for use at run-time, through (a) Context, a partial state which is relevant to the status and intention actor, which connects the context of variation points, such as OR-decomposition to G1 including contextual decomposition link for G2, the achievement of these goals is necessary to achieve G1 only if context C1 implemented. While the G3 become a legitimate alternative, only if C2 implemented.

In addition C3, C4 and C5, a context that applies to AND-decomposition to G2. (b) Activation consists of triggering event rules and precondition. A goal that is activated when a trigger event occurs, and the precondition implemented, for example, activation rules for top-level goals include G3 class system is activated as send messages automatically, and G4 delegated to the actor of majors if C2 implemented. While the G5, G6, G7 activated as the send a warning message before the specified time or exceeds the time. (c) Declarative goals are goals that are met only if the achievement of the conditions is met, the satisfaction of these goals independent of satisfaction sub-goal or plan that is connected by means-end decomposition. Achievement of the above conditions is expressed as state context models, such conditions are likely to G4 achievement is to determine the replacement schedule. This goal declarative face of uncertainty, for example, the collection goal lecturer time and find an empty room that had an OR-decomposition. (d) Time limits define the maximum amount of time that an agent can achieve a goal (goal timeouts), for example, G3 must be reached within 60 min prior to the time schedule of lectures, while the G5, G6, G7 must be achieved no more than 10 min from the time schedule. (e) The plan is a collection of actions connected with the goals that were executed by agents, each action performed correctly if post condition achieved within a certain time limit, and at that time precondition done, if not, then the plan includes failed.

3.3 Adaptive Requirements Modeling Language (ARML)

ARML approach [6, 7] model each goal and plan to use a domain ontology, through inference rule are representing. The models were developed using the Techno language through the addition of a new concept (context and resources), with two rules of

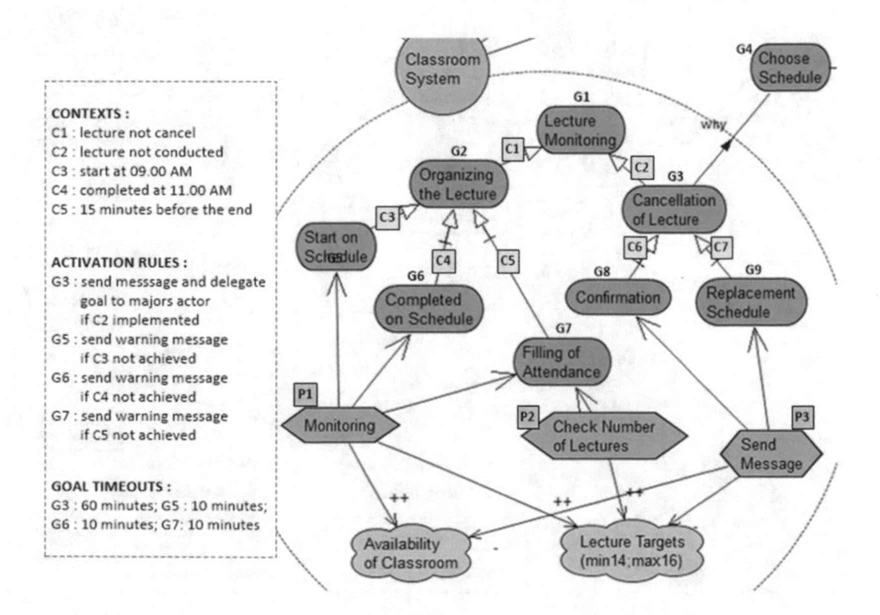

Fig. 3. Modeling of Adaptive STSs in lecture monitoring

relations, namely relegation and influences. In addition, a requirement can be mandatory (M) or optional (O) through inference relations and conflict, to represent how the element of satisfaction can affect the satisfaction of others.

In Fig. 4, the goal "determine of the replacement schedule" is modeled as a mandatory nodes (M node, an unary relation), decomposed through inference relations to mandatory goal, namely the collect time lecturers, find free classroom, choose schedule (black I node, binary relation) as the fact that the goal of "determining of the replacement schedule" will be met through mutual satisfaction of the three goals. Quality constraints (done if the number of lecture < 14 sessions) is placed in inference relation. Influence relations between the three decomposition occurs goal, to describe the context of the prevailing conditions and the availability of resources that may affect the achievement of the "choose schedule" goal.

Requirements analysis for the "manually" goal, connected through inference node (I) which is decomposed into several sub-task as a candidate solution. Each candidate's solutions include mobile phone resource and the domain assumptions (lecturers have a mobile phone and a laptop), connected by symbols preferences are used to compare the requirements in the candidate solutions, for example, send SMS preferable to send the email. While other issues, for instance, related to the lecturer does not have access so did not get the message, then this may be identified through the relegation relation, this

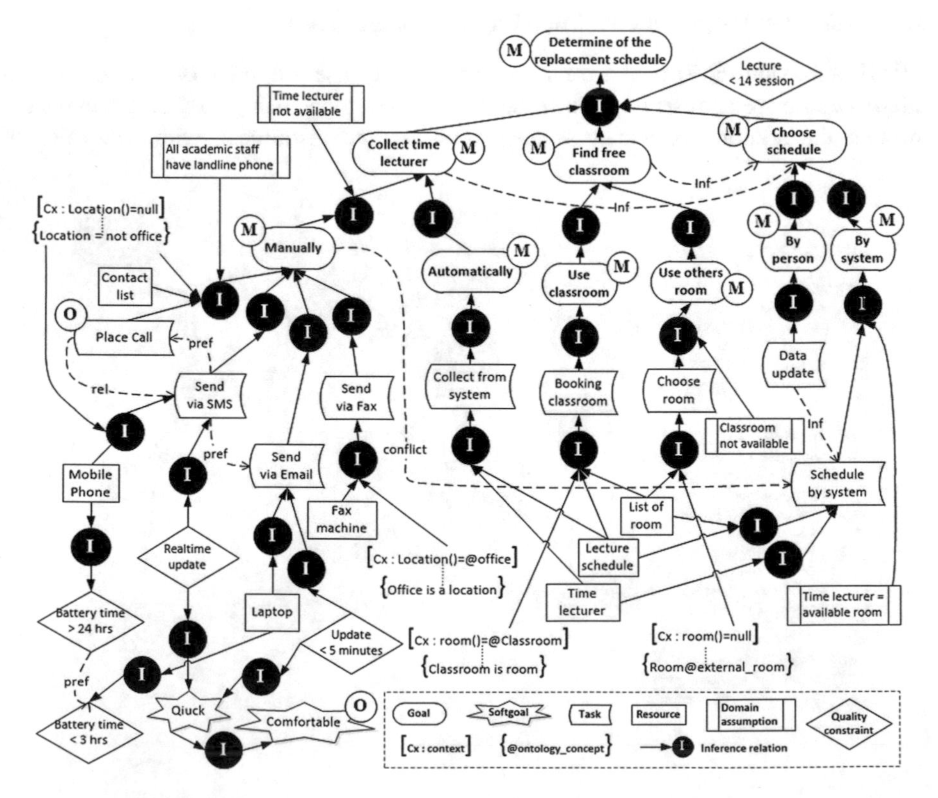

Fig. 4. Modeling of ARML in college scheduling of a replacement

condition makes it possible to take into account situations where a lecturer context changed, so did not get the message. In determining the choice of solution, this can be done through the plan "place call" with the domain assumptions e.g. academic staff have a land-line phone and a contact list resource. Here, the candidate solution is rated and evaluated by quality constraints.

3.4 ZANSHIN Framework

Zanshin framework [8, 9] consists of two approaches, namely awareness requirements (AwReqs) and evolution requirements (EvoReqs). AwReqs an indicator in the convergence of run-time requirements, or constraint for another state requirements (goal elements). While EvoReqs evolve the model automatically, in response to AwReqs failure to determine the change requirements when certain conditions apply. Modeling begins by identifying AwReqs, through the perspective that the system is able to adapt and the importance of determining the level of failure can be tolerated. In Fig. 5, there are seven AwReqs (AwReqs-1 to AwReqs-7). AwReqs also can represent trends success rate requirements, for example, AwReqs-7 empty classroom should not always be available, if at a certain time classroom situation really busy. So that the requirements (for domain assumptions) are not always treated as invariants that must always be accomplished (or should always true). It means that the system may fail to achieve one of the initial requirements (or assumptions become incompatible). Therefore, through a feedback loop provided a way to determine the level of each critical requirements, and monitor the system so aware of the failure. After AwReqs obtained, identification parameters (variation points (VP) and the control variable (CV)), which in the event of a change can have an impact on the indicator (AwReqs) relevant.

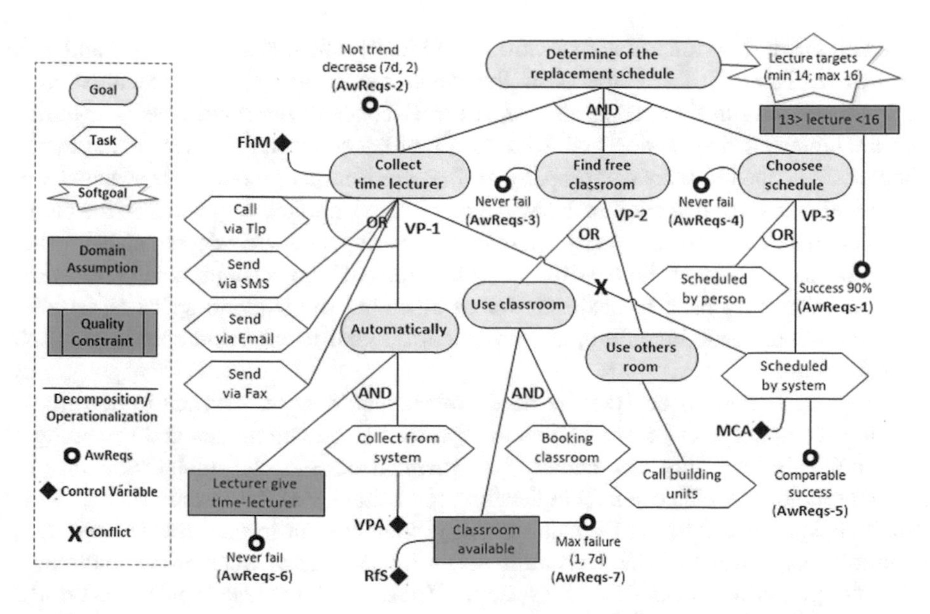

Fig. 5. Modeling of AwReqs. and EvoReqs. in college scheduling of a replacement.

In the case, for example if there is a negative trend in the success rate of collect time lecturer for 2 weeks (AwReqs-2: 7 days (d), 2), such as the agenda of other work, it can be tolerated at most occurs 3 times per semester, thus it (relaxing) reducing constraints to 3 weeks. So variable FHM (From how Many), determine how much time the willingness of lecturer to be collected, changes in the value of these variables can affect the goal itself and the satisfaction of other requirements became clear. Then if classrooms are not available (AwReqs-7: 1, 7d), the system cannot determine a new room, then it does is increase the RFS (Rooms for schedules). As another example, if we know the domain assumptions willingness time lecturers (AwReqs-6) is invalid, then replace (replace it) with other tasks that would verify the validity of the use of willingness time lecturers, it is associated variables VPA (View Private Appointments) that can be worth a yes or no. MCA (Maximum Conflicts allowed) variable associated with (AweReqs-5: Comparison of success rate), requires the system to choose the date on which the number of scheduling conflicts does not exceed the value specified in this variable.

After identifying the parameters that influence AwReqs, then modeling the impact of the differential relations, for example, Δ(AR8 /RFS) [0, maxRooms] > 0, represents the fact that by increasing the number of classrooms, then the domain assumptions classrooms are available to be fulfilled more often. Figures in brackets indicate the interval in which this relationship applies (maxRooms: the qualitative value that should, and can be replaced by a corresponding amount). Then after doing modeling impact activities for each pair of indicators and parameters individually, activities for the improvement of the impact analysis of the indicator to a specified combination, whether the cumulative and if possible, the effects are sorted from the largest to the smallest.

4 Discussion and Future Work

Based on the description of the four works, we identify the essential elements and main features in Table 1. Two aspects of the basic needs, namely the specification of design-time and run-time, consisting of, (a) goal concept: the requirements specification are represented as models goal with details of the concept, (b) environment model: concept developed to represent property system environment (context) associated with the model goals, (c) behavior analysis: the mechanism applied to determine the choice of system behavior to meet every goal, (d) run-time dependencies: dependency artifacts requirements at design time with the operation of the system during run-time, (e) adaptation strategies: mechanisms for determining the solution space as an alternative solution (reconfiguration), and modification goal to represent space new problems (evolution).

Tropos4AS able to analyze the needs of stakeholders and system-to-be through defining the concept of its goal, so as to represent the requirements and reasoning at run-time. This capability can actually be enhanced through the implementation of a domain ontology, which can help in detailing the behavior of the system, such as in the Adaptive STSs and ARML. Domain ontology management in goal models requires a specific strategy if the model decision in selecting candidates wants more solutions to fulfill the properties of self-adaptive systems. Zanshin offers these capabilities through

Table 1. Comparison of self-adaptation requirements specification.

Model	Design-time			Run-time	
	Goal concept	Environment model	Behavioral analysis	Run-time dependency	Adaptation strategy
Tropos 4AS (2011, 2015)	Goal, softgoal, plan, resources, goal type (achieve, maintain, perform)	UML Class, Condition relation (pre, context, creation, achieve, failure)	Variability design for failures model (goal satisfaction)	Transition rules, run-time goal state (<<sequence>> <<inhibits>>)	Inference rules: L/R [rule-name], elicitation and recovery model for failure
Adaptive STSs (2013)	Goal, softgoal, plan, resources, domain assumptions	Context model (context sensor, agent, context actuator), plan specification	Reconcile and compensation strategy, cost and contribution to soft-goals	Activations rules, context, time limits, declarative goals, plan	MDRC, DLV-complex reasoner, spesific thread for failure model
ARML (2011, 2012)	Goal, task, quality constraints, domain assumptions	Domain ontology, context concept, relation concept, resources	Linking domain ontology to goal model, and ECA rules modeling	Inference relation, conflict, preferences, relegation, influence rels	Run-time requirements artifact, SALmon, CARE App., ECA rules
Zanshin (2012, 2013)	Goal, task, quality constraints, domain assumptions	AwaReqs. in goal models, parameter: variation points, control variable	Diferential relations (Δ (AwReqs-"n"/ CV) > 0), ECA rules modeling	EvoReqs., controller and system target dependency, MAPE loop	Qualitative reasoning, and diferential relations, ECA rules

the centralization of feedback control loop, but related domain model that represents the problem domain as requirements, have not been included. While Tropos4AS have this capability, in addition to the ability of the high variability and the concept of independence, so that integration Tropos4AS and Zanshin to the attention of our work in the future.

Adaptive STSs also proposed a self-adaptation capabilities similar to Zanshin, based on the architecture approach of a model driven requirements, but through a compensation strategy for the behavior of multi-agent systems. While the new Tropos4AS developing behavior based on the single-agent system. Comparison of both the architectural concept is also our concern. When viewed from the run-time requirements specification, Tropos4AS and Adaptive STSs ability to realize the goal through agent executable models, while Zanshin and ARML using ECA rules as primitive operations on the model of the goal. Thus, we assume, design-based independent software agents can exploit human-oriented abstractions such as agent and goals so that construction of this language suitable for representing real-world requirements and reasoning at run-time. While the centralized feedback loop approach can actually be integrated to enhance the adaptability functions at run-time, through the development of data mining algorithm.

5 Conclusion

Self-adaptation requirements is a concept that can address the needs in translating real-world conditions, related to the diversity of the elements involved and the anticipation of amendments. The four works that are discussed in this paper enough to give an idea of the variety offered alternative solutions to achieve self-adaptive requirements capability. Based on these studies, we see two main strengths that are considered to be a great opportunity to leverage the power of self-adaptation requirements, namely (a) the goals oriented approach through Tropos/i * model, can capture and represent the variability in context and behavior of the system (domain models), as the concept of requirements, (b) models of dynamic systems through a feedback loop, can be developed to establish assurance criteria for management system and mechanisms of adaptation, as the concept of self-adaptation. Based on these assumptions, our next job is to formulate a formal framework that can bridge the integration of both approaches, and how the data mining technique could facing uncertain environment at run-time, through determining inference context algorithm based model of sensor and its data history.

References

1. Aradea, D., Supriana, I., Surendro, K.: Roadmap dan area penelitian self-adaptive systems. Prosiding Seminar Nasional Teknik Informatika dan Sistem Informasi (SeTISI), Universitas Maranatha Bandung (2015)
2. Bresciani, P., Perini, A., Giorgini, P., Giunchiglia, F., Mylopoulos, J.: TROPOS: an agent-oriented software development methodology. J. Auton. Agent. Multi-agent Syst. **8**(3), 203–236 (2004)
3. Morandini, M.: Goal-oriented development of self-adaptive systems. Ph.D. thesis, University of Trento (2011)
4. Morandini, M., Penserini, L., Perini, A., Marchetto, A.: Engineering requirements for adaptive systems. J. Requirements Eng. **21**, 1–27 (2015). Springer
5. Dalpiaz, F., Giorgini, P., Mylopoulos, J.: Adaptive socio-technical systems: a requirements-based approach. J. Requirements Eng. **18**(1), 1–24 (2013)
6. Qureshi, N.A.: Requirements engineering for self-adaptive software: bridging the gap between design-time and run-time. Ph.D. thesis, University of Trento (2011)
7. Qureshi, N.A., Jureta, I.J., Perini, A.: Towards a requirements modeling language for self-adaptive systems. In: Regnell, B., Damian, D. (eds.) REFSQ 2012. LNCS, vol. 7195, pp. 263–279. Springer, Heidelberg (2012). doi:10.1007/978-3-642-28714-5_24
8. Souza, V.E.S.: Requirements-based software system adaptation. Ph.D. thesis, University of Trento (2012)
9. Souza, V.E.S., Lapouchnian, A., Angelopoulos, K., Mylopoulos, J.: Requirements-driven software evolution. J. Comput. Sci. Res. Dev. **28**(4), 311–329 (2013). Springer

Dynamic Trackback Strategy for Email-Born Phishing Using Maximum Dependency Algorithm (MDA)

Isredza Rahmi A. Hamid(✉), Noor Azah Samsudin, Aida Mustapha, and Nureize Arbaiy

Faculty Computer Science and Information Technology, University Tun Hussein Onn, Parit Raja, Johor, Malaysia
{rahmi, azah, aidam, nureize}@uthm.edu.my

Abstract. Generally, most strategy prefers to use fake tokens to detect phishing activity. However, using fake tokens is limited to static feature selection that needs to be pre-determined. In this paper, a tokenless trackback strategy for email-born phishing is presented, which makes the strategy dynamic. Initially, the selected features were tested on the trackback system to generate phishing profile using Maximum Dependency Algorithm (MDA). Phishing emails are split into group of phishers constructed by the MDA algorithm. Then, the forensic analysis is implemented to identify the type of phisher against already assumed group of attacker either single or collaborative attacker. The performance of the proposed strategy is tested on email-born phishing. The result shows that the dynamic strategy could be used for tracking and classifying the attacker.

Keywords: Phishing · Trackback strategy · Forensic · Maximum Dependency Algorithm · Clustering

1 Introduction

Organized crime is defined as illegal activities done by collaborative links which usually pretend to be legal businesses. The escalating number of organized crime causes a vital risk to financial organization and has been studied by many researchers from several grounds of taught [1–5]. According to U.S national incidence report, 1 in 90 users lost their money due to phishing which totaled to $483 million between year 2004 to 2009 [6]. Generally, highly income returns are the key motivation of the organized crime to launch attack. Though, some of them may also be inspired by their group ideology and political view. Moreover, the continuous development of electronic commerce opens new insight for the criminals to exploit free resources toward establishing their illegal activities. Due to this, cybercrime cause modest distress to Internet user thru phishing attack by luring their victims to a bogus website and deceived them to reveal credential information.

At present, tracking the phisher who uses identity concealment techniques is very challenging. Phisher always covers their tracks by impersonating using fake identities,

T. Herawan et al. (eds.), *Recent Advances on Soft Computing and Data Mining*,
Advances in Intelligent Systems and Computing 549, DOI 10.1007/978-3-319-51281-5_27

redirecting users to another server or modify the email header information. In this paper, a framework of trackback phishing emails based on clustering and similarity measurement approach to track email-born phishing activities is proposed. Current clustering phishing attack related research usually concentrated on the structural of email properties based on significant features such as hyperlink [5, 7, 8] and Domain Name Server (DNS) [9, 10]. However, we follow a different approach. We tested for email's header and email's body dataset. The key ideas are to track the phishing activity in email and profile type of phisher either single or collaborative attacker. In the proposed trackback framework, we apply clustering techniques using Maximum Dependency Algorithm (MDA) presented in [11, 12] to generate group of phishers from selected feature vector. Phishing emails are clustered based on phisher email's body features. Email's body-based feature can be extracted from characteristics of email where phishers normally have their own signatures [4, 13, 14]. Therefore, a phisher's class can be expected to show a collection of different activities. Ten-fold cross validation is used to evaluate the precision of phisher group that have been identified.

The remainder of this paper is organized as follows. Section 2 describes related research regarding phishing trackback approaches proposed in recent year. Section 3 examines the phishing email's body-based feature set used in the experiment and forensic filtering algorithms as well. Section 4 gives the performance analysis result and the effectiveness of the proposed forensic trackback approach. Section 5 concludes the work and direction for future work.

2 Related Work

Recently, the concept of tracking phishing email using honeypot to track the phisher activities have been discussed in [3, 9, 15, 16]. Honeypots is a trap set to detect unauthorized access by using fake token to bait the attacker. Therefore, it is limited to detect activity that interacts with the fake token. In other words, the system cannot capture the attack activity against the systems unless the attacker interacts with the honeypot.

Chandrasekaran et al. [15] submitted false credential to phishing sites as phoneytokens. The main idea is to identify phishing sites based on the response of fake input. The PHONEY prototype lies between a user's Mail Transfer Agent (MTA) and Mail User Agent (MUA) where it processes each arriving email for phishing attacks. They tested on 20 different phishing emails focusing on Uniform Resource Locator (URL) and form features. However, many phishing email attacks are created without form and hyperlinks which show that this technique still have flaws in classification. Our work differs from [15] in such a way that we focused on structural of email for forensic analysis and did not use any phoneytoken to track phishing activity.

Gajek et al. [3] discussed forensic framework for profiling and tracing phishing activity in phishing network. The main idea is to populate their database with fingerprinted credentials (phoneytokens) which could lure phishers to a fake system that simulate the original service. Then, phoneypot pretends to be the original service in order to profile phishers' behaviour. This approach causes phishers to spend more time and resource to acquire financial benefit and increases the risks to track phishers.

They are interested in tracing the phisher's agents and not in the technical means used by phishers. Our approach is different where we did not use any phoneytoken. Work by [3] derived three classes of phishing profiles; non-phisher, definite phisher and potential phisher. Meanwhile, we categorized phisher into two types; single attack and collaborative attack. This paper proposed a new trackback strategy that includes clustering the phisher's activity and conducting forensic analysis on the produced cluster. The ultimate aim is to distinguish between single and collaborative attack.

3 Phishing Email Trackback Approach

A high level description of the implementation of tracking phishing email is shown in Fig. 1. The first phase is to perform feature selection approach to select a subset of relevant features to be used in the construction of the model. The feature selection approach has been discussed in [13]. The most informative features are selected using a learning model and a classification algorithm. Then, the dataset is split into selected training and testing ratios. The training data was used to train the clustering algorithm while the testing data to estimate the error rate of the training classifier. The next phase is profiling the attacker using the clustering algorithm. Based on the training data of the feature selection phase, the profiling algorithm phase is generated to create phishing profiles. The profile produced is used to train the classification algorithm where it is supposed to forecast the unidentified class label into normal or phishing email [17].

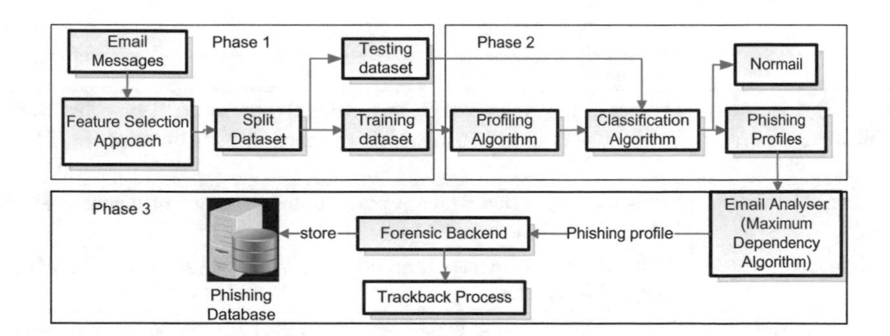

Fig. 1. Trackback process

The trackback framework based on clustering algorithm and similarity measurement is proposed in this paper. The first step is to extract email features from the collection of the phishing email. In this step, the phishing email is already selected by the profiling phase. The next step is the email analyzer phase. We propose different phishing trackback model by implementing clustering algorithm using the phishing email as feature vectors. Then, we generate the group of attacker based on clustering prediction. This group of attacker is sent to the next step which is forensic backend. We then applied similarity measurement in forensic backend step to smooth the phishing group in order to categorize the attacker into single or collaborative attacker.

The experiment focused on email's body features only. Eight email's body-based features were selected to evaluate the proposed strategy.

3.1 Feature Extraction and Selection

In this section, feature set description and feature selection approach are explained.

3.1.1 Feature Set Description

We extracted eight email's body feature from the Nazario phishing email dataset [18] as listed in Table 1. Phishers have diverse ways of attack. In one circumstance, phishers may lure their victim by inserting images, malicious file or link in the form which can safely detour anti-phishing techniques. When the user clicked the hyperlink, this will lead them to a bogus site. We believed that the phisher's profiles should be able to distinguish between different groups [17, 19]. Different group of phisher might insert a different fake link which will redirect the user to a phishing site. Others may send email from different email server or different attachment file. Thus, we want to identify groups based on email's body which could be extracted from the emails characteristics. These characteristics are then reflected as features that will correspond to the proposed trackback framework. Based on the characteristics collected, the data clustering approach would provide the relationships between the features in the phishing emails and their pre-specified classes.

Table 1. Body-based features

Features	Description
Body (8) Subjectblacklist (BSB)	Defines binary value which represents the presence of blacklist words in the subject of an email which included in bags of words in [94]
Hypertextblacklist (BHB)	Defines binary value which represents the presence of blacklist words in hypertext at the email's body
Htmlcontent (BHC)	Define if the emails have html content. Value '1' if the email contains html, '0' otherwise
Textcontent (BTC)	Defines binary feature which represents the presence of blacklist words in the hypertext at the email's body. The words included: click here, bank, renew, dispute, login, sign, update, resolution, activate, unsuspended [94]
Imagecontent (BIC)	Define if the emails have image content. Value '1' if the email contains images, '0' otherwise
Appcontent (BAC)	Define if the emails have application content. Value '1' if the email contains an application, '0' otherwise
Multipartcontent (BMC)	Define if the emails have multipart content. Value '1' if the email contains multipart, '0' otherwise
Language (BL)	Text value which defines language used in the emails. i.e. English, Spanish

3.1.2 Constructing Feature Vector

In this section, we discuss the process used for constructing phishing feature vector for the datasets. Let E and F denote the emails and feature vector space respectively:

$$E = \{e_1, e_2, \ldots, e_N\} \tag{1}$$

$$F = \{f_1, f_2, \ldots, f_M\} \tag{2}$$

Where N is the total number of emails and M refers to the size of the feature vector. Let x_{ij} be the value of jth feature of ith document. Therefore, the presentation of each dataset D is given as follows:

$$D_d = \{x_{ij}\} | 1 \leq i \leq N; 1 \leq j \leq M; d \in body \tag{3}$$

For example, the feature vector for the *body* dataset consists of eight features $F_i, 1 \leq i \leq 8$ for all phishing emails. Note that *body* dataset consists of binary and text data. So, the dataset for D_{body} consists of:

$$D_{body} = \{SBB, BHB, BHC, BTC, BIC, BAC, BMC, BL\}.$$

The D_{body} feature set is loaded into the email analyser module. This configuration is dynamic where the rule sets could be configured with whatever features (rule) desired for a given email data. Upon receiving these datasets, the email analyser will apply a clustering algorithm approach on the generated analysis data to produce a group of phisher. Then, the phisher groups are submitted to the forensic backend for further action. However, we did not carry out the actual forensic filtering algorithm in this paper.

3.2 Dynamic Trackback Strategy

In this section, the description of the Dynamic Trackback Strategy is given. The MDA [11] is incorporated as the clustering algorithm. A schematic representation of the MDA is shown in Fig. 2. MDA is used to generate group of phishing profile, Q_{pp} based on the maximum dependency degree of attributes. The accurate selection of clustering attributes denotes by the higher degree of dependency. The phishing profiles generated based on clustering algorithm's predictions are set to be the elements of profiles. These predictions are further utilized to generate complete profiles of phishing emails. Note that this work differs from [11] where we incorporated forensic backend phase for tracking attacker purposes. Let X_i be the corresponding input value for the algorithm such that,

$$X_i = (x_{i1}, x_{i2}, \ldots, x_{iN}) \tag{4}$$

where x_i is the value of attribute i in the email and N is the total number of attributes in the unsupervised datasets. The steps for the email analyser phases are as follows:

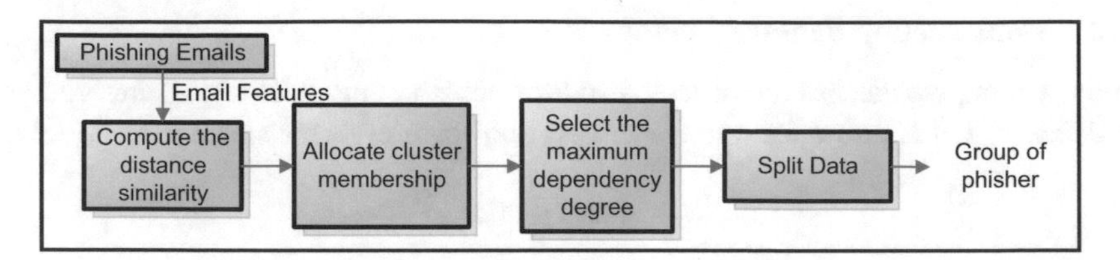

Fig. 2. Maximum Dependency Algorithm (MDA) components.

Step 1: (Compute the distance similarity). Compute the distance similarity using the maximum dependency degree of attribute.
Step 2: (Allocate cluster membership). Allocate cluster membership according to the maximum dependencies degree of attributes a_i with respect to all a_j, where $i \neq j$.
Step 3: (Select maximum dependency degree). Select maximum dependency degree of each attribute based on Marczeweski–Steinhaus (MZ) metric [20–22].
Step 4: (Split data). Split data to decide the final clusters based on maximum dependencies degree value.
Step 5: (Forensic analysis). To classify attacker into single or collaborative attack using email header information.

Figure 3 shows the generalized forensic filtering algorithm which consists of two steps that are compute the similarity measurement and produce prediction. First, the similarity measurement is calculated using Euclidean Distance. The Euclidean distance between two attacker $Phisher_1$ and $Phisher_2$ is defined as $e(phisher_1, phisher_2) = \sqrt{\sum_{i \in N} \left(x_{phisher_1} - x_{phisher_2}\right)^2}$, where N is the number of data. The numerical difference for each corresponding attributes of attacker $Phisher_1$ and $Phisher_2$ is measured. Then, combine the square of differences in each data into a whole distance. This square function normalizes the distance similarity score to a value between 0 and 1, where a value of 1 show that two attackers might cooperate together while value 0 means that two attackers do not work together. Finally, we refined each group of attacker into two categories: single attack or collaborative attacker. Moreover, we are able to track the origin of the attacker by looking into detail the email header information. Thus, actions are taken based on the forensic backend result where may comprise having trackback server to alert the email system regarding the violation.

```
Algorithm: Forensic Filtering Algorithm
Input: phishing profiles with class attributes
Output : Classify group of attacker
Begin:
   Step 1: Compute the similarity measurement, αij of
item, N
   Step 2: Produce prediction
End
```

Fig. 3. Forensic filtering algorithm

4 Performance Analysis

In this section, we discussed the analysis and performance evaluation of the proposed framework and test it on a series of dataset of various features.

4.1 Experimental Setup

We extracted eight body-based features from the Nazario phishing email dataset [18]. Then, Maximum Dependency Degree (MDA) [12] is used as the clustering algorithm because it performed well for categorical data. For forensic backend phases, we adapted the Euclidean Distance similarity measurement to classify attacker into single or collaborative attacker. We used the default setting for both algorithms in our initial preliminary testing. The *body-based* dataset consists of 3185 phishing email which has been pre-classified into groups of attacker to evaluate the predicted cluster by the clustering algorithm. Ten-fold cross validation is used for evaluation purposes. In order to measure the proposed trackback framework, we run the following tests:

(a) Maximum dependency value: To assess the proposed framework systematically, we tested on the proposed *body-based* features. Then, we grouped the phishing email for each the dataset into fix number of splits.
(b) Split size selection: To test the performance of the clustering algorithm to various numbers of data splitting. We used one to five data splitting information to determine the performance of clustering algorithm to body based features.
(c) Forensic analysis: With the aim to classify attacker into single or collaborative attack, we proposed forensic features extracted from email headers. We tested the datasets with the proposed forensic email features. Note that, the dataset has been pre-classified into groups of phisher by the clustering algorithm.

We ran a series of tests using classification algorithms to *body-based* features with pre-determined class by the clustering algorithm. AdaBoost and Hidden Markov Model (HMM) algorithm are selected because these are the frequently used classification methods. The default setting is set for the classification algorithms in the initial preliminary testing. In order to measure the performance of the proposed trackback framework, we used the following two performance metrics. These metrics are: (1) *False Positive Rate (FPR):* How many normal emails are misclassified as phishing email?, and (2) *False Negative Rate (FPR):* How many negative emails are misclassified as normal phishing email?

4.2 Result and Discussion

Comprehensive experiments of the proposed trackback framework on body-based phishing email features are carried out. The results of the experiments will be discussed in this section.

4.2.1 Maximum Dependency Degree Value

The degree dependency attributes of body-based features can be summarized in Table 2. The body features are body_subjectblacklist (BSB), body_hypertextblacklist (BHB), body_textcontent (BTC), body_htmlcontent (BHC), body_imagecontent (BIC), body_appcontent (BAC), body_multipartcontent (BMC), body_language (BL). Table 3 summarized the split information for body-based features while Fig. 4 shows five groups of attacker based on four split for body-based features.

Table 2. Maximum dependency matrix *of body-based* features

	BHB	BTC	BHC	BIC	BAC	BMC	BL
BSB	0	0	0	0	0	0	0
	BSB	BTC	BHC	BIC	BAC	BMC	BL
BHB	0	0	0	0	0.01	0	0
	BSB	BHB	BHC	BIC	BAC	BMC	BL
BTC	0	0	0.16	0	0.01	0	0
	BSB	BHB	BTC	BIC	BAC	BMC	BL
BHC	0	0	0.84	0	0.01	0	0
	BSB	BHB	BTC	BHC	BAC	BMC	BL
BIC	0	0	0	0	0.01	0	0.01
	BSB	BHB	BTC	BHC	BIC	BMC	BL
BAC	0	0.35	0.84	0.84	0.12	0	0.01
	BSB	BHB	BTC	BHC	BIC	BAC	BL
BMC	0	0.35	0.84	0.84	0.12	0.01	0.03
	BSB	BHB	BTC	BHC	BIC	BAC	BMC
BL	0	0	0	0	0	0	0

4.2.2 Split Size Selection

Table 4 shows the results of our analysis in detail where *body-based* features achieved False Positive (FP) and False Negative (FN) value tested on Adaboost and Hidden Markov Model (HMM) algorithm. Feature vector for this analysis were prepared for *body-based* datasets. It was interesting to notice that *body-based* feature achieved 99% True Positive value tested on Adaboost (split number 4) and HMM (slit number 2) algorithm. In general, the proposed framework capable to track *body-based* email features. However, the selection of features to be track is crucial in order to identify group of attacker.

The distribution of single and collaborative attack for body-based features are shown in Fig. 5 The result shows that the proposed trackback framework manages to distinguish type of attacker either single or collaborative attack by implementing the Euclidean Distance as the similarity measurement algorithm.

Table 3. Split group of phishing attacker based on *body* feature

Split	MDA	Degree attribute
1	0.31	Is a degree of attribute 7
2	0.307	Is a degree of attribute 6
3	0.121	Is a degree of attribute 4
4	0.025	Is a degree of attribute 3
5	0.002	Is a degree of attribute 5
6	0.001	Is a degree of attribute 2
7	0	Is a degree of attribute 1
8	0	Is a degree of attribute 8

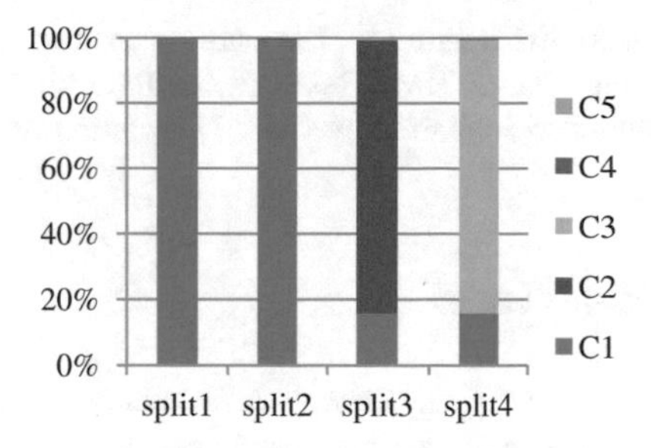

Fig. 4. Numbers of split groups for *body-based* features

Table 4. FN and FP rate for various numbers of split

Split number		1		2		3		4		5	
Feature	Algorithm	FN	FP	FN	FP	FN	FP	FN	FP	FN	FP
Body-based	AdaBoost	0.00	0.00	0.00	0.00	0.01	0.00	1.00	0.99	0.97	0.91
	HMM	0.00	1.00	0.01	0.99	0.16	0.84	0.16	0.84	0.17	0.83

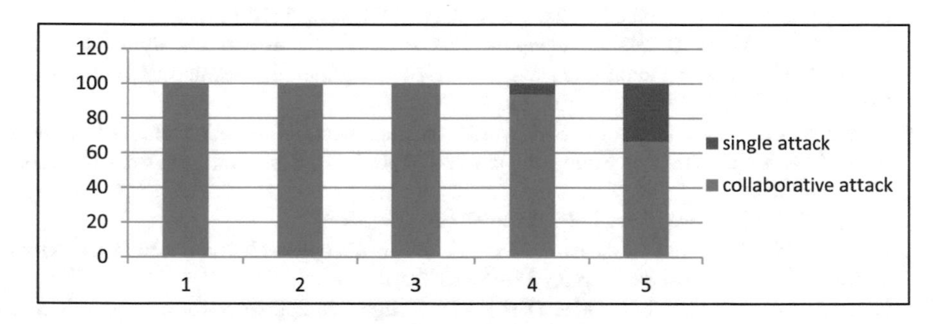

Fig. 5. Distribution of single and collaborative attack for five splits

5 Conclusion and Future Direction

Phishing email is an evolving problem and solving this problem has proven to be very challenging. In this paper, we proposed a dynamic phishing trackback strategy by performing MDA clustering and forensic analysis. First, we formulate the trackback framework as a clustering problem using email-born features in the phishing emails as feature vectors. The proposed framework is then tested on proposed body-based feature set. The experiment results shows that the body-based features achieved highest precision value when tested using the AdaBoost algorithms. Then, the distance similarity for each attribute in class is calculated to classify the attacker either single or collaborative attacker. We plan to investigate other clustering algorithm and similarity measurement that could be integrated with our proposed trackback framework.

Acknowledgement. The authors express appreciation to the University Tun Hussein Onn Malaysia (UTHM), Research and Innovation Fund (ORICC Fund), Short Term Grant Vot U653 and also supported by Gates IT Solution Sdn. Bhd. under its publication scheme.

References

1. Abawajy, J., Kelarev, A.: A multi-tier ensemble construction of classifiers for phishing email detection and filtering. Cyberspace Saf. Secur. **7672**, 48–56 (2012)
2. Fette, I., Sadeh, N., Tomasic, A.: Learning to detect phishing emails. In: Proceedings of the 16th International Conference on World Wide Web, WWW 2007, p. 649 (2007)
3. Gajek, S., Sadeghi, A.-R.: A forensic framework for tracing phishers. In: Fischer-Hübner, S., Duquenoy, P., Zuccato, A., Martucci, L. (eds.) Privacy and Identity 2007. ITIFIP, vol. 262, pp. 23–35. Springer, Heidelberg (2008). doi:10.1007/978-0-387-79026-8_2
4. Hamid, I.R.A., Abawajy, J., Kim, T.: Using feature selection and classification scheme for automating phishing email detection. Stud. Inf. Control **22**(1), 61–70 (2013)
5. Ma, L., Ofoghi, B., Watters, P., Brown, S.: Detecting phishing emails using hybrid features. In: Symposia and Workshops on Ubiquitous, Autonomic and Trusted Computing, UIC-ATC 2009, pp. 493–497 (2009)
6. State of the Net 2010: Consumer Reports National Research Center (2010)
7. Yearwood, J., Mammadov, M., Webb, D.: Profiling phishing activity based on hyperlinks extracted from phishing emails. Soc. Netw. Anal. Min. **2**(1), 5–16 (2012)
8. Yearwood, J., Webb, D., Ma, L., Vamplew, P., Ofoghi, B., Kelarev, A.: Applying clustering and ensemble clustering approaches to phishing profiling. In: 8th Australasian Data Mining Conference, AusDM 2009, vol. 101, pp. 25–34 (2009)
9. Garera, S., Provos, N., Chew, M., Rubin, A.D.: A framework for detection and measurement of phishing attacks. In: Proceedings of the 2007 ACM Workshop on Recurring Malcode, pp. 1–8 (2007)
10. Wong, M.W.: SPF overview. Linux J. **2004**(120), 2 (2004)
11. Herawan, T., Deris, M.M., Abawajy, J.H.: A rough set approach for selecting clustering attribute. Knowl.-Based Syst. **23**(3), 220–231 (2010)
12. Herawan, T., Yanto, I.T.R., Mat Deris, M.: Rough set approach for categorical data clustering. In: Ślęzak, D., Kim, T.-h., Zhang, Y., Ma, J., Chung, K.-i. (eds.) DTA 2009. CCIS, vol. 64, pp. 179–186. Springer, Heidelberg (2009). doi:10.1007/978-3-642-10583-8_21
13. Hamid, I.R.A., Abawajy, J.: Hybrid feature selection for phishing email detection. In: Xiang, Y., Cuzzocrea, A., Hobbs, M., Zhou, W. (eds.) ICA3PP 2011. LNCS, vol. 7017, pp. 266–275. Springer, Heidelberg (2011). doi:10.1007/978-3-642-24669-2_26
14. Hamid, I.R.A., Abawajy, J.: Phishing email feature selection approach. In: IEEE 10th International Conference on Trust, Security and Privacy in Computing and Communications (TrustCom), pp. 916–921 (2011)
15. Chandrasekaran, M., Chinchani, R., Upadhyaya, S.: PHONEY: mimicking user response to detect phishing attacks. In: 2006 International Symposium on a World of Wireless, Mobile and Multimedia Networks, WoWMoM 2006, vol. 2006, pp. 668–769 (2006)
16. Li, S., Schmitz, R.: A novel anti-phishing framework based on honeypots. In: eCrime Researchers Summit, eCRIME 2009, pp. 1–13 (2009)
17. Hamid, I.R.A., Abawajy, J.H.: An approach for profiling phishing activities. Comput. Secur. **45**, 27–41 (2014)
18. Nazario, J.: Phishing corpus. http://monkey.org/~jose/wiki/doku.php

19. Hamid, I.R.A., Abawajy, J.H.: Profiling phishing email based on clustering approach. In: 12th IEEE International Conference on Trust, Security and Privacy in Computing and Communications (TrustCom), pp. 628–635 (2013)
20. Yao, Y.Y.: Two views of the theory of rough sets in finite universes. Int. J. Approx. Reason. **15**(4), 291–317 (1996)
21. Yao, Y.Y.: Constructive and algebraic methods of the theory of rough sets. Inf. Sci. (Ny) **109** (1–4), 21–47 (1998)
22. Yao, Y.Y.: Information granulation and rough set approximation. Int. J. Intell. Syst. **16**, 87–104 (2001)

A Case Based Methodology for Problem Solving Aiming at Knee Osteoarthritis Detection

Marisa Esteves[1], Henrique Vicente[2,3], José Machado[3], Victor Alves[3], and José Neves[3(✉)]

[1] Departamento de Informática, Universidade do Minho, Braga, Portugal
marisa.araujo.esteves@gmail.com

[2] Departamento de Química, Escola de Ciências e Tecnologia, Universidade de Évora, Évora, Portugal
hvicente@uevora.pt

[3] Centro Algoritmi, Universidade do Minho, Braga, Portugal
{jmac,valves,jneves}@di.uminho.pt

Abstract. Knee osteoarthritis is the most common type of arthritis and a major cause of impaired mobility and disability for the ageing populations. Therefore, due to the increasing prevalence of the malady, it is expected that clinical and scientific practices had to be set in order to detect the problem in its early stages. Thus, this work will be focused on the improvement of methodologies for problem solving aiming at the development of *Artificial Intelligence* based decision support system to detect knee osteoarthritis. The framework is built on top of a *Logic Programming* approach to *Knowledge Representation* and *Reasoning*, complemented with a *Case Based* approach to computing that caters for the handling of incomplete, unknown, or even self-contradictory information.

Keywords: Knee osteoarthritis · Knee X-ray image feature extraction · Knowledge representation and reasoning · Logic programming · Case-based reasoning

1 Introduction

At the moment more than a hundred of different types of arthritis are known [1]. *OsteoArthritis* (*OA*) is the most common type of arthritis or degenerative joint disease, and it is a major cause of impaired mobility and disability for the ageing populations. It occurs when the cartilage or cushion between joints breaks down and due to the degradation of cartilage, joint space width is reduced [2]. This may result in bone rubbing on bone, leading to pain, stiffness and swelling pain. *OA* develops slowly and the pain it causes worsens over time.

The disease may affect any joint in the body. Nonetheless, the knee is one of the most commonly affected areas. In fact, knee joint is a key piece of the human body, since it is where the femur (thigh bone) and the tibia (shin bone) meet, allowing the bones to move freely (however, within certain limits) [2]. The whole weight of the body lies on the knee whether we are running, walking, or even standing. When the

T. Herawan et al. (eds.), *Recent Advances on Soft Computing and Data Mining*,
Advances in Intelligent Systems and Computing 549, DOI 10.1007/978-3-319-51281-5_28

knee has osteoarthritis, its surface becomes damaged and it does not move as well as it should do. Knee osteoarthritis has no specific causes and several factors may lead to the development of the disease, like age, gender, heredity, weight, repetitive stress injuries, and other illnesses [1, 2]. It should be noted that age is the strongest predictor of the development and progression of knee *OA*. Indeed, the disease most commonly affects the middle-aged and elderly, although it may begin earlier because of injury or overuse. It is estimated to affect over 630 million people worldwide, i.e., 15% of all the people on the globe [1]. The prevalence of *OA* is increasing and this places a globally major burden on individuals, health and social care systems. Hence, due to the increasing prevalence of osteoarthritis, it is mandatory the development of clinical and scientific tools that may detect *OA* in an early stage of the disease.

The diagnosis of knee *OA* includes a medical history and a physical examination. Usually conventional radiographic knee images are used to detect and distinguish various forms of *OA* (Fig. 1). *X-rays* of an arthritic knee may show a narrowing of the joint space and changes in the bone. These images can also be used to calculate a few parameters of utmost importance in the diagnosis of this disease, like *Knee Joint Space Width* (*KJSW*), *Knee Joint Space Perimeter* (*KJSP*), and *Knee Joint Space Area* (*KJSA*). For instance, the measurement of *KJSW*, between the distal femur and the proximal tibia is an indirect way of measuring the tibia-femoral cartilage thickness, which is considered a valuable parameter in assessing knee cartilage disease such as knee *OA* [3].

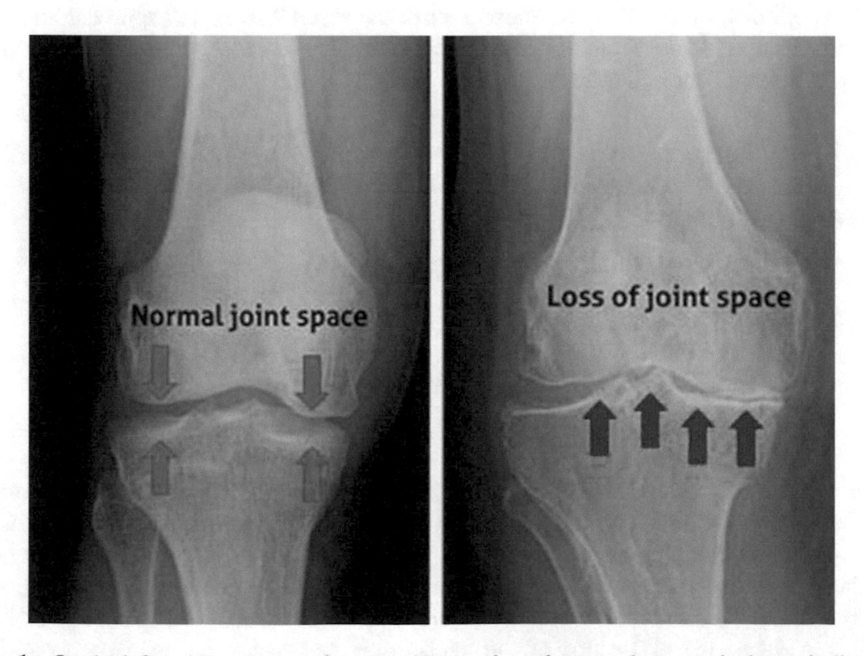

Fig. 1. On the left an *X-ray image* of a normal knee where the space between the bones indicates healthy cartilage. On the right an *X-ray image* of an arthritic knee showing severe loss of joint space.

Solving problems related to the diagnosis of knee *OA* requires a proactive strategy able to take into account all these factors. This work is focused on the development of a hybrid methodology for problem solving, aiming at the elaboration of clinical decision support systems to detect knee osteoarthritis based on knee joint space parameters obtained from knee *X-ray* images, according to a historical dataset, under a *Case Based Reasoning* (*CBR*) approach to problem solving [4, 5]. Indeed, *CBR* provides the ability of solving new problems by reusing knowledge acquired from past experiences [4], i.e., *CBR* is used especially when similar cases have similar terms and solutions, even when they have different backgrounds [5]. Indeed, its use may be found in different arenas, namely in *The Law* [6], *Online Dispute Resolution* [7, 8], or *Medicine* [9]. In this work it will be emphasized the handling of self-contradictory data, information or knowledge, which it is a major development with respect to [10].

This paper develops along five sections. In a former one a brief introduction to the problem is made. Then the proposed approach to knowledge representation and reasoning is introduced. In the third and fourth sections it is assumed a case study and presented a solution to the problem. Finally, in the last section, the most relevant conclusions are described and possible directions for future work are outlined.

2 Knowledge Representation and Reasoning

Many approaches to knowledge representation and reasoning have been proposed using the *Logic Programming* (*LP*) epitome, namely in the area of *Model Theory* [11, 12], and *Proof Theory* [13, 14]. In the present work the *Proof Theoretical* approach in terms of an extension to the *LP* language is followed. An *Extended Logic Program* is a finite set of clauses, given in the form:

$$
\begin{array}{l}
\{ \\
\quad \neg p \leftarrow not\, p, not\, exception_p \; (that\, stand\, for\, predicate's\, clousure) \\
\quad p \leftarrow p_1, \cdots, p_n, not\, q_1, \cdots, not\, q_m \\
\quad ?(p_1, \cdots, p_n, not\, q_1, \cdots, not\, q_m) \; (n, m \geq 0) \\
\quad exception_{p_1} \\
\quad \cdots \\
\quad exception_{p_j} \; (0 \leq j \leq k), \; being\, k\, and\, integer\, number \\
\} :: scoring_{value}
\end{array}
$$

where "*?*" is a domain atom denoting falsity, the p_i, q_j, and p are classical ground literals, i.e., either positive atoms or atoms preceded by the classical negation sign $\neg$ [13]. Under this formalism, every program is associated with a set of *abducibles* [11, 12], given here in the form of exceptions to the extensions of the predicates that make the program. The term $scoring_{value}$ stands for the relative weight of the extension of a specific *predicate* with respect to the extensions of the peers ones that make the inclusive or global program.

Indeed, and in order to evaluate the knowledge that may be associated to a logic program, an assessment of the *Quality-of-Information* (*QoI*), given by a truth-value in the interval [0, 1], that branches from the extensions of the predicates that make a program, inclusive in dynamic environments, is set [15, 16]. On the other hand, a measure of one's confidence that the argument values or attributes of the terms that make the extension of a given predicate, with relation to their domains, fit into a given interval, is also considered, and labeled as *DoC* (*Degree of Confidence*) [17]. The *DoC* is evaluated as described in [17] and computed using $DoC = \sqrt{1 - \Delta l^2}$, where Δl stands for the argument interval length, which was set to the interval [0, 1]. Thus, the universe of discourse is engendered according to the information presented in the extensions of such predicates, according to productions of the type:

$$predicate_i - \bigcup_{1 \leq j \leq l} clause_j((QoI_{x_1}, DoC_{x_1}), \cdots, (QoI_{x_l}, DoC_{x_l})) :: QoI_j :: DoC_j \quad (1)$$

where $\cup$, m and l stand, respectively, for *set union*, the *cardinality* of the extension of $predicate_i$ and number of attributes of each clause [17]. The subscripts of *QoIs* and *DoCs*, $x_1, \ldots, x_l$, stand for the attributes values ranges.

2.1 Case Based Reasoning

The *CBR* methodology for problem solving stands for an act of finding and justifying the solution to a given problem based on the consideration of similar past ones, by reprocessing and/or adapting their data or knowledge [4, 5]. In *CBR* – the cases – are stored in a *Case Base*, and those cases that are similar (or close) to a new one are used in the problem solving process. The typical *CBR* cycle presents a mechanism that should be followed, where the former stage entails an initial description of the problem. The new case is defined and it is used to retrieve one or more cases from the *Case Base*. At this point it is important to identify the characteristics of the new problem and retrieve cases with a higher degree of similarity to it. Thereafter, a solution for the problem emerges, on the *Reuse* phase, based on the blend of the new case with the retrieved ones. The suggested solution is reused (i.e., adapted to the new case), and a solution is provided [4, 5]. However, when adapting the solution it is crucial to have feedback from the user, since automatic adaptation in existing systems is almost impossible. This is the *Revise* stage, in which the suggested solution is tested by the user, allowing for its correction, adaptation and/or modification, originating the test repaired case that sets the solution to the new problem. The test repaired case must be correctly tested to ensure that the solution is indeed correct. Thus, one is faced with an iterative process since the solution must be tested and adapted while the result of applying that solution is inconclusive. During the *Retain* (or *Learning*) stage the case is learned and the knowledge base is updated with the new case [4, 5].

Despite promising results, the current *CBR* systems do not cover all areas, and in some cases, the user cannot choose the similarity(ies) method(s) and is required to follow the system defined one(s), even if they do not meet their needs. But, worse than that, in real problems, access to all necessary information is not always possible, since

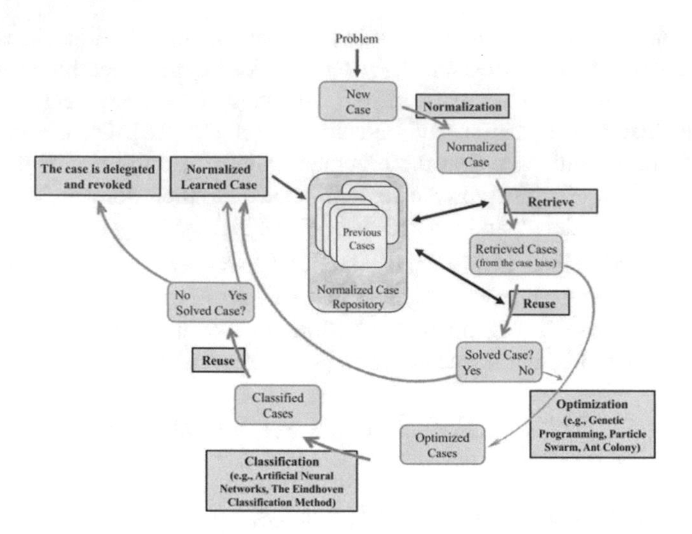

Fig. 2. An extended view of the *CBR* cycle [18].

existent *CBR* systems have limitations related to the capability of dealing, explicitly, with unknown, incomplete, and even self-contradictory data, information or knowledge. To make a change, a different *CBR* cycle was induced (Fig. 2). It takes into consideration the case's *QoI*s and *DoCs* [18]. It deals not only with unknown, incomplete, and even contradictory information or knowledge, in an explicit way, but also contemplates the cases optimization in the *Case Based*, whenever they do not comply with the terms under which a given problem as to be addressed (e.g., the expected degree of confidence on the diagnostic was not attained), either using particle swarm optimization procedures [19], or genetic algorithms [14], just to name a few.

2.2 Methods

In order to develop an intelligent support system aiming to detect knee osteoarthritis, a database was set. The data was taken from the health records of patients at a major health care institution in the north of Portugal. This section sets (briefly) the process of data set creation and how it is processed. Usually the information contained on medical images is too large and unorganized. Therefore, some steps must be followed in order to extract the most relevant features for the study and eliminate some artefacts that can cause some impairment on the results that one is trying to obtain. A set of knee *X-ray DICOM* images, corresponding to antero-posterior views of knees of different patients, was collected in order to accomplish this study. All the images were classified as *knee without OA* or *knee with OA* based on a comparison effectuated with already classified knee X-ray images. The attributes considered for the detection of knee *OA* include the age, gender, three measures of *Knee Joint Space Width* (*KJSW*) and its average, the *Knee Joint Space Perimeter* (*KJSP*) and the *Knee Joint Space Area* (*KJSA*).

A *Java-based* image-processing framework named *imageJ* [20] was used to extract the necessary features from the *X-ray* images. Firstly, the patient's age and gender were extracted from the *DICOM* header with the *Show Info* option. After that, the region of interest, corresponding to the knee joint space, was cropped with the *Crop* option in order to measure the desired features. Then, with the option *Measure*, the *KJSW* was measured vertically with recourse to the *straight-line* command in three different points of the image, i.e., left (*KJSW1*), center (*KJSW2*) and right (*KJSW3*) and the correspondent average (*KJSWA*) was, afterwards, calculated. Finally, once again with the option *Measure*, the *KJSP* and *KJSA* were calculated using the *freehand selection*. It should be noted that the unit of all measures was the pixel.

3 A Logic Programming Approach to Knowledge Representation and Reasoning

It is now possible to build up a knowledge database given in terms of the extension of the relation showed in Fig. 3, which stand for a situation where one has to detect knee osteoarthritis based on knee *X-ray* images. Under this scenario, some incomplete and/or unknown data is present. For instance, the age of patient 3 is unknown, which is represented by the symbol $\perp$.

The values presented in the column *Age* range in the interval [22, 97]. In column *Gender*, 0 and 1 stand for *female* and *male*, respectively. The values that populate the columns *KJSW1*, *KJSW2*, *KJSW3* and *KJSWA* range in the interval [0, 50], while the values displayed in the columns *KJSP* and *KJPA* range in the intervals [800, 1400] and [4000, 12000], respectively. The *Descriptions* column stands for free text fields that allow for the registration of relevant patient features.

Applying the algorithm presented in [17] to all the fields that make the knowledge base for *Knee Osteoarthritis Detection* (Fig. 3), excluding of such a process the *Description* ones, and looking to the DoC_s values obtained as described in [17], it is possible to set the arguments of the predicate ***k**nee **o**steo**a**rthritis **detec**tion* (koa_{detec}) referred to below, that also denotes the objective function with respect to the problem under analyze:

$$koa_{detec} : Age, Gender, KJSW1, KJSW2, KJSW3, KJSWA, \mathrm{KJSP}, KJSA \rightarrow \{0, 1\}$$

where 0 (zero) and 1 (one) denote, respectively, the truth values *false* and *true*.

	Knee Osteoarthritis Detection									
Attributes of the Feature Vector:	#	Age	Gender	KJSW1	KJSW2	KJSW3	KJSWA	KJSP	KJSA	Description
Feature Vector Attributes:	1	72	2	22	20	21	21	992	6678	Description 1
	2	87	2	19	7	12	13	944	5085	Description 2
	3	$\perp$	1	31	36	35	34	1269	10768	Description 3
	...	...	...	...	...	...	...	...	...	...
	n	87	2	34	18	12	21	1193	8863	Description *n*
Feature Vector Domains:		[22, 97]	[0, 1]	[0, 50]	[0, 50]	[0, 50]	[0, 50]	[800, 1400]	[4000, 12000]	

Fig. 3. A fragment of the knowledge base for Knee Osteoarthritis Detection.

Exemplifying the application of the algorithm presented in [17], to a term (patient) that presents feature vector (*Age* = 69, *Gender* = 1, *KJSW1* = 27, *KJSW2* = 26, *KJSW3* = 27, *KJSWA* = 26.7, *KJSP* = ⊥, *KJSA* = [5724, 5966]), one may get:

$$
\begin{array}{l}
\{ \\
\quad \neg\, koa_{detec}\left((QoI_{Age}, DoC_{Age}), \cdots, (QoI_{KJSP}, DoC_{KJSP}), (QoI_{KJSA}, DoC_{KJSA})\right) \\
\quad \leftarrow not\; koa_{detec}\left((QoI_{Age}, DoC_{Age}), \cdots, (QoI_{KJSP}, DoC_{KJSP}), (QoI_{KJSA}, DoC_{KJSA})\right) \\
\quad koa_{detec} \underbrace{((1,1), \cdots, (1,0), (1,0.999))}_{attribute's\ quality-of-information\ and\ respective\ confidence\ values} :: 1 :: 0.88 \\
\qquad \underbrace{[0.63, 0.63] \cdots [0,1][0.22, 0.24]}_{attribute's\ values\ ranges\ once\ normalized} \\
\qquad \quad \underbrace{[0,1] \cdots [0,1][0,1]}_{attribute's\ domains\ once\ normalized} \\
\}
\end{array}
$$

4 A Case Based Approach to Computing

A soft computing approach to model the universe of discourse based on a *CBR* methodology for problem solving is now set. Indeed, contrasting with other problem solving methodologies (e.g., *Decision Trees* or *Artificial Neural Networks*), in a *CBR* based one relatively little work is done offline. Undeniably, in almost all the situations the work is performed at query time. The main difference between this new approach and a typical *CBR* relies on the fact that not only all the cases have their arguments set in the interval [0, 1], but it also caters for the handling of incomplete, unknown, or even self-contradictory data or knowledge [18]. Thus, the classic *CBR* cycle was changed (Fig. 2), being the *Case Base* given in terms of triples that follow the pattern:

$$Case = \{ < Raw_{data}, Normalized_{data}, Description_{data} > \}$$

where Raw_{data} and $Normalized_{data}$ stand for themselves, and $Description_{data}$ is made on a set of strings or even in free text, which may be analyzed with string similarity algorithms but, as it was stated before, will not be object of attention in this work. The objective here is to present a perspective of the way one's handle self-contradictory data, information or knowledge.

Indeed, when confronted with a new case the system is able to retrieve all cases that meet such a structure, i.e., it considers the attributes *DoC*'s value of each case or of their optimized counterparts when analysing similarities among them. Thus, under the occurrence of a new case, the goal is to find similar cases in the *Case Base*. Having this in mind, the algorithm given in [17] is applied whenever one is faced a new case, now with feature vector (*Age* = ⊥, *Gender* = 0, *KJSW1* = 22, *KJSW2* = 24, *KJSW3* = 24, *KJSWA* = 23.7, *KJSP* = [1001, 1023], *KJSA* = 6708, *Description* = *Description new*). One may get:

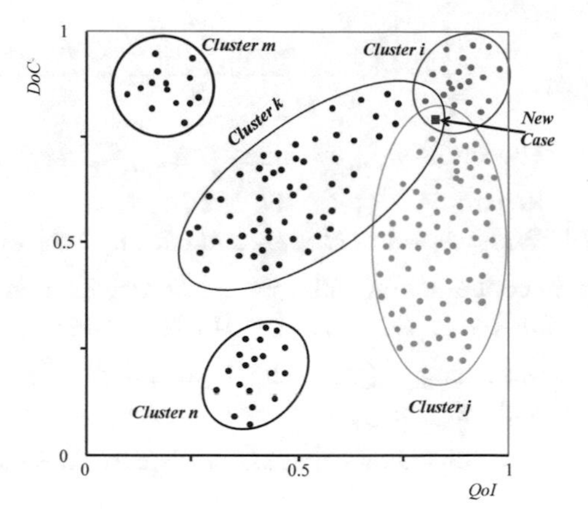

Fig. 4. A case's set divided into clusters.

$$\frac{koa_{detec_{new}}((1,0),\cdots,(1,0.999),(1,1)) :: 1 :: 0.875}{new\ case}$$

Then, the *new case* can be painted on the Cartesian plane in terms of its *QoI* and *DoC*, and by using clustering techniques, it is feasible to identify the clusters that intermingle with the new one (symbolized as a square in Fig. 4). The *new case* is compared with every retrieved case from the cluster using a similarity function *sim*, given in terms of the average of the modulus of the arithmetic difference between the arguments of each case of the selected cluster and those of the *new case* (*description's* will not be object of study in this work). Thus, one may have:

$$\frac{\begin{array}{c} koa_{detec_{i,1}}((1,1),\cdots,(1,1),(1,0.97)) :: 1 :: 0.871 \\ koa_{detec_{i,2}}((1,0),\cdots,(1,0.95),(1,1)) :: 1 :: 0.869 \\ \vdots \\ koa_{detec_{j,1}}((1,1),\cdots,(1,0),(1,0.98)) :: 1 :: 0.372 \\ koa_{detec_{j,2}}((1,0),\cdots,(1,0.87),(1,0.98)) :: 1 :: 0.481 \\ \vdots \\ koa_{detec_{k,1}}((1,1),\cdots,(1,0),(1,0.95)) :: 1 :: 0.619 \\ koa_{detec_{k,2}}((1,1),\cdots,(1,1),(1,0.98)) :: 1 :: 0.748 \\ \vdots \end{array}}{cases\ from\ retrieved\ clusters}$$

With respect to $cluster_i$, and assuming that every attribute has equal weight, the dissimilarity between $koa_{detec_{new}^{DoC}}$ and $koa_{detec_{i,1}^{DoC}}$, i.e., $koa_{detec_{new \to i,1}^{DoC}}$, is computed as follows:

$$dissim\_koa_{detec^{DoC}_{new \to i,1}} = \frac{\|1-1\| + \cdots + \|0-1\| + \|0.999 - 0.97\|}{8} = 0.254$$

and the similarity between $koa_{detec^{DoC}_{new}}$ and $koa_{detec^{DoC}_{i,1}}$, i.e., $sim\_koa_{detec^{DoC}_{new \to i,1}}$ is $1 - 0.254 = 0.746$. Regarding *QoI* the procedure is similar, returning $sim\_koa_{detec^{DoC}_{new \to i,1}} = 1$. With respect to $cluster_j$ and $cluster_k$, whose data, information or knowledge is in part in contradiction with that of $cluster_i$, the computational processes are similar. Thus $sim\_koa_{detec^{DoC}_{new \to j,1}}$ is $1 - 0.752 = 0.248$ and $sim\_koa_{detec^{DoC}_{new \to k,1}}$ $1 - 0.506 = 0.494$. Regarding *QoI,* the procedure is similar, returning $sim\_koa_{detec^{DoC}_{new \to j,1}} = 1$, and $sim\_koa_{detec^{DoC}_{new \to k,1}} = 1$.

A brief look to these results shows that $cluster_i$'s cases are closer to the *new one,* and therefore the problem solving process must continue based on these facts, i.e., once all the $cluster_i$*'s* cases are analysed, a possible solution to the problem may be found in the historical case that presents the best similarity value.

5 Conclusions

This work presents a new methodology for problem solving aiming at the development of intelligent decision support systems to detect knee osteoarthritis, among other illnesses, in which all steps are subject to formal proof, which in this specific work is highlighted with the handling of self-contradictory data, information or knowledge. Indeed, it is centred on a formal framework based on *LP* for *Knowledge Representation and Reasoning*, complemented with a *CBR* approach to computing. Under this approach the cases' retrieval and optimization phases were heightened, when compared with existing systems. Additionally, under this scenery the users may define the weights of the cases' attributes on the fly, letting them to choose the most appropriate strategy to address the problem (i.e., it gives the user the possibility to narrow the search space for similar cases at runtime).

In future work, the knowledge base will be extended in order to include the patient clinical history, namely the factors that can lead to the development of knee *OA*, like heredity, weight and repetitive stress injuries. Additionally, instead of simply detecting knee *OA*, the method must detect the different stages of *OA* severity, based, for example, on the Kellgren-Lawrence classification grades [22]. Moreover, string similarity descriptions will be considered and implemented, either in terms of free text or as logical *formulae*, being therefore context dependent and subject to formal proof.

Acknowledgments. This work has been supported by COMPETE: POCI-01-0145-FEDER-007043 and FCT – Fundação para a Ciência e Tecnologia within the Project Scope: UID/CEC/00319/2013.

References

1. Tanna, S.: Osteoarthritis: opportunities to address pharmaceutical gaps. In: Warren Kaplan, W., Laing, R., (eds.) Priority Medicines for Europe and the World, pp. 6.12-1–6.12-25. World Health Organization Editon, Geneva (2004). http://archives.who.int/prioritymeds/report/index.htm
2. Arya, R.K., Jain, V.: Osteoarthritis of the knee joint: an overview. J. Indian Acad. Clin. Med. **14**, 154–162 (2013)
3. Anas, I., Musa, T.A., Kabiru, I., Yisau, A.A., Kazaure, I.S., Abba, S.M., Kabir, S.M.: Digital radiographic measurement of normal knee joint space in adults at Kano, Nigeria. Egypt. J. Radiol. Nucl. Med. **44**, 253–258 (2013)
4. Aamodt, A., Plaza, E.: Case-based reasoning: foundational issues, methodological variations, and system approaches. AI Commun. **7**, 39–59 (1994)
5. Richter, M.M., Weber, R.O.: Case-Based Reasoning: A Textbook. Springer, Berlin (2013)
6. Balke, T., Novais, P., Andrade, F., Eymann, T.: From real-world regulations to concrete norms for software agents – a case based reasoning approach. In: Poblet, M., Schild, U., Zeleznikow, J. (eds.) Proceedings of the Workshop on Legal and Negotiation Decision Support Systems (LDSS 2009), pp. 13–28. Huygens Editorial, Barcelona (2009)
7. Carneiro, D., Novais, P., Andrade, F., Zeleznikow, J., Neves, J.: Using case based reasoning to support alternative dispute resolution. In: Carvalho, A.F., Rodríguez-González, S., Paz-Santana, J.F., Corchado-Rodríguez, J.M. (eds.) Distributed Computing and Artificial Intelligence. Advances in Intelligent and Soft Computing, vol. 29, pp. 123–130. Springer, Berlin (2010)
8. Carneiro, D., Novais, P., Andrade, F., Zeleznikow, J., Neves, J.: Using case-based reasoning and principled negotiation to provide decision support for dispute resolution. Knowl. Inf. Syst. **36**, 789–826 (2013)
9. Ping, X.-O., Tseng, Y.-J., Lin, Y.-P., Chiu, H.-J., Feipei Lai, F., Liang, J.-D., Huang, G.-T., Yang, P.-M.: A multiple measurements case-based reasoning method for predicting recurrent status of liver cancer patients. Comput. Ind. **69**, 12–21 (2015)
10. Chandrasekaran, B.: On evaluating artificial intelligence systems for medical diagnosis. AI Mag. **4**, 34–48 (1983)
11. Kakas, A., Kowalski, R., Toni, F.: The role of abduction in logic programming. In: Gabbay, D., Hogger, C., Robinson, I. (eds.) Handbook of Logic in Artificial Intelligence and Logic Programming, vol. 5, pp. 235–324. Oxford University Press, Oxford (1998)
12. Pereira, L., Anh, H.: Evolution prospection. In: Nakamatsu, K. (ed.) New Advances in Intelligent Decision Technologies – Results of the First KES International Symposium IDT 2009. Studies in Computational Intelligence, vol. 199, pp. 51–64. Springer, Berlin (2009)
13. Neves, J.: A logic interpreter to handle time and negation in logic databases. In: Muller, R., Pottmyer, J. (eds.) Proceedings of the 1984 Annual Conference of the ACM on the 5th Generation Challenge, pp. 50–54. Association for Computing Machinery, New York (1984)
14. Neves, J., Machado, J., Analide, C., Abelha, A., Brito, L.: The halt condition in genetic programming. In: Neves, J., Santos, M.F., Machado, J.M. (eds.) EPIA 2007. LNCS (LNAI), vol. 4874, pp. 160–169. Springer, Heidelberg (2007). doi:10.1007/978-3-540-77002-2_14
15. Lucas, P.: Quality checking of medical guidelines through logical abduction. In: Coenen, F., Preece, A., Mackintosh, A. (eds.) Research and Development in Intelligent Systems XX, pp. 309–321. Springer, London (2003)
16. Machado, J., Abelha, A., Novais, P., Neves, J., Neves, J.: Quality of service in healthcare units. In: Bertelle, C., Ayesh, A. (eds.) Proceedings of the ESM 2008, pp. 291–298. Eurosis – ETI Publication, Ghent (2008)

17. Fernandes, F., Vicente, H., Abelha, A., Machado, J., Novais, P., Neves J.: Artificial neural networks in diabetes control. In: Proceedings of the 2015 Science and Information Conference (SAI 2015), pp. 362–370. IEEE Edition (2015)
18. Neves, J., Vicente, H.: Quantum approach to Case-Based Reasoning (in preparation)
19. Mendes, R., Kennedy, J., Neves, J.: Watch thy neighbor or how the swarm can learn from its environment. In: Proceedings of the 2003 IEEE Swarm Intelligence Symposium (SIS 2003), pp. 88–94. IEEE Edition (2003)
20. Rasband, W.S.: ImageJ. U. S. National Institutes of Health, Bethesda, Maryland, USA (1997–2015). http://imagej.nih.gov/ij/
21. Shamir, L., Ling, S.M., Scott, W.W., Bos, A., Orlov, N., Macura, T., Eckley, D.M., Ferrucci, L., Goldberg, I.G.: Knee x-ray image analysis method for automated detection of osteoarthritis. IEEE Trans. Biomed. Eng. **56**, 407–415 (2009)

Preliminary Study for the Implementation of Electrical Capacitance Volume Tomography (ECVT) to Display Fruit Content

Riza Agustiansyah[1], Rohmat Saedudin[1](✉), and Mahfudz Al Huda[2]

[1] Information System Department, Industrial Engineering Faculty, Telkom University, Bandung, Indonesia
{rizaagustiansyah, rdrohmat}@telkomuniversity.ac.id
[2] CTECH Laboratories EDWAR Technology, Tangerang, Indonesia
huda1126@gmail.com

Abstract. There are some problems in the Indonesian fruit export that need a solution which is a non-destructive tool on fruit to distinguish the conditions of raw, ripe and rotten fruits. In this preliminary research, the first step is to measure the electrical characteristic which is the capacitance of fruit. The result shows, in general, the value of the capacitance decreases when frequency is enlarged. It also can be concluded that the differences in capacitance seen more clearly at high frequencies. On the use of multi channel ECVT scanner, the resulting image shows only the outside of the fruit, so it is difficult to distinguish the condition of each fruits. Further studies are to make a sensor that can wrap the fruit inside so that the inner of the fruit can be more clearly seen. The algorithm used for the image reconstruction in this research is Linear Back Projection (LBP).

Keywords: ECVT · Non-destructive test on fruit · Image reconstruction technique

1 Introduction

Fruit is one of export commodities in Indonesia. Hasanuddin Ibrahim, Director-General horticulture of the Ministry of agriculture, refers to the value of the export of pineapple can reach US $ 250 million, meanwhile mangosteen reach US $ 20 million. The export value of Mango reaches US $ 3 million and bananas are US $ 2 million. In 2013 the temporary numbers from the Central Bureau of statistics, mango export-oriented fruit production reaching 2.05 million tonnes per year, bananas reaching 5.3 million tonnes, salak reaching 991,762 tons and Orange 1.4 million tons. However, there are some countries which still refused a number of commodity export fruit from Indonesia. During the last five years, Indonesian fruit having difficulty to penetrates the Japanese and Chinese market. The two countries refused Indonesian fruit because they said it not hygienic and exposed with fruit fly pests. As is known, fruit flies are pests that attack fruits and vegetables. The fruits affected with fruit flies diseases become rotten. Japan and China countries fear when eggs or larvae in rotten fruit will breed in their country and ruining their fruit commodities [1].

T. Herawan et al. (eds.), *Recent Advances on Soft Computing and Data Mining*,
Advances in Intelligent Systems and Computing 549, DOI 10.1007/978-3-319-51281-5_29

To address these trade barriers, the Government of Indonesia held a Mutual Recognition Agreement (MRA) with several countries. In the near future Indonesia will sign a MRA with New Zealand, Australia, and the United States. With the MRA between Indonesia and the three countries, the trusts are built in the import-export of agricultural products. According to Hasanuddin Ibrahim, Director-General horticulture of the Ministry of agriculture, the three countries above have believed in professionalism of our exporters, so now this is the task from the quarantine agency in all respective countries to tighten the inspection process [2].

To perform the non destructive test on fruits, there are already some research conducted, such as the using of ultrasonic [3], the acoustic wave [4], and high-frequency electromagnetic waves [5]. Meanwhile, current research for the non destructive test on fruits uses Impedance spectrometry because it's a low-cost and relatively simple technology [6, 7]. However, there are shortcomings of that technology such as it cannot give an image output inside the fruit. The data remains in the form of a graph that illustrates the level of fruit maturity and acidity.

One of the alternatives of imaging techniques that are safe, and inexpensive, which is being developed at this moment is Electrical Capacitance Volume Tomography (ECVT), which is the development of Electrical Capacitance Tomography (ECT). The basic principle of ECVT is the reconstruction of permittivity distribution in the areas reviewed. The permittivity distribution is based on the value of the capacitance of the electrode pairs located on the sensor [8]. ECVT is capable of imaging the human body because within the human body there are conductive cells (neurons) that serve to transmit information. Potential arising out of brain activity will exert influence on the permittivity value around the active parts so they can be imaged with ECVT [9].

In this study, it expected to produce a tool that can display an image of the fruit content with ECVT technology. Furthermore, the ECVT for imaging the fruit called ECVT Fruit Scanner. This paper aims to examine the basic concepts used in the ECVT for Fruit scanner with different conditions and types of fruit. Fruit that will be tested on this research is orange and mango with the kind of raw, ripe and rotten conditions.

In the next section will be discussed the basis of the ECVT technology, then continued in part 3 with previous research about ECVT. Section 4 will discuss the results of the data retrieval which aims at finding the electrical characteristics of the fruit that will be applied to the ECVT fruit scanner to get the optimum results. Conclusion section contains an analysis of the measurement results and discussion.

2 Theoretical Background

2.1 The Basic Principle of ECVT

The basic principle of ECVT is basically the same as the ECT, namely collecting data capacitance of electrodes mounted on a wall outside the vessel (forward problem) and reconstruct the image data from the measurement of capacitance (inverse problem). Poisson's equation for the electric potential is written in Eq. (1) [8].

$$\nabla \cdot \varepsilon(x, y, z,)\nabla\phi(x, y, z) = -\rho(x, y, z), \tag{1}$$

with ε (*x*, *y*, *z*) is the permittivity distribution, while the ϕ (*x*, *y*, *z*) is the potential distribution of the electric field, and ρ (*x*, *y*, *z*) is the charge density. The value of the measured capacitance Ci from pair of source and electrode i determined by integrating Poisson equation

$$C_i = -\frac{1}{\Delta Vi} \oiint_{Ai} \varepsilon(x, y, z)\nabla\emptyset(x, y, z)dA, \tag{2}$$

The value of the measured capacitance Ci from pair of source and electrode i determined with Δ Vi is the voltage difference between the electrode pair and Ai is the surface area covering electrode detector. The relationship of permittivity distribution and the measured capacitance Ci seems not linear. To solve the Eq. (2), linearization techniques used with sensitivity model. Linear and discrete forms of forward problem are then written in Eq. (3) [8].

$$C_{Mx1} = S_{MxN} G_{Nx1}, \tag{3}$$

C is the capacitance data with M × 1dimensional, S is sensitivity matrix with M × N dimensional, and G is an image vector (the permittivity distribution) with N × 1 dimensional. N is the number of voxel and M is the combination of electrode pairs. Inverse problem on image reconstruction from ECVT is capacitance data, namely finding a vector image of g. There is some image reconstruction algorithm was ever used on the ECVT scanners, such as Linear Back Projection (LBP), Iterative Linear Back Projection (ILBP), and Neural Network Multicriterion Optimization of Image Reconstruction Technique (NN-MOIRT). Until now, NN-MOIRT Algorithms on reconstruction techniques produce the best image [8].

ECVT Scanner System consists of three main devices, as shown by Fig. 4, namely (a) computer, (b) capacitance sensor such as in the form of a tube, and (c) Data Acquisition System (DAS). The tube sensor consists of electrodes that mounted on the inner side of tube. On a measurement, electrode pairs can function as a transmitter (source electrode) and the receiver (receiver electrodes). The measurement is repeated until all of the electrodes in turns into a source electrode and receiver electrode. Thus, the number of combination electrode pair M on the sensor can be expressed in Eq. (4) [8].

$$M = \frac{n(n-1)}{2}, \tag{4}$$

n is the number of electrodes. DAS is used to measure the voltage signal and converts it into capacitance. DAS control the input voltage and measure the output of each pair of electrodes. The data obtained are then processed by computer to permittivity before it mapped and becomes the image [9].

2.2 The Development of ECVT Technology

Basically, ECVT (Electrical Capacitance Volume Tomography) is similar to USG, CT Scan and MRI which are widely used in the medical world. But unlike the CT Scan and MRI which are used only to see what happens in the human body, ECVT is more sophisticated because patients do not have to enter into a tube like an MRI to have an output displays a two-dimensional image. ECVT system consists of sensors system, data acquisition systems and a computer for controlling and processing, data reconstruction and display. With this technology, the scanning can be done from the outside, without touching the object even nano-scale objects and objects moving at high speed can be seen. In a related development, ECVT technology already recognized even by NASA, Exxon Mobil, Shell, BP, ConocoPhillips, Dow Chemical, Mitsubishi Chemicals including the Us Department of energy (Morgantown National Laboratory). In Indonesia, this technology is used for scanning a high-pressure gas cylinder, such as gas-fueled vehicle Bus TransJakarta. Up to now, CTECH Edwar Technology Labs still continue to develop ECVT technology for various applications. The research that is being developed currently at Ctech Labs includes Breast Cancer scanner, Brain Cancer Scanner and the healing therapy of cancer [10].

Table 1. Researches comparison of non-destructive tests on fruits.

Last research	Method	Low cost	Simple technique	Data/ graph	3D image
Determination of avocado and mango fruit properties by ultrasonic techniques [3]	Ultrasonic test			√	√
Development of an automated monitoring device to quantify changes in firmness of apples during storage [4]	Ultrasonic test			√	√
Nondestructive measurement for mango inspection [5]	High-frequency electromagnetic waves			√	√
Assessment of quality of fruits using impedance spectroscopy [6]	Impedance spectrometry	√	√	√	
Electrical behavior of garut citrus fruits during ripening changes in resistance and capacitance models of internal fruits [7]	Impedance spectrometry	√	√	√	
Multi-spectral prediction of unripe tomatoes [11]	Spectral and colour measurements	√	√	√	

3 Related Works

In Table 1, there are some researches that have been done in the field of nondestructive test to measure the quality of fruits.

From previous studies related to non-destructive test on fruit, there is no research that using ECVT. ECVT is technology that is relatively inexpensive because it uses low voltage energy source. The voltage used in the ECVT technology ranging from 5 to 12 V, very low if compared to other Tomography technology such as MRI or CT Scan. In addition, when compared with previous studies about nondestructive test for fruit, by using ECVT it is expected that the output not only electrical characteristics data can be obtained to distinguish the conditions and types of fruit, but also image depicting the inside of the fruit.

4 Data Retrieval

The preliminary stages on this research were to measure the electrical characteristic from the fruit which is the capacitance. The aim of the measurement of capacitance is to compare electrical characteristics for each condition and type of fruit. In addition, from the results of these measurements, it is expected to know the optimal frequency as preparations for creating a Data Acquisition System (DAS) unit as a part of ECVT Scanner to display a good image inside the fruit.

There are 2 ways of characteristic measurements are performed in this research as follows:

1. Using LCR Meter
2. Using a signal Generator and Oscilloscope

Measurement of electrical characteristics is done using juice from orange and mango fruits that are filled in the tube with 2 sensors as shown in Fig. 1. The measurement is done to citrus fruits and mango on some fruit condition such as raw, ripe and rotten with the level of acidity (pH) that measured by the pH meter pH-2011 with specs:

Range: 0.00–14.00 pH
Resolution: 0.01 pH
Accuracy: ± 0.1 pH

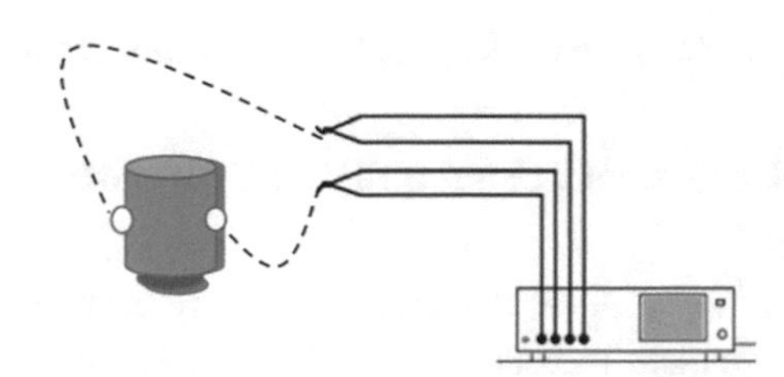

Fig. 1. Capacitance measurement scheme using LCR Meter.

Further tests are performed using existing sensor-sized tubular ECVT with 32 sensors to get a preliminary overview of the resulting image.

4.1 Capacitance Measurement Using LCR Meter

LCR meter that are used: GW Instek LCR821 from Good Will Instrument Co., Ltd. Taiwan. Juice from citrus fruits and mango with various conditions placed on the tube with 2 sensors connected to an LCR Meter as shown in Fig. 1.

For calibration of measurement, we conducted a reading on capacitors 1, 10 and 100pF-sized. In Table 2 the following are the results of the reading of the 3 capacitors with input voltage of 10 V and frequency of 1 kHz.

Table 2. The results of the measurement of capacitor using LCR Meter with an input voltage of 10 V and frequency of 1 kHz.

Capacitor size	LCR meter reading
1 pF	2.25 pF
10 pF	11.97 pF
100 pF	97.05 pF

Following are the results of the measurements which performed using LCR Meter with input voltage 10 V. Frequency measurements varied between 100 Hz to 100 kHz. The result in Fig. 2 is the average value from 5 times measurements of the same sample. The difference on each specimen is very small and almost invisible when plotted on a graph with a logarithmic scale.

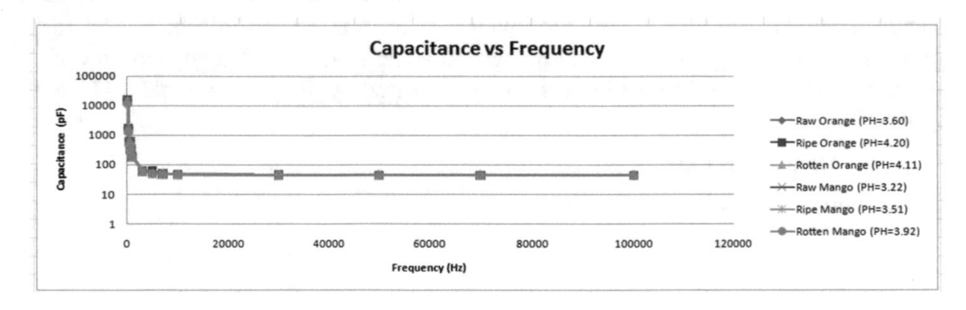

Fig. 2. Measurement results using LCR Meter with input voltage 10 V.

4.2 Capacitance Measurement Using a Signal Generator and Oscilloscope

The measurement is using a signal Generator Tetronix AFG2021 and Oscilloscope Tetronix TBS1064. Complete measurement guide can be seen on the manual from Tetronix about the Capacitance and Inductance Measurements Using an Oscilloscope

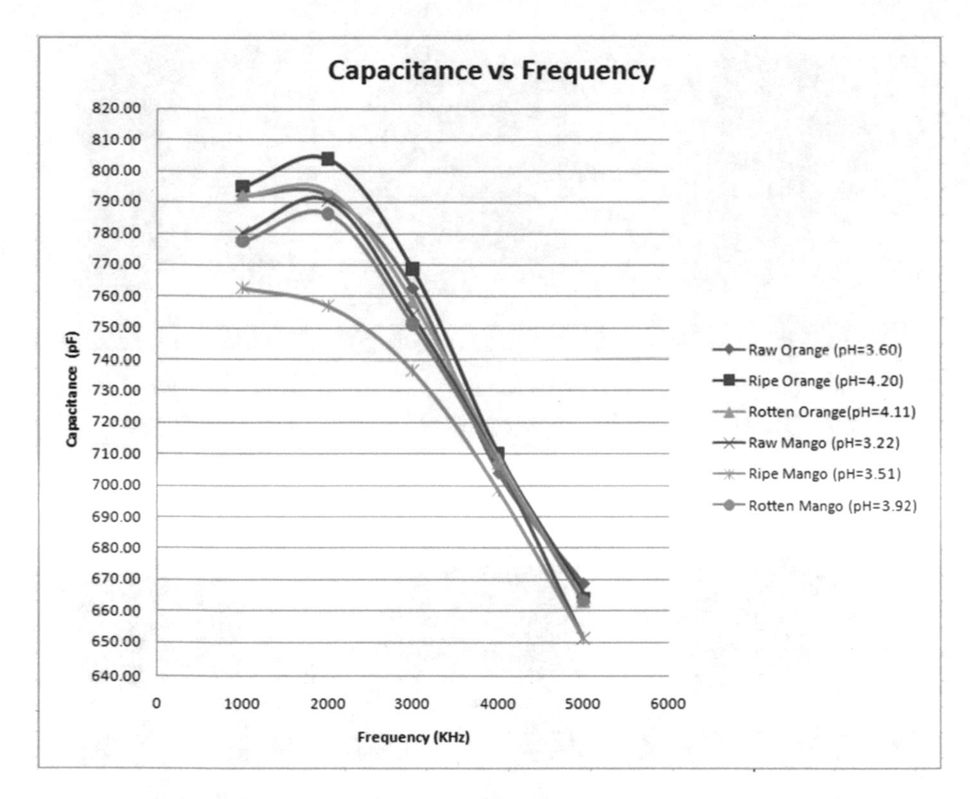

Fig. 3. Measurement result of the capacitance measurements which performed using a signal generator and oscilloscope with input voltage 10 V.

and a Function Generator [12]. The measurement is done at tool measurement range tool at frequency 1 to 5 MHz.

Measurement results are shown in Fig. 3 where the value of the capacitance decreases with the rise of frequency. The difference in the value of capacitance in this result is more clearly seen when compared with measurements using LCR Meter on the frequency 100 Hz to 100 kHz in Fig. 2.

4.3 Image Processing of Orange and Mango Fruits Using ECVT Scanner with 32 Electrode Sensor Tube

Further tests are performed using existing sensor-sized tubular ECVT with 32 sensors to get a preliminary overview of the resulting image. The algorithm used for the image reconstruction in this research is Linear Back Projection (LBP). The instrument used in the measurement is at frequency MHz. The difference between each specimen is not clearly visible. This is most likely due to the large number of air cavity between fruit and the tube-shaped measurement tool. There still need to adjust the design the tool to be fit with mangoes and oranges (Fig. 4).

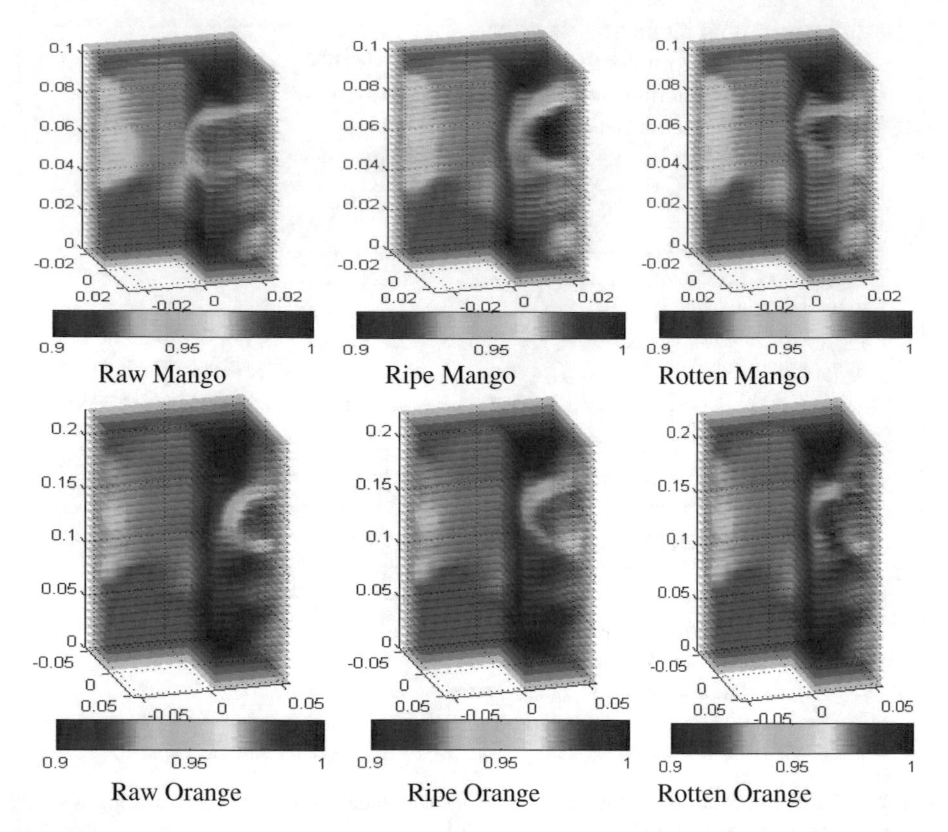

Fig. 4. The results of image processing of citrus and mango fruits with sensor tube 32 electrodes ECVT.

5 Results and Discussion

From the results of the electrical characteristics measurement in the form of capacitance data on orange and mango fruits in some different conditions, in general, the value of the capacitance decreases when frequency is enlarged. On measurements using the RCL Meters in the range of 100 Hz–100 kHz, the difference between some fruit condition looks very small. While on measurements using a signal generator and oscilloscope with 1 MHz frequency range 1–5 MHz, the capacitance difference between fruit conditions more clearly visible. From the above cases it can be concluded that the differences in capacitance seen more clearly at high frequencies so DAS ECVT needs to be designed at high frequencies.

On the use of multi channel ECVT with sensor tube that consists of 32 electrodes, the resulting image shows only the outside of the fruit, so it is difficult to distinguish the condition of each fruits. This is due to the cavity between the tube and the fruit that is still too large.

Further studies need to be made a sensor which can wrap the fruit inside so that the inside of the fruit can be more clearly seen. The algorithm used for the image reconstruction in this research is the Linear Back Projection (LBP). Further research will be used other algorithm such as Linear Iterative Back-Projection (ILBP), and Neural Network Multicriterion Optimization of Image Reconstruction Technique (NN-MOIRT) to produce a better image.

References

1. Tribunnews.com: Indonesia Berencana Ekspor Buah-buahan ke Jepang dan Tiongkok. http://www.tribunnews.com/bisnis/2014/08/22/indonesia-berencana-ekspor-buah-buahan-ke-jepang-dan-tiongkok. News date: 22 Agustus 2014, Accessed 12 Dec 2014
2. Republika.co.id: Upaya Buah Lokal Tembus Pasar Asia. http://www.republika.co.id/berita/koran/ekonomi-koran/14/09/04/nbd7g84-upaya-buah-lokal-tembus-pasar-asia. News date: 04 September 2014, Accessed 12 Dec 2014
3. Mizrach, A.: Determination of avocado and mango fruit properties by ultrasonic techniques. Ultrasonic **38**(1–8), 717–722 (2000)
4. Belie, N.D., Schotte, S., Coucke, P., Baerdemaeker, J.D.: Development of an automated monitoring device to quantify changes in firmness of apples during storage. Postharvest Biol. Technol. **18**, 1–8 (2000)
5. Krairiksh, M., Mearnchu, A., Phongcharoenpanich, C.: Nondestructive measurement for mango inspection. In: International Symposium on Communications and Information Technologies (ISCIT 2001), Sapporo, Japan, 26–29 October 2001
6. Rehman, M., Basem, A.J.A., Izneid, A., Abdullah, M.Z., Arshad, M.R.: Assessment of quality of fruits using impedance spectroscopy. Int. J. Food Sci. Technol. **46**, 1303–1309 (2011)
7. Juansah, J., Budiastra, I.W., Dahlan, K.: Electrical behavior of garut citrus fruits during ripening changes in resistance and capacitance models of internal fruits. Int. J. Eng. Technol. IJET-IJENS **12**(4), 1–8 (2012)
8. Taruno, W.P., Marashdeh, Q., Fan, L.S.: Electrical capacitance volume tomography (ECVT). IEEE Sens. J. **7**, 525–535 (2007)
9. Taruno, W.P., Ihsan, M.F., Baidillah, M.R., et al.: Electrical capacitance volume tomography for human brain motion activity observation. In: Middle East Conference on Biomedical Engineering (MECBME), 17–20 February 2014, pp. 147–150 (2014)
10. CTECH Laboratories EDWAR Technology Website. http://www.c-techlabs.com
11. Hahn, F.: Multi-spectral prediction of unripe tomatoes. Biosyst. Eng. **81**, 147–155 (2002)
12. Tetronix: Capacitance and Inductance Measurements Using an Oscilloscope and a Function Generator, Application Note (2011). http://www.tetronix.com/afg2000

Dependency Scheme for Revise and Reasoning Solution

Wiwin Suwarningsih[1,2](✉), Ayu Purwarianti[1], and Iping Supriana[1]

[1] School of Electronic Engineering and Informatics, Institute Technology Bandung, Bandung, Indonesia
wiwin.suwarningsih@students.itb.ac.id
[2] Research Center for Informatics, Indonesian Institute of Science, Bandung, Indonesia

Abstract. Revising the solution is one step in the cycle of Case-Based Reasoning (CBR). Revising the process is a task to make any improvements to a solution that cannot be reused by a question and answer system. This paper presents a new scheme to improve the solution in question-answer system using the dependency approach between words or phrases. The results of the repair cases should be initially tested to ensure that the obtained solution has met the criteria of a problem. The testing process will be conducted using the search dependency structure of the solution. By using the data of 135 Indonesian sentence English-premises solution in the form of a sentence, the scheme is built to produce an accuracy of 80,74%.

Keywords: Revise solution · CBR · Graph dependency · Indonesian medical sentences · Question-answering pairs

1 Introduction

Question-Answer system with the medical domain is rooted in Evidence Based Medicine (EBM). Case Based Reasoning (CBR) in this case becomes one approach that can be used to support the application of EBM. The process flow of CBR in solving problems is defined into 4 steps [1]: (i) Retrieve – it is to take the most similar issue/case, (ii) Reuse – it is to reuse the issues/cases to solve the problem, (iii) Revise – it is to revise, the solutions submitted if necessary, and (iv) Retain – it is to maintain/save the new solution as part of the problem/new cases. In medical, CBR has been applied in a diagnosis and becomes a part of the therapy. A number of related methods have been used for field orientation method for control engineering learning [2], retrieval method for developing mentor [3]. On the other hand, in a medical decision support system using CBR such as [4]. This application is regarding the misdiagnosis of heart disease. This system has three functions: looking for similar cases; determining the difference, and transferring the diagnosis similar to the current case.

Based on the description above, this paper is designed to explain a new scheme in revise solutions to support medical debriefing system in Indonesian language. Overall, this paper is aimed to provide the following contributions: (i) to produce a tuple relation

T. Herawan et al. (eds.), *Recent Advances on Soft Computing and Data Mining*,
Advances in Intelligent Systems and Computing 549, DOI 10.1007/978-3-319-51281-5_30

between words on sentence answer, and (ii) to create a new strategy for the extraction solutions by determining the relationship dependencies between words in a sentence.

The rest of the paper is organized as follows: Sect. 2 is designed to present some related work; Sect. 3 to describe the proposed method; Sect. 4 for Analysis and Discussion and Sect. 5 as last part to present our conclusion.

2 Related Work

In common, revising a solution concept is not dealt with specifically because this stage is a choice revision when some necessary corrections are made if not required and ignored. Several case studies for an improvement used an ontology matching, such as [5] to find some correspondences between the entities associated with the semantic knowledge representation. Ontology, in this case, allows for matching of data revealed knowledge and is able to adapt and produce a number of appropriate solutions. In [6] stated that the case of significant improvements can be obtained by matching ontology that refers to the existing knowledge base. Matching ontologies is the solution to the problem of semantic heterogeneity.

In addition, there are some studies focused on the use of words such as dependency approach to research conducted by [7] establishing a dependency relationship between the words/phrases that can enhance the error correction and error positions sentence written by the language in which POS-tag and the language model are not sufficiently powerful to detect a preposition, verb and noun types of errors. In [8] it was stated that Indonesian Language is rich of the morphological phenomena such as word prefixes, substituting word, and nonverbal clauses. Thus, a way is required to explain how the scheme accommodates the phenomena in determining the dependency relationship between the two words.

Another related work, there are several research studies using Semantic Role Labeling (SRL) to make any improvements such solution as [9, 10] stating that the semantic roles of arguments will aid the error detection and the introduction of the entities in the sentence. [11, 12] asserts that SRL can identify the semantic relationship predicate-argument in a natural text by analyzing the target verbs and some of its constituents. In [12–14] writes that SRL can be used to identify and label an accompanying argument predicate in a sentence depending upon their semantic relationships. From these studies, we obtained some things that can be explored further and improved to produce strategy using SRL for revising and reasoning solution.

3 Propose Method

We propose the new scheme for revise the solution. In this scheme, the improvement process was focused on the refinement solutions in the form of single or compound Indonesian sentence (The flowchart of this study is illustrated in Fig. 1). The method to be used was the extraction of relationships between words with dependency graph based element of PICO SRL. Stages performed consisted of: (i) the extraction solution of selected cases, the extraction was performed to predict the dependencies structure

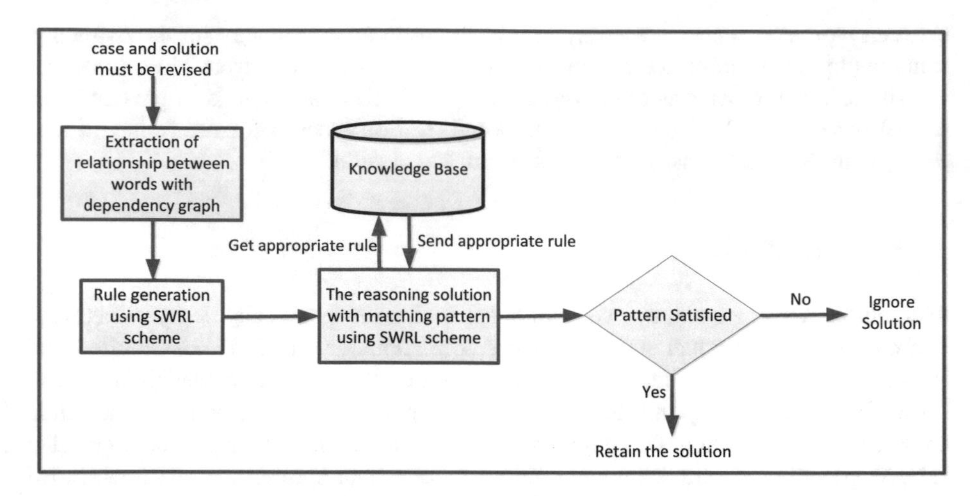

Fig. 1. Proposed method for revise solution

among words (ii) specifying the dependencies between words based SRL, (iii) generating the rule by the scheme SWRL (semantic web rule language) and a test of solution improvement results in a way reasoning to the knowledge base.

4 Result and Discussion

This section explains the results and discussion of the proposed method. Regarding those going to cover the extraction solution of selected cases, it can be done by determining a dependency between words on the solution, generation scheme rules SWRL and test solution improved results.

4.1 Relation Extraction Between Words Using Dependency Graph

Dependency graph is defined as a directed graph consisting of a set of nodes, bows and the priority order on the node. Node was labeled with the forms of words while the arc was labeled with the dependency type. The Stanford representation of dependencies [15] was designed to provide a simple overview of grammatical relations in English sentences that could be easily understood and effectively used by people without any requirement of linguistic skills in taking a textual relationship [16]. When compared with the representation of sentence structure, the dependency graph represented all contexts as a relationship of dependency. Considering that Indonesian language has a number of similarities with English sentence structure, it is expected that it could represent these dependencies Stanford dependency grammar in Indonesian.

The extraction step was performed to predict the dependencies' structure among words. This was performed as the words becoming the head could be used to carry information [17], and focuses on the use of important words (ex. based on SRL) [18]. The strategies undertaken to determine dependency graph (see Fig. 2) are presented as follows:

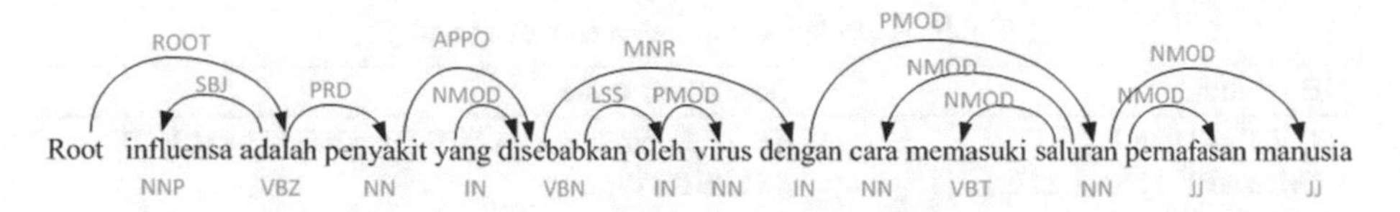

Fig. 2. Words dependency graph

- Evaluating dependence by calculating dependency parsing suitable: for a given word, Head and labeling must comply with the reference and the hypothesis.
- Evaluating SRL by calculating the semantic matching bow: bow in accordance with the predicate, the word labeled as Head of the argument and the argument labeled.
- Evaluating these two tasks as a special case to consider the roots of the tree of dependencies and identity predicate.
- Assumption: that the reference and the hypothesis of words and sentences that are evaluated are identical.

Here is an example of a sentence to be repaired.

sentence-1: influensa adalah penyakit yang disebabkan oleh virus dengan cara memasuki saluran pernafasan manusia.

(in English: Influenza is a disease caused by the virus by entering human respiratory tract.)

By using the strategies above dependency graph, then for sentence-1 form dependency graph can be seen in Fig. 2.

Referring to Fig. 2, it is assumed that the relation between the words will be extracted only among the entities contained in the same sentence. This means that information is obtained only from the same sentence including two entities that have a relevance to do an extraction process relation. Each sentence can be attributed to its dependence graph, where the word or phrase is described as nodes and dependencies between words or phrases as illustrated by arrows. To obtain information from the unstructured sentences, it is necessary to determine the definition of target information as structured information to be extracted. This information may include any entities or relationships between entities. This phase was carried out purposely to define the semantic relationships in a way of entities into the arguments of the relation. Name Relation produced are: "*DISEBABKAN-OLEH*" (CAUSED-BY), "*INDIKASI*" (INDICATION), "*MENGATASI*" (FIGHT), "*BERADA-DI*" (STAY-IN), "*MEMASUKI*" (ENTER), "*UNTUK-MENGOBATI*" (TO-TREAT), "*MENGAKIBATKAN*" (EFFECT), "*BERPOTENSI*" (POTENTIAL), "*MENGENDALIKAN*" (CONTROLLING). A number of sample sentences and relationships can be seen in Table 1.

Table 1. Relationship between names entity

NE relation	Generation tuple
[POPULATION] **DISEBABKAN-OLEH** [PROBLEM]	([POPUL:'influenza'] **'DISEBABKAN-OLEH'** [PROBL:'Virus'])
[CONTROL] **MENGATASI** [POPULATION]	([CONTR:'konsumsi madu'] **'MENGATASI'** [POPUL:'batuk berdahak'])
[OUTCOME] **INDIKASI** [POPULATION]	([OUTCO:'hemoglobin tinggi'] **'INDIKASI'** [POPUL:'penyakit jantung'])
[DRUG] **UNTUK-MENGOBATI** [POPULATION]	([DRUG:'teosol'] **'UNTUK-MENGOBATI'** [POPUL:'asma'])
[DRUG] **MENGAKIBATKAN** [POPULATION]	([DRUG:'teosol'] **'MENGAKIBATKAN'** [POPUL:'pusing'])
[CONTROL] **MENGAKIBATKAN** [OUTCOME]	([CONTR:'konsumsi madu'] **'MENGAKIBATKAN'** [OUTCO:'badan sehat'])
[CONTROL] **MENGENDALIKAN** [POPULATION]	([CONTR:'konsumsi madu'] **'MENGENDALIKAN'** [POPUL:'batuk berdahak'])
[DRUG] **MENGENDALIKAN** [POPULATION]	([DRUG:'teosol'] **'MENGENDALIKAN'** [POPUL:'asma'])
[PROBLEM] **BERPOTENSI** [POPULATION]	([PROBL:'asap kendaraan'] **'BERPOTENSI'** [POPUL:'asma'])
[DRUG] **BERPOTENSI** [OUTCOME]	([DRUG:'paracetamol'] **'BERPOTENSI'** [OUTCO:'panas menurun'])

4.2 Generation Rules with SWRL Scheme

Extraction dependent relationship between words was then used to generate a number of rules with SWRL scheme. The generation of this rule is used for the search process in the knowledge base when a new solution is generated. The strategy used to generate the rules included (i) creating a relationship with an adaptation of the references and (ii) generation rule of instantiation relations.

4.2.1 Creating a Relationship with an Adaptation of the References

References used for the manufacture of a relation are the relation NE rules (Table 1). The adaptation of this reference means to use historical data that had been made before the study was conducted. An adaptation reference was made to validate the relationship built between the words. Hence, the prediction of the rules of the relationship between words can be done with the correct data. Results relation to the example sentence-1 can be seen in Table 2.

Table 2. Relation of instantiation for example sentences-1

Template of relation	Form of instantiation relationships
<POPULATION> **DISEBABKAN-OLEH** <PROBLEM>	*influensa* **DISEBABKAN-OLEH** *virus*
<POPULATION> **DISEBABKAN-OLEH** <PROBLEM>	*Penyakit* **DISEBABKAN-OLEH** *virus*
<ILLNES> **DIINDIKASIKAN** <PROBLEM>	*Influensa* **DIINDIKASIKAN** *virus*
<PROBLEM> **BERADA-DI** <ORGANS>	*virus* **BERADA-DI** *saluran pernafasan*
<PROBLEM> **MEMASUKI** <ORGANS>	*virus* **MEMASUKI** *saluran pernafasan*

4.2.2 Generation Rule of Instantiation Relation

Referring to the research that has been conducted by [18] regarding the information extraction using SWRL, then at the stage of generating rules from the instance of these relations, a same scheme (SWRL) would be used. The stages comprised (i) the grammatical analysis: this stage to labeling with NE Tagger words and grouping words into phrases. (Ii) Semantic Analysis: This stage of the word ontology mapping any medical knowledge. (Iii) Generation rules: This stage is the establishment of rules with scheme of SWRL to be used for pattern matching rules to the data in the knowledge base. Below, the stages of generation rules are briefly outlined.

4.2.2.1 Grammatical Analysis

Grammatical structure of sentences in the text was analyzed to identify how this relationship was used to describe the concept of a domain. Recognizing the relationship between words resembles to an atom on ontology scheme such as the concept, instances, part-of is an important step for the rules of acquisition of the text. Ideally, each domain concept recognized in the text is linked with the concept of another domain via a relationship to determine how the ontological atoms are linked to each other. Based on the example sentence-1, text analysis was done to convert the text by using NE Tagger to establish a set of words. The results from the conversion of text to sentence-1 are:

Influensa/NNP adalah/VBZ penyakit/NN yang/IN disebabkan/VBN oleh/IN virus/NN dengan/IN cara/NN memasuki/VBT/ saluran/NN pernafasan/JJ manusia/JJ

The identification of grammatical next is the establishment of example sentences-1 to determine the sentence separation into body and head. The identification of grammatical formed by the phrase is [NP IN VP]. Because the sample sentence-1 is a compound sentence, the separation of sentences separated by conjunctions (conjunctions) that the word 'with'. Hence, the separation of words formed by the phrase (see Fig. 3) is a Body Part: *influensa* (influenza), *penyakit* (disease), *disebabkan* (caused), virus (virus) and Section Head: *memasuki* (entering), *saluran pernafasan* (respiratory), *manusia* (human).

influensa	NNP	influensa	body
adalah	VBZ	*adalah : stop word*	
penyakit	NN	penyakit	
yang	IN	*yang : stop word*	
disebabkan	VBN	disebabkan	
oleh	IN	*oleh : stop word*	
virus	NN	virus	
dengan	IN	***dengan : Konjungsi***	
cara	NN	*cara : stop word*	head
memasuki	VBG	memasuki	
saluran	NN	saluran	
pernafasan	JJ	pernafasan	
manusia	JJ	manusia	

Fig. 3. Gramatical analysis result

4.2.2.2 Semantic Analysis

Semantic analysis was performed to see the role of the word served to bridge a semantic gap between phrases and vocabulary ontology terminology. This step made it possible to find words into phrases in the ontology. The first step was to identify the entity of ontologies that correspond to each term of phrase input. When the input did not belong ontology term, we then must look for any synonyms in Indonesian corpus (for medical entities) and among those selected in accordance with the existing data in the ontology. The results of the mapping of each word in the input sentence with the phrase contained in ontology of P3K: influensa → Class ∈ OntologiP3K; *penyakit* → object property ∈ OntologiP3K; *disebabkan* → object property ∈ OntologiP3K; *virus* → Class ∈ OntologiP3K; *memasuki* → object property ∈ OntologiP3K; *saluran pernafasan* → Subclass ∈ OntologiP3K; *manusia* → Class ∈ OntologiP3K.

The second step of semantic analysis was pairing to the shape argument <argument, relation, argument> for validation to the ontology P3K. All possible pairs of arguments combined with relation to obtain an optimal partner. The result of a combination of a couple arguments and these relations selected by the selection strategy are as follows: (i) the couple's argument containing "relationship" is ignored, (ii) all of the pairs argument combined with all relationships (*"DISEBABKAN-OLEH", "INDIKASI", "MENGATASI", "UNTUK-MENGOBATI", "MENGAKIBATKAN", "BERPOTENSI", "MENGENDALIKAN", "BERADA-DI", "MEMASUKI"*), (iii) selection argument the couple and relatives using NE relation rules. All possible pairs of input text argument can be seen in Fig. 4.

4.2.2.3 Generation Rules

At this stage the step was to describe the generation SWRL rules. This stage was needed to obtain some formal rules of written language SWRL by utilizing a sequence of semantic link existing between the components of each pair-argument at the previous phase. Rule generation algorithm is as follows:

Part of Body	Part of Head
(influensa, penyakit)	(cara, memasuki)
(influensa, disebabkan)	(cara, saluran)
(influensa, virus)	(cara, pernafasan)
(penyakit, disebabkan)	(cara, manusia)
(penyakit, virus)	(memasuki, saluran)
(disebabkan, virus)	(memasuki, pernafasan)
	(memasuki, manusia)
	(saluran, pernafasan)
	(saluran, manusia)

Fig. 4. Relationship form <argument, relation, argument> from input text

Algorithm Rule generation

```
Getting all the combinations of pair relations
Instantiation relation to form pairs (argument1, relationships,
argument2).
{The process of looping through a combination of pair relations}
While (the end of the combination partner relationship) do
   {Generate Rule for Body}
   IF find format (t1 argument, relationship, argument-2) THEN
      Change Argument1 be scheme-1-> argument1(?X)
      Change Relationships and argument-2 a scheme-2-
        >relation(?X,argument2)
      Create Relationships scheme-1 and scheme-2 with relation:
        and (^)
      IF find format(arguments, relationship)OR (relation,
      argument) THEN ignore

      {Generate Rule for Body}
       IF found form (relation, argument)
                THEN transform to scheme-2 -> relation (?X,
                     argument).
      IF found form (argument1, relationships, argument2)
                THEN transform to scheme-2 -> relation (?X,
                    argument)
            IF found form (arguments, relationship) THEN ignore
      end-while

{Looping process ends}
Selection Rule of the generation reference table
```

Encryption refers to the generation rule, generate rule results with SWRL schemes have variations as shown in Table 3.

4.2.3 Testing of Revise Solutions Results

The solution improved results were tested to determine whether the solution could be reused. The test scenario was the generation of pattern matching rules as an example in Table 5.7 to the pattern of the rules derived from the knowledge base. If the patterns

Table 3. Rule generation result

Pair of (argument1, relation, argument2)	Generating rule
(influensa, disebabkan_oleh, virus)	influensa (?x) ^ disebabkan_oleh (?x, virus)
(penyakit, disebabkan_oleh, virus)	penyakit (?x) ^ disebabkan_oleh (?x, virus)
(influensa, diindikasikan, virus)	influensa (?x) ^ diindikasikan (?x, virus)
(virus, berada-di, saluran_pernafasan)	virus (?x) ^ berada-di (?x, saluran_pernafasan)
(virus, memasuki, saluran_pernafasan)	virus (?x) ^ memasuki (?x, saluran_pernafasan)

were matching, it means that the search was successful and the solution could be reused and stored in the database of QA pairs. If the pattern did not match, then the solution may be ignored or manual analysis with the support of experts.

The test on solutions was made by taking 135 examples of solutions in the form of a sentence to answer some cases. Accuracy obtained was 80.74% (109 correct out of 135). From the test results 26 sentences were obtained to do the search. Error in searching this knowledge base was caused by such things as the following definition of partner relations between argument and imprecise and ambiguous occurs when rules are raised because of the lack of specifications defining arguments and relationships.

5 Conclusion

This study has presented a new scheme to revise and reasoning cases. As for the method of use and the generation rule, dependency graph with SWRL scheme produce significant improvement solutions. By using the data 135 solution in the form of a sentence, the scheme was built to produce an accuracy of 80.74%. However, there were still a few of errors at the time of the search to the current knowledge base improved yields tested cases. It was caused by a pair of defining relations between argument and imprecise, ambiguity when rules are raised for the lack of specifications defining arguments and relationships. Future work to resolve ambiguities and errors in the definition of a couple of arguments and relationships can be conducted using a semantic approach template to generate the value of optimum accuracy.

Acknowledgments. We would also thank the anonymous reviewers for their detailed comments, which have helped us to improve the quality of this work.

References

1. Aamodt: cased-based reasoning: foundation issues. In: AICOM, pp. 39–597 (1994)
2. Acevedo, P.R., Martinez, J.J.F.: Case-based reasoning and system identification for control engineering learning. IEEE Trans. Educ. **51**(2), 271–281 (2008)
3. Shimazu, H., Shibata, A., Nihei, K.: Expert guide a conversational case-based reasoning tool for developing mentors in knowledge spaces. Appl. Intell. **14**(1), 33–48 (2001)

4. Salem, A.-B.M., El Bagoury, B.M.: A case-based adaptation model for thyroid cancer diagnosis using neural networks. In: Proceedings of the Sixteenth International FLAIRS Conference, pp. 155–159. AAAI Press (2003)
5. Euzenat, J., Valtchev, P.: Similarity-based ontology alignment in OWL-lite. In: de Mantaras and Saitta, pp. 333–337 (2004)
6. Pavel, S.: Ontology matching: state of the art and future challenges. IEEE Trans. Knowl. Data Eng. **25**(1), 158–176 (2013)
7. Chatterji, S., Sarkar, T.M., Dhang, P., Deb, S., Sarkar, S., Chakraborty, J., Basu, A.: A dependency annotation scheme for Bangla Treebank. Lang. Resour. Eval. **48**(3), 443–477 (2014). Springer, Netherland
8. Irmawati, B., Shindo, H., Matsumoto, Y.: A Dependency annotation scheme for Indonesian, the association for natural language processing, pp: 740–743 (2015)
9. Punyakanok, V., Roth, D., Yih, W.: The necessity of syntactic parsing for semantic role labeling. In: Proceedings of CoNLL, pp. 1117–1123 (2004)
10. Palmer, M., Gildea, D., Kingsbury, P.: The proposition bank: an annotated corpus of semantic roles. Comput. Linguist. **31**(1), 71–106 (2005)
11. Liu, H., Che, W., Liu, T.: Feature engineering for Chinese semantic role labeling. J. Chin. Inf. Process. **21**(1), 79–84 (2007)
12. Bi, Y., Chen, J.: A study on semantic role labeling of Korean sentence. In: International Conference on Asian Languages Processing, pp. 102–107 (2009)
13. Gildea, D., Lurafsky, D.: Automatic labeling of semantic roles. Comput. Linguist **28**(3), 245–288 (2002)
14. Zhang, M., Che, W., Zhou, G., Aw, A., Tan, C.L., Liu, T., Li, S.: Semantic role labeling using a grammar-driven convolution tree kernel. IEEE Trans. Audio Speech Lang. Process. **16**(7), 1315–1329 (2008)
15. de Marneffe, M.-C., Manning, C.D.: The Stanford typed dependencies representation (2008). http://nlp.stanford.edu/pubs/dependencies-coling08.pdf
16. Levy, R., Manning, C.D.: Deep dependencies from context-free statistical parsers: correcting the surface dependency approximation. ACL **42**, 328–335 (2004)
17. Zouaq, A., Gagnon, M., Ozell, B.: Semantic analysis using dependency-based grammars and upper-level ontologies. Int. J. Comput. Linguist. Appl. **1**(1–2), 234–240 (2010)
18. Boufrida, A., Boufaida, Z.: Automatic rules extraction from medical texts. In: International Workshop on Advanced Information Systems for Enterprises (IWAISE), pp. 29–33 (2014)

A New Binary Similarity Measure Based on Integration of the Strengths of Existing Measures: Application to Software Clustering

Rashid Naseem(✉) and Mustafa Mat Deris

Faculty of Computer Science and Information Technology,
Universiti Tun Hussein Onn Malaysia, 86400 Parit Raja, Batu Pahat, Johor, Malaysia
rnsqau@gmail.com

Abstract. Different binary similarity measures have been explored with different agglomerative hierarchical clustering approaches for software clustering, to make the software systems understandable and manageable. Similarity measures have strengths and weakness that results in improving and deteriorating clustering quality. Determine whether strengths of the similarity measures can be used to avoid their weaknesses for software clustering. This paper presents the strengths of some of the well known existing binary similarity measures. Using these strengths, this paper introduces an improved new binary similarity measure. A series of experiments, on five different test software systems, is presented to evaluate the effectiveness of our new binary similarity measure. The results indicate that our new measure show the combined strengths of the existing similarity measures by reducing the arbitrary decisions, increasing the number of clusters and thus improve the authoritativeness of the clustering.

Keywords: Binary similarity measures · Improved measure · Agglomerative hierarchical clustering · Software clustering

1 Introduction

Clustering is an approach that makes clusters of similar entities in the data. In the software domain, an important application of clustering is to modularize a software system or to recover the module architecture or components of the software systems by clustering the software entities, e.g. functions, files or classes, in the source code. Recovery is very important when no up-to-date documentation of a software system is available [1].

Agglomerative Hierarchical Clustering (AHC) algorithms have commonly used by researchers for software clustering [3,4]. AHC comprises of two main factors, a similarity measure to find the association between two entities and a linkage method to update the similarity values between entities in each iteration. However, selection of a similarity measure is an important factor in AHC [5], that has a major influence on the clustering results [7]. For software clustering

T. Herawan et al. (eds.), *Recent Advances on Soft Computing and Data Mining*,
Advances in Intelligent Systems and Computing 549, DOI 10.1007/978-3-319-51281-5_31

the comparative studies has reported that Jaccard binary similarity measure produced better clustering results [7]. In our previous study [6], we proposed a new binary similarity measure, called JaccardNM, which could overcome some deficiencies of Jaccard binary similarity measure.

In this paper, we explore the integration of the existing binary similarity measures for AHC algorithms using linkage methods (e.g. Complete Linkage (CL) and Singel Linkage (SL)). For example, we select the Jaccard similarity measure, which produces a relatively large number of clusters [8,9] and the JaccardNM binary similarity measure which takes less number of arbitrary decisions [6,10]. Creating large number of clusters means that a clustering approach may creates compact clusters, hence improving the quality of clustering results [3]. Arbitrary decision is the arbitrary clustering of two entities when there exist more than two equally similar entities, hence arbitrary decisions create problems and reduce the quality of clustering results [3,6]. This analysis leads us to introduce better binary similarity measures by combining the Jaccard and JaccardNM measures, i.e. "Jaccam".

The paper is organized as follows: Sect. 2 illustrate the software clustering using AHC algorithm. Section 3 shows and analyze the strengths of the existing similarity measures and the new proposed similarity measure. Section 4 gives the experimental results and discussion on comparing our new similarity measure with existing similarity measures by using arbitrary decisions, number of clusters and authoritativeness as evaluation criteria. Section 5 concludes this paper.

2 Software Clustering Using AHC

Algorithm 1 presents the main steps of AHC, which starts by grouping the entities into small clusters in a bottom up fashion. In every iteration, AHC clusters the most similar entities until the targeted number of clusters is reached or a final large cluster that contains all entities is formed. When AHC is employed for the software clustering, the first step that occurs is the selection of the entities to be clustered where each entity is described by different features.

ALGORITHM 1. Agglomerative Hierarchical Clustering (AHC) Algorithm

Input: Feature (*F*) matrix
Output: Hierarchy of Clusters (Dendrogram)

initialization;
1 Create a similarity matrix by calculating similarity using a **Similarity Measure** between each pair of entities;
repeat
  2 Group the most similar (singleton) clusters into one cluster (using maximum value of similarity in similarity matrix);
  3 Update the similarity matrix by recalculating similarity using a **Linkage Method** between newly formed cluster and existing (singleton) clusters;
until *the required number of clusters or a single large cluster is formed*;

2.1 Selection of Entities and Features

Selecting the entities and features associated with entities depends on the type of software system to be clustered. Researchers have used different types of entities

Table 1. An example feature ($E \times F$) matrix

	f1	f2	f3	f4	f5	f6	f7
E1	1	1	0	0	0	0	0
E2	1	1	0	0	0	0	0
E3	1	0	1	1	0	0	0
E4	0	0	1	1	1	0	0
E5	0	0	0	0	0	1	0

Table 2. The similarity matrix derived from the matrix in Table 1 by using the Jaccard similarity measure

	E1	E2	E3	E4	E5
E1					
E2	1				
E3	0.25	0.25			
E4	0	0	0.5		
E5	0	0	0	0.2	

e.g. files [11], classes [12] and methods [9]. Researchers have also used different types of features to describe the entities such as global variables used by an entity [2], procedure calls [11]. Features may be in binary or non-binary format. A binary feature represents the presence or absence of a feature, while non-binary features are weighted features, to demonstrate the strength of the relationship between entities. Binary features are widely used for software clustering [5,13].

When entities and features are extracted from a software system, it results in a feature matrix of size $E \times F$, where E is the total number of entities and F is the total number of features. AHC takes $E \times F$ as input, as shown in Algorithm 1. Table 1 shows an example 0–1 feature matrix *ExF*, which contains 5 entities (E1–E5) and 7 binary features (f1–f7). In Table 1, for example, f1 is present in entities E1, E2, and E3 while absent in entities E4 and E5.

2.2 Selection of Similarity Measure

The first step of the AHC process is to calculate the similarity between the entities to obtain a similarity matrix. For this purpose a similarity measure can be used. Some of the well known binary similarity measures:

$$Jaccard = a/(a+b+c) \tag{1}$$

$$JaccardNM = a/(2(a+b+c)+d) \tag{2}$$

All the existing binary similarity measures are expressed as combinations of the four quantities associated with the pair of entities (Ei, Ej): (1) the number of features common to both entities, denoted by a; (2) the number of features present in Ei, but not in Ej, denoted by b; (3) the number of features present in Ej, but not in Ei, denoted by c; (4) the number of features absent in both entities, denoted by d. It is important to note that a+b+c+d is equal to the total number of features F.

To illustrate the calculation of Jaccard measure as defined in Eq. 1, Table 2 gives the corresponding similarity matrix of the feature matrix shown in Table 1. The similarity between E1 and E2 is calculated using the quantities defined by a, b, c, and d, and in this case a = 2, b = 0, c = 0, and d = 5. Putting all these values in Jaccard similarity measure, we get similarity value '1' (shown in Table 2). Likewise, similarity values are calculated for each pair of entities and are presented in Table 2. Now AHC will group the most similar entities in Table 2, according to the Step 2 in Algorithm 1. E1 and E2 have the highest similarity value, so AHC groups these entities in a single cluster (E1E2). A new cluster is therefore formed, and AHC will update the similarity values of E1E2 and all other (singleton) clusters, i.e., E3, E4, and E5. To update these similarity values different linkage methods can be used, which are described in the next subsection.

2.3 Selection of the Linkage Method

When a new cluster is formed, the similarities between new and the existing clusters are updated using a linkage method (Step 3 of Algorithm 1). There exist a number of linkage methods which update similarities differently. However, in this study we only discuss those linkage methods which are widely used for software clustering. They are listed below, where (EmEn) represents a new cluster and Eo represents an existing singleton cluster.

- *CL(EmEn, Eo)= min(similarity(Em, Eo), similarity(En, Eo))*
- *SL(EmEn, Eo)= max(similarity(Em, Eo), similarity(En, Eo))*

In the illustrative example, we update similarity values between a new cluster (E1E2) and existing singleton clusters using CL method. The updated similarity matrix is shown in Table 3. For example, the CL method returns the minimum similarity value between E1 and E3 (i.e., 0.25) and E2 and E3 (i.e., 0.25). Both of the returned values are the same (if there was a minimum, that would be selected), therefore, AHC selects this similarity value as the new similarity between (E1E2) and E3, as shown in Table 3. Similarly, all similarity values are updated between (E1E2) and E4 and E5.

Table 3. The updated similarity matrix from the values in Table 2 using CL linkage method

	E1E2	E3	E4	E5
E1E2				
E3	0.25			
E4	0	0.5		
E5	0	0	0.2	

AHC repeats Steps 2 and 3 until all entities are merged in one large cluster, or the desired number of clusters is obtained. At the end AHC results in a hierarchy of clusters, also known as dendrogram. The obtained hierarchy is then evaluated to assess the performance of similarity measures and linkage methods.

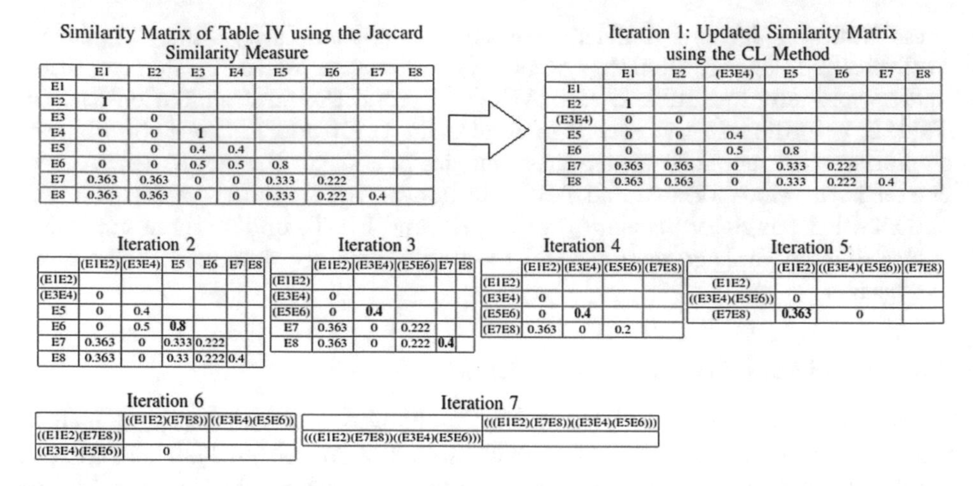

Similarity Matrix of Table IV using the Jaccard Similarity Measure

	E1	E2	E3	E4	E5	E6	E7	E8
E1								
E2	1							
E3	0	0						
E4	0	0	1					
E5	0	0	0.4	0.4				
E6	0	0	0.5	0.5	0.8			
E7	0.363	0.363	0	0	0.333	0.222		
E8	0.363	0.363	0	0	0.333	0.222	0.4	

Iteration 1: Updated Similarity Matrix using the CL Method

	E1	E2	(E3E4)	E5	E6	E7	E8
E1							
E2	1						
(E3E4)	0	0					
E5	0	0	0.4				
E6	0	0	0.5	0.8			
E7	0.363	0.363	0	0.333	0.222		
E8	0.363	0.363	0	0.333	0.222	0.4	

Iteration 2

	(E1E2)	(E3E4)	E5	E6	E7	E8
(E1E2)						
(E3E4)	0					
E5	0	0.4				
E6	0	0.5	**0.8**			
E7	0.363	0	0.333	0.222		
E8	0.363	0	0.33	0.222	0.4	

Iteration 3

	(E1E2)	(E3E4)	(E5E6)	E7	E8
(E1E2)					
(E3E4)	0				
(E5E6)	0	**0.4**			
E7	0.363	0	0.222		
E8	0.363	0	0.222	**0.4**	

Iteration 4

	(E1E2)	(E3E4)	(E5E6)	(E7E8)
(E1E2)				
(E3E4)	0			
(E5E6)	0	**0.4**		
(E7E8)	0.363	0	0.2	

Iteration 5

	(E1E2)	((E3E4)(E5E6))	(E7E8)
(E1E2)			
((E3E4)(E5E6))	0		
(E7E8)	**0.363**	0	

Iteration 6

	((E1E2)(E7E8))	((E3E4)(E5E6))
((E1E2)(E7E8))		
((E3E4)(E5E6))	0	

Iteration 7

	(((E1E2)(E7E8))((E3E4)(E5E6)))
(((E1E2)(E7E8))((E3E4)(E5E6)))	

Fig. 1. The similarity matrix and iterations in clustering process using Jaccard measure and CL method

3 The Jaccam Similarity Measure

As discussed in Sect. 1, we define a new similarity measure which has the combined strengths of the Jaccard and JaccardNM defined in Eqs. 1 and 2, respectively. To highlight the strengths of these existing measures we first show a small example case study, and then define our new similarity measure.

3.1 An Example Case Study

To illustrate the strengths of existing similarity measures, we take an example feature matrix (see Table 4). Feature matrix contains 8 entities (E1–E8) and 13 features (f1–f13). Using feature matrix shown in Table 4, we illustrate the strengths of Jaccard and JaccardNM similarity measures and CL method is used to updated the similarity matrix.

Jaccard with CL Clustering Process. First we illustrate the Jaccard measure with the CL method. The first step of AHC is to create the similarity matrix using a similarity measure. After applying Jaccard measure to feature matrix in Table 4, we get the similarity matrix shown in Fig. 1. In the first iteration of AHC, a maximum similarity value from the similarity matrix is selected to make a new cluster or to update a cluster. So, AHC searches for a maximum similarity value in the similarity matrix but it finds maximum similarity value '1' two times. Hence, there are two arbitrary decisions as (E1E2) has similarity value equal to 1, meanwhile (E3E4) also has the same similarity value. At this stage, AHC arbitrarily selects similarity value of (E3E4), so (E3E4) cluster is made (see Iteration 1 in Fig. 1).

Table 4. Feature matrix

	f1	f2	f3	f4	f5	f6	f7	f8	f9	f10	f11	f12	f13
E1	0	0	0	0	0	1	1	1	1	1	1	1	1
E2	0	0	0	0	0	1	1	1	1	1	1	1	1
E3	1	1	0	0	0	0	0	0	0	0	0	0	0
E4	1	1	0	0	0	0	0	0	0	0	0	0	0
E5	1	1	1	1	1	0	0	0	0	0	0	0	0
E6	1	1	1	1	0	0	0	0	0	0	0	0	0
E7	0	0	1	1	1	1	1	0	0	1	0	1	0
E8	0	0	1	1	1	1	0	1	1	0	1	0	0

The CL method is used to update the similarity values between the new cluster i.e. (E3E4) and all existing singleton clusters, and the updated similarity matrix (see Iteration 1 in Fig. 1). In the second iteration, AHC searches again for the maximum value in updated similarity matrix i.e. matrix in Iteration 1. This time it makes (E1E2) as a new cluster and updates its similarity values with all other existing clusters, as shown in Iteration 2 of Fig. 1. In iterations 3 and 4 it makes clusters of (E5E6) and (E7E8), respectively. In Iteration 3 it can be seen that there are two maximum values (i.e. 0.4), hence AHC may select either again. As stated before, AHC will select value that occurs later, therefore it makes cluster (E7E8). In the remaining iterations, AHC makes clusters of ((E3E4) (E5E6)), ((E1E2) (E7E8)) and (((E1E2) (E7E8)) ((E3E4) (E5E6))), as shown in Fig. 1

JaccardNM with CL Clustering Process. Now we apply the JaccardNM measure on the feature matrix given in Table 4, and get similarity matrix which can be seen in Fig. 2. The process for making clusters is the same as discussed in Subsect. 3.1. As per the AHC, the first cluster formed is (E1E2), second is (E5E6), third is (E7E8), fourth is ((E1E2) (E7E8)), fifth is (E3E4), sixth is ((E3E4) (E5E6)), and the last is (((E1E2) (E7E8)) ((E3E4) (E5E6))). The similarity matrices during iterations, i.e. from the first iteration to the seventh (n−1) iteration, are given in Fig. 2. In each iteration, the CL method is used to update the similarity between newly formed and existing (singleton) clusters.

Discussion on the Results of Jaccard and JaccardNM Measures. In the previous two Subsects. 3.1 and 3.2, we observed that the Jaccard measure results in more number of clusters as compared to the JaccardNM measure. JaccardNM creates less number of arbitrary decisions as compared to the Jaccard. The JaccardNM produces results as expected because the main intuition of introducing this measure is to reduce the arbitrary decisions [6]. Hence, from these results we can easily conclude that the Jaccard has the strength to create more number of clusters, while the JaccardNM has the strength to reduce the number of arbitrary decisions.

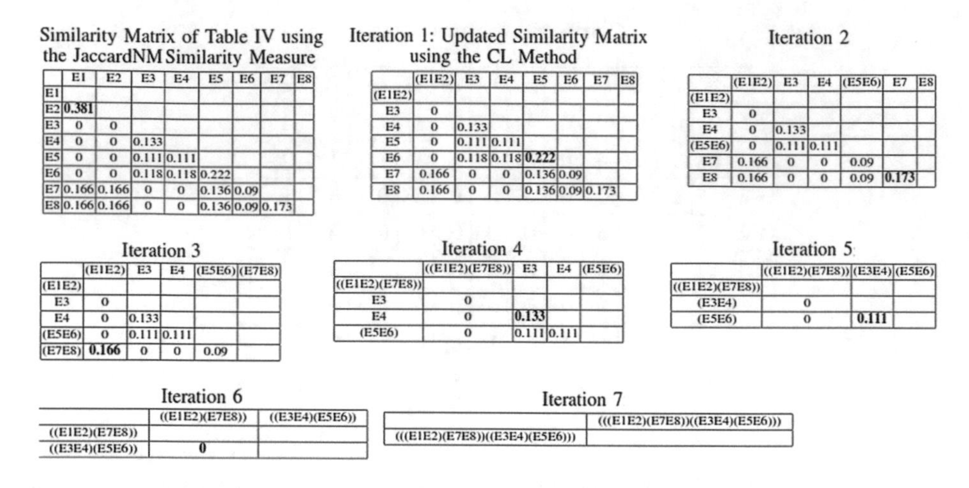

Similarity Matrix of Table IV using the JaccardNM Similarity Measure

	E1	E2	E3	E4	E5	E6	E7	E8
E1								
E2	**0.381**							
E3	0	0						
E4	0	0	0.133					
E5	0	0	0.111	0.111				
E6	0	0	0.118	0.118	0.222			
E7	0.166	0.166	0	0	0.136	0.09		
E8	0.166	0.166	0	0	0.136	0.09	0.173	

Iteration 1: Updated Similarity Matrix using the CL Method

	(E1E2)	E3	E4	E5	E6	E7	E8
(E1E2)							
E3	0						
E4	0	0.133					
E5	0	0.111	0.111				
E6	0	0.118	0.118	**0.222**			
E7	0.166	0	0	0.136	0.09		
E8	0.166	0	0	0.136	0.09	0.173	

Iteration 2

	(E1E2)	E3	E4	(E5E6)	E7	E8
(E1E2)						
E3	0					
E4	0	0.133				
(E5E6)	0	0.111	0.111			
E7	0.166	0	0	0.09		
E8	0.166	0	0	0.09	**0.173**	

Iteration 3

	(E1E2)	E3	E4	(E5E6)	(E7E8)
(E1E2)					
E3	0				
E4	0	0.133			
(E5E6)	0	0.111	0.111		
(E7E8)	**0.166**	0	0	0.09	

Iteration 4

	((E1E2)(E7E8))	E3	E4	(E5E6)
((E1E2)(E7E8))				
E3	0			
E4	0	**0.133**		
(E5E6)	0	0.111	0.111	

Iteration 5

	((E1E2)(E7E8))	(E3E4)	(E5E6)
((E1E2)(E7E8))			
(E3E4)	0		
(E5E6)	0	**0.111**	

Iteration 6

	((E1E2)(E7E8))	((E3E4)(E5E6))
((E1E2)(E7E8))		
((E3E4)(E5E6))	**0**	

Iteration 7

	(((E1E2)(E7E8))((E3E4)(E5E6)))
(((E1E2)(E7E8))((E3E4)(E5E6)))	

Fig. 2. The similarity matrix and iterations in clustering process using JaccardNM measure and CL method

3.2 The New Jaccam Measure

To combine the strengths of these existing similarity measures, the add operation is used to combine the existing similarity measures. The following subsections introduce our new measure and its analysis.

Addition of the Jaccard and JaccardNM Measures. The strengths of the Jaccard and JaccardNM measures can be combined by adding both the similarity measures to get the "Jaccam" measure. "Jaccam" is defined as:

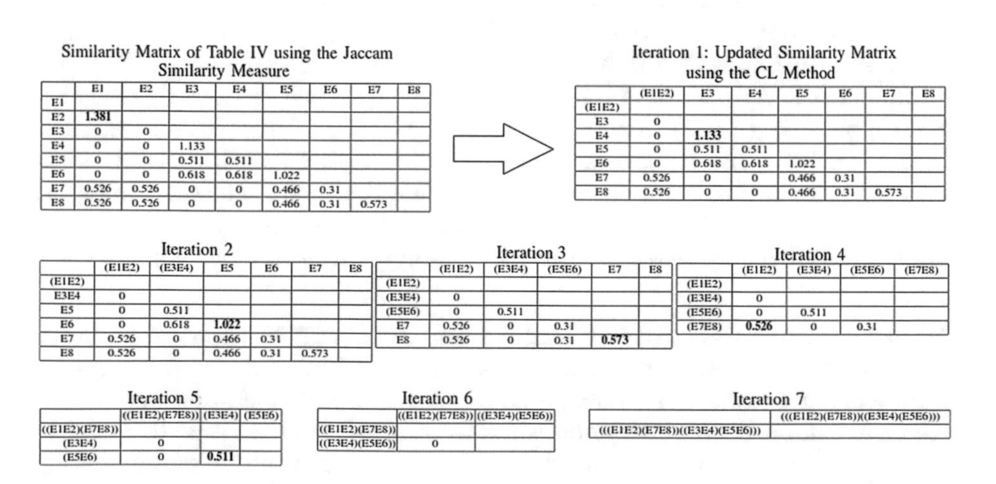

Similarity Matrix of Table IV using the Jaccam Similarity Measure

	E1	E2	E3	E4	E5	E6	E7	E8
E1								
E2	**1.381**							
E3	0	0						
E4	0	0	1.133					
E5	0	0	0.511	0.511				
E6	0	0	0.618	0.618	1.022			
E7	0.526	0.526	0	0	0.466	0.31		
E8	0.526	0.526	0	0	0.466	0.31	0.573	

Iteration 1: Updated Similarity Matrix using the CL Method

	(E1E2)	E3	E4	E5	E6	E7	E8
(E1E2)							
E3	0						
E4	0	**1.133**					
E5	0	0.511	0.511				
E6	0	0.618	0.618	1.022			
E7	0.526	0	0	0.466	0.31		
E8	0.526	0	0	0.466	0.31	0.573	

Iteration 2

	(E1E2)	(E3E4)	E5	E6	E7	E8
(E1E2)						
E3E4	0					
E5	0	0.511				
E6	0	0.618	**1.022**			
E7	0.526	0	0.466	0.31		
E8	0.526	0	0.466	0.31	0.573	

Iteration 3

	(E1E2)	(E3E4)	(E5E6)	E7	E8
(E1E2)					
(E3E4)	0				
(E5E6)	0	0.511			
E7	0.526	0	0.31		
E8	0.526	0	0.31	**0.573**	

Iteration 4

	(E1E2)	(E3E4)	(E5E6)	(E7E8)
(E1E2)				
(E3E4)	0			
(E5E6)	0	0.511		
(E7E8)	**0.526**	0	0.31	

Iteration 5

	((E1E2)(E7E8))	(E3E4)	(E5E6)
((E1E2)(E7E8))			
(E3E4)	0		
(E5E6)	0	**0.511**	

Iteration 6

	((E1E2)(E7E8))	((E3E4)(E5E6))
((E1E2)(E7E8))		
((E3E4)(E5E6))	0	

Iteration 7

	(((E1E2)(E7E8))((E3E4)(E5E6)))
(((E1E2)(E7E8))((E3E4)(E5E6)))	

Fig. 3. The similarity matrix and iterations in clustering process using Jaccam measure and CL method

$$Jaccam = Jaccard + JaccardNM = \frac{a(3(a+b+c)+d)}{(a+b+c)(2(a+b+c)+d)} \quad (3)$$

The Example Feature Matrix and Jaccam Measure. To demonstrate the strengths of our new measure, we now apply the Jaccam similarity measure to the example feature matrix shown in Table 4. The corresponding similarity matrix using the Jaccam similarity measure is shown in Fig. 3. The CL linkage method is used to update the similarity matrix during the clustering process. We can see from similarity matrix that the Jaccam prioritizes the similarity values between pair of entities (E1E2) and (E3E4), as done by the JaccardNM in the similarity matrix given in the similarity matrix in Fig. 2. Hence, the decision to cluster the entities is no longer arbitrary. Entities E1 and E2 have a high value of similarity and are grouped first (see Iteration 1 in Fig. 3). Then in the subsequent iterations, the AHC makes clusters of (E3E4), (E5E6) and (E7E8). Note that in Iteration 3 the Jaccard measure creates arbitrary decisions while our new measure Jaccam does not, as shown in Figs. 1 and 3, respectively.

It is interesting to note that the Jaccam measure creates clusters as created by the Jaccard measure (i.e. 4 clusters) and similar to the JaccardNM measure, takes no arbitrary decisions. It can be inferred that our new measure has the strength to create larger number of clusters while reducing the arbitrary decisions taken by the AHC during the clustering process, so the Jaccam outperforms the existing similarity measure.

4 Experimental Setup, Results and Analysis

In this section, we present the test software systems used for experimental purposes and the setup of clustering process including the selection of assessment criteria. The assessment criteria are used to compare our new similarity measures with the well-established similarity measures for software clustering.

4.1 Datasets

To conduct the experiments we have used four software systems developed in Java, C and C++. These test software systems are different in their source code sizes and application areas. We use an open source software system, i.e. (1) Weka, an open source data mining software system used for data pre-processing, clustering, regression, classification, and visualization. In current study we use Weka version 3.4. All the proprietary software systems are developed using C++ programming language and they are: (1) FES, a fact extractor software system to extract the entities and features of software systems developed in Visual C++ programming language; (2) PLC, is a printer language converter into intermediate language; and (3) PLP, a parser software system used to parse a printer language. We obtained the extracted feature matrices of all these proprietary software systems from Muhammad et al. [2]. For all test systems classes are selected as entities.

4.2 Algorithms and Evaluation Criteria

For experiments we used CL and SL methods with Jaccam, Jaccard and JaccardNM measures. To evaluate the results, we consider authoritativeness (MoJoFM [14]), arbitrary decisions [13] and the number of clusters [3].

4.3 Arbitrary Decision

Table 5 presents the experimental results for all similarity measures. This table lists the average number of arbitrary decisions which are taken by the AHC using different similarity measures in each iteration. The first column in Table 5 shows the linkage methods. Similarity measures are shown in the second column while the arbitrary decision values for all test systems are given in the next four columns. The last column shows the average values for each similarity measure. The bold face values enclosed in parentheses indicate best values, while only bold face values represent the better values.

Table 5. Experimental results using arbitrary decisions

Algorithm	Measure	FES	PLC	PLP	Weka	Average
CL	Jaccam	10.28	**37.59**	**9.24**	695.42	94.07
	Jaccard	10.43	72.10	10.72	700.39	99.21
	JaccardNM	**10.26**	37.63	9.25	**695.13**	**94.03**
SL	Jaccam	3.00	**(35.90)**	**(1.85)**	374.69	51.93
	Jaccard	3.20	70.47	3.30	384.05	57.63
	JaccardNM	**(2.98)**	**(35.90)**	**(1.85)**	**(372.77)**	**(51.69)**

As can be seen from Table 5, our proposed similarity measure has reduced the arbitrary decisions for each linkage method. It is very interesting that Jaccam in most cases produce the similar number of arbitrary decisions as obtained by JaccardNM and also less than the Jaccard. The fact that both similarity measures, i.e. Jaccam and JaccardNM, have the ability to count all features, i.e. a, b, c and d. Therefore, both measures can clearly more distinguish the entities. Jaccard measure does not count d therefore can not distinguishes the entities and thus creates a large number of arbitrary decisions.

4.4 Number of Clusters

Table 6 shows the maximum number of non-singleton clusters, created by AHC during all iterations. The values enclosed in parentheses indicate best values, while only bold face values shows the better values. As can be seen from Table 6, that number of clusters created by Jaccam measure is higher than that created by JaccardNM measure and similar to that created by Jaccard measure.

Table 6. Experimental Results using Number of Clusters

Method	Measure	FES	PLC	PLP	Weka	Average
CL	Jaccam	**(10)**	10	**(12)**	**(55)**	**(10.88)**
	Jaccard	**(10)**	10	**(12)**	**(55)**	**(10.88)**
	JaccardNM	9	**(11)**	11	35	8.25
SL	Jaccam	**8**	**6**	**8**	**31**	**6.63**
	Jaccard	**8**	**6**	**8**	**31**	6.63
	JaccardNM	4	5	4	12	3.13

It can also be seen that for all software systems, our new measure substantially increased the number of clusters similar to the Jaccard measure. It is very interesting to note that the new measure integrating the Jaccard similarity measure as achieved equally large number of clusters as the Jaccard measure.

4.5 Authoritativeness

Authoritativeness finds the similarity between automated results (AR) and the authoritative decomposition (AD) prepared by a human expert. The AR should resemble the AD as much as possible for the better clustering results. In this study, the widely used MoJoFM [14], is utilized. MoJoFM finds the move and join operations to convert the AR into AD.

$$MoJoFM(AR, AD) = \left(1 - \frac{mno(AR, AD)}{max(mno(\forall AR, AD))}\right) * 100 \qquad (4)$$

where $mno(AR, AD)$ is the minimum number of 'move' and 'join' operations required to translate AR in to AD and $max(mno(\forall AR, AD))$ is the maximum of $mno(\forall AR, AD)$. MoJoFM results into a percentage of the similarity between two decompositions. A higher percentage indicates greater similarity between the AR and AD.

The MoJoFM values for the series of experiments are given in Table 7. This table shows the maximum MoJoFM values selected during the iterations of clustering process. The bold face values indicate the better values for a test system/method. The values enclosed in parentheses indicate best values in the Table 7. The average values for each similarity measure is shown in the last column of Table 7.

As can be seen from Table 7 that, in most of the cases our new measure outperform the existing ones. This is because in previous Subsects. 4.3 and 4.4, we shown that our new similarity measure results in smaller number of arbitrary decisions and larger number of clusters.

Table 7. MoJoFM Results for all Similarity Measures

Method	Measure	FES	PLC	PLP	Weka	Average
CL	Jaccam	**45.00**	61.54	(**65.67**)	**30.45**	(**40.53**)
	Jaccard	43.00	61.00	51.00	**30.45**	37.09
	JaccardNM	43.00	(**65.00**)	60.00	30.13	39.63
SL	Jaccam	**47.50**	**63.08**	**59.70**	22.12	**38.48**
	Jaccard	35.00	55.00	28.00	**23.08**	28.22
	JaccardNM	43.00	42.00	28.00	17.31	26.06

5 Conclusion

This paper presents a new binary similarity measures (namely Jaccam) for software clustering. This measure integrates the strengths of the following existing binary similarity measures: Jaccard and JaccardNM. An example case study is used to show how our new measure integrates strengths of the existing similarity measures, i.e., reducing the arbitrary decisions and increasing the number of clusters. The Jaccam and existing binary similarity measures are assessed using four different software systems implemented in different programming languages.

One of the most remarkable strengths from the integration of the existing binary similarity measures is that Jaccam results in the large number of clusters which results in improving the authoritativeness. The new measure also reduces arbitrary decisions to lessen the complications of making clusters during the clustering process.

Acknowledgment. The authors would like to thank Office of Research, Innovation, Commercialization and Consultancy Office (ORICC), Universiti Tun Hussein Onn Malaysia (UTHM) for financially supporting this research under the Postgraduates Incentive Grant (GIPS) vote no. U063.

References

1. Shtern, M., Tzerpos, V.: Methods for selecting, improving software clustering algorithms. Softw. Pract. Exp. **44**(1), 33–46 (2014)
2. Muhammad, S., Maqbool, O., Abbasi, A.Q.: Evaluating relationship categories for clustering object-oriented software systems. IET Softw. **6**(3), 260 (2012)
3. Maqbool, O., Babri, H.: Hierarchical clustering for software architecture recovery. IEEE Trans. Softw. Eng. **33**(11), 759–780 (2007)
4. Shtern, M., Tzerpos, V.: On the comparability of software clustering algorithms. In: 2010 IEEE 18th International Conference on Program Comprehension, pp. 64–67. IEEE, June 2010
5. Cui, J.F., Chae, H.S.: Applying agglomerative hierarchical clustering algorithms to component identification for legacy systems. Inf. Softw. Technol. **53**(6), 601–614 (2011)

6. Naseem, R., Maqbool, O., Muhammad, S.: An improved similarity measure for binary features in software clustering. In: 2010 Second International Conference on Computational Intelligence, Modelling and Simulation, pp. 111–116. IEEE, September 2010
7. Shtern, M., Tzerpos, V.: Clustering methodologies for software engineering. Adv. Softw. Eng. **2012**, 1–18 (2012)
8. Maqbool, O., Babri, H.: The weighted combined algorithm: a linkage algorithm for software clustering. In: Eighth European Conference on Software Maintenance and Reengineering, pp. 15–24. IEEE (2004)
9. Saeed, M., Maqbool, O., Babri, H., Hassan, S., Sarwar, S.: Software clustering techniques and the use of combined algorithm. In: Seventh European Conference on Software Maintenance and Reengineering, pp. 301–306. IEEE Computer Society (2003)
10. Naseem, R., Maqbool, O., Muhammad, S.: Improved similarity measures for software clustering. In: 2011 15th European Conference on Software Maintenance and Reengineering, pp. 45–54. IEEE, March 2011
11. Andritsos, P., Tzerpos, V.: Information-theoretic software clustering. IEEE Trans. Softw. Eng. **31**(2), 150–165 (2005)
12. Bauer, M., Trifu, M.: Architecture-aware adaptive clustering of OO systems. In: Eighth European Conference on Software Maintenance and Reengineering, 2004, CSMR 2004, Proceedings, pp. 3–14 (2004)
13. Naseem, R., Maqbool, O., Muhammad, S.: Cooperative clustering for software modularization. J. Syst. Softw. **86**(8), 2045–2062 (2013)
14. Wen, Z., Tzerpos, V.: An effectiveness measure for software clustering algorithms. In: Proceedings, 12th IEEE International Workshop on Program Comprehension, 2004, pp. 194–203. IEEE (2004)

A Comparative Study of Linear and Nonlinear Regression Models for Outlier Detection

Paul Inuwa Dalatu[1,3](✉), Anwar Fitrianto[1], and Aida Mustapha[2]

[1] Department of Mathematics, Universiti Putra Malaysia, Serdang, Selangor, Malaysia
dalatup@gmail.com, anwar@upm.edu.my
[2] Soft Computing and Data Mining Centre, Universiti Tun Hussein Onn Malaysia, Serdang, Selangor, Malaysia
aidam@uthm.edu.my
[3] Department of Mathematics, Adamawa State University, Mubi, Nigeria

Abstract. Artificial Neural Networks provide models for a large class of natural and artificial phenomena that are difficult to handle using classical parametric techniques. They offer a potential solution to fit all the data, including any outliers, instead of removing them. This paper compares the predictive performance of linear and nonlinear models in outlier detection. The best-subsets regression algorithm for the selection of minimum variables in a linear regression model is used by removing predictors that are irrelevant to the task to be learned. Then, the ANN is trained by the Multi-Layer Perceptron to improve the classification and prediction of the linear model based on standard nonlinear functions which are inherent in ANNs. Comparison of linear and nonlinear models was carried out by analyzing the Receiver Operating Characteristic curves in terms of accuracy and misclassification rates for linear and nonlinear models. The results for linear and nonlinear models achieved 68% and 93%, respectively, with better fit for the nonlinear model.

Keywords: Variable selection · Best-subsets regression · Linear regression · Artificial Neural Network · Nonlinear regression

1 Introduction

An outlier is defined as an observation or measurement that does not conform to the other values contained in a given dataset [20] or an observation that significantly deviates from the bulk of data [2]. Outliers may also resulted from abnormal operation, badly calibrated or faulty instruments, electrical faults or problems in communication channels or master servers where measurements are recorded [9]. An outlier can be identified when the observation value exceeds the values of other observations in the sample by a large amount, perhaps three or four standard deviations away from the mean of all the observations [4].

The detection of outliers can be the key discovery to be made from very large databases since the discovery often leads to more interesting and useful results

T. Herawan et al. (eds.), *Recent Advances on Soft Computing and Data Mining*,
Advances in Intelligent Systems and Computing 549, DOI 10.1007/978-3-319-51281-5_32

than the discovery of inliers [5]. This means, the detection can be translated into actionable information in a wide variety of applications [10].

One active approach to solving outlier detection is by using an Artificial Neural Network (ANN) [1]. ANNs are abstract representations of biological neural systems that are able to capture many of their characteristics [7]. ANN is able to perform calculations from time to time with it units and does not involve complex processes, which makes it very suitable for finding outliers [17]. Most importantly, ANN offers a potential solution to fit all the data including any outliers instead of removing the outliers [6].

In general, the widespread popularity of ANNs in many fields, particularly in real-life problems in engineering and manufacturing, is mainly due to their ability to approximate complex multivariate nonlinear functions directly from the input samples [21]. ANNs provide models for a large class of natural and artificial phenomena that are difficult to handle using classical parametric techniques [8], thus, the main motivation for this research. The remainder of this paper is structured as follows. Section 2 starts with the related works on linear and nonlinear regression. Section 3 presents the proposed comparative methodology for both regression models in a classification task. Section 4 presents the experiments and results, which cover the results of the best subsets regression, ANN, and ROC analysis. Finally, Sect. 5 concludes the paper.

2 Related Work

Contrary to the conventional model-based techniques, ANNs are data-driven and self-adaptive, whereby the technique requires very few a priori hypotheses underlying distributions of the problem under study. The network allows the inclusion of a large number of variables with only a small set of assumptions before the models can be constructed [15]. ANNs are good in universal functional approximations, where they can represent a wide variety of interesting functions given appropriate parameters to any desired degree of accuracy and high precision [12]. Because ANNs are primarily inherent with nonlinear rule, they have the ability to model multi-dimensional systems with maximum flexibility under the consideration to learn [23]. They are capable of learning information that is apparently hidden in the data. The number of hidden layers are determined through trial and error. To date, much research has been carried out to combine and compare techniques such as linear or nonlinear regression, and ANNs. Several studies have shown that ANN models are much more important better than the traditional linear model and their outputs are more accurate [13,19,23].

Recent efforts have shown developments in the combination of ANN with statistical models such as nonlinear multiple regression, nonlinear logistic regression, and nonlinear multinomial logistic regression [14]. The motivation behind these efforts lies in the fact that linear models are not sufficient to capture real-world phenomena. [3] used nonlinear regression in the environmental sciences by using the Support Vector Machine (SVM) combined with the evolutionary strategy. Meanwhile, [7] produced an efficient self-organizing multilayer ANN for modeling nonlinear systems. Nonetheless, the ANN techniques have a number of inherent weaknesses, it is a black-box implementation and therefore, to describe how an ANN system solves a given problem.

3 Methodology

The aim of this paper is to compare the predictive results of linear and nonlinear regression models in outlier detection. To achieve this, two sets of experiments are carried out. The first experiment is to measure the performance of a standard linear model using best subsets regression to choose a subgroup of variables from an overall set of variables that produced the highest classification accuracy. The second experiment is to measure the performance of a nonlinear regression model based on an ANN using the selected explanatory variables and the dependent variable produced by the best subsets regression. Finally, both the linear and nonlinear models will be compared in a standard supervised learning approach via a classification experiment. The chosen variables will be taken as input for four well-known classification algorithms, which are the K-Nearest Neighbor (k-NN) (for $k = 1$ to $k = 5$), ANN, SVM with Linear and Radial Basis Kernel, and kernel-based Naive Bayes [1].

3.1 Best Subsets Regression

The procedure is to explore multiple variable models for classification, a standard multiple linear regression model [16] based on a best-subsets algorithm [1,3] was determined.

The most widely accepted method of detecting the presence of multicollinearity is by using Variance Inflation Factors (VIFs) [13]. Therefore, to avoid multicollinearity among the independent variables, the VIF variable with highest degree of values are removed. The VIF index is calculated using Eq. 1.

$$\frac{1}{1 - R_m^2} \tag{1}$$

3.2 ANN-based Nonlinear Regression

The ANN model in this research is using the Multi-Layer Perceptron (MLP) [1,23]. In this research, to identify the best correlation among the independent variables and the dependent variable, two feed-forward MLPs are chosen: one for nonlinear mapping of independent variables (x) into a single predicted score u and the other for linear transferring of the dependent variable (y) into a predicted score named v [1].

Figure 1 shows the architecture for the MLP model in calculating the nonlinear score of u. Finally, output layer has one output unit.

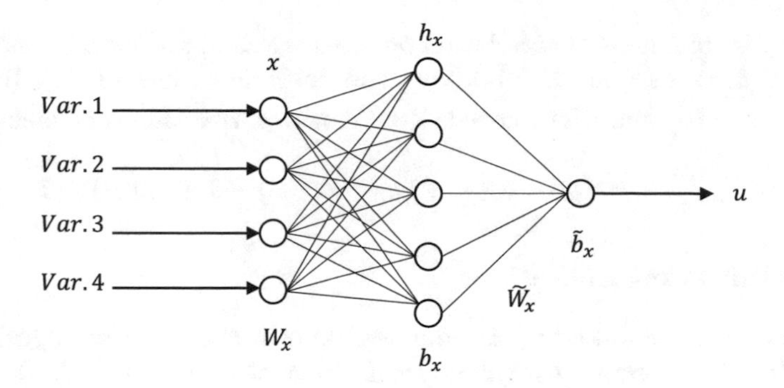

Fig. 1. Structure of MLP model for computing nonlinear score u

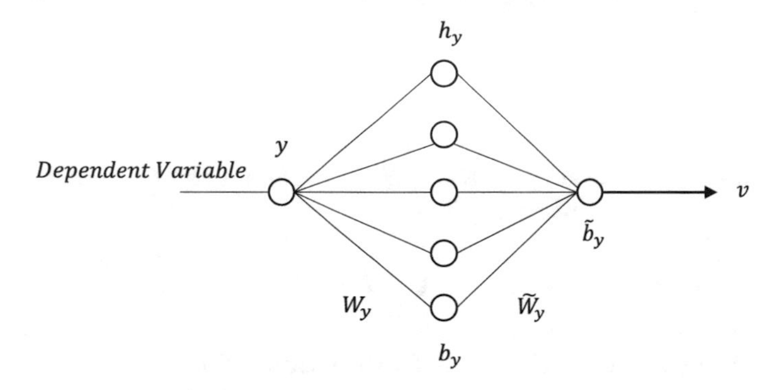

Fig. 2. Structure of MLP model for computing linear score v

Figure 2 shows linear transferring of the dependent variable is estimated. The MLPs uses hidden neurons to determine the number of neurons through trial-and-error evaluation; due to the fact that there is no established distinct theory to carry out the optimum number in the hidden unit [1].

The MLPs, which consists of input variable vectors x and y were transferred to the hidden layer u and v as shown in Eq. 2.

$$\begin{cases} h_x = tanh(W_x \cdot x + b_x) \\ h_y = W_y \cdot y + b_y \end{cases} \tag{2}$$

where W_x and W_y are the weight matrices between the input and the hidden unit, while b_x and b_y are the bias parameter of hidden units. The values u and v are estimated through linear combination of the hidden neurons' h_x and h_y, as shown in Eq. 3.

$$\begin{cases} u = \widetilde{W_x} \cdot h_x + \widetilde{b}_x \\ u = \widetilde{W_y} \cdot h_y + \widetilde{b}_y \end{cases} \tag{3}$$

In order to gain highest correlation between u and v, the specific cost function $J = -corr(u, v)$ is gained by obtaining the optimum values of W_x, W_y, b_x, b_y, $\widetilde{W}_x$, $\widetilde{W}_y$, $\widetilde{b}_x$, and $\widetilde{b}_y$ and where $corr$ is the Pearson correlation coefficient formula.

$$J_m = -corr(u, v) + \langle u \rangle + \langle v \rangle + \sqrt{\langle u^2 \rangle} - 1 + \sqrt{\langle v^2 \rangle} - 1 \quad (4)$$

3.3 Classification Algorithms

In this research, the following [1], four well-known classification algorithms are considered. The proposed classifiers are K-Nearest-Neighbor (k-NN), Artificial Neural Networks (ANNs), Support Vector Machines (SVMs) with linear and Radial Basis Function (RBF), and Naive Bayes (NB) in a kernel configuration.

K-Nearest-Neighbor: The conventional distance metric used in $k - NN$ is in a high-dimensional feature space, whereby the norm distance in a space is computed using the most common distance function such as Euclidean distance [12]. The Euclidean distance among test points f_t and training points f_s with n attributes are estimated using Eq. 5.

$$d = [(f_{t1} - f_{s1})^2 + (f_{t2} - f_{s2})^2 + \ldots + ({}_{tn} - f_{sn})^2] \quad (5)$$

Artificial Neural Networks with Radial Basis Function: The ANN classifier calculates the sum of inputs in a computational neuron and uses an activation sigmoidal function for outputs. The output layer is linear, supplying the response of the network to the activation pattern applied to the input layer. Figure 3 shows an RBF network.

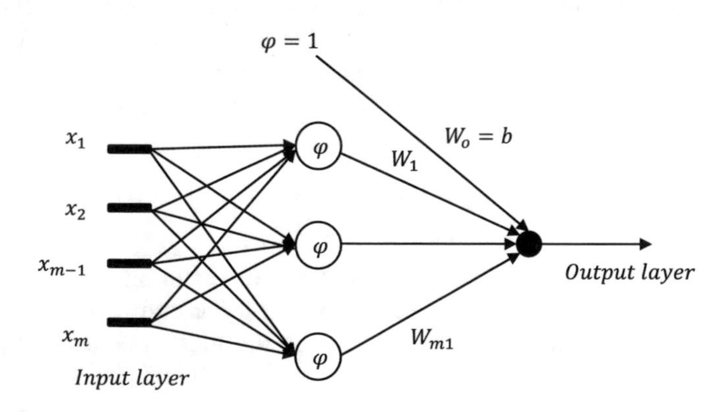

Fig. 3. Radial-basis function network

The RBF method initiates N basis functions, one for each data point, takes the form $\phi(\|X - X^n\|)$, where $\phi(.)$ is nonlinear function. The result is transferred by using linear combination of the basis function as shown in Eq. 6.

$$f(X) = \sum_{n=1}^{N} W_n \phi(\|X - X^n\|) \tag{6}$$

Naive Bayesian: To classify a test instance x, one approach is to formulate a probabilistic model to estimate the posterior $P(Y|X)$ of different Y and predict the one with the largest posterior probability [18] using the conditional independent assumption in Eq. 7.

$$P(X|Y = y) = \prod_{k=1}^{n} P(X_k|Y = y) \tag{7}$$

where each attribute set $X = (X_1, X_2, ..., X_N)$.

3.4 Evaluation Metrics

The predictive performance of linear and nonlinear models were computed by evaluating true positive rate and the true negative rate [1]. Sensitivity is the numbers correctly identified to be in the positive group over those actually belonging to the positive group. Specificity is the proportion of numbers correctly identified to be negative group over those actually belonging to the negative group [1,22].

The interpretation of both sensitivity and specificity with values evaluated from the linear and nonlinear models from actual values is plotted for possible value cutoff of the Receiver Operating Characteristics (ROC) curves and the Area Under each ROC curve (AUC).

4 Experiments and Results

Both linear and nonlinear regression models are evaluated in terms of the prediction rate in a standard classification task. Training applied to 10-fold cross-validation technique, which provides insight for considering the generalization capability on future as-yet-unseen data. The experiments used the Glass identification dataset, which was motivated by criminological investigation [11].

4.1 Best-Subsets Regression

From the dataset, four independent variables (var. 1 to var. 4) were selected. The four independent variables accounted for 73% of the variance among the dependent variables in the multiple linear model. The Pearson coefficient of correlation r, coefficient of determination R^2, and p-value are shown in Table 1. The VIF value results are shown in Table 2.

Table 1. Summary of linear and nonlinear models

Models	r	R^2	Std. error	F	Sig.
Linear	0.856	0.733	1.09798	143.23(0.67)	p = 0.000
Nonlinear	0.944	0.891	0.9389	177.62(0.83)	p = 0.000

Table 2. Summary of Linear and Nonlinear Models for Independent Variables

Ind. variables	β Coeff	Sigma	VIF
var. 1	+2.1219	0.0000	1.493
var. 2	−2.1462	0.000	3.213
var. 3	+1.3657	0.000	2.216
var. 4	−1.3175	0.000	2.630

From Table 2, the VIF value for this model ranged between 3.213 for var. 2 and 1.493 for var. 1. However, these values were below the recommended level [4]. The calculation of the predicted standardized score is based on the standardized β of multiple regression models in Eq. 8.

$$2.1219(var.1) - 2.1462(var.2) + 1.3657(var.3) - 1.3175(var.4) \tag{8}$$

The variance of the true error rate estimator is approximately 67.76% and the bias of the true error rate estimator is 32.24% (less accurate), while for the nonlinear model the variance of the true error rate estimator is approximately 92.52% and the bias of true error rate estimator is 7.48% (very accurate). Therefore, the nonlinear model prediction score is better than the linear model.

4.2 ANN-based Nonlinear Regression

The explanatory and response data were taken for the explanatory of the MLP. Pearson correlation coefficient between the nonlinear value u and the linear value v is 0.944, is higher than that obtained with the linear model. The value of R^2 also showed an increase from 0.733 to 0.891, while the standard error of the estimate decreased from 1.09798 to 0.9389. The results has shown better fit for the nonlinear model.

4.3 Classification Results

The classification methods results based on the best subset of the four selected independent variables (var. 1 to 4) are shown in Table 3. The results shows that individual classifier showed very good performance rate, especially k-NN with $k = 2$. The lowest performance is obtained with the ANN.

Table 3. Classification results with selected independent variable subsets

Technique	Prediction rate (%)
k-NN (k=1)	66.7
k-NN (k=2)	93.3
k-NN (k=3)	86.7
k-NN (k=4)	86.7
k-NN (k=5)	80.0
ANNpr	60.0
SVM (Linear)	80.0
SVM (RBF)	80.0
Naive Bayes	80.0

4.4 ROC Analysis

Fig. 4 illustrates the ROC curves for both linear and nonlinear models. Table 4 shows the results for the nonlinear and linear model (in parenthesis). The accuracy and misclassification rates are 92.5% (155+43) and 7.5% (8+8) respectively. The accuracy and misclassification rates are 71.5% (127+26) and 28.5% (36+25) respectively.

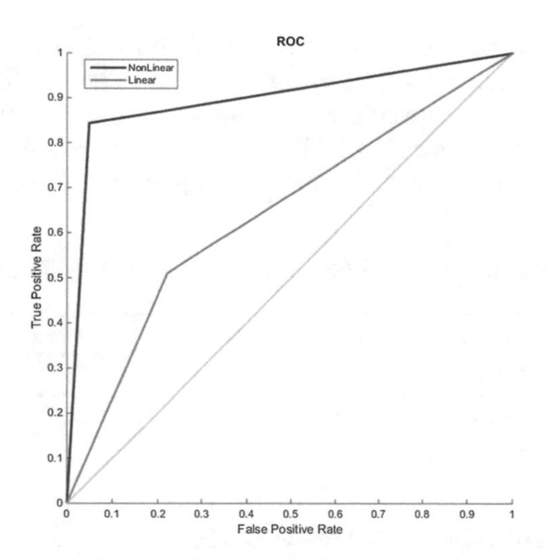

Fig. 4. ROC curve for nonlinear and linear models

Table 5 shows the results of ROC analysis in computing both the linear and nonlinear models. AUC estimated for the probability of the area under the curves are approximately 64% and 90% for the models, respectively. The standard error

Table 4. Results for nonlinear (linear) model

		Predicted (Model)	
		Positive	Negative
Actual (Target)	Positive	155 (127)	8 (36)
	Negative	8 (24)	43 (26)

for nonlinear (linear) is 0.5289 (0.3378), and the confidence interval for linear is 0.6375 (lower bound) and 1.9617 (upper bound), with range of 1.3242. Meanwhile, the confidence interval for the nonlinear is 3.6084 (lower bound) and 5.6816 (upper bound), with the range of 2.0732. The nonlinear model shows better performance to linear model, as it has much wider range of variation.

Table 5. Results of ROC analysis

Model	AUC	Std. error	Sig.	Confidence interval	
				Lower bound	Upper bound
Linear	0.6445	0.3378	$p<0.05$	0.6375	1.9617
Nonlinear	0.8970	0.5289	$p<0.05$	3.6084	5.6816

Table 6 shows the predictive performance for both linear and nonlinear models considering sensitivity, specificity, and accuracy. The sensitivity for nonlinear (linear) is 0.9509(0.7791), specificity for nonlinear (linear) is 0.8431(0.5098) and accuracy for nonlinear (linear) is 0.9252(0.7150). The results showed that nonlinear model achieved higher performance as opposed to the linear model.

Table 6. Predictive performance at best cut-off points

Model	Sensitivity	Specificity	Accuracy
Linear	0.7791	0.5098	0.7150
Nonlinear	0.9509	0.8431	0.9252

5 Conclusions and Future Works

ANN plays an important role in many classification tasks in term of predictive decisions in information processing. The objective of this research is to compare the linear and nonlinear regression models via a classification experiment using ANN for evaluations. Nonetheless, the outcomes were not able to discriminate the real achievement of each variable in the final prediction due to the black box approach of ANN and the fact that the number of hidden neurons in the ANNs is usually estimated by trial-and-error method. In future, this research will explore ways of making each variable give its numerical contributions in model building, we may adopt some methods such as cross-validation and some basic contents for guidance on the selection of the appropriate ANN architecture.

Acknowledgments. This project is sponsored by Universiti Tun Hussein Onn Malaysia under the Short Term Grant (STG) Scheme Vot U129.

References

1. Landi, A., Piaggi, P., Laurino, M., Menicucci, D.: Artificial neural networks for nonlinear regression and classification. In: 10th International Conference on Intelligent Systems Design and Application (ISDA), pp. 115–120 (2010)
2. Rusiecki, A.: Robust LTS backpropagation learning algorithm. In: Sandoval, F., Prieto, A., Cabestany, J., Graña, M. (eds.) IWANN 2007. LNCS, vol. 4507, pp. 102–109. Springer, Heidelberg (2007). doi:10.1007/978-3-540-73007-1_13
3. Lima, A.R., Cannon, A.J., Hsieh, W.W.: Nonlinear regression in environmental sciences by support vector machines combined with evolutionary strategy. Comput. Geosci. **50**, 136–144 (2012)
4. Gujarati, D.N., Porter, C.: Basic Econometrics. McGraw Hill Education, Singapore (2009)
5. Williams, G., Baxter, R., He, H., Hawkins, S., Gu, L.: A comparative study of RNN for outlier detection in data mining. CSIRO Math. Inf. Sci. **1**, 709–712 (2002). ISSN: 0–7695-1754-4
6. Motulsky, H.J., Brown, R.E.: Detecting outliers when fitting data with nonlinear regression - a new method based on robust nonlinear regression and the false discovery rate. BMC Bioinform. **7**(123), 20 (2006)
7. Han, H.-G., Wang, L.-D., Qiao, J.-F.: Efficient self-organizing multilayer neural network for nonlinear system modeling. Neural Netw. **43**, 22–32 (2013)
8. Muzhou, H., Lee, M.H.: A new constructive method to optimize neural network architecture and generalization, 1–8 (2013). CoRR abs/1302.0324
9. Garces, H., Sbarbaro, D.: Outliers detection in industrial databases: an example sulphur recovery process. In: World Congress, vol. 18(1) (2011)
10. Singh, K., Upadhyaya, S.: Outlier detection: applications and techniques. Int. J. Comput. Sci. Issues (IJCSI) **9**(3), 307–323 (2012)
11. Lichman, M.: UCI Machine Learning Repository. University of California, Irvine (2013)
12. Khashei, M., Hamadani, A.Z., Bijari, M.: A novel hybrid classification model of ANN and Multiple linear regression models. Expert Syst. Appl. **39**(3), 2696–2720 (2012)
13. Kutner, M.H., Nachtsheim, C.J., Neter, J.: Applied Linear Regression Models. McGraw Hill, New York (2008)
14. Fallah, N., Gu, H., Mohammed, K., Seyyedsalehi, S.A., Nourijelyani, K., Eshraghian, M.R.: Nonlinear poisson regression using neural networks: simulation study. Neural Comput. Appl. **18**(8), 939–943 (2009)
15. Husin, N.A., Salim, N.: A comparative study for backpropagation neural network and nonlinear regression models for predicting dengue outbreak. Junal Teknologi Maklumat Bil **20**(4), 97–112 (2008)

16. Maliki, O.S., Agbo, A.O., Maliki, A.O., Ibeh, L.M., Agwu, C.O.: Comparison of regression model and artificial neural network model for the prediction of electrical power generated in Nigeria. Adv. Appl. Sci. Res. **2**(5), 329–339 (2011). ISSN: 0976–8610
17. Yang, P., Zhu, Q., Zhong, X.: Subtractive clustering based RBF neural network model for outlier detection. J. Comput. **4**(8), 755–761 (2009)
18. Liu. Q., Lu, J., Chen, S., and Zhao, K.: Multiple naïve bayes classifiers ensemble for traffic incident detection. Math. Probl. Eng. **2014** (2014)
19. Tiryaki, S., Aydin, A.: An ANN for predicting compression strength of heat treated woods and comparison with a multiple linear regression model. Constr. Build. Mater. **62**, 102–108 (2014)
20. Cateni, S., Colla, V., Vannucci, M.: Outlier detection methods for industrial applications. INTECH Open Access Publishers (2008)
21. Koncsos, T.: The application of neural networks for solving complex optimization problems in modeling. In: Conference of Junior Researchers in Civil Engineering (2012)
22. Cherkassky, V., Mulier, F.M.: Learning from Data: Concepts, Theory and Methods, 2nd edn., pp. 1538–1550 (2007). ISSN-13: 978–0471681823
23. Bo, Z.: A prediction model based on linear regression and artificial neural network analysis of the hairiness of polyester cotton winding yarn. Adv. Multimedia Softw. Eng. Comput. **128**, 97–103 (2012)

Clustering Based on Classification Quality (CCQ)

Iwan Tri Riyadi Yanto[1(✉)], Rd Rohmat Saedudin[2], Dedy Hartama[3], and Tutut Herawan[4]

[1] Department of Information System, Ahmad Dahlan University, Yogyakarta, Indonesia
yanto.itr@is.uad.ac.id

[2] Department of Industrial Engineering, Telkom University, Bandung, West Java, Indonesia
rdrohmat@telkomuniversity.ac.id

[3] Department of Information System, Tunas Bangsa AMIK and STIKOM, Pematangsiantar, Indonesia
dedyhartama@amiktunasbangsa.ac.id

[4] Department of Information System, University of Malaya, Kuala Lumpur, Malaysia
tutut@um.edu.my

Abstract. Clustering a set of objects into homogeneous classes is a fundamental operation in data mining. Categorical data clustering based on rough set theory has been an active research area in the field of machine learning. However, pure rough set theory is not well suited for analyzing noisy information systems. In this paper, an alternative technique for categorical data clustering using Variable Precision Rough Set model is proposed. It is based on the classification quality of Variable Precision Rough theory. The technique is implemented in MATLAB. Experimental results on three benchmark UCI datasets indicate that the technique can be successfully used to analyze grouped categorical data because it produces better clustering results.

Keywords: Clustering · Rough set · Variable precision rough set model · Classification quality

1 Introduction

Cluster analysis is a data analysis tool used to group data with similar characteristics. It has been used in data mining tasks such as unsupervised classification and data summation, as well as segmentation of large heterogeneous data sets into smaller homogeneous subsets that can be easily managed, separately modeled and analyzed [1]. The basic objective in cluster analysis is to discover natural groupings of objects [2].

A variety of clustering algorithms exists to group objects having similar characteristics. But the implementations of many of those algorithms are challenging in the process of dealing with categorical data. While some of the algorithms cannot handle categorical data, others are unable to handle uncertainty within categorical data in

T. Herawan et al. (eds.), *Recent Advances on Soft Computing and Data Mining*,
Advances in Intelligent Systems and Computing 549, DOI 10.1007/978-3-319-51281-5_33

nature [3]. Several clustering analysis techniques for categorical data exist to divide similar objects into groups. Some are able to handle uncertainty in the clustering process, whereas others have stability issues [4].

Recently, many attentions have been put on categorical data clustering, where data objects are made up of non-numerical attributes. For categorical data clustering, a new trend has become in algorithms which can handle un-certainty in the clustering process. One of the well-known techniques is based on rough set theory [5–7]. The first attempt on using rough set theory for selecting a clustering (partitioning) attribute was proposed by Mazlack et al. [8]. Mazlack proposed a technique called TR (Total Roughness) which is based on accuracy of approximation of a set [5], where the highest value is the best selection of attribute. One of the successful pioneering rough clustering for categorical data techniques is Minimum-Minimum Roughness (MMR) proposed by Parmar et al. [9]. The algorithm for selecting a clustering attribute is based on the opposite of accuracy of approximation of a set [5]. To this, TR and MMR possibly provide the same result on selecting a clustering attribute. The algorithms are based on lower and upper approximations of a set [5–7]. However, the original rough set model is quite sensitive to noisy data [10] and some limitation was reported in [11]. There are drawbacks, particularly losing more useful information for demanding the inclusion of the absolutely precision in the classical definition of rough set. In order to overcome the drawback, Ziarko [11] proposed the VPRS model to deal with noisy data and uncertain information by introducing an error parameter β, where $0 \leq \beta < 0.5$ as a new way to deal with the noisy data.

Inspired VPRS for handling noisy data, in this paper, we propose an alternative technique for categorical data clustering that there are addresses above issue. For selecting the clustering attribute, it is based on the classification quality of Variable Precision Rough theory.

2 Variable Precision Rough Set

Variable precision rough set (VPRS) extends rough set theory by the relaxation of the subset operator [11]. It was proposed to analyze and identify data patterns which represent statistical trends rather than functional. The main idea of VPRS is to allow objects to be classified with an error smaller than a certain pre-defined level. This introduced threshold relaxes the rough set notion of requiring no information outside the dataset itself.

Definition 1. Let a set U as a universe and $X, Y \subseteq U$, where $X, Y \neq \phi$. The error classification rate of X relative to Y is denoted by $e(X, Y)$, is defined by

$$e(X,Y) = \begin{cases} 1 - \frac{|X \cap Y|}{|X|} & , |X| > 0 \\ 0 & , |X| = 0 \end{cases}. \tag{1}$$

Definition 2. Let U be a finite set and a set $X \subseteq U$. Given β be a real number within the range $0 \leq \beta < 0.5$. The B_β-lower approximation of X, denoted by $\underline{B}_\beta(X)$ and B_β-upper approximation of X, denoted by $\overline{B}_\beta(X)$, respectively, and are defined by

$$\underline{B}_\beta(X) = \{x \in U : e([x]_B, X) \leq \beta\}\ \overline{B}_\beta(X) = \{x \in U : e([x]_B, X) < 1 - \beta\}. \tag{2}$$

The set $\underline{B}_\beta(X)$ is called the positive region of X. It's the set of object of U that can be classified into X with error classification rate not greater than β. Then we have $\underline{B}_\beta(X) \subseteq \overline{B}_\beta(X)$ if only if $0 \leq \beta < 0.5$, which means that β be restricted in an interval $[0, 0.5)$ in order to keep the meaning of the "upper" and "lower" approximations.

The attributes dependency degree of rough set model in Variable Precision Rough Set Model is called the measure of classification quality. Based on Ziarko's notions, it is given in the following definition.

Definition 3. The accuracy of approximation variable precision (accuracy of variable precision roughness) of any subset $X \subseteq U$ with respect to $B \subseteq A$ is denoted by $\alpha_{B_\beta}(X)$. It is presented as

$$\alpha_{B_\beta}(X) = \frac{|\underline{B}_\beta(X)|}{|\overline{B}_\beta(X)|} \tag{3}$$

where $|X|$ denotes cardinality of X. If $\beta = 0$, it is the traditional rough set model of Pawlak.

Proposition 4. Let $S = (U, A, V, f)$ be an information system, $\alpha_B(X)$ be an accuracy of roughness and $\alpha_{B_\beta}(X)$ is an accuracy of variable precision roughness given β the error factor of variable precision. $(0 \leq \beta < 0.5) \Rightarrow \alpha_B(X) \leq \alpha_{B_\beta}(X)$.

Proof. Based on Definition 5, if $\beta \geq 0.5$, then $\underline{B}_\beta(X) \not\subset \overline{B}_\beta(X)$. Thus, for $0 \leq \beta < 0.5$, we have $\underline{B}_0(X) \supseteq \underline{B}_\beta(X)$ and $\overline{B}_0(X) \subseteq \overline{B}_\beta(X)$. Consequently $|\underline{B}_0(X)| \leq |\underline{B}_\beta(X)|$ and $|\overline{B}_0(X)| \geq |\overline{B}_\beta(X)|$.

For $\beta = 0$, based on Definition 5, $\alpha_B(X) = \alpha_{B_\beta}(X)$.
For $0 < \beta < 0.5$, we have $|\underline{B}(X)| \leq |\underline{B}_\beta(X)|$ and $|\overline{B}_\beta(X)| \leq |\overline{B}(X)|$. Hence

$$\frac{|\underline{B}(X)|}{|\overline{B}(X)|} \leq \frac{|\underline{B}_\beta(X)|}{|\overline{B}_\beta(X)|}.$$

Therefore, $\alpha_B(X) \leq \alpha_{B_\beta}(X)$. □

Definition 5. Let $S = (U, A, V, f)$ be an information system and let D and C be any subsets of A. Given β be a real number within the range $0 \leq \beta < 0.5$. The measure of classification quality of attribute C on attributes D, denoted by $D \Rightarrow_\gamma C$, is defined by

$$\gamma = \frac{\sum_{X \in U/D} |\underline{C}_{\beta}(X)|}{|U|}, \tag{4}$$

Obviously, $0 \le \gamma \le 1$. Attribute D is depends on C with the classification error not greater than β if elements of the universe U can be classified to equivalence classes of the partition U/D, employing C.

3 Classification Quality for Selecting Clustering Attribute

In this section, we will present the proposed technique, which is clustering based on classification quality of Variable Precision Rough Set (CCQ). The technique uses the classification quality in variable precision of attributes of rough set theory.

Proposition 9. Let $S = (U, A, V, f)$ be an information system and let D and C be any subsets of A. Given β be a real number within the range $0 \le \beta < 0.5$. If D depends on C with the classification error not greater than β, then $\alpha_{D_\beta}(X) \le \alpha_{C_\beta}(X)$, for every $X \subseteq U$.

Proof. Let D and C be any subsets of A in information system. From the hypothesis, we have the portioning U/D is finer that U/C. Therefore, for every $x \in X \subseteq U$, $e([X]_C, X) \le e([X]_D, X)$. And hence, for every $X \subseteq U$, we have

$$\underline{D}_{\beta}(X) \subseteq \underline{C}_{\beta}(X) \subseteq \overline{C}_{\beta}(X) \subseteq \overline{D}_{\beta}(X).$$

Consequently

$$\alpha_{D_\beta}(X) = \frac{|\underline{D}_{\beta}(X)|}{|\overline{D}_{\beta}(X)|} \le \frac{|\underline{C}_{\beta}(X)|}{|\overline{C}_{\beta}(X)|} = \alpha_{C_\beta}(X).$$

□

The attribute with highest average of classification quality is selected as the clustering decision.

Definition 10. Suppose $a_i \in A$, $V(a_i)$ has k-different values, say y_k, $k = 1, 2, \cdots, n$. Let $X(a_i = y_k)$, $k = 1, 2, \cdots, n$ be a subset of the objects having k-different values of attribute a_i. The measure of classification quality of the set $X(a_i = y_k)$, $k = 1, 2, \cdots, n$ for given β error factor, with respect to a_j, where $i \neq j$, can be generalized as follows

$$\begin{aligned} \gamma_{a_i}(a_j) &= \frac{|X_{\beta a_j}(a_i = y_1)|}{|U|} + \frac{|X_{\beta a_j}(a_i = y_1)|}{|U|} + \cdots + \frac{|X_{\beta a_j}(a_i = y_n)|}{|U|} \\ \gamma_{a_i}(a_j) &= \frac{\sum_{U/a_j} |X_{\beta}(a_i = y_k)|}{|U|}, \quad k = 1, 2, \cdots, n. \end{aligned} \tag{5}$$

Definition 11. Given n attributes, mean classification quality of attribute $a_j \in A$ with respect to $a_j \in A$, where $i \neq j$, denoted as $CCQ(a_i)$ is obtained by following formula

$$CCQ(a_i) = mean\left(\gamma_{a_i}(a_j)\right), i,j < n. \tag{6}$$

3.1 Example

The following table is a Discretized of supplier information system containing 15 objects with 4 categorical-valued conditional attributes; Demand Delivery WH, Production Plan, Sales forecast, and Supply. Then, we will select a clustering attribute among all candidates.

The procedure to find CCQ value is described here. To obtain the values of CCQ, firstly, we must obtain the equivalence classes induced by indisceribility relation of singleton attribute.

$X(\text{Demand}=1)=\{1,2,3,4,5\}$, $X(\text{Demand}=2)=\{6,7,8,9,15\}$, $X(\text{Demand}=3)=\{10,11,12,13,14\}$,

$$U/\text{Demand}=\{\{1,2,3,4,5\},\{6,7,8,9,15\},\{10,11,12,13,14\}\}.$$

$X(\text{DeleveryWH}=1)=\{1,4,6,10,12,14\}$, $X(\text{DeleveryWH}=2)=\{2,3,7,9\}$,

$X(\text{DeleveryWH}=3)=\{5,8,11,13,15\}$,

$$U/\text{DeleveryWH}=\{\{1,4,6,10,12,14\},\{2,3,7,9\},\{5,8,11,13,15\}\}.$$

$X(\text{ProductionPlan}=1)=\{1,5,8,9,11\}$, $X(\text{ProductionPlan}=2)=\{2,3,4,6,7,10,12,13,14,15\}$

$$U/\text{ProductionPlan}=\{\{1,5,8,9,11\},\{2,3,4,6,7,10,12,13,14,15\}\}.$$

$X(Salesforcast=1)=\{1,2,6,8,10,11\}$, $X(Salesforcast=2)=\{3,4,5,7,9,12,13,14,15\}$

$$U/Salesforcast=\{\{1,2,6,8,10,11\},\{3,4,5,7,9,12,13,14,15\}\}.$$

$X(Supply=1)=\{1,5,6,7,8,9,12,13,14\}$; $X(Supply=2)=\{2,3,4,10,11,15\}$

$$U/Supply=\{\{1,5,6,7,8,9,12,13,14\},\{2,3,4,10,11,15\}\}$$

Based on Definition 1, the error classification attribute Production Plan with respect to Demand is calculated as follow.

$$c(Demand = 1, ProductionPlan = 1) = 1 - \frac{|\{1,5\}|}{|\{1,2,3,4,5\}|} = 1 - \frac{2}{5} = \frac{3}{5},$$

$$c(Demand = 2, ProductionPlan = 1) = 1 - \frac{|\{8,9\}|}{|\{6,7,8,9,15\}|} = 1 - \frac{2}{5} = \frac{3}{5},$$

$$c(Demand = 3, ProductionPlan = 1) = 1 - \frac{|\{11\}|}{|\{10,11,12,13,14\}|} = 1 - \frac{1}{5} = \frac{4}{5},$$

$$c(Demand = 1, ProductionPlan = 2) = 1 - \frac{|\{2,3,4\}|}{|\{1,2,3,4,5\}|} = 1 - \frac{3}{5} = \frac{2}{5},$$

$$c(Demand = 2, ProductionPlan = 2) = 1 - \frac{|\{6,7,15\}|}{|\{6,7,8,9,15\}|} = 1 - \frac{3}{5} = \frac{2}{5},$$

$$c(Demand = 3, ProductionPlan = 2) = 1 - \frac{|\{10,12,13,14\}|}{|\{10,11,12,13,14\}|} = 1 - \frac{4}{5} = \frac{1}{5}.$$

Table 1. A discretized supplier information system

	D	DWH	PP	SF	S
1	1	1	1	1	1
2	1	2	2	2	2
3	1	2	2	1	2
4	1	1	2	1	2
5	1	3	1	1	1
6	2	1	2	2	1
7	2	2	2	1	1
8	2	3	1	2	1
9	2	2	1	1	1
10	3	1	2	2	2
11	3	3	1	2	2
12	3	1	2	1	1
13	3	3	2	1	1
14	3	1	2	1	1
15	2	3	2	1	2

By given $\beta = 0.2$, the quality of classification of the set of attribute Production Plan with respect to Demand as follows

$$\begin{aligned}\gamma &= \frac{\sum_{U/a_j} \left| X_\beta (a_i = y_k) \right|}{|U|} \\ &= \frac{|\{\phi\}| + |\{10, 11, 12, 13, 14\}|}{|15|} \\ &= \frac{5}{15} = \frac{1}{3}.\end{aligned}$$

Following the same procedure, the quality of classification on all attributes with respect each to the other are computed. These calculations are summarized in Table 2.

Table 2. The measure of classification quality of Table 1

Attribute (with respect to)	The quality of classification				Mean
Demand	DWH	PP	SF	S	0
	0	0	0	0	
Delivery WH	D	PP	SF	S	0
	0	0	0	0	
Production Plan	D	DWH	SF	S	0.283
	0.333	0.4	0	0.4	
SalesForcast	D	DWH	PP	S	0.083
	0.333	0	0	0	
Supply	D	DWH	PP	SF	
	0.333	0	0.333	0	0.1665

With CCQ technique, From Table 2, the highest quality of classification of attributes is Production Plan. Thus, attribute Production Plan is selected as a clustering attribute.

For objects splitting, we use a divide-conquer method. For example, in Table 2 we can cluster (partition) the objects based on the decision attribute selected, i.e., Production Plan. Notices that, the partition of the set of animals induced by attribute Production Plan is

$$U/PP = \{\{1,5,8,9,11\},\{2,3,4,6,7,10,12,13,14,15\}\}.$$

To this, we can split the objects using the hierarchical tree as follows (Fig. 1).

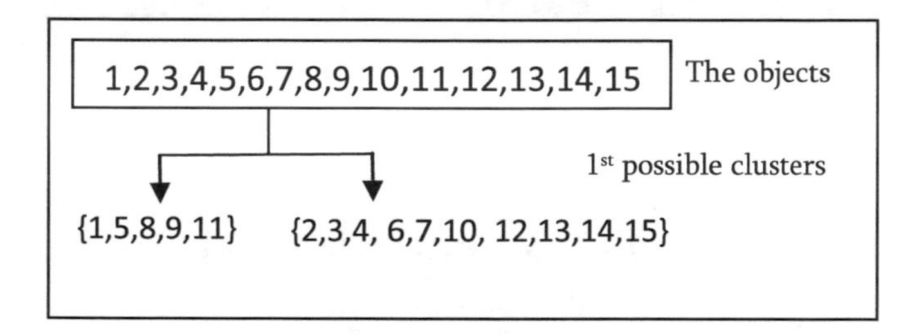

Fig. 1. The objects splitting.

The technique is applied recursively to obtain further clusters. At subsequent iterations, the leaf node having more objects is selected for further splitting. The algorithm terminates when it reaches a pre-defined number of clusters. This is subjective and is pre-decided based either on user requirement or domain knowledge.

4 Experiment Results

We elaborate the proposed technique through the three UCI benchmark datasets taken from: Http:/kdd.ics.uci.edu. Balloon dataset contains 16 instances and 4 categorical attributes; Color, Size, Act and Age. Tic-Tac-Toe Endgame dataset The data contains 958 of instances and 9 categorical-attributes; top left square (TLS), top middle square (TMS), top right square (TRS), middle left square (MLS), middle middle square (MMS), middle right square(MRS), bottom left square (BLS), bottom middle square (BMS), bottom right square (BRS) and a class attribute. Hayes-Roth dataset contains 132 training instances, 28 test instances and 4 attributes; hobby, age, educational level and marital status. The algorithms of TR, MMR, and CCQ are implemented in MATLAB version 7.6.0.324 (R2008a). They are executed sequentially on a processor Intel Core 2 Duo CPUs. The total main memory is 1G and the operating system is Windows XP Professional SP3. The experiment results is summarized in Table 3.

The TR, MMR and CCQ use different techniques in selecting clustering attribute. TR uses the total average of mean roughness, MMR uses the minimum of mean roughness and CCQ uses the measure of classification quality of Variable Precision Rough Set to select a clustering attribute. Based on Table 3 the decision cannot be

Table 3. The experiment results

Technique	Data set		
	Ballon	Tic tac toe	Hayes-Roth
TR	0	0	0
Attribute selected	All	All	All
MMR	1	1	1
Attribute selected	All	All	All
CCQ ($\beta = 0.4$)	0.8667	0.4541	0.3535
Attribute selected	3 dan 4	5	3

obtained using TR and MMR, because the value of TR and MMR of attributes in all datasets are same (for TR is 0 and for MMR is 1, respectively). But, the clustering attribute can be selected based on the highest values using CCQ.

The purity of clusters was used as a measure to test the quality of the clusters [9] The purity of a cluster and overall purity are defined as

$$Purity(i) = \frac{\begin{array}{c}\textit{The number of data in both the ith cluster}\\ \textit{and its corresponding class}\end{array}}{\textit{The number of data in the data set}} \tag{7}$$

$$Overal\,Purity = \frac{\sum_{i=1}^{\#\,of\,cluster} Purity(i)}{\#\,of\,cluster}$$

We also use Rand Measure which is external validity to analyze the cluster The adjusted Rand index [12] is the corrected for chance version of the Rand index that computes how similar the clusters (returned by the clustering algorithm) are to the benchmark classifications. The Adjusted Rand Index is as follows

$$RI = \frac{\sum_{i=1}^{m}\sum_{j=1}^{K}\binom{n_{ij}}{2} - \binom{n}{2}^{-1}\sum_{i=1}^{m}\binom{n_i}{2}\sum_{j=1}^{K}\binom{n_j}{2}}{\frac{1}{2}\left[\sum_{i=1}^{m}\binom{n_i}{2} + \sum_{j=1}^{K}\binom{n_j}{2}\right] - \binom{n}{2}^{-1}\sum_{i=1}^{m}\binom{n_i}{2}\sum_{j=1}^{K}\binom{n_j}{2}} \tag{8}$$

where n_{ij} represents the number of objects that are in predefined class i and cluster j, n_i indicates the number of objects in a priori class i, n_j indicates the number of objects cluster j, and n is the total number of objects in the data set.

CR index takes its values from the interval $[-1,1]$, in which the value 1 indicates perfect agreement between partitions, whereas values near 0 correspond to cluster agreement found by chance (Table 4).

Table 4. The cluster validity

	Data set		
	Ballon	Tic tac toe	Hayes-Roth
Purity	0.83	0.69	0.63
Rank index	66.3158	60.4557	54.0806

5 Conclusion

In this paper, we have proposed an alternative technique for categorical data clustering using Variable Precision Rough Set model. For selecting the clustering attribute, it is based on the classification quality of variable precision of attributes in the rough set theory. We present an example how our technique run. Further, we compare our technique on three benchmark datasets; Balloon and Tic-Tac-Toe Endgame and Hayes-Roth taken from UCI ML repository. The results show that our technique provides better performance in selecting the clustering attribute. Since TR and MMR are based on the traditional definition of rough set theory, thus our technique is different from TR and MMR.

References

1. Huang, Z.: Extensions to the k-means algorithm for clustering large data sets with categorical values. Data Min. Knowl. Disc. **2**(3), 283–304 (1998)
2. Johnson, R., Wichern, W.: Applied Multivariate Statistical Analysis. Prentice Hall, New York (2002)
3. Park, I.-K., Choi, G.-S.: Rough set approach for clustering categorical data using information-theoretic dependency measure. Inf. Syst. **48**, 289–295 (2015). ISSN 0306-4379
4. Li, M., Deng, S., Wang, L., Feng, S., Fan, J.: Hierarchical clustering algorithm for categorical data using a probabilistic rough set model. Knowl. Based Syst. **65**, 60–71 (2014). ISSN 0950-7051
5. Pawlak, Z.: Rough sets. Int. J. Comput. Inf. Sci. **11**, 341–356 (1982)
6. Pawlak, Z.: Rough Sets: A Theoretical Aspect of Reasoning About Data. Kluwer Academic Publisher, Dordrecht (1991)
7. Pawlak, Z., Skowron, A.: Rudiments of rough sets. Inf. Sci. **177**(1), 3–27 (2007)
8. Mazlack, L.J., He, A., Zhu, Y., Coppock, S.: A rough set approach in choosing partitioning attributes. In: Proceedings of the ISCA 13th International Conference, CAINE-2000, pp. 1–6 (2000)
9. Parmar, D., Wu, T., Blackhurst, J.: MMR: an algorithm for clustering categorical data using rough set theory. Data Knowl. Eng. **63**, 879–893 (2007)
10. Gong, Z.T., Shi, Z.H., Yao, H.Y.: Variable precision rough set model for incomplete information systems and its B-reducts. Comput. Inf. **31**(2012), 1385–1399 (2012)
11. Ziarko, W.: Variable precision rough set model. J. Comput. Syst. Sci. **46**, 39–59 (1991)
12. Hubert, L., Arabie, P.: Comparing partitions. J. Classif. **2**(1), 193–218 (1985)

A Framework of Clustering Based on Chicken Swarm Optimization

Nursyiva Irsalinda[1(✉)], Iwan Tri Riyadi Yanto[2], Haruna Chiroma[3], and Tutut Herawan[3]

[1] Mathematics Department, Universitas Ahmad Dahlan, Yogyakarta, Indonesia
nursyiva.irsalinda@gmail.com
[2] Information System Department, Universitas Ahmad Dahlan, Yogyakarta, Indonesia
yanto.itr@is.uad.ac.id
[3] Department of Information System, University of Malaya, Kuala Lumpur, Malaysia
tutut@um.edu.my

Abstract. Chicken Swarm Optimization (CSO) algorithm which is one of the most recently introduced optimization algorithms, simulates the intelligent foraging behaviour of chicken swarm. Data clustering is used in many disciplines and applications. It is an important tool and a descriptive task seeking to identify homogeneous groups of objects based on the values of their attributes. In this work, CSO is used for data clustering. The performance of the proposed CSO was assessed on several data sets and compared with well known and recent metaheuristic algorithm for clustering: Particle Swarm Optimization (PSO) algorithm, Cuckoo Search (CS) and Bee Colony Algorithm (BC). The simulation results indicate that CSO algorithm have much potential and can efficiently be used for data clustering.

Keywords: Clustering · Metaheuristic · Chicken swarm optimization · Optimization

1 Introduction

Data clustering is the process of grouping together similar multi-dimensional data vectors into a number of clusters or bins. Clustering algorithms have been applied to a wide range of problems, including data mining [4, 5], exploratory data analysis, mathematical programming [3, 7] and image segmentation [8]. Clustering techniques have been used successfully to address the scalability problem of machine learning and data mining algorithms, where prior to, and during training, training data is clustered, and samples from these clusters are selected for training, thereby reducing the computational complexity of the training process, and even improving generalization performance [1, 2, 3, 6].

Many clustering methods have been proposed. They are classified into several major algorithms: hierarchical clustering, partitioning clustering, density based clustering and graph based clustering.

T. Herawan et al. (eds.), *Recent Advances on Soft Computing and Data Mining*,
Advances in Intelligent Systems and Computing 549, DOI 10.1007/978-3-319-51281-5_34

K-means is one of the most popular partitioning algorithm because simple and efficient [4]. Unfortunately, it has disadvantages such as needs to define the number of clusters before starting, its performance depends strongly on the initial centroids and may get trapped in local optimal solutions. To avoid the inconvenience of K-means, several metaheuristic were developed. Most of them are evolutional and populatin based such as Genetic Algorithm (GA), Particle Swarm Optimization (PSO), Bee Colony Algorithm (BCA) and Cuckoo search (CS).

The previous metaheuristic algorithms are single swarm optimization algorithm. Their common essence is to simulate and reveal some natural phenomena and processes developed according to the system initializing a set of initial solution, the operation iterative rules specific for a group of solutions combined with the search mechanism itself are iterative, and finally get the optimal solution [7]. Algorithm to obtain better performance is still being developed. Therefore, in 2014 Xianbing Meng et al. proposed multi swarm optimization algorithm called Chicken Swarm Optimization (CSO). CSO can achieve optimization results both accuracy and robustness optimization in terms compared to previous single swarm optimization algorithms. So that, to obtain the better performance of clustering, we propose in this paper to use CSO algorithm. The CSO mimicking the hierarchal order in the chicken swarm and the behaviors of the chicken swarm, including roosters, hens and chicks. CSO can efficiently extract the chickens' swarm intelligence to optimize problems. It is a population based and this algorithms overcomes the problem of local and global optimum.

The remainder of this paper is organized as follows: Sect. 2 presents the related works. Section 3 provides the cluster analysis. In Sect. 4 the basics of CSO is presented. The proposed approach for data clustering is explained in Sect. 5. The detailed experimental results and comparisons are proved in Sect. 6. Finally, the conclusion of this study and the future work are drawn in Sect. 7.

2 Related Works

To overcome the disadvantage of K-means, several metaheuristic were developed. For instance the Genetic algorithm (GA) is evolutionary population optimization based; it uses natural genetics and evolution: selection, mutation and crossover [10]. It is still suffers from the difficulty of coding modelling and the operation of crossover and mutation are too expensive. More over it needs to much parameter to handle. Particle Swarm Optimization (PSO) incorporates swarming behaviours observed in flocks of bird and school of fish. Like GA, it needs much parameter to manipulate. The ant colony algorithm is one another metaheuristic inspired from the behaviour of the real ants to find the shortest path from the nest to the food sources [11, 12]. Artificial Bee Colony (ABC) algorithm mimicking the foraging behaviour of honey bee colony. In ABC algorithm, the position of a food source represents a possible solution to the optimization problem and the nectar amount of a food source corresponds to the quality (fitness) of the associated solution. The number of the employed bees or the onlooker bees is equal to the number of solutions in the population. ABC algorithm has many advantages but it has two major weaknesses: one is slower convergence speed; the other is getting trapped in local optimal value early [8]. Another algorithm for data clustering is based on Cuckoo Search (CS) optimization. Cuckoo is generic and robust for many optimization problems and it has attractive features like easy implementation, stable convergence characteristic and good computational efficiency. All of the previous metaheuristic algorithms for clustering are single swarm optimization. In 2014 Xianbing Meng et al. proposed multi swarm optimization algorithm called Chicken Swarm Optimization (CSO). CSO can achieve optimization results both accuracy and robustness optimization in terms compared to previous single swarm optimization algorithms. So that, to obtain the better performance of clustering, we propose in this paper to use CSO algorithm. In this metaheuristic no much parameters is used. We only need to define the group of rooster, hen and chick in the chicken swarm which does not really affect in the results of clustering. More over, the research of the optimal solution is done by mathematical function. In each generation we select the best solution and the next generation calculated by Chicken Swarm Optimization formula. Thereby, we always obtain the optimal solution.

3 Cluster Analysis

The main goal of the clustering process is to group the most similar objects in the same cluster or group. Each object is defined by a set of attributes or measurements. To determine the similar objects, we use the measure of similarity between them. In this paper we use the Euclidian distance to calculate the similarity between the objects. It is the most popular metric done by this formula:

$$istance(o_i, o_j) = \left(\sum_{p=1}^{m} |o_{ip} - o_{jp}|^{\frac{1}{2}}\right)^2 \tag{1}$$

where: m is the number of attributes and o_{ip} is the value of the attribute number p of the object number $i(o_i)$.

4 Basic of CSO

Chicken Swarm Optimization (CSO) based on the chicken behavior was proposed by Meng et al. [9]. As in [9], there are at least four rules in the chicken behavior, as follows

(1) In the chicken swarm, there exist several groups. Each group comprises a dominant rooster, a couple of hens, and chicks.
(2) How to divide the chicken swarm into several groups and determine the identity of the chickens (roosters, hens and chicks) all depend on the fitness values of the chickens themselves. The chickens with best several fitness values would be acted as roosters, each of which would be the head rooster in a group. The chickens with worst several fitness values would be designated as chicks. The others would be the hens. The hens randomly choose which group to live in. The mother-child relationship between the hens and the chicks is also randomly established.

```
Pseudo code of the CSO

  Initialize a population of N chickens and define
  the related parameter;
Evaluate the N chicken`s fitness values, t = 0;
While (t < Max Generation)
    If (t == 0)
       Rank the chicken`s fitness values and
       establish a hierarchal order in the swarm;
       Divide the swarm into different groups, and
       determine the relationship between the chicks
       and hens in a groups; End if.
    For i = 1:N
       If i == the rooster Update its solution using
rooster formula
       If i == the hen Update its solution using hen
formula
       If i == the chick Update its solution using
chick formula
       Evaluate the new solution
       If the new solution is better than its
previous one, update it.
    End for
End While
```

Fig. 1. Pseudocode of standard CSO

(3) The hierarchal order, dominance relationship and mother-child relationship in a group will remain unchanged. These status only update every several (G) steps.
(4) Chickens follow their group-mate rooster to search for food, while they may prevent the ones from eating their own food. Assume chickens would randomly steal the good food already found by others. The chicks search for food around their mother (hen). The dominant individuals have advantage in competition for food.

Based on the four rules, the basic steps of the CSO can be summarized by the pseudo code as follows (Fig. 1).

5 Clustering Based on CSO

To solve data clustering problem, the basic CSO is adapted to reach the centroids of the clusters. For doing this, we suppose that we have n objects and every objects is defined by m attributes. In this study, the main goal of the CSO is to find k centroids of clusters which minimize the Eq. (1). The data set must be represented by a matrix (n, m), such as the row-i corresponds to the object number.

In the CSO mechanism, the solution are the chicken and each chicken is represented by a matrix with k rows and m columns, where the matrix rows are the centroids of cluster.

We propose a CSO algorithm for data clustering through the following steps:

1. Generate randomly Initialize a population of N chickens.
2. Evaluate the N chicken's fitness values.
3. Determine the relationship between the rooster, chicks and hens in a groups.
4. Calculate the fitness value of the new solutions.
5. Compare the new solutions with the old one, if the new solution is better than its previous one, replace the old solution by the new one.
6. Find the best solution
   End While;
7. Print the best solution.

6 Experimental Result

In order to test the validity and the efficiency of the proposed approach, We elaborate the four approaches through the UCI benchmark datasets. The result of a clustering algorithms are be evaluated and validated by internal and external validity [22]. The external is used to analyze the cluster in this study is Rand Measure. The adjusted Rand index [11] is the corrected-for-chance version of the Rand index that computes how similar the clusters (returned by the clustering algorithm) are to the benchmark classifications. The Adjusted Rand Index as follows

$$RI = \frac{\sum_{i=1}^{m}\sum_{j=1}^{K}\binom{n_{ij}}{2} - \binom{n}{2}^{-1}\sum_{i=1}^{m}\binom{n_{i.}}{2}\sum_{j=1}^{K}\binom{n_{.j}}{2}}{\frac{1}{2}\left[\sum_{i=1}^{m}\binom{n_{i.}}{2} + \sum_{j=1}^{K}\binom{n_{.j}}{2}\right] - \binom{n}{2}^{-1}\sum_{i=1}^{m}\binom{n_{i.}}{2}\sum_{j=1}^{K}\binom{n_{.j}}{2}}$$

where n_{ij} represent the number of objects that are in predefined class i and cluster j, $n_{i}.$ indicates the number of objects in a priori class i, $n_{.j}$ indicates the number of objects cluster j, and n is the total number of objects in the data set.

Davies Bouldin index and Dunn index are used to assess the quality of clustering algorithms based on internal criterion. Davies Bouldin index attempts to minimize the average distance between each cluster and the one most similar to it [12]. It is defined as follows

$$DB = \frac{1}{K}\sum_{k=1}^{K}\max_{\mathrm{k}\neq\mathrm{m}}\left(\frac{\sigma_k + \sigma_m}{d(c_k, c_m)}\right)$$

where K is the number of clusters, σ_k is the average distance of all elements in cluster k and $d(c_k, c_m)$ is the distance between cluster k and cluster m. The clustering algorithm that produces a collection of clusters with the smallest Davies–Bouldin index is considered the best algorithm based on this criterion. Dunn's Validity Index [13] attempts to identify those cluster sets that are compact and well separated. The Dunn's validation index can be calculated with the following formula

$$Dn = \min_{1 \le k \le K}\left(\min_{k+1 \le m \le K}\left(\frac{d(c_k, c_m)}{\max\limits_{1 \le n \le k} d'(n)}\right)\right)$$

where $d(c_i, c_j)$ represents the inter cluster distance between cluster k and cluster m. It may be any number of distance measure, such as the distance between the centroids of the cluster. $d'(n)$ is called the intra cluster distance of cluster n that may be measured in variety mays, such as the maximal distance between any pair of element in cluster n.

6.1 Iris Dataset

The iris dataset contains 150 objects and 4 attributes which are unscrewed into 3 classes of 50 instances representing a type of iris plant. Table 1. shows that the proposed algorithm obtained better performance in terms of accuracy and Rand index. Based on internal validation CSO also has good performance in term of the Dunn's validity index and Davies Bouldin index.

6.2 Ecoli Dataset

The Ecoli dataset contains 336 numbers of instances and 7 attributes condition, The data set is classified into 8 classes where each class represent the localization site. The evaluation result is summarized in Table 2. The CSO Algorithm has better performance based on Davies Boulding index, and also still gives good results analyzed by other internal and external validity index.

6.3 Ionospehere Dataset

The Ionosphere dataset contains 351 number of instances and 34 predictor attributes. The data is classified into binary classes either "good" or "bad". The computation result shows that the CSO has better performance in term of internal validity and external validity. The summary of the experiment result is illustrated in Table 3.

Table 1. The evaluation for iris data set

Method	External validation		Internal validation	
	Accuracy	Rand index	Dunn index	Davies Boludin index
GA	0.6	0.66971	0.029031	1.0648
CS	0.68667	0.75374	0.020419	**1.5088**
PSO	0.66667	0.75982	**0.15604**	0.34151
CSO	**0.92667**	**0.90971**	0.10281	0.3595

Table 2. The evaluation for Ecoli data set

Method	External validation		Internal validation	
	Accuracy	Rand index	Dunn index	Davies Bouldin index
GA	**0.63988**	**0.82637**	**0.057473**	0.41989
CS	0.47321	0.74053	0.03733	0.40288
PSO	0.47917	0.7599	0.031671	**0.53819**
CSO	0.61905	0.76041	0.050097	0.37858

Table 3. The evaluation for ionosphere data set

Method	External validation		Internal validation	
	Accuracy	Rand index	Dunn index	Davies Boludin index
GA	0.62393	0.52938	0.026323	1.6593
CS	0.59829	0.51795	0.030362	0.95699
PSO	0.66952	0.55621	**0.047005**	1.1032
CSO	**0.7208**	**0.59635**	0.0356	**1.6748**

Table 4. The evaluation for cancer data set

Method	External validation		Internal validation	
	Accuracy	Rand index	Dunn index	Davies Boludin index
GA	0.87848	0.78618	0.039498	**1.4807**
CS	0.58419	0.51346	0.039014	2.2236
PSO	0.90483	0.82752	**0.056254**	1.2584
CSO	**0.93851**	**0.88441**	0.039778	0.95345

6.4 Cancer Data Set

The Cancer data set represent the Wisconsin breast cancer dataset. The data contains 683 instances with 9 features. Each instance has one of two possible classes benign or malignant. The experiment results are summarized in the Table 4 which is showing that the CSO give better performance.

7 Conclusion

In this paper, we have presented a new approach for solving the data clustering problem. The approach principally based on the chicken swarm optimization. The proposed algorithm is applied to four different data sets. Simulation experiments show that the proposed approach obtains the better results in term of the internal and external validity. In order to improve the performance and as a future works, we plan to hybridize to other algorithm and also we still consider to apply the algorithm into the other real dataset.

References

1. Engelbrecht, A.P.: Sensitivity analysis of multilayer neural networks. Ph.D. thesis, Department of Computer Science, University of Stellenbosch, Stellenbosch, South Africa (1999)
2. Fisher, D.: Knowledge acquisition via incremental conceptual clustering. Mach. Learn. **2**, 139–172 (1987)
3. Potgieter, G.: Mining continuous classes using evolutionary computing. M. Sc. thesis, Department of Computer Science, University of Pretoria, Pretoria, South Africa (2002)
4. Evangelou, I.E., Hadjimitsis, D.G., Lazakidou, A.A., Clayton, C.: Data mining and knowledge discovery in complex image data using artificial neural networks. In: Workshop on Complex Reasoning an Geographical Data, Cyprus (2001)
5. Quinlan, J.R.: C4.5: Programs for Machine Learning. Morgan Kaufmann, San Mateo (1993)
6. Premalatha, K.: A new approach for data clustering based on PSO with local search. Comput. Inf. Sci. **1**(4), 139–145 (2008)
7. Rao, M.R.: Cluster analysis and mathematical programming. J. Am. Stat. Assoc. **22**, 622–626 (1971)
8. Lillesand, T., Keifer, R.: Remote Sensing and Image Interpretation. Wiley, Hoboken (1994)
9. Meng, X., Liu, Yu., Gao, X., Zhang, H.: A new bio-inspired algorithm: chicken swarm optimization. In: Tan, Y., Shi, Y., Coello, C.C. (eds.) ICSI 2014. LNCS, vol. 8794, pp. 86–94. Springer, Heidelberg (2014). doi:10.1007/978-3-319-11857-4_10
10. Yang, X.-S., He, X.: Bat algorithm: literature review and applications. Int. J. Bio-Inspired Comput. **5**(3), 141–149 (2013)
11. Hubert, L., Arabie, P.: Comparing partitions. J. Classif. **2**(1), 193–218 (1985). http://doi.org/10.1007/BF01908075
12. Davies, D.L., Bouldin, D.W.: A cluster separation measure. IEEE Trans. Pattern Anal. Mach. Intell. **1**(2), 224–227 (1979)
13. Dunn†, J.C.: Well-separated clusters and optimal fuzzy partitions. J. Cybern. **4**(1), 95–104 (1974). http://doi.org/10.1080/01969727408546059

Histogram Thresholding for Automatic Color Segmentation Based on *k*-means Clustering

Adhi Prahara[1(✉)], Iwan Tri Riyadi Yanto[2], and Tutut Herawan[3]

[1] Informatics Department, Universitas Ahmad Dahlan, Yogyakarta, Indonesia
adhi.prahara@tif.uad.ac.id

[2] Information System Department, Universitas Ahmad Dahlan, Yogyakarta, Indonesia
yanto.itr@is.uad.ac.id

[3] Department of Information Systems, University of Malaya, Kuala Lumpur, Malaysia
tutut@um.edu.my

Abstract. Color segmentation method has been proposed and developed by many researchers, however it still become a challenging topic on how to automatically segment color image based on color information. This research proposes a method to estimate number of color and performs color segmentation. The method initiates cluster centers using histogram thresholding and peak selection on CIE L*a*b* chromatic channels. *k*-means is performed to find optimal cluster centers and to assign each color data into color labels using previously estimated clusters centers. Finally, initial color labels can be split or merge in order to segment black, dark, bright, or white color using luminosity histogram. The final cluster is evaluated using silhouette to measure the cluster quality and calculate the accuracy of color label prediction. The result shows that the proposed method achieves up to 85% accuracy on 20 test images and average silhouette value is 0.694 on 25 test images.

Keywords: Automatic color segmentation · Histogram thresholding · Cluster centers initialization · *k*-means clustering

1 Introduction

One of the most difficult in image processing is segmentation step. Segmentation is a process that partitions image into segments [1]. In color image segmentation, to segment image into its appropriate color label is difficult especially when there is no provided information on how many number of color labels should be made. The problem also comes from image dimension. Many digital color images have three dimensions of color channels that each dimension is often related to each other e.g. in RGB (red-green-blue) color space or independent to each other e.g. HSV (hue-saturation-value) and CIE L*a*b* (*Commission Internationale de l'Eclairage*-Lab) color space. The purpose of color image segmentation is that each segment will have pixel that similar to pixel in its own segment and different to pixel in other segment.

T. Herawan et al. (eds.), *Recent Advances on Soft Computing and Data Mining*,
Advances in Intelligent Systems and Computing 549, DOI 10.1007/978-3-319-51281-5_35

There are many ways to segment color in digital images. One can segment using three-dimensional color data [2]. Two-dimensional correlation of each channel can also be used, e.g. RG (red-green), GB (green-blue), or RB (red-blue) [3]. More efficient way, color image is transformed into color histogram [3, 4] or transform to other color space that has independent color channels [5, 6] such as CIE L*a*b* and HSV.

Color segmentation can be done by clustering method. Many of clustering methods need to be provided with number of cluster and points to start the initial cluster centers. Some methods have been developed to estimate the number of cluster or initiate cluster centers [7–9] and showing good results. However, choosing the clustering methods also depend on the data that will be used. Color data can be separated well using specific distance metric e.g. Euclidean, squared Euclidean, or city block.

This research proposes a method to estimate number of color and performs color segmentation. In summary, the contribution of this work is given as follow:

a. The proposed segmentation method initiates cluster centers using histogram thresholding and peak selection on CIE L*a*b* chromatic channels.
b. *k*-means is performed to find optimal cluster centers and to assign each color data into color labels using previously estimated clusters centers.
c. Finally, initial color labels can be split or merge in order to segment black, dark, bright, or white color using luminosity histogram.
d. The final cluster is evaluated using silhouette to measure the cluster quality and calculate the accuracy of color label prediction.

The rest of this paper is organized as follow: Sect. 2 presents related work. Section 3 presents the proposed segmentation method. Section 4 presents the results and discussion. Finally, the conclusion of this work is described in Sect. 5.

2 Related Works

Many color segmentation methods have been proposed. Jassim and Altaani proposed a method for color image segmentation by combining Otsu method applied in each channel of RGB image and median filter to smoothen the distorted image caused by formulation of new color image from previous Otsu method [2]. The median filter was also used to increase the segmented regions. The result of their experiment showed that the method was fast, easy to be implemented and good for medical image processing.

Kurugollu *et al.* proposed histogram multi-thresholding on bands of color and fusion the resulting segmentation [3]. The method divided the RGB color images into subsets of pair RB (red-blue), RG (red-green), and BG (blue-green), resulting bands of two-dimensional histogram. Histogram peaks was selected using multi thresholding scheme. The result of each band-pair's segmentation would be fusion into segmentation map. Result showed that using two-dimensional histogram for segmentation was superior than using one-dimensional histogram.

Cheng and Sun proposed hierarchical approach to segment color images using homogeneity [4]. Homogeneity histogram via multi-level thresholding was used in order to find uniform regions. Peak finding algorithm was applied to select significant peak from the histogram. Hue component was used in histogram analysis to segment the image. After that, a region merging was performed to avoid over-segmentation then CIE L*a*b* color space was used to measure the color difference. The result showed effectiveness and superiority of the proposed method on tested color image.

Severino and Gonzaga used color mixture to segment color image [5]. They proposed Hue, Saturation, Mixture color space as planes that described the RGB cube. They used the method in skin classification. The proposed method was surpassed the performance of all compared methods.

Angulo and Serra proposed color segmentation method by ordered merging [6]. At the first step, they determined color space which suitable for morphological operation. Next, they compared other color segmentation methods such as pyramid of watersheds and different color gradients with the proposed method, multi-scale color segmentation using merging chromatic-achromatic partitions ordered by saturation components. The result showed that saturation component plays an important role in order to merge the chromatic and achromatic information.

Tan *et al.* proposed a color segmentation method using Fuzzy C-Means (FCM) with initialization scheme to determine the number of cluster and cluster centers called Hierarchical Approach (HA) [10]. Color image was split into multiple regions and merging technique was used to determine number of cluster. Cluster centers were obtained from FCM. Result showed the HA initialization was superior to the state-of-the-art initialization for FCM and could be applied to segment any color images.

Wang *et al.* proposed color image segmentation method using support vector machine (SVM) and fuzzy C-means (FCM) [11]. The pixel was used as input of SVM which trained by FCM with the extracted pixel-level features. They combined the advantages of local information of color images with ability of SVM classifier to segment the image. The result showed that the method was effective for decreasing computational time and increasing the quality of color image segmentation.

This research proposes automatic color segmentation method using histogram thresholding and *k*-means clustering. The method estimates number of cluster and centers using histogram thresholding and peak selection on CIE L*a*b* chromatic channels. The clustering method uses *k*-means to find optimal cluster centers and to assign each color data into color labels. The initial color labels will be split or merge in order to avoid over-segmentation and to segment black, dark, bright, or white color using luminosity histogram. The final segmentation is enhanced using median filter.

3 Proposed Method

In this section, we present the proposed method which is automatic segmentation on color images based on histogram thresholding and *k*-means clustering. The proposed method consists of four main steps, i.e. extracting color data, find cluster centers using histogram thresholding and peak selection, calculate initial segmentation, and perform final segmentation. The general steps of the method are shown in Fig. 1.

From Fig. 1, the first step consists of procedure to extract color data from input image. Input color image is enhanced using median filter to remove noise and smoothing the color image. Enhanced image is converted into CIE L*a*b* color space.

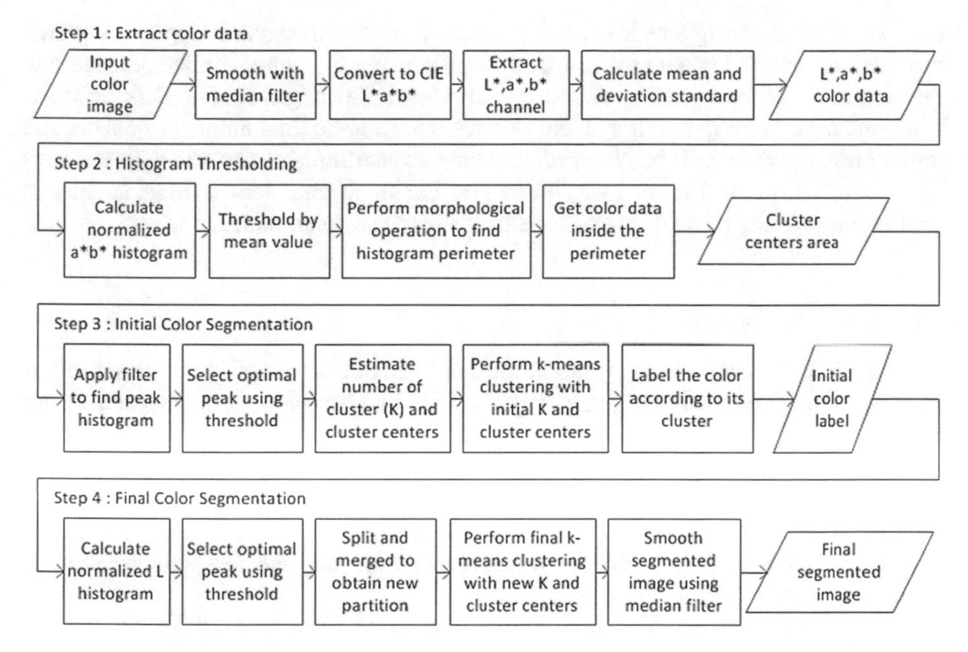

Fig. 1. The general steps of automatic color segmentation using histogram thresholding and *k*-means clustering.

Every channel of CIE L*a*b* color space is extracted and calculated its mean and deviation standard to be used as threshold.

At the second step, histogram thresholding is performed to localize cluster centers area. This step constructs a two-dimensional histogram using a* and b* chromatic channels of CIE L*a*b* color space. The histogram value is divided by its maximum value to normalize the range into 0–1. A mean threshold is applied to separate high and low-density value. In order to localize the cluster centers area, a morphological operation is performed. The morphological operation localizes the high-density value by dilating and filling the hole to find the perimeter of cluster centers area. Data inside the perimeter is likely a cluster centers area.

At the third step, initial number of cluster and cluster centers will be estimated on cluster centers area. A peak filter is applied to find all the peaks that represent the highest density around its neighborhood. All of peaks will be sorted by descending order. An iterative procedure will compare the higher value of peak with other peak according to the order. Lower peak that has squared distance higher than variance will be kept as superior peak otherwise it will be merged with the higher peak. The final number of peaks will become initial number of cluster and its value will become the initial cluster centers. The *k*-means clustering with this defined initial number of cluster and cluster centers is performed to assign every color data into its optimal color label.

At the last step, the initial color label will be refined to differentiate the black, dark, bright, and white color using luminosity channel of CIE L*a*b* color space. One- dimensional normalized histogram is constructed using luminosity channel for each color label. To ignore low-density value, a mean threshold is applied. Optimal peak will be selected among the histogram's peak using the same iterative procedure in the step three. If there is more than one peak, the cluster might contain dark or bright color and need to be split. After all clusters are updated, the final number of cluster and initial cluster centers will be obtained. *K*-means clustering is performed with this new number of cluster and cluster centers to find the final color labels. Median filter is applied on final segmented image to remove noise and to smooth the image.

3.1 Color Histogram

The CIE L*a*b* color space is one of the most popular to quantify the visual differences of color. It is consists of luminosity layer L*, and two chromaticity layers, a* and b*. The chromaticity layer a* indicating where color falls along the red-green axis and the chromaticity layer b* indicating where the color falls along the blue-yellow axis. The color information is stored in a* and b* layers and can be measured using Euclidean distance. By using two chromaticity layers a* and b*, the histogram dimension can be reduced into two-dimensional from original three-dimensional of RGB color space.

If the size of image is $N \times M$, I_A and I_B is a* and b* intensity data, p and q is the index of histogram bins for a* and b* channel, i and j is the index of a* and b* coordinate intensity data in the image. The two-dimensional a*–b* normalized histogram (*hist_AB*) can be calculated using (1).

$$hist\_AB(p,q) = \frac{\sum_{i=0,j=0}^{M,N} (I_A(i,j) = p, I_B(i,j) = q)}{\max(hist\_AB)} \tag{1}$$

If the size of image is $N \times M$, I_L is luminosity channel intensity data, p is the index of histogram bins for luminosity channel, i and j is the index of luminosity coordinate intensity data in the image, then one-dimensional luminosity normalized histogram (*hist_L*) can be calculated using (2).

$$hist\_L(p) = \frac{\sum_{i=0,j=0}^{M,N} (I_L(i,j) = p)}{\max(hist\_L)} \tag{2}$$

3.2 Histogram Thresholding

Histogram thresholding uses statistical color data such as mean, deviation standard and variance value as threshold. Mean is used as threshold to ignored low-density value of histogram, deviation standard is used to determine the length of peak filter kernel, and variance is used as threshold to select optimal peak by its squared distance. Morphological operation is applied on two-dimensional normalized histogram to localize the high-density values as cluster centers area. The morphological operations are dilation and filling. Dilation is used to expand the binary image formed by normalized histogram and filling is used to fill the hole of color data distribution. The result of morphological operations is cluster centers area's perimeter. Color data that lies inside the perimeter will be kept otherwise will be ignored from the calculation of initial cluster centers. The area inside perimeter will become cluster centers area.

3.3 Number of Cluster and Cluster Centers Estimation

This method uses statistic color information to obtain the threshold, performs histogram thresholding, and peak selection to estimate number of cluster and cluster centers. Peak is a color data with value superior to other peak inside the radius of filter kernel. The length of filter kernel is determined from the value of deviation standard. A peak filter with 3 × 3 kernel is shown in Fig. 2.

1	1	1
-1	8	-1
-1	-1	-1

Fig. 2. A peak filter with 3 × 3 kernel

From Fig. 2, the kernel ensures that the value will be compared to its neighborhood and if the result is higher than 0 then the value is considered as a peak. The procedure to find optimal peak after peak extraction is explained below:

a. All peak is sorted by descending order. This step ensures that higher value of peak will be prioritized than lower value of peak.
b. An iterative procedure is performed from the higher peak to compare with other lower peak below it.
c. Squared distance is used to measure the distance of current peak with other peak using variance as a threshold. If distance is higher than variance then the lower peak will be promoted to higher peak otherwise it is merged with current peak.
d. The iterative procedure is done until all peak have been evaluated.

The result of the iterative procedure is optimal number of peak and peak coordinate that will be used as initial number of cluster and cluster centers.

3.4 Color Segmentation

Color segmentation uses *k*-means clustering to assign every color data to its optimal color label. *K*-means is one of the unsupervised clustering methods, which usually uses Euclidean distance to assign data into its optimal cluster. The data within cluster are close as possible to other data and as far as possible from data in other cluster. This research provides *k*-means with initial number of cluster (*k*) and cluster centers. Color segmentation procedure using *k*-means will be explained in these steps:

a. Initialize number of cluster and cluster centers using statistical color information, histogram thresholding and peak selection.
b. Calculate Euclidean distance of each color data to each cluster centers.
c. Assign each color data to the nearest cluster centers using (3) where $c_{(i)}$ is the *i-th* distance data, x_i is the *i-th* color data, and μ_j is the *j-th* cluster center.

$$c_{(i)} = \arg \min_{1 \leq j \leq K} \left\| x_i - \mu_j \right\|^2 \tag{3}$$

d. After all color data assign to its nearest cluster, calculate new cluster centers.
e. Repeat step b to d until converge or all cluster member are not changing label.
f. Get the final number of cluster, cluster centers, and cluster label.

3.5 Evaluation

The performance of proposed method is evaluated by two methods. Segmentation accuracy is determined by comparing actual number of color in supervised color image with segmentation result. Accuracy is calculated using (4).

$$accuracy = \frac{number\ of\ right\ color\ prediction}{number\ of\ test\ color\ images} \times 100\% \tag{4}$$

Segmentation quality is evaluated using silhouette method. The silhouette measure how similar the data is to data in its own cluster when compared to data in other cluster [12]. If *a(i)* is the average distance from the *i-th* data to the other data in the same cluster as *i* and *b(i)* is the minimum average distance from the *i-th* data to data in different cluster then the silhouette value for *i-th* data *s(i)* can be calculated using (5).

$$s(i) = \frac{b(i) - a(i)}{\max\{a(i), b(i)\}} \tag{5}$$

4 Result and Discussion

The proposed method is tested using Matlab 2010b and running on laptop with Intel i5 processor and 8 GB of RAM. Test is conducted on 20 supervised color images and 25 natural color images. Supervised color image contain mostly 1–12 primary and secondary color as shown in Fig. 3a–e. Natural color image is obtained from the internet and McGill Calibrated Color Image Database [13], which consists of objects in nature. Some of natural color image are shown in Fig. 4a–e.

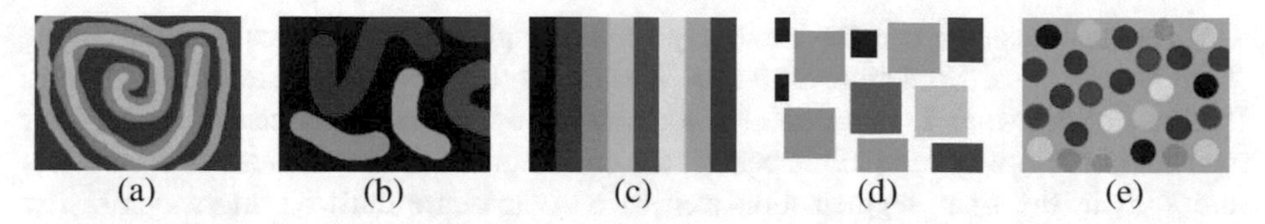

Fig. 3. Supervised color image. (a) 4 colors image (b) 6 colors image (c) 8 colors image (d) 10 colors image and (e) 12 colors image

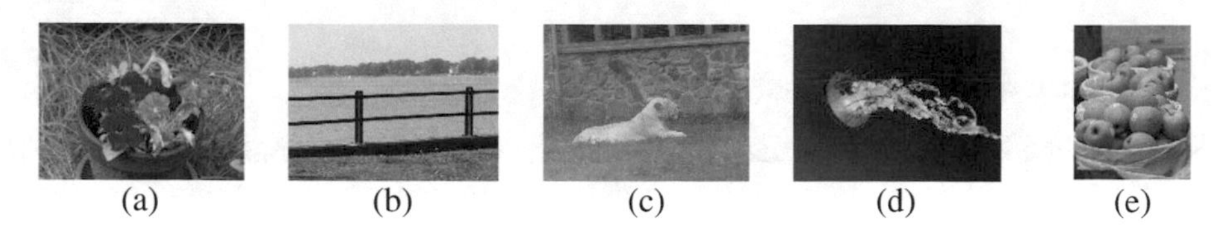

Fig. 4. Natural color image. (a) flowers (b) sea (c) animal (d) jellyfish (e) apple

4.1 Cluster Initialization Result

Proposed method uses histogram thresholding and peak selection to estimate number of cluster and cluster centers. In initial segmentation step, the method uses mean threshold to ignore low-density of a*b* histogram and 0.8*variance as threshold to select optimal peak. In final segmentation step, the method uses mean threshold to ignore low-density of L* histogram and variance as distance threshold to select optimal peak to split the cluster. Figure 5 shows the distribution of selected peak, initial cluster centers, and final cluster centers. From Fig. 5, the green mark is the perimeter, cyan mark is distribution of color data, and yellow mark is the high-density area. The blue cross marker is the initial peaks, red circle marker is the initial cluster centers, and the red cross marker is the final cluster centers.

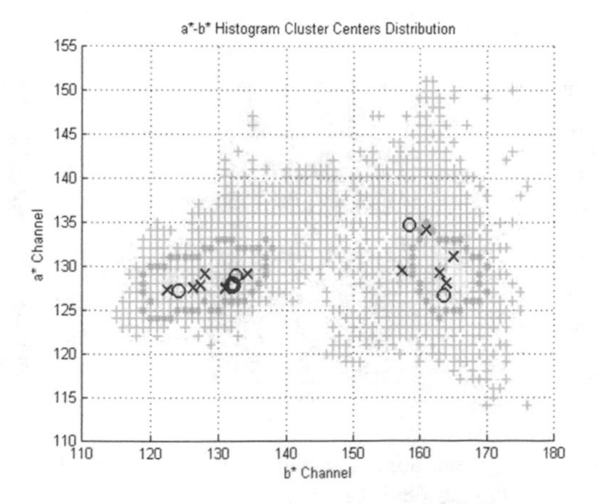

Fig. 5. Cluster centers localization

The final cluster centers are likely to be inside the perimeter and lies in the high-density area. Sometimes the final cluster centers are lies outside the perimeter because of the wide distribution of color data. Notice that the number of final cluster centers is seven while the initial centers are only four. This means some of the clusters are split in the final segmentation step to accommodate dark or bright color. The distance of final cluster centers is close to each other because it is plotted on a*b* histogram while the main difference is on the L* channel that represent the dark or bright intensity.

4.2 Segmentation Result

Segmentation is done by *k*-means clustering method using initial number of cluster and cluster centers obtained from histogram thresholding and peak selection. The small population of color data might be preserved if the average value of all data is low and can be merged if the average value is high. Color image are shown in Fig. 6a and d, the initial segmentation are shown in Fig. 6b and e, and the final segmentation is shown in Fig. 6c and f.

The result between initial segmentation and final segmentation are looks different because of the different number of cluster applied in the segmentation. In Fig. 6e, the lotus flower image is segmented with $k = 4$ and the final segmentation with $k = 8$. In final segmentation step, initial cluster might be split to deal with dark or bright color. However, sometimes it is same as the initial cluster if the cluster is already optimal. The number of cluster becomes the number of predicted color by the method.

Sometimes, there are color data that not assign to their right color label. Figure 7a shows color image and Fig. 7b shows the false color label (red square marker). The number of color should be 12 but the method is predicted 10. The distance to differentiate color can become large depend on the variance, resulting the low-density peak to be merged to the higher density peak around its variance length and the color data represent by that peak is also merged to another color label. The variance threshold is enough to differentiate the color label because smaller threshold may result in over segmentation.

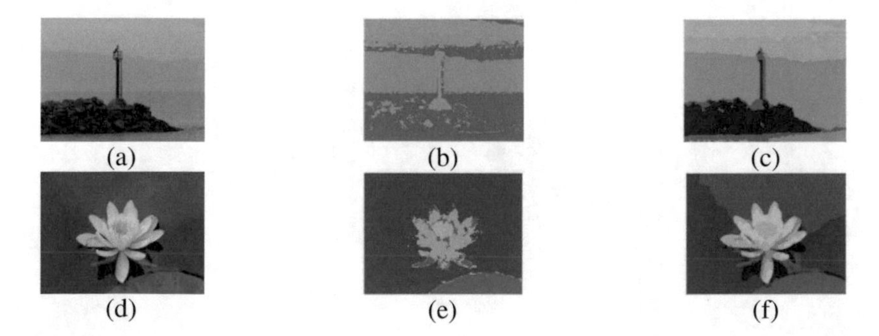

Fig. 6. Result of color segmentation. (a) lighthouse color image (b) lighthouse initial segmentation, $k = 4$ (c) lighthouse final segmentation, $k = 5$; (d) lotus original image (e) lotus initial segmentation, $k = 4$ and (f) lotus final segmentation, $k = 8$

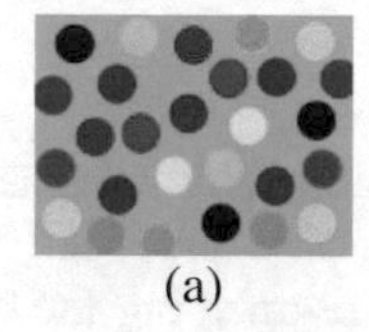
(a)

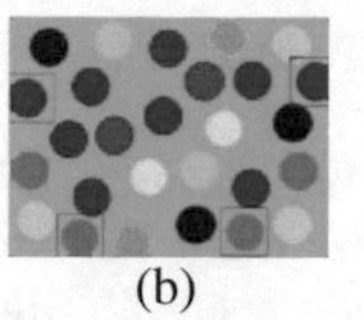
(b)

Fig. 7. False color label after final segmentation of supervised color image. (a) color image (b) false segmentation result (red square marker)

4.3 Evaluation

Supervised color image is used to measure how well the proposed method segments color image by comparing the result of segmentation with actual number of color. Accuracy of color prediction with 20 test images is 85%. With that accuracy, the method can predict number of color from color image quite well. Performance of segmentation is measure using silhouette method. Silhouette value is ranged from -1 to $+1$. The higher value means quality of segmentation is good and the lower value means poor quality segmentation that can be caused by too many or too few cluster. The result of average silhouette value tested on 25 natural image is 0.694. Figure 8a and d show the original image, Fig. 8b and e show the final segmentation and Fig. 8c and f show the silhouette visualization.

For color image in Fig. 8a, the predicted number of color is 6 and the average silhouette value is 0.758 and for color image in Fig. 8d, the predicted number of color is 7 and the average silhouette value is 0.853. As the result shows, the method produces appropriate cluster in color segmentation. Although, there are some cluster that only have small member (thin shape of the silhouette) and some part of the cluster that close to another cluster (the sharp edge of the silhouette), most of the cluster have medium-high value. The overall high value proves that this method produces appropriate number of cluster and good quality segmentation.

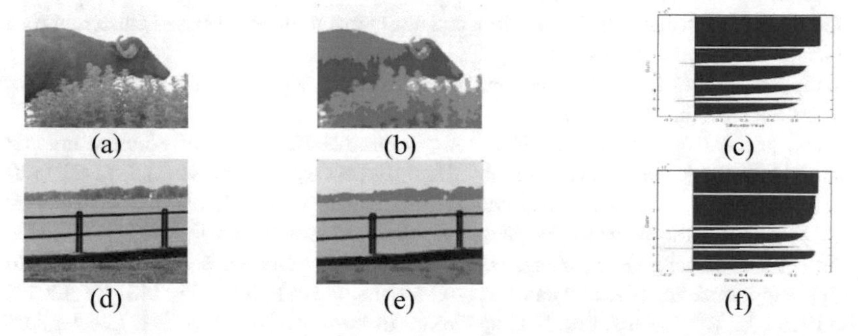

Fig. 8. Cluster performance measurement using silhouette visualization. (a) buffalo: color image (b) buffalo: final segmentation (c) buffalo: silhouette visualization (d) sea: color image (e) sea: final segmentation (f) sea: silhouette visualization

5 Conclusion

This paper has discussed *k*-means clustering-based method with emphasizes on histogram thresholding for automatic color segmentation. The proposed method has obtained good result in estimating number of color and segmenting the color image. This research shows that final cluster centers is likely inside the perimeter of estimated cluster centers area and lies inside the high-density of histogram value. The final cluster centers also not far from initial cluster centers obtained from peak selection. Therefore, the further development of this method is expected to localize the exact location of cluster centers without or with some minimal iterative procedure. The *k*-means clustering method is suitable for color segmentation and resulted in good quality of cluster. For future works, the method can be combined with spatial color information in order to produce good segmentation of color and even can be used to separate objects.

Acknowledgement. This research is supported by University of Malaya research grant UMRG.

References

1. Gonzalez, R.C., Woods, R.E.: Digital Image Processing, 3rd edn. Prentice Hall, Upper Saddle River (2008)
2. Jassim, F.A., Altaani, F.H.: Hybridization of Otsu method and median filter for color image segmentation. Int. J. Soft Comput. Eng. (IJSCE) **3**, 69–74 (2013)
3. Kurugollu, F., Sankur, B., Harmanci, A.E.: Color image segmentation using histogram multithresholding and fusion. Image Vis. Comput. **19**(13), 915–928 (2001)
4. Cheng, H.-D., Sun, Y.: A hierarchical approach to color image segmentation using homogeneity. IEEE Trans. Image Process. **9**(12), 2071–2082 (2000)
5. Severino Jr., O., Gonzaga, A.: A new approach for color image segmentation based on color mixture. Mach. Vis. Appl. **24**(3), 607–618 (2013)
6. Angulo, J., Serra, J.: Color segmentation by ordered mergings. Proc. Int. Conf. Image Process. (ICIP) **2**, 125–128 (2003)
7. Meilă, M., Heckerman, D.: An experimental comparison of model-based clustering methods. Mach. Learn. **42**(1), 9–29 (2001)
8. Khan, S.S., Ahmad, A.: Cluster center initialization algorithm for K-modes clustering. Expert Syst. Appl. **40**(18), 7444–7456 (2013)
9. Celebi, M.E., Kingravi, H.A., Vela, P.A.: A comparative study of efficient initialization methods for the k-means clustering algorithm. Expert Syst. Appl. **40**(1), 200–210 (2013)
10. Tan, K.S., Lim, W.H., Isa, N.A.M.: Novel initialization scheme for Fuzzy C-Means algorithm, on color image segmentation. Appl. Soft Comput. **13**(4), 1832–1852 (2013)
11. Wang, X.-Y., Zhang, X.-J., Yang, H.-Y., Bu, J.: A pixel-based color image segmentation using support vector machine and Fuzzy C-Means. Neural Netw. **33**, 148–159 (2012)
12. Kaufman, L., Rousseeuw, P.J.: Finding Groups in Data: An Introduction to Cluster Analysis. Wiley, Hoboken (1990)
13. McGill Vision Research: McGill calibrated colour image database. http://tabby.vision.mcgill.ca/html/browsedownload.html

Does Number of Clusters Effect the Purity and Entropy of Clustering?

Jamal Uddin(✉), Rozaida Ghazali, and Mustafa Mat Deris

Faculty of Computer Science and Information Technology,
Universiti Tun Hussein Onn Malaysia, Batu Pahat, Malaysia
jamal_maths@yahoo.co.uk

Abstract. Cluster analysis automatically partitioned the data into a number of different meaningful groups or clusters using the clustering algorithms. Every clustering algorithm produces its own type of clusters. Therefore, the evaluation of clustering is very important to find the better clustering algorithm. There exist a number of evaluation measures which can be broadly divided internal, external and relative measures. Internal measures are used to assess the quality of the obtained clusters like cluster cohesion and number of clusters (NoC). The external measures such as purity and entropy find the extent to which the clustering structure discovered by a clustering algorithm matches some external structure while the relative measures are used to assess two different clustering results using internal or external measures. To explore the effect of external evaluations specifically the NoC on internal evaluation measures like purity and entropy, an empirical study is conducted. The idea is taken from the fact that the NoC obtained in the clustering process is an indicator of the successfulness of a clustering algorithm. In this paper, some necessary propositions are formulated and then four previously utilized test cases are considered to validate the effect of NoC on purity and entropy. The proofs and experimental results indicate that the purity maximizes and the entropy minimizes with increasing NoC.

Keywords: Clustering · Purity · Entropy · Number of clusters

1 Introduction

Clustering is the process of dividing into clusters that are meaningful, useful and depict the natural inherited structure of the data. These conceptually meaningful clusters of objects shares common characteristics that helps to analyze and describe the data. Clustering has played an important role in a wide variety of fields: psychology and other social sciences, biology, statistics, pattern recognition, information retrieval, machine learning, and data mining [1]. In general, major clustering methods can be classified into the following categories: Partitioning algorithms, Hierarchy algorithms, Density-based, Grid-based, and

T. Herawan et al. (eds.), *Recent Advances on Soft Computing and Data Mining*,
Advances in Intelligent Systems and Computing 549, DOI 10.1007/978-3-319-51281-5_36

Model-based [2]. For clustering based on Pawlaks rough set model and its associated data type, the information system is a basic idea of many papers [3].

Clusters obtained as a result of clustering process must be assessed to evaluate their quality. Cluster validation or cluster evaluation is important and should be a part of any clustering. A key motivation is that due to the number of different kinds of possible clusterings, each clustering algorithm defines its own type of clusters in a data set. The performance of these clustering algorithms can be judged by evaluating them comparatively. The evaluation may be carried out by using external, relative and internal assessments [4].

For external assessment, the expert decomposition or gold standard, which has been obtained by a manual inspection of the system are compared with obtained clusters. Researchers have proposed different measures for performing the external assessment such as entropy, purity, precision, recall, and the F-measure [1]. These measures evaluate the extent to which a cluster contains objects of a single class. The relative assessment compares the results of same algorithm when using different features or similarity measures or the results of two different clustering algorithms. The measures used for comparison during external assessment may be employed to compare clustering results for relative assessment [4]. Internal assessment is an intrinsic evaluation of clustering results and it refers to measuring the goodness of a clustering structure without external information. Different measures have been adopted to evaluate cluster quality internally. Anquetil and Lethbridge [5] use cohesion and coupling of clusters within a decomposition to evaluate its quality. Moreover, [5–7] used the NoC and cluster size for evaluating clusters internally.

The purity measure evaluates the coherence of a cluster [8] and as a part of internal evaluation the large NoC means more cohesive and low coupled clusters are created [9]. The NoC obtained at each step of the clustering process is an indicator of the success fullness of a clustering algorithm [4] and high purity is easy to achieve when the NoC is large [10]. In this work, we explore the effect of the NoC on purity and entropy of clustering. The rough set approach has ability to deal with vagueness and uncertainty and can characterize set of objects in terms of attribute values [11,12] hence it can be used to obtain clusters. To explore the effects of NoC on cluster purity and entropy, four(4) test cases from previously utilized research examples are considered. The propositions and experimental results proves that the more the NoC are, the higher cluster purity is and lesser the cluster entropy.

Organization of rest of this article includes: Sect. 2 defines some preliminaries with experimental setup. Section 3 gives the detail of proposed methodology and propositions whereas, Sect. 4 explains the results of different examples case studies. Finally, Sect. 5 concludes the article.

2 Preliminaries

The mostly used external cluster evaluation measures are purity and entropy. A perfect clustering solution will be the one that leads to clusters that contain objects from only a single expert cluster, in which case the purity is 1 and entropy will be zero. Rough clustering approach [12] is used to cluster the data sets. Motivation behind using rough set theory is that it can represent subsets of a universe in terms of equivalence classes of a clustering of the universe as an information system [13]. Each part of our experimental setup is illustrated in subsequent paras.

2.1 Purity

Purity measures the extent to which a cluster contains objects of a single class [1]. In general, the larger the values of purity, the better the clustering solution is. For each cluster, the expert cluster distribution of the data is calculated first, i.e., for cluster j we compute P_{ij}, the probability that an element of cluster i belongs to class j as $P_{ij} = e_{ij}/e_i$, where e_i is the number of elements/objects in cluster i and e_{ij} is the number of elements/objects of class j in cluster i. The purity of cluster i is $P_i = max_j P_{ij}$ and the overall purity for e number of total data points of a clustering is,

$$P = \sum_{i=1}^{k} \frac{e_i}{e} P_i \tag{1}$$

2.2 Entropy

Entropy is another measure of the degree to which each cluster consists of objects of a single class [1]. Hence, smaller the entropy values, better the clustering solution is [14]. Using the class distribution and previous terminology, the entropy of each cluster i is calculated using the standard formula,

$$E_i = -\sum_{j=1}^{L} P_{ij} log_2 P_{ij} \tag{2}$$

Where L is the number of classes. The total entropy for a set of clusters is calculated as the sum of entropies of each cluster weighted by the size of each clusters, that is,

$$E = \sum_{i=1}^{k} \frac{e_i}{e} E_i \tag{3}$$

where K is the NoC and e is the total number of objects.

2.3 Pawlak's Rough Set

Zdzislaw Pawlak in the early 1980s introduced a mathematical tool that is rough set theory to deal with vagueness and uncertainty to aid decision making [12].

Definition 1. An information system is a 4-tuple (quadruple) $IS = (O, A, V, \alpha)$, where O is a non-empty finite set of objects, A is a non-empty finite set of attributes, $V = \bigcup_{a \in A} V_a$, V_a is the domain(value set) of attribute a, $\alpha : U \times A \rightarrow V$ is a function such that $\alpha(u, a) \in V_a$ for every $(u, a) \in U \times A$, called information function [12].

Definition 2. Let B be any subset of A. Two elements $x, y \in U$ is said to be B-indiscernible(i.e. indiscernible by the set of attribute $B \subseteq A$ in *IS*) if and only if $\alpha(x, b) = \alpha(y, b)$ for every $b \in B$. Obviously, every subset of A induces unique indiscernibility relation. Notice that, an indiscernibility relation induced by the set of attributes B, denoted by $IND(B)$, is an equivalence relation. It is well known that, an equivalence relation induces unique clustering. The clustering of U induced by *IND(B)* in *IS* denoted by U/B and the equivalence class in the clustering U/B containing $x \in U$, denoted by $[x]_B$. The cardinality of indiscernibility relation of an attribute(s) will show the NoC obtained by that attribute and can be evaluated as,

$$card(IND(B)) = |IND(B)| \tag{4}$$

3 Methodology

In the clustering process, if a clustering algorithm produces a larger NoC then it represent small size clusters, hence more similar objects or more specific information is combined which automatically improve the purity of clusters. That means more compact clusters has been obtained in this case so, clusters obtained are more cohesive and low coupled [9]. Moreover, as the granularity of the partitioning becomes coarser, our knowledge or information about a particular value also decreases [15]. The proof of by increasing NoC can increase the purity is illustrated in Proposition 1.

Proposition 1. *High purity is easy to achieve when the NoC is large.*

Proof: *Equation 1 of purity can be simplified to,*

$$purity = \sum_{i=1}^{k} \frac{e_i}{e} P_i = \sum_{i=1}^{k} \frac{e_{ij}}{e} \tag{5}$$

If we consider the worst possible case i.e. in selected best attributes each cluster in them has just single object correctly classified to particular class, than Eq. 5 gives,

$$\begin{aligned} purity(k\text{clusters}) &= \frac{1}{e} + \frac{1}{e} + \ldots + \frac{1}{e} = \frac{k}{e} \\ purity(k-1\text{clusters}) &= \frac{1}{e} + \frac{1}{e} + \ldots + \frac{1}{e} = \frac{k-1}{e} \\ purity(k-2\text{clusters}) &= \frac{1}{e} + \frac{1}{e} + \ldots + \frac{1}{e} = \frac{k-2}{e} \end{aligned}$$

Which shows that reducing the NoC minimize the purity of clusters because, $\frac{k}{e} > \frac{k-1}{e} > \frac{k-2}{e}$ is always true.

Moreover, if each cluster contains elements from only one class then its ideal case and purity is 1 [1]. Hence, if we consider best possible case i.e. all objects of clusters of clustering solution correctly classified to particular class, than Eq. 5 shows,

$$purity(k\text{clusters}) = \frac{a}{e} + \frac{b}{e} + ... + \frac{x}{e} = \frac{a+b+..+x}{e} = 1$$

$$purity(k-1\text{clusters}) = \frac{c}{e} + \frac{d}{e} + ... + \frac{y}{e} = \frac{c+d+..+y}{e} = 1$$

$$purity(k-2\text{clusters}) = \frac{f}{e} + \frac{g}{e} + ... + \frac{z}{e} = \frac{f+g+..+z}{e} = 1$$

In this case we always get purity 1 which means reducing the NoC has no effect on the purity of clusters as long as all objects of clusters are correctly classified to particular expert cluster. The entropy decreases as the classes become smaller through finer partitioning [15]. The effect of NoC on clustering entropy is explained in Proposition 2.

Proposition 2. *Entropy minimizes with increasing NoC.*

Proof: Equation 2 of entropy can be simplified to,

$$E = \sum_{i=1}^{k} \frac{e_i}{e} E_i = -KL\frac{e_{ij}}{e} log_2 \frac{e_{ij}}{e_i} \tag{6}$$

If in selected group of clusters, each cluster in them has just single object correctly classified to particular class then $e_{ij} = e_{sj} = e_{tj} = 1$. For this worst possible case for any clustering solution, the Eq. 6 results,

$$Entropy(k\text{clusters}) = -kL\frac{1}{e}log_2\frac{1}{e_i} = -\frac{kL}{e}log_2\frac{1}{e_i}$$

$$Entropy((k-1)\text{clusters}) = -(k-1)L\frac{1}{e}log_2\frac{1}{e_s} = -\frac{(k-1)L}{e}log_2\frac{1}{e_s}$$

$$Entropy((k-2)\text{clusters}) = -(k-2)L\frac{1}{e}log_2\frac{1}{e_t} = -\frac{(k-2)L}{e}log_2\frac{1}{e_t}$$

For inequality $k > k-1 > k-2$ we have $e_i < e_j < e_t$, because by reducing NoC will increase the size of each cluster. Hence, $-\frac{kL}{e}log_2\frac{1}{e_i} < -\frac{(k-1)L}{e}log_2\frac{1}{e_s} < -\frac{(k-2)L}{e}log_2\frac{1}{e_t}$ is always true. Hence, it shows that the entropy will minimize for increasing NoC.

Ideally, if each cluster will contain elements from only one class then entropy is 0 [1]. Considering this best possible case for any clustering solution, where all objects inside clusters are correctly classified to a particular expert cluster then $e_{ij} = e_i, e_{sj} = e_s, e_{tj} = e_t$. Hence, Eq. 6 simplified to,

$$Entropy(k\text{clusters}) = -kL\frac{1}{e}log_2\frac{e_{ij}}{e_i} = -\frac{kLe_{ij}}{e}log_2 1 = 0$$

$$Entropy((k-1)\text{clusters}) = -(k-1)L\frac{1}{e}log_2\frac{e_{sj}}{e_s} = -\frac{(k-1)Le_{sj}}{e}log_2 1 = 0$$

$$Entropy((k-2)\text{clusters}) = -(k-2)L\frac{1}{e}log_2\frac{e_{tj}}{e_t} = -\frac{(k-2)Le_{tj}}{e}log_2 1 = 0$$

In this case we always get entropy 0 because reducing the NoC has no effect on the entropy of clusters as long as all objects of clusters are correctly classified to only one expert cluster.

4 Experiments and Discussion

Four examples are considered to explore the effects of NoC on cluster purity and entropy. Indiscernibility relations using rough set theory will be used to produce clusters. Obviously, each attribute or subset of attributes induces unique indiscernibility relation which is an equivalence relation and hence it induces unique clustering. The cardinality of indiscernibility relations gives the NoC produced by each attribute.

Example 1: First considering the Suraj's flu patients information system from [16] as shown in Table 1. Six patients are considered with three possible symptoms i.e. Headache, Muscle-pain and Temperature.

Following are the indiscernibility relations induced by each attribute:

U/Headache $= \{(1,4,6),(2,3,5)\}$, U/Muscle-pain $= \{(1,3,4,6),(2,5)\}$,
U/Temperature $= \{(1,2,5),(3,6),(4)\}$.

After calculating the NoC as cardinality of indiscernibility relations, purity and entropy of each clustering attribute, the results are summarized in Fig. 1. The results shows that the clusters produced by attribute *Temperature* are larger in number than *Headache* and *Muscle − pain*. Also, the attribute *Temperature* clusters have comparatively better purity and lesser entropy.

Example 2: Table 2 contains the characterization of six stores in terms of six factors E, Q, S, R, L and P from [17].

Table 1. Suraj's flu patients data [16]

Patient	Headache	Muscle-pain	Temperature	Cluster label
1	No	Yes	High	C_1
2	Yes	No	High	C_1
3	Yes	Yes	Very high	C_1
4	No	Yes	Normal	C_2
5	Yes	No	High	C_2
6	No	Yes	Very high	C_1

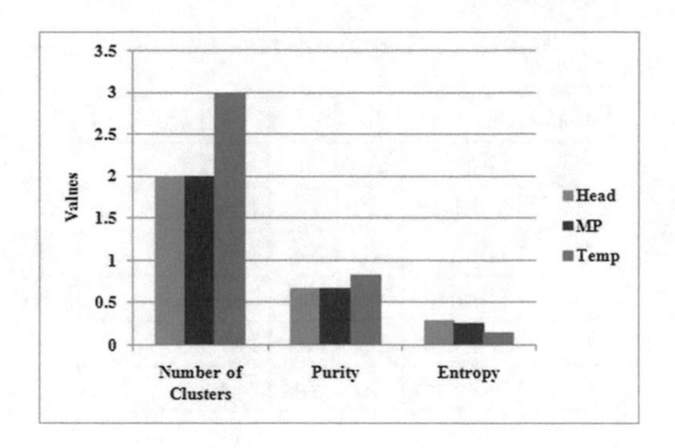

Fig. 1. Evaluation performance Suraj's flu data set

Table 2. Stores characterization data set [17]

Case	Q	S	R	L	P
S1	High	Good	Yes	Yes	No
S2	High	Good	No	Yes	No
S3	Medium	Good	Yes	Yes	Yes
S4	Low	Avg	Yes	Yes	Yes
S5	Low	Good	Yes	Yes	Yes
S6	High	Avg	No	No	Yes

Each attribute results following indiscernibility relations:

$U/Q=\{(1,2,6),(3),(4,5)\}$, $U/S=\{(1,2,3,5),(4,6)\}$,
$U/R=\{(1,3,4,5),(2,6)\}$, $U/L=\{(1,2,3,4,5),(6)\}$.

Figure 2 shows that the maximum clusters, high purity and lesser entropy was produced by attribute Q as compare to remaining attributes.

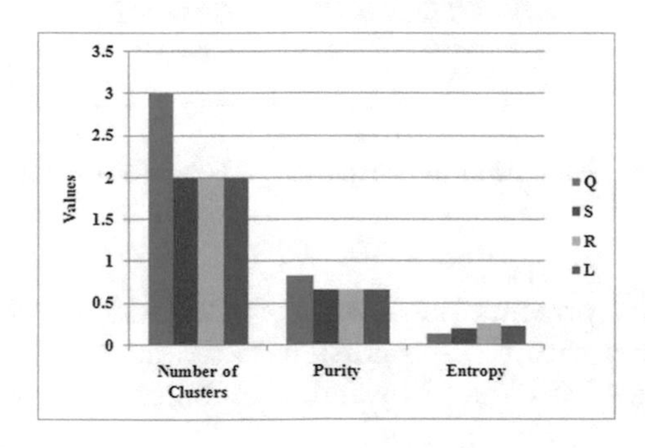

Fig. 2. Evaluation performance of stores characterization data set

Table 3. Grzymala data set [18]

Case	A1	A2	A3	A4	Decision
1	High	Yes	No	Yes	1
2	V.high	Yes	Yes	No	1
3	High	No	No	No	0
4	High	Yes	Yes	Yes	1
5	Normal	Yes	No	No	0
6	Normal	No	Yes	Yes	0

Example 3: Now considering another information system from [18] shown in Table 3. This is a patient data set having a decision attribute and four conditional attributes, i.e. A1, A2, A3 and A4 expressing certain symptoms of a disease.

Each attribute results following indiscernibility relations:

$U/A1=\{(1,3,4),(2),(5,6)\}$, $U/A2=\{(1,2,4,5),(3,6)\}$,
$U/A3=\{(1,3,5),(2,4,6)\}$, $U/A4=\{(1,4,6),(2,3,5)\}$,

Results of Grzymala data set are presented in Fig. 3. This figure illustrates that the clusters produced by attribute A1 are maximum as compare to remaining attributes with better purity and lesser entropy.

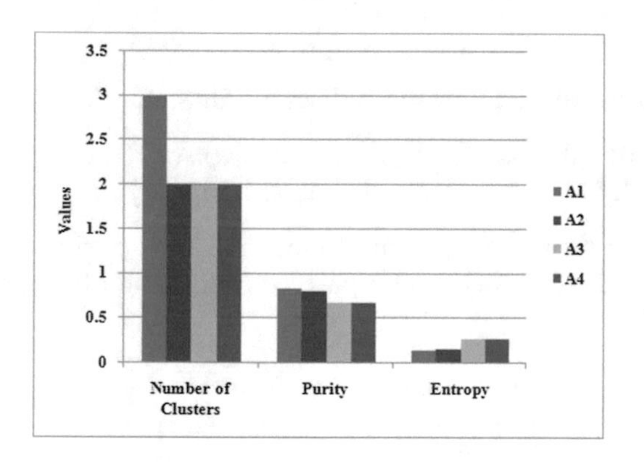

Fig. 3. Evaluation performance of Grzymala's data set

Example 4: Table 4 presents the Pawlak's modified data set taken from [19]. There are six objects with three conditional attributes (i.e. a, b and c). There are three partitions of U induced by indiscernibility relation on each attribute that are,

$U/a=\{(O1,O2,O3),(O4,O5,O6)\}$, $U/b=\{(O1,O2,O3,O4,O6),(O5)\}$,
$U/c=\{(O1,O4),(O2,O5),(O3,O6)\}$.

Table 4. Pawlak's modified data set [19]

U	a	b	c	Cluster label
O1	1	Yes	Fair	C_1
O2	1	Yes	Good	C_2
O3	1	Yes	Excellent	C_2
O4	0	Yes	Fair	C_1
O5	0	No	Good	C_1
O6	0	Yes	Excellent	C_2

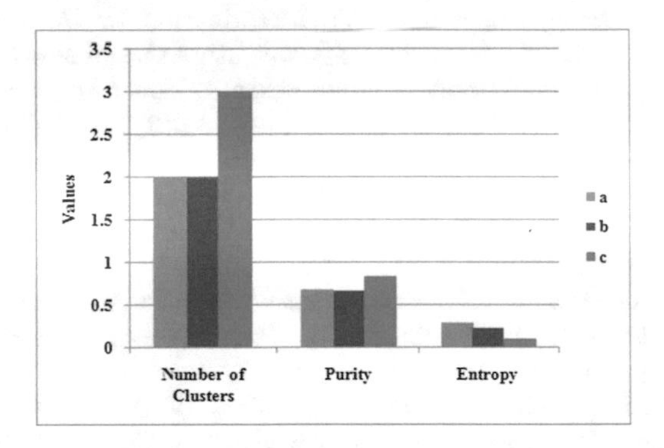

Fig. 4. Evaluation performance of Pawlak's modified data set

These results illustrated in Fig. 4 shows that the clusters produced by attribute c are larger in number than attributes a and b. Moreover, this attribute c also shows better purity and lesser entropy as compare to other attributes.

Hence, for all considered cases, the selected attribute with maximum NoC has always better purity and entropy of obtained clusters. This significant effect of the NoC on purity and entropy validate the Propositions 1 and 2. Finally, based on above results we can infer that the purity maximizes whereas the entropy minimizes with increasing NoC.

5 Conclusion

Various clustering algorithms are used for automatically partitioning of the data into meaningful clusters and each clustering algorithm results into different combinations of clusters. Therefore, it is necessary to evaluate and validate a clustering algorithm by traditional types of the evaluation measures like internal, external and relative measures. The relationship and effect of internal measures like NoC on external measures such as purity and entropy is studied in this work. Two important propositions are formulated to explore the effect. Moreover, four test cases were considered to validate this effect. A series of experimental results reveal that NoC has the significant effect on purity and entropy in a way that purity maximizes and entropy minimizes with the increasing NoC. This study can be useful for the selection of better clustering algorithm in terms of NoC.

Acknowledgment. The authors would like to thank Universiti Tun Hussein Onn Malaysia (UTHM) and Ministry of Higher Education (MOHE) Malaysia for financially supporting this research under the Fundamental Research Grant Scheme (FRGS), Vote No. 1235.

References

1. Tan, P.-N., Steinbach, M., Kumar, V.: Introduction to Data Mining. Addison-Wesley, Boston (2006). Chap. 8. http://www-users.cs.umn.edu/~kumar/
2. Verma, S., Nagwani, N.K.: Software bug classification using suffix tree clustering (STC) algorithm. Int. J. Comput. Sci. Technol. **4333**, 36–41 (2011)
3. Düntsch, I., Gediga, G.: Rough set clustering. Brock University, Department of Computer Science, Rough, Ontario, Canada, Technical report (2015)
4. Maqbool, O., Babri, H.A.: Hierarchical clustering for software architecture recovery. IEEE Trans. Softw. Eng. **33**(11), 759–780 (2007)
5. Anquetil, N., Lethbridge, T.C.: Experiments with clustering as a software remodularization method. In: Proceedings of the Sixth Working Conference on Reverse Engineering, pp. 235–255 (1999)
6. Davey, J., Burd, E.: Evaluating the suitability of data clustering for software remodularisation. In: Proceedings Seventh Working Conference on Reverse Engineering, pp. 268–276. IEEE Computer Society (2000). http://ieeexplore.ieee.org/lpdocs/epic03/wrapper.htm?arnumber=891478
7. Wu, J., Hassan, A.E., Holt, R.C.: Comparison of clustering algorithms in the context of software evolution. In: IEEE International Conference on Software Maintenance, ICSM, vol. 2005, pp. 525–535 (2005)
8. Huang, A.: Similarity measures for text document clustering. In: Proceedings of the Sixth New Zealand, pp. 49–56, April 2008. http://nzcsrsc08.canterbury.ac.nz/site/proceedings/Individual_Papers/pg049_Similarity_Measures_for_Text_Document_Clustering.pdf
9. Wang, Y., Liu, P., Guo, H., Li, H., Chen, X.: Improved hierarchical clustering algorithm for software architecture recovery. In: International Conference on Intelligent Computing and Cognitive Informatics, pp. 1–4 (2010)
10. Christopher, P.R., Manning, D., Schütze, H.: Introduction to Information Retrieval, April 2009
11. Uddin, J., Ghazali, R., Deris, M.M., Naseem, R., Shah, H.: A survey on bug prioritization. Artif. Intell. Rev. **46**, 1–36 (2016). doi:10.1007/s10462-016-9478-6
12. Pawlak, Z., Skowron, A.: Rudiments of rough sets. Inf. Sci. **177**(1), 3–27 (2007)
13. Herawan, T., Deris, M.M., Abawajy, J.H.: A rough set approach for selecting clustering attribute. Knowl. Based Syst. **23**(3), 220–231 (2010). doi:10.1016/j.knosys.2009.12.003
14. Zhao, Y.: Criterion functions for document clustering: experiments and analysis (Technical report), pp. 1–30 (2001). http://scholar.google.com/scholar?hl=en&btnG=Search&q=intitle:Criterion+Functions+for+Document+Clustering+?+Experiments+and+Analysis#4
15. Beaubouef, T., Petry, F.E., Arora, G.: Information-theoretic measures of uncertainty for rough sets and rough relational databases. J. Inf. Sci. **5**, 185–195 (1998)

16. Suraj, Z.: An introduction to rough set theory and its applications. In: ICENCO 2004, Cairo, Egypt, 27–30 December 2004
17. Pawlak, Z.: Rough Sets Theoretical Aspects of Reasoning about Data (1991)
18. Grzymala-Busse, J.W.: Rough set theory with applications to data mining. In: Negoita, M.G., Reusch, B. (eds.) Real World Applications of Computational Intelligence. Studies in Fuzziness and Soft Computing, vol. 179, pp. 221–244. Springer, Heidelberg (2005). doi:10.1007/11364160_7
19. Pawlak, Z., et al.: Rough sets. Commun. ACM **38**(11), 88–95 (1995). http://portal.acm.org/citation.cfm?doid=219717.219791

Text Detection in Low Resolution Scene Images Using Convolutional Neural Network

Anhar Risnumawan[1(✉)], Indra Adji Sulistijono[2], and Jemal Abawajy[3]

[1] Mechatronics Engineering Division, Politeknik Elektronika Negeri Surabaya (PENS), Kampus PENS, Surabaya, Indonesia
[2] Graduate School of Engineering Technology, Politeknik Elektronika Negeri Surabaya (PENS), Kampus PENS, Surabaya, Indonesia
{anhar,indra}@pens.ac.id
[3] School of Information Technology, Deakin University, Geelong, Australia
jemal.abawajy@deakin.edu.au

Abstract. Text detection on scene images has increasingly gained a lot of interests, especially due to the increase of wearable devices. However, the devices often acquire low resolution images, thus making it difficult to detect text due to noise. Notable method for detection in low resolution images generally utilizes many features which are cleverly integrated and cascaded classifiers to form better discriminative system. Those methods however require a lot of hand-crafted features and manually tuned, which are difficult to achieve in practice. In this paper, we show that the notable cascaded method is equivalent to a Convolutional Neural Network (CNN) framework to deal with text detection in low resolution scene images. The CNN framework however has interesting mutual interaction between layers from which the parameters are jointly learned without requiring manual design, thus its parameters can be better optimized from training data. Experiment results show the efficiency of the method for detecting text in low resolution scene images.

Keywords: Scene text detection · Low resolution images · Cascaded classifiers · Convolutional Neural Network · Mutual interaction · Joint learning

1 Introduction

Detecting text on scene images has attracted wide interest due to its usefulness in varieties of real world applications, such as robots navigation, assisting visually impaired people, tourists navigation, enhancing safe vehicle driving, etc. [2,4]. Differing from the problems of printed documents analysis where the characters are typically monotone on uniform backgrounds, detecting text on scene images

The original version of this chapter was revised: Co-author name has been deleted. The erratum to this chapter is available at DOI: 10.1007/978-3-319-51281-5_65

T. Herawan et al. (eds.), *Recent Advances on Soft Computing and Data Mining*, Advances in Intelligent Systems and Computing 549, DOI 10.1007/978-3-319-51281-5_37

is highly complicated due to complex background and high variation of fonts, size, and color. Thus, text on scene images has to be robustly detected, which is usually indicated by forming bounding boxes on the detected text.

In most embedded-systems-based platforms such as robotics and mobile devices, low resolution images are often acquired from low resolution camera since it has small size, modularity with the platforms, and its compatibility with low processing power of most embedded systems. Ranging from robot localization applications using road signs as a GPS alternative, wearable-devices-based scene text detection such as Google Goggles[1], vOICe[2], and Sypole[3] to help visually impaired people.

Dealing with low resolution images for detection is generally solved using many features which are cleverly integrated followed by cascaded classifiers to get better discriminative system as compared to the high-resolution-images-based systems [5,11–13,18–20]. Many features integration and cascaded classifier are extremely important to reduce misclassification rate due to the noise caused by low resolution images. The features are however prone to error due to imperfection from manual design, while the cascaded classifiers require tedious jobs of hand-tuning before preparing another classifier to be cascaded, as shown in Fig. 1.

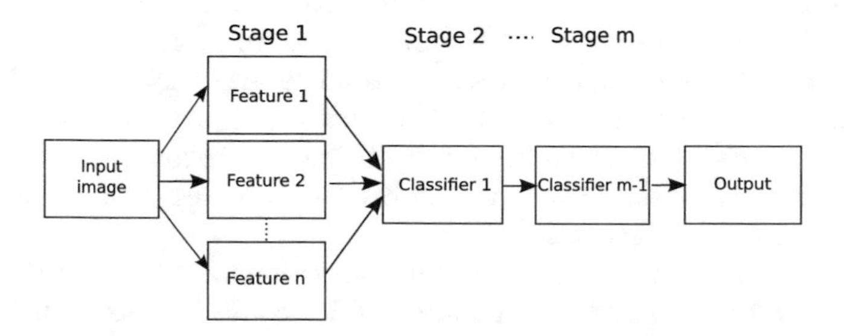

Fig. 1. General cascaded framework for detection in low resolution images, where there is no mutual interaction between stages. Each stage usually needs careful design and tune before working on another stage to be cascaded.

The cascaded framework as shown in Fig. 1 has no mutual interaction between stages. More specifically, the parameter on each stage is individually designed and tuned during learning, which is then the parameters become fixed when a new stage is designed. For example, the Histogram of Oriented Gradient (HOG) feature is individually designed with its parameters manually tuned given the linear SVM classifier being used in [1]. Then HOG feature become fixed when new classifiers are designed [6].

[1] http://www.google.com/mobile/goggles/#text.

[2] http://www.artificialvision.com/android.htm.

[3] http://tcts.fpms.ac.be/projects/sypole/index.php?lang=en.

In this paper, we show that the most notable cascaded framework that is using many integrated features and cascaded classifiers is equivalent to a deep convolutional neural network (CNN) [3], to deal with text detection on low resolution scene images. The method mainly differs from the existing manually hand-crafted features and explicitly learn each classifier, the method implicitly integrates many features and has mutual interaction between stages. This mutual interaction can be seen as the stages are connected and jointly learned for which the back-propagation technique is employed to obtain parameters simultaneously.

2 Text Detection Using CNN

Overview of the system is shown in Fig. 2. Given a low resolution input scene image, response map is computed from CNN, and bounding boxes are formed to indicate the detected text lines. One can see the noise caused by low resolution from the zoomed image in Fig. 2b. These for example, the characters are not smooth as in high resolution images, touching between characters may be classified as non-text, missing characters due to imperfect geometrical properties, and non-text may be classified as text due to noise.

Notable works which solve text detection on low resolution images are Mirmehdi et al. [7] and Nguyen et al. [10], but it only works for document which is differ from scene images. Sanketi et al. [14] used several stages to localize text on low resolution scene images. However, the method is just a proving concept. It is also noted that several stages system could hardly be maintained as the error from the first stage could be propagated due to imperfect tuning, and requires manually hand-crafted features.

Classifier is the main part of scene text detection system. It has been proven that increasing the performance of the classifier could increase the accuracy of detection, for example the works by [8,9] applied cascaded classifiers to boost the accuracy.

Conventionally, in order to classify an image patch u whether it is text or non-text, a set of features $\mathbf{Q}(u) = (Q_1(u), Q_2(u), \ldots, Q_N(u))$ are extracted and a binary classifier k_l for each text and non-text label l is learned. From the probabilistic point of view, this means classifiers are learned to yield a posterior probability distribution $p(l|u) = k_l(\mathbf{Q}(u))$ over labels given the inputs. The objective is then to maximize to recognize the labels l contained in image patch u such that $l^* = \text{argmax}_{l \in L} p(l|u)$, where $L = \{\text{text}, \text{non-text}\}$. Those features $\mathbf{Q}$ often require hand-crafted manual design and optimized through a tedious jobs of trial-and-error cycle from which adjusting the features and re-learning the classifiers are needed. In this work, CNN is applied instead to learn the representation, jointly optimizing the features as well as the classifiers.

Multiple layers of features are stacked in CNN. A convolutional layer consist of N linear filters which is then followed by a non-linear activation function h. A feature map $f_m(x, y)$ is an input to a convolutional layer, where $(x, y) \in \mathcal{S}_m$ are spatial coordinates on layer m. The feature map $f_m(x, y) \in \mathbb{R}^C$ contains

(a) Low resolution image, 250x138 pixels

(b) Zoomed image patch

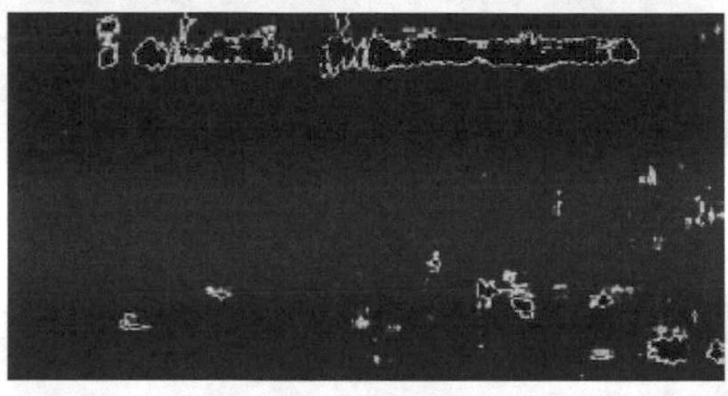

(c) Response map from CNN

(d) Bounding boxes formation

Fig. 2. Overview of the system using low resolution input image. Response map is shown as the probability output [0, 1] from CNN, ranging from blue(lowest) - yellow - red(highest). Note that most of the probability values spread around the detected text line. Best viewed in color.

C channels or $f_m^c(x, y)$ to indicate c-th channel feature map. The output of convolutional layer is a new feature map f_{m+1}^n such that,

$$f_{m+1}^n = h_m(W_{mn} \bigotimes f_m + b_{mn}) \tag{1}$$

where W_{mn} and b_{mn} denote the n-th filter kernel and bias, respectively and $\bigotimes$ denotes convolution operator. An activation layer h_m such as the Rectified Linear Unit (ReLU) $h_m(f) = \max\{0, f\}$ is used in this work. In order to build translation invariance in local neighborhoods, convolutional layers can be intertwined with normalization, subsampling, and pooling layers. Pooling layer is obtained by performing operation such as taking maximum or average over local neighborhood contained in channel c of feature maps. The process starts with f_1 equal to input image patch u, performs convolution layer by layer, and ends by connecting the last feature map to a logistic regressor for classification to get the probability of the correct label. All the parameters of the model are jointly learned from the training data. This is achieved by minimizing the classification loss over a training data using Stochastic Gradient Descent (SGD) and back-propagation.

We apply CNN structure as shown in Fig. 3. An image patch $u \in \mathbb{R}^{32\times32}$ is used as input. The input u is firstly normalized using its mean and variance.

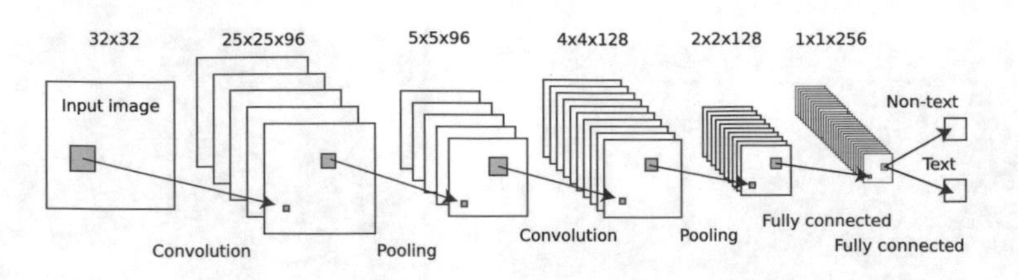

Fig. 3. CNN structure for this work.

The normalized input then becomes an input to two convolutional and pooling layers. We found that the normalization is important to reduce the loss as the input is now centered and bounded to $[0, 1]$. We use the number of first filters $N_1 = 96$ and the second $N_2 = 128$. After each convolutional layer and the first fully-connected (FC) layer we intertwine with ReLu activation layer. ReLu is simply passing all the feature maps if it is not less than zero $h_m(f) = \max\{0, f\}$. In practice we found that ReLu will increase the convergence rate of learning process instead of using sigmoid function as it could easily deteriorate the gradient.

After the last fully-connected layer we apply sigmoid activation function followed by softmax loss layer. In our experiments, it was found empirically that sigmoid function after last fully-connected layer yields superior performance. This can be thought as bounded the output to be within the labels which are $\{0, 1\}$ for non-text and text labels, respectively. During testing, the output is the probability value of each label.

In order to form bounding boxes from the response maps, we apply non-maximal suppression (NMS) as in [15]. In particular, for each row in the response maps, we scan the possible overlapping line-level bounding boxes. Each bounding box is then scored by the mean value of probability contained in response maps. Then NMS is applied again to remove overlapping bounding boxes.

3 Relationship to the Existing Cascaded Framework

We show that the cascaded framework can be seen as a Convolutional Neural Network. Figure 4 shows an illustration.

In the cascaded framework, many features are extracted from the input image. Those features are used to provide rich information to the classifier for better input representation. The classifier then decide the correct label of the input. The classifiers generally contain non-linear transformation such as in Support Vector Machine (SVM) from which the input features space are projected or mapped onto higher dimension to apply linear operations. In order to make better discriminative system, several classifiers are usually cascaded.

The above discussion shows that the cascaded framework can be viewed as a kind of CNN, where the stages are now becomes layers. Many features are represented by feature maps with its channels. And a classifier's non-linear transformation as a convolutional operation as in Eq. 1 containing non-linear

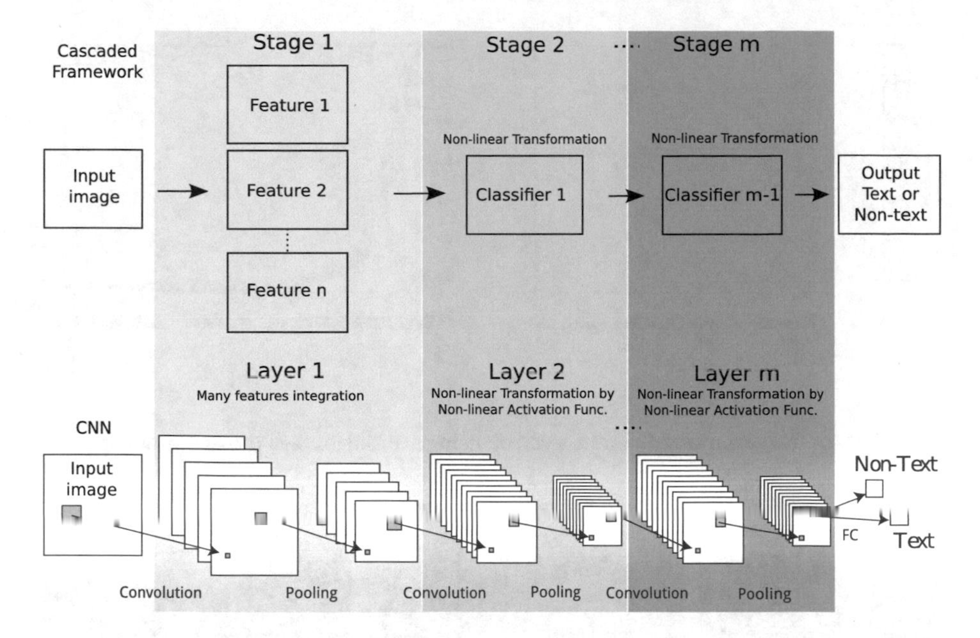

Fig. 4. Analogy between cascaded framework and Convolutional Neural Network structure.

activation function of convolution between the filter kernel and its bias with feature maps. However, in the CNN, many features integration, non-linear transformation, together with maximum or average pooling, are all involved in a layer. It is fully feed-forward and can be computed efficiently. So the method able to optimize an end-to-end mapping that consists of all operations. In addition, the last fully-connected layer (FC) can be thought as pixel-wise non-linear transformation. A new stage or layer could be easily cascaded to make better discriminative system.

4 Experimental Results

We train the CNN using ICDAR 2003 training set and synthetic data from [16]. Some of the training data is used as validation set. The learning rate of SGD parameters is set to 0.05 and decreases by multiplication of 0.1 for every 100,000 iterations. Maximum iteration is set to 450,000. Momentum parameter is 0.9. The validation set is tested for every 1000 iterations. Using a standard PC core i5 4 Gb RAM running GPU 1 Gb memory the training takes about 1 week. These validation test curves for every 1000 iteration during training as shown in Fig. 5.

The performance of our method is investigated further by reducing the resolution image, as shown in Fig. 6. Our method shows able to detect the text line which is indicated by the probability values contained in the response map spreading around the text line, while removing a lot of non-text. The bounding

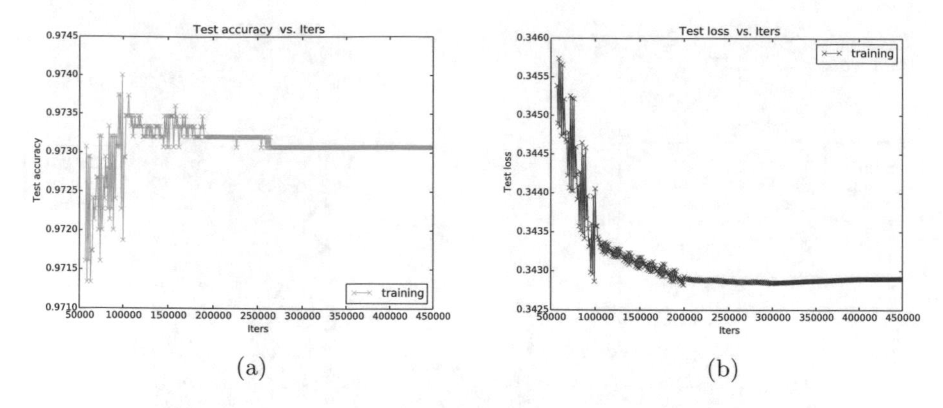

(a) (b)

Fig. 5. Test accuracy (a) and loss (b) using validation set during training.

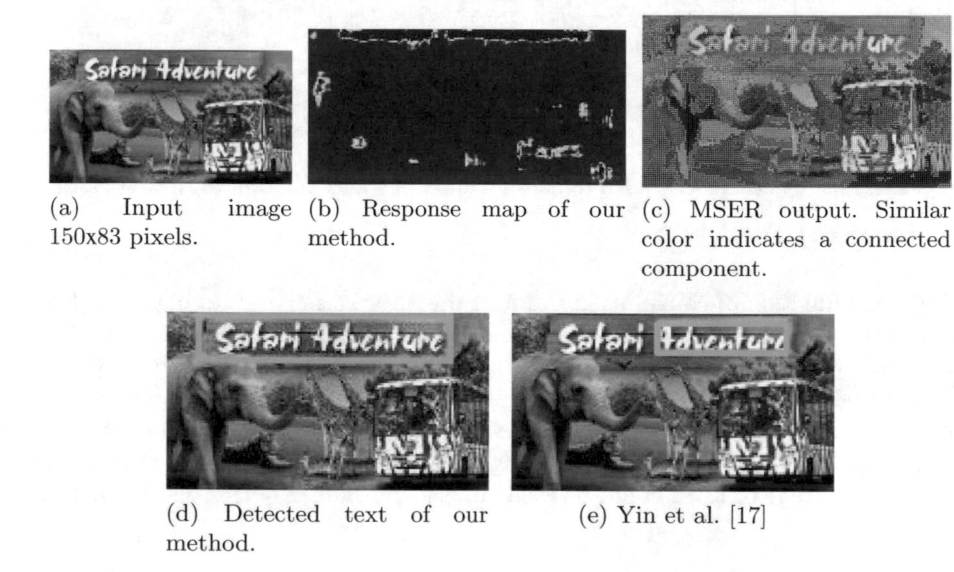

(a) Input image 150x83 pixels. (b) Response map of our method. (c) MSER output. Similar color indicates a connected component.

(d) Detected text of our method. (e) Yin et al. [17]

Fig. 6. Result and comparison with low resolution input image. Best viewed in color.

boxes could be easily formed using NMS as the non-text responses do not have regular pattern (such as horizontal alignment).

On the contrary, the well known methods based on connected components analysis such as Maximally Stable Extremal Regions (MSER)-based show many components are extracted as shown in Fig. 6c including non-text and fews are text. Moreover, note that some of the characters are touching its neighborhood, e.g., character S-a, r-i in the word 'Safari', and the characters stroke width are relatively small. These could hardly be classified as text since most of the connected components analysis investigate geometrical structure of each component. We compare with the result from Yin et al. [17] as a representation from other

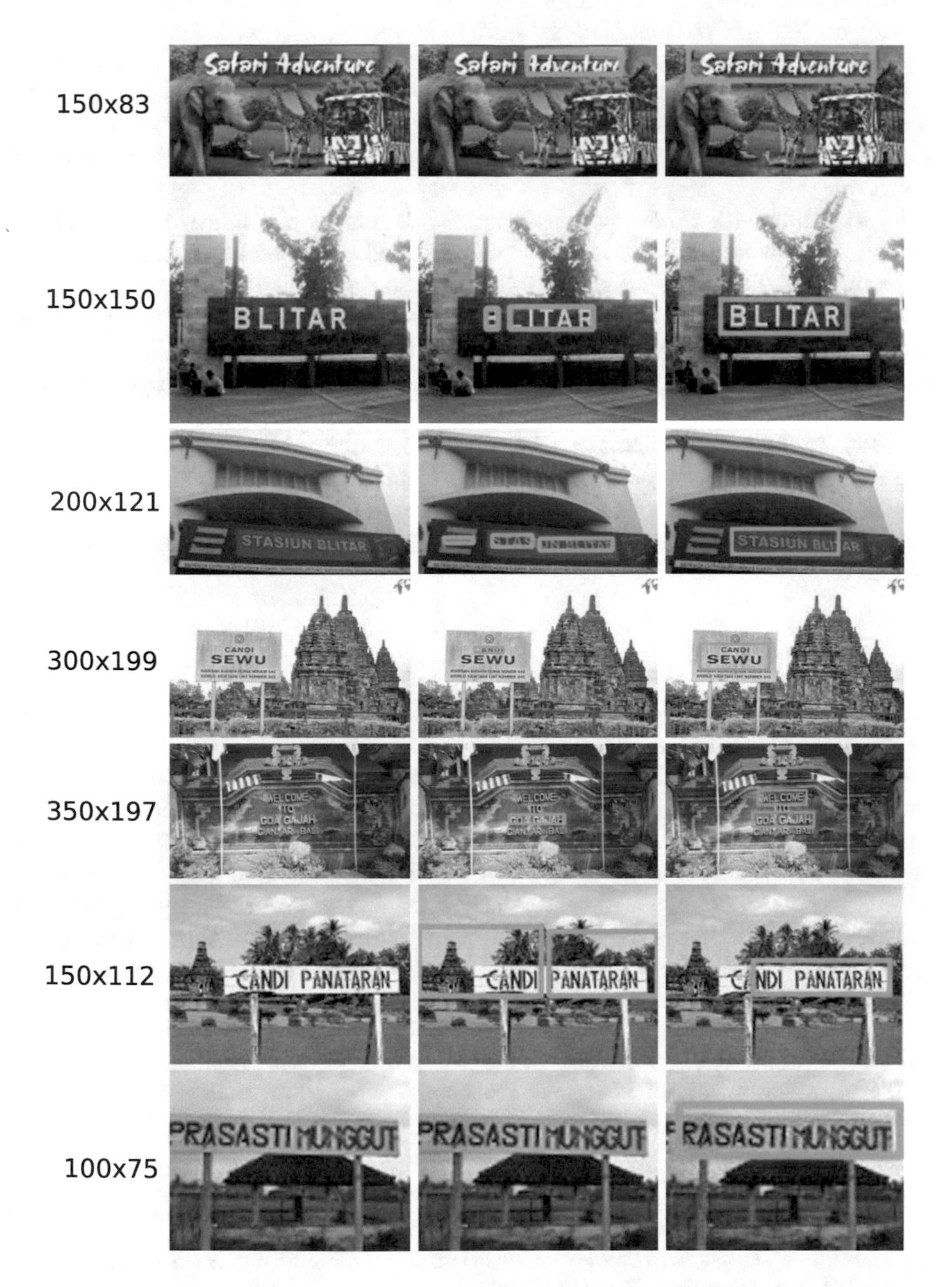

Fig. 7. Results in low resolution scene images input (left-side) between our method (right-side) and Yin et al. [17] (middle-side). Dimension values at the left most side indicate resolution of the corresponding images. Best viewed in color.

cascaded-framework-like methods, since their method achieved the best performance on ICDAR competition dataset, it is also similar with the definition of the cascaded framework in this work, online demo is provided thus easy to make comparison, and more importantly it used connected components analysis using MSER-based. As can be seen the whole text line could not be detected as in Fig. 6e due to imperfect geometrical structure of the components because of noise.

Test result with other low resolution scene images are shown in Fig. 7. One can observe the background are very complex and characters are very small with a lot of noise. Our method shows fairly well to detect the text lines, though it misses some very small characters. Those characters in such away too small for even human could not recognize correctly. It is noted that the best performance result on ICDAR could not correctly detect the text lines as shown in the middle Fig. 7. This could be attributed to the imperfect geometrical structure of the components due to low resolution images, or it could be the framework not deep enough to create better discriminative system. In our experiments, the limit of the lowest characters resolution using our method seems to be constrained by half of the image patch size. Below this value, the text lines may partly be detected.

5 Conclusion

In this paper, we have presented a framework that has mutual interaction between stages or layers using CNN to deal with text detection problems in low resolution images. The framework can be viewed as the conventional cascaded framework to build better discriminative system. Interestingly, many features integration, non-linear transformation, together with maximum or average pooling, are all involved in a layer. It is fully feed-forward and can be computed efficiently. Thus it is able to jointly optimize an end-to-end mapping that consists of all existing cascaded operations. The experiments show encouraging results that it would be beneficial for the future works on text detection in low resolution scene images.

Acknowledgements. The authors would like to thank Pusat Penelitian dan Pengabdian Masyarakat (P3M) of Politeknik Elektronika Negeri Surabaya (PENS) for supporting this research by Local Research Funding FY 2016.

References

1. Dalal, N., Triggs, B.: Histograms of oriented gradients for human detection. In: CVPR, vol. 1, pp. 886–893. IEEE (2005)
2. Jung, K., Kim, K.I., Jain, A.K.: Text information extraction in images and video: a survey. Pattern Recogn. **37**(5), 977–997 (2004)
3. LeCun, Y., Boser, B., Denker, J.S., Henderson, D., Howard, R.E., Hubbard, W., Jackel, L.D.: Backpropagation applied to handwritten zip code recognition. Neural Comput. **1**(4), 541–551 (1989)

4. Liang, J., Doermann, D., Li, H.: Camera-based analysis of text and documents: a survey. Intl. J. Doc. Anal. Recogn. (IJDAR) **7**(2–3), 84–104 (2005)
5. Mählisch, M., Oberländer, M., Löhlein, O., Gavrila, D., Ritter, W.: A multiple detector approach to low-resolution fir pedestrian recognition. In: Proceedings of the IEEE Intelligent Vehicles Symposium (IV2005), Las Vegas, NV, USA (2005)
6. Maji, S., Berg, A.C., Malik, J.: Classification using intersection kernel support vector machines is efficient. In: CVPR, pp. 1–8. IEEE (2008)
7. Mirmehdi, M., Clark, P., Lam, J.: Extracting low resolution text with an active camera for OCR. In: Spanish Symposium on Pattern Recognition and Image Processing IX, pp. 43–48 (2001)
8. Neumann, L., Matas, J.: Real-time scene text localization and recognition. In: CVPR, pp. 3538–3545. IEEE (2012)
9. Neumann, L., Matas, J.: On combining multiple segmentations in scene text recognition. In: ICDAR (2013)
10. Nguyen, M.H., Kim, S.-H., Lee, G.: Recognizing text in low resolution born-digital images. In: Jeong, Y.-S., Park, Y.-H., Hsu, C.-H.R., Park, J.J.J.H. (eds.) Ubiquitous Information Technologies and Applications. LNEE, vol. 280, pp. 85–92. Springer, Heidelberg (2014). doi:10.1007/978-3-642-41671-2_12
11. Risnumawan, A., Chan, C.S.: Text detection via edgeless stroke width transform. In: ISPACS, pp. 336–340. IEEE (2014)
12. Risnumawan, A., Shivakumara, P., Chan, C.S., Tan, C.L.: A robust arbitrary text detection system for natural scene images. Expert Syst. Appl. **41**(18), 8027–8048 (2014)
13. Sahli, S., Ouyang, Y., Sheng, Y., Lavigne, D.A.: Robust vehicle detection in low-resolution aerial imagery. In: SPIE Defense, Security, and Sensing, p. 76680G. International Society for Optics and Photonics (2010)
14. Sanketi, P., Shen, H., Coughlan, J.M.: Localizing blurry and low-resolution text in natural images. In: 2011 IEEE Workshop on Applications of Computer Vision (WACV), pp. 503–510. IEEE (2011)
15. Wang, K., Babenko, B., Belongie, S.: End-to-end scene text recognition. In: ICCV, pp. 1457–1464. IEEE (2011)
16. Wang, T., Wu, D.J., Coates, A., Ng, A.Y.: End-to-end text recognition with convolutional neural networks. In: ICPR, pp. 3304–3308. IEEE (2012)
17. Yin, X.-C., Yin, X., Huang, K., Hao, H.-W.: Robust text detection in natural scene images. IEEE Trans. Pattern Anal. Mach. Intell. **36**(5), 970–983 (2014)
18. Zhang, J., Gong, S.: People detection in low-resolution video with non-stationary background. Image Vis. Comput. **27**(4), 437–443 (2009)
19. Zhao, T., Nevatia, R.: Car detection in low resolution aerial images. Image Vis. Comput. **21**(8), 693–703 (2003)
20. Zhu, J., Javed, O., Liu, J., Yu, Q., Cheng, H., Sawhney, H.: Pedestrian detection in low-resolution imagery by learning multi-scale intrinsic motion structures (mims). In: CVPR, pp. 3510–3517 (2014)

Handling Imbalanced Data in Churn Prediction Using RUSBoost and Feature Selection (Case Study: PT.Telekomunikasi Indonesia Regional 7)

Erna Dwiyanti(✉), Adiwijaya, and Arie Ardiyanti

School of Computing, Telkom University, Bandung, Indonesia
emailnya.erna@gmail.com,
{adiwijaya,ardiyanti}@telkomuniversity.ac.id

Abstract. Solving imbalance problems is a challenging tasks in data mining and machine learning. Most classifiers are biased towards the majority class examples when learning from highly imbalanced data. In practice, churn prediction is considered as one of data mining application that reflects imbalance problems. This study investigates how to handle class imbalance in churn prediction using RUSBoost, a combination of random under-sampling and boosting algorithm, which is combined with feature selection for better performance result. The datasets used are broadband internet data collected from a telecommunication industry in Indonesia. The study firstly select the important features using Information Gain, and then building churn prediction model using RUSBoost with C4.5 as the weak learner. The result shows that feature selection and RUSBoost improve 16% of the performance of prediction and reduce 48% of the processing time.

Keywords: Information gain · RUSBoost · Imbalance problems · Churn prediction

1 Introduction

In practice, there are many problems associated with the imbalanced data, which make the data mining algorithm lose its power. The data is said to be imbalance if the sample from one class is in greater number than other [1]. The class that has more numbers of instances is called as majority class, while the one that has relatively less number of instances is called minority class [1]. In the context of data mining, the minority class is often become the primary interest and has a great value. In such situation, most of the classifier show poor classification and biased towards the major classes [2]. For example, a binary classification model on training dataset contains only 1% of positive class, a model can achieve 99% accuracy by classifying all examples as belonging to the majority class and simply ignores the minority class.

In the telecommunication industry, the term "churn" refers to the loss of subscribers who switch from one provider to another during a given period [3]. Churn prediction is considered as one of data mining application that reflecting imbalance problems. Based on data from PT. Telkom Indonesia, the average churn rate for internet broadband

T. Herawan et al. (eds.), *Recent Advances on Soft Computing and Data Mining*,
Advances in Intelligent Systems and Computing 549, DOI 10.1007/978-3-319-51281-5_38

customer of PT.Telkom Indonesia from January until June 2015 is about 2.8% per month. It indicates that one in thirty five subscribers of a given company discontinues their services every month. As it is more profitable for the company to retain the existing customers than to constantly attract new customers [3], it is very important to build an accurate churn prediction model to identify those customers who are most prone to churn. It is not an easy task for machine learning algorithm to predict churn with a very small sample of positive class. Therefore this research attempts to investigate how to handle the imbalance problems in churn prediction so a better churn prediction model can be developed.

Several solutions to the imbalance data problem are previously proposed both at the data and algorithmic levels [4]. At the data level, the solutions include sampling and feature selection. At the algorithmic level, the solutions include threshold method, one-class learning, and cost sensitive method. Both data and algorithmic approach will be examined in this work.

Feature selection is often used as a preprocessing step in machine learning. It is a process of selecting a subset of original features so that the feature space is optimally reduced according to a certain evaluation criterion [5]. Various type of feature selection has been studied on different datasets of imbalance data and Information Gain gives best result dealing with imbalance data [6]. Another technique at data level approach is data sampling, which has received much attention to handle imbalance problem. Data sampling tries to overcome imbalanced class distribution problems by adding or removing samples from the data [4]. Process of adding new sample in existing is known as over-sampling and process of removing a sample known as under-sampling. Both oversampling and undersampling method have their benefits and drawbacks. The main drawback of undersampling is the loss of important information that comes with deleting examples from training data [7]. While the benefit of undersampling is that it takes less time to train the data, since the training dataset is reduced. On the other hand, oversampling results in no lost information, but it can lead to overfitting when it is performed by duplicating examples [8].

Boosting is a technique for dealing with imbalanced data that works on algorithmic level and AdaBoost is the most commonly used boosting algorithm which has been shown to improve the performance of any weak classifier. A number of variations has been proposed to make AdaBoost cost sensitive or to improve performance on imbalanced data. The new novel hybrid data sampling and boosting algorithm is RUSBoost which combines RUS and AdaBoost [9]. In this paper, we present RUSBoost combined with Information Gain as feature selection at data preprocessing method for handling imbalance problems on churn prediction.

RUSBoost is a hybrid method that combines data sampling and boosting algorithm [10]. RUSBoost algorithm is simillar with SMOTEBoost algorithm, both algorithms combine sampling with AdaBoost as boosting algorithm. The difference between the two algorithms is the sampling technique used. RUSBoost uses random undersampling, while SMOTEBoost uses SMOTE as an over sampling method. Random under sampling is a sampling technique which removes examples from the majority class randomly. While SMOTE is more complex and consumes a lot of time since it is adding synthetic examples to minority class by extrapolating between existing examples. Both over sampling and under sampling have their own drawbacks. Over sampling increases the training time, and tends to overfitting. While the drawback of under sampling is loss of important information when removing examples from majority class. However, the under sampling drawback is greatly overcome by combining it with AdaBoost and feature selection.

2 Methodology

The proposed method is described in the design process flowchart in Fig. 1 below:

Data Collection is a process to collect data. In our research, data from October 2014 until November 2015 were collected using SQL from oracle server provided by PT. Telkom Indonesia. Data preparation included data integration, data cleaning, and data

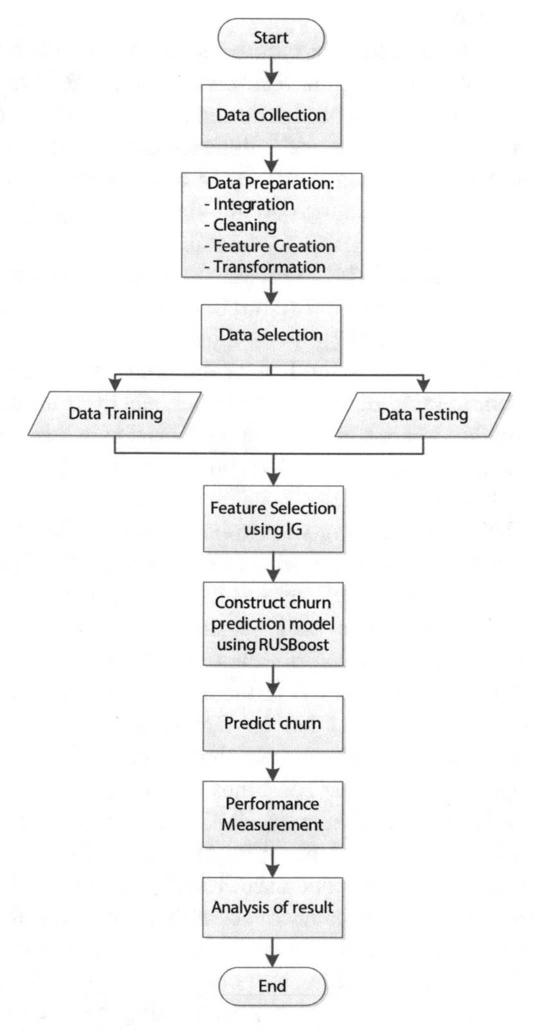

Fig. 1. Design process of proposed scheme

transformation. Data integration merged data from many tables and sources to become one integrated table. The data that the value was missed were deleted from data cleaning process. Total data after data cleaning was 97.530 rows with 1527 churn and the rest was not churn. After data cleaning, new features were created based on information of selected feature in previous step. After feature creation, there were 53 attributes that had been used for this research. And for the last step in data preparation was data transformation, which transformed the data types from nominal to numerical attributes.

Data Selection process was conducted in order to find how many months of customer data histories used to get the highest F-Score. Customer data histories used for this step were 3 months, 6 months, 9 months, and 12 months. After finding how long customer histories used in previous step, the next process is Feature Selection, i.e. selecting the important features that have high impact on predicting churn. Feature selection was conducted by Information Gain algorithm. The output of the feature selection will be used in the Churn prediction using RUSBoost Algorithm [9]. The result of churn prediction was analyzed using F-Score measurement. The F Score conveyed the balance between the precision and the recall.

RUSBoost Algorithm
Input:
Dataset S $\{(x_1,y_1), \ldots, (x_m,y_m)\}$ $x_i \in X$, with labels $y_i \in Y = \{1, \ldots, C\}$, where C_m, ($C_m <$ C), corresponds to a minority class, and C_{mj} ($C_{mj} \neq C_m$, $C_{mj} <$ C) corresponds to majority class.
T Number of iteration
N desired percentage of total instances to be represented by minority class

1. Initialize $B = \{(i,y) ; i = 1, \ldots, m ; y \neq y_i\}$
2. Initialize weight D_1 over the examples, such that $D_1(i,y) = 1/m$
3. **For** $t = 1, 2, 3, 4, \ldots, T$ **do**
4. Create temporary dataset S'_t with distribution D'_t by using random under sampling
5. Call *WeakLearner,* input : S'_t with distribution D'_t , output: prediction
6. Weak hypothesis $h_t \leftarrow$ prediction
7. Compute the pseudo loss of hypothesis h_t:

$$\varepsilon_t = \sum_{(i,y)\epsilon B} D_t(i,y)(1 - h_t(x_i,y_i) + h_t(x_i,y))$$

8. Set update weight parameter β_t and w_t

$$\beta_t = \frac{\varepsilon_t}{(1-\varepsilon_t)}$$

$$w_t = 1/2(1 - h_t(x_i,y) + h_t(x_i,y_i))$$

9. Update weight of D_{t+1}

$$D_{t+1}(i,y) = D_t(i,y).\beta_t^{w_t}$$

10. Normalize D_{t+1}

$$D_{t+1}(i,y) = \frac{D_{t+1}(i,y)}{\sum_m D_{t+1}(i,y)}$$

11. **end for**
12. **return** the final hypothesis h_{fin}

$$h_{fin} = \arg\max \sum_{t=1}^{T} \left(log\frac{1}{\beta_t}\right).h_t(x,y)$$

First, initialize D1(i) = 1/m for all i as the weights of each example, where m is the number of examples in the training dataset. And for T iteration, the dataset is being sampled and trained to get the hypothesis. For each iteration, the dataset is randomly undersampling and train using weaklearner (this study used C4.5), the pseudo loss is calculated and the weight is updated. And the final result is *H(x)* as the average of the weak hypothesis.

3 Experiments

3.1 Data Collection

The data was collected using SQL from oracle server provided by PT. Telkom Indonesia. The data was collected from October 2014 until November 2015. The data sets contained 53 attributes with 52 attributes as predictors and 1 attribute for label. Total records of training data after preprocessing process were 97.530 rows with 1.57% churn rate or 96.063 Not Churn (negative samples) and 1527 Churn (positive samples). While total records of testing data were 98.573 rows with 96.860 Not Churn and 1713 Churn. The data is represented in Fig. 2.

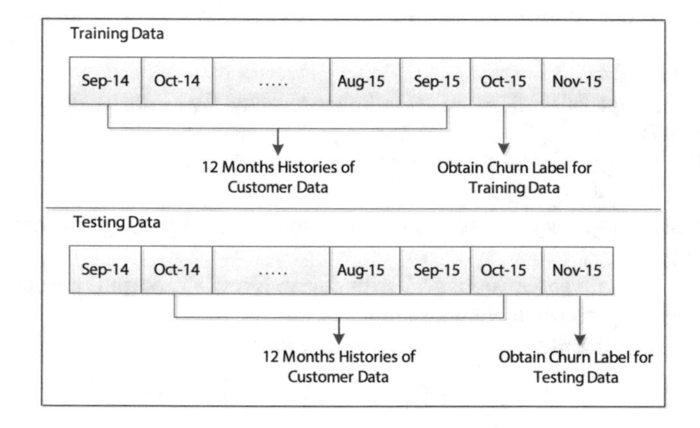

Fig. 2. Data collection

Based on broadband Internet service information, the following features were selected churn prediction:

1. Account information, is the information of customer account. It is include service packages, area, and age
2. The historical information of payments and bills
3. Broadband monthly usage information
4. The historical information of service fault.

3.2 Experimental Design Summary

The experiment design consists of 3 stages:

1. The first experiment aimed to get the best dataset for data training. How many periode of historical data are needed to get the best churn prediction? This experiment will process 3, 6, 9, and 12 months of customer historical data.
2. The second experiment aimed to find out the mayor factor that influenced churn prediction. Information Gain was used for selecting the important attributes.
3. The third experiment performed RUSBoost for building churn prediction model with different "target class distribution" (5%, 10%, 15%, 20%, 25%, 35%, 40%, 45%, and 50%) and evaluated the result.

4 Result

4.1 Experiment 1

From the first experiment, the best F-Score is obtained from 3 months(1), 9 months(1), and 12 months dataset. This 3 datasets were examined in the next experiment in order to get the comprehensive result. The 6 months data was excluded for the next experiments because it did not have enough information to predict churn (Table 1).

Table 1. Result of experiment 1

Exp.	Datasets	Att.	F-Score
1.1	**3 months (1)**	**16**	**0.344**
1.2	3 months (2)	16	0.199
1.3	3 months (3)	16	0.133
1.4	3 months (4)	16	0.129
1.5	3 months (5)	16	0.186
1.6	3 months (6)	16	0.161
1.7	3 months (7)	16	0.200
1.8	3 months (8)	16	0.233
1.9	3 months (9)	16	0.211
1.10	3 months (10)	16	0.212
1.11	6 months (1)	28	0.298
1.12	6 months (2)	28	0.172
1.13	6 months (3)	28	0.093
1.14	6 months (4)	28	0.090
1.15	6 months (5)	28	0.142
1.16	6 months (6)	28	0.164
1.17	6 months (7)	28	0.164
1.18	**9 months (1)**	**40**	**0.343**
1.19	9 months (2)	40	0.174
1.20	9 months (3)	40	0.082
1.21	9 months (4)	40	0.091
1.22	**12 months**	**52**	**0.331**

4.2 Experiment 2

The second experiment aimed to find out the important attributes that were used. Experiment 2 used the results from the first experiment, 3 months (1), 9 months (1), and 12 months datasets were used in second experiment.

The thresholds used for deviding attributes that were used, were divined by

a. IG values that have almost the same magnitude.
   To obtain the attributes that had almost the same value, the thresholds were based on the first number of decimal values
b. Ratio IG
   The high ratio were considered as threshold. Higher ratio indicated a larger difference between IG(n) and IG(n + 1).

The selected attributes for each datasets are:

Table 2. Selected Attributes

3 months (1)	9 months (1)	12 months
1.USAGE_N-1	1.USAGE_N-4	1.USAGE_N-4
2.USAGE_N	2.USAGE_N-3	2.USAGE_N-3
3.TAG_N	3.USAGE_N-1	3.USAGE_N-1
	4.USAGE_N	4.USAGE_N
	5.USAGE_N-7	5.USAGE_N-7
	6.USAGE_N-8	6.USAGE_N-9
	7.USAGE_N-6	7.USAGE_N-8

From Table 2, the total selected attributes for 9 months (1) and 12 months were the same, 7 attributes, which was USAGE. It showed that the total bandwidth that customer used were the most important factor for predicting churn. While 3 months (1) dataset gave the best prediction when 3 attributes were selected. Besides USAGE, TAG or amount of customer bill was selected as the important factor for indicating churn. From this experiment, it was proven that service usage and billing were the mayor factors influencing customer churn in telecommunication industry as stated in previous literature study [11].

4.3 Experiment 3

From the third experiment, the best F-Score was obtained from 10% churn rate for 9 months dataset (Table 3).

The comparison result for RUSBoost that using IG and without IG are in Table 4.

From Table 4, 9 months (1) dataset gave the highest F-Score when using undersampling, RUSBoost and IG. It showed that 9 months (1) dataset with 7 selected attributes provided the best information for churn prediction. The use of information gain improved the overall performance, with 16% improvement average. While the use of undersampling improved the overall performance significantly with 56% improvement average.

Table 3. Result of experiment 3

Exp.	Data history	Attr.	% Churn	F-Score
3.1	3 months (1)	3	1.57%	0.449
3.2	3 months (1)	3	5%	0.448
3.3	**3 months (1)**	**3**	**10%**	**0.461**
3.4	3 months (1)	3	15%	0.425
3.5	3 months (1)	3	20%	0.396
3.6	3 months (1)	3	25%	0.383
3.7	3 months (1)	3	30%	0.360
3.8	3 months (1)	3	35%	0.358
3.9	3 months (1)	3	40%	0.364
3.1	3 months (1)	3	45%	0.321
3.11	3 months (1)	3	50%	0.320
3.12	9 months (1)	7	1.57%	0.232
3.13	9 months (1)	7	5%	0.344
3.14	**9 months (1)**	**7**	**10%**	**0.531**
3.15	9 months (1)	7	15%	0.403
3.16	9 months (1)	7	20%	0.410
3.17	9 months (1)	7	25%	0.392
3.18	9 months (1)	7	30%	0.387
3.19	9 months (1)	7	35%	0.381
3.2	9 months (1)	7	40%	0.348
3.21	9 months (1)	7	45%	0.343
3.22	9 months (1)	7	50%	0.348
3.23	12 months	7	1.57%	0.234
3.24	12 months	7	5%	0.344
3.25	**12 months**	**7**	**10%**	**0.413**
3.26	12 months	7	15%	0.405
3.27	12 months	7	20%	0.403
3.28	12 months	7	25%	0.390
3.29	12 months	7	30%	0.386
3.3	12 months	7	35%	0.379
3.31	12 months	7	40%	0.384
3.32	12 months	7	45%	0.363
3.33	12 months	7	50%	0.377

Table 4. Comparison result

Dataset	Without sampling		Undersampling (10% churn)	
	RUSBoost	RUSBoost & IG	RUSBoost	RUSBoost & IG
3 months (1)	0.413	0.449	0.425	**0.461**
9 months (1)	0.202	0.232	0.410	**0.531**
12 months	0.165	0.234	0.399	**0.413**
Average f-score	0.260	0.305	0.411	**0.468**

By using the 10% churn rate, the two methods were compared in terms of processing time for learning the dataset. The comparison of execution time for RUSBoost with IG and RUSBoost without IG is in Table 5.

Table 5. Comparison of execution time

Dataset	Time (sec)	
	RUSBoost	RUSBoost & IG
3 months (1)	439.69	230.43
9 months (1)	687.17	302.45
12 months	701.23	410.56
Average time	609.36	314.48

From Table 5, RUSBoost and IG decreased the processing time. The average reduction in time was -48%. It showed that IG improved the performance of churn prediction and also speed up the processing time.

5 Conclusion

This research proposed the combination of feature selection technique (this research used Information Gain) and RUSBoost. The performance of prediction models developed by RUSBoost without any feature selection was compared with RUSBoost with feature selection. Information Gain can assist RUSBoost to improve the churn prediction performance. By using F-Score as the performance measurement, the combination method can improve the F-Score 16% compared to RUSBoost without IG. Although the improvement of F-Score is not significant, it is proven that IG and RUSBoost can handle imbalanced data in churn prediction. Information Gain reduced the attributes by selecting the important ones based on entropy. By reducing attributes, it gives faster time for learning the data. And from the result in this research experiments, RUSBoost with IG can reduce 48% of processing time compared to RUSBoost without IG. The implementation of under sampling method to "rebalance" the data can improve the performance significantly. Based on the experiment, the best proportional of churn rate is 10%.

For the future works, additional learner for RUSBoost is needed to be evaluated, such as Logistic Regression, SVM, Neural Network, and many more. Parameter selection for boosting iteration is also needed to be investigated in attempt to identify potentially favorable parameter values.

References

1. Chawla, N.V., Japkowicz, N., Kotcz, A.: Editorial: special issue on learning from imbalanced data sets. ACM SIGKDD Explor. Newsl. **6**(1), 1–6 (2004)
2. Weiss, G.M.: Mining with rarity: a unifying framework. ACM SIGKDD Explor. Newsl. **6** (1), 7–19 (2004)
3. Berson, A., Smith, S.J., Thearling, K.: Building Data Mining Applications for CRM. McGraw-Hill Osborne (2000)
4. Kotsiantis, S., Kanellopoulos, D., Pintelas, P., et al.: Handling imbalanced datasets: a review. GESTS Int. Trans. Comput. Sci. Eng. **30**(1), 25–36 (2006)
5. Tang, J., Alelyani, S., Liu, H.: Feature selection for classification: a review. In: Data Classification: Algorithms and Applications, p. 37 (2014)
6. Jamali, I., Bazmara, M., Jafari, S.: Feature Selection in Imbalance data sets. Int. J. Comput. Sci. Issues **9**(3), 42–45 (2012)
7. Batista, G.E., Prati, R.C., Monard, M.C.: A study of the behavior of several methods for balancing machine learning training data. ACM SIGKDD Explor. Newsl. **6**(1), 20–29 (2004)
8. Drummond, C., Holte, R.C., et al.: C4. 5, class imbalance, and cost sensitivity: why under-sampling beats over-sampling. In: Workshop on Learning from Imbalanced Datasets II, vol. 11 (2003)
9. Seiffert, C., Khoshgoftaar, T.M., Van Hulse, J., Napolitano, A.: RUSBoost: A hybrid approach to alleviating class imbalance. IEEE Trans. Syst. Man Cybern. Part A Syst. Hum. **40**(1), 185–197 (2010)
10. Chawla, N.V., Lazarevic, A., Hall, L.O., Bowyer, K.W.: SMOTEBoost: improving prediction of the minority class in boosting. In: Lavrač, N., Gamberger, D., Todorovski, L., Blockeel, H. (eds.) PKDD 2003. LNCS (LNAI), vol. 2838, pp. 107–119. Springer, Heidelberg (2003). doi:10.1007/978-3-540-39804-2_12
11. Effendy, V., Adiwijaya, Z., Baizal, A.: Handling imbalanced data in customer churn prediction using combined sampling and weighted random forest. In: 2014 2nd International Conference on Information and Communication Technology (ICoICT), pp. 325–330 (2014)

Extended Local Mean-Based Nonparametric Classifier for Cervical Cancer Screening

Noor Azah Samsudin(✉), Aida Mustapha, Nureize Arbaiy, and Isredza Rahmi A. Hamid

Faculty of Computer Science and Information Technology, University Tun Hussein Onn Malaysia, 86400 Parit Raja, Johor, Malaysia
{azah, aida, nureize, rahmi}@uthm.edu.my

Abstract. Malignancy associated changes approach is one of possible strategies to classify a Pap smear slide as positive (abnormal) or negative (normal) in cervical cancer screening procedure. The malignancy associated changes (MAC) approach acquires analysis of the cells as a group as the abnormal phenomenon cannot be detected at individual cell level. However, the existing classification algorithms are limited to automation of individual cell analysis task as in rare event approach. Therefore, in this paper we apply extended local-mean based nonparametric classifier to automate a group of cells analysis that is applicable in MAC approach. The proposed classifiers extend the existing local mean-based nonparametric techniques in two ways: voting and pooling schemes to label each patient's Pap smear slide. The performances of the proposed classifiers are evaluated against existing local mean-based nonparametric classifier in terms of accuracy and area under receiver operating characteristic curve (AUC). The extended classifiers show favourable accuracy compared to the existing local mean-based nonparametric classifier in performing the Pap smear slide classification task.

Keywords: Nonparametric classifier · Malignancy associated changes · Cancer screening

1 Introduction

In cervical cancer screening procedure, a patient will undergo a procedure known as Pap smear test which involves analysis of cells on a slide, collected from a woman's cervix [1]. The essential step in the Pap test is the microscopic examination of the cells for abnormal signs. There are two common approaches in detecting the abnormal signs on the slide: rare event (RE) and malignancy associated changes (MAC) [2]. The RE approach acquires cell-by-cell scrutinisation to be performed with each slide carries thousands of cells. Consequently, the RE approach is often commented as prone to human weaknesses such as fatigue, inexperience and time consuming. In addition, some diagnostic cells may fail to appear on the slide due to sampling error, Fortunately, MAC is an alternative approach that solves the problem with sampling error [2]. The primary advantage of the MAC approach is that abnormality can be detected by analysing intermediate cells, which lessen the dependency on availability of diagnostic cells.

T. Herawan et al. (eds.), *Recent Advances on Soft Computing and Data Mining*,
Advances in Intelligent Systems and Computing 549, DOI 10.1007/978-3-319-51281-5_39

However, most existing classification algorithms are developed to label individual object or instance which accommodates the RE approach [1, 2]. Different from the RE strategy, the MAC approach uses the summary statistics of the features in intermediate cells. Our interest is to automate the slide classification problem using the MAC approach. We bear in mind, there are two challenges in automating the MAC approach. First, the classification result shall be based on the analysis on a group cells rather than on individual cell. Second, the use of summary statistics as features in the existing MAC approach is prone to 'curse of dimensionality' problem [3] that is the increasing number of features will make the slide labelling task more difficult.

In this study, we describe application of extended local mean-based nonparametric classifiers [4] to automate the MAC approach in cervical cancer screening procedure. The extended local mean based nonparametric classifiers will also be evaluated against the existing conventional MAC approach using two performance measures: classification accuracy and the area under receiver operating characteristic curve (AUC) [5].

This paper is organised as follows: Sect. 2 describes the classifiers used in the experiment. Section 3 presents the data set and discusses the experiment conducted using the classifiers on a real data set. Section 4 presents the results and finally, Sect. 5 offers some conclusions.

2 Extended Local Mean-Based Nonparametric Classifier

Generally, in a classification problem, we are given a data set consisting of N samples and their associated class labels l, such as $\mathbf{X} = \{(\mathbf{x}_1,c_1), (\mathbf{x}_2,c_2), \ldots, (\mathbf{x}_N,c_N)\}$. Let $\mathbf{X}_{\mathrm{TE}}$ be the test set and $\mathbf{X}_{\mathrm{TR}}$ be the training set such that $\mathbf{X}_{\mathrm{TR}} \in \mathbf{X}, \mathbf{X}_{\mathrm{TE}} \in \mathbf{X}$ where $\mathbf{X} = \mathbf{X}_{\mathrm{TR}} \cup \mathbf{X}_{\mathrm{TE}}$ and $\mathbf{X}_{\mathrm{TR}} \cap \mathbf{X}_{\mathrm{TE}} = \varnothing$. Let $\mathbf{X}_{\mathrm{TE}}$ consists of N_{TE} instances that is, $\mathbf{X}_{\mathrm{TE}} = \{\mathbf{x}^1, \ldots, \mathbf{x}^{N_{TE}}\}$. Each instance $\mathbf{x}$ is represented by n-dimensional measurements, which are also known as feature vectors—that is, $\mathbf{x} = (f_1, f_2, \ldots, f_n)$. c_N is a class label l, for $\mathbf{x}_N$, where l belongs to a set of class labels, such that $l = 1, \ldots, L$, $L > 1$. While in the general classification problem a classifier is presented with an individual instance $\mathbf{x}$, in automating the MAC approach, a classifier shall be presented with multiple instances that is multiple feature vectors. In other words, our proposed classifiers are designed to label the multiple feature vectors in the $\mathbf{X}_{\mathrm{TE}}$ as a group.

We have applied three classification algorithms in our experiment. First, we have automated the conventional MAC approach using existing local mean-based nonparametric classifier in which we refer to as LNP. Then we have demonstrated two strategies to extend the existing LNP classifier: pooling and voting schemes. We refer to the extended LNP using pooling scheme as ELNP(P) and the extended LNP using voting scheme as ELNP(V). Tables 1 and 2 are to be used to describe the ELNP(V) and ELNP (P) respectively. Note that the LNP classifier requires us to select a value for parameter k. The procedure to select k value for our experiments will be presented in Sect. 3.

As shown in Tables 1 and 2, let $\mathbf{X}_{\mathrm{TE}}$ denotes a test set of N_{TE} cells, such that $\mathbf{X}_{\mathrm{TE}} = \{\mathbf{x}^1, \ldots, \mathbf{x}^{N_{TE}}\}$. Note that, each ith cell, $\mathbf{x}^i$ is represented by multiple features. Let v_l^i denotes votes for each class based on the number of nearest neighbours from class l for i^{th} sample. Note that v_l^v denotes total votes for class l based on v_l^i and, v_l^p denotes

Table 1. Extended LNP classifier using voting scheme (ELNP(V)) [4]

Test set, $\mathbf{X}_{TE}$	Class l Metric	Class label for every $\mathbf{x}^i$
$\mathbf{x}^1$	y_1^1	$c_l^1 = \underset{l}{\mathrm{argmin}}(y_l^1)$
⋮	⋮	⋮
$\mathbf{x}^{N_{TE}}$	$y_1^{N_{TE}}$	$c_l^{N_{TE}} = \underset{l}{\mathrm{argmin}}(y_l^{N_{TE}})$
		$y_l^V = \sum_{i=1}^{N_{TE}} c_l^i$ voting scheme

Table 2. Extended LNP classifier using pooling scheme (ELNP(P)) [4]

Test set, $\mathbf{X}_{TE}$	Class l Metric
$\mathbf{x}^1$	y_1^1
⋮	⋮
$\mathbf{x}^{N_{TE}}$	$y_1^{N_{TE}}$
	$y_l^P = \sum_{i=1}^{N_{TE}} y_l^i$ pooling scheme

total votes for class l. Let c_l^i denotes class label for each ith cell and $\boldsymbol{c}_l$ denotes a class label for $\mathbf{X}_{TE}$. A discussion on how the ELNP(V) and ELNP(P) will be used to classify the $\mathbf{X}_{TE}$ is briefly described in Sects. 2.2 and 2.3:

2.1 Local Mean-Based Nonparametric Classifier

The existing local mean-based nonparametric (LNP) classifier is a variant of the *k*-nearest neighbour (*k*-NN) rule [6]. Let $\mathbf{X}_{KNN}$ be the set of k-nearest neighbour for a sample $\mathbf{x}$. Each jth training sample $\mathbf{x}_j \in \mathbf{X}_{KNN}$ contributes a single vote towards the votes for each class based on their corresponding class labels. The aim of the classification problem is to determine the class membership of a single sample, $\mathbf{x}$ based on majority votes. On the other hand, the LNP classifier differs from the *k*-NN in that the *k*-NNs are determined for each class, l. In this way, the local mean vector of every class, $\boldsymbol{\mu}_l$ is determined for the *k*-NNs of each class, such that $\boldsymbol{\mu}_l = \frac{1}{k}\sum_{r=1}^{k} \mathbf{x}_l^r$, where $\mathbf{x}_l^r$ denotes the rth neighbour among the *k*-NNs with class label c_l. Next, the distance, d_l, between the test sample, $\mathbf{x}$, and each class local mean vector, $\boldsymbol{\mu}_l$, is calculated. Finally, $\mathbf{x}$ is given a class label based on minimum distance between $\mathbf{x}$ and the respective local mean vector, i.e., $c_l = \underset{l}{\mathrm{argmin}}(y_l)$.

In this study, we have applied the LNP to automate the existing MAC data classification approach that is the LNP classifier is presented with the slide summary statistics of the cells' features, such as mean and/or variance of $\mathbf{X}_{TE}$. A slide is assigned a class label that has minimum distance with its mean vector determined by the k neighbours. The use of LNP allows us to compare the existing MAC classification approach with the proposed ELNP(V) and ELNP(P). Note that, our study aims to extend the use of the LNP to improve classification accuracy in labelling the cells as a group. Thus, we choose to accumulate the voting results from LNP in two ways: voting and pooling.

2.2 Extending LNP Classifier Using Voting Scheme

We have demonstrated the possibility of extending LNP classifier using accumulated information gained from voting processes as in Table 1. The formulae applied in the voting schemes and the base of class label decisions are presented in Table 1. In our experiment, ELNP(V) is presented with the raw measurements for a group of cells that represent a slide. Using the existing LNP classifier, these cells will be assigned class labels. After all cells in the $\mathbf{X}_{TE}$ have been labelled, the number of cells designated normal and abnormal will be counted, and the majority vote will be used to determine the final class label.

2.3 Extending LNP Classifier Using Pooling Scheme

In ELNP(P), the total votes of the neighbours from every class are accumulated for the group. The group is then labelled in accordance with the largest total votes. The formulae applied in the pooling scheme and the base of class label decisions are presented in Table 2.

All of the classifiers have the same ultimate aim that is to determine a slide's class membership. However, they differ in their approach. The LNP classifier uses summary statistics to classify each slide as in the conventional MAC approach. The ELNP(V) and ELNP(P) classifiers use the raw measurements of the cells as a group to classify each slide. In principal, the ELNP(V) and ELNP(P) classifiers accumulate information from the group of cells and classify the slide as a whole. The ELNP(V) requires sample cells to be labelled individually prior to labelling the slide. The next section will explain how the experiments are conducted in the study.

3 Experimental Setup

This study aims to evaluate the performance of the ELNP(V) and ELNP(P) using real MAC data set. In this section, we describe the attributes of the MAC data set, feature selection phase, k value estimation, and experimentation procedure.

3.1 Pap Smear Data Set

The data set used in our experiment consists of MACs cell measurements for a set of Papanicolaou-stained cervical smear slides obtained from the Cytology Department, Queensland Medical Laboratory (QML). According to the QML's diagnosis, 99 slides were classified as normal (negative) and the other 40 slides were classified as abnormal (positive). For our experiments, we randomly selected data for 1,000 cells from each slide. Each measured cell presented a total of 29 features, referred to as $\{F_1, \ldots, F_{29}\}$. The descriptions of these features can be found in [7]. We applied a normalisation transform [8] to ensure that all of the measurements of the features for the 139,000 cells (1,000 cells $\times$ 139 slides) had a zero mean and unit variance such that all measurements were scaled to the range $\sim N(0, 1)$. The transformation was class label independent; therefore, the mean and variance of each class may be different.

3.2 Feature Selection

There are two main tasks in the feature-selection phase of our experiments. First is to determine the optimal number of features needed to represent each slide and second is to determine the optimal subset of features to represent the slides.

As noted earlier, the MAC data consist of 139 slides, with each comprising 1,000 cells. Each cell contains a set of 29 features, $\{F_1, \ldots, F_{29}\}$. We determined mean, μ_i and standard deviation, σ_i for every feature, F_i of each slide. As a result, each slide is represented by a set of summary statistics, $\{\mu_1 \ldots \mu_{29}, \sigma_1, \ldots, \sigma_{29}\}$ as features. We used μ_i and σ_i for the feature-selection phase because the conventional MAC approach uses summary statistics as features to classify each slide. Note that to estimate the effectiveness of our feature-selection method, we also included an additional random feature, F_{30}. We generated 139 observations using a uniform distribution function in MATLAB for feature F_{30}. The μ_{30} and σ_{30} of F_{30} were also determined.

To determine the optimal number of features for our experiments, we used the feature set of summary statistics, $\{\mu_1 \ldots \mu_{30}, \sigma_1, \ldots, \sigma_{30}\}$. The approach used to determine the optimal number of features was to apply an inner hold-out within 10-fold cross-validation. Our data consist of 99 slides from the normal class and 40 slides from the abnormal class. The data are divided into a validation set and a training set, and in every fold, approximately 90 per cent of the slides are left for training. In effect, the training set contains about 90 normal slides and 36 abnormal slides for each fold. Therefore, applying the inner hold-out, means that in every fold, 30 per cent of the training set slides were held as an inner test set partition and 70 per cent of the training set slides were assigned to an inner training set.

The slides were then used in the inner training set to select 10 feature subsets from the feature set $\{\mu_1 \ldots \mu_{30}, \sigma_1, \ldots, \sigma_{30}\}$ using the 'plus-l-take-away-r' algorithm [8]. Each selected feature subset was evaluated using the Mahalanobis criterion function [8]. The 10 training set partitions of the cross-validation resulted in 10 iterations of the inner hold-out approach. Ten feature subsets were therefore obtained, with each containing up to 10 features. Then, we used a logistic regression classifier [9] to evaluate combinations of the features from size 1 to 10 for each subset. This means that for every training set partition, the inner training set is used to train the logistic regression classifier, while the inner test set is used to estimate the area under receiver operating characteristics curve (AUC) of the logistic regression classifier. At the end of the 10-fold cross-validation, 10 sets of the AUC for every feature subset size are obtained. The results of the mean AUC were then plotted against the feature subset size. We then considered the number of features with the maximum mean AUC and identified eight features. It should be noted that determining the feature subsets of size eight within the 10-fold cross-validation is likely to result in different feature subsets to be used by different validation sets in the classifier performance-evaluation phase.

The LNP classifier was presented with the selected features originally from the feature set $\{\mu_1 \ldots \mu_{30}, \sigma_1, \ldots, \sigma_{30}\}$. However, for the proposed classifiers, the aim is to classify the slide using the cells' raw measurements instead of the slide summary statistics. Therefore, to present the slides to the ELNP(V) and ELNP(P), the selected features are traced to the original feature set $\{F_1, \ldots, F_{30}\}$. For example, if σ_{12}, μ_{15} and σ_{20} are selected for summary statistics-based classifiers, then F_{12}, F_{15} and F_{20} are the features presented to the proposed classifiers. Using the selected features from the summary statistics to trace the features for the proposed classifiers may allow us to directly compare the classifiers independently of the features used. Therefore, we believe that evaluating the performance of the proposed classifiers with a potentially (pessimistically) biased feature subset is acceptable. The results of our feature-selection phase are presented in Table 3.

Table 3. Feature subsets selected in every fold for the eight-feature experiments. M refers to the mean of the feature and S refers to the standard deviation of the feature. F refers to the raw measurement of the feature.

Cross-validation	Eight-feature subsets
1	[S_6 S_7 S_8 S_{22} M_{27} S_{28} S_{29}] [F_6 F_7 F_8 F_{22} F_{27} F_{28} F_{29}]
2	[M_1 M_4 S_4 M_5 M_6 M_8 S_{19} S_{21}] [F_1 F_4 F_5 F_6 F_8 F_{19} F_{21}]
3	[S_3 S_8 S_{14} M_{21} M_{22} S_{26} M_{29} S_{29}] [F_3 F_8 F_{14} F_{21} F_{22} F_{26} F_{29}]
4	[S_6 S_{16} S_{19} M_{23} M_{24} S_{25} S_{26} S_{27}] [F_6 F_{16} F_{19} F_{23} F_{24} F_{25} F_{26} F_{27}]
5	[S_8 S_9 S_{14} M_{24} M_{26} S_{28} M_{29} S_{29}] [F_8 F_9 F_{14} F_{24} F_{26} F_{28} F_{29}]
6	[S_3 M_5 M_6 S_{13} M_{18} S_{22} S_{25} M_{28}] [F_3 F_5 F_6 F_{13} F_{18} F_{22} F_{25} F_{28}]
7	[S_2 S_3 S_4 S_6 S_7 S_{16} S_{23} S_{25}] [F_2 F_3 F_4 F_6 F_7 F_{16} F_{23} F_{25}]
8	[S_2 S_9 S_{11} S_{17} S_{25} M_{26} M_{28} S_{28}] [F_2 F_9 F_{11} F_{17} F_{25} F_{26} F_{28}]
9	[S_{12} M_{17} M_{20} M_{22} S_{26} M_{28} M_{29} S_{29}] [F_{12} F_{17} F_{20} F_{22} F_{26} F_{28} F_{29}]
10	[S_3 M_{12} S_{13} S_{14} M_{15} M_{24} M_{25} S_{29}] [F_3 F_{12} F_{13} F_{14} F_{15} F_{24} F_{25} F_{29}]

3.3 Estimation of k Value

When designing the LNP classifier, it is important to select an optimal number of neighbours, which are referred to as $k*$. In our experiments, for the LNP classifier the same inner partition of the training set for feature selection is used to determine $k*$ in the 10-fold cross-validation. In this way, each fold may utilise a different $k*$ to evaluate the LNP classifier on its corresponding validation set.

In the $k*$ selection, the inner test set is used to evaluate the classifier's performance on a set of k values in the range of 1 to 26—for example, $k = \{1, \ldots, 26\}$. The AUC for the classifier is evaluated for every k value. In each fold, the $k*$ is selected based on the maximum AUC on the inner test set. It should also be noted that the same $k*$ is used to implement the ELNP(V) and ELNP(P) to enable the direct comparison of performance with the LNP classifier independent of the $k*$ value. Therefore, evaluating the performance of the ELNP(V) and ELNP(P) with a potentially (pessimistically) biased $k*$ value is acceptable.

3.4 Performance Evaluation

To evaluate the performance of the classifiers, stratified 10-fold cross-validation [10] was conducted. The stratified 10-fold cross-validation requires the data set to be divided into 10 subsets of about equal size. Each subset is assigned a representative proportion of slides from every class in the data set. Thus, each subset contains approximately 10 normal slides and four abnormal slides. In every fold, one subset is assigned as a validation set, and another nine subsets are assigned as a training set.

The training set is used to design the classifiers, and the validation set is used to evaluate the performance of the classifiers. The LNP classifier was developed to demonstrate the conventional MAC approach. Therefore, the summary statistics of the cells are applied to represent each slide in the validation set. For the ELNP(V) and ELNP(P) classifiers, the raw measurements of the cells are used instead of slide summaries. However, in our experiments, only 100 cells (selected at random) were used from every slide of the validation set rather than all 1,000 cells in order to reduce the computational complexity.

One of the performance measures used in evaluating a classifier's performance is their accuracy that is the probability that the slides in the validation set are classified correctly by the classifiers. We will obtain 10 sets of accuracy measures for each classifier from the 10-fold cross-validation and then use the mean accuracy and standard deviation of the accuracy for each classifier for performance evaluation. In

Table 4. Confusion matrix

True class label	Predicted class label		
	Negative	Positive	
Negative	TN	FP	Total_Negative
Positive	FN	TP	Total_Positive
	Total negative prediction	Total positive prediction	N

addition to the mean accuracy and its standard deviation, we will also use the analysis on the area under the ROC curve (AUC), which was initially presented in [5]. Using the ROC analysis, four possible outcomes are obtained as illustrated in Table 4: True Negative (TN), False Positive (FP), False Negative (FN), and True Positive (TP).

On each validation set partition, a set of the four outcomes will be obtained at various decision threshold values. The decision thresholds are varied based on the scores assigned to each slide. Table 5 indicates how these scores are assigned to each slide by each classifier. The probability of true positive and probability of false positive were then plotted as a ROC curve.

Table 5. Scoring system of LNP, ELNP(V) and ELNP(P) classifiers

Classifiers	Scoring system
LNP	Distance from class 1 mean vector against distance from class 2 mean vector
ELNP(V)	Number of cells assigned to Class 1 against total number of cells in the test slide
ELNP(P)	Total distances of neighbouring cells that belong to Class 1 against total distances of neighbouring cells that belong to Class 2

We will consider two ways of combining the ROC curves from the 10 different validation partitions in presenting the experiment results: pooling and averaging [5]:

Averaging: We will calculate the ROC curve (AUC) at each successive point of (P(TP), P(FP)) pair for every validation set. The AUC is to be estimated using the formula of trapezoidal numerical integration. There will be 10 sets of AUC to be estimated for the 10 validation set partitions. We will use the mean and standard deviation of the 10 AUC values to evaluate the performance of the classifiers.

Pooling: We will also present an average or 'group' ROC curve for each classifier. Thus, we will pool the frequencies of TPs and FPs at varying decision thresholds, as defined in Table 5, for every classifier. At the end of the experiments, we will obtain 10 sets of these frequencies. These frequencies will then be averaged to plot a single ROC curve for every classifier.

4 Results and Discussions

Table 6 shows the mean accuracy (Mean Acc) and mean AUC for the experiments with eight features. It also presents the plot of the average ROC curves for the classifiers in Fig. 1. Overall, the extended local mean- based nonparametric classifiers show higher mean accuracy and mean AUC than the LNP classifier. Figure 1 shows that ELNP(V) and ELNP(P) classifiers perform better than the LNP classifier. ELNP(P) has the best performance among the LNP based classifiers. In fact, ELNP(P) has the maximum AUC for the experiment using eight features.

Table 6. Results of mean accuracy (Mean Acc) and mean area under the receiver operating characteristics (ROC) curve (AUC).

Classifiers	Eight features	
	Mean Acc ± std dev	Mean AUC ± std dev
LNP	0.736 ± 0.107	0.780 ± 0.136
ELNP(V)	0.843 ± 0.125	0.910 ± 0.124
ELNP(P)	0.864 ± 0.137	0.925 ± 0.101

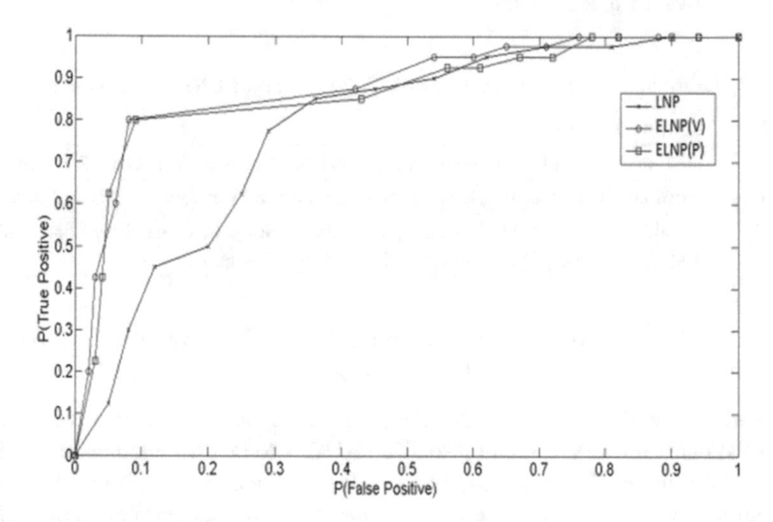

Fig. 1. ROC curve of LNP, ELNP(V) and ELNP(P) classifiers for eight features.

5 Conclusions

The study has shown that the extended LNP techniques can potentially improve cervical cancer screening procedure. Two approaches can be considered: voting scheme and pooling scheme. The study clearly demonstrates that the ELNP(V) and ELNP(P) techniques can be effectively utilised to reduce Pap smear slide misclassifications in cervical cancer screening procedure. The apparent finding from the results is that the pooling scheme of ELNP(P) is more effective than the voting scheme of ELNP(V). This finding is in agreement with our previous results [4]. In conclusion, both extended LNP classification techniques are applicable to detecting the MAC phenomenon using the raw measurements of the cells as a group, thus potentially improve cervical cancer screening procedure as a whole.

Acknowledgments. This work was supported in part by a grant from the Ministry of Education of Malaysia, Research Acculturation Grant Scheme (RAGS), Vot R045 and in part by a grant from Research Gates IT Solution Sdn. Bhd.

References

1. Ushizima, D.M., Gomes, A.H., Bianchi, A.G.C.: Automated Pap smear cell analysis: optimizing the cervix cytological examination. In: 12th International Conference on Machine Learning and Applications, Miami (2013)
2. Moshavegh, R., Bejnordi, B.E., Mehnert, A., Sujathan, K., Malm, P., Bengtsson, E.: Automated segmentation of free-lying cell nuclei in Pap semars for malignancy-associated change analysis. In: 34th Annual International Conference of the IEEE EMBS, San Diego, California, USA (2012)
3. Jain, A.K., Duin, R.P.W., Mao, J.: Statistical pattern recognition: a review. IEEE Trans. Pattern Anal. Mach. Intell. **22**(1), 4–37 (2000)
4. Samsudin, N.A., Bradley, A.P.: Nearest neighbour group-based classification. Pattern Recogn. **43**(10), 3458–3467 (2010)
5. Bradley, A.P.: The use of the area under the ROC curve in the evaluation machine learning algorithms. Pattern Recogn. **30**(7), 1149–1155 (1997)
6. Yamamoto, M., Hamamoto, Y.: A local mean-based nonparametric classifier. Pattern Recogn. Lett. **27**(10), 1151–1159 (2006)
7. Doudkine, A., MacAulay, C., Poulin, N., Palcic, B.: Nuclear texture measurements in image cytometry. Pathologica **87**(3), 286–299 (1995)
8. Webb, A., Copsey, K.: Statistical Pattern Recognition. Wiley, London (2011)
9. Kleinbaum, D.G., Klein, M.: Logistic Regression: A Self-Learning Text. Springer, New York (2010)
10. Breiman, L., Friedman, J., Stone, C.J., Olshen, R.A.: Classification and Regression Trees. Taylor and Francis, Belmont (1984)

Adaptive Weight in Combining Color and Texture Feature in Content Based Image Retrieval

Ema Rachmawati(✉), Mursil Shadruddin Afkar, and Bedy Purnama

Telkom University, Bandung, Indonesia
emarachmawati@telkomuniversity.ac.id

Abstract. Low-level image feature extraction is the basis of content based image retrieval (CBIR) systems. In that process, the usage of more than one descriptors has tremendous impact on the increasing of system accuracy. Based on that fact, in this paper we combined color and texture feature in the feature extraction process, namely Color Layout Descriptor (CLD) for color feature extraction and Edge Histogram Descriptor (EHD) for texture feature extraction. We measure the system performance on retrieving top-5, top-10, top-15, and top-20 relevant images. We successfully demonstrated in the experiment, that the combination of color and texture descriptor might be improved the performance of retrieval system, significantly. In our proposed system, the combination of CLD and EHD reaches 72.82% in accuracy, using adaptive weight in Late Fusion Method.

Keywords: CBIR · Color layout descriptor · Edge histogram descriptor · Adaptive weight · Late fusion method

1 Introduction

"Any technology that in principle helps to organize digital picture archives by their visual content" is categorized as Content Based Image Retrieval (CBIR) system [1]. In CBIR system, the user usually submits an example image, and the system will search for the most similar images in the database, then retrieved those similar images to the user. It is called *query by image*. In order to provide the most similar/relevant images to the query image, extracting proper features from the images should be conducted. Further, a suitable distance function must be defined in the selected feature space that will measure the similarity between query image and the images in the database.

Related to this, a proper feature vector is commonly used to represent the visual content of an image. Those features are usually extracted using some image processing techniques [2]. CBIR systems often use more than one type of features [3], such as color, texture, spatial information, etc. Color is one of the most widely used visual features in image and video retrieval. It has special characteristics that is relatively robust to changes in the background colors and is also independent of image size and orientation. On the other hand, the visual patterns that have properties of homogeneity or not that result from the presence of multiple colors or intensities in the image, is

T. Herawan et al. (eds.), *Recent Advances on Soft Computing and Data Mining*,
Advances in Intelligent Systems and Computing 549, DOI 10.1007/978-3-319-51281-5_40

usually referred as texture. Describing textures in images by appropriate texture descriptors provides powerful means for similarity matching and retrieval. According to that facts, since the visual appearance of object in image can be determined using color and texture in a different way, hence, it makes sense trying to join them together.

In this paper we proposed a combination of color and texture feature in the feature representation of CBIR system. We applied color and texture feature representation standardized by MPEG-7 [4], namely Color Layout Descriptor (CLD) and Edge Histogram Descriptor (EHD). MPEG-7 visual description tools contains basic structure and descriptor that includes basic visual features: color, texture, shape, motion, and localization [5].

Color Layout Descriptor was recommended as one of good color descriptor [6]. Reference [7] used color layout descriptor as a feature description for high-speed image/video segment retrieval. The image retrieval system introduced by [8] is based on a query by layout method using CLD and EHD. While [9] combine CLD with texture descriptor (Gabor filters) to construct robust feature set in CBIR system. While Edge Histogram Descriptor (EHD) is defined in MPEG-7 for describing nonhomogeneous texture [5].

In this paper, we demonstrated how feature vectors from CLD and EHD is combined. However, CLD and EHD has different distance function, hence a function to calculate distance between combined feature vectors is needed. Related to that, we adapted the Late Fusion Method in [10] to calculate final similarity value of using combined color and texture feature. We proposed the use of adaptive weight instead of fixed weight as used in [10]. The aim of the use of adaptive weight is to give better proportion to the color and texture feature representation.

The remainder of the paper is organized as follows. Section 2 gives a brief overview of the proposed system. Section 3 describes the experimental result. Finally, Sect. 4 provides a conclusion to this paper.

2 Proposed System

The pipeline of our proposed CBIR system consists of several processes as illustrated in Fig. 1. Feature extraction were applied to color images of our dataset to extract color and texture feature. We use Color Layout Descriptor [7] for extracting color feature and Edge Histogram Descriptor [11] for extracting texture feature. This process yielded feature vectors which will be saved in internal storage. Furthermore, we conduct linear combination of color and texture feature into a compact feature vector. These feature vectors further used as feature representation of color images in the content based image retrieval system.

The retrieval process is depicted in the right side of Fig. 1. Starting by giving particular query image to the system, the feature extraction process will extract the color and texture feature of the query image. Further, the similarity value will be calculated between query image and all images in the feature storage using Late Fusion Method [10].

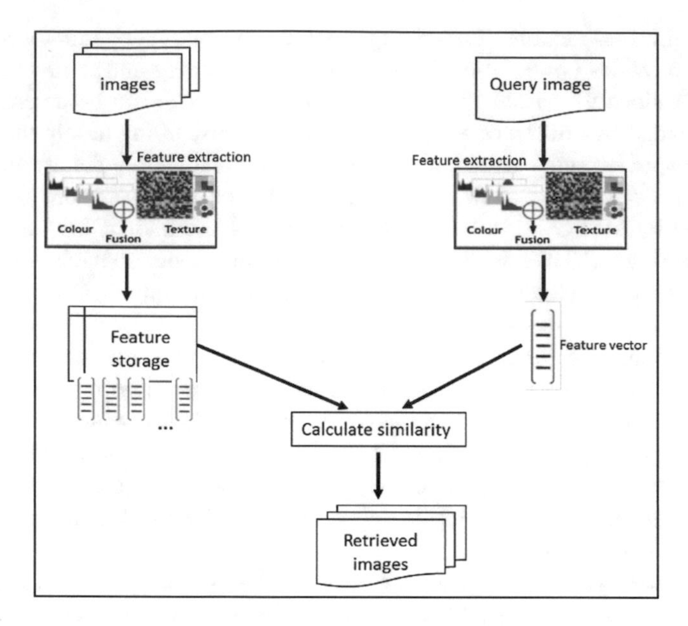

Fig. 1. Proposed system

2.1 Color Feature Extraction

In this paper, Color Layout Descriptor (CLD) is used in the color feature extraction process. CLD is the color descriptor in Content-Based Image Retrieval which extracts spatial color information in image [7]. The pipeline of extracting color feature using CLD is illustrated in Fig. 2.

The input image is divided into 64 non-overlapping blocks. For each block, we extract the representative color by averaging the values of all pixels in each block. This will result in 8×8 matrix for each color component (Red, Green, and Blue), that will further be converted into YCbCr color space. Further, each matrix is transformed by 8×8 DCT (Discrete Cosine Transform) and is quantized to obtain 8×8 DCT matrices. Zigzag scanning is then performed to obtain color feature vector. This feature vector at least consist of 12 elements and 192 elements for the maximum (64 elements for each color component).

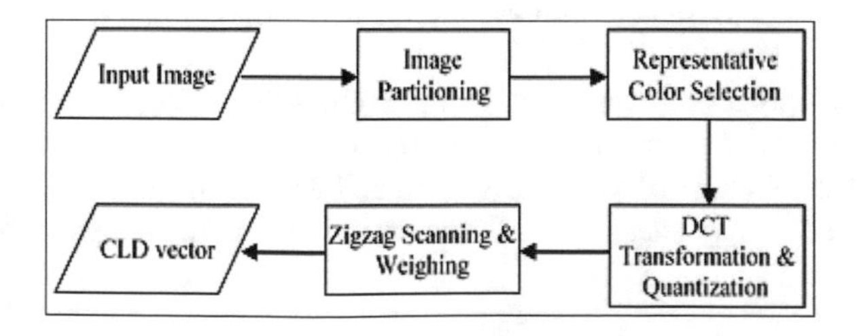

Fig. 2. CLD extraction process

2.2 Edge Histogram Descriptor

For extracting texture feature, we use Edge Histogram Descriptor (EHD) as texture feature extraction. EHD is the texture descriptor in Content-Based Image Retrieval which extracts spatial edge distribution in image [11]. The main phases in the extraction process of Edge Histogram Descriptor can be seen in Fig. 3.

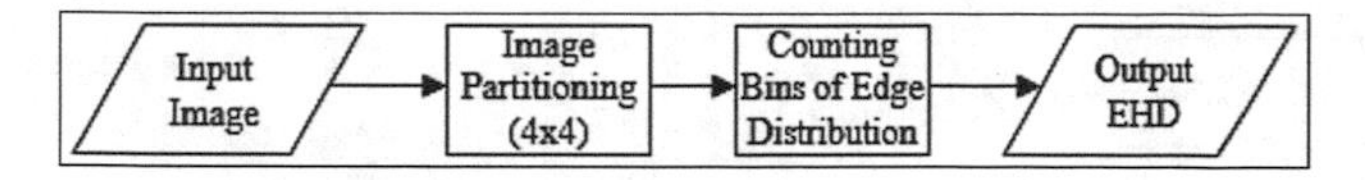

Fig. 3. EHD extraction process

In this section, we described the implementation of EHD [11], briefly. The input image is divided into 4 × 4 non-overlapping blocks (16 sub images). Further, each sub image is divided into several image blocks. The next step is counting the edge distribution. This process is aimed to obtain local, semi-global, and global edge distributions. The final feature vector is created by concatenating elements of local, semi-global, and global edge distribution.

Edge orientation of each sub image is categorized into five different types: vertical, horizontal, 45°, 135°, and non-directional. Local edge distribution can be obtained by calculating the distribution of those five edge orientation types in each sub image. It resulted in 80 elements for local edge distribution (16 sub images, each sub image results in 5 elements for each edge distribution type).

Semi-global edge distribution can be obtained from the value of each edge distribution in combination of sub images, called segments. It resulted in 65 elements for semi-global edge distribution (13 segments, each segment results in 5 elements). Moreover, the global edge distribution is obtained by calculating the edge distribution for the whole image. It resulted in 5 elements of global edge distributions. The total elements for EHD vector is 150 elements (80 local + 65 semi-global + 5 global).

The calculation of edge distribution was conducted in each image block of sub image. Each image block was divided into 2 × 2 sub blocks to calculate the magnitude of five edge orientations using (1). The magnitude $m(i,j)$ is calculated for each image block in sub image. $B(i,j)$ is sub block located at i^{th} row and j^{th} column, whereas $f(k)$ is a filter coefficient of edge orientation.

$$m(i,j) = \left| \sum_{k=0}^{3} B(i,j) \times f(k) \right| \tag{1}$$

2.3 Combination of CLD and EHD

The combination of CLD and EHD feature vector was obtained by concatenating the elements of CLD with the elements of EHD. For example, if there is a CLD vector which contains 12 elements and an EHD vector contains 80 elements, the combined vector of CLD and EHD has 92 elements. The first 12 elements was obtained from CLD vector and the rest was obtained from EHD vector. The illustration of this process can be seen in Fig. 4.

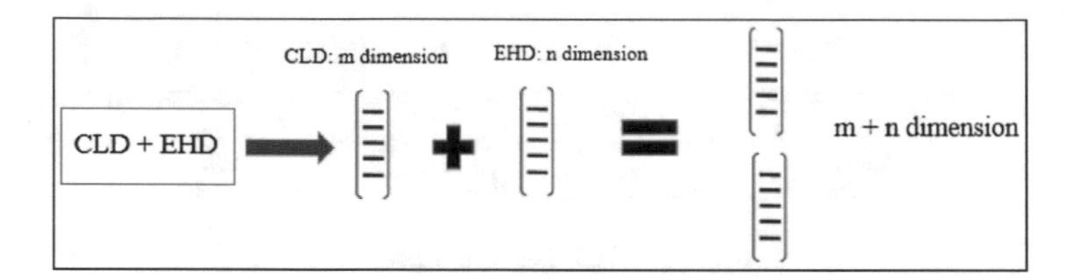

Fig. 4. Illustration of the combined feature vector

2.4 Distance Function

There were several distance functions we used in the experiment to calculate similarity of feature vectors. We use Euclidean distance to calculate similarity of color feature vector, as can be seen in (2). $D(A, B)$ is distance between image A and B, where Y_i, Cb_i, and Cr_i are values in each channel Y, Cb, and Cr at i^{th} element obtained from DCT process and zigzag scanning.

$$D(A, B) = \sum_{i=0}^{j} \sqrt{(Y_{A_i} - Y_{B_i})^2 + (Cb_{A_i} - Cb_{B_i})^2 + (Cr_{A_i} - Cr_{B_i})^2} \quad (2)$$

To calculate similarity of texture feature vector, we use Manhattan distance, as can be seen in (3). $D(A, B)$ in (3) is a distance between image A and B if feature vector of EHD contains local, semi-global, and global elements. If feature vector of EHD contains only local edge distribution (80 elements), the distance can be calculated as in (4).

$$D(A,B) = \sum_{i=0}^{79} |Loc_{A_i} - Loc_{B_i}| + \sum_{j=0}^{64} |SG_{A_j} - SG_{B_j}| + 5 \times \sum_{k=0}^{4} |Glo_{A_k} - Glo_{B_k}| \quad (3)$$

$$D(A,B) = \sum\nolimits_{i=0}^{79} |Loc_{A_i} - Loc_{B_i}| \quad (4)$$

2.5 Late Fusion Method

Having different distance function, the distance of two combined feature vector could not be calculated by either Euclidean Distance or Manhattan Distance. Therefore, the distance calculation of combined feature vector was conducted using Late Fusion Method [10]. In this method, we combine algorithms after the matching scores are calculated. It takes the normalized value of texture descriptor distance (5) and color descriptor distance (6) and weighs both of the value to obtain the distance (7).

$$\mathrm{n_T} = \frac{\mathrm{T}}{\max(\mathrm{T})} \tag{5}$$

$$\mathrm{n_C} = \frac{\mathrm{C}}{\max(\mathrm{C})} \tag{6}$$

$$\mathrm{Q} = \mathrm{w.n_C} + (1 - \mathrm{w})\mathrm{n_T} \tag{7}$$

T is a texture distance between query image and database image, C is a color distance between query image and database image, n_T is a normalized value of texture distance between query image and database image, n_C is a normalized value of color distance between query image and database image, and w is the weight for color distance. The weight for texture distance is set so that the summation of weight of color and texture distance always 1.

In this paper we define the value of w so that it can be adaptive. Meaning, the value of w is determined by the characteristics of data being examined. The value of $w_{adaptive}$ is defined in (8), with h_C is the correct images retrieved using CLD only and h_E is the correct images retrieved using EHD only.

$$\mathrm{w_{adaptive}} = \frac{\mathrm{h_C}}{(\mathrm{h_C} + \mathrm{h_E})} \tag{8}$$

3 Experimental Result

In this section, we explain the experiment in order to measure the performance of CBIR system. The effect of feature extraction was measured by calculating the system accuracy when using color feature only, texture feature only, and combined color and texture feature.

3.1 Dataset

The dataset used in this experiment is Wang dataset [12], which contains 1000 images, categorized into 10 classes. The classes in database are Africa, beach, monument, bus, dinosaur, elephant, flower, horse, mountain, and food. Images in the dataset had either portrait or landscape layout. Portrait images had dimension 256 × 384 pixels and the landscape ones had dimension 384 × 256 pixels. The example of image in each class is shown in Fig. 5.

Fig. 5. Example of image in each class of WANG database

3.2 Analysis on Using CLD and EHD as Feature Vector

The length of feature vector extracted using CLD used in the experiment is 12, 18, 22, and 192. A feature vector has length of 12 consists of (6, 3, 3), meaning, 6 elements of Y component + 3 elements of Cb component and 3 elements of Cr component. While the length of 18 consists of (6, 6, 6), the length of 22 consists of (10, 6, 6), and the length of 192 consists of (64, 64, 64).

From Fig. 6(a), system get highest accuracy when it used feature vector of CLD of length 192, because it stored more data from image (all of spatial information) to count than the other vector with different length. Furthermore, we also conduct experiment on using feature vector of EHD. Feature vector length of EHD tested in this proposed system was 80 and 150. The length of 80 consists of 80 local elements of the image whereas the length of 150 consists of 80 local elements, 65 semi-global elements, and 5 global elements. The accuracy achieved is shown in Fig. 6(b). It can be seen that the highest accuracy is achieved when it used feature vector of EHD with length of 150. This kind of feature vector stored not only local edge distribution but also semi-global and the edge distribution of the image itself.

Moreover, we conduct experiment on using the feature vector combination of CLD and EHD. There was eight combination of combined feature vector of CLD and EHD, as can be seen in Fig. 6(c). The weight (w) was set at 0.5, with the aim to get the proportional effect of color and texture feature. From Fig. 6(c), it can be inferred that the accuracy is increasing significantly.

We also conduct experiment on retrieving several images on our system, namely top-5, top-10, top-15, and top-20 images. From the result shown in Fig. 7, it can be seen that when only top-5 images retrieved, the accuracy achieved is 72.82%. While when top-20 images retrieved, the accuracy is decreased to 59.535%.

On measuring the effect of using adaptive weight as proposed in (8), we compare the system accuracy of using combined CLD and EHD using adaptive weight and pre-defined weight, namely 0.2, 0.3, 0.4, 0.5, 0.6, 0.7, and 0.8. In this case, we use feature vector CLD18 and EHD150. The result of this process can be seen in Fig. 8.

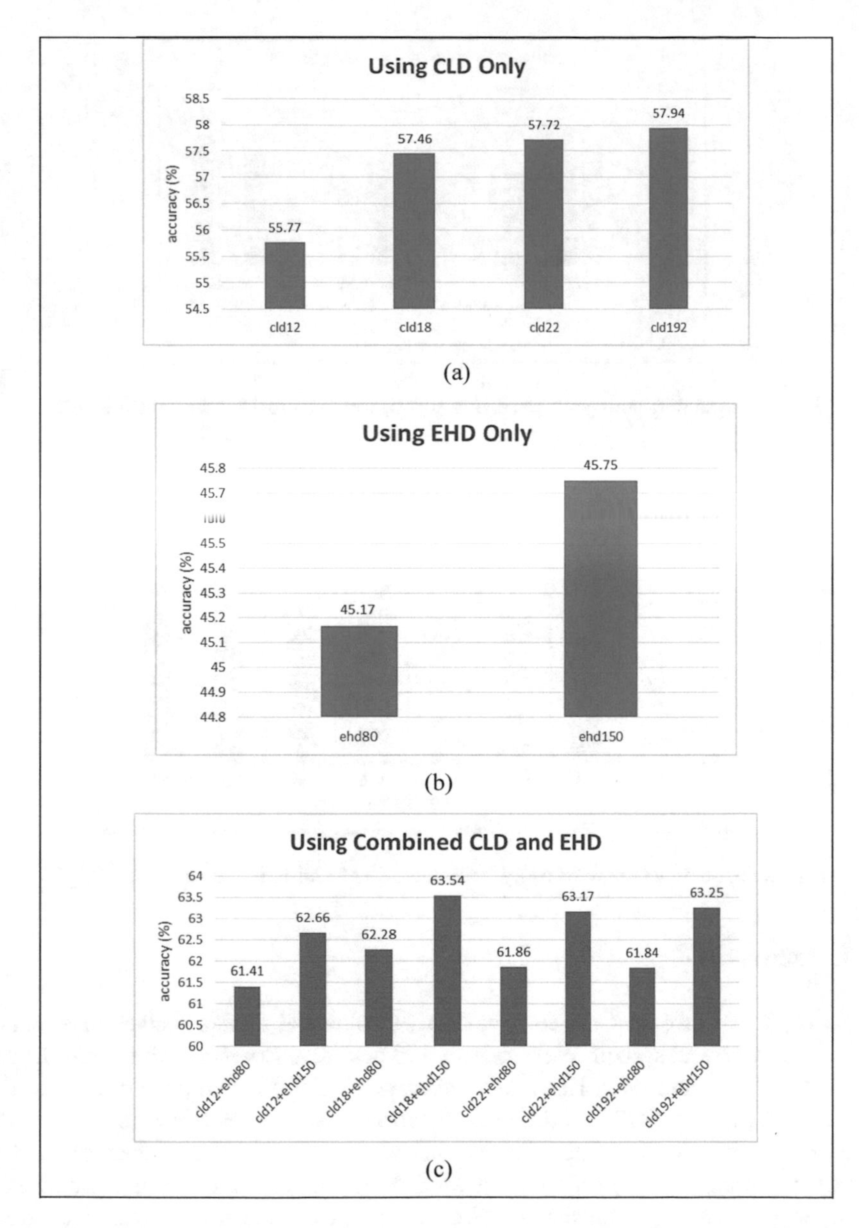

Fig. 6. System accuracy of using feature vector: (a) CLD only; (b) EHD only; and (c) combined CLD and EHD

The adaptive weight would change according to characteristics of retrieval result. In this case, if the accuracy result on retrieving images using CLD is larger than using EHD, the value of w become larger. This applies vice versa. If the accuracy result on retrieving images using EHD is larger than using CLD, the value of w become smaller.

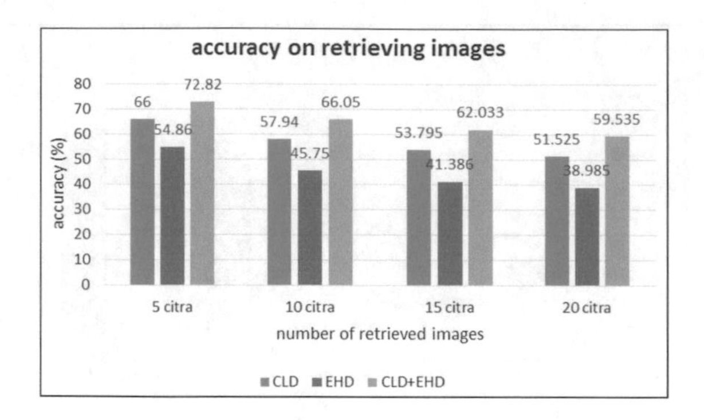

Fig. 7. System accuracy on retrieving top-5, top-10, top-15, and top-20 images

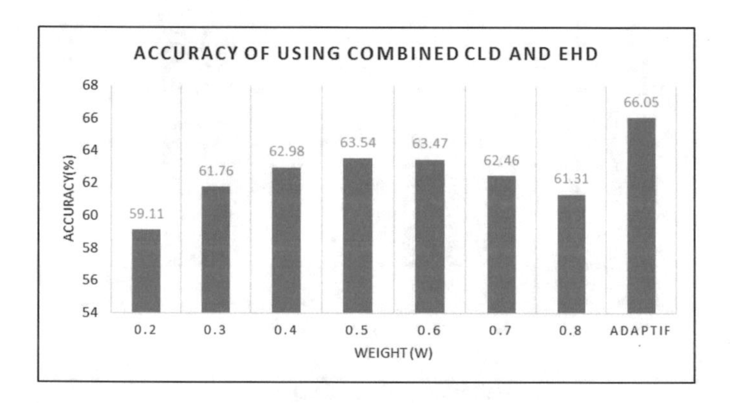

Fig. 8. System accuracy of using combined CLD and EHD on different weights

4 Conclusion

In this paper we proposed the combination of color and texture feature in a content based image retrieval system. We demonstrated how feature vectors from Color Layout Descriptor (CLD) and Edge Histogram Descriptor (EHD) are combined. As stated in this paper, CLD and EHD has different distance function, hence we need a different distance function to calculate distance between those combined feature vectors. Moreover, we adapted the Late Fusion Method in the similarity calculation of using combined color and texture feature. We has proposed the use of adaptive weight instead of fixed weight in the feature representation and proved that it has a significant impact on increasing system accuracy. The purpose of using adaptive weight is to give better proportion to the color and texture feature representation.

The best accuracy achieved in the system we build is 72.82% in using the combination of cld18 and ehd150. From the experiments we have conducted, it can be concluded that CBIR using combination more than one descriptor can increase the performance of the system rather than using only one descriptor.

References

1. Datta, R., Joshi, D., Li, J., Wang, J.Z.: Image retrieval: ideas, influences, and trends of the new age. ACM Comput. Surv. **40**(2), 1–60 (2008)
2. Penatti, O.A.B.B., Valle, E., Torres, R.: Comparative study of global color and texture descriptors for web image retrieval. J. Vis. Commun. Image Represent. **23**(2), 359–380 (2012)
3. Choraś, R.S., Andrysiak, T., Choraś, M.: Integrated color, texture and shape information for content-based image retrieval. Pattern Anal. Appl. **10**(4), 333–343 (2007)
4. Chang, S.F., Sikora, T., Puri, A.: Overview of the MPEG-7 standard. IEEE Trans. Circuits Syst. Video Technol. **11**(6), 688–695 (2001)
5. Sikora, T.: The MPEG-7 visual standard for content description-an overview. IEEE Trans. Circuits Syst. Video Technol. **11**(6), 696–702 (2001)
6. Eidenberger, H.: How good are the visual MPEG-7 features? In: SPIE Visual Communications and Image Processing Conference, pp. 476–488 (2003)
7. Kasutani, E., Yamada, A.: The MPEG-7 color layout descriptor: a compact image feature description for high-speed image/video segment retrieval. In: International Conference on Image Processing, vol. 1, pp. 674–677 (2001)
8. Kim, S.M., Park, S.J., Won, C.S.: Image Retrieval via query-by-layout using MPEG-7 visual descriptors. ETRI J. **29**(2), 246–248 (2007)
9. Jalab, H.A.: Image retrieval system based on color layout descriptor and Gabor filters. In: 2011 IEEE Conference on Open Systems, pp. 32–36 (2011)
10. Bleschke, M., Madonski, R., Rudnicki, R.: Image retrieval system based on combined MPEG-7 texture and colour descriptors. In: 2009 MIXDES International Conference Mixed Design of Integrated Circuits & Systems, pp. 0–4 (2009)
11. Won, C.S., Park, D.K., Park, S.J.: Efficient use of MPEG-7 edge histogram descriptor. ETRI J. **24**(1), 23–30 (2002)
12. Li, J., Wang, J.Z.: Automatic linguistic indexing of pictures by a statistical modeling approach. IEEE Trans. Pattern Anal. Mach. Intell. **25**(9), 1075–1088 (2003)

WordNet Gloss for Semantic Concept Relatedness

Moch Arif Bijaksana(✉) and Rakhmad Indra Permadi

School of Computing, Telkom University, Bandung, Indonesia
arifbijaksana@telkomniversity.ac.id, permadi.rakhmadindra@gmail.com

Abstract. Semantic lexical similarity and relatedness are important issues in natural language processing (NLP). Similarity and relatedness are not the same, while they are very closely related. To date, in many works these two issues are mixed up which harm system's effectiveness. A popular approach to measure semantic similarity and relatedness is utilizing WordNet, a lexical database. This paper shows that Wordnet's gloss is a potential source for measuring semantic relatedness. Experiment result using WordSim353 relatedness database confirms the effectiveness of the approach.

Keywords: Semantic textual relatedness · Semantic textual similarity · Lexical relatedness · Gloss · WordNet

1 Introduction

A meaningful text (such as document and paragraph) is constructed of meaningful smaller unit text (such as sentence and word). Word is the smallest unit representing semantic. Concept is a set of synonymous words (synset). For example word *oil* and *petroleum* are synonym, therefore these two words are in the same concept. An effective semantic textual similarity and relatedness (STSR), thus depend on effective semantic similarity and relatedness of concepts.

Many NLP applications could benefit from effective STSR. STSR are dealing with processing of natural language processing application such as textual entailment [5], word sense disambiguation [1,13,15,29], information extraction, information retrieval [12,19,22,30], text clustering [7], spelling error [10], identifying the discourse structure [17], text summarization, plagiarism detection (text reusable) [24], pharaphase detection, community question answering [18], automatic essay grading, and ontology construction [4,9].

While semantic similarity used in many NLP cases, semantic relatedness is needed in more cases [10]. Semantic relatedness can be defined as any kind of lexical or functional association that may exist between two concepts [8]. For example, *shampoo* and *hair* have a close semantic relationship, while they are not similar. *Shampoo* is a cleansing agent, while *hair* a covering for the body. Other examples of high related pairs with low similarity are *bank-money*, *planet-astronomer*, *computer-software*, and *car-journey*. Semantic relatedness has broad

T. Herawan et al. (eds.), *Recent Advances on Soft Computing and Data Mining*,
Advances in Intelligent Systems and Computing 549, DOI 10.1007/978-3-319-51281-5_41

notion [28]. It can defined as any kind of lexical or functional association that may exist between two words [8].

WordNet is a lexical database of English [21]. Nouns, verbs, adjectives and adverbs are grouped into sets of cognitive synonyms (synsets), each expressing a distinct concept. In WordNet, at least two knowledge can be exploited to measure STSR, taxonomic structure and gloss. A concept contains a gloss (a brief definition). For example, word *shampoo* has three synsets, two nouns and one verb. Their glosses are *cleansing agent consisting of soaps or detergents used for washing the hair* for *shampoo#n#1* that is the first concept of *shampoo* as noun, *the act of washing your hair with shampoo* for *shampoo#n#2*, and *use shampoo on (hair)* for *shampoo#v#1* [1].

We conducted experiments on popular semantic relatedness database WordSim353 [2], compared to existing similarity and relatedness measurements to evaluate our gloss-based approach. The results show that gloss-based approach is more effective.

The rest of this paper is structured as follows: Sect. 2 discusses related work. Section 3 presents gloss-based semantic lexical relatedness. The experiment design is described in Sect. 4, whereas the results are discussed in Sect. 5. Finally, Sect. 5 gives concluding remarks.

2 Related Work

Several approaches can be used in STSR, *path-based*, *information content-based*, *gloss-based*, and *vector-base* [33]. These approaches can utilize language resource such as WordNet.

In a *path-based* approach (also known as *edge-based*) [23], taxonomic structure of WordNet is exploited [3,11,14,25,32]. Wang and Hirst in [31] proposed a new approach for measuring the length and depth of path. The most influence taxonomic structure for path-based is *is-a* relationship, consist of *hypernym* and *hyponym*. Assume C_1 and C_2 are concepts. C_2 is a *hypernym* (or *superordinate*) of C_1 if C_1 is a (kind of) C_2. And, C_1 is a *hyponym* (or *subordinate*) of C_2 if C_1 is a (kind of) C_2. *Path-based* is an efficient approach, however it is suffered from two weakness of WordNet structure. First, paths can have different weight, which WordNet does not have information about the weight. Second, a concept has one or more hypernym, while in WordNet just only single hypernym for each concept.

In *information content* (IC) approach, STSR calculated based on content of information of text. The content of information of two WordNet concepts are derived from probability of these concepts from a large corpus. The IC approach was pioneered by Resnik [26]. Followed by several work including [11,16,20,27]. IC approach overcome the first drawback of WordNet structure (no weight for path), however the second drawback (single hypernym) still persists.

[1] A list of WordNet 3.0 glosses in logical forms with XML forms is available in http://wordnetcode.princeton.edu/standoff-files/wn30-lfs.zip.

3 Gloss-Based Semantic Concept Relatedness

Two closely related concepts can have a distance in WordNet structure. For example *hair* and *shampoo.*

- *hair* upward structure:
  entity#n#1 < physical_entity#n#1 < object#n#1 < whole#n#2 < natural_object#n#1 < covering#n#1 < body_covering#n#1 < hair#n#1
- *shampoo* upward structure:
  entity#n#1 < physical_entity#n#1 < matter#n#3 < substance#n#1 < material#n#1 < chemical#n#1 < compound#n#2 < formulation#n#1 < cleansing_agent#n#1 < shampoo#n#1

As shown in the above structure, *hair* and *shampoo* has lowest common subsumer (LCS) *physical entity.* This LCS is just one below the root, i.e. *entity.* However, with gloss of each concept:

- $gloss_{hair}$ = *"a covering for the body (or parts of it) consisting of a dense growth of threadlike structures (as on the human head); helps to prevent heat loss"*
- $gloss_{shampoo}$ = *"shampoo (cleansing agent consisting of soaps or detergents used for washing the hair)"*

It is clear from glosses above that the two concepts closely related. The definition of *shampoo* is *...for washing hair* that is $(C_{hair} \cup gloss_{hair}) \cap (C_{shampoo} \cup gloss_{shampoo}) = \{\text{"}hair\text{"}\}$. In which concept of *hair* and gloss of *shampoo* has an overlap word. Where C_{hair} is the set of synonym of *hair.*

Semantic lexical relatedness (SLR) between word w_1 and word w_2 is squared of overlap between gloss of w_1 and w_2.

$$SLR_1(w_1, w_2) = |(C_{w_1} \cup gloss_{w_1}) \cap (C_{w_2} \cup gloss_{w_2})|^2 \tag{1}$$

The same case as concepts *bank* and *money.*

- *bank* upward structure:
  abstraction#n#6 < group#n#1 < social_group#n#1 < organization#n#1 < institution#n#1 < financial_institution#n#1 < bank#n#2
- *money* upward structure:
  abstraction#n#6 < measure#n#2 < system_of_measurement#n#1 < standard#n#1 < medium_of_exchange#n#1 < money#n#1

- $gloss_{bank}$ = *"a financial institution that accepts deposits and channels the money into lending activities"*
- $gloss_{money}$ = *"the most common medium of exchange; functions as legal tender"*

However, in some cases of concept pairs like *car* and *journey*

- $gloss_{car,auto,automobile,machine,motorcar}$ = *"a motor vehicle with four wheels; usually propelled by an internal combustion engine"*
- $gloss_{journey,journeying}$ = *"the act of traveling from one place to another"*

$$(C_{w_1} \cup gloss_{w_1}) \cap (C_{w_2} \cup gloss_{w_2}) = \{\}, \text{ where } w_1 \text{ is } car \text{ and } w_2 \text{ is } journey.$$

Therefore in our gloss-based approach we use taxonomic structure of WordNet hyponym, hypernym, and holonym (and meronym). Holonymy is the opposite of meronymy. In a meronym, the name of a constituent part of, the substance of, or a member of something. For example *car* has meronym such as *air bag.*

$$\begin{aligned} SRL_2(C_1, C_2) = {} & SRL(gloss_{C_1}, gloss_{C_2}) \\ & + SRL_1(gloss_{C_1}, hype_{C_2}) \\ & + SRL_1(gloss_{C_1}, hypo_{C_2}) \\ & + SRL_1(gloss_{C_1}, holo_{C_2}) \\ & + SRL_1(hype_{C_1}, gloss_{C_2}) \\ & + SRL_1(hype_{C_1}, hype_{C_2}) \\ & + SRL_1(hype_{C_1}, hypo_{C_2}) \\ & + SRL_1(hype_{C_1}, holo_{C_2}) \\ & + SRL_1(hypo_{C_1}, gloss_{C_2}) \\ & + SRL_1(hypo_{C_1}, hype_{C_2}) \\ & + SRL_1(hypo_{C_1}, hypo_{C_2}) \\ & + SRL_1(hypo_{C_1}, holo_{C_2}) \\ & + SRL_1(holo_{C_1}, gloss_{C_2}) \\ & + SRL_1(holo_{C_1}, hype_{C_2}) \\ & + SRL_1(holo_{C_1}, hypo_{C_2}) \\ & + SRL_1(holo_{C_1}, holo_{C_2}) \end{aligned} \tag{2}$$

In SLR_2 from Eq. 2, the range of the result is 0 until infinity. To normalize the result with range 0 until 1, SLR_2 is divided by the number maximum overlapping words. It can be obtained from gloss of structure with minimum number of word.

The sense of word chosen influences the relatedness score, for example word *tiger* dan *cat.*

- Sense 1 - *tiger (a fierce or audacious person).*
- Sense 2 - *tiger, Panthera tigris (large feline of forests in most of Asia having a tawny coat with black stripes; endangered).*

- Sense 1 - *cat, true cat (feline mammal usually having thick soft fur and no ability to roar: domestic cats; wildcats)*

We can see that sense 2 of *tiger* and sense 1 of *cat* are closely related. However it doesn't for pair of sense 2 of *tiger* and sense 1 of *cat.*

4 Evaluation

In this section, we first discuss the gold standard data collection used for our experiments. We also describe the baseline models, experiments and discussions.

In this study we used a 252 pair of words from WordSim353 [2] as gold standard to test our system. Table 1 presents sample of database used. This database is selected from original WordSim353 [6].

Table 1. Sample of WordSim353 related pairs

Word1	Word2	Score	Word1	Word2	Score
Maradona	Football	8.62	Treatment	Recovery	7.91
OPEC	Oil	8.59	Baby	Mother	7.85
Money	Bank	8.50	Money	Deposit	7.73
Computer	Software	8.50	Television	Film	7.72
Lawyer	Evidence	6.69	Development	Issue	3.97
Fertility	Egg	6.69	Day	Summer	3.94
Precedent	Law	6.65	Theater	History	3.91
Minister	Party	6.63	Situation	Isolation	3.88
Cup	Substance	1.92	Professor	Cucumber	0.31
Forest	Graveyard	1.85	King	Cabbage	0.23

Three popular baseline STSR measurements are used: WUP a path-based [32], Lin [16] and JCN [11] both are information content based.

We used Pearson Correlation to evaluate our system. The purpose is to show how precisely our system in measuring semantic relatedness. We compare gold standards relatedness pairs of WordSim353 that comes from human annotators with the score of semantic relatedness generated in our system.

$$r(x, y) = \frac{\sum(x - \bar{x})(y - \bar{y})}{\sqrt{\sum(x - \bar{x})^2}\sqrt{\sum(y - \bar{y})^2}}$$

where x is human judgement and y is SRL of a pair of concept C_1 and C_2. While $\bar{x}$ and $\bar{y}$ are average values.

Table 2 shows that *gloss-based* approach is much more effective that other approaches for measuring semantic lexical relatedness.

Not all concepts in WordNet have holonym. We can see that the usage of holonym increases correlation score slightly (Table 3).

Table 2. Experiment result

Method	Correlation
Gloss (our method)	23.6 %
WUP	4.9 %
JCN	−2.5 %
Lin	4.2 %

Table 3. Gloss-based, with and without holonym

Method	Correlation
With holonym	23.0 %
Without holonym	23.6 %

5 Conclusions

Measurement of semantic similarity and relatedness of concepts or lexical are used in many natural language applications. While similarity and relatedness are close, they should be treated differently. In this paper, we show that *gloss* of WordNet, supported by taxonomy structure, is potentially used for semantic lexical relatedness. It is much more effective than *path* based and *information content* based. While utilization of the main structure of WordNet (hypernym and hyponym) boost performance, holonym degrade performance slightly.

References

1. Agirre, E., Rigau, G.: Word sense disambiguation using conceptual density. In: Proceedings of the 16th Conference on Computational Linguistics (COLING), pp. 16–22. Association for Computational Linguistics (1996)
2. Agirre, E., Alfonseca, E., Hall, K., Kravalova, J., Paşca, M., Soroa, A.: A study on similarity, relatedness using distributional, wordnet-based approaches. In: Proceedings of Human Language Technologies: The 2009 Annual Conference of the North American Chapter of the Association for Computational Linguistics, pp. 19–27. Association for Computational Linguistics (2009)
3. Bhattacharya, A., Bhowmick, A., Singh, A.K.: Finding Top-k similar pairs of objects annotated with terms from an ontology. In: Gertz, M., Ludäscher, B. (eds.) SSDBM 2010. LNCS, vol. 6187, pp. 214–232. Springer, Heidelberg (2010). doi:10.1007/978-3-642-13818-8_17
4. Caraballo, S.A.: Automatic construction of a hypernym-labeled noun hierarchy from text. In: Proceedings of the 37th Annual Meeting of the Association for Computational Linguistics on Computational Linguistics, pp. 120–126. Association for Computational Linguistics (1999)

5. Dzikovska, M.O., Nielsen, R.D., Brew, C., Leacock, C., Giampiccolo, D., Bentivogli, L., Clark, P., Dagan, I., Dang, H.T.: Semeval-2013 task 7: the joint student response analysis and 8th recognizing textual entailment challenge. In: Proceedings of the 9th International Workshop on Semantic Evaluation (SemEval 2013), pp. 263–274. Association for Computational Linguistics (2013)
6. Finkelstein, L., Gabrilovich, E., Matias, Y., Rivlin, E., Solan, Z., Wolfman, G., Ruppin, E.: Placing search in context: the concept revisited. ACM Trans. Inf. Syst. **20**(1), 116–131 (2002)
7. Gad, W.K., Kamel, M.S.: New semantic similarity based model for text clustering using extended gloss overlaps. In: Perner, P. (ed.) MLDM 2009. LNCS (LNAI), vol. 5632, pp. 663–677. Springer, Heidelberg (2009). doi:10.1007/978-3-642-03070-3_50
8. Gurevych, I.: Using the structure of a conceptual network in computing semantic relatedness. In: Dale, R., Wong, K.-F., Su, J., Kwong, O.Y. (eds.) IJCNLP 2005. LNCS (LNAI), vol. 3651, pp. 767–778. Springer, Heidelberg (2005). doi:10.1007/11562214_67
9. Hearst, M.A.: Automatic acquisition of hyponyms from large text corpora. In: Proceedings of the 14th Conference on Computational Linguistics (COLING), pp. 539–545. Association for Computational Linguistics (1992)
10. Hirst, G., Budanitsky, A.: Correcting real-word spelling errors by restoring lexical cohesion. Nat. Lang. Eng. **11**(01), 87–111 (2005)
11. Jiang, J.J., Conrath, D.W.: Semantic similarity based on corpus statistics and lexical taxonomy. In: Proceedings of International Conference Research on Computational Linguistics (ROCLING X) (1997)
12. Khoo, C.S.G., Na, J.-C.: Semantic relations in information science. Ann. Rev. Inf. Sci. Technol. **40**, 157 (2006)
13. Leacock, C., Chodorow, M.: Combining local context, WordNet similarity for word sense identification. In: Fellbaum, C., (ed.) WordNet: An Electronic Lexical Database, pp. 265–283 (1998)
14. Lee, J.H., Kim, M.H., Lee, Y.J.: Information retrieval based on conceptual distance in IS-A hierarchies. J. Documentation **49**(2), 188–207 (1993)
15. Lin, D.: Using syntactic dependency as local context to resolve word sense ambiguity. In: Proceedings of the 35th Annual Meeting of the Association for Computational Linguistics and Eighth Conference of the European Chapter of the Association for Computational Linguistics (ACL-EACL), pp. 64–71. Association for Computational Linguistics (1997)
16. Lin, D.: An information-theoretic definition of similarity. In: Proceedings of the 5th International Conference on Machine Learning, (ICML '98), vol. 98, pp. 296–304 (1998)
17. Manabu, O., Takeo, H.: Word sense disambiguation and text segmentation based on lexical cohesion. In: Proceedings of the 15th Conference on Computational Linguistics (COLING), pp. 755–761. Association for Computational Linguistics (1994)
18. Màrquez, L., Glass, J., Magdy, W., Moschitti, A., Nakov, P., Randeree, B.: Semeval-2015 task 3: Answer selection in community question answering. In: Proceedings of the 9th International Workshop on Semantic Evaluation (SemEval 2015) (2015)
19. Meij, E., IJzereef, L., Azzopardi, L., Kamps, J., de Rijke, M.: Combining thesauri-based methods for biomedical retrieval. In: Proceeding of The Fourteenth Text REtrieval Conference (TREC) (2005)

20. Meng, L., Junzhong, G., Zhou, Z.: A new model of information content based on concept's topology for measuring semantic similarity in wordnet. Intl. J. Grid Distrib. Comput. **5**(3), 81–94 (2012)
21. Miller, G.A.: WordNet: a lexical database for English. Commun. ACM (CACM) **38**(11), 39–41 (1995)
22. Myaeng, S.H., Khoo, C., Li, M.: Linguistic processing of text for a large-scale conceptual information retrieval system. In: Tepfenhart, W.M., Dick, J.P., Sowa, J.F. (eds.) ICCS-ConceptStruct 1994. LNCS, vol. 835, pp. 69–83. Springer, Heidelberg (1994). doi:10.1007/3-540-58328-9_5
23. Pesquita, C., Faria, D., Falcão, A.O., Lord, P., Couto, F.M.: Semantic similarity in biomedical ontologies. PLoS Comput. Biol. **5**(7), e1000443 (2009)
24. Potthast, M., Hagen, M., Beyer, A., Busse, M., Tippmann, M., Rosso, P., Stein, B.: Overview of the 6th international competition on plagiarism detection. In: Cappellato, L., Ferro, N., Halvey, M., Kraaij, W. (eds.) Working Notes Papers of the CLEF 2014 Evaluation Labs, CEUR Workshop Proceedings, CLEF and CEUR-WS.org, September 2014. http://www.clef-initiative.eu/publication/working-notes
25. Rada, R., Mili, H., Bicknell, E., Blettner, M.: Development and application of a metric on semantic nets. IEEE Trans. Syst. Man Cybern. **19**(1), 17–30 (1989)
26. Resnik, P.: Using information content to evaluate semantic similarity in a taxonomy. In: Proceedings of the 14th International Joint Conference on Artificial Intelligence (IJCAI '95), pp. 448–453 (1995)
27. Sánchez, D., Batet, M.: A new model to compute the information content of concepts from taxonomic knowledge. Intl. J. Semant. Web Inf. Syst. (IJSWIS) **8**(2), 34–50 (2012)
28. Vede, C.: Understanding semantic relationships. VLDB J. **2**(4), 455–488 (1993)
29. Sussna, M.: Word sense disambiguation for free-text indexing using a massive semantic network. In: Proceedings of the Second International Conference on Information and Knowledge Management (CIKM '93), pp. 67–74. ACM (1993)
30. Voorhees, E.M.: Query expansion using lexical-semantic relations. In: Proceeding of The Seventeenth Annual International ACM/SIGIR Conference on Research and Development in Information Retrieval (SIGIR), pp. 61–69. Springer, London (1994)
31. Wang, T., Hirst, G.: Refining the notions of depth and density in wordnet-based semantic similarity measures. In: Proceedings of the Conference on Empirical Methods in Natural Language Processing, pp. 1003–1011. Association for Computational Linguistics (2011)
32. Wu, Z., Palmer, M.: Verbs semantics and lexical selection. In: Proceedings of the 32nd Annual Meeting on Association for Computational Linguistics, pp. 133–138. Association for Computational Linguistics (1994)
33. Zhang, Z., Gentile, A.L., Ciravegna, F.: Recent advances in methods of lexical semantic relatedness-a survey. Nat. Lang. Eng. **19**(04), 411–479 (2013)

Difference Expansion-Based Data Hiding Method by Changing Expansion Media

Tohari Ahmad(✉), Diksy M. Firmansyah(✉), and Dwi S. Angreni

Department of Informatics, Institut Teknologi Sepuluh Nopember (ITS), Surabaya 60111, Indonesia
tohari@if.its.ac.id,
{diksy14, shinta14}@mhs.if.its.ac.id

Abstract. In this era, protecting secret data has played an important role since such data may be transmitted over public networks or stored in public storages. One possible method to protect the data is by implementing steganography/data hiding algorithms, such as Difference Expansion (DE). It works by embedding a secret message on the difference value of two pixels, in the case the cover is an image. Because the data changes directly, Difference Expansion has a problem on the limit values which are called overflow and underflow. This affects the amount of the secret message and quality of the resulted stego data. In this paper, we propose to change the embedding method on a matrix which is generated from an LSB image. Therefore, there is no restriction on media value where the data is embedded. The experimental result shows that this proposed method is able to improve the performance of stego data.

Keywords: Data hiding · Difference Expansion · Expansion media · Data security

1 Introduction

The concept of "What You See Is What You Get (WYSIWYG)" which we often face when viewing images or other media is no more appropriate and is not able to deceive an attacker because this concept is not always true [1]. An image is not only what we see with the Human Visual System (HVS); it can represent thousands words. Since pre-digital era, people have been designing excellent methods to secretly communicate. There are three inter-related techniques in the topic of data security: steganography, watermarking and cryptography.

As a concept, steganography and cryptography have a very close relationship [1–3, 5]. Although both have the same goal, their way and use have significant differences. Steganography is hidden writing which hides the presence of the message while cryptography is secret writing which provides security in the message content such as that implemented in [1] and [4], respectively. Different from steganography, steganalysis tries to break the information which has been embedded in the respective media.

Generally, steganography and steganalysis are an important area of research within the paradigm of data hiding [5]. It is believed that digital steganography provides secure communications that have become the needs of most applications in the world today.

T. Herawan et al. (eds.), *Recent Advances on Soft Computing and Data Mining*,
Advances in Intelligent Systems and Computing 549, DOI 10.1007/978-3-319-51281-5_42

Various media types such as text, image, audio, and video can act as a cover to bring confidential information.

There are various methods which have been introduced to hide a secret message, such as [8–17]. The existing algorithms, however, have two common problems: the capacity of the secret which can be embedded into the cover, and the quality of the resulted stego data. In this paper, we propose a method which is able to work on those problems by changing the expansion media.

The rest of the paper is organized as follows. Section 2 describes the methods which relate to our proposed technique. Section 3 presents our method whose experimental result is provided in Sect. 4. This is followed by the conclusion which is drawn in Sect. 5.

2 Related Works

Some research which relate with data hiding are provided in this section. This includes the Difference Expansion method that has relatively good performance.

2.1 Data Hiding

Steganography or data hiding is generally divided into two categories: reversible and irreversible [6]. The irreversible techniques can extract the secret message only. The cover data (e.g., an image) changes or becomes damaged, which leads to be unreadable to the human. Reversible technique is in the opposite. Here, both the secret message and the cover data can be extracted, back to their original forms [7].

An example of irreversible technique is the method proposed by Battisti et al. [8] which proposes an approach to hide data by using p-Fibonacci number sequence. Dey et al. [9, 10] and Nosrati et al. [11] propose a method which is an improvement of data hiding approaches using the p-Fibonacci numbers sequence. Then, Dey et al. [12] introduce a method of embedding using a linked-list in 24 bit RGB color image. An example of reversible techniques of data hiding is a method presented by Ni et al. [13], which hides data by using a histogram shifting of the original image. It is then refined by Kuo et al. [14] which propose a technique based on the division of the blocks to hide data in images.

In further development, Holil and Ahmad [15] and Ahmad et al. [16] embed the secret data in a medical image. They believe that it is useful for maintaining the confidentiality of the patient medical record. A medical image certainly has information about the patient, such as a disease they suffer, the identity of the patient, and other sensitive information. Due to its nature, this information needs to be hidden. So it will not be known easily by unauthorized parties. In a data hiding method, this information can be embedded in a medical image so that they are not aware of the presence of the secret message. By using the reversible method, not only the secret message, but also the medical image can be restored. The research in [16] is actually an improvement of the previous Difference Expansion (DE) [17]. This method is able to present a simple and efficient reversible technique of embedding data for digital images. Additionally, it has advantages in terms of the amount of secret data can be hidden or the quality of the stego image. However, this paper changes directly the pixel values, which can reduce the quality of the image.

Difference Expansion (DE) is a reversible method of steganography which is introduced by Tian [17]. This technique provides an additional space with a few redundancies on the image content. This method also provides the best capacity limits of payloads and quality of the embedded image. Furthermore, DE has also low computational complexity. This method begins with reversible integer transform by using an 8-bit grayscale image (x, y), where x and y are the value of pixels in a pair. Here, $0 \leq (x, y) \leq 255$. The average value l and the difference h are calculated as in (1) whose inverse transform is provided in (2).

$$l = \left\lfloor \frac{x+y}{2} \right\rfloor, \quad h = x - y. \tag{1}$$

$$x = l + \left\lfloor \frac{h+1}{2} \right\rfloor; \; y = l - \left\lfloor \frac{h}{2} \right\rfloor. \tag{2}$$

Difference Expansion embeds the secret message by adding it to the binary representation of h. Mathematically, this process can be represented by using the following equation:

$$h' = 2h + b. \tag{3}$$

where h' is the value of the difference after being embedded by a message, and b is the secret message itself. In order to extract the secret message and to recover the value of difference, the Eq. (4) is applied.

$$b = LSB(h'); h = \left\lfloor \frac{h'}{2} \right\rfloor. \tag{4}$$

The implementation of the Eq. (2) is done directly on the pixel values. This poses a problem in the process of embedding of data because the range of pixel values is only from 0 to 255. This can cause either overflow or underflow. The former is a condition in which the modified pixel value exceeds the value of 255, while the latter is a condition in which the modified pixel value is less than 0. In order to overcome this problem, the value of the embedded data h' must satisfy the Eq. (5) for $b = 0$ or $b = 1$.

$$|h'| \leq 2\lfloor 255 - l \rfloor; |h'| \leq 2l + 1. \tag{5}$$

2.2 Difference Expansion (DE) Embedding Process

The stage of embedding data is performed by changing all changeable difference values. This is done by either adding new or changing the LSB. In order to make the process able to recover the data, the image (i.e., the cover data) is embedded by the original value of the modified LSB. Briefly, the data embedding process of DE consists of some phases: counting the value of difference, partitioning the difference values into four sets, shaping the location map, collecting the original value of the LSB, embedding the data to perform the replacement, and finally, calculating the integer inverse transform value. In more details, those phases can be described as follows:

(1) The original images are grouped into a number of pixel pair values. Each pair of pixel consists of two adjacent pixels. This process can be done by constructing a pair of horizontal pixels with the same row and column sequentially: $(i, 2j - 1)$ and $(i, 2j)$ where i and j are the number of row and column respectively; a pair of vertical pixels: $(2i - 1, j)$ and $(2i, j)$; or specific key patterns. This pairing can be applied to either all pixels or only certain pixels of the image. Additionally, integer transformation is implemented to each pair of pixels, and the value of difference h is to be sorted into one dimensional list.

(2) Four different sets of difference values are generated: EZ, EN, CN and NC:
   i. *EZ*: consists of all the expandable values and expandable with $h = 0$ and $h = -1$.
   ii. *EN*: comprises all the expandable h values that are not members of the *EZ*.
   iii. *CN*: contains all the changeable h values that are not members of the combined *EZ* and *EN*.
   iv. *NC*: comprises all non-changeable value h.

(3) The location map is developed from the chosen expandable difference value. Each value of h in the *EZ*, is used in DE. Depending on the size of its payload, some difference values in *EN* are selected for the DE. In details, the selected subsets from *EN* are denoted by $EN1$ and $EN2$.

(4) The original LSB of the difference value in $EN2$ and *CN* are collected. For every h in $(EN2 \cup CN)$, $LSB(h)$ will be collected into a bit stream C, except that with $h = 1$ or $h = -2$ in $(EN2 \cup CN)$, where its LSB is ignored, so that the values 1 and 0 in sequence can be determined by the location map.

The payload P which includes the hash value of the original image, is embedded along with the location map L and the original value of LSB C. Those three values are combined into a single bit stream $B = L \cup C \cup P = b1, b2, \ldots, bm$ where $bi \in \{0, 1\}$, $1 \leq i \leq m$, and m is the bit length of B. The value of C is added to the end of L whose result is appended by P. The pseudo code of the data embedding process can be represented as follows:

```
Set i = 1 dan j = 0
While (i ≤ m)
     j = j + 1
if hj ∈ (EZ U EN1)
     hj = 2 x hj + bi
     i = i + 1
Else if hj ∈ (EN2UCN)
     hj = 2 x hj + bi
     i = i + 1
```

3 Changing Expansion Media

In our proposed method, we enhance the existing DE method [17], such that its capability increases. Even though our proposed method can be applied to any media, in this paper we use an image as the cover.

3.1 Data Embedding

The algorithm consists of five steps, namely LSB extraction, matrix generation, secret message embedding with DE, payloads generation, and payloads substitution to the cover image to get the stego image. This process can be depicted as follows.

(1) Extract LSB of the pixels from the cover image. The extraction process can be carried out horizontally, vertically or any other pattern. Assume that the LSB extraction process is done horizontally, we obtain the binary matrix with dimension $m \times (n - (n \bmod 16))$ as previously described.

(2) Group every 8 bits and represents it into integer format to obtain a matrix of integers with dimensions $\frac{1}{8} \times m \times (n - (n \bmod 16))$. Because the values of the matrix is the representation of the 8-bit binary, then it is likely that its value is from 0 to 255.

(3) Embed the secret message to the matrix using DE formula. Then, the transformation is performed by using (6). In this process, the resulted matrix has the same dimensions as before, however, its values may different. This is because the previous may have values between 0 and 511.

$$x' = l + \left\lfloor \frac{h+1}{2} \right\rfloor + 128; \; y' = l - \left\lfloor \frac{h}{2} \right\rfloor + 128. \tag{6}$$

(4) After obtaining the matrix, change its values back into binary representation which is to be the payloads. Since the values in the matrix are from 0 to 511, they are represented by 9 bits whose dimension is $m \times (n - (n \bmod 16))$. Next, this payload is used for replacing the LSB of the cover image. As this larger than the cover, the payload is firstly compressed. In the case that the length of payload is less than the required, the remaining spaces are filled with the original LSB values.

(5) Finally, replace the LSB of the cover with the payloads to obtain stego image. The overall process of this embedding step is provided in Fig. 1.

3.2 Data Extraction

The extraction step comprises five steps: extraction LSB to get payloads, decompression payloads, the generation matrix, extraction secret message with DE, and substitution matrix results with stego image to obtain the original cover image. These steps are as follows.

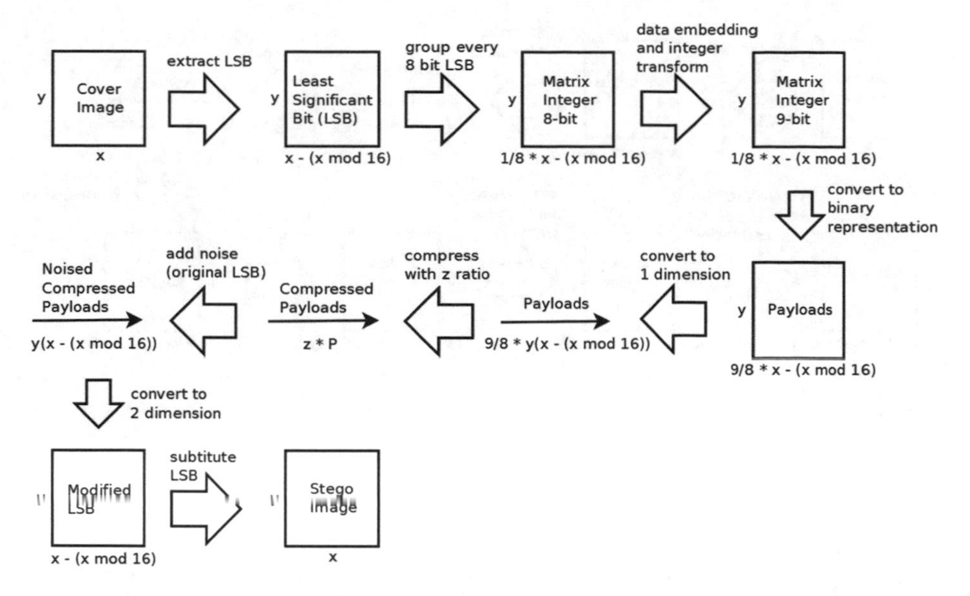

Fig. 1. Data embedding process

(1) Extract LSB of the pixels in the stego image. The order of the extraction process is same as that of embedding, which can be horizontally, vertically or other patterns. Let it is done horizontally. We have a binary matrix whose dimension is $m \times (n - (n \bmod 16))$. This is a binary representation of payloads compression results in embedding algorithm.

(2) Decompress the payload. Here, we get a binary matrix with dimensions $\frac{9}{8} \times m \times (n - (n \bmod 16))$.

(3) Group every 8 bits in the binary matrix. Next, convert them to an integer format to obtain a matrix of integers with dimension $\frac{1}{8} \times m \times (n - (n \bmod 16))$. This matrix is that used for embedding the secret message.

(4) Extract the secret message of the integers matrix. The transformation is done by using (7) to have the original matrix values. Once all values are recovered, convert them to binary to have a binary matrix with dimension $m \times (n - (n \bmod 16))$.

$$x' = l + \left\lfloor \frac{h+1}{2} \right\rfloor - 128; \; y' = l - \left\lfloor \frac{h}{2} \right\rfloor - 128. \tag{7}$$

(5) The resulting binary matrix is the original LSB from the cover image. To get the complete cover image, substitute LSB stego image with the value of the binary matrix. Extraction and recovery process are depicted in Fig. 2.

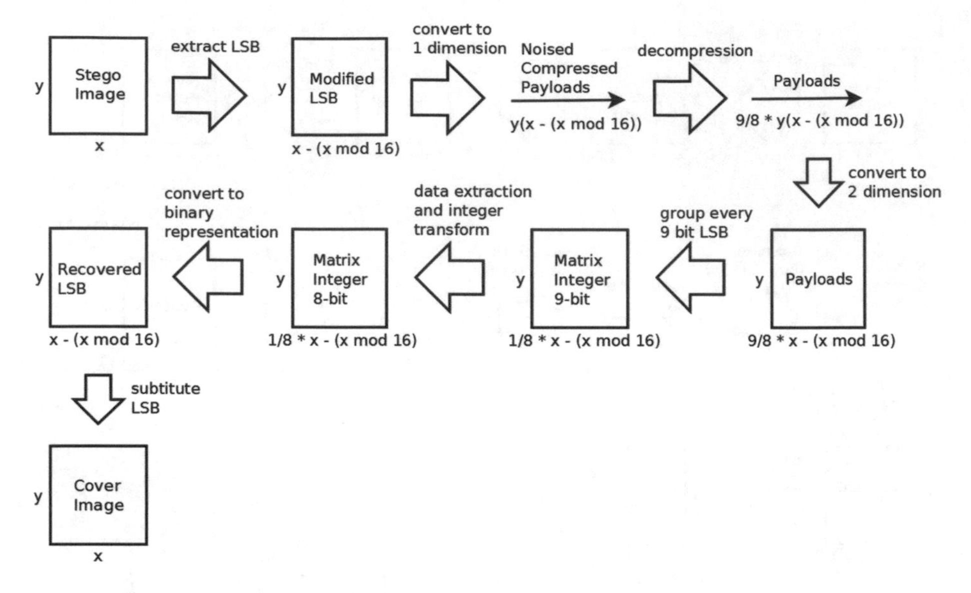

Fig. 2. Data extraction process

4 Experimental Results

The proposed method is implemented in the Java programming language with the Java Development Kit (JDK) version 1.8.0-25 which runs on top of NetBeans IDE 8.0.2. The program runs on Intel Celeron Processor 1000 M (1.8 GHz), 2048 MB DDR3 SDRAM with Windows 8 Pro 64 bit. In addition, we also implement the DE algorithm [17] for a comparison purpose. For the experiment, we use medical images taken from [18] for the cover or carrier. Those images are firstly converted from their original *.jpg format into *.png which we work on. The secret message is from lorem ipsum paragraphs of length 1500 bytes generated by lorem ipsum paragraphs generator [19]. Lorem ipsum itself is an example of texts which is often used by a wide variety of software industry as an example of text-based content for their products. The original cover image and the stego images generated by DE and proposed methods are provided in Fig. 3.

The experiment is conducted for evaluating the capacity and quality. Capacity is calculated by using the bits per pixel (bpp). It is done by dividing the maximum number of secret message that can be accommodated by the cover image with the dimensions of the cover image itself. The quality is calculated by using Peak Signal-to-Noise Ratio (PSNR).

The experimental results are presented in Table 1. It is shown that the quality of the stego image significantly goes up, where on average, the PSNR rises about 14 points. The highest increase happen in abdominal image, from 34.140 to 51.763. This has a positive effect on the user perception, considering that minimum recommended PSNR of stego image is 35, Nevertheless, as a trade-off, the capacity of the secret message slightly decreases as represented by the bpp values. This problem however, may be overcome by enlarging the cover image. Consequently, the size of the stego image is

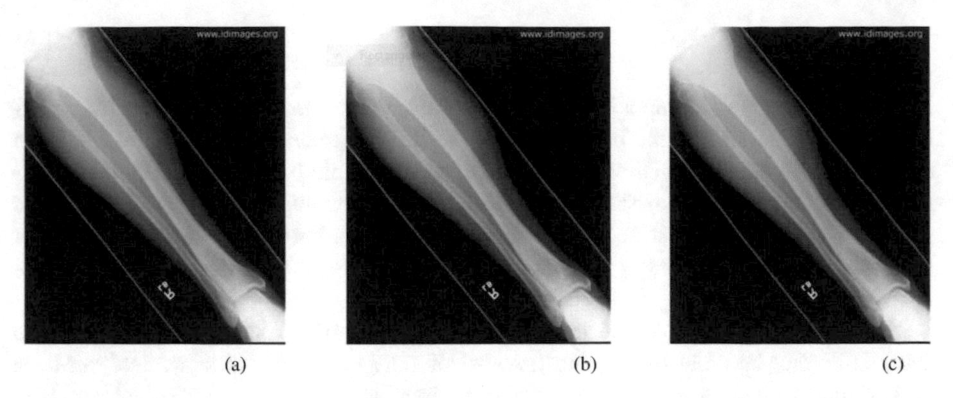

Fig. 3. (a) Original cover image [18], (b) stego image with DE, (c) stego image with the proposed method

affected. We believe that this method is appropriate to use for hiding smaller secret data with higher level of confidentiality. This is because people may not be aware of the hidden data if the stego image is looked normally, which is represented by its relatively high PSNR value. Overall, the users may select an appropriate data hiding method to be used according to their purpose. This can more focus either on the quality or capacity.

Table 1. Performance of DE and the proposed method.

Image	abdominal.png		leg.png		lung.png		hand.png	
Performance	DE	Proposed	DE	Proposed	DE	Proposed	DE	Proposed
PSNR (dB)	34.140	51.763	37.980	52.175	41.850	52.361	35.230	-
Capacity (bpp)	0.275	0.061	0.418	0.060	0.437	0.060	0.471	-

4.1 Discussion

In this experiments we use the class java.util.zip Deflater for the compression method. By using this class, we find that our proposed method can not work on a certain image (i.e., hand.png). This is based on the assumption that in order to have this method runs properly, the minimum compression ratio is 8/9. This is because every 9 bits of matrix resulted from the embedding process should be able to be compressed into 8 bit at most. In this case, the compression method is not able to achieve that condition.

A better compression ratio means better the quality of stego image (higher PSNR) since more salt is to be used. In this proposed method, the salt is taken from the original LSB of the cover image. A possible compression method which is appropriate to use is the arithmetic coding, even though other methods may also be applicable.

Selection of a suitable compression method may be one of the problems that can be raised in the future. In addition, putting the compression on the other step is also possible to increase the quality of the stego image.

5 Conclusion

In this paper, we have proposed a DE-based data hiding method. Some improvements have been made so that the performance is relatively better than the original DE. In addition, this proposed method also removes the possible overflow or underflow values. Consequently, the possibility of successful embedding process is higher. This proposed method, however, relies on the characteristic of the compression technique. If the compression ratio does not reach 8/9, then the embedding process does not work because every 8 bit LSB of the cover changes to 9 bit.

In the future, we would like to work further on increasing the quality of the stego data. This may be done by varying the size of the generated matrix, which can be a bigger matrix. This size may also result to increasing the capacity of secret data. Furthermore, the method should work on any compression ratio. In this case, the matrix generation process should be explored.

References

1. Cheddad, A., Condell, J., Curran, K., McKevitt, P.: Digital image steganography: survey and analysis of current metods. Sig. Process. **90**(3), 727–752 (2010)
2. Johnson, N.F., Jajodia, S.: Steganalysis of images created using current steganography software. In: Aucsmith, D. (ed.) IH 1998. LNCS, vol. 1525, pp. 273–289. Springer, Heidelberg (1998). doi:10.1007/3-540-49380-8_19
3. Chandramouli, R., Kharrazi, M., Memon, N.: Image steganography and steganalysis: concepts and practice. In: Kalker, T., Cox, I., Ro, Y.M. (eds.) IWDW 2003. LNCS, vol. 2939, pp. 35–49. Springer, Heidelberg (2004). doi:10.1007/978-3-540-24624-4_3
4. Ahmad, T., Hu, J., Han, S.: An efficient mobile voting system security scheme based on elliptic curve cryptography. In: 3rd International Conference on Network and System Security, pp. 474–479 (2009)
5. Subhedar, M.S., Mankar, V.H.: Current status and key issues in image steganography: a survey. Comput. Sci. Rev. **13**, 95–113 (2014)
6. Sarkar, T., Sanyal, S.: Reversible and irreversible data hiding technique (2014). https://arxiv.org/abs/1405.2684v2
7. Sarkar, T., Sanyal, S.: Steganalysis: detecting LSB steganographic techniques (2014). https://arxiv.org/abs/1405.5119v1
8. Battisti, F., Carli, M., Neri, A., Egiaziarian, K.: A generalized fibonacci LSB data hiding technique. In: IEEE 3rd International Conference on Computers and Devices for Communication (CODEC-2006) TEA, pp. 18–20 (2006)
9. Dey, S., Abraham, A., Sanyal, S.: An LSB data hiding technique using natural numbers. In: 3rd International Conference on Intelligent Information Hiding and Multimedia Signal Processing, vol. 2, pp. 473–476 (2007)
10. Dey, S., Abraham, A., Sanyal, S.: An LSB data hiding technique using prime numbers. In: 3rd International Symposium on Information Assurance and Security, pp. 101–108 (2007)
11. Nosrati, M., Karimi, R., Nosrati, H., Nosrati, A.: Embedding stego-text in cover images using linked list concepts and LSB technique. J. Am. Sci. **7**, 97–100 (2011)
12. Dey, S., Abraham, A., Bandyopadhyay, B., Sanyal, S.: Data hiding techniques using prime and natural numbers. J. Digit. Inf. Manag. **6**, 463–485 (2008)

13. Ni, Z., Shi, Y.Q., Ansari, N., Su, W.: Reversible data hiding. IEEE Trans. Circ. Syst. Video Technol. **16**(3), 354–362 (2006)
14. Kuo, W.C., Jiang, D.J., Huang, Y.C.: A reversible data hiding scheme based on block division. In: Congress on Image and Signal Processing, vol. 1, pp. 365–369 (2008)
15. Holil, M., Ahmad, T.: Secret data hiding by optimizing general smoothness difference expansion based method. J. Theoret. Appl. Inf. Technol. **72**, 155–163 (2015)
16. Ahmad, T., Holil, M., Wibisono, W., Ijtihadie, R.M.: An improved quad and RDE-based medical data hiding method. In: IEEE International Conference on Computational Intelligence and Cybernetics, pp. 141–145 (2013)
17. Tian, J.: Reversible data embedding using a difference expansion. IEEE Trans. Circ. Syst. Video Technol. **13**(8), 890–896 (2003)
18. Partners Infectious Disease Images - eMicrobes Digital Library (2002). http://www.idimages.org/images/
19. Lorem Ipsum - All the facts - Lipsum generator. http://www.lipsum.com/

Ensemble Methods and their Applications

A New Customer Churn Prediction Approach Based on Soft Set Ensemble Pruning

Mohd Khalid Awang[1(✉)], Mokhairi Makhtar[1],
Mohd Nordin Abd Rahman[1], and Mustafa Mat Deris[2]

[1] Faculty of Informatics and Computing, Universiti Sultan Zainal Abidin,
22200 Besut, Terengganu, Malaysia
{khalid,mokhairi,mohdnabd}@unisza.edu.my
[2] Faculty of Computer Science and Information Technology,
Universiti Tun Hussein Onn, Batu Pahat, Johor, Malaysia
mmustafa@uthm.edu.my

Abstract. Accurate customer churn prediction is vital in any business organization due to higher cost involved in getting new customers. In telecommunication businesses, companies have used various types of single classifiers to classify customer churn, but the classification accuracy is still relatively low. However, the classification accuracy can be improved by integrating decisions from multiple classifiers through an ensemble method. Despite having the ability of producing the highest classification accuracy, ensemble methods have suffered significantly from their large volume of base classifiers. Thus, in the previous work, we have proposed a novel soft set based method to prune the classifiers from heterogeneous ensemble committee and select the best subsets of the component classifiers prior to the combination process. The results of the previous study demonstrated the ability of our proposed soft set ensemble pruning to reduce a substantial number of classifiers and at the same time producing the highest prediction accuracy. In this paper, we extended our soft set ensemble pruning on the customer churn dataset. The results of this work have proven that our proposed method of soft set ensemble pruning is able to overcome one of the drawbacks of ensemble method. Ensemble pruning based on soft set theory not only reduce the number of members of the ensemble, but able to increase the prediction accuracy of customer churn.

Keywords: Ensemble pruning · Customer churn prediction · Ensemble selection · Soft set · Ensemble methods

1 Introduction

Customer churn prediction becomes the most significant activities that form the foundation for all Customer Relationship Management (CRM) in any business operations. The cost of acquiring new customers is relatively high, thus maintaining current customers is vital. In the telecommunications industry, customer churn becomes more apparent because of the fast growth of wireless technology. Customers have more choices and able to select and change from one package to another package offered by the different service providers. Therefore, to remain competitive in an increasingly

T. Herawan et al. (eds.), *Recent Advances on Soft Computing and Data Mining*,
Advances in Intelligent Systems and Computing 549, DOI 10.1007/978-3-319-51281-5_43

saturated market, companies will strive to retain their customers and try to reduce the cost of acquiring new customers. Many researchers have proposed various models of customer churn prediction, for example the neural network, decision tree, genetic algorithm, and regression analysis [28–32]. However, current approaches for churn prediction are not effective and need to be improved because there are many uncertainty factors that contribute to customer churn.

Ensemble methods or multiple classifiers are known as learning algorithms that train a set of classifiers and combine them to achieve the best prediction performance [1]. Ensemble methods such as bagging [2], boosting [3], stacking [4], Bayes optimal classifier [5], rotation forest [5], ensemble selection [6] and hybrid intelligent system [7] have proven to improve the performance of the classification.

The most essential concepts of ensemble methods consist of two main stages which is the construction of multiple base classifier models and their combination. One of the obvious disadvantages of ensemble methods is the production of a large number of individuals or base classifiers which sometimes referred as overproduce. Recent work [8, 9] suggested another step prior to ensemble combination, which is known as ensemble pruning, ensemble selection or ensemble thinning [8–12]. The purpose of pruning is to find the minimal or the optimum number of base classifiers from a repository of classifiers and at the same time maintaining the classification or prediction performance. However, despite of the importance of the pruning phase, only a few researches focused on the selection of ensemble's classifiers.

The paper proposed a novel approach for an ensemble pruning method based on the soft set theory for customer churn prediction [32]. Our approach aims to solve the problem of representing less redundant ensemble classifiers based on the dimensionality reduction of soft set theory and at the same time producing highest customer churn prediction accuracy.

The rest of this paper is organized as follows. Section 2 describes the ensemble methods and ensemble pruning. Section 3 discusses the soft set and its reduction algorithm. Section 4, soft set pruning method describes the soft set theoretical analysis of the granular metadata generated by the decisions of base classifiers. Section 5 describes the experimental setting and results. Finally, Sect. 6 summarizes this work.

2 Ensemble Methods and Ensemble Pruning

Previous researchers have proposed various ensemble methods as learning algorithms in data mining to increase the classifier performance and accuracy [1–7]. Most of the previous studies focus on the ensemble construction and ensemble combination in improving the accuracy and performance of classification, but rarely consider the ensemble pruning algorithms. Nevertheless, there are few researches focusing on ensemble methods [8–13].

2.1 A Taxonomy of Ensemble Pruning Methods

Previous ensemble pruning techniques can be categorized into three branches, the ordering-based pruning, the clustering-based pruning and the optimization-based pruning. Tsoumakas et al. [8] provided a brief taxonomy on ensemble pruning. Order-Based Pruning ranks the individual classifiers according to some criterion. The classifiers in the front-part of the rank will be considered as the best candidate to form the final ensemble. Reduce-Error Pruning [14], Kappa Pruning [15] and Boosting-Based Pruning [16] are belongs category. On the hand, the Clustering-Based Pruning identifies a number of representatives of individual classifiers to construct the final ensemble. Ensemble pruning groups together the individual classifiers into a number of clusters based on their similarities. Some of the works in this method including Hierarchical Agglomerative Clustering [17], k-means Clustering [18] and Deterministic Annealing [19]. The last category is the Optimization-Based Pruning, which aims to select the subset of individual classifiers that maximizes or minimizes an objective related to the final ensemble. Some researchers under this category proposed Mathematical Programming Pruning [20] and Probabilistic Pruning [21].

3 Soft Set Theory

3.1 Basic Concepts of Soft Set

Let U be initial universal set and let E be a set of parameters. Let $P(U)$ denote the power set of U. A pair (F, E) is called a soft set over U, if only if F is a mapping given by $F: E \rightarrow P(U)$ [22, 23].

3.2 Soft Set Based Reduction Based on Discernibility

The most fundamental concept in rough set is set approximation and it is carried out by discernibility matrix and discernibility function. Based on [24] that every rough set is a soft set, we proposed a similar concept of discernibility function in rough set [25, 26] to reduct and discern the soft set ensemble of classifiers [32]. Our proposed soft set ensemble reduction is based on the work done by Skrowron [25]. He introduced the representation of the decision table into the discernibility matrix to compute reduct.

4 Proposed Soft Set Pruning Ensemble Method

A reduct in ensemble methods can be defined as the irreducible subset of classifiers, which keeps the same discernibility as the original set of classifiers. The process of reducing the number of classifiers is known as pruning. The first step in the soft set pruning ensemble methods is to generate the decision table of the testing data set. The decision table is then transformed into a soft set representation. The next step is to apply the reduction algorithm on the soft set table. Based on [24] that every rough set is a soft set, we proposed a similar concept of discernibility function in rough set [25] to reduct and discern the data sets. Then the table is transformed into discernibility matrix. The next step is to perform the discernibility function on the discernibility matrix. The discernibility function will produce set of reducts. Finally, we apply the distributive law on the reduct to generate reduct teams.

4.1 A New Soft Set Ensemble Selection Algorithm

Input: Decision tables of the testing dataset
Output: Team/teams of ensemble

```
1. Start
2. Construct the decision table of the testing data set.
3. Transform the decision table into softest representation
4. Remove any redundant representations
5. Transform the softest representation into discernibility
   matrix
6. Transform the discernibility matrix into discernibility
   function
7. Apply the absorption law to get the set of the reduct.
8. Apply the distributive law to construct the reduct teams
9. End
```

4.2 Soft Ensemble Representation

Suppose that there are ***M*** *instances* in the test data set which consist of (r_1, r_2, r_3, r_4, r_5, r_6, r_7) and ***N*** number of *classifiers* in our pool of classifiers such as (c_1, c_2, c_3, c_4). Each instance of test data set is mapped against each type of classifiers to produce *N* numbers of *prediction output*.

Step 2: The $M \times N$ matrix is considered as the *decision table* representing the *M* numbers of instances and *N* number of classifiers (Table 1).

Table 1. An example of prediction output.

U	c_1	c_2	c_3	c_4	c_5
r_1	yes	no	yes	no	**no**
r_2	yes	yes	no	no	**no**
r_3	yes	yes	yes	yes	**yes**
r_4	no	yes	no	yes	**yes**
r_5	no	no	yes	no	**no**
r_6	no	no	no	no	**no**
r_7	no	no	yes	no	**no**

A *Prediction Output* is defined as a 7-tuple S = (U, A, V, f), where $U = \{u_0, u_1, \cdots, u_{|U-1|}, u_{|U|}\}$ is a non-empty finite set of objects, $A = \{a_0, a_1, \cdots, a_{|A-1|}, a_{|A|}\}$ is a non-empty finite set of attributes, $V = \bigcup_{e_i \in A} V_{e_i}$, where V_a is the domain (value set) of attribute a, $f : U \times A \rightarrow V$ is an information $f(x, a) \in V_a$. Function, such that, for ever $f(x, a) \in V_a$.

Step 3: Thus, we can make one-to-one corresponding between a Boolean-valued prediction results and a soft set.

Proposition 1. If (F, E) is a soft set over the universe U, then (F, E) is a Boolean-valued information system S = (U, A, V{0,1}, f).

Step 4: Remove any redundant representations. Based on our previous algorithm, we notice that the number of classifiers will have a significant effect on discernibility function and soft set reduction. Therefore, we proposed reduction of redundancy prior to ensemble matrix construction using Euclidean Distance function [31]. From Table 2, we could notice that $F(c_4) = F(c_5)$. $F(c_4) = \{r_3, r_4\}$, $F(c_5) = \{r_3, r_4\}$. Therefore, we eliminate all the redundant representation by taking only the first representation.

Table 2. Boolean-valued of classifier's prediction

U	c_1	c_2	c_3	c_4	c_5
r_1	1	0	1	0	0
r_2	1	1	0	0	0
r_3	1	1	1	1	1
r_4	0	1	0	1	1
r_5	0	0	1	0	0
r_6	0	0	0	0	0
r_7	0	0	1	0	0

Step 5: For the information system S from Table 2 we obtain the discernibility matrix presented in Table 3, and the following discernibility functions:

Table 3. Boolean-valued of classifier's prediction

	r_1	r_2	r_3	r_4	r_5	r_6	r_7
r_1	0						
r_2	c_2, c_3	0					
r_3	c_2, c_4	c_3, c_4	0				
r_4	c_1, c_2, c_3, c_4	c_1, c_4	c_1, c_3	0			
r_5	c_1	c_1, c_2, c_3	c_1, c_2, c_4	c_2, c_3, c_4	0		
r_6	c_1, c_3	c_1, c_2	c_1, c_2, c_3, c_4	c_2, c_4	c_3	0	
r_7	c_1	c_1, c_2, c_3	c_1, c_2, c_4	c_2, c_3, c_4	0	c_3	0

Step 6: The discernibility functions are as follows:

$$f(r1) = \{c2 \vee c3\} \wedge \{c2 \vee c4\} \wedge \{c1 \vee c2 \vee c3 \vee c4\} \wedge \{c1\} \wedge \{c1 \vee c3\} \wedge \{c1\};$$
$$f(r2) = \{c3 \vee c4\} \wedge \{c1 \vee c4\} \wedge \{c1 \vee c2 \vee c3\} \wedge \{c1 \vee c2\} \wedge \{c1 \vee c2 \vee c3\};$$
$$f(r3) = \{c1 \vee c3\} \wedge \{c1 \vee c2 \vee c4\}' \wedge \{c1 \vee c2 \vee c3 \vee c4\} \wedge \{c1 \vee c2 \vee c4\};$$
$$f(r4) = \{c2 \vee c3 \vee c4\} \wedge \{c2 \vee c4\} \wedge \{c2 \vee c3 \vee c4\};$$
$$f(r5) = \{c3\};$$
$$f(r6) = \{c3\};$$

Step 7: Generating the reduct based on the following indiscernibility functions:

$$f(r_i) = f(r1) \wedge f(r2)\ f(r3) \wedge f(r4) \wedge f(r5) \wedge f(r6) \wedge f(r7)$$

by applying the absorption law for each $f(r_i)$, we obtain:

$$f(R) = \{c2 \vee c4\} \wedge \{c1\} \wedge \{c3\}$$

Step 8: At the end of the discernibility function, we applied the distributive law to gain the final reducts. By applying the distributive law for each of f(R), we obtain the 2 reduct sets; R1 = {c2, c1, c3}, R2 = {c4, c1, c3}. Based on the reduction method, we can reduce the ensemble size and select the team to produce a good and efficient ensemble.

5 Experimental Evaluation

In order to validate the performance of the proposed soft set ensemble selection algorithm, we construct our ensemble on two (2) customer churn datasets.

5.1 Data Set

The first dataset [**DataSet1**] is based on the actual customer dataset taken from one of the leading telecommunications company in Malaysia. The total number of datasets is 272 records, which later divided into training (217) and testing (55). The second dataset [**DataSet2**] is based on the customer churn taken from the UCI Repository of Machine Learning Databases. The dataset consists of 3,333 cleaned objects and 20 instances along with one indicator whether or not to churn [27]. The dataset is divided into training and testing dataset with the proportion of 2978 and 355 respectively.

5.2 Learning Algorithm for Classifiers

We create our heterogeneous ensemble by selecting ten different classifiers which are as in Table 4.

Table 4 displays the prediction accuracy of each of the classifiers with the highest accuracy of an individual classifier is 0.71% for **DataSet1** while 0.85% for **DataSet2**.

Table 4. Classifier prediction

Classifiers	Team representation	Prediction accuracy	
		Set1	Set2
meta.EnsembleSelection	0000000001	0.42	0.84
rules.DecisionTable	0000000010	0.71	0.85
meta.StackingC	0000000100	0.71	0.85
meta.AdaBoostM1	0000001000	0.44	0.85
meta.Bagging	0000010000	0.65	0.77
rules.ZeroR	0000100000	0.71	0.85
bayes.NaiveBayesUpdateable	0001000000	0.64	0.82
rules.JRip	0010000000	0.64	0.85
trees.J48	0100000000	0.45	0.85
lazy.IBk	1000000000	0.69	0.85

Table 5. Number of classifier before and after pruning

	Original ensemble size	After soft set pruning algorithm	
		DataSet1	DataSet2
Number of classifiers in ensemble	10	7	5
Number of all possible combinations of classifiers (2^n)	$2^{10} = 1024$	$2^7 = 128$	$2^5 = 32$
Percentage of reduction		87.5%	96.9%

Based on the number of classifiers in the ensemble, we could end up with 1024 combination of different classifiers teams for both dataset as illustrated in Table 5.

Table 5 shows the size of ensemble before and after the soft set pruning algorithm. The soft set pruning algorithms for **DataSet1** pruned 3 classifiers which are {`c5, c8, c9`}. The actual size of the ensemble and the number of all possible combination of classifiers are significantly reduced from 1024 teams in 128, which is a 87.5% reduction.

Meanwhile the soft set pruning algorithms for **DataSet2** drops 5 classifiers which are {`c2, c5, c7, c8, c9`}. The actual size of the ensemble and the number of all possible combination of classifiers are significantly reduced by 96.9%, which drop from 1024 teams to only 32 teams.

Tables 6 and 7 shows some of the possible combinations of classifiers in the ensemble methods that produce the best prediction accuracy. Based on the experiment, we could conclude that the performance of the ensemble classifiers is better than single classifiers. Furthermore, the soft set pruning algorithm on ensembles produce the minimum number of classifiers team. The experimental result shows that the performance of the proposed soft set based pruning is as good as the full ensemble.

Table 6. DataSet1 - ensemble combination with the highest accuracy

Best team of ensemble classifiers	FULL ensemble or pruned	Prediction accuracy	Number of classifiers in the team ensemble
1000110000	Full [1024 teams]	0.75	3
1000010000	**Pruned** [128 teams]	0.75	**2**

Table 7. DataSet2 - ensemble combination with the highest accuracy

Best team of ensemble classifiers	FULL ensemble or pruned	Prediction accuracy	Number of classifiers in the team ensemble
1111110000	Full [1024 teams]	0.86	6
0011000001	**Pruned** [32 teams]	**0.86**	**3**

The best ensemble team for **DataSet1** only consists of two (2) classifiers; which is 1000010000 = {c1, c6}, while the **DataSet2** produce the best ensemble team of three (3) classifiers; which is 0011000001 = {c3, c4, c10}.

6 Conclusion

In this paper, a new customer churn prediction approach based on soft set ensemble pruning method is proposed. Heterogeneous ensemble is generated based on ten different classifier algorithms. It's acknowledged that the most significant advantage of soft set theory is its great ability of dimensionality reduction. Based on this soft set reduction algorithm, the ensemble is pruned and only a subset of the classifiers is considered prior to ensemble combination. From the experiment, we could claim that soft set ensemble pruning algorithm able to produce the highest customer churn prediction accuracy with the minimum number of classifiers. Nevertheless, there could be several directions to explore in the future works. One of our future works will be on discovering an algorithm for ensemble optimization and combination based on soft set theory.

Acknowledgement. This work is partially supported by UniSZA and KPT (Grant No. FRGS/2/2013/ICT07/UniSZA/02/2).

References

1. Dietterich, Thomas, G.: Ensemble methods in machine learning. In: Kittler, J., Roli, F. (eds.) MCS 2000. LNCS, vol. 1857, pp. 1–15. Springer, Heidelberg (2000). doi:10.1007/3-540-45014-9_1
2. Breiman, L.: Bagging predictors. Mach. Learn. **24**(2), 123–140 (1996)
3. Freund, Y., Schapire, R.E.: Experiments with a new boosting algorithm. In: ICML, vol. 96, pp. 148–156 (1996)
4. Breiman, L.: Stacked regressions. Mach. Learn. **24**(1), 49–64 (1996)
5. Wang, H., Fan, W., Yu, P.S., Han, J.: Mining concept-drifting data streams using ensemble classifiers. In: Proceedings of the Ninth ACM SIGKDD International Conference on Knowledge Discovery and Data Mining, ACM, pp. 226–235, August 2003
6. Rodriguez, J.J., Kuncheva, L.I., Alonso, C.J.: Rotation forest: a new classifier ensemble method. IEEE Trans. Pattern Anal. Mach. Intell. **28**(10), 1619–1630 (2006)
7. Caruana, R., Niculescu-Mizil, A., Crew, G., Ksikes, A: Ensemble selection from libraries of models. In: Proceedings of the Twenty-First International Conference on Machine Learning, p. 18. ACM, July 2004
8. Tsoumakas, G., Partalas, I., Vlahavas, I.: A taxonomy and short review of ensemble selection. In: Workshop on Supervised and Unsupervised Ensemble Methods and Their Applications, July 2008
9. Partalas, I., Tsoumakas, G., Katakis, I., Vlahavas, I.: Ensemble pruning using reinforcement learning. In: Antoniou, G., Potamias, G., Spyropoulos, C., Plexousakis, D. (eds.) SETN 2006. LNCS (LNAI), vol. 3955, pp. 301–310. Springer, Heidelberg (2006). doi:10.1007/11752912_31
10. Martinez-Muoz, G., Hernández-Lobato, D., Suarez, A.: An analysis of ensemble pruning techniques based on ordered aggregation. IEEE Trans. Pattern Anal. Mach. Intell. **31**(2), 245–259 (2009)
11. Caruana, R., Munson, A., Niculescu-Mizil, A.: Getting the most out of ensemble selection. In: Sixth International Conference on Data Mining, ICDM 2006, pp. 828–833. IEEE (2006)
12. Cruz, R.M., Sabourin, R., Cavalcanti, G.D., Ren, T.I.: META-DES: a dynamic ensemble selection framework using meta-learning. Pattern Recogn. **48**(5), 1925–1935 (2015)
13. Taghavi, Z.S., Sajedi, H.: Ensemble pruning based on oblivious Chained Tabu Searches. Int. J. Hybrid Intell. Syst. **12**(3), 131–143 (2016)
14. Fürnkranz, J., Widmer, G.: Incremental reduced error pruning. In: Proceedings of the 11th International Conference on Machine Learning (ML-1994), pp. 70–77 (1994)
15. Margineantu, D.D., Dietterich, T.G.: Pruning adaptive boosting. In: ICML, vol. 97, pp. 211–218, July 1997
16. Schapire, R.E., Singer, Y.: BoosTexter: a boosting-based system for text categorization. Mach. Learn. **39**(2), 135–168 (2000)
17. Strehl, A., Ghosh, J.: Cluster ensembles – a knowledge reuse framework for combining multiple partitions. J. Mach. Learn. Res. **3**, 583–617 (2003)
18. Topchy, A., Jain, A.K., Punch, W.: Clustering ensembles: models of consensus and weak partitions. IEEE Trans. Pattern Anal. Mach. Intell. **27**(12), 1866–1881 (2005)
19. Bakker, B., Heskes, T.: Clustering ensembles of neural network models. Neural Netw. **16**(2), 261–269 (2005)
20. Zhang, Y., Burer, S., Street, W.N.: Ensemble pruning via semi-definite programming. J. Mach. Learn. Res. **7**, 1315–1338 (2005)

21. Chen, H., Tino, P., Yao, X.: A probabilistic ensemble pruning algorithm. In: Sixth IEEE International Conference on Data Mining Workshops, ICDM Workshops 2006, pp. 878–882. IEEE (2006)
22. Molodtsov, D.: Soft set theory-first results. Comput. Math Appl. **37**(4), 19–31 (1999)
23. Maji, P.K., Biswas, R., Roy, A.: Soft set theory. Comput. Math Appl. **45**(4), 555–562 (2003)
24. Herawan, T., Deris, M.M.: A direct proof of every rough set is a soft set. In: 2009 Third Asia International Conference on Modelling and Simulation, pp. 119–124. IEEE, May 2009
25. Skowron, A., Rauszer, C.: The discernibility matrices and functions in information systems. In: Słowiński, R. (ed.) Intelligent Decision Support, pp. 331–362. Springer, Netherlands (1992)
26. Kong, Z., Gao, L., Wang, L., Li, S.: The normal parameter reduction of soft sets and its algorithm. Comput. Math Appl. **56**(12), 3029–3037 (2008)
27. Hall, M., Frank, E., Holmes, G., Pfahringer, B., Reutemann, P., Witten, I.H.: The WEKA data mining software: an update. ACM SIGKDD Explor. Newsl. **11**(1), 10–18 (2009)
28. Christopher, M., Peck, H.: Marketing Logistics. Routledge, London (2012)
29. Ismail, M.R., Awang, M.K., Rahman, M.N.A., Makhtar, M.: A multi-layer perceptron approach for customer churn prediction. Int. J. Multimed. Ubiquit. Eng. **10**(7), 213–222 (2015)
30. Awang, M.K., Rahman, M.N.A., Ismail, M.R.: Data mining for churn prediction: multiple regressions approach. In: Kim, T.-h., Ma, J., Fang, W.-c., Zhang, Y., Cuzzocrea, A. (eds.) FGIT 2012. CCIS, vol. 352, pp. 318–324. Springer, Heidelberg (2012). doi:10.1007/978-3-642-35603-2_47
31. Gower, J.C.: Properties of Euclidean and non-Euclidean distance matrices. Linear Algebra Appl. **67**, 81–97 (1985)
32. Awang, M.K., Makhtar, M., Rahman, M.N.A., Deris, M.M.: A new soft set based pruning algorithm for ensemble method. J. Theor. Appl. Inf. Technol. **88**(3), 384–391 (2016)

An Association Rule Mining Approach in Predicting Flood Areas

Mokhairi Makhtar(✉), Nur Ashikin Harun, Azwa Abd Aziz, Zahrahtul Amani Zakaria, Fadzli Syed Abdullah, and Julaily Aida Jusoh

Faculty of Informatics and Computing, Universiti Sultan Zainal Abidin, Tembila Campus, 22200 Besut, Terengganu, Malaysia
{mokhairi, azwaaziz, zahrahtulamani, fadzlihasan, julaily}@unisza.edu.my, shikin_harun91@yahoo.com

Abstract. This study focuses on the application of Association rules mining for the flood data in Terengganu. Flood is one of the natural disasters that happens every year during the monsoon season and causes damage towards people, infrastructure and the environment. This paper aimed to find the correlation between water level and flood area in developing a model to predict flood. Malaysian Drainage and Irrigation Department supplied the dataset which were the flood area, water level and rainfall data. The association rules mining technique will generate the best rules from the dataset by using Apriori algorithm which had been applied to find the frequent itemsets. Consequently, by using the Apriori algorithm, it generated the 10 best rules with 100% confidence level and 40% minimum support after the candidate generation and pruning technique. The results of this research showed the usability of data mining in this field and can help to give early warning towards potential victims and spare some time in saving lives and properties.

Keywords: Data mining · Association rule · Apriori algorithm · Flood disaster

1 Introduction

Data mining is the processes of knowledge discovery where patterns, correlations and trends in data will be discovered; hence, the new knowledge will help in developing new ideas and solutions. In fact, the purpose of data mining is to expedite understanding in vast volumes of data by ascertaining interesting regularities or exceptions [1, 8]. The size of database with the data stored expands through the passage of time, where different industries and enterprises do not have any idea that data actually can evolve into a boastful decision-making. For example, market basket analysis aims to find interesting rules that have the minimum transactional support and minimum confidence [2] in which the analysis can lead to the applications of association rules in many fields.

Associations rules are widely used in many kind of areas, ranging from healthcare [6, 11], sports [12], biomechanical [14], manufacturing industry [7], and hydrological [3, 4]. Data mining when applied in a domain of natural disaster can help and aid in disaster control measures. In a tropical cyclone development, the roles of multiple

T. Herawan et al. (eds.), *Recent Advances on Soft Computing and Data Mining*, Advances in Intelligent Systems and Computing 549, DOI 10.1007/978-3-319-51281-5_44

associated physical processes help in attaining the solution, besides improving intensity forecasting [15]. Moreover, in using data mining techniques, the disaster recovery planning and management have been developed in a collaboration of public and private sectors, where the information retrieval help communities better apprehend the present disaster circumstances and how to act for it [16]. In fact, forest fires can also create ecological and economical destruction. By using data mining methods and meteorological data, they are capable in predicting the burned area and the model is also useful in improving the firefighting resource management [5].

This study was conducted in Terengganu because it obligates flood problems during the North East monsoon between Novembers till March every year. When flood happens in a village, other surrounding villages can actually get an early warning in order to get ready for any circumstances. However, the existing warning that has been established is unreachable and too sudden. Therefore, in order to seek which village is associated to another village when flood happens, association rules algorithm has been adopted for hydrological data. The association rules are normally applied in the transaction of a supermarket; then, the rules successfully made the sales rise rapidly. Based on this motivation, the association rules are used in predicting flood area to find the association between places. In addition, this research aimed to identify the correlation between river flow and flood area during the monsoon season. Besides, this study will design a model to predict the flood area in Terengganu and also to implement association rule mining for flood area prediction. The flood data in Terengganu had not been tested with the data mining techniques to predict flood area yet. Therefore, this study will help to create a model for improving flood management; hence, the damage can be reduced and lives can be saved if there is early warning for potential victims.

2 Materials and Methods

This section explains the process of the research and also the data used and the algorithm that had been applied.

2.1 Research Workflow

Knowledge Discovery in Database (KDD) process is identifying valid, novel and potential pattern from vast sets of data. The core process in KDD is the data mining, where algorithm is applied to extract many valuable information.

In phase 1, data was selected from the database. This research collected secondary data from the Malaysian Drainage and Irrigation Department. The department supplied the dataset for rainfall data and water level data from 2009 until 2015. The data that had been selected were for three months, which were November, December and January. Besides, the records were also collected from the department's report yearly. Then, from the raw data, the target data that had been selected and undergone the pre-processing step. At this phase, the data had been cleaned from noise, redundancy and any incompleteness. The transformation phase ensured the reliability and accuracy of datasets. Then, for the core phase, which was data mining, was the analytical process. In this paper, association rules, which was the Apriori Algorithm, were applied in order to generate vital rules and patterns. Then, the interpretation of the results will end with knowledge discovery.

2.2 Apriori Algorithm

The Apriori Algorithm is a prominent algorithm for mining frequent itemsets for boolean association rules. It is deliberated to be functional on a transaction database in order to ascertain the pattern in transaction made by customers in a store [13]. The flow of Apriori Algorithm starts with scanning transaction the database to get the support (S) of each *I*-itemset, and *S* will be compared with minimal support (min_sup) and will receive a set of frequent *I*-itemset, which is denoted as *Lk*. It will use the Apriori property to prune the unfrequent k-itemset from the set. The candidate set will be checked, either it is equal to NULL or not. If not, the process continues by generating the next candidate sets. If it is equal to NULL, the process stops.

Figure 1 shows the pseudocode of the Apriori Algorithm, where the originality of the algorithm is applied in supermarket transactions [2].

1: $k = 1$.
2: $F_k = \{\, i \mid i \in I \wedge \sigma(\{i\}) \geq N \times minsup\}$. {**Find all frequent 1-itemsets**}
3: **repeat**
4: $\quad k = k + 1$.
5: $\quad C_k =$ apriori-gen(F_{k-1}). {**Generate candidate itemsets**}
6: $\quad$ **for each transaction** $t \in T$ **do**
7: $\quad\quad C_t =$ subset(C_k, t). {**Identify all candidates that belong to** t}
8: $\quad\quad$ **for each candidate itemset** $c \in C_t$ **do**
9: $\quad\quad\quad \sigma(c) = \sigma(c) + 1$. {**Increment support count**}
10: $\quad\quad$ **end for**
11: $\quad$ **end for**
12: $\quad F_k = \{\, c \mid c \in C_k \wedge \sigma(c) \geq N \times minsup\}$. {**Extract the frequent** k**-itemsets**}
13: **until** $F_k = \emptyset$
14: Result $= \bigcup F_k$.

Fig. 1. Pseudocode of Apriori algorithm

The association rule generation was applies in two steps [13]; first, minimum support was applied to find all frequent itemsets in database. Second, the frequent itemset and the minimum confidence constraint were used to form rules.

3 Results and Discussion

For our test, we considered 27 attributes which were the villages in the Marang district in Terengganu. From the pre-processing process, only data that correlated with the selected water level were considered as transaction sets. Table 1 presents an example of the information regarding the area for a district that had flood. In the Marang District of Terengganu, there are 27 villages. The name of the village was considered as the items. This particular village will have flood during the monsoon seasons. Table 1 shows the list of itemsets. For this experiment, the Apriori Algorithm generated strong rules from the frequent itemsets. We set up the minimum support which was 40%, and the minimum confident was 90%.

Table 1. The list of villages (items)

Item_ID	Item name	Denoted as
1	KG. SERATING	A
2	KG. PULAU RUSA	B
3	KG. ALOR WAN SYED	C
4	KG. PADANG LEBAN	D
5	KG. TANJUNG MELOR	E
6	KG. TEMALA	F
7	KG. BUKIT KECIL	G
8	KG. LAMA	H
9	KG. SEBERANG MARANG	I
10	KG. GONG BALAI	J
11	KG. GONG NANGKA	K
12	TAMAN MAS	L
13	KG. PAYA	M
14	KG. KUBU	N
15	KG. SELAWA	O
16	TUN TELANAI	P
17	TAMAN TASIK	Q
18	TAMAN BUKIT PAYONG	R
19	KG. BATANGAN	S
20	KG. GONG BERIS	T
21	KG. PASIR PUTIH	U
22	KG. BARAT	V
23	KG. PENGKALAN BERANGAN	W
24	KG. JENANG	X
25	KG. BUKIT GASING	Y
26	KG. WAKAF TAPAI	Z

This research aimed to find the association rules that satisfied the requirements of 40% support and 90% confidence. Table 2 shows 9 transactions which were based on the items from Table 1.

To find the frequent set f L_1, we counted how many times the village appeared in the transaction or we called it as frequency like in Table 3.

The support that we set earlier was 40%, so, it diminished the items that were less than 40% of their support. Table 4 showed large itemset for L_1 after pruning the itemset which less that minimum support.

Based on L_1, we now can form candidate pairs or C_2 by joining the items. Table 5 indicates the joining of items and its frequency which then we calculate the supports.

Table 2. Transaction storage

TID	Items
1	A, B, C, D, E, F, G, H, I, J, K, L, M, N
2	C, G, H, I, J, K, O, P, Q, R, S, T
3	C, D, G, H, I, J, K, M, O, R, S, T, U, V
4	J, U, W, X, Y
5	J, U, W, X, Y
6	J, U, W, X, Y
7	C, F, N, U, X, Z
8	C, F, N, U, X, Z
9	C, F, N, U, X, Z

Table 3. Computing L_1

Item_ID	Item name	Frequency	Support(X) (Number of time X appears)/N(no. of transactions) = P (X)
1	A	1	0.11
2	B	1	0.11
3	C	6	0.67
4	D	2	0.22
5	E	1	0.11
6	F	4	0.44
7	G	3	0.33
8	H	3	0.33
9	I	3	0.33
10	J	6	0.67
11	K	3	0.33
12	L	1	0.11
13	M	2	0.22
14	N	4	0.44
15	O	2	0.22
16	P	1	0.11
17	Q	1	0.11
18	R	2	0.22
19	S	2	0.22
20	T	2	0.22
21	U	7	0.77
22	V	1	0.11
23	W	3	0.33
24	X	6	0.66
25	Y	3	0.33
26	Z	3	0.33

Table 4. The frequent 1-itemset of L_1

Item_ID	Item name	Frequency	Support(X) (Number of time X appears)/N(no. of transations) = P (X)
3	C	6	0.67
6	F	4	0.44
10	J	6	0.67
14	N	4	0.44
21	U	7	0.77
24	X	6	0.66

Table 5. The candidate 2-itemsets or C_2

C2	Frequency	Support(X) (Number of time X appears)/N(no. of transactions) = P (X)
{C,F}	4	0.44
{C,J}	3	0.33
{C,N}	4	0.44
{C,U}	4	0.44
{C,X}	3	0.33
{F,J}	1	0.11
{F,N}	4	0.44
{F,U}	3	0.33
{F,X}	3	0.33
{J,N}	1	0.11
{J,U}	4	0.44
{J,X}	3	0.33
{N,U}	3	0.33
{N,X}	3	0.33
{U,X}	6	0.67

Table 6 presented the process of L_2 after the next process of joining of the sets. Meanwhile Table 7 displayed the process of pruning from L_2 to C_3. There were three itemset which satisfied the minimum threshold.

Based on Table 8, the itemset of L_3 which is {C,F,N} was the only member of L_3. When we generated C4 with L3, it turned out to be empty; therefore, the process was terminated. The support of these 6 rules was 4.

{C,F,N}

C→FN FN→C
F→CN CN→F
N→CF CF→N

According to Table 9, the confidence of all these rules, except the first one which was 67%, was 100%; so, the last five association rules qualified as the best rules. Rules

Table 6. Computing L_2

L2	Frequency	Support(X) (Number of time X appears)/N(no. of transactions) = P (X)
{C,F}	4	0.44
{C,N}	4	0.44
{C,U}	4	0.44
{F,N}	4	0.44
{J,U}	4	0.44
{U,X}	6	0.67

Table 7. Computing C_3

C3	Frequency	Support(X) (Number of time X appears)/N(no. of transactions) = P (X)
C,F,N	4	0.44
C,F,U	3	0.33
J,U,X	3	0.33

Table 8. Computing L_3

L3	Frequency	Support(X) (Number of time X appears)/N(no. of transactions) = P (X)
C,F,N	4	0.44

Table 9. Computing C_3

C3	Support of CFN	Frequency of LHS	Confidence
C → FN	4	6	0.67
F → CN	4	4	1
N → CF	4	4	1
FN → C	4	4	1
CN → F	4	4	1
CF → N	4	4	1

L_2 which were not included in L_3 should be checked first, so that any rules which did not overlap will be also qualified as the best rule.

The full list of all the rules are shown in the Table 10 below.

The summary of the association rules generated for Marang District is shown in Fig. 2.

Figure 2 shows the best rule generated from the datasets. There were two main steps of this algorithm, which were the join step and the pruning step. The join step created the generation of the candidates, Ck (C2 and C3) from the Lk-1 (L1, L2, L3). Meanwhile, the pruning step will select the frequent itemsets from the Lk.

Association rule has been applied in many field these recent years. The major challenge in mining data is the efficacy of the mining results. Usually, the produce results are either well-known already to the science community or are problematic to interpret and use [15]. Nevertheless, most of the researches that were carried out

Table 10. The best rules generated

All association rules	Denoted as
KG. JENANG → KG. PASIR PUTIH	X → U
KG. ALOR WAN SYED → KG. TEMALA	C → F
KG. ALOR WAN SYED → KG. KUBU	C → N
KG. KUBU → KG. TEMALA	N → F
KG. TEMALA → KG. KUBU	F → N
KG. ALOR WAN SYED → KG.TEMALA, KG. KUBU	C → FN
KG. TEMALA → KG. ALOR WAN SYED, KG,KUBU	F → CN
KG. KUBU → KG. ALOR WAN SYED, KG.TEMALA	N → CF
KG. TEMALA,KG. KUBU → KG. ALOR WAN SYED	FN → C
KG. ALOR WAN SYED,KG. KUBU → KG. TEMALA	CN → F
KG. ALOR WAN SYED, KG. TEMALA → KG. KUBU	CF → N

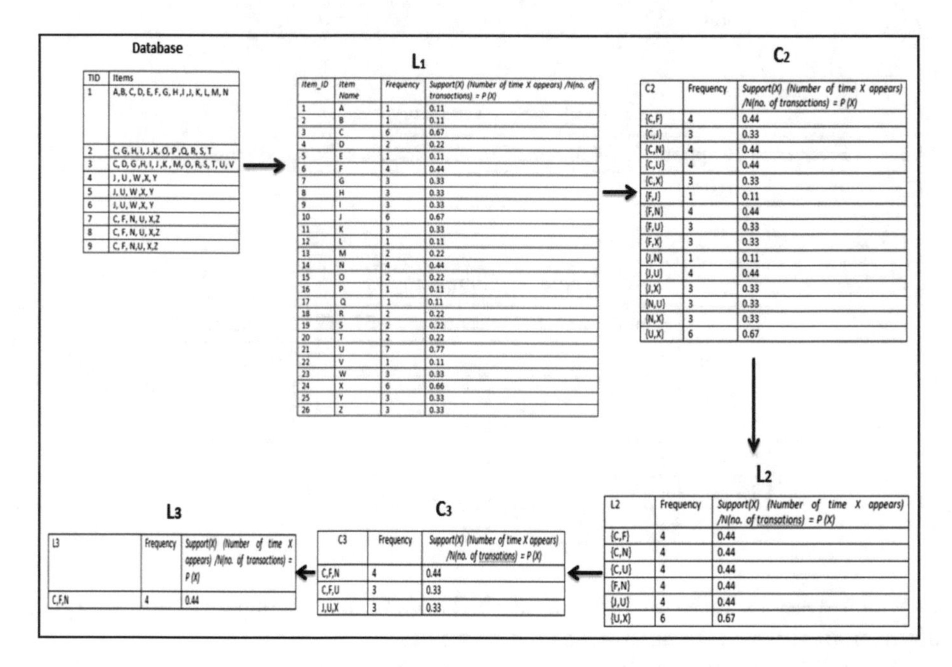

Fig. 2. Flow of generating rules based on the Apriori algorithm

produced such great results and that were comprehensive [1, 3, 7]. In fact, in this study, apriori algorithm had been used and it generated frequent itemsets where the correlation of the attributes was created [2, 10]. The results indicated that the association among vilages can be used as an early warning in the area. Although the rules generated were vital, in order to generate the best rule, the minimum threshold needed to be satisfied, where it will sometimes generate unnecessary itemsets. Therefore, the selection of support and confidence level are very crucial and tactical. In fact, mining the least itemsets which is a practice of finding the rare set of item may also divulge a valuable knowledge, for example, in educational data [1] and many more. As a matter of fact, there are lots of other researches that can be done; finding frequent itemsets is probably more useful and precise, such as applying it to the data streams which discard the non-frequent set and obtain frequent itemsets [9], finding hidden patterns in weather datasets [3], increasing the effectiveness of response and efficiency of time and cost for the wooden door industry [7] and many more.

4 Conclusion

In this paper, we presented an approach by applying Apriori Algorithm to predict flood area. The 10 best rules were generated and the rule of Kg. Jenang implied that Kg. Pasir Putih received 100% confidence and 40% support. This algorithm is usually applied in market transaction, but this research revealed that the application of association rules in disaster management is a powerful tool, which is predicting flood and thus avoiding adverse impact in risk areas. The future works should consider integrating more variables to make a better and more precise prediction.

Acknowledgements. The presented work was funded by the Ministry of Higher Education Malaysia under the Research Acculturation Grant Scheme (RAGS) reference code RR095 and UniSZA. The authors would like to thank the Malaysian Drainage and Irrigation Department for supplying the data of flood in Terengganu and to all those who had participated in this research.

References

1. Abdullah, Z., Herawan, T., Noraziah, A., Deris, M.M.: Mining least association rules of degree level programs selected by students. Int. J. Multimedia Ubiquit. Eng. **9**(1), 241–253 (2014)
2. Agrawal, R., Imielinski, T., Swami, A.: Mining association in large databases. In: Proceedings of 1993 ACM SIGMOD International Conference Management Data - SIGMOD 1993, pp. 207–216 (1993)
3. Athiyaman, B., Sahu, R.: Hybrid data mining algorithm: an application to weather data. J. Indian Res. **1**(4), 71–83 (2013)
4. Aziz, A.A., Harun, N., Makhtar, M., Hassan, F., Jusoh, J.A., Zakaria, Z.A.: A conceptual framework for predicting flood area in Terengganu during monsoon season using association rules. J. Theor. Appl. Inf. Technol. **87**(3), 512–519 (2016)
5. Cortez, P., Morais, A.: A data mining approach to predict forest fires using meteorological data. In: Neves, M.F.S.J., Machado, J. (eds.) New Trends in Artificial Intelligence, Proceedings of the 13th EPIA 2007, Portuguese Conference on Artificial Intelligence, Guimarães, Portugal, pp. 512–523 (2007)
6. Cremaschi, P., Carriero, R., Astrologo, S., Coli, C., Lisa, A., Parolo, S., Bione, S.: An association rule mining approach to discover lncRNAs expression patterns in cancer datasets. Biomed Res. Int. 146250 (2015)
7. Djatna, T., Alitu, I.M.: An application of association rule mining in total productive maintenance strategy: an analysis and modelling in wooden door manufacturing industry. Procedia Manuf. **4**, 336–343 (2015)

8. Klemettinen, M., Mannila, H.: Finding interesting rules from large sets of discovered association rules. In: Proceedings of Third International Conference Information Knowledge Management, pp. 401–407 (1994)
9. Nath, N.D., Meena, M.J., Syed Ibrahim, S.P.: Mining frequent itemsets in real time. In: Vijayakumar, V., Neelanarayanan, V. (eds.) ISBCC – 2016. SIST, vol. 49, pp. 325–334. Springer, Heidelberg (2016)
10. Agrawal, R., Srikant, R.: Fast algorithms for mining association rules. Ann. Pharmacother. **42**(1), 62–70 (2008)
11. Rameshkumar, K., Sambath, M., Ravi, S.: Relevant association rule mining from medical dataset using new irrelevant rule elimination technique. In: 2013 International Conference on Information Communication and Embedded Systems, ICICES 2013, pp. 300–304 (2013)
12. Sun, J., Yu, W., Zhao, H.: Study of association rule mining on technical action of ball games. In: 2010 International Conference Meas Technology Mechatronics Automation ICMTMA 2010, vol. 3, pp. 539–542 (2010)
13. Tanna, P., Ghodasara, Y.: Using Apriori with WEKA for frequent pattern mining. arXiv Preparation arXiv1406.7371, vol. 12, no. 3, pp. 127–131 (2014)
14. Vannozzi, G., Croce, U.D., Starita, A., Benvenuti, F., Capozzo, A.: Knowledge discovery in databases of biomechanical variables: application to the sit to stand motor task. J. Neuroeng. Rehabil. **1**(7), 1–10 (2004)
15. Yang, R., Tang, J., Sun, D.: Association rule data mining applications for Atlantic tropical cyclone intensity changes. Weather Forecast. **26**(3), 17 (2011)
16. Zheng, L., Shen, C., Tang, L., Zeng, C., Li, T., Luis, S., Chen, S.C.: Data mining meets the needs of disaster information management. IEEE Trans. Hum.-Mach. Syst. **43**(5), 451–464 (2013)

The Reconstructed Heterogeneity to Enhance Ensemble Neural Network for Large Data

Mumtazimah Mohamad(✉), Mokhairi Makhtar, and Mohd Nordin Abd Rahman

Faculty Informatics and Computing, University Sultan Zainal Abidin, Besut Campus, 22200 Besut, Terengganu, Malaysia
{mumtaz,mokhairi,mohdnabd}@unisza.edu.my

Abstract. This paper present an enhanced approach for ensemble multi classifier of Artificial Neural Networks (ANN). The motivation of this study is to enhance the ANN capability and performance using reconstructed heterogeneous if the homogenous classifiers are deployed. The clusters set are partitioned into two sets of cluster; clusters of a same class and clusters of multi class which both of them were using different partition techniques. Each partitions represented by an independent classifier of highly correlated patterns from different classes. Each set of clusters are compared and the final decision is voted by using majority voting. The approach is tested on benchmark large dataset and small dataset. The results show that the proposed approach achieved almost near to 99% of accuracy which is better classification than the existing approach.

Keywords: Multiple classifier · Diversity · Parallel heterogeneity · Ensemble · Neural network

1 Introduction

In the recent years, the has been a substantial computing power that has led to high dimensional data structures, either a high number of samples or the number of features or high number of classes. The enormous size of this kind of data involves very challenging problem especially in recognition and classification task. A large number of datasets can cause the training procedure unfeasible for many types of classifiers. Furthermore, it also affected the complexity of a classification process. Applications such as handwritten character recognition or any other problem containing thousands of different classes still remain a challenge [1]. The challenge of training a classifier with a large number of samples lies in the computational complexity of the training task. This is because the variability of the categories is high and several examples from each category are necessary for a correct data representation. The methods and the current methodology still have issues with higher time [2, 3]. Some methods can only deal with categorical attributes and some only deals with numerical [4]. Most methods do not use all the important features that are related to the data. Moreover, data elimination cannot be used, as every category's samples must be used for complete training.

In addition, the enhancement in recognition method with high efficiency and accuracy to support high dimensional data access has become the choice of researchers

T. Herawan et al. (eds.), *Recent Advances on Soft Computing and Data Mining*,
Advances in Intelligent Systems and Computing 549, DOI 10.1007/978-3-319-51281-5_45

[5]. The challenge of training a classifier with a large number of samples lies in the computational complexity of the training task. This is because the variability of the categories is high and several examples from each category are necessary for a correct data representation. The methods and the current methodology have the disadvantage of higher time [6]. Some methods can only deal with categorical attributes and some only deals with numerical [4]. Most methods do not use all the important features that are related to the data. Moreover, data elimination cannot be used, as every category's samples must be used for complete training.

In this paper, the discussion focused on model of heterogeneity of a reconstructed ensemble multi classifier of neural network. The model has been implemented and tested on large data and performance of the classifier.

2 Related Works

The large datasets classification using Artificial Neural Network (ANN) is rarely being discussed. It might due to available modern techniques that seem to have complicated algorithm and not tested to the real environment. However, research by [7] shows a bright side on large dataset in ANN classification tasks after many other researcher ignoring the universal ANN approximation. ANN first known as 'connectionist model' emerged from the introduction of simplified neuron by McCulloch and Pitts in 1943. These neurons presents as biological model that attempt to mimic cognitive capabilities of human beings. ANN has the ability to learn complex nonlinear input-output relationships for sufficiently large data in training times that scales linearly with data size [8]. In addition, an enhancement alternative is encouraged in ANN that transform a traditional optimization algorithms to complex classification tasks that can be formulated as optimization problems [9]. Other method such as SVM is not suitable for large datasets classification because it needs to solve the quadratic programming problem in order to find a separation hyper-plane, which causes an intensive computational complexity [2].

The use of a single ANN usually for large datasets leads to unstable learner and it is sensitive to the initial conditions, however works differently for different training data [10]. Therefore, an ensemble technique is practically employed in order to preserve the capability of ANN classifiers. Previous researchers found the performance of multiple classifier can be better than their base model since individual ANN tends to make errors on different examples [11]. It is a significant research challenge to develop ensemble heterogeneous of the different set of instances to get better classification performance. It is because when different areas of input spaces have been learned by some classifiers, they become specialized in specific areas of the input spaces, and consequently have fewer errors in those areas.

The existing ensemble algorithm designed for single source cannot be applied directly to the result of a different heterogeneous schema as they often vary greatly from traditional single classifier [12]. Since ANN observes the target data perceptually, this will suggest that there are learning algorithms that imposed requirement scale for computing time approximately and linearly with total volume of data.

3 Proposed ANN Ensemble Classifiers on Clusters

3.1 Reconstructed Clusters and Ensemble Classifier

In our case, the diversity of classifier is generated using the diverse errors on different samples strategy [13]. This is to make sure the classifier has the ability to find the extent of heterogeneity among classifiers and estimate the improvement or deterioration inaccuracy of individual classifiers when they have been combined [14]. The proposed ensemble classifier creation approach is based on the notion of heterogeneity of the classifier and its notion of clustering. The objective is to partition the data set into multiple clusters and deploy a different set of classifier to learn the decision boundaries from all clusters and within the clusters.

The process of designing an ensemble strategies consists of two main steps [15]. The first step is related to data set partitioning process into two categories of layer. A first set of clusters composed of patterns of a same class of geometrically close and the second set of clusters with pattern of multiple classes. The first set of cluster $C_{1,1}$ to $C_{1,n}$ is an atomic cluster that is composed of patterns of a same class while other clusters $(C_{2,1}\ to\ C_{2,n})$ composed of patterns from multiple classes. First set of clustering is used to partitioned data with its classes into geometrically close. Both clustering set are presented based on distributed reordering techniques. The distributed reordering technique alters the sequence of pattern order in certain training in each cluster by deploying the pattern neighbour samples are not independent in sequence. This can be meant that first pattern will not affect the second samples. In this case, the co-variance between two samples is not zero and it will complicate the computations. All clusters network is randomly reordered at the beginning for each classifier cycle with the probability of clusters of $\frac{C_i 1}{N}, i \in [1, \ldots, N]$.

Classifiers are trained dependently in first clustering set and independently in the second clustering set. Both sets will overlap where the same patterns are included in the training of multiple classifiers. The test patterns also belong to different clusters at both categories and this achieving heterogeneity rules. We deploy the idea of creating ensemble and fusing their decisions. The contribution of the proposed approach lies in introduction of different fusing at each level in achieving heterogeneity.

3.2 Theoretical Modeling

Let h_i be the ith classifier on an ensemble of L classifiers, $S = (x_1, y_1), (x_2, y_2), \ldots, (x_N, y_N)$ the training data set, S_i a version of the available dataset where each pattern is described by a vector of n continuous valued features $x_j = < x_{j1}, x_{j2}, \ldots x_{jn} >$ and a class label y_j with $y_j \in \{class_1, class_2, \ldots class_{Nclass}\}$. Set of cluster is denoted by ι and K clusters at level ι are denoted by $class_1, c_{l,1}, c_{l,2}, \ldots c_{l,k}$ where $1 \le l \le N_{sets}$. $H(x)$ a function that combines/selects the decisions of various learners about a new input pattern x.

A pattern in training set can be considered as a point in K-Mean space of dimension n in order to cluster data point into groups that are geometrically similarly close. This algorithm aims at minimizing an objective function, in this case a squared error function. The objective function:

$$J = \sum_{j=1}^{k} \sum_{i=1}^{n} \left\| x_i^j - C_j \right\|^2 \quad (1)$$

where $\left\| x_i^j - C_j \right\|^2$ is a chosen distance measure between a data point and the cluster centre cj, is an indicator of the distance of the n data points from their respective cluster centres. The other set of cluster is denoted as given *k* clusters from *K* classifiers and the random weight initialization for each of cluster using sampling without replacement with the probability of cluster $\frac{K_i 1}{N}, i \in [1, \ldots, N]$. The multiclass set of cluster with x = <x_{i1}, x_{i2}, ... x_{in}> that drawn randomly using probability *m*. A set of K clusters $\{\Omega_{l1}, \Omega_{l2}, \ldots \Omega_{lk},\}$ at set cluster *l*, with the associated cluster centre $Y_l = \{k_{l1}, k_{l2}, \ldots k_{lk},\}$ are initialized using stochastic learning and the clustering algorithm aims to minimize ANN objective function.

Consider a ANN function learning setup where each example z is a pair (x; y) composed of an arbitrary input x and a scalar output y. The loss function l(ŷ; y) that measures the cost of predicting ŷ when the actual answer is y. The F group functions fw(x) parametrized by a weight vector w. Function f ∈ F shall minimizes the loss Q(z; w) = l(fw(x), y) averaged on the examples. The calculation for computation of $z_1 \ldots z_n$ for an unknown distribution dP(z) is based on:

$$E(f) = \ell((f_x), y) dP(z) E_n(f) = \frac{1}{n} \sum_{x=i}^{n} \ell(f(x_i), y_i) \quad (2)$$

The error risk *En*(*f*) measures the training set performance. The expected risk *E*(*f*) measures the generalization performance and the expected performance on future examples. It has often been proposed to minimize the error risk *En*(*fw*) using gradient descent. Each iteration updates the weights w on the basis of the gradient of $E_n(f_w)$),

$$w_t + 1 = w_t - \gamma \frac{1}{n} \sum_{r=1}^{n} \nabla w Q(z_i w_i) \quad (3)$$

where γ is an chosen gain value. Under sufficient regularity assumptions, when the initial estimate w0 is close enough to the optimum, and when the gain γ is sufficiently small, this algorithm achieves linear convergence. *Q* is surrogate loss of the labeled example $(z_i w_i)$.

First cluster process each pattern (x_1, y_1) belong to a cluster $C_{l,k}$, where $1 \le k \le K$. The first set cluster comprised of single class clusters while the second set is multiclass set clusters. The class of first set of clusters is saved. The distributed portioning deployed by second set of clusters generates well distributed of multiclass clusters. After that, Back Propagation ANN is set up to for the second set of clusters in order to learn the decision boundaries on its pattern. The weight matrices is initialized by randomly select pattern sequence by using stochastic methods where single data point from the portioned data and compute error function to the direction of the gradient with

Table 1. Learning algorithm

```
Begin
Set k clusters, L number of set, l=1, weight random
for all class, l<2
Generate cluster using K-mean c_{l,1}, c_{l,2}, ... c_{l,k}
   Save weights and configuration.
end for
for t=1 to J
    Generate C_{l,k} using sampling without replacement
Draw X_i from C_{l,k} using probability Pthreads
  for i=1 to N pattern
Tr_i = x_i (i, allColumns)= C(randRow, allColumns)
      Adjust probabilities P
      end for
   end for
Train ANN
Compare result and adjust weight of configuration of c_{l,1}
End
Vote the best cluster.
Test
```

Table 2. Data set and ANN parameters

Data sets	N	BPNN configuration							
		Inputs	Outputs	Hidden nodes	Hidden layer number	Weights	Learning rate	Momentum	Square error
Iris	150	4	1	5	1	761	0.01	0.8	0.001
MNIST	60000	784	1	30	1	23581	0.01	0.8	0.001

respect to that data point. The learning and prediction model that has been described is presented in Table 1.

A test pattern x is classified by finding the appropriate cluster at each set of clusters. The distance measure between x and the center of K is computed using (1) and the appropriate set of clusters at l is selected as P_l at set of clusters l. During prediction P_l from N_{sets} sets, the decisions are fused into a final majority voting selection rule. The experiment has been conducted upon several numbers on benchmark datasets to verify the capability of proposed heterogeneity of ensemble classifier. The dataset is compiled from National Institute Science & Technology and from UCI Iris. A summary data is presented in Table 2. The dataset have been pre-processed using data normalization and de-normalization and then divided into training set, testing set and validation set. Note that the objective of the experiment was to improve the performance of classifier on ensemble learning, therefore, the clusters are restricted to parameter settings that perform best on all data found by trial and error basis.

4 Result and Discussion

The experimental results are presented here to demonstrate the performance and the effectiveness of the proposed approach. The cluster set created is based on the heterogeneity concept deploy for ensemble network where changes in cluster content at different initialization and pattern selection. The classification result on benchmark data sets of a proposed ensemble classifiers are compared against the large and small data sets. Figure 1 represents the class cluster in co-occurrences matrices for cluster set of a same class for dataset Iris and MNIST (Fig. 2). The first figure and the second figure for each dataset were different for each class. It is because the content of the cluster is changes in both cases as the weight are initialized randomly. Each cluster has an identical patterns showed difference learning that make errors that uncorrelated.

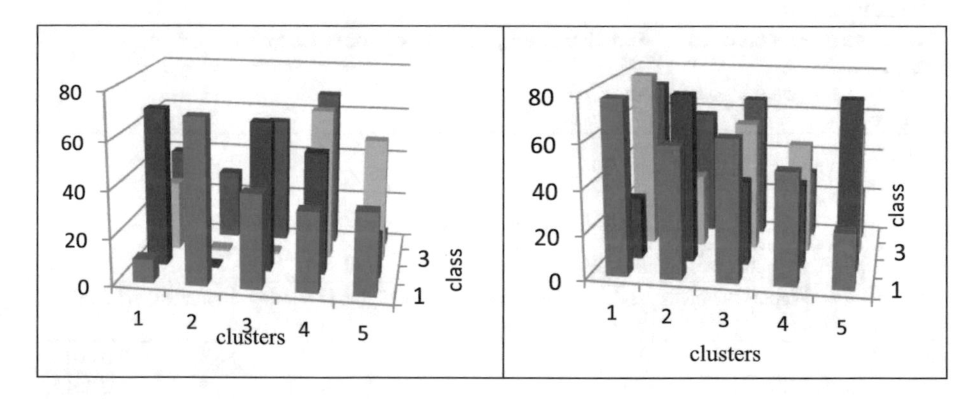

Fig. 1. Different clustering set of Iris datasets

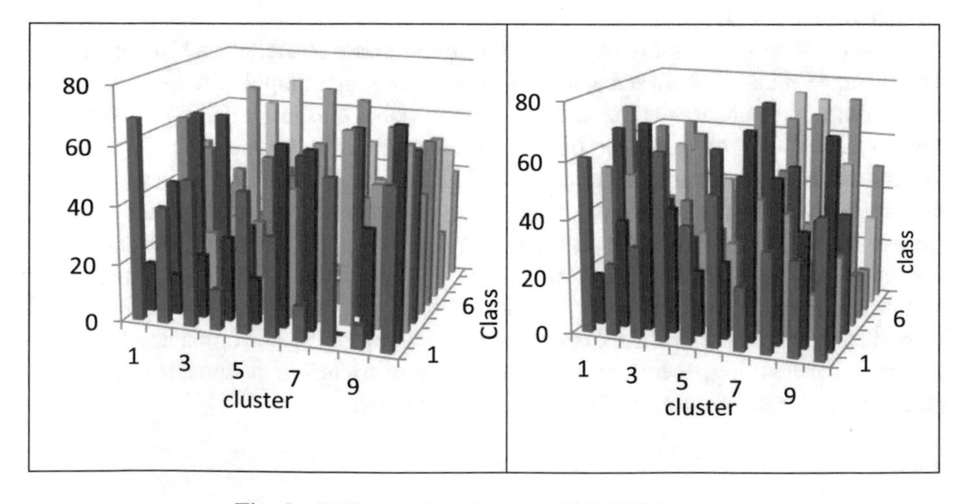

Fig. 2. Different clustering set of MNIST datasets

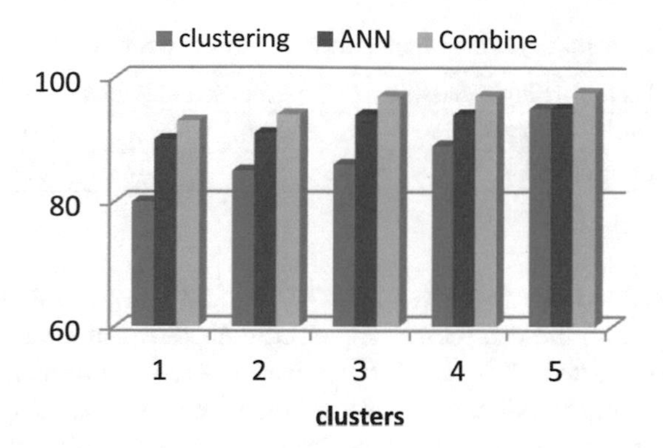

Fig. 3. Performance accuracy for Iris dataset

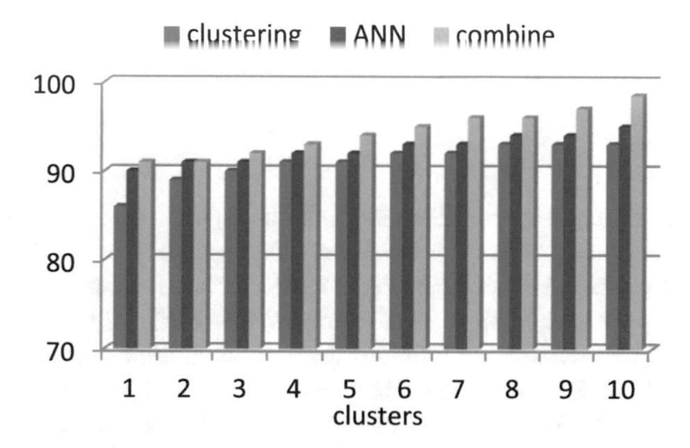

Fig. 4. Performance accuracy for MNIST dataset

Figure 3 represents the class the classification performance achieved for first set of clusters, second set and clusters and combined clusters for Iris dataset, as well as in Fig. 4 for MNIST dataset. All of the clusters were selected using majority voting. The graph in Fig. 3 indicated that proposed ensemble multi classifier improved a significant accuracy from classifiers that use only clustering or ANN method. Note that, this an effect of heterogeneity element for cross comparison between clusters with same classes and cluster with different classes.

Figure 4 shows the major improvement of element of heterogeneity in each decision boundary of clusters of a similar class and clusters with different classes. It is found that, the large dimension of data takes good advantages in heterogeneity where it trained on identical classes and mix with the ANN training that lead to higher classification accuracy.

Table 3 shows the standard deviation of the classification of the beast accuracy for clusters of classifier as shown in Fig. 4. In general, the best reconstructed heterogeneity gave better classification accuracy. The variation for Iris data improved from 1.81 to 2.10 for clusters set of consider both clustering and ANN. Dataset MNIST shows the

Table 3. Standard deviation of classification accuracy for small and large data

Dataset	Single classifier	Std	Proposed classifier	Std
Iris	80.2	1.81	97.6	2.10
MNIST	85.4	4.2	98.5	5.33

opportunity of best variation improved from 4.2 to 5.33 point of standard of deviation. This is regarding to all the dataset has improved their decision boundary then classify better for many of clusters. It is found that it is easy to learn decision boundaries with clustering and multiple classifier of ANN are given the clustered boundaries that learns other set cluster that similarly close of a same classes in addition to their reordering technique to generate heterogeneity elements. For a smaller data set; the proposed technique shows positive impact in accuracy and variation. This result is somewhat similar to what have found in [12] where an integrated the trained component classifiers shows significant enhancement to the performance.

5 Conclusion

In this paper, an improved approach towards creating and training of an ensemble multi classifier. The proposed approaches strengthen the classifiers of a same class on different partition by using k-mean clustering. In order to have better decision, two group of clusters created represented as weak clusters and ANN-trained clusters. The decision boundaries is robust because the multiple decision is obtained, hence improved the overall performance. The result shows the use of more than one variant of cluster; one set with identical of similarly class and one set multiple class, contribute to an improved of accurate decision. The proposed approach has been compared to the use of heterogeneity created by single clustering or classification only and it has outperformed the commonly used classifier approach. This is due to the fact that heterogeneity of patterns is important in order to enhance the capability of classifier. When the heterogeneity level is enhanced the decision boundary seems easy to draw. The study shows positive impact occurred in large dataset which shows the significant sight in modeling large data classification.

Acknowledgments. This work supported by Fundamental Research Grant Scheme under Malaysia Ministry of Higher Education (MOHE) and Center of Research and Innovation Management of Universiti Sultan Zainal Abidin, Terengganu, Malaysia.

References

1. Bottou, L.: Large-scale machine learning with stochastic gradient descent. In: Lechevallier, Y., Saporta, G. (eds.) 19th International Conference on Computational Statistics, pp. 177–186. Physica-Verlag HD, Paris (2010)
2. Vijayalakshmi, M., Devi, M.R.: A survey of different issues of different clustering algorithms used in large data sets. Int. J. Adv. Res. Comput. Sci. Softw. Eng. **3**, 137–141 (2012)
3. Wen, Y.-M., Wang, Y.-N., Liu, W.-H.: Using parallel partitioning strategy to create diversity for ensemble learning. In: 2nd IEEE International Conference on Computer Science and Information Technology, ICCSIT 2009, Beijing, pp. 585–589 (2009)
4. Seiffertt, J., Wunsch. D.C.: Back propagation on time scales. In: Unified Computational Intelligence for Complex Systems, vol. 6, pp. 77–89. Springer, Heidelberg (2010)
5. Parvin, H., Minaei, B., Alizadeh, H., Beigi, A.: A novel classifier ensemble method based on class weightening in huge dataset. In: Liu, D., Zhang, H., Polycarpou, M., Alippi, C., He, H. (eds.) ISNN 2011. LNCS, vol. 6676, pp. 144–150. Springer, Heidelberg (2011). doi:10.1007/978-3-642-21090-7_17
6. Jing, Y., Xiaoqin, Z., Shuiming, Z., Shengli, W.: Effective neural network ensemble approach for improving generalization performance. IEEE Trans. Neural Netw. Learn. Syst. **24**, 878–887 (2013)
7. Ciresan, D.C., Meier, U., Gambardella, L.M., Schmidhuber, J.: Deep big simple neural nets excel on handwritten digit recognition. Neural Comput. **22**, 3207–3220 (2010)
8. Bishop, C.M.: Neural Networks for Pattern Recognition. Clarendon, Oxford (1995)
9. Yao, Y.: On complexity issues of online learning algorithms. IEEE Trans. Inf. Theory **56**, 6470–6481 (2010)
10. Windeatt, T.: Accuracy diversity and ensemble MLP classifier design. IEEE Trans. Neural Netw. **17**, 1194–1211 (2006)
11. Sospedra, J.T.: Ensembles of artificial neural networks: analysis and development of design methods. Ph.D. doctoral dissertation, Department of Computer Science and Engineering, Universitat Jaume I, Castellon (2011)
12. Peng, K., Obradovic, Z., Vucetic, S.: Towards efficient learning of neural network ensembles from arbitrarily large datasets. In: ECAI, p. 623 (2004)
13. Wang, S., Yao, X.: Relationships between diversity of classification ensembles and single-class performance measures. IEEE Trans. Knowl. Data Eng. **25**, 206–219 (2013)
14. Fernández, C., Valle, C., Saravia, F., Allende, H.: Behavior analysis of neural network ensemble algorithm on a virtual machine cluster. Neural Comput. Appl. **21**, 535–542 (2012)
15. Polikar, R.: Ensemble based systems in decision making. IEEE Circ. Syst. Mag. **6**, 21–45 (2006)

A New Mobile Malware Classification for SMS Exploitation

Nurzi Juana Mohd Zaizi(✉), Madihah Mohd Saudi, and Adiebah Khailani

Faculty of Science and Technology, Universiti Sains Islam Malaysia (USIM), Bandar Baru Nilai, 71800 Nilai, Negeri Sembilan, Malaysia
{njuana,madihah}@usim.edu.my, a_diebah3@yahoo.com

Abstract. Mobile malware is ubiquitous in many malicious activities such as money stealing. Consumers are charged without their consent. This paper explores how mobile malware exploit the system calls via SMS. As a solution, we proposed a system calls classification based on surveillance exploitation system calls for SMS. The proposed system calls classification is evaluated and tested using applications from Google Play Store. This research focuses on Android operating system. The experiment was conducted using Drebin dataset which contains 5560 malware applications. Dynamic analysis was used to extract the system calls from each application in a controlled lab environment. This research has developed a new mobile malware classification for Android smartphone using a covering algorithm. The classification has been evaluated in 500 applications and 126 applications have been identified to contain malware.

Keywords: Mobile malware · Data classification · Android · Big data · Covering algorithm

1 Introduction

Technology advancement has made mobile phones powerful and useful in everyday life. Attacker will always find a way to exploit this device for their gain. Vulnerable devices are exploited to gain root privilege to harm the target. Android is one of the top operating systems for smartphones. Android operating system can be exploited via SMS. Recently, there is a new Android Trojan that specializes in stealing mobile banking information by intercepting SMS [1]. Each day, there are better enhancements of the devices either in detecting, classifying or deleting suspicious thing behind the operating system. Unfortunately, smartphones running Android are increasingly targeted by attackers and infected with malicious software. Figure 1 depicts the android malware samples according to the GData Mobile Malware Report [2]. According to the GData Mobile Malware Report [2], a total of 440,267 new malware files were identified in Android in the first quarter of 2015. This represents an increase of 6.4 percent compared to the fourth quarter of 2014 which is 413,871. On average, the experts discovered almost 4,900 new Android malware files every day in the first quarter of 2015, an increase of almost 400 more new malware files per day compared to the second half of 2014. In the first quarter of 2015, the analysts identified a new malware

T. Herawan et al. (eds.), *Recent Advances on Soft Computing and Data Mining*,
Advances in Intelligent Systems and Computing 549, DOI 10.1007/978-3-319-51281-5_46

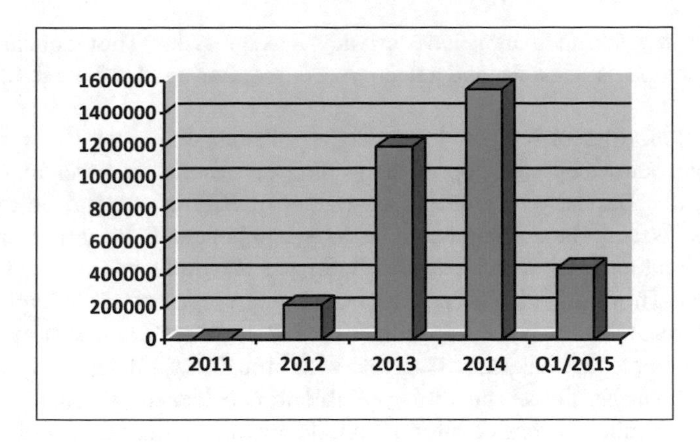

Fig. 1. New android malware samples

sample every 18 s, per hour this makes about 200 new Android malware. Meanwhile, financially motivated Android malware makes up around half of the malware analyzed which is 50.3 percent. This type of malware includes banking Trojans, ransomware, SMS Trojans and the like samples [2].

This findings show that there is a need for stopping the increasing of malware on Android markets and smartphones before it gets worse considering that the Android OS is known as the worst platform for malware.

2 Related Works

Malware analysis is the process of identifying the instances of malware based on different classification schemes by using the attributes of known malware characteristics. There are two ways to analyze malware in Android devices; static and dynamic program analysis of the malware binary.

Static analysis has good coverage on the coding but lack of run-time information leading to imprecise output. Whilst, dynamic analysis takes a long time to find and analyze the small amount of code implementing the malicious functionality as Android malware is often hidden in applications that have legitimate functionality [3].

According to Abhijit et al. [4], the static analysis uses requested permission to check the risk of an application. However, it is difficult to obtain high detection rate by using permission based methods. In contrast, dynamic analysis usually deals with features like dynamic code loading and system calls that can be collected while an application is running.

Qian et al. [5] proposed a two steps model which combines static and dynamic analysis approaches. First, the static analysis uses a permission combination matrix to determine whether an application has potential risks. Then, the suspicious applications are sent into the dynamic monitoring module to track the call information of the sensitive APIs while it is running. The experimental results show that almost 26%

applications in Android market have privacy leakage risks. They conclude that the proposed method is feasible and effective for monitoring these kind of malicious behaviour.

Octeau et al. [6] describe how to overlay a probabilistic model which is trained by using domain knowledge, on top of static analysis results, in order to triage static analysis results. Basically, a probabilistic model of ICC is overlaid on top of static analysis results and since computing the inter-component links is a prerequisite to inter-component analysis, a formalism for inferring ICC links based on set constraints is introduced. They computed all potential links in a corpus of 11,267 applications in 30 min and triage them using a probabilistic approach. It was found that over 95.1% of all 636 million potential links are associated with probability values below 0.01 and are thus likely unfeasible links. Thus, it is possible to consider only a small subset of all links without significant loss of information. This work is the first significant step in making static inter-application analysis more tractable, even at large scales.

Park et al. [7] presented a novel security model RGBDroid, which serves as a new approach to protect the system against privilege escalation attacks in Android system. To protect a system from the adverse actions attainable through privilege escalation, the RGBDroid acts in responding to the attacks instead of trying to prevent privilege escalation attacks. It has the capability to determine whether an application illegally acquires root-level privilege and not allowing an illegal root-level process to access protected resources regarding to the principle of least privilege. Even when malware obtains root-level privilege, the RGBDroid can still protects the Android system against malicious applications by exploiting vulnerabilities of the Android platform.

Weichselbaum et al. [8] proposed the ANDRUBIS, a completely automated, publicly available and comprehensive analysis system for Android applications. In order to increase code coverage, this standalone tool performs a combination of static analysis techniques and dynamic analysis on both Dalvik VM and system level, together with several stimulation techniques. The results of the static analysis are used to perform more effective and efficient dynamic analysis. In analyzing and detecting Android malware, ANDRUBIS follows the same basic principle that research on x86 malware relies on and each application passes through three stages; static analysis (yields information immediately by just looking at a sample's application package and code), dynamic analysis (executes the sample in a sandbox and provides details on its behaviour during runtime) and post-processing (produce more meaningful results if they are applied to a rich feature set). The results are consistent with previous research and the system's soundness and effectiveness is verified for analyzing Android applications.

3 Methodology

The objective of this research is to develop system calls classification based on surveillance exploitation system calls for SMS. Figure 2 depicts the framework to achieve the objective. This research uses two types of dataset: (1) training dataset, and (2) testing dataset. These datasets include Drebin dataset and applications from Google Play Store. A total of 5560 malware applications from Drebin dataset was tested and

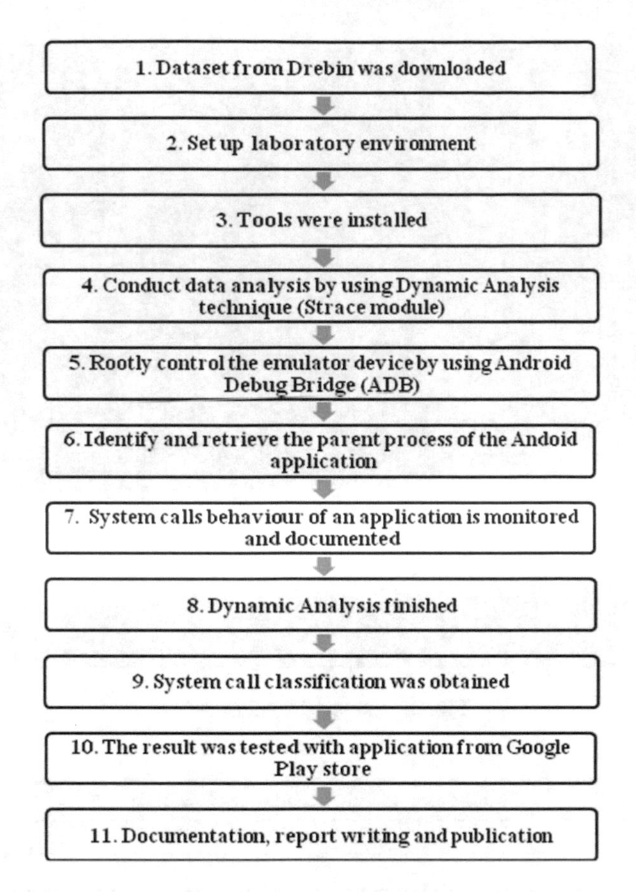

Fig. 2. The research process

the patterns of system calls were classified. For the purpose of data analysis, 500 random applications from Google Play Store were downloaded and analyzed.

A controlled laboratory environment was set up to conduct the experiment. Drebin dataset was first downloaded into a laptop. By using the Strace tool, system call generated from Android samples was examined to outline the system call that may exploit SMS message features. All system calls generated from the dataset were collected and tabulated. Malicious dataset was installed in the Genymotion emulator. Dynamic analysis was used to capture all system calls of each application. The system calls can only be generated based on the user interaction with the applications. The behaviour of the applications then was monitored through system calls. Figure 3 depicts an example of system calls captured in a running application.

Figure 4 depicts the steps during the dynamic analysis was carried out.

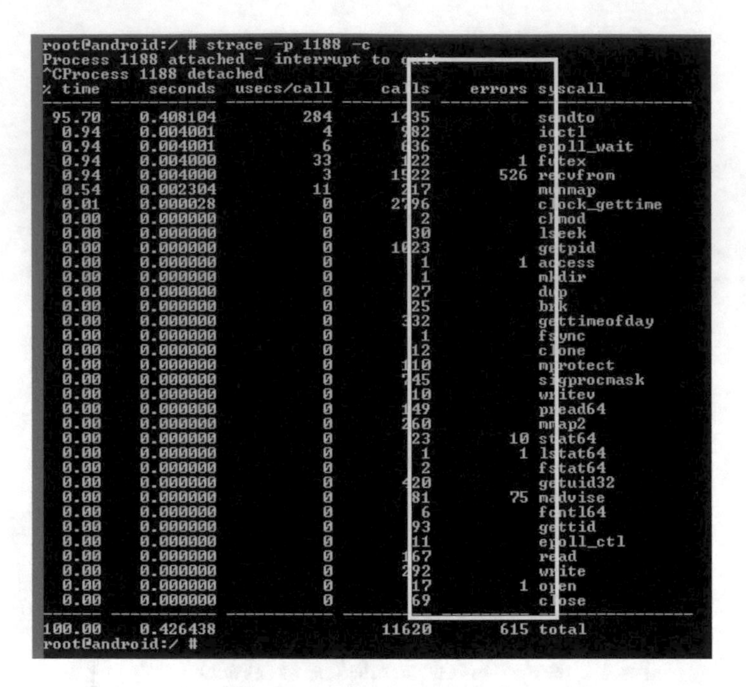

```
root@android:/ # strace -p 1188 -c
Process 1188 attached - interrupt to quit
^CProcess 1188 detached
% time     seconds  usecs/call     calls    errors syscall
------ ----------- ----------- --------- --------- ----------------
 95.70    0.408104         284      1435           sendto
  0.94    0.004001           4       982           ioctl
  0.94    0.004001           6       636           epoll_wait
  0.94    0.004000          33       122         1 futex
  0.94    0.004000           3      1522       526 recvfrom
  0.54    0.002304          11       217           munmap
  0.01    0.000028           0      2796           clock_gettime
  0.00    0.000000           0         2           chmod
  0.00    0.000000           0        30           lseek
  0.00    0.000000           0      1023           getpid
  0.00    0.000000           0         1         1 access
  0.00    0.000000           0         1           mkdir
  0.00    0.000000           0        27           dup
  0.00    0.000000           0        25           brk
  0.00    0.000000           0       332           gettimeofday
  0.00    0.000000           0         1           fsync
  0.00    0.000000           0        12           clone
  0.00    0.000000           0       110           mprotect
  0.00    0.000000           0       745           sigprocmask
  0.00    0.000000           0        10           writev
  0.00    0.000000           0       149           pread64
  0.00    0.000000           0       260           mmap2
  0.00    0.000000           0        23        10 stat64
  0.00    0.000000           0         1         1 lstat64
  0.00    0.000000           0         2           fstat64
  0.00    0.000000           0       420           getuid32
  0.00    0.000000           0        81        75 madvise
  0.00    0.000000           0         6           fcntl64
  0.00    0.000000           0        93           gettid
  0.00    0.000000           0        11           epoll_ctl
  0.00    0.000000           0       167           read
  0.00    0.000000           0       292           write
  0.00    0.000000           0        17         1 open
  0.00    0.000000           0        69           close
------ ----------- ----------- --------- --------- ----------------
100.00    0.426438                 11620       615 total
root@android:/ #
```

Fig. 3. Captured system calls

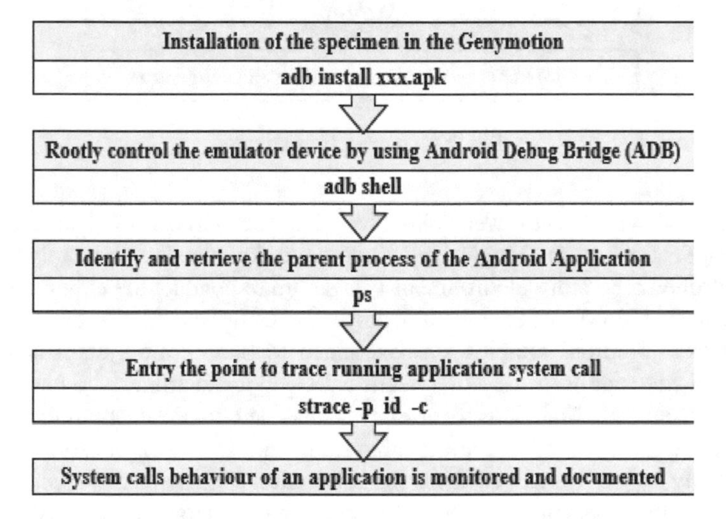

Fig. 4. Dynamic analysis process

3.1 Percentage of Occurrence

The result of occurrence of each system call is tabulated. Next, the presence of the system calls in the application sample is indicated as 1 while system call labelled 0 indicates that the system call is absent. The percentage of occurrence of each system call is recorded to compare their existence in each application sample. This approach makes it easier to summarize and analyse the data because their patterns can be seen clearly.

From the data analysis, four system calls related with the message were noted. First is recvfrom() which read incoming data or message from the remote side locally or via network. Second is recv() which function to receive message from socket. Third are sendmsg() and sendto() with the same function which is sending message on socket. However, sendto() system call seems to have high rate of occurrence. This shows that sendto() is a basic system call that uses in an application to perform non-malicious activity. Thus, sendto() was eliminated as the bad system call that exploit through SMS.

3.2 Similarity Analysis Procedure Based on System Calls Pattern

To determine normal and suspicious system calls which might contribute in the execution of malicious actions, a method of detecting similarity system calls between the malicious and legitimate application is used.

By comparing system calls generated from both malicious and legitimate applications, suspicious and normal system calls can be determined after taking into account their similarity and percentage of occurrence. From the list of suspicious system calls, the one that attacker might use in exploiting SMS could be detected easily.

3.3 Covering Algorithm

To collect the system calls from the dataset, the Covering Algorithm has been used as the basis for the formation of the new classification for SMS exploitation. The following describes the general to specific rules used in the pseudocodes:

Each class C initializes E to the instance set. While E contains an instance in class C. Create a rule R with an empty left-hand side (LHS) that predicts class C. Therefore, R covers for each attribute A that is not mentioned in R, and each value v, considers as

- *Adding the condition E = v to the (LHS) of R*
- *Select E and v to maximize the accuracy*
- *Finally add (E = v) to R* [9].

4 Results

Based on the experiments, thousands of system call have been retrieved. It is not feasible to enumerate all system calls but the focus is on the system calls that generate bad activities in the background. There are 59 system calls captured from the execution of the dataset. By examining the list, there are several system calls which could be used by the attackers to exploit the SMS messages. From this exploitation, attackers can gain money as victim might send SMS message without noticing it. The victim would be subscribed to the premium-rate SMS message which could lead to losing money. By reviewing each function of the system calls (Linux documentation), four system calls

Table 1. List of new patterns of system calls that exploit through SMS

Pattern representation	Pattern
P1	c27 + c28 + c29 + c30 + c31
P2	c27 + c28 + c29 + c30 + c31 + c1
P3	c27 + c28 + c29 + c30 + c31 + c3
P4	c27 + c28 + c29 + c30 + c31 + c1 + c3
P5	c27 + c28 + c29 + c30 + c31 + c1 + c6
P6	c27 + c28 + c29 + c30 + c31 + c1 + c3 + c6
P7	c27 + c28 + c29 + c30 + c31 + c1 + c3 + c16
P8	c27 + c28 + c29 + c30 + c31 + c1 + c3 + c35
P9	c27 + c28 + c29 + c30 + c31 + c1 + c6 + c16
P10	c27 + c28 + c29 + c30 + c31 + c1 + c6 + c35
P11	c27 + c28 + c29 + c30 + c31 + c1 + c16 + c35
P12	c27 + c28 + c29 + c30 + c31 + c1 + c3 + c6 + c16 + c35
P13	c27 + c28 + c29 + c30 + c31 + c9
P14	c27 + c28 + c29 + c30 + c31 + c51
P15	c27 + c28 + c29 + c30 + c31 + c44
P16	c27 + c28 + c29 + c30 + c31 + c9 + c51
P17	c27 + c28 + c29 + c30 + c31 + c9 + c44
P18	c27 + c28 + c29 + c30 + c31 + c51 + c44
P19	c27 + c28 + c29 + c30 + c31 + c9 + c51 + c44
P20	c27 + c28 + c29 + c30 + c31 + c2
P21	c27 + c28 + c29 + c30 + c31 + c23
P22	c27 + c28 + c29 + c30 + c31 + c13
P23	c27 + c28 + c29 + c30 + c31 + c7
P24	c27 + c28 + c29 + c30 + c31 + c2 + c23
P25	c27 + c28 + c29 + c30 + c31 + c2 + c13
P26	c27 + c28 + c29 + c30 + c31 + c2 + c7
P27	c27 + c28 + c29 + c30 + c31 + c23 + c13
P28	c27 + c28 + c29 + c30 + c31 + c23 + c7
P29	c27 + c28 + c29 + c30 + c31 + c13 + c7
P30	c27 + c28 + c29 + c30 + c31 + c2 + c13 + c7
P31	c27 + c28 + c29 + c30 + c31 + c2 + c23 + c7
P32	c27 + c28 + c29 + c30 + c31 + c2 + c23 + c13
P33	c27 + c28 + c29 + c30 + c31 + c2 + c23 + c13 + c7

were used to send and receive message from a socket: (1) *sendto()*, (2) *sendmsg()*, (3) *recvfrom()* and (4) *recv()*.

From the data analysis, *sendmsg(), recvfrom()* and *recv()* system calls have low rate of occurrence. These system calls are the most likely system calls that attacker could use to exploit a device through SMS. Thus, all patterns that consist of those 3 system calls were analyzed. From the 5560 testing applications executed, this research has found 33 patterns of malicious system calls that exploit through SMS. Table 1 depicts

Table 2. Percentage of match system calls pattern

Pattern	Google play	Percentage
P1	23	4.60%
P2	8	1.60%
P3	3	0.60%
P4	4	0.80%
P5	1	0.20%
P6	0	0%
P7	4	0.80%
P8	3	0.60%
P9	7	1.40%
P10	1	0.20%
P11	4	0.80%
P12	6	1.20%
P13	1	0.20%
P14	0	0%
P15	3	0.60%
P16	5	1%
P17	0	0%
P18	8	1.60%
P19	1	0.20%
P20	0	0%
P21	3	0.60%
P22	7	1.40%
P23	3	0.60%
P24	5	1%
P25	4	0.80%
P26	6	1.20%
P27	2	1.40%
P28	4	0.80%
P29	1	0.20%
P30	3	0.60%
P31	1	0.20%
P32	3	0.60%
P33	2	1.40%
Total	**126**	**500**

the 33 new classifications that have been extracted. These patterns become a new format dataset that are used in next step which is data evaluation.

The new mobile malware classifications were evaluated with the applications from Google Play Store. Over 500 applications installed. This research has found that there are 126 applications matched with the new mobile malware classification for SMS exploitation proposed in this research. Table 2 depicts the percentage of pattern that matches the system calls pattern that have been proposed.

5 Conclusions

Android-based applications can be developed freely in an open-source form. Applications that contain malicious codes may disguise as a legitimate application and be distributed through Android official market or other third-party markets. This research targeted 5560 Android malware samples from Drebin dataset to analyze suspicious and normal system calls. This research has successfully developed 33 new system call classifications for exploiting SMS. These are beneficial in identifying mobile malware exploitation in application. This research has proven that the mobile malware could exploit through SMS feature on Android from the patterns obtained.

Acknowledgment. The authors would like to express their gratitude to Universiti Sains Islam Malaysia (USIM) and Islamic Science Institute (ISI), USIM for the support and facilities provided. This research is supported by grants [FRGS/1/2014/ICT04/USIM/02/1] and [PPP/USG-0116/FST/30/13216].

References

1. Brook, C.: New Banking Trojan Targets Android, Steals SMS. Threat Post. https://threatpost.com/new-banking-trojan-targets-android-steals-sms/110819/. Accessed 25 Feb 2016
2. GData Mobile Malware Report, Threat Report: Q1/2015. https://public.gdatasoftware.com/Presse/Publikationen/Malware_Reports/G_DATA_MobileMWR_Q2_2015_US.pdf. Accessed 25 Feb 2016
3. Zhang, D., Kong, W.K., You, J., Wong, M.: On-line palmprint identification. In: Proceedings of IEEE Transactions on Pattern Analysis and Machine Intelligence, pp. 1041–1050. IEEE Computer Society, Washington DC (2003)
4. Abhijit, B., Xin, H., Kang G.S., Taejoon, P.: Behavioral detection of malware on mobile handsets. In: ACM 978-1-60558-139-2/08/06, Breckenridge (2008)
5. Qian, Q., Cai, J., Xie, M., Zhang, R.: Malicious behavior analysis for android applications. Int. J. Netw. Secur. **18**(1), 182–192 (2015)
6. Octeau, D., Jha, S., Dering, M., McDaniel, P., Bartel, A., Li, L., Klein, J., Traon, Y.L.: Combining static analysis with probabilistic models to enable market-scale android inter-component analysis. In: Proceedings of the 43rd Annual ACM SIGPLAN-SIGACT Symposium on Principles of Programming Languages (POPL 2016), pp. 469–484. ACM, New York (2016)
7. Park, Y., Lee, C.H., Lee, C., Lim, J.H., Han, S., Park, M., Cho, S.J.: RGBDroid: a novel response-based approach to android privilege escalation attacks. In: Proceedings of the 5th USENIX Conference on Large-Scale Exploits and Emergent Threats. USENIX, Berkeley (2012)
8. Weichselbaum, L., Neugschwandtner, M., Lindorfer, M., Fratantonio, Y., Veen, V., Platzer, C.: Andrubis: android malware under the magnifying glass. Vienna University of Technology, Technical report TR-ISECLAB-0414-001 (2014)
9. Witten, I.H., Frank, E.: Data Mining: Practical Machine Learning Tools and Techniques, (Morgan Kaufmann Series in Data Management Systems), 2nd edn. Morgan Kaufmann Publishers Inc., San Francisco (2005)

Data Mining Techniques for Classification of Childhood Obesity Among Year 6 School Children

Fadzli Syed Abdullah[1(✉)], Nor Saidah Abd Manan[1], Aryati Ahmad[1], Sharifah Wajihah Wafa[1], Mohd Razif Shahril[1], Nurzaime Zulaily[1], Rahmah Mohd Amin[1], and Amran Ahmed[2]

[1] Universiti Sultan Zainal Abidin, 21300 Kuala Terengganu Terengganu, Malaysia
{fadzlihasan, aryatiahmad, sharifahwajihah, razifshahril, rahmahamin}@unisza.edu.my, {sl0901, sl0831}@putra.unisza.edu.my

[2] Universiti Malaysia Perlis, 02600 Arau, Perlis, Malaysia
amranahmed@unimap.edu.my

Abstract. Today, data mining is broadly applied in many fields, including healthcare and medical fields. Obesity problem among children is one of the issues commonly explored using data mining techniques. In this paper, the classification of childhood obesity among year six school children from two districts in Terengganu, Malaysia is discussed. The data were collected from two main sources; a *Standard Kecergasan Fizikal Kebangsaan untuk Murid Sekolah Malaysia*/National Physical Fitness Standard for Malaysian School Children (SEGAK) Assessment Program and a set of distributed questionnaire. From the collected data, 4,245 complete data sets were promptly analyzed. The data preprocessing and feature selection were implemented to the data sets. The classification techniques, namely Bayesian Network, Decision Tree, Neural Networks and Support Vector Machine (SVM) were implemented and compared on the data sets. This paper presents the evaluation of several feature selection methods based on different classifiers.

Keywords: Bayesian network · Childhood obesity · Classification · Decision tree · Feature selection · Neural network · Support Vector Machine

1 Introduction

Today, data mining is broadly applied in many fields, including healthcare and medical fields. The application of data mining is to improve the decision making in medical issues [1]. The increase of biomedicine data has led to the importance of data mining in handling uncertainties [2, 3]. Due to the enormous number of data sets, the application of data mining allows many possibilities for unseen patterns in medical data sets [4]. The application of data mining also may help the medical practitioners in order to focus their approaches from population-based into individual-based [5].

T. Herawan et al. (eds.), *Recent Advances on Soft Computing and Data Mining*, Advances in Intelligent Systems and Computing 549, DOI 10.1007/978-3-319-51281-5_47

In this century, the prevalence of childhood obesity is increasing at an alarming rate. This makes it one of the most critical public health problems around the world [6]. The prevalence of overweight and obesity among school aged children in North-eastern Romania were considerably high; 24% and 7%, respectively [7]. In 2014, a finding had stated a high percentage in the prevalence of overweight and obese among Iraqi secondary school students; 20.6% and 11.3%, respectively [8]. In the United States, the prevalence of obesity among 2 to 19 year old children remained high, 16.9% [9]. In Jena (Germany), the prevalence of obesity for both boys and girls had increased from 10.0% and 11.7% in 1975 to 16.3% and 20.7% in 1995 [10]. While in Malaysia, the overall overweight Malaysians aged 7 to 12 years old was considered high (19.9%) [11]. Another study indicated the prevalence of obesity among school children in Kota Kinabalu, Sabah, Malaysia was 2.5% [12]. A study also reported that the prevalence of overweight children in Malaysia had increased by 5%, while obese children increased by 2.5% from 2002 to 2008 [13].

The aim of this study was to identify the factors that influence the childhood obesity using various feature selection techniques. Other than that, this paper reports the application of various classifiers for the classification of childhood obesity.

2 Data Collection

2.1 Study Sample

The target population was 12 year old (year 6) students from two districts of Terengganu; Kuala Terengganu and Besut. A total of 153 primary schools were selected with 81 in Kuala Terengganu and 54 in Besut. The data were collected from two main sources; *Standard Kecergasan Fizikal Kebangsaan untuk Murid Sekolah Malaysia* (SEGAK) Assessment Program and the study questionnaire (i.e. consist of sociodemographic, physical activity and dietary assessment questionnaire). The total of SEGAK data of Kuala Terengganu and Besut students were 5,222, and 1,374, respectively. Otherwise, the total of completed questionnaire from Kuala Terengganu and Besut were 3,385 and 1,692, respectively. Overall, only 4,245 data were completely matched between SEGAK and questionnaire.

2.2 SEGAK Assessment

SEGAK assessment is coordinated for all students during *Pendidikan Jasmani dan Pendidikan Kesihatan* (PJPK) class. The data obtained from SEGAK were Body Mass Index (BMI) value and category, SEGAK score and grade.

Equation (1) shows the calculation of BMI. The chart of BMI-for-age was used for reference, in which BMI less than the 5th percentile is considered as “underweight”, the BMI greater than the 5th but less than the 85th percentile is considered as “normal”. Based on the chart, a BMI greater than the 85th percentile is considered as “overweight” while a BMI greater than 95th percentile is considered as “obese”.

$$BMI = \frac{weight\ in\ kilograms}{(height\ in\ meters)^2} \quad (1)$$

The SEGAK assessment consists of four tests, including step-up test (NTB), push-up test (TT/S), partial curl-up test (RTS) and sit and reach test (JM). Equation (2) shows the calculation of total score for SEGAK. If the total score greater than 18 the student is considered as "best/A" while if greater than 15 the student is considered as "good/B". Other than that, the student is considered as "normal/C" if the total score is greater than 12 and considered as "below normal/D" if the total score is greater than 8 but if the total score is less than 7 the student is considered as "not active/E".

$$\text{NTB score} + \text{TT/S score} + \text{RTS score} + \text{JM score} = \text{Total Score} \quad (2)$$

2.3 Questionnaire

The questionnaire was distributed to the students through the teachers. The questionnaire was divided into three sections; personal information, physical activity and dietary. The personal information section was filled by the parents/guardians of the students, whilst the physical activity and dietary section were filled by the students under the supervision of PJPK teachers. The data obtained from the questionnaire are listed in Table 1.

Table 1. Data collected from distributed questionnaire

Section	Description
Socio-demographic	Children information, parent and household information, children's health information, family health history
Physical activity	Children daily and weekend activity, activity during PJPK class and recess time, sedentary activity
Dietary	Food Frequency Questionnaire (FFQ) – the frequency of food consumption and the portion size over a defined period

3 Data Processing

Figure 1 shows the flow of data processing. It consists of three main processes; data preprocessing, feature selection and classification. In data preprocessing, two processes were involved; data cleaning and discretization.

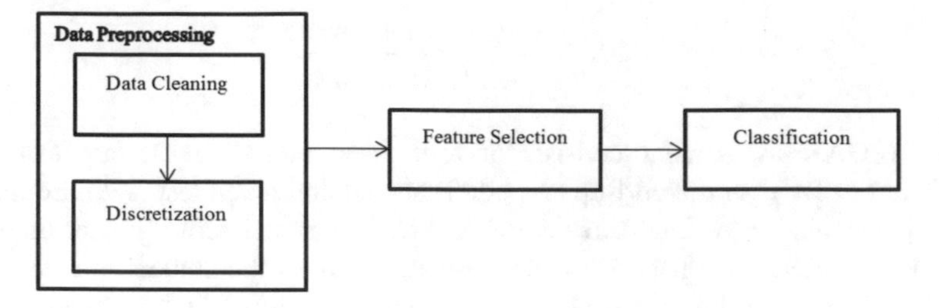

Fig. 1. The flow of data processing

3.1 Data Cleaning

Data cleaning is a process of remove incomplete and inconsistent data also the outliers and fill in missing values [14]. There was a constraint that the age of a student cannot be less or more than 12 years old, otherwise the data was regarded as abnormal. *Mykid* was used as student id, there was a constraint that the student must have the *Mykid*, otherwise the data is regarded as abnormal. The abnormal data were discarded.

3.2 Discretization

Discretization is a method that converts a continuous variable into a discrete variable [15]. The discretization was preferred in this study since almost of the data sets are continuous or numerical data type. In order to get better classification performance, the continuous values in the data sets was changed into nominal values using the discretization method in Weka.

3.3 Feature Selection

Feature selection is a process of reduction of feature in order to increase the accuracy and performance of the classification [16, 17]. Liu in [18] concluded that the less the number of features the shorter the training time will be taken. In addition, the final models can be simplified. Using Weka, the feature selection can be implemented by specifying a search method and an attribute evaluator. The search method is a searcher algorithm while the attribute evaluator is an algorithm to evaluate the selected attributes [19]. A study proved that using CfsSubsetEval as an attribute selector along with BestFirst, genetic search, greedy stepwise and linear forward search as search methods the classification can perform better [20]. Since different classifiers inherit different learning bias, the effect of feature selection implementation also can be seen on different classifiers [21]. On heart disease data set, information gain, ReliefF, CFS and the Wrapper give better performance of naïve Bayes [22].

3.4 Classification

Many studies had been conducted in predicting childhood obesity using various classification techniques. The techniques are Bayesian classifiers [23–27], decision tree [23–25], neural network [28] and Support Vector Machine (SVM) [25].

Bayesian Classifier. This technique predicts class by probabilities [26]. Bayesian network and Naïve Bayes are the fundamental methods in Bayes' algorithms. The probabilistic relationships among classes depicts by a graphical model using Bayesian networks [29]. The graphical model depicts by a directed graph with nodes that represent attributes and the arcs represent the dependencies [23]. The following formula is the Bayesian Theorem to calculate $P(H|X)$, the probability of hypothesis H holds given the observed sample with unknown class X:

$$P(H|X) = \frac{P(X|H)P(H)}{P(X)} \quad (3)$$

Where $P(H)$ is the prior probability of H and $P(X)$ is the prior probability of X. While $P(X|H)$ is a posteriori probability of X conditioned on H [26]. In [27], the results showed that Naïve Bayes classifier may produce better prediction result with a good variable selection technique.

Decision Tree. This algorithm has been applied for multi-stage decision making [26]. The model builds by this algorithm is tree-like structure. The model comes with three types of nodes; starts from the root node, internal node and ends with a terminal node [25]. Splitting algorithm in decision tree techniques are used to split the input into subgroups by identifying the variable and its corresponding threshold. The splitting process is repeated until a complete tree is built [23]. There are many well-known algorithms for decision tree; CART, ID3, C4.5, CHAID and J48 [30]. In our study, we implemented C4.5 (weka.classifiers.trees.J48) due to [31, 32] where C4.5 was the most applicable classifier for predicting childhood obesity.

Neural Network. This technique is based on the weight connection between a collection of neuron-like [33]. A biological neuron in human brain is a cell that consists of a nucleus, axon and dendrite branches. The axon is a transmitter that transmits signals to other neurons while dendrite is a receiver that receive signals from other neurons [23]. Unlike decision tree, neural network has three main layers; input layer, output layer and hidden layers. During the learning process, the weights can be adjusted to satisfy the links of input and output [34]. This algorithm is well-known in classification and prediction because it has a high tolerance to noise and possible to classify unseen patterns in data sets [25]. Multi-layer perceptron (MLP) is the most powerful algorithm with back-propagation for prediction and classification [35].

Support Vector Machine (SVM). This machine learning technique is based on statistical learning theory. SVM can be solved easily since it can be expressed in the quadratic optimization programming standard form. The performance of SVM is very good because the ability of its kernel mapping techniques to deal with the similarity in a high-dimensional Hilbert space. Other than that, the SVM also can simplify the testing procedure by compressed a large data set into a comparably small data set. The SVM was the best algorithm for childhood obesity prediction [25].

4 Experiments and Results

Under the data exploration mode, we explored almost all feature selection techniques in Weka to collect optimal subset of attributes. The feature selection is implemented to find more related attributes for all possible combinations of attribute evaluators and search method. Table 2 shows the group of feature selection based on selected features.

Table 2. Feature selection grouped based on selected features

Attribute evaluator	Search method	Number of selected features	Group
CfsSubsetEvaluator	Best first	22	A
	Genetic search	20	B
	Greedy stepwise	22	A
	Linear forward	22	A
Consistency	Best first	17	C
	Genetic search	26	D
	Greedy stepwise	17	C
	Linear forward	29	E

The feature selection results were grouped based on the similarities of feature selected. The CfsSubsetEvaluator using best first, greedy stepwise and linear forward search method select similar attributes. The Consistency using best first and greedy stepwise search method can be grouped together since the selected features are alike. The CfsSubsetEvaluator using genetic search method and Consistency using genetic search also linear forward search method was unable to be grouped with another feature selection method because the selected features were unique. Table 3 shows the selected features by each group.

The feature selection then was evaluated based on the various classifiers. Table 4 shows the sensitivity and accuracy of various classifiers based on group. Based on the results, it can be seen that all the classifiers have comparable sensitivity and accuracy. J48 is marginally better than other classifiers on every group. SMO gives better accuracy than other classifiers except J48. MLP gives the lower sensitivity and accuracy on every group. On each group, the sensitivity and accuracy given by BayesNet is equal to Naïve Bayes.

Other than that, it can be seen that the best result is from Group E; the Consistency using linear forward search method which improves J48. Group B; the CfsSubsetEvaluator using genetic search method is improves BayesNet, Naïve Bayes and SMO. Group A; the CfsSubsetEvaluator using best first or greedy stepwise or linear forward search method improves MLP.

Table 3. Selected features

Group	Number of selected features	List of selected features
A (CfsSubsetEvaluator + Best First/Greedy Stepwise/Linear Forward)	22	Gender, SEGAK grade, CHO2, Pro2, Niacin, C, birth order, birthweight cat, breastfeed age, edulevel mother, GDM, birthweight mother, healthprob mother, maritalstatus father, edulevel father, underweight prob, obes history, heart history2, highbloodpressure history2, underweight history2, cancer history2
B (CfsSubsetEvaluator + Genetic Search)	20	SEGAK grade, CHO2, Fibre2, B2, birth order, birthweight cat, breastfeed, breastfeed age, edulevel mother, GDM, birthweight mother, healthprob mother, maritalstatus father, age father, underweight prob, obes history, highbloodpressure history2, underweight history2, cancer history2
C (Consistency + Best First/Greedy Stepwise)	17	District, school location, SEGAK grade, CPAQ cat, Pro2, Fat2, PUFA2, C2, birth order, household number, birthweight cat, edulevel mother, age mother, age father, birthweight father, familyincome cat, highbloodpressure history2
D (Consistency + Genetic Search)	26	District, school location, SEGAK grade, Pro2, Fat2, Fibre, Fibre2, PUFA2, RE2, household number, blood group, birthweight cat, edulevel mother, age mother, occupation mother, familyincome cat, birthweight father, epilepsy, blood disease, heart disease, diabetes, kidney history, highbloodpressure history2, diabetes history2, cancer history
E (Consistency + Linear Forward)	29	District, SEGAK grade, Fibre2, SFA, Sugar, blood group, edulevel mother, age mother, naritalstatus mother, GDM, healthprob mother, familyincome cat, birthweight father, healthprob father, age father, maritalstatus father, edulevel father, epilepsy, blood disease, eyesight prob, learning prob, obes history, heart history, highbloodpressure history, highbloodpressure history2, diabetes history, diabetes history2, cancer history

Table 4. The sensitivity and accuracy of classifiers based on group

Classifiers	Group									
	A		B		C		D		E	
	Sen. (%)	Acc. (%)	Sen. (%)	Acc. (%)	Sen. (%)	Acc. (%)	Sen. (%)	Acc. (%)	Sen. (%)	Acc. (%)
BayesNet	64.12	82.10	64.45	82.22	63.19	81.96	61.37	80.68	61.58	80.80
J48	***65.25***	**82.63**	***65.11***	**82.56**	***65.40***	**82.70**	***65.30***	**82.65**	***65.44***	**82.72**
Naive bayes	64.08	82.04	64.52	80.14	63.79	81.90	61.37	80.68	61.63	80.81
MLP	63.25	81.63	62.43	79.81	60.85	80.42	62.43	81.21	61.86	80.93
SMO	64.00	82.00	64.17	82.08	63.93	82.00	63.82	81.91	64.15	82.07

5 Conclusion

In conclusion, this study presents the comparison of performance between four classifiers in the classification of childhood obesity; Bayes Net, J48, Naïve Bayes, MLP and SMO. Based on the result, J48 and SMO appear to be the best classifiers for predicting childhood obesity on these data sets. Other than that, the CfsSubsetEvaluator using genetic search method was good overall performer by improving the accuracy of all classifiers. Consistency using best first or greedy stepwise search method chose fewer features but there were weak attribute interactions on BayesNet, Naïve Bayes, MLP and SMO. Other integrated feature selection methods and classifiers can be implemented in future to improve the evaluation and the accuracy of childhood obesity prediction.

Acknowledgments. This study was funded by the Ministry of Education Malaysia, grant no. [FRGS/2/2013/SKK/UNISZA/01/1].

References

1. Adnan, M.H.M., Husain, W., Rashid, N.A.: Data mining for medical systems: a review. ACIT **2012**, 978–981 (2012)
2. Lucas, P.J., Lucas, P.: Bayesian analysis, pattern analysis, and data mining in health care health care, March 2016
3. Jacob, S.G., Ramani, R.G.: Data mining in clinical data sets: a review. Int. J. Appl. Inf. Syst. **4**(6), 15–26 (2012)
4. Milovic, B., Milovic, M.: Prediction and decision making in health care using data mining **1**(2), 69–76 (2012)
5. Bellazzi, R., Zupan, B.: Predictive data mining in clinical medicine: Current issues and guidelines. Int. J. Med. Inform. **77**, 81–97 (2008)
6. WHO, Childhood overweight and obesity. http://www.who.int/dietphysicalactivity/childhood/en/. Accessed 03 Apr 2016
7. Mocanu, V.: Prevalence of overweight and obesity in urban elementary school children in Northeastern Romania: its relationship with socioeconomic status and associated dietary. BioMed Res. Int. **1**, 2013 (2013)
8. Qadir, M.S., Rampal, L., Sidik, S.M., Said, S., Ramzi, Z.S.: Prevalence of obesity and associated factors among secondary school students in Slemani City Kurdistan Region, Iraq. Malaysia J. Med. Heal. Sci. **10**(2), 27–38 (2014)
9. Ogden, C.L., Carroll, M.D., Kit, B.K., Flegal, K.M.: Prevalence of childhood and adult obesity in the United States, 2011–2012. JAMA **311**(8), 806–814 (2014)
10. Kromeyer-Hauschild, K., Zellner, K., Jaeger, U., Hoyer, H.: Prevalence of overweight and obesity among school children in Jena (Germany). Int. J. Obes. **23**, 1143–1150 (1999)
11. Naidu, B.M., Mahmud, S.Z., Ambak, R., Sallehuddin, S.M., Mutalip, H.A., Saari, R., Sahril, N., Hamid, H.A.A.: Overweight among primary school-age children in Malaysia. Asia Pac. J. Clin. Nutr. **22**(May), 408–415 (2013)
12. Chong, H.L., Soo, T.L., Rasat, R.: Childhood obesity – prevalence among 7 and 8 year old primary school students in Kota Kinabalu. Med. J. Malaysia **67**(2), 147–150 (2012)

13. Ismail, M., Ruzita, A., Norimah, A., Poh, B., Shanita, S.N., Mazlan, M.N., Roslee, R., Nurunnajiha, N., Wong, J., Zakiah, M.N., Raduan, S.: Prevalence and trends of overweight and obesity in two cross-sectional studies of Malaysian children, 2002–2008. In: MASO 2009, pp. 26–27, August 2009
14. Han, J., Kamber, M.: Data preprocessing. In: Kamber, M. (ed.) Data Mining Concepts and Techniques, 3rd edn, pp. 83–124. Elsevier, Amsterdam (2014)
15. Lustgarten, J.L., Gopalakrishnan, V., Grover, H., Visweswaran, S.: Improving classification performance with discretization on biomedical datasets. AMIA Annu. Symp. Proc. **2008**, 445–449 (2008)
16. Janecek, A., Gansterer, W.N.W., Demel, M., Ecker, G.: On the relationship between feature selection and classification accuracy. FSDM **4**, 90–105 (2008)
17. Singhi, S.K., Liu, H.: Feature subset selection bias for classification learning. In: Proceedings of the 23rd International Conference on Machine Learning, pp. 849–856 (2006)
18. Liu, Y., Schumann, M.: Data mining feature selection for credit scoring models. J. Oper. Res. Soc. **56**(9), 1099–1108 (2005)
19. Tetko, I., Baskin, I., Varnek, A.: Descriptor selection bias. In: Tutorial on Machine Learning, pp. 1–12 (2013)
20. Vasantha, M., Bharaty, V.S.: Evaluation of attribute selection methods with tree based supervised classification-a case study with mammogram images. Int. J. Comput. Appl. **8**(12), 35–38 (2010)
21. Liu, H., Motoda, H.: Feature Selection for Knowledge Discovery and Data Mining. Springer, Heidelberg (1998)
22. Hall, M.A., Holmes, G.: Benchmarking attribute selection techniques for data mining. IEEE Trans. Knowl. Data Eng. **15**(6), 1437–1447 (2003)
23. Adnan, M.H.M., Husain, W., Damanhoori, F.: A survey on utilization of data mining for childhood obesity prediction (2010)
24. Pochini, A., Wu, Y., Hu, G.: Data mining for lifestyle risk factors associated with overweight and obesity among adolescents (2014)
25. Zhang, S., Tjortjis, C., Zeng, X., Qiao, H., Buchan, I., Keane, J.: Comparing data mining methods with logistic regression in childhood obesity prediction. Inf. Syst. Front. **11**(4), 449–460 (2009)
26. Soni, S., Pillai, J.: Usage of nearest neighborhood, decision tree and Bayesian classification techniques in development of weight management counseling system. In: Proceedings - 1st International Conference on Emerging Trends in Engineering Technology, ICETET 2008, pp. 691–694 (2008)
27. Adnan, M.H.M., Husain, W., Rashid, N.A.: Hybrid approaches using decision tree, naïve bayes, means and euclidean distances for childhood obesity prediction. Int. J. Softw. Eng. Appl. IJSEIA **6**(3), 99–106 (2012)
28. Novak, B., Bigec, M.: Application of artificial neural networks for childhood obesity \prediction. In: IEEE, pp. 377–380 (1995)
29. Chen, J., Xing, Y., Xi, G., Chen, J., Yi, J., Zhao, D., Wang, J.: A comparison of four data mining models: bayes, neural network, SVM and decision trees in identifying syndromes in coronary heart disease. In: Liu, D., Fei, S., Hou, Z.-G., Zhang, H., Sun, C. (eds.) ISNN 2007. LNCS, vol. 4491, pp. 1274–1279. Springer, Heidelberg (2007). doi:10.1007/978-3-540-72383-7_148
30. Dangare, C.S., Apte, S.S.: Improved study of heart disease prediction system using data mining classification techniques. Int. J. Comput. Appl. **47**(10), 44–48 (2012)
31. Suca, C., Cordova, A., Condori, A., Sulla, J.C.Y.J.: Comparison of Classification Algorithms for Prediction of Cases of Childhood Obesity, April 2016

32. Suguna, M.: Childhood obesity epidemic analysis using classification algorithms. Int. J. Mod. Comput. Sci. **4**(1), 22–26 (2016)
33. Kaur, H., Wasan, S.K.: Empirical study on applications of data mining techniques in healthcare. J. Comput. Sci. **2**(2), 194–200 (2006)
34. Yoo, I., Alafaireet, P., Marinov, M., Pena-Hernandez, K., Gopidi, R., Chang, J.-F., Hua, L.: Data mining in healthcare and biomedicine: a survey of the literature. J. Med. Syst. **36**, 2431–2448 (2012)
35. Xing, Y., Wang, J., Zhao, Z., Gao, A.: Combination data mining methods with new medical data to predicting outcome of coronary heart disease. In: ICCIT 2007, pp. 868–872 (2007)

Multiple Criteria Preference Relation by Dominance Relations in Soft Set Theory

Mohd Isa Awang[1(✉)], Ahmad Nazari Mohd Rose[1],
Mohd Khalid Awang[1], Fadhilah Ahmad[1], and Mustafa Mat Deris[2]

[1] Faculty of Informatics and Computing,
Universiti Sultan Zainal Abidin, Terengganu, Malaysia
{isa, anm, khalid, fad}@unisza.edu.my
[2] Faculty of Computer Science and Information Technology,
Universiti Tun Hussein Onn Malaysia, Johor, Malaysia
mmustafa@uthm.edu.my

Abstract. This paper presents the applicability of soft set theory for discovering the preference relation in multi-valued information systems. The proposed approach is based on the notion of multi-soft sets. An inclusion of objects into value set of decision class in soft set theory is used to discover the relation between objects based on preference relation. Results from the experiment shows that dominance relation based on soft theory for preference relation is able to produce a finer object classification by eliminating inconsistencies during classification process as opposed to the expert judgement classification.

Keywords: Multiple criteria preference relation · Soft set theory · Soft-dominance relation · Multi-valued information system

1 Introduction

Many decision-making problems are characterized by the ranking of objects according to a set of criteria with pre-defined preference-ordered decision classes, such as credit approval [18], stock risk estimation [1], mobile phone alternatives estimation [10]. Models and algorithms were proposed for extracting and aggregating preference relations based on distinct criteria. The underlying objectives are to understand the decision process, to build decision models and to learn decisions rules from data.

Rough set theory provides an effective tool for dealing with imprecise and vague information system [14, 19]. It has been widely applied in feature evaluation [17], attribute reduction and rule extraction [9]. Pawlak's rough set model was constructed based on equivalence relations. These relations are viewed by many to be one of the main limitations when employing the model involving complex decision tasks. However, [6] has proposed the idea of dominance rough set model in overcoming the limitations in dealing with complex decision tasks.

In multiple criteria preference relation, there are preference structures between conditions and decisions. Greco et al. [6] introduced a dominance rough set model that is suitable for preference analysis. In [6], the decision-making problem with multiple attributes and multiple criteria were examined, where dominance relations were

T. Herawan et al. (eds.), *Recent Advances on Soft Computing and Data Mining*,
Advances in Intelligent Systems and Computing 549, DOI 10.1007/978-3-319-51281-5_48

extracted from multiple criteria and similarity relations were constructed from numerical attributes and equivalence relations were then constructed from nominal features. An extensive review of multi-criteria decision analysis based on dominance rough sets is given in [7]. Rough dominance relation has also been applied to ordinal attribute reduction and multi-criteria classification. While rough set theory is well-known and often useful approach in describing uncertainty, it certainly has some inherent difficulties as pointed by Molodtsov [13].

Soft set theory that was proposed by Molodtsov [13] provides an effective tool in dealing with inconsistencies which is free from the difficulties and limitations in existing methods. As reported in [13], a wide range of applications of soft sets have been developed in many different fields. There has been a rapid growth of interest in soft set theory and its application especially in decision making in recent years. Maji et al. [12] discussed the application of soft set theory in a decision making. Based on fuzzy soft sets, Roy and Maji [15] presented a method of object recognition from an imprecise multi-observer data and applied it to decision making problems. Chaudhuri et al. [2] define the concepts of soft relation and fuzzy soft relation then apply them to solve a number of decision making problems. Feng et al. [3] introduce an adjustable approach to fuzzy soft set and investigate the application of the weighted fuzzy soft set in decision making. Feng et al. [4] also present the application of level soft sets in decision making based on interval-valued fuzzy soft sets. Jiang et al. [11] present an adjustable approach to intuitionistic fuzzy soft sets based decision making by using level soft sets of intuitionistic fuzzy soft sets.

Although the Molotosv's proposal of soft set has been studied and applied by several authors in the cases of uncertainty in decision making, the presence of multiple criteria preference were not been considered. Only recently, multiple criteria decision making analysis under soft set theory has been discussed in [20]. Isa et al. had introduced a soft-dominance relation in decision making analysis in the presence of multiple criteria evaluation [20]. Thus, in this paper, it is proposing the general framework for preference relation by applying dominance relation based on soft set theory. In the proposed scheme, the decision system is transformed into the equivalent multi-soft set where in each soft set, the predicates are ordered according to the preference order, and then the approximations will be obtained using dominance-based soft set approach (DSSA) to classify objects into groups similarity preferences.

The rest of this paper is organized as follows. Section 2 describes the concept of soft set theory for multi-valued information systems. The preference relation based on soft set theory is introduced in Sect. 3. An illustrative example is given in Sect. 4 followed by the conclusion of our work is described in Sect. 5.

2 Soft Set Theory

Throughout this section U refers to an initial universe, E is a set of parameters, $P(U)$ is the power set of U and $A \subseteq E$.

Definition 1 (See [13]). *A pair (F,A) is called a soft set over U, where F is a mapping given by*

$$F : A \rightarrow P(U).$$

In other words, a soft set over U is a parameterized family of subsets of the universe U. For $\varepsilon \in A$, $F(\varepsilon)$ may be considered as a set of ε-elements of the soft set (F,A) or as the set of ε-approximate elements of the soft set. Clearly, a soft set is not a (crisp) set. As for illustration, Molodtsov has considered several examples in [13]. The example shows that, soft set (F,A) can be viewed as a collection of approximations, where each approximation has two parts:-

(i) A predicate p; and
(ii) An approximate value-set v (or simply to be called value-set v).

We denote $(F,A) = \{p_1 = v_1, p_2 = v_2, \ldots, p_n = v_n\}$, where n is a number of predicates.

Based on the definition of an information system and soft set, we then show that a soft set is a special type of information systems, i.e., a binary-valued information system.

Proposition 2. *A pair $(F,A) = \{(F,a_i) : 1 \leq i \leq |A|\}$ is a multi-soft set over the universe U, then (F,A) is a multi-valued information system $S = (U,A,V,f)$.*

Proof. The above proposition has been proven in [8].

Since the definition of soft sets is based on the mapping of value sets to the set of objects, it can then only handle one kind of inconsistency of decision - the one related to the inclusion of objects into different value-set of attributes, i.e., predicates. While this is sufficient for classification of taxonomy type, the classical soft set approach fails in case of ordinal classification with monotonicity constraints [16], where the value-sets of attributes are preference ordered. In this case, decision examples may be inconsistent in the sense of violation of the dominance principle which requires that an object x dominating object y on all considered criteria (i.e., x having evaluations at least as good as y on all considered criteria).

3 Preference Relation Based on Soft Set Theory

In this section, we present the principle of preference relation based on soft set theory which can be used in decision making analysis. As mention the previous section, multi-valued information system $S = (U,A,V,f)$ can be represented by a multi-soft set $(F,A) = \{(F,a_i) : 1 \leq i \leq |A|\}$, where A is a finite set of parameters representing the set of attributes in multi-valued information system. The set A is, in general, divided into set C of condition attributes and set D of decision attributes.

Condition attributes with value sets ordered according to decreasing or increasing preference of a decision maker are called criteria. Since for each criterion $c \in A$ is represented by criterion soft set $C = (F,c)$ in multi-soft set (F,A), the value sets for criterion c is equivalence to the value sets of the soft set when the predicates are ordered according to decreasing or increasing preference. For soft set $C = (F, c)$, $\succeq_c$ is an outranking relation on U with reference to soft set $C \in (F, A)$ such that $x \succeq_c y$ means "x is at least as good as y with respect to soft set C".

Definition 3 *(See [20]). For criterion soft set $C = \{p_i = v_i, i = 1, 2, \ldots, n\}$, for all $r, s \in \{1, .., n\}$, such that $r > s$, predicate p_r is dominance than predicate p_s with respect to C, if the value of p_r is preferred to p_s, and we denote that by $p_r \succeq_c p_s$.*

Definition 4 *(See [20]). For criterion soft set $C = \{p_i = v_i, i = 1, 2, \ldots, n\}$, the set of objects in value-set v_r are preferred to the set of objects in value-set v_s with respect to C, if predicate p_r is dominance than predicate p_s, we denote it by $v_r \succeq_c v_s$ iff $p_r \succeq_c p_s \forall r, s \in \{1, \ldots, n\}$.*

Furthermore, let suppose that the set of decision attributes D is a singleton $\{d\}$ and is represented by decision soft set $D = (F, d)$. The values of predicate in D make a partition of universe U into a finite number of decision classes, $Cl = \{Cl_t, t = 1, \ldots, n\}$, such that each $x \in U$ belongs to one and only one class $Cl_t \in Cl$. It is supposed that the classes are preference-ordered, i.e. for all $r, s \in \{1, .., n\}$, such that $r > s$, the objects from Cl_r are preferred to objects from Cl_s. If $\succeq$ is a comprehensive weak preference relation on U, i.e. if for all $x, y \in U$, $x \succeq y$ means "x is comprehensively at least as good as y", it is supposed: $[x \in Cl_r, y \in Cl_s, r > s] \Rightarrow [x \succeq y$ and not $y \succeq x]$. The above assumptions are typical for consideration of ordinal classification problems (also called multiple criteria sorting problems).

The set to be approximated are called *upward union* and *downward union* of classes, respectively:

$$Cl_t^{\succeq} = \bigcup_{s \geq t} Cl_s, \quad Cl_t^{\preceq} = \bigcup_{s \leq t} Cl_s, \quad t = 1, \ldots, n.$$

The statement $x \in Cl_t^{\succeq}$ means "x belongs to at least class Cl_t", while $x \in Cl_t^{\preceq}$ means "x belongs to at most class Cl_t". Let us remark that $Cl_1^{\succeq} = Cl_n^{\preceq} = U$, $Cl_n^{\succeq} = Cl_n$ and $Cl_1^{\preceq} = Cl_1$. Furthermore, for $t = 2, .., n$,

$$Cl_{t-1}^{\preceq} = U - Cl_t^{\succeq} \text{ and } Cl_t^{\succeq} = U - Cl_{t-1}^{\preceq}.$$

The key idea of the multiple criteria preference relation is representation (approximation) of knowledge generated by decision soft set, using granules of knowledge generated by criterion soft sets. In preference relation, the knowledge to be represented is a collection of upward and downward unions of classes, and the granules of knowledge are sets of objects *defined using a soft dominance* relation [20].

x dominates y with respect to $P \subseteq C$(shortly, xP_{soft}-*dominates* y), denoted by $xD_P y$, if for every criterion soft set $q \in P$ and $x \in v_i, y \in v_j$: $v_i \succeq_q v_j$. The relation P_{soft}-dominance is reflexive and transitive, i.e. it is a partial order.

Given a set of soft sets $P \subseteq C$ and $x \in U$, the "granules of knowledge" used for approximation are:

- a set of objects dominating x, called P_{soft}*-dominating* set, $D_P^+(x) = \{y \in U : yD_Px\}$,
- a set of objects dominated by x, called P_{soft}*-dominated* set, $D_P^-(x) = \{y \in U : xD_Py\}$.

From the above, it can be seen that the "*granules of knowledge*" have the form of upward (positive) and downward (negative) dominance cones in the evaluation space. Recall that the dominance principle requires that an object x dominating object y on all considered criterion soft sets or criteria (i.e., x having evaluation at least as good as y on all considered soft set) should also dominate y on decision soft set (i.e., x should be assigned to at least as good decision class as y). This is the only principle widely agreed upon in the multiple criteria comparisons of objects.

Given $P \subseteq C$, the inclusion of an object $x \in U$ to the upward union of classes $Cl_t^{\succeq}(t = 2, \ldots, n)$ is inconsistent with the dominance principle if one of the following conditions holds:

- x belongs to class Cl_t or better, but it is P_{soft}-dominated by an object y belonging to a class worse than Cl_t, i.e. $x \in Cl_t^{\succeq}$ but $D_P^+(x) \cap Cl_{t-1}^{\preceq} \neq \varphi$,
- x belongs to a worse class than Cl_t but it P_{soft}-dominates an object y belonging to class Cl_t or better, i.e. $x \notin Cl_t^{\succeq}$ but $D_P^-(x) \cap Cl_{t-1}^{\succeq} \neq \varphi$.

If, given a set of soft set $P \subseteq C$, the inclusion of $x \in U$ to $Cl_t^{\succeq}(t = 2, \ldots, n)$ is inconsistent with the dominance principle, then x belongs to $Cl_t^{\succeq}$ with some ambiguity. Thus, x belongs to $Cl_t^{\succeq}$ without any ambiguity with respect to $P \subseteq C$, if $x \in Cl_t^{\succeq}$ and there is no inconsistency with the dominance principle. This means that all objects P_{soft}-dominating x belong to $Cl_t^{\succeq}$, i.e., $D_P^+(x) \subseteq Cl_t^{\succeq}$.

Furthermore, x possibly belongs to $Cl_t^{\succeq}$ with respect to $P \subseteq C$ if one of the following conditions holds:

- according to decision soft set $D = (F, d)$, object x belongs to $Cl_t^{\succeq}$,
- according to decision soft set $D = (F, d)$, object x does not belong to $Cl_t^{\succeq}$, but it is inconsistent in the sense of the dominance principle with an object y belonging to $Cl_t^{\succeq}$.

In terms of ambiguity, x possibly belongs to $Cl_t^{\succeq}$ with respect to $P \subseteq C$, if x possibly belongs to $Cl_t^{\succeq}$ with or without ambiguity. Due to the reflexivity of the soft dominance relation D_P, the above conditions can be summarized as follows: x possibly belongs to class Cl_t or better, with respect to $P \subseteq C$, if among the objects P_{soft}-dominated by x there is an object y belonging to class Cl_t or better, i.e., $D_P^-(x) \cap Cl_t^{\succeq} \neq \varphi$.

The P_{soft}-lower approximation of $Cl_t^{\succeq}$, denoted by $\underline{P}(Cl_t^{\succeq})$, and the P_{soft}-upper approximation of $Cl_t^{\succeq}$, denoted by $\overline{P}(Cl_t^{\succeq})$, are defined as follows (t = 1,…,n):

$$\underline{P}(Cl_t^{\succeq}) = \left\{x \in U : D_P^+(x) \subseteq Cl_t^{\succeq}\right\},$$

$$\bar{P}(Cl_t^{\succeq}) = \left\{x \in U : D_P^-(x) \cap Cl_t^{\succeq} \neq \varphi\right\}.$$

Analogously, one can define the P_{soft}-lower approximation and the P_{soft}-upper approximation of $Cl_t^{\preceq}$ as follows (t = 1,…,n):

$$\underline{P}(Cl_t^{\preceq}) = \left\{x \in U : D_P^-(x) \subseteq Cl_t^{\preceq}\right\},$$
$$\overline{P}(Cl_t^{\preceq}) = \left\{x \in U : D_P^+(x) \cap Cl_t^{\preceq} \neq \varphi\right\}.$$

The P_{soft}-lower and P_{soft}-upper approximation defined above, satisfy the following properties for each $t \in \{1,\ldots,n\}$ and for any $P \subseteq C$:

$$\underline{P}(Cl_t^{\succeq}) \subseteq Cl_t^{\succeq} \subseteq \overline{P}(Cl_t^{\succeq}),\ \underline{P}(Cl_t^{\preceq}) \subseteq Cl_t^{\preceq} \subseteq \overline{P}(Cl_t^{\preceq}).$$

The P_{soft}-lower and P_{soft}-upper approximations of $Cl_t^{\succeq}$ and $Cl_t^{\preceq}$ have an important complementary property, according to which,

$$\begin{array}{c}\underline{P}(Cl_t^{\succeq}) = U - \overline{P}(Cl_{t-1}^{\preceq}) \text{ and } \overline{P}(Cl_t^{\succeq}) = U - \underline{P}(Cl_{t-1}^{\preceq}), t = 2,\ldots,n \\ \underline{P}(Cl_t^{\preceq}) = U - \overline{P}(Cl_{t+1}^{\succeq}) \text{ and } \overline{P}(Cl_t^{\preceq}) = U - \underline{P}(Cl_{t+1}^{\succeq}), t = 1,\ldots,n-1\end{array}.$$

The P_{soft}-boundaries of $Cl_t^{\succeq}$ and $Cl_t^{\preceq}$, denoted by $Bn_P(Cl_t^{\succeq})$ and $Bn_P(Cl_t^{\preceq})$ respectively, and defined as follows (t = 1,…,n):

$$Bn_P(Cl_t^{\succeq}) = \overline{P}(Cl_t^{\succeq}) - \underline{P}(Cl_t^{\succeq}), \text{ and } Bn_P(Cl_t^{\preceq}) = \overline{P}(Cl_t^{\preceq}) - \underline{P}(Cl_t^{\preceq}).$$

Due to complementary property, $Bn_P(Cl_t^{\succeq}) = Bn_P(Cl_{t-1}^{\preceq})$, for t = 2,…,n.

For any criterion soft set $P \subseteq C$, we define the accuracy of approximation of $Cl_t^{\succeq}$ and $Cl_t^{\preceq}$ for all $t \in T$ respectively as

$$\alpha_P(Cl_t^{\succeq}) = \frac{\left|\underline{P}(Cl_t^{\succeq})\right|}{\left|\overline{P}(Cl_t^{\succeq})\right|},\ \alpha_P(Cl_t^{\preceq}) = \frac{\left|\underline{P}(Cl_t^{\preceq})\right|}{\left|\overline{P}(Cl_t^{\preceq})\right|}.$$

The quality of approximation of the ordinal classification Cl by a set of soft set P is defined as the ration of the number of objects P_{soft}-consistent with the dominance principle and the number of all objects in U. Since the P_{soft}-consistent objects are those which do not belong to any P_{soft}-boundary $Bn_P(Cl_t^{\succeq})$, t = 2,…,n, or $Bn_P(Cl_t^{\preceq})$, t = 1,…, n-1, the quality of approximation of the ordinal classification Cl by a set of soft set P, can be written as

$$\gamma_P(Cl) = \frac{\left|U - \left(\left(\bigcup_{t \in T} Bn_P(Cl_t^{\succeq})\right) \cup \left(\bigcup_{t \in T} Bn_P(Cl_t^{\preceq})\right)\right)\right|}{|U|}.$$

$\gamma_P(Cl)$ can be seen as a degree of consistency of the objects from U, where P is the set of criterion soft set and Cl is the considered ordinal classification. Every minimal subset $P \subseteq C$ such that $\gamma_P(Cl) = \gamma_C(Cl)$ is called a *reduct* of Cl and is denoted by RED_{Cl}.

Moreover, for a given set of U one may have more than one *reduct*. The intersection of all *reducts* is known as the core, denoted by $CORE_{Cl}$.

The dominance-based soft approximations of upward and downward unions of classes can serve to classify objects based on similarity preference into different clusters. It is therefore meaningful to formalize object classification by the following definition.

Definition 5. *Let $(F,A) = \{(F,a_i) : 1 \leq i \leq |A|\}$ be a multi-soft set representing information system, $S = (U,A,V,f)$ and $Cl_t^{\succeq}(t = 1,\ldots,n)$ are positive dominance classes in decision soft set (F,d). The inclusion of object $x \in U$ into cluster Cr_t, for $t = 1,2,..n$, is defined as follow.*

$$Cr_t = \left\{ \begin{array}{l} x \in \underline{P}\left(Cl_t^{\succeq}\right),\ t = n \\ x \in \underline{P}\left(Cl_t^{\succeq} \backslash \underline{P}\left(Cl_{t+1}^{\succeq}\right),\ t = 1,..,n-1\right) \end{array} \right\}$$

4 Experiments

On the basis of data from Grabisch [5] which has been modified in [16], the same data is used here as a test case for an ordinal classification problem. Students of a college must obtain an overall evaluation on the basis of their achievements in Mathematics, Physics and Literature. The director of the college wants to assign students to three classes: bad, medium and good. To fix the classification rules, the director is asked to present some examples. The examples concern with eight students described by means of four attributes (see Table 1 below):

The components of the multi-valued information table S are:

$$U = \{1,2,3,4,5,6,7,8\}$$

Table 1. Multi-valued information table (decision table)

Student	A_1 (Mathematics)	A_2 (Physics)	A_3 (Literature)	A_4 (Evaluation)
1	good	medium	bad	bad
2	medium	medium	bad	medium
3	medium	medium	medium	medium
4	good	good	medium	good
5	good	medium	good	good
6	good	good	good	good
7	bad	bad	bad	bad
8	bad	bad	medium	bad

$$A = \{A_1, A_2, A_3, A_4\}$$

$$V_1 = V_2 = V_3 = V_4 = \{bad, medium, good\}$$

the information function $f(x, q)$, taking values $f(1, A_1)$ = good, $f(1, A_2)$ = medium, and so on.

Within this approach, we approximate the class $Cl_t^{\succeq}$ of "(at least) good", "(at least) medium" and "(at least) bad" students. Since information table as in Table 1 above consist of three decision classes, we have $Cl_1^{\preceq} = Cl_1$, and $Cl_3^{\geq} = Cl_3$. Moreover, $C = \{A_1, A_2, A_3\}$ and $D = \{A_4\}$. In this case, however, A_1, A_2 and A_3 are criteria and the classes are preference-ordered. Furthermore, the multi-soft set equivalent to the multi-valued information table as in Table 1 is given below:

$$(F, A) = \begin{cases} (F, a_1) = \{bad = \{7, 8\}, medium = \{2, 3\}, good = \{1, 4, 5, 6\}\} \\ (F, a_2) = \{bad = \{7, 8\}, medium = \{1, 2, 3, 4, 5\}, good = \{4, 6\}\} \\ (F, a_3) = \{bad = \{1, 2, 7\}, medium = \{3, 4, 8\}, good = \{5, 6\}\} \\ (F, d) = \{bad = \{1, 7, 8\}, medium = \{2, 3\}, good = \{4, 5, 6\}\} \end{cases}.$$

Our experiment obtained the following results. The P_{soft}-lower approximations, P_{soft}-upper approximations and the P_{soft}-boundaries of classes $Cl_1^{\succeq}$, $Cl_2^{\succeq}$ and $Cl_3^{\succeq}$ are equal to:

$$\underline{P}(Cl_1^{\succeq}) = \{U\},\ \overline{P}(Cl_1^{\succeq}) = \{U\},\ Bn_P(Cl_1^{\succeq}) = \{\varphi\},$$

$$\underline{P}(Cl_2^{\succeq}) = \{3, 4, 5, 6\},\ \overline{P}(Cl_2^{\succeq}) = \{1, 2, 3, 4, 5, 6\},\ Bn_P(Cl_2^{\succeq}) = \{1, 2\},$$

$$\underline{P}(Cl_3^{\succeq}) = \{4, 5, 6\},\ \overline{P}(Cl_3^{\succeq}) = \{4, 5, 6\},\ Bn_P(Cl_3^{\succeq}) = \{\varphi\}.$$

Therefore, the accuracy of the approximation for $Cl_1^{\succeq} = Cl_3^{\succeq} = 1$, and $Cl_2^{\succeq} = 0.67$, while the quality of approximation is equal to 0.8. Using Definition 5, the clusters are obtained and given as follows: $Cr_3 = \{4, 5, 6\}, Cr_2 = \{3\}$, and $Cr_1 = \{1, 2, 7, 8\}$. Please take note that object (student) 2 is assigned to cluster #1 together with students #2, #7 and #8, although the overall evaluation for student #2 was better than the other three students within the cluster. It can be seen from Table 1 that student #2 do have not better evaluation than student #1 on all the considered criteria, however, the overall evaluation of student #2 is better than the overall evaluation of student #1. Therefore, this can be seen as an inconsistency revealed by the approximation based on soft-dominance that cannot be captured by the normal approximation by mapping of parameters to the set of objects under consideration.

5 Conclusion

In this paper we have presented the applicability of soft set theory for discovering preference relation in multi-valued information system. Dominance relation based on soft set theory is used to deal with typical inconsistencies during the consideration of criteria or preference-ordered decision classes. Based on the approximations obtained through the soft-dominance relation, it is possible to classify objects based on the preferential information contained in the decision system. The classifications or clusters obtained by this approach are able to eliminate inconsistency in the process of classifying object according to their similarity preferences. Our experiment shows that the quality of approximation obtained by the proposed approach, i.e., DSSA is equal to 0.8 which is better the the quality of approximation obtained by [16].

References

1. Albadvi, A., Chaharsooghi, S., Esfahanipour, A.: Decision making in stock trading: an application of PROMETHEE. Eur. J. Oper. Res. **177**(2), 673–683 (2007)
2. Chaudhuri, A., De, K., Chatterjee, D.: Solution of decision making problems using fuzzy soft relations. Int. J. Inf. Technol. **15**(1), 78–107 (2009)
3. Feng, F., Jun, Y.B., Liu, X., Li, L.: An adjustable approach to fuzzy soft set based decision making. J. Comput. Appl. Math. **234**, 10–20 (2010)
4. Feng, F., Li, Y., Violeta, L.-F.: Application of level soft sets in decision making based on interval-valued fuzzy soft sets. J. Comput. Math. Appl. **60**, 1756–1767 (2010)
5. Grabisch, M.: Fuzzy integral in multicriteria decision making. Fuzzy Sets Syst. **89**, 279–298 (1994)
6. Greco, S., Matarazzo, B., Slowinski, R.: Rough approximation of a preference relation in a pairwise comparison table. Rough Sets in Knowledge Discovery: Applications, case studies, and software systems. p. 13 (1998)
7. Greco, S., Matarazzo, B., Slowinski, R.: Rough sets theory for multicriteria decision analysis. Eur. J. Oper. Res. **129**(1), 1–47 (2001)
8. Herawan, T., Deris, M.M.: On multi-soft sets construction in information systems. In: Huang, D.-S., Jo, K.-H., Lee, H.-H., Kang, H.-J., Bevilacqua, V. (eds.) ICIC 2009. LNCS (LNAI), vol. 5755, pp. 101–110. Springer, Heidelberg (2009). doi:10.1007/978-3-642-04020-7_12
9. Hu, Q., Xie, Z., Yu, D.: Hybrid attribute reduction based on a novel fuzzy-rough model and information granulation. Pattern Recogn. **40**(12), 3509–3521 (2007)
10. IsIklar, G., Buyukozkan, G.: Using a multi-criteria decision making approach to evaluate mobile phone alternatives. Comput. Stand. Interfaces. **29**(2), 265–274 (2007)
11. Jiang, Y., Tang, Y., Chen, Q.: An adjustable approach to intuitionistic fuzzy soft sets based decision making. J. Appl. Math. Modell. **35**, 824–836 (2011)
12. Maji, P.K., Roy, A.R., Biswas, R.: An application of soft sets in a decision making problem. Comput. Math Appl. **44**, 1077–1083 (2002)
13. Molodtsov, D.: Soft set theory-first results. Comput. Math Appl. **37**, 19–31 (1999)
14. Pawlak, Z.: Rough Sets: Theoretical Aspects of Reasoning About Data. Springer, Heidelberg (1991)
15. Roy, A.R., Maji, P.K.: A fuzzy soft set theoretic approach to decision making problem. J. Comput. Appl. Math. **203**, 412–418 (2007)
16. Slowinski, R.: Rough set approach to knowledge discovery about preferences. In: Nguyen, N.T., Kowalczyk, R., Chen, S.-M. (eds.) ICCCI 2009, LNAI 5796, pp. 1–21. Springer-Verlag, Berlin Heidelberg (2009)
17. Wang, X., Yang, J., Teng, X., Xia, W., Jensen, R.: Feature selection based on rough sets and particle swarm optimization. Pattern Recogn. Lett. **28**(4), 459–471 (2007)

18. Yu, L., Wang, S., Lai, K.: An intelligent-agent-based fuzzy group decision making model for financial multicriteria decision support: the case of credit scoring. Eur. J. Oper. Res. **195**(3), 942–959 (2009)
19. Zdzislaw, P.: Rough sets. Int. J. Comput. Inf. Sci. **11**(5), 341–356 (1982)
20. Isa, A.M., Rose, A.N.M., Deris, M.M.: Dominance-based soft set approach in decision-making analysis. In: Tang, J., King, I., Chen, L., Wang, J. (eds.) ADMA 2011. LNCS (LNAI), vol. 7120, pp. 299–310. Springer, Heidelberg (2011). doi:10.1007/978-3-642-25853-4_23

Reduce Scanning Time Incremental Algorithm (RSTIA) of Association Rules

Iyad Aqra[1(✉)], Muhammad Azani Hasibuan[2], and Tutut Herawan[1]

[1] Department of Information Systems, University of Malaya, 50603 Pantai Valley, Kuala Lumpur, Malaysia
i_aqra@siswa.um.edu.my, tutut@um.edu.my
[2] Faculty of Industrial Engineering, Telkom University, Bandung, Indonesia
muhammadazani@telkomuniversity.ac.id

Abstract. In the real world where large amounts of data grow steadily, some old association rules can become stale, and new databases may give rise to some implicitly valid patterns or rules. Hence, updating rules or patterns is also important. A simple method for solving the updating problem is to reapply the mining algorithm to the entire database, but this approach is time-consuming. This paper reuses information from old frequent itemsets to improve its performance and addresses the problem of high cost access to incremental databases in which data are very changing by reducing the number of scanning times for the original database. A log file has been used to keep track of database changes whenever, a transaction has been added, deleted or even modified, a new record is added to the log file. This helps identifying the newly changes or updates in incremental databases. A new vertical mining technique has been used to minimize the number of scanning times to the original database. This algorithm has been implemented and developed using C#.net and applied to real data and gave a good result comparing with pure Apriori.

Keywords: Association rule mining (ARM) · Incremental learning · Itemset · Vertical layout

1 Introduction

On the other hand, depending on the classes of knowledge derived, the mining approaches may be classified as finding association rules, classification rules, clustering rules, and sequential patterns 4, among others. Among them, finding association rules in transaction databases is most commonly seen in data mining [1–3].

All frequent itemsets are first found based on a user-defined minimum support threshold and then the association rules are derived from the discovered frequent itemsets based on the user-defined minimum confidence threshold. In association rule mining, each item is treated as a binary variable to discover relationships among itemsets or products.

Continuously, works are yet going on improvement and development of data mining algorithms, and this works have produced a variety of efficient techniques. Generally, the data mining has high attractive field, and it is becomes popular hotspot

T. Herawan et al. (eds.), *Recent Advances on Soft Computing and Data Mining*,
Advances in Intelligent Systems and Computing 549, DOI 10.1007/978-3-319-51281-5_49

field for researcher. since the ARM have introduced in [1, 4], it has one of the important nook of data mining field particularly, In many mining techniques and learning [2], ARM is the most popular method used for knowledge discovery to find the relationships among items or products. Moreover, where is data? That mean we can use data mining technique. The structure of data is not matter, the mining technique work over many database type like: transaction databases, temporal databases, relational databases, and multimedia databases, among others (we need reference) [5]. The most common implementations use the Apriori algorithm [1] for generating and testing the candidate itemsets level by level. So recently, many researchers going on development of data mining algorithm and it has attractive attention, while there are wide convergence between data mining and other practice fields.

Where, data mining defined as complex process on order to extract useful pattern from huge database, moreover, it has capability of looking for knowledge from huge database, with different database form for real application. On other word, data mining is processes goal to produce accommodating pattern called knowledge from enormous data on varies databases. There are numerous tasks or strategy in data mining, for example [3]: association rules discovery, clustering rules, classification rules, associative technique, sequential patterns, and decision tree. Among them, discovery association rules in databases is most commonly process seen in data mining, and most of the data mining techniques have discovery association rules. Where ARM acquainted as iterative statistical process, it aims to discovery correlation between collection features on database.

Assumption for discovery itemset is going to became immensely colossal volumes, it is difficult to manage and used. But association frequent itemset is the essential step in some of data mining technique like associative mining [6], classification mining [1], clustering mining [14], and this technique arrange the frequent itemset and rule with opportune way to use it, either arrange it a group by clustering, or build classifier to select suitable rule in associative and classification mining.

In general, all data mining techniques and algorithms have output named, as mention before, knowledge, particularly it is represent or called rule, this knowledge has difference characteristic for each, but totally, this knowledge form different technique are cores in basic common: the knowledge discovered is statistical terms, since the rules hold significant representation for data in dataset is better compatible with other rule has less representation data.

Indeed, it is hard to get one rule to reflect all data, in other words; it is insubstantial for one rule to represent for all the data. For that, the learned knowledge usually contains a lot of rule. Other characteristic of extracted Knowledge, and this has more essential for our research work in this search project, the extraction knowledge is valid for instance time, or the knowledge represent the present state for mined database, while there are no critical and significant changing in database, either insightful addition, updated, or deletion transaction. For that, there are two methodology to keep legitimacy for knowledge, one of them when state of database change rediscovery knowledge from scratch, the second perspective, mined knowledge by do impact of change in database in knowledge if there are addition represent this addition to knowledge, or if there are deletion do a reversal for effect erased transaction record in extracted knowledge.

There are many proposed algorithm, which manage association rule discovery, the primary one is apriori [4], Eclat [7], DHP [8], TIDAPriori [9], McEclat [10]. And MsApriori proposed in [11] making enhancement to collect rare frequent item, and as we mention many data mining technique use association rule as essential step in the technique step.

In general, the association rule mining problem, we can decomposed it to two sub problems:

1. Discovery the large itemset set called frequent itemset, which have occurrence in database more than minimum support. Where the minimum support is input threshold.
2. The second problem, using the yield from frequent itemset to generate association rule.

There are different data mining problems, but the mining of association rules is an important one. A well-known illustration or case for association rules is about basket market analysis. Where, a record in the sales data depicts all the items that are purchased in a single transaction. Together with other information such as the transaction time, customer-id, etc. mining association rules from such a database is to discover from the huge amount of past transactions, all the rules like "A customer who buys item A and item B is most likely to buy item C in the same transaction". Where A, B and C are initially unknown. Such rules are very useful for marketers to develop and implement customized promoting marketing programs and strategies.

A feature of data mining problems is that in order to have stable and reliable results, a huge amount of data has to be collected and analyzed. The large amount of input data and mining results poses a maintenance problem. While new transactions are being appended to a database and obsolete ones are being removed. Rules that already discovered also have to be updated. In this paper we examine the problem of maintaining discovered association rules. We propose a new incremental algorithm which can handle all the update cases including insertion, deletion and modification of transactions.

However, in Apriori [1], the discovery of frequent itemsets from transactional databases is proficient in a step savvy style. The frequent itemsets have found in a particular step (n), they will be used to produce potential frequent itemsets – known as candidate itemset - at next step (n + 1). During each step, a database scan is vital to perform support counting of the new candidate itemsets. After the presentation of Apriori, various association rule algorithms [7, 12] have concentrated on improving Apriori candidate generation step by reducing the number of database passes, main memory usage and other CPU costs.

In this paper, a new incremental technique in light of vertical mining has been introduced which points with reduce the number of original database scanning time. This technique build the itemset and every mining time update this set without rebuild itemset. The other thing, it is dealing with all manipulation operations.

The rest of this paper in next section, Sect. 2, some of the literature review have been stated and reviewed. In the Sect. 3, the proposed approach has been explained. The conclusion has been stated in Sect. 4.

2 Literature Review

In the real world where large amounts of data grow steadily, some old association rules can become stale, and new databases may give rise to some implicitly valid patterns or rules. Hence, updating rules or patterns is also important. A simple method for solving the updating problem is to reapply the mining algorithm to the entire database, but this approach is time-consuming. The algorithm in this paper reuses information from old frequent itemsets to improve its performance. Several other approaches to incremental mining have been proposed.

Although many mining techniques for discovering frequent itemsets and associations have been presented, the process of updating frequent itemsets remains trouble for incremental databases. The mining of incremental databases is more complicated than the mining of static transaction databases, and may lead to some severe problems, such as the combination of frequent itemsets occurrence counts in the original database with the new transaction database, or the rescanning of the original database to check whether the itemsets remain frequent while new transactions are added.

This work proposes an algorithm for incremental mining, which can discover the latest rules and doesn't need to rescan the original database. In (D.W. Cheung) the authors have proposed an algorithm called Fast Update algorithm (FUP) to efficiently generate associations in the updated database.

The FUP algorithm relies on Apriori and considers only these newly added transactions. Let db be a set of new transactions and DB be the updated database (including all transactions of DB and db). An itemset x is either frequent or infrequent in DB or db. Therefore, x has four possibilities, as shown in Table 1. In the first pass, FUP scans db to obtain the occurrence count of each 1-itemset. Since the occurrence counts of Fk in DB are known in advance, the total occurrence count of arbitrary x is easily calculated if x is in Case 2. If x is unfortunately in Case 3, DB must be rescanned. Similarly, the next pass scans db to count the candidate 2-itemsets of db. If necessary, DB is rescanned. The process is reiterated until all frequent itemsets have been found. In the worst case, FUP does not reduce the number of the original database must be scanned.

Furthermore, a recently many algorithms have proposed [7, 13, 14], they adopt the tid-list intersection methods of [7] from association rule to discover classification rules in a single training data scan. A tid-list of an item is the transaction numbers (tids) in the database that contain that item. Experimental results on real world data and synthetic data revealed that algorithms that employ tid-lists fast intersection method outperform Apriori-like ones with regards to processing time and memory usage. In spite of the advantage of tid-list intersection approach, when the cardinality of the tid-list

Table 1. Four scenarios associated with an itemset in DB

DB	db	
	Frequent itemset	Infrequent itemset
Frequent itemset	Case 1: Frequent	Case 2
Infrequent itemset	Case 3	Case 4: Infrequent

becomes very large, intersection time gets larger as well. This happens particularly for large and correlated transactional databases.

Finally, developing efficient frequent rule items discovery methods, which decrease the number of database scans and minimize the use of complex data structure objects during the learning step is vital. This is because that most of the time is spent during the training phase.

3 Basic Concept and Terminology

ARM involves looking for relation between item in transactions, detecting items which tend to occur together in transactions and the association rules that relate them [14].

Several key terms are utilized in frequent itemset mining and have been specified in the Introduction. In this section, we clarify and formulate these expressions to present the fundamental concepts of frequent itemset mining. For a clearer depiction, we employ market basket as an example to exhibit meaning in a significant manner. The following definition describes the notion of item set.

Consider $I = \{i_1, i_2, i_3, \cdots, i_m\}$ as a set of items, in the other word the set I refer to things or products in supermarket according to popular case study for association rule mining. A transaction T is over I is defined as a pair $T = (tid, I')$, where *tid* is the transaction identifier and $I' \subseteq I$. The large amount from transaction, set of {T}, collected in one collection, and it called transactional database.

The association rule mining process compound by two step: the first one is to discover the frequent patterns, they named itemset, $Itemset = \{A, B, C, \cdots\}$.

Let D, the task relevant data, is a set of database transactions where each transaction T is a set of items such that T is a subset of I. Each transaction is associated with an identifier, called TID. Let A be a set of items. A transaction T is said to contain A if and only if A $\subseteq$ T. An Association Rule is an implication of the form A $\Rightarrow$ B, where A $\subset$ I, B $\subset$ I, and A $\cap$ B = φ.

3.1 Apriori Algorithm

The Apriori algorithm concentrates primarily on the discovery of frequent itemsets according to a user-defined minSup. The algorithm relies on the fact that an itemset could be frequent only when each of its subset is frequent; otherwise, the itemset is infrequent. In the first pass, the Apriori algorithm constructs and counts all 1-itemsets. (A k-itemset is an itemset that includes k items.) After it has found all frequent 1-itemsets, the algorithm joins the frequent 1-itemsets with each other to form candidate 2-itemsets. Apriori scans the transaction database and counts the candidate 2-itemsets to determine which of the 2-itemsets are frequent. The other passes are made accordingly. Frequent (k − 1)-itemsets are joined to form k-itemsets whose first k-1 items are identical. If k 3, Apriori prunes some of the k-itemsets; of these, (k − 1)-itemsets have at least one infrequent subset. All remaining k-itemsets constitute candidate k-itemsets. The process is reiterated until no more candidates can be generated.

The following example is showing the procedure of association rule discovery according to Apriori perception.

Example 1: Consider the database presented in Table 2 with a minimum support requirement is 50%. The database includes 11 transactions. Accordingly, the supports of the frequent itemsets are at least six. The first column "TID" includes the unique identifier of each transaction, and the "Items" column lists the set of items of each transaction.

Let Ck be the set of candidate k-itemsets and Fk be the set of frequent k-itemsets. In the first pass, the database is scanned to count C1. If the support count of a candidate exceeds or equals six, then the candidate is added to F1. The outcome is shown in the Tables 3, 4 and 5. Then, F1 & F1 forms C2 (Apriori-gen function is used to generate C2) after the database has been scanned for a second time, Apriori examines which itemset of C2 exceeds the predetermined threshold.

Moreover, C3 is generated from F2 as follows. Table 4 presents two frequent 2-itemsets with identical first item, such as {BC} and {BE}. Then, Apriori tests whether the 2-itemset [15] is frequent. In the Table 5, the itemset {BCE} a candidate in 3-itemset is infrequent, so {BCE} must be pruned. Apriori stops to look for frequent itemsets when no candidate 4-itemset can be joined from F3. Apriori scans the database k times when candidate k-itemsets are generated.

Table 2. An example of a transaction database

TID	Items
001	A C D E
002	A C D
003	B C E
004	A B C E
005	A B E
006	B C E
007	A B E
008	B C D E
009	A B C D
010	C E F
011	A B C F

The following picture capture the database Passes.

Table 3. First pass in Apriori algorithm

Scan **DB**

C_1	Count
{A}	7
{B}	8
{C}	9
{D}	4
{E}	8
{F}	2

Apply MinSup ⇨

F_1	Count
{A}	7
{B}	8
{C}	9
{E}	8

Table 4. Second pass in Apriori algorithm

F_1	C_2		C_2	Count	Apply MinSupp ⇨	F_2	Count
	{AE}		{AB}	5		{BC}	6
	{AC}	Scan	{AC}	5		{BE}	6
	{AE}	DB	{AE}	4		{CE}	6
	{BC}	⇨	{BC}	6			
	{BE}		{BE}	6			
	{CE}		{CE}	6			

Table 5. Third pass in Apriori algorithm

	C_2	Scan DB ⇨	C_2	Count	Apply MinSupp ⇨	F_2	Count
$F_2 \infty F_2$	{BCE}		{BCE}	4		{}	

4 Reduce Scanning Time Incremental Algorithm (RSTIA)

Apriori algorithm is based on finding large itemsets from database transactions by keeping a count for every itemset. However, since the number of possible itemsets is exponential to the number of items in the database, it is impractical to count every subset we encounter in the database transactions. The Apriori algorithm tackles the combinatorial explosion problem by using an iterative approach to count the itemsets.

The iterative nature of the Apriori algorithm implies that at least n database passes are needed to discover all the large itemsets if the biggest large itemsets are of size n. since database passes involve slow access, to increase efficiency, we should minimize the number of database passes during the mining process. One solution is to generate bigger-sized candidate itemsets as soon as possible, so that their supports can be counted early.

This paper addresses the problem of high cost access to incremental databases in which data are very changing and time-varying by reducing the number of scanning times for the original database.

In this approach, a log file has been used to keep track of database changes, the log file contains three columns which are the transactionID, ActionID and finally ActionDate. Whenever, a transaction has been added, deleted or even modified, a new record is added to the log file. This helps identifying the newly changes or updates in incremental databases by avoiding scanning database to locate newly updates. On the other hand, an xml file has been used to store the date of the last time the algorithm has been executed to ensure that on the next time the algorithm should be executed from the last time it executed.

This algorithm is Apriori-based but, tries to solve the shortcoming of multi scan to the original database by generating bigger-sized candidate itemsets as soon as possible, so that their supports can be counted early.

The algorithm instruction and procedure step list as:

Step 1: check the xml file, if it exists then go step 2, else call vertical_ Apriori.
Step 2: read the log file and compare the ActionDate attribute values with the stored value in the xml file, if they are greater than the value of xml file, then go to step 3.
Step 3: for each record in the log file that satisfy step 2, check the value of ActionID, if one then go to step 4, if two got step 5, if 3 go to steps 5, 4 respectively.

Only two differences are, that a new information is added to frequent times which is the transactions where they lay, and the other difference is that vertical_Apriori scans database fir.

Step 4: one means a new transaction has been added to the database, so, for each item that involved in such transaction, increment its support by one, and concatenate the transactionID to each item involved in it.
Step 5: two means a transaction has been deleted from the database, so, for each item that involved in such transaction, decrement its support by one, and remove the transactionID from each item involved in it.

The coming example capture the methodology of association rule discovery from purposed algorithm perspective.

Example 2: Consider the database presented in Table 6 with a minimum support requirement is 50% and xml file doesn't exists. The history of manipulation over the database shown in log file (Table 7). The database includes 4 transactions. Accordingly, the supports of the frequent itemsets are at least 2. The first column "TID" includes the unique identifier of each transaction, and the "Item" column lists the set of items of each transaction.

Table 6. An example of a transaction database

TID	Item
1	ab
2	bc
3	ac
4	abc

Table 7. LOG file

TID	ActionID	Date
1	1	2011/05/18 12:30
2	1	2011/05/18 12:35
3	1	2011/05/18 12:40
4	1	2011/05/18 12:42

First, there is no rule-generation has been done yet so the algorithm scans the original database and counts all frequent 1-itemsets and store them into FrequentItemsL1 file with their support and transaction ID they located to make it easily finding such transactions, see table below (Tables 8, 9, 10, 11 and 12).

Table 8. First intermediate itemset

ListIndex	Itemset	Support	TransactionList
1	a	3	1,3,4
2	b	3	1,2,4
3	c	3	2,3,4

Table 9. Second intermediate itemset

ListIndex	Itemset	Support	TransactionList
1	ab	2	1,4
2	ac	2	3,4
3	bc	2	2,4

Table 10. Third intermediate itemset

ListIndex	Itemset	Support	TransactionList
1	abc	1	4

Table 11. All frequent itemset

ListIndex	Itemset	Support
1	a	3
2	b	3
3	c	3
4	ab	2
5	ac	2
6	bc	2

Table 12. All frequent itemset

ListIndex	Itemset	Confidence
1	a->b	66.6%
2	a->c	66.6%
3	b->a	66.6%
4	b->c	66.6%
5	c->a	66.6%
6	c->b	66.6%

In this step, we use FrequentItemsL1 file to generate Candidate itemset of size 2, C2 and count their support by scanning the original database and store them to AllFrequentItem file as it is in previous step.

In the next step, C3 is generated and their support is counted again by scanning the original database and then the resultant C3 with support are appended to the AllFrequentitmem file.

In this step, Apriori Stops, since there is no more candidate itemsets, so we use the All Frequent item file to generate the final_all_frequent_itemset by matching the required support with their support, and discover the most frequent itemsets in the database.

And finally, the rules are generated as follows:

5 Comparison

In this work, the incremental ARM problem has been studied, and a new approach (RSTIA) has been introduced. The RSTIA has improved and included many features over existing techniques. The RSTIA handle the changing in the threshold. The RSTIA does not require scanning database under any case.

In next table, the enhancement features have been summarized, and brief comparison between proposed algorithm and existing approaches state in the next table (Table 13).

Table 13. Comparison between INAP and incremental algorithms

Algorithm	Add	Update	Delete	Rescan solved	The support accuracy	Threshold changes
FUP	✓	×	×	×	×	×
NFUP	✓	×	×	×	×	
IMSC	✓	×	×	×	×	✓
MAAP	✓	×	×	×	×	×
RSTIA	✓	✓	✓	✓	✓	✓

6 Conclusion

This paper reuses information from old frequent itemsets to improve its performance and addresses the problem of high cost access to incremental databases in which data are very changing by reducing the number of scanning times for the original database. A log file has been used to keep track of database changes.

Whenever, a transaction has been added, deleted or even modified, a new record is added to the log file. This helps identifying the newly changes or updates in incremental databases. This algorithm has been implemented and developed using C#.net and applied to real data and gave a good result comparing with pure Apriori.

References

1. Agrawal, R., Imieliński, T., Swami, A.: Mining association rules between sets of items in large databases. ACM SIGMOD Rec. **22**, 207–216 (1993)
2. Bose, I., Mahapatra, R.K.: Business data mining - a machine learning perspective. Inf. Manag. **39**(3), 211–225 (2001)
3. Fayyad, U., Piatetsky-Shapiro, G., Smyth, P.: From data mining to knowledge discovery in databases. AI Mag. **17**(3), 37 (1996)
4. Agrawal, R., Srikant, R.: Fast algorithms for mining association rules. In: Proceedings of 20th International Conference Very Large Data Bases, VLDB (1994)
5. Lin, C.-W., et al.: Efficient updating of discovered high-utility itemsets for transaction deletion in dynamic databases. Adv. Eng. Informat. **29**(1), 16–27 (2015)
6. Chen, F., et al.: Principal association mining: an efficient classification approach. Knowl. Based Syst. **67**, 16–25 (2014)
7. Zaki, M.J., et al.: New algorithms for fast discovery of association rules. In: KDD (1997)
8. Park, J.S., Yu, P.S., Chen, M.-S.: Mining association rules with adjustable accuracy. In: Proceedings of the Sixth International Conference on Information and Knowledge Management. ACM (1997)
9. Li, Z.-C., He, P.-L., Lei, M.: A high efficient AprioriTid algorithm for mining association rule. In: Proceedings of 2005 International Conference on Machine Learning and Cybernetics. IEEE (2005)
10. Schlegel, B., et al.: Scalable frequent itemset mining on many-core processors. In: Proceedings of the Ninth International Workshop on Data Management on New Hardware. ACM (2013)
11. Lee, Y.-C., Hong, T.-P., Lin, W.-Y.: Mining association rules with multiple minimum supports using maximum constraints. Int. J. Approx. Reason. **40**(1), 44–54 (2005)
12. Li, W., Han, J., Pei, J.: CMAR: accurate and efficient classification based on multiple class-association rules. In: Proceedings IEEE International Conference on Data Mining, ICDM 2001. IEEE (2001)
13. Zaki, M.J.: Scalable algorithms for association mining. IEEE Trans. Knowl. Data Eng. **12**(3), 372–390 (2000)
14. Ibrahim, H.M., Marghny, M., Abdelaziz, N.M.: Fast vertical mining using Boolean algebra. Editor. Pref. **6**(1) (2015)
15. Cheung, D.W., et al.: Maintenance of discovered association rules in large databases: an incremental updating technique. In: Proceedings of the Twelfth International Conference on Data Engineering. IEEE (1996)

A New Multi Objective Optimization to Improve Growth Domestic Produce of Economic Using Metaheuristic Approaches: Case Study of Iraq Economic

Ahmed Khalaf Zager Al Saedi[1(✉)], Rozaida Ghazali[2], and Mustafa Mat Deris[2]

[1] Ministry of Higher Education and Scientific Research (MOHESR), Misan University, Amarah, Iraq
ahmedkhalafalsady@yahoo.com

[2] Faculty of Computer Science and Information Technology, Universiti Tun Hussein Onn Malaysia, Parit Raja, Malaysia
{rozaida,mmustafa}@uthm.edu.my

Abstract. Currently, optimization problems are some of the immediate concern in economics. Peoples' need is fast diversifying, while resources remain limited. This phenomenon is called the Multi-Objective Optimization (MOO) problem. Current techniques are mostly grounded in redundancy, large size path, long processing time. At this point in time, economic problems can be solved by utilizing mathematical principles, and one of the most common and effective approach include metaheuristics as soft computing techniques approaches in the context of the development of significance based plan reduction in the growth domestic product (GDP). The indicators in this model can be utilized to assess the state of a nation's economy. This paper will discuss metaheuristics as soft computing techniques such Ant Colony Optimization (ACO) and Artificial Bees Colony (ABC) in order to propose an effective solution in the reduction of the complexity of MOO in the economy via the determination of an efficient strategy (plan). Experimental results proved that the usage of metaheuristics as soft computing techniques approaches is effective and more promising that current techniques, while ABC is superior to ACO in the context of search time and the exploration of an efficient global strategy (plan).

Keywords: Economic · Gross domestic product · Ant Colony Optimization · Artificial Bees Colony

1 Introduction

We are proposing that four objectives be realized in the quest of maximizing the Growth Domestic Produces (GDP), which are the minimization of the consumption of energy resources, maximization of investments, the minimization of federal expenditure, and the selection of alternative versions pertaining to the real sector of the national economy. Metaheuristic approaches can be used for analyses and economic processes due to the fact that it shows a high complexity, a large body of input information,

T. Herawan et al. (eds.), *Recent Advances on Soft Computing and Data Mining*,
Advances in Intelligent Systems and Computing 549, DOI 10.1007/978-3-319-51281-5_50

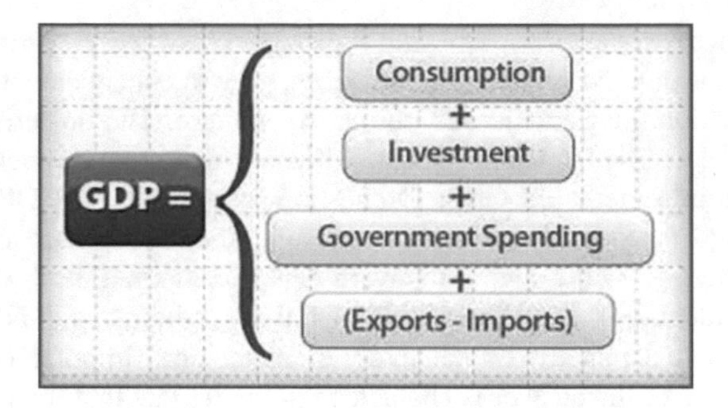

Fig. 1. Domain effective

and the availability of indefinite and hard of formulate factors, alongside a few criteria based on the selection of the best version to confirm the direction of the development of the economy

In this paper choosing best strategies to overcome the problem by remove the not promising plan (strategy) and applying possible resources in boarder area to increase the Gross Domestic Product (GPD).

The most important measure of economic activity in a country, GPD is the crossing point of three sides of the economy: expenditure, output, and income. To measure economic activity, one needs a meaningful aggregation of all kinds of productions. The territory's productions are the crossing result of (1) effective demand, (2) production capabilities, and (3) income. Income arises from payments distributed to production factors and it provides the necessary finance for demand shows in Fig. 1 [2].

We would like to draw your attention to the fact that it is not possible to modify a paper in any way, once it has been published. This applies to both the printed book and the online version of the publication. Every detail, including the order of the names of the authors, should be checked before the paper is sent to the Volume Editors.

2 Multi Objective Optimization in Economic

It is the field of multi-criteria decision, which is interested in the problems of mathematical examples that involve more than one objective function to be examples of these functions simultaneously making. The application of multi-objective examples in many scientific fields, including economics, where you need the optimal decisions that must be taken in the presence of a trade-off between two or more conflicting goals. Reduce costs while maximizing comfort while buying a car, maximizing performance while reducing fuel consumption and emission of pollutants from a car, and those examples of problems examples multiple objectives that include two and three goals respectively In practical problems, there could be more than three goals. It will not look for an optimal solution, however, an official solution is indeed possible. Mathematics is a tool that can be used by economists to create meaningful, testable propositions on wide-ranging and complex subjects that could be expressed in an informal manner. The multi-objective Optimization Problem (MOP) (also called multi-criteria optimization, multi-performance or vector optimization problem) can be regarded as a finding problem [3]. In the context of economic sciences, growth enhancement should result in increased production, investment, and public expenditure. The objectives of this work could result in decreased unemployment, inflation, and low balanced trade deficit, which translates into increased Gross Domestic Product (GDP). GDP = C + G + I + NX [3], where C is equal to all private consumption, or consumer spending, in a nation's economy, G is the sum of government spending, I is the sum of all the country's investment, including businesses capital expenditures, and NX is the nation's total net exports, calculated as total exports minus total imports (NX = Export − Imports). Multiples of the answer is characterized by the fact that the objectives do not agree with one another. This is exemplified in the case of when a person is intending to purchase a second-hand car, which needs to be in a good condition and also cost effective. It should also be pointed out that the Pareto frontier in the context of economics is that it is a Pareto efficient allocation, where the marginal rate of substitution is similar across all consumers.

3 Metaheuristic Approaches

In computer science, a metaheuristic is heuristic create to determine a heuristic that might produce an enough great answer to an optimization problem [4]. Metaheuristics create some opinion about the optimization problem being solved, and so they may be useful for a different kind of problems [5].

3.1 Ant Colony Optimization

This algorithm is based on the inside interactions of an ant colony when the ants are foraging (looking for food). Ant Colony Optimization (ACO) is a probability based approach [6]. This algorithm requires a graph before computation of the optimal solution. The algorithm has many variants like the elitist ant system, max-min ant system, rank based ant system and recursive ACO. The algorithm offers an inherent parallelism. It is efficient for Optimization problems and can be used in real time situations. The optimal solution while using this algorithm is guaranteed. Ant Colony System (ACS) is an agent-based system, which is based on the biological ants and their social behaviour [7]. The Ants can choose any path to reach its destination. In MOO-ACO also parametrized by the number of ant colonies number (CN) pheromone structure (PS). Algorithm 1 describe the generic framework of MOO-ACO (CN, PS). Basically, the algorithm follows the Max-Min ant system scheme [8]. First, pheromone triples are initialized to a given upper bound T_{max}. Then, at each cycle every ant constructed absolution, and pheromone trail are updated. To prevent premature convergence, pheromone trails are bounded within two given bounds $T_{min}\ and\ T_{max}$ such that $0 < T_{min} < T_{max}$. The algorithm stops iterating when a maximum number of cycles has been performed.

```
Algorithm 1 ACO for MOO (CN, N):
   Initialize all pheromone trails to T max
   Repeat
      For each colony c in 1.......CN (CN = colony number)
         For each ant k in 1.........ant number
            Construct a solution
      For I in 1..............N (N=number of phonemic structure)
         Update the ith pheromone structure trails
         If trails is lower than T min then set it
         to T min (T min= lower bound)
         If a trail is greater than T max then set
         it to T max (T max= upper bound)
      Until a maximal number of cycles reached.
```

```
Algorithm 2 Construct of solution S:
  S ← Ø    (S=solution)
  Us ← V    (candidate vector)
     While us ≠ Ø do
        Choose VI ∈ us with probability PS (vi)
        Add VI at the end of S
        Remove from us non promising. (us candidate
        vector )
  End wile
```

Algorithm 2 describes the algorithm used by ants to construct solutions in a construction graph G = (V, E) the definition of which depends on the problem (X, D, C, and F) to solve. At each iteration, a vertex of G is chosen within a set of candidate vertices us; it is added to the solution S and the set of candidate vertices is updated by removing vertices that violate constraints of C. The vertex VI to be added to the solution S by ants of the colony c is randomly chosen with the probability pc S (VI) defined as follows:

$$p_s(v_i) = \begin{cases} \frac{[T_s(\mathbf{v}_i)]^{\alpha}.[\eta_s(\mathbf{v}_i)]^{\beta}}{\sum_{vj \in US}[T_s(\mathbf{v}_i)]^{\alpha}[\eta_s(\mathbf{v}_i)]^{\beta}}, & \\ 0, & otherwise \end{cases} \quad (1)$$

Where $[T_s(\mathbf{v}_i)]$ and $[\boldsymbol{\eta}_s(\mathbf{v}_i)]$ respectively are the pheromone and the heuristic factors of the candidate vertex are two parameters that determine their relative importance. The definition of these two factors depends on the problem to be solved and on the parameters NA and PS; they will be detailed later.

3.2 Artificial Bees Colony (ABC)

Many functions using with proposed algorithm has important rules in metaheuristic approaches based algorithms, which can affect the quality of objective in economics.

3.2.1 Subset Function

An input objective graph with a set of resources R = {R0,⋯, Rn−1}. Indicated by Eq. 2:

$$S = \{R_0\} + \bigcup_{i=0}^{n-1} N(\{R_i\}) \tag{2}$$

Where n = Total a number of food source. $R = \{R_0, \ldots, R_{n-1}\}$ positions of food source.

N $(\{R_i\})$ = Neighbors of (R_i). The objective of (Eq. 2) was converted objective graph with a set of resources contains R_0, to subset.

3.2.2 Plan Function

In this algorithm convert connected subset of partitions, each partition denoted by plan in Eq. 3:

$$P_s = \bigcup_{j=1}^{k} S_{j,} \tag{3}$$

and each path that the bees are vested in represents the plan, where p_s *is set partition* (Plan), J is no plan, while S set of the plan this objective function (Eq. 3) converts the set of relation to the partition or the plan. Each path the bees follow when visiting food source represents a given partition or plan.

3.2.3 Cost Estimation

To conclude the cost for each plan, we use Eqs. 4 and 5:

$$plan\ size = \frac{n(a) * n(b)}{\prod_{Cj \in C} max(v(c_j, a), v(c_j, b))} \tag{4}$$

Where C is public attribute over a, b, V(c_j, a) = distinct values for attribute C in relation a, N(plane) = size of resultant resource to the operation of two resources a and b.

$$\text{Cost} = \sum_{i=1}^{n-1} N(Plan_i) \tag{5}$$

Here N ($plan_i$) is the number of tuples in the relation t_i. ($plan_i$) is an internal node of the join processing tree. Each nectar (food source) becomes a possible solution for optimization problem based on the amount of nectar (resource in economic) representation the efficiency of the solution by (Eq. 6).

$$p_i = \frac{f(x_i)}{\sum_{k=1}^{S} f(x_k)} \tag{6}$$

Where: $f(x_i)$ is a mount of nectar for the food source located at x_i. S is the total number of food sources. Assume the number of employed equal to the number of food source mainly one employed to every food source. For each iteration every employed bee select a food source in the neighborhood of the current food source and evaluate its nectar by Eq. 6.

3.2.4 Time Function

One contribution of this research work is to reduce the responding time of MOO. In RDBMS to minimize the total execution time and cost of parallel tasks, each task shows its cost and execution time in function with certain parameters and variables. Since the capacity of the obtained part for each task is in proportion with the tasks in the certain resource is the same, the execution time of objective n on r is equal to Eq. 7.

$$\mathrm{T} = \frac{2n^3 - 2n}{3} \tag{7}$$

$$\mathrm{P} = \frac{n^3 - n}{3} \tag{8}$$

Where N is the number table (resources) in the economic, T = complex time.

4 Example of Methodology

There are many researchers who implement some sort of multi-optimization, meaning that the solution found depends on the set of random variables being generated [9]. In combinatorial optimization looking over a large set of feasible solutions, metaheuristics is often capable of looking for excellent solutions at lower computational effort compared to the optimization algorithm, iterative methods, or simple heuristics [10]. This makes them useful towards optimization problems [11]. There are several books and papers published on the topic [12, 13, 14].

Consider all factor of GDP is population or location or site or public each site have multi objective and limited resources in the country for year then works with each public as set of solution then attempt to find sufficient solution by using stochastic estimation such as in Iraq contain more one pupations like Baghdad, Basra and Mosul, etc. shows in Figs. 2 and 3. Need to optimize cost and complexity time for each public in one country for one year.

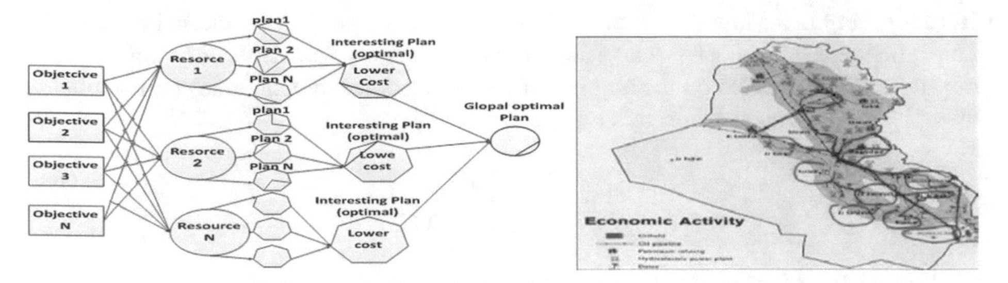

Fig. 2. MOO based on metaheuristic approaches

Fig. 3. Iraq resources

We have land there more than one objective to investing this land such as use it as stadium or manufacture or agricultural land should choose efficient strategy as promising and can reuse in anther resources in future and leave not interesting strategies. In another hand the majority of connected subsets of objective within the batch possess many plans. The bees eschew food source that has previously been visited. This makes optimization unnecessarily redundant. For example, take into account the linked subset of connected resources R0 ⋈ R1 ⋈ R2 for objectives 1 and 2, after the optimization of objective 1, the interesting plan associated with the linked subsets are kept in objective 1. The results from objective 1 is reusable in objective 2, and by using this approach, optimization can be shared across the linked subsets.

5 Experimental Result

To calculate the GPD for each public in the country for only final goods for example bread and butter calculates contribution (quantity * price) for bread and butter to before latest year and apply the same thing for the latest year then use this formula $\frac{final-intial}{intial} \times 100\%$ to calculate the change in GDP for 2-years (Table 1).

The percent of change of GDP of from 2015 to 2016 equal to 62.5%. The different of quantity of beard from 100 in 2015 to 500 in 2016 the rate of change id = s25% and in butter from 50 to 60 the rate percentage is 20% and change the price effect to the GDP for bread 33% and butter is change 25%. In economic the change in GDP in consumption effective with the change of quantity and price.

Table 1. GDP for two years based on final goods bread and butter

Years	Final goods	Price	Quantity	Contribution to GDP
2015	Bread	3 $	100	300
	Butter	2$	50	100
Total				$400
2016	Final goods			
	Bread	4$	125	500
	Butter	2.5$	60	150
Total				$650

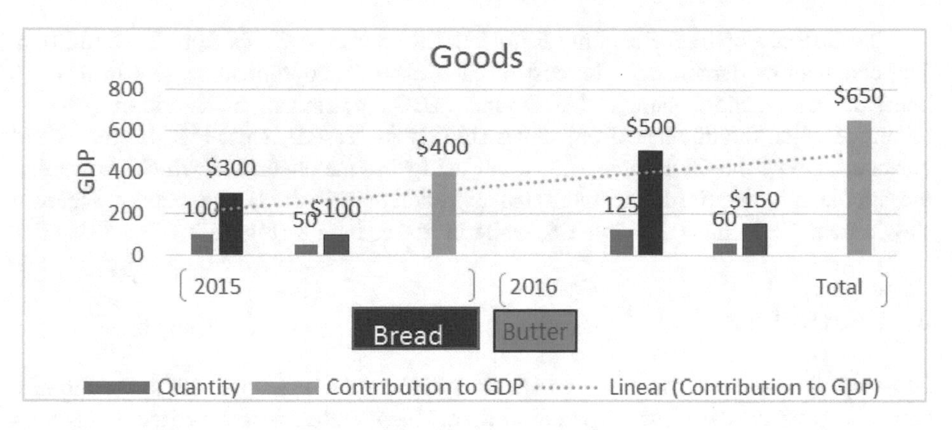

Fig. 4. GDP for two years based on final goods bread and butter

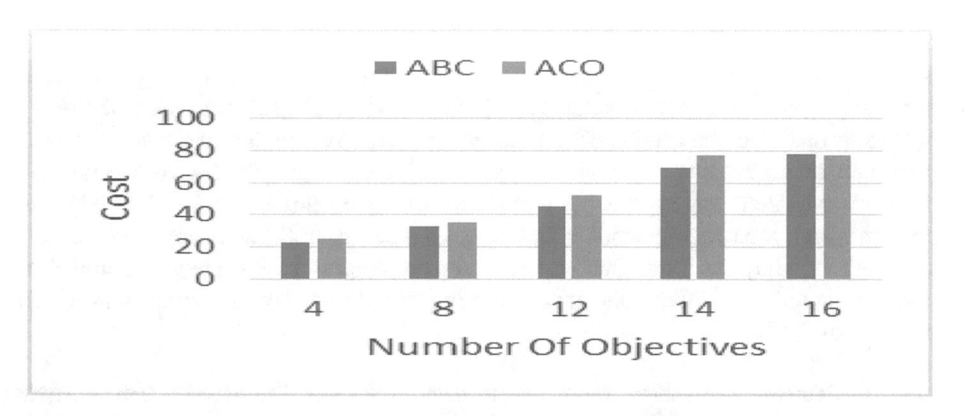

Fig. 5. The effective number of objective

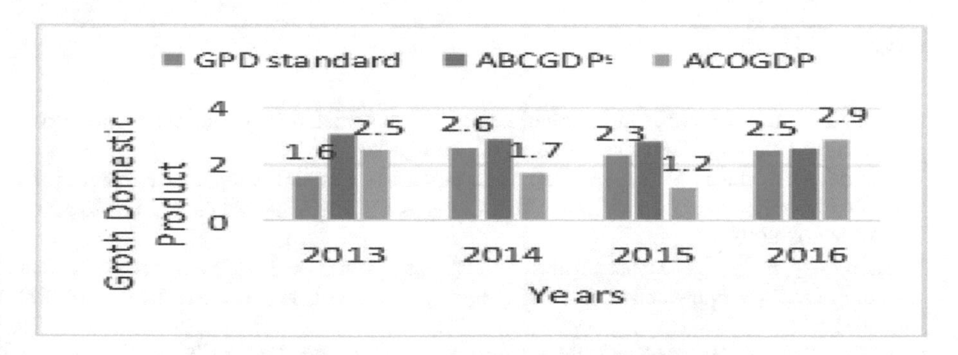

Fig. 6. Improver GDP based on ABC and ACO

The Institute of Iraqi Economy Refold (IIER) reported GDPs from 2013 to 2016. The chart below demonstrate the cost of each objective optimization based on meta-heuristic approaches, namely (ACO and ABC) optimization algorithm, can be increased when the number of objectives slightly increased, as per Fig. 4. The second bar chart shows the GDP increasing when using the metaheuristic, which shows that our approach is superior to the traditional approach used in the Iraqi economy. Figure 6 shows Iraqi GDP in 2013 being 1.6, while in 2016, its 2.8 (Figure 5).

6 Conclusions

GDP represents one of the primary indicators that can be used to assess the health of a nation's economy for a limited period of time. The metaheuristic approach is regarded as a very effective algorithm in solving MOO and reducing the complexity time in economics. An excellent optimization algorithm is not only capable of enhancing the plans' efficiency, but it could also decrease the query time. The successful application of any macroeconomic models relies on its exploitation of the objective models. MOO is crucial towards research in economics. Metaheuristics as soft Computing techniques approaches provide effective solutions for MOP NP-hard problem, while the ABC outperform ACO in the context of exploration and exploitation. Solving this problem is important in economics due to the direct effect on the people that face economic problems. Metaheuristic approaches provide an effective solution for the MOP NP-hard problem, and it is recommended that Iraq change its traditional approach to this approach and rely on more than one resource to support its economy. Examples of resource include agriculture and manufacturing-based industry, the product of which can be exported.

Acknowledgment. The authors would like to thank Universiti Tun Hussein Onn Malaysia (UTHM) and Ministry of Higher Education (MOHE) Malaysia for financially supporting this research under the Fundamental Research Grant Scheme (FRGS), Vote No. 1235.

References

1. Ulungu, E.L., Teghem, J.M.: Multi-objective combinatorial, optimization problems. A survey. J. Multi-Criteria Dec. Anal. **3**, 83–104 (1994)
2. Bianchi, L.M., Luca, M.G.: A survey on metaheuristics as soft computing techniques for stochastic combinatorial optimization. Nat. Comput. Int. J. **8**, 239–287 (2009). doi:10.1007/s11047-008-9098-4
3. Blum, C., Roli, A.: Metaheuristics as soft computing techniques in combinatorial optimization: overview and conceptual comparison. ACM Comput. Surv. **35**(3), 268–308 (2003)
4. Stützle, T., Hoo, H.: MAX − MIN ant system. J. Fut. Gener. Comput. Syst. **16**, 889–914 (2000)
5. Unler, A.: Improvement of Energy Demand Forecast Using Swarm Intelligent. Elsevier

6. Alsaedi, A.K.Z., Ghazali, R., Deris, M.M.: An efficient Multi Join Query Optimization for relational database management system using two phase Artificial Bess Colony algorithm. In: Badioze Zaman, H., Robinson, P., Smeaton, A.F., Shih, T.K., Velastin, S., Jaafar, A., Mohamad Ali, N. (eds.) IVIC 2015. LNCS, vol. 9429, pp. 213–226. Springer, Heidelberg (2015). doi:10.1007/978-3-319-25939-0_19
7. Alsaedi A.K.Z., Ghazali, R., Deris, M.M.: Materialize view selection for objective optimization in data warehouse system using heuristic approaches. J. Next Gener. Inf. Technol. **6**(3) (2015)
8. Alsaedi, A.K.Z., Ghazali, R., Deris, M.M.: Materializing multi join query optimization for RDBMS using swarm intelligent approach. Int. J. Comput. Inf. Syst. Ind. Manag. Appl. **7**(1), 74–83 (2014). https://doaj.org/article/6075444fdcb14d689551d93d0e56eccf. ISSN 2150 7988
9. Mladinero, M.: Single-objective and multi objective optimization using the HUMANT algorithm. CRORR **6**(2) (2015)
10. Chande, S.V., Sinha, M.: Optimization of relational database queries using genetic algorithms. In: Proceedings of the International Conference on Data Management, IMT Ghaziabad (2010)
11. Steinbrunn, M., Moerkotte, G., Kemper, A.: Heuristic and randomized optimization for the join ordering problem. Very Large Data Bases J. **6**(3), 191–208 (1997). doi:10.1007/s007780050040
12. Almery, M., Farahad, A.: Application of bees algorithm in multi join objective optimization. Indexing and retrieval. ACSIJ Int. J. Comput. Sci. **1**(1) (2012)
13. Kadkhodaei, H., Mahmoud, F.: A combination method for join ordering problem in relational databases using a genetic algorithm and an ant colony. In: Proceedings of the 2011 IEEE International (2011)
14. Li, N., Liu, Y., Dong, Y., Gu, J.: Application of ant colony optimization algorithm to multi-join objective optimization. In: Proceedings of the 3rd International Symposium on Advances in Computation and Intelligence (ISICA 2008), pp. 189–197 (2008). http://www.springer.com/computer/information+systems+and+applications/book/978-3-40-92136-3
15. Mukul, J., Praveen, S.: Objective optimization: an intelligent hybrid approach using cuckoo and tabu search. Int. J. Intell. Inf. Technol. **9**(1), 40–55 (2013)
16. Pandao, M., Isalkar, A.D.: Multi objective optimization using a heuristic approach (2012)
17. Chande, S.V., Snik, M.: Genetic optimization for the join ordering problem of database queries. Department of Computer Science International School of Informatics and Management, Jaipur, India (2007)
18. Pandao, M., Isalkar, A.: Multi objective optimization using a heuristic approach. Int. J. Comput. Sci. Netw., Hardware Compon. RDBMS. J. Comput. Eng. Inf. Technol. (2013). ISSN: 2277-5420

The Algorithm Expansion for Starting Point Determination Using Clustering Algorithm Method with Fuzzy C-Means

Edrian Hadinata[1(✉)], Rahmat W. Sembiring[2], Tien Fabrianti Kusumasari[3], and Tutut Herawan[4]

[1] Sekolah Tinggi Teknik Harapan, Medan, North Sumatera, Indonesia
edrianhadinata@gmail.com
[2] Politeknik Negeri Medan, Medan, North Sumatera, Indonesia
rahmatws@yahoo.com
[3] Faculty of Industrial Engineering, Telkom University, Bandung, Indonesia
tienkusumasari@telkomuniversity.ac.id
[4] University of Malaya, Kuala Lumpur, Malaya, Malaysia
tutut@um.edu.my

Abstract. The starting point determination in Fuzzy C-Means algorithm (FCM) is taken by random. Thus, the algorithm for starting point determination was developed with Hierarchical Agglomerative Clustering approach as a substitution of membership degree randomization process in the early iteration. It is expected that the clustering process will produce fewer iteration. The process contained on this algorithm is the incorporation of a number of clusters based on the approach contained in complete linkage. Then it will calculate the difference in the objective function for each iterations after the clustering process has been conducted on the FCM. The iteration process will be stopped after the difference of objective function is smaller than the prescribed limit. In this research, analysis of variance from the obtained cluster produces a good homogeneity and heterogeneity value. In addition, the number of iteration is getting fewer.

Keywords: Fuzzy C-Means · Starting point determination · Complete Linkage

1 Introduction

The development of group analysis is started from a hierarchical method which its development will form a tree diagram based on the distance to form a group. While a non- hierarchical method do partition in determining the number of groups first which then adapted to the objectives of the research. For example the k-means algorithm is included in the exclusive category because of its data can be ascertained be a group only and not being a member of other groups, but if the data have two or more entering the members of the group, then it is called the overlapping which is expressed with degrees of membership. The Fuzzy C-Means (FCM) is included in it as well as included in Fuzzy – Based non-hierarchical clustering method [1].

T. Herawan et al. (eds.), *Recent Advances on Soft Computing and Data Mining*,
Advances in Intelligent Systems and Computing 549, DOI 10.1007/978-3-319-51281-5_51

In the calculation of a of Fuzzy C-Means Clustering Algorithm method, the optimum limit for this algorithm is often causing problems. The determination of a different central point produces different cluster because of the value of initial μ membership degree which is formed distributed randomly [6]. That will produce looping calculation until the difference between objective function P reached smaller than ξ. Because of that situation, researchers usually take the alternatives i.e., taking the iteration boundary (maxlter) based on a specified value. The Calculation of the optimum objective function as a stopping point in Fuzzy C-Means is based on the calculations of $P_t - P_{t-1} < \xi$ with the un-known numbers of looping. If ξ has fairly small value as 0.05, the number of looping to reach $P_t - P_{t-1} < \xi$ on certain cases with the high level of data scattering will increase the number of iteration calculation, if the value of ξ is fairly large then the data will not be entering the proper clusters.

2 Basic Concept of Clustering Method

Clustering is a method of data analysis, which is often included as one of the data mining methods, in which its objective is to group the data with the same characteristics belongs to one area while the data with different characteristics belongs to another area.

There are several approaches used in developing a clustering method. Two main approaches are Partitioning Clustering and hierarchical clustering approach [7]. The Partitioning Clustering, or often referred to partition-based clustering, classifies data by sorting through the data to be analyzed into the existing clusters. The Hierarchical Clustering classifies the data by creating a hierarchical in a form of dendogram in which the similar data will be placed on adjacent hierarchy and the remote hierarchy.

2.1 Clustering with the Hierarchy Approach

Hierarchical Clustering that makes hierarchy on cluster depicted in a tree structure is called Dendogram [8]. One way to facilitate the dendogram development for Hierarchical Clustering is to create a similarity matrix containing the level of similarity between the grouped data. The level of similarity can be calculated in various ways such as the Euclidean Distance Space. Departing from the similarity matrix, we can choose the type of linkage which will be used to classify the analyzed data [3].

2.2 Complete Linkage

The method of Complete Linkage Agglomerative Clustering (CL) is commonly called the method of *Furthest Neighbor Technique*. The process this method is generally almost the same as the single linkage method, but in a partner searching, the complete linkage method is looking for a partner which distance is farthest away from the observation values.

$$d_{CL}(G,H) = max(d_{ii'}); i \in G; i' \in H \tag{1}$$

Inequality between the G, H is the dissimilarity between two points on the opposing groups. Results of Complete Linkage algorithm is created in a dendogram that is commonly called a tree diagram. Each branch will be met and put together. Furthermore, this process will cut the tree branches and then $d_{[CL]}$ will generate the farthest point.

2.3 Making the Centroid Data

Making the centroid data or data center is based on paper multistage random sampling FCM Algorithm which states that a small group of vector can be used for approximating the center cluster for the whole large group of data [2].

For that reason, the approach of complete linkage algorithm which seek the cluster center based on the farthest partner is expected to be accurate to predict the value of cluster centers which are studied. However, in complete linkage algorithm which choose the cluster center with a ratio of maximum distances A to B will cause that cluster center is still skew in the most maximum distance, so that the cluster center is less accurate to represent a group of values.

At the cluster center searching modeling using complete linkage algorithm is changed to the midpoint of the comparison of two minimum and maximum distances.

$$V_{ik}(A,B) = max\ (d(A,B)) - \frac{1}{2}|max\ (d(A,B)) - min\ (d(A,B))| \quad (2)$$

Whereas, the calculation of the distance for searching the new membership function on Fuzzy C-Means algorithm is conducted when the initialization process of the starting point has been done.

$$d_i = \sum\nolimits_{j=1}^{c} \sum\nolimits_{k=1}^{s} \left(X_{ij} - V_{jk}\right)^2 \quad (3)$$

2.4 Fuzzy C-Means

The Fuzzy clustering is the process of determining the degree of membership, and then uses it by inserting it into the element of the data in one or more cluster groups.

This will give information about the similarity of each object. One of several Fuzzy Clustering Algorithms used is fuzzy C-Means algorithm. The Vector of fuzzy clustering, *V* = *{v1, v2, v3,…, vc}*, is an objective function that can be defined by the degree of membership of the data *Xij* and cluster center V_{jk}.

$$J_i(X,\mu,V) = \sum\nolimits_{j=1}^{n} \sum\nolimits_{k=1}^{c} (\mu_{ik})^m d^2(X_{ij}, V_{kj}) \quad (4)$$

As such μ_{ik} is the degree of membership of *Xj* and center cluster is a part of the membership of matrix $[\mu_{ij}]$. d^2 is the root of the Euclidean distance and *m* is the fuzzy parameter that the fuzziness degree average of each data whose membership degree is not greater than 1.0. [9].

In the early stages or $t = 0$, the fuzzy C-Means does the randomization process with the condition of $\mu_{ik} \in [0, 1]$ as such:

$$\sum\nolimits_{k=1}^{c} \mu_{ik} = 1, i = 1, 2, \ldots, N \quad (5)$$

Searching a new membership degree candidate can be obtained by using Eq. (6)

$$W_k = \frac{\sum\nolimits_{i=1}^{N} (\mu_{ik})^m * X_i}{\sum\nolimits_{i=1}^{N} (\mu_{ik})^m} \quad (6)$$

Furthermore, the function (4) describes the constraint Optimation problems, but if it is converted into unconstraint Optimation problem by using a lagrange multiplier technique, it will formed the Eq. (7) which is used to search a new membership degree.

$$u_{ik} = \frac{1}{\sum\nolimits_{l=1}^{c} \left(\frac{u_{ik}}{d_{il}}\right)^{\frac{2}{m-1}}} \quad (7)$$

In which i = 1,2,…N, j = 1,2,…,n dan k = 1,2,..c for $d_{ik} = \|X_{ij} - V_{jk}\| > 0$, $\forall$ *i and k*.

Then calculate the objective function at the iteration t, P_t

$$P_t = \sum_{i=1}^{n} \sum_{k=1}^{c} \left(\left[\sum_{j=1}^{m} (X_{ij} - V_{kj})^2 \right] (\mu_{ik})^w \right) \quad (8)$$

Looping will be done simultaneously with the addition of the value of t with 1 if the condition $\|P_t - P_{t-1}\| > \xi$ and will be back in the calculation of the Eq. (6).

2.5 Analysis Cluster

In cluster analysis, the object grouping is performed based on the similarity and dissimilarity. Each object which is incorporated in one or more groups in the Fuzzy C-Means has a higher degree of homogeneity than other objects. For that reason, testing can be carried out by looking at the value of the variance or the distribution of data. Cluster variance can be determined by the equation.

$$V_c^2 = \frac{1}{n_c - 1} \sum\nolimits_{i=1}^{n_c} (x_i - \overline{x_c})^2 \quad (9)$$

Based on the Eq. (9) which resulted in the variance of each cluster, the density of a cluster can be obtained by analysis of *variance within cluster*, according to the Eq. (10).

$$V_w = \frac{1}{N - c} \sum\nolimits_{i=1}^{c} (n_i - 1).V_i^2 \quad (10)$$

Another analysis is carried out to look at the distribution of data between clusters (*variance between clusters*) that can be calculated by Eq. (11) below.

$$V_b = \frac{1}{k-1}\sum\nolimits_{i=1}^{k} n_i(\overline{x_i} - \overline{x}) \tag{11}$$

Cluster with a minimum value of V_w could represent Internal *Homogenity* so that the cluster is closer to the ideal. While V_b with the largest value represent the *External Homogenity*. The next equation can express the restriction variance.

$$V = \frac{V_w}{V_b} \tag{12}$$

3 Methodology

3.1 Introduction

Clustering method uses the objective function as a condition of looping to find the exact cluster center. Thus, the tendency data is obtained to enter the appropriate cluster at the last step [5].

Therefore, to calculate the distance for each point and analyze movement value as well as the changes of the center point for each its looping and to determine the value of the early center point generated from the previous calculation of the initial μ value which is expected to be able to find an alternative or addition for other models to limit the number of finite iterations in order to achieve more efficient iteration value.

3.2 Research Framework

There are some stages carried out in this Study as followed:

1. *Data Collection*, Data were taken from the UCI Machine Learning Repository which has been prepared by Fereitas et al. (accessed on December 5th, 2014) which will be used as a test. The data set taken is a data set grammatical Facial Expression. This data set is taken to develop the appropriate model for interpreting the grammatical form through facial expressions of Brazilian Sign Language (Libras). [4]
2. *Determination of the starting point*, Having the results of the data initialization was known, the process of determining the starting point by using a Hierarchical Clustering Method with Agglomeratif Complete Linkage as the algorithm that determine the formation of the initial cluster centers in the previous step.
3. *The establishment of a New Cluster Center and a Membership Function Searching*, The resulting initial cluster centers from step 4 will be used to establish a new center cluster and a new membership function.
4. *Algorithm Making*, The establishment of new rules on Fuzzy C-Means that has been developed in accordance with some of the processes taken for each steps.

5. *Cluster Analysis*, This section will discuss the value of homogeneity and heterogeneity of each cluster, by noticing and analyzing the cluster if it was ideal.
6. *Conclusion*, The conclusion was drawn by making documentation from the supporting theory, design, testing and test results as well as the suggestions that can be carried out in further research (Fig. 1).

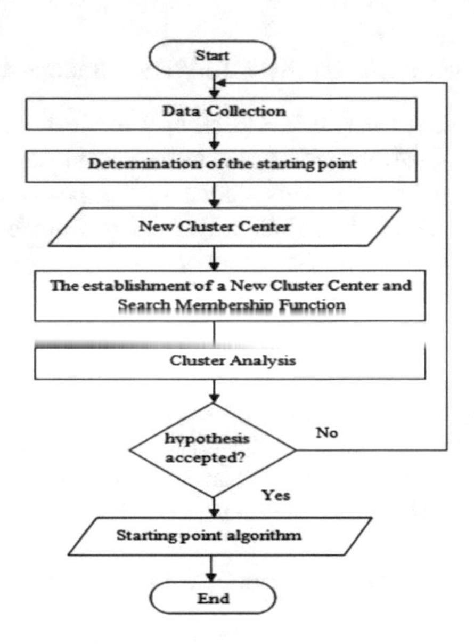

Fig. 1. Research framework flowchart

4 Experimental Result and Discussion

4.1 Introduction

In this stage, the results and discussion of the development of starting point on clustering method with Fuzzy C-Means algorithm was presented, i.e., the starting point which is previously formed by random is being changed by using complete linkage method on agglomerative clustering.

Tests on both algorithms will be carried out using the same parameters, namely: the smallest error which is expected $\xi = 0{:}35$, for weight m = 8. The example of the analyzed data is affirmative testing comparing the number of iterations and the results of the segmentation between Fuzzy C-Means with the random starting point and Fuzzy C-Means developed by producing the starting point.

4.2 Testing on Fuzzy C-Means

On the first step, randomization process of membership degree is carried out, this process is carried out when $t = 0$ with the condition of the Eq. (5). Then the searching process for degree of membership candidate is conducted as well as the calculation of objective function to determine the limit stop for each looping. If P_t-P_{t-1} is still greater than the specified value, the new membership degree on t = 1 will be formed.

The last objective function difference for P_{11}-P_{10} is still greater than ξ, namely 46156.3–51400.78 = 5244.497217, since the difference is still large enough to reach ξ = 0.35, the calculations of looping process will still be carried out.

4.3 Starting Point Determination by Complete Linkage Approach

Complete Linkage algorithm search in searching the cluster center to determine the starting point on Fuzzy C-Means (FCM) is starting from determining the number of clusters = number of rows or $c = n$. This algorithm is expected to be able to reduce the number of iterations contained in ordinary FCM algorithm by replacing the central cluster produced randomly on the first iteration.

Suppose $0x_1$ ... $0x_n$ as X_i ... X_n and X_j...X_n. Perform calculation of distance for each data object is carried out. Calculate every distance of d_{ij} between each pair of X_i and X_j. for every $X_i \in R^p$ *dan* $d_{ij} = \|X_i - X_j\|^2$. Then, for the processing step give index on each cluster, clustering calculations can be expressed by $C = \{i_1, i_2, i_3, \ldots, i_n\}$. For each level, the cluster contains the i value which has been united. To the lowest level contains $C = \{i\}$ for each cluster, on the last level or the top level there is only one cluster $C = \{1, \ldots, n\}$ in which a reduction in the number of clusters for each iteration is occurred.

Starting by treating each data as a group, we choose the distance of two smallest groups.

$$min_{u,v}(d_{uv}) = d_{9,8} = 0.000120999$$

On searching of the cluster center value of the complete linkage algorithm, the minimum value is also determined. But in this case, Based on [5] in which it divides a group of data into several blocks with the difference of a minimum value and the maximum value divided by the standard deviation. In other cases [2], it expressed in multistage random sampling using small samples.

To draw the data center on a small sample, the maximum value is added with the difference between maximum and minimum values of the selected group and divided into two based on the Eq. (2) in dendogram, see Fig. 2.

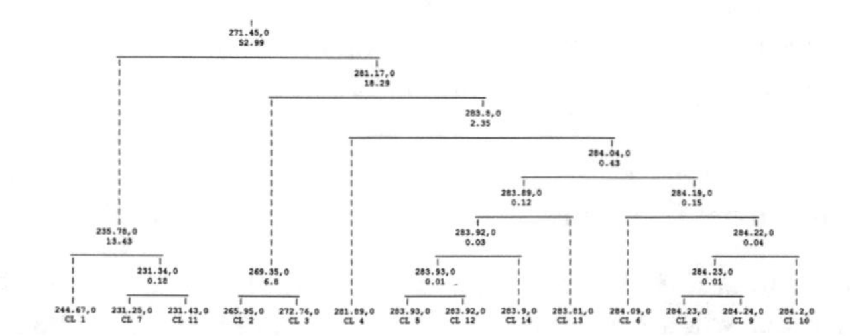

Fig. 2. The objective function graphic

4.4 Initialization of the Initial Value

The algorithm result was inserted as a replacement cluster center of the early iteration. This starting point determination replace some steps in the early iterations of the Fuzzy C-Means algorithm of which is the producing the random values shown in Eq. (5), and the searching calculation of the vector V_k in Eq. (2).

The cluster center included in Eq. (3), in which $V = \{v_1, v_2, v_3, \ldots, v_c\}$ as such each v_j is the cluster center for column $-k$.

$$V_{kj} = \begin{bmatrix} 282.9503 & 281.17 & 277.34 & 273.83 \\ 269.354 & 269.48 & 261.81 & 257.72 \\ 238.0038 & 240.95 & 235.99 & 231.25 \end{bmatrix}$$

Then, calculate the objective function based on the Eq. (8), the objective function is calculated as a measurement to limit the number of iterations, as the result is:

$$P_{t=1} = 231802.309;\ P_{t=0} = 0;$$
$$|231802.309 - 0| > \xi\ ;\ \xi = 0.35$$

For $P_{t=0} = 0$, the searching calculation of the new membership degrees is carried out until it reaches the difference of the objective function < 0:35. Therefore, the t value is added by 1.

Thus, the number of objective function for several iterations can be seen in Fig. 3 below. The Calculation of objective function is carried out until the 5th iteration.

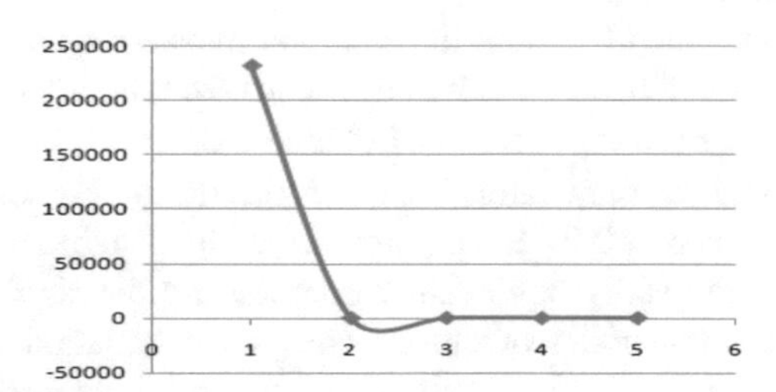

Fig. 3. The objective function graph on $t = 5$

Because of the difference in the objective function $\|P_t - P_{t\text{-}1}\| < 0.35$, the looping of the searching calculation for membership degree value will stop at the 4th iteration or $t = 4$.

Here we can see that by defining a starting point instead of random values in fuzzy C-Means algorithm can help reduce the number of iterations that occur.

4.5 Creating the Algorithm

The approach used for producing clusters from the previous rules resulted in new rules, as follows:

Step 1. Insert the observation data
Step 2. Calculate the number of data *(n)* and initialization of cluster number = n-data;; $c = n;\ \xi = 0.35,\ m = 8,\ t = 0;\ P_0 = 0$
Step 3. Specify the number of required cluster, *cMin* = *num_cluster*;
Step 4. Calculate the distance between X_j and X_i, in which the value of $X_i = Data_i$ and $X_j = Data_i$
Step 5. Combine column $min(d(X_j,X_i))$, in which $merge(d_i,d_j) = max(d(X_i,X_j))$
Step 6. Replace the Cluster center $c_i = data_{i\text{-}baru}$; with the Eq. 2.3; *n-1;numCi* = *n*
Step 7. Repeat the Step 5 if $numC_i \neq cMin;$
Step 8. Specify $V_{ik} = ci;$ If *numCi* = *cMin; t* +=*1;*
Step 9. Produce V_{ik}

4.6 Accuracy and Testing

The testing of homogeneity value of each object in one group could also describe whether the clusterization process is running well or whether the object is included in the exact cluster or not. Besides that, the testing of heterogeneity among the objects in one group with other groups can figure out whether the object in one group is different from other objects in different groups [10].

Then, they are calculated by using the equation in one-way analysis of variance to test whether the existing population is homogeneous. In which the homogeneity of the population stated $\sigma_1^2 = \sigma_2^2 = \sigma_3^2 = \ldots = \sigma_k^2$ (Table 1).

From the F list with *df* numerator 2 and *Df* (*degree of freedom*) denominator 12. While the odds 0.95 or $\alpha = 0.05$ is acquired from the F table = 3.88, if F result is smaller than F table then H_0 is significantly accepted by the significance level 0.05.

Then, to see the comparison of cluster analysis calculation on Fuzzy C-Means without using the starting point for iteration $t = 15$ can be obtained in Table 2, which states that the value of F results on iteration $t = 15$ is larger than the F results on the iteration $t = 4$ by using starting point.

Table 1. Ratio the comparison of *Between Group* and *Within Group* with the starting point on $t = 4$

Source	Df	SS	MS	F result
Mean	1	1031568	1031568	0.992012
Between Group	2	758.2668	379.1334	
Within Group	12	4586.237	382.1864	

Table 2. The comparison ratio of *Between Group* and *Within Group* without starting point on $t = 15$

Source	Df	SS	MS	F result
Mean	1	738155.6	738155.6	27.88809
Between Group	2	2860.863	1430.431	
Within Group	12	410.3348	51.29185	

5 Conclution

In the determination of starting point on Fuzzy C-Means algorithm using this *clustering* method, the use of agglomerative algorithm with complete linkage *clustering* has significant affect on the reduction of the iterations number. The importance of iteration number efficiency on this algorithm can reduce the running process that speed up the computer performance in analysis process.

The use of random data in this case is not included since it can generate the arbitrary value of center point that causing the distance which is multiplied by the random value does not significantly represent the existing data. It resulting the difference of the objective function is still large in the early iteration and will be able to increase the number of iterations to an extent that cannot be determined. Therefore, the researcher usually use iteration limit by inserting the maximum value for grouping by using this algorithm.

In the test criteria, it is described that the value of the obtained F results are smaller than F table value in the significance level of 0.05 with a level of truth 0.95. While the value for the variance between clusters is 379.1334 which is smaller than the variance within clusters with a value of 382.1864. This value proves the value of the data distribution on each cluster in this research is ideal.

References

1. Agarwal, C.C., Reddy, C.K.: Data Clustering Algorithms and Applications. Data Mining and Knowledge Discovery Series. Chapman & Hall/CRC Book, CRC Press, Taylor and Francis Group, Boca Raton (2014). ISBN 978-1-4665-5821-2
2. Cheng, T.W., Goldgof, D.B., Hall, L.O.: Fast fuzzy clustering. J. Fuzzy Sets Syst. **93**(1), 49–56 (1998)
3. Everitt, B.S., Landau, S., Leese, M., Stahl, D.: Cluster Analysis. Willey Series in Probability and Statistics, 5th edn. John Willey & Sons, Ltd, UK (2011). King's College, London
4. Freitas, F.D.A., Perez, S.M.: Grammatical Facial Expressions Data Set. Machine Learning Repositori, UCI, 5 December 2014. https://archive.ics.uci.edu/ml/datasets/Grammatical+Facial+Expressions
5. Hung, M.C., Yang, D.L.: An efficient fuzzy C-Means clustering algorithm. In: Proceedings of IEEE International Conference, Department of Information Engineering pp. 225 – 232 (2001)

6. Karlina, T., Afrida, H., Arifin, F.: Pengembangan Algoritma K-Modes pada Penentuan Titik Pusat Awal untuk Mengelompokkan Penyakit pada Kacang Kedelai. Prosiding IES-2006-Politeknik Elektronika Negeri Surabaya-ITS, pp. 253–259 (2006)
7. Oliveira, J.V.D., Pedrycz, W.: Advance in Fuzzy Clustering. Willey, England (2007). The Atrium, Southern Gate, Chichester. British Library Cataloguing in Publication Data
8. Sembiring, R.W., Zain, J.M., Embong, A.: A comparative agglomerative hierarchical clustering method to cluster implemented course. J. Comput. **2**(12), 33–38 (2010)
9. Valarmathie, P., Srinath, M.P., Ravichandran, T., Dinakaran, K.: Hybrid fuzzy C-Means clustering technique for gene expression data. Int. J. Res. Rev. Appl. Sci. **1**(1), 33–37 (2009)
10. Yatracos, Y.G.: Variance and Clustering. Proc. Am. Math. Soc. **126**(4), 1177–1179 (1998)

On Mining Association Rules of Real-Valued Items Using Fuzzy Soft Set

Dede Rohidin[1,2,3(✉)], Noor A. Samsudin[1,2,3], and Tutut Herawan[1,2,3]

[1] Faculty of Computer Science, Telkom University, Bandung, Indonesia
rohidin@tass.telkomuniversity.ac.id
[2] Faculty of Computer Science and Information Technology, Universiti Tun Hussein Onn Malaysia, Parit Raja, Malaysia
[3] Faculty of Computer Science and Information Technology, University of Malaya, Kuala Lumpur, Malaysia

Abstract. Association rules is s one of data mining method that have been implemented in many discipline areas. This rule is able to find interesting relation between the data in a large data set. The traditional association rule has been employed to handle crisp set of items. However, for real-valued items, the traditional association rules fail to handle them. This paper introduces an alternative method for mining association rules for real-valued items. It is based on the concept of hybridization between fuzzy and soft sets. This combination is called fuzzy soft association rules. The results show that the introduced concept was able to mine an interesting association rules among the real number of items where they are represented in fuzzy soft set. Furthermore, it has the ability in dealing with uncertainty or vague data.

Keywords: Data mining · Association rules · Soft set · Fuzzy soft set

1 Introduction

As an approach to discover cooperative example from market basket data, Agrawal *et al.* introduced association rules mining for the first time [1]. The problem on designing the catalog and determining layout of store can be solved by learning of customers' buying pattern. Since then, many problems in other domains (e.g. credit card fraud, network intrusion detection, genetic data analysis) were implemented by association rules meaning for finding solution. How to generate all of association rules from a collection of transactions data is a focus of association rules mining. All rules must have support value greater or equal than minimum support and confidence value must greater or equal than minimum confidence. Capacity to discover interesting relationship between the data in a large data item is advantages of association rules [2].

There are many association of rule mining algorithms that have been proposed in the data mining literature, as example is Apriori [3] and FP-growth [34]. Apriori follows a breadth-first-search strategy while FP-growth follows a depth-first search strategy. Many researcher implemented association rules in classification [2, 4–6], medical [7–10], finance [11, 12], etc. Class association rules (CARs) is a framework which integrated classification and association rule mining [6]. Many problems in

T. Herawan et al. (eds.), *Recent Advances on Soft Computing and Data Mining*,
Advances in Intelligent Systems and Computing 549, DOI 10.1007/978-3-319-51281-5_52

current classification system that can be solved by CARs. Kamruzzaman reported that combination of association rule, naive Bayes and genetic algorithm was performed in larger data set and more classes [13]. However, the fuzzy-valued transactions that are common in the real worlds. One of big challenge for researchers in this field is how to develop an algorithm that is able to handle various types of data

Soft set as a collection of approximate descriptions of an object was introduced in 1999 by Molodstov [14]. The theory of soft set is convenient and easily applicable because there is no restriction on the approximate description [15]. For parameters with binary numbers, the soft set theory can perform in Boolean dataset [16, 17], but in real number parameters it was still difficult to work with. For that purpose, to handle parameters which have real value, R.K. Maji introduced the concept of fuzzy soft set based on soft set [18]. This theory was growing very fast and have been implemented in various field such as medical diagnosis [7–10, 17, 19], decision making problem [20–26], and data classification [27]. On medical field, fuzzy soft set has been applied for many purposes [28], fuzzy soft set approach is better than KNN and Fuzzy KNN for gene classification as stated in [28].

Up to present, much of research on fuzzy soft sets have been achieved in theoretical and implementation aspect. As a first result, D. Molodtsov proposed the basic notions of soft sets theory and discussed some problems of the future [14]. Maji presented several algebraic operations on soft set such as subset, super set, complement [18], AND, OR, union and intersection. T. Hong and Y. Lee [29] compared conventional crisp-set mining and fuzzy-mining approaches, and the outcome demonstrate that fuzzy-mining is smoother because off the fuzzy membership. The idea of a multi-soft sets representing a multivalued information system was presented by T. Herawan and M.M. Deris [15]. T. Herawan, M.M. Deris [17] Discussed subjective perspective against recent soft decision making method through an artificial patients dataset. N. Kalaiselvi used a classifier based on fuzzy similarity to improve the precision to anticipate inter- stage cancer genes [28]. H. Ma *et al.* [16] introduced the weighted fuzzy soft set and investigated its application to decision making. Y. Jiang and H. Liu proposed implementation of ontology-based soft set in decision making [30]. N. Çağman [31] defined Fuzzy Parameterized of Fuzzy Soft (FPFS) sets and their operations that allows constructing more efficient decision processes. S.E. Glu introduced ways to make decision-making more efficient through use of soft fuzzy soft aggregation operator [32]. Recognition strategy based on multi observer input parameter data set was presented by [20] Roy and Maji. N. Çağman and S. Enginoğlu [23] defined matrices of soft set and their operations. B. Handaga, T. Herawan and M.M. Deris [27] designed Fuzzy Soft Set Classifier (FSSC) as new classification algorithm based on fuzzy soft set for handle a numerical data. A. Kharal presented a new conjecture for finding the optimum choice between two parties [24]. S. Alkhazaleh introduced M-IVFSS, the concept of multi-interval-valued-fuzzy-soft-set and examined its application to decision making [22].

From the analysis above, we observe that the study which combines association rules and hybrid fuzzy soft set is still an outstanding challenge. Therefore, this paper proposes a fuzzy soft association rules as a new concept of combination between hybrid fuzzy soft set and association rules. In summary, the contributions of this work are described as follow.

a. We propose the notion of fuzzy soft association rules.
b. We illustrate trough examples on how to mine fuzzy association rules as proof of concepts.
c. We show that the presented definition is able to mine an interesting association between the real number items that represented in fuzzy soft set.

The rest of this paper is structured as follows. Section 2 explorers the fundamental of association rules, soft set and hybrid fuzzy soft sets. The proposed method, fuzzy soft association rules will be discussed in Sect. 3. Finally, conclusions and future work are presented in Sect. 4.

2 Rudimentary

In this section, we reviewed some theory that are useful in developing our proposed method such as association rules [1], soft set [14], fuzzy set theory [33], and hybrid fuzzy soft set [32].

2.1 Association Rules

Let $I = \{i_1, i_2, \ldots, i_{|A|}\}$, for $|A| > 0$ as a *set of items* and the set $D = \{t_1, t_2, \ldots, t_{|U|}\}$, for $|U| > 0$ refers to the transactional dataset, where each transaction $t \in D$ is a list of distinct items $t = \{i_1, i_2, \ldots, i_{|M|}\}$, $1 \leq |M| \leq |A|$ and each transaction can be identified by a distinct identifier *TID*. Let, a set $X \subseteq t \subseteq I$ called an *itemset*. An *itemset* with *k-items* is called a *k-itemset*. The *support* of an *itemset* X, denoted sup(X) is defined as a number of transactions contain X. An *association rule* between sets X and Y, *denoted* by the form $X \Rightarrow Y$, where $X \cap Y = \phi$ The itemsets X and Y are called *antecedent* and *consequent*, respectively. The *support* of an association rule $X \Rightarrow Y$, denoted sup *(X Y)*, is defined as a number of transactions in D contain $X \cup Y$. The *confidence* of an association rule $X \Rightarrow Y$, denoted conf($X \Rightarrow Y$) is defined as a ratio of the numbers of transactions in D contain $X \cup Y$ to the number of transactions in D contain X. Thus, $conf(X \Rightarrow Y) = \frac{\sup(X \Rightarrow Y)}{Sup(X)}$.

2.2 Soft Set Theory

Let U be an initial universe of objects and E is the set of parameters in relation to objects in U. Parameters are often attributes, characteristics, or properties of objects. Let *P(U)* denote the power set of U and $A \subseteq E$. The notion of soft set in presented in Definition 2.1 as follow.

Definition 2.1 (Molodtsov, 1999). *Let U be non-empty universal set, E be a set of parameters, and P(U) denote the power set of U. A pair (F,E) is called a soft set over U, if only if F is a mapping given by F : E→P(U). Thus, a soft set over U can be represented by the collection of ordered pairs,*

$$F_E = \{(x, f_E(x)) | x \in E, f_E(x) \in P(U)\}$$

From Definition 2.1, a soft set *(F,E)* over the universe *U* can be regarded as a parameterized family of subsets of the universe *U*, which gives an approximate (soft) description of the objects in *U*. A hybridization of fuzzy and soft set theory called fuzzy soft set is given in the following sub-section.

2.3 Fuzzy Soft Set Theory

The notion of a fuzzy soft set is given in the following definition.

Definition 2.2. *Let U be a universe. a fuzzy set X over U is a set defined by a function* μ_x *representing the mapping* $\mu_x : U \to [0, 1]$

μ_x is function that represents membership of *X*, and the value $\mu_x(u)$ is a grade of membership of $u \in U$. The value represents the degree of *u* belongs to the fuzzy set *X*. Thus, we can represent the fuzzy set *X* over U as follows.

$$X = \{(\mu_x(u)/u) : u \in U, \mu_x(x) \in [0, 1]\}$$

Note that. *F(U)* denoted the set of all the fuzzy sets over U. We can define a fuzzy soft set (*fuzzy soft set*) F_A over U as a function f_A that represent a mapping.

$f_A : A \to F(U)$ and $f_A(x) = 0$, if $x \notin A$, $A \subseteq E$. In order pair notations, the *fuzzy soft set* F_A over U can be write as follows:

$$F_A = \{(x, f_A(x))/x \in A, f_A(x) \in F(U)\}$$

Let us consider an example of a fuzzy soft as follow.

Example 1. Suppose $U = \{u_1, u_2, u_3, u_4, u_5, u_6\}$ is the set of houses under consideration, $E = \{x_1, x_2, x_3, x_4, x_5\}$ is the set of parameters where each parameter is a fuzzy word or a sentence involving fuzzy words, E = {expensive (x_1), beautiful (x_2), wooden (x_3), cheap (x_4), in the green surroundings (x_5)}. In the case, to define a fuzzy soft set means to point out expensive house, beautiful house, and so on. The *fuzzy soft set* F_A describes the "attractive-ness of the houses" which Mr. X is going to buy. Suppose that

$$\begin{aligned}
f_A(x_1) &= \{(u_1, 0.5), (u_2, 1), (u_3, 0.4), (u_4, 1), (u_5, 0.3), (u_6, 0)\} \\
f_A(x_2) &= \{(u_1, 1), (u_2, 0.4), (u_3, 1), (u_4, 0.4), (u_5, 0.6), (u_6, 0.8)\} \\
f_A(x_3) &= \{(u_1, 0.2), (u_2, 0.3), (u_3, 1), (u_4, 1), (u_5, 1), (u_6, 0)\} \\
f_A(x_4) &= \{(u_1, 1), (u_2, 0), (u_3, 1), (u_4, 0.2), (u_5, 1), (u_6, 0.2)\} \\
f_A(x_5) &= \{(u_1, 0.8), (u_2, 0.1), (u_3, 0.5), (u_4, 0.3), (u_5, 0.2), (u_6, 0.3)\}
\end{aligned}$$

then the F_A can be written as,

$$F_A = \{(x_1, \{(u_1, 0.5), (u_2, 1), (u_3, 0.4), (u_4, 1), (u_5, 0.3), (u_6, 0)\}),$$
$$(x_2, \{(u_1, 1), (u_2, 0.4), (u_3, 1), (u_4, 0.4), (u_5, 0.6), (u_6, 0.8)\}),$$
$$(x_3, \{(u_1, 0.2), (u_2, 0.3), (u_3, 1), (u_4, 1), (u_5, 1), (u_6, 0)\}),$$
$$(x_4, \{(u_1, 1), (u_2, 0), (u_3, 1), (u_4, 0.2), (u_5, 1), (u_6, 0.2)\}),$$
$$(x_5, \{(u_1, 0.8), (u_2, 0.1), (u_3, 0.5), (u_4, 0.3), (u_5, 0.2), (u_6, 0.3)\})\},$$

In matrix representation, the F_A can be presented by Table 1 as follow.

Table 1. Table representation of F_A

U	x_1	x_2	x_3	...	x_n
u_1	$\mu_{f_{A(x_1)}}(u_1)$	$\mu_{f_{A(x_2)}}(u_1)$	$\mu_{f_{A(x_3)}}(u_1)$	...	$\mu_{f_{A(x_n)}}(u_1)$
u_2	$\mu_{f_{A(x_1)}}(u_2)$	$\mu_{f_{A(x_2)}}(u_2)$	$\mu_{f_{A(x_3)}}(u_2)$	...	$\mu_{f_{A(x_n)}}(u_2)$
⋮	⋮	⋮	⋮	⋮	⋮
u_m	$\mu_{f_{A(x_1)}}(u_m)$	$\mu_{f_{A(x_2)}}(u_m)$	$\mu_{f_{A(x_3)}}(u_m)$	...	$\mu_{f_{A(x_n)}}(u_m)$

where $\mu_{f_{A(x_i)}}$ is the membership function of f_A.

If $a_{ii} = \mu_{f_{A(x_i)}(u_j)}$ i = 1, 2, 3,…, n and j = 1, 2, 3…m then the f_A can be represented in form of matrix,

$$[a_{ij}] = \begin{bmatrix} a_{11} & a_{12} & \cdots & a_{1n} \\ a_{21} & a_{22} & \cdots & a_{2n} \\ \vdots & \vdots & \vdots & \vdots \\ a_{m1} & a_{m2} & \cdots & a_{mn} \end{bmatrix}$$

The matrix is called $m \times n$ of the f_A over U. The matrix of the f_A from Example 1 is,

$$F_A = \begin{vmatrix} 0,5 & 1 & 0,2 & 1 & 0,8 \\ 1 & 0,4 & 0,3 & 0 & 0,1 \\ 0,4 & 1 & 1 & 1 & 0,5 \\ 1 & 0,4 & 1 & 0,2 & 0,3 \\ 0,3 & 0,6 & 1 & 1 & 0,2 \\ 0 & 0,8 & 0 & 0,2 & 0,3 \end{vmatrix}$$

The following theorem presents a property the cardinal set of F_A.

Theorem 2.1. *The cardinal set of* F_A *denoted by* $cF_A = \left\{\mu_{f_A(x)}/x : x \in E\right\}$ *is a fuzzy soft set over E.*

Proof. Let $F_A \in FS(U)$ with cardinal set, where $cF_A = \left\{\mu_{f_A(x)}/x : x \in E\right\}$ is a membership function of cF_A and

$$\mu_{cF_A} = \frac{|f_A(x)|}{|U|} \in [0, 1]$$

$|U|$ is the cardinal value of the universe, and $|f_A(x)|$ is the scalar cardinality of $f_A(x)$. Thus, it is clear that $\mu_{cF_A} : E \to [0,1]$. □

The following definition describes the representation of cF_A.

Definition 2.3. *Let* $F_A \in$ FS(U) *and* $cF_A \in$ FS(U). *Assume tha* $E = \{x_1, x_2, \ldots, x_n\}$ *t and* $A \subseteq E$. *The* cF_A *can be presented by Table* 2 *as follow*From Definition 2.3, if $a_{ii} = \mu_c F_A(x_j)$ where j = 1,2,…,n then we can be represented the cardinal set cF_A in form of matrix,

Table 2. The table representation of cF_A

E	x_1	x_2	…	x_n
$\mu_c F_A$	$\mu_c F_A(x_1)$	$\mu_c F_A(x_2)$		$\mu_c F_A(x_n)$

$$\lfloor a_{ij} \rfloor = [a_{11} \quad a_{12} \quad \cdots \quad a_{1n}]$$

From Example 1 the cardinal set of F_A over E is as follows;

$$cF_A = |0,53 \quad 0,70 \quad 0,58 \quad 0,57 \quad 0,37|$$

The following definition presents the notion of AND operation in fuzzy soft sets.

Definition 2.4. (AND operation on two fuzzy set). *If* F_A *and* G_A *are two fuzzy soft sets, then "*F_A *AND* F_B*" denoted by* $F_A \wedge F_B$ *is defined by* $F_A \wedge F_B = F_C$ *where*

$$F_C = \{(xy, f_C(xy)/xy \in A\,x\,B, f_C(xy) \in FS(U)\} \text{ and } f_C(xy) = f_A(x) \cap f_B(y)$$

Let us consider the following example to illustrate AND operation in fuzzy soft sets.

Example 2. From Example 1, let A = $\{x_1, x_4\}$ and B = $\{x_2, x_3\}$, where A, B $\subseteq$ E. The F_A and F_B *are* two *fuzzy soft set* over U that defined by a function f_A and f_B, respectively.

$$f_A(x_1) = \{(u_1, 0.5), (u_2, 1), (u_3, 0.4), (u_4, 1), (u_5, 0.3), (u_6, 0)\}$$
$$f_A(x_4) = \{(u_1, 1), (u_2, 0), (u_3, 1), (u_4, 0.2), (u_5, 1), (u_6, 0.2)\}$$

$$f_B(x_2) = \{(u_1,1),(u_2,0.4),(u_3,1),(u_4,0.4),(u_5,0.6),(u_6,0.8)\}$$
$$f_B(x_3) = \{(u_1,0.2),(u_2,0.3),(u_3,1),(u_4,1),(u_5,1),(u_6,0)\}$$

$F_A \wedge F_B = F_C$ where

$C = \{x_1x_2, x_1x_3, x_4x_2, x_4x_3\}$
$=$ {expensive-beautiful, expensive-wooden, cheap-beautiful, cheap-wooden}

$$\begin{aligned} f_C(x_1x_2) &= f_A(x_1) \cap f_B(x_2) \\ &= \{min((u_1,0.5),(u_1,1)), \min((u_2,1),(u_2,0.4)), \min((u_3,0.4),(u_3,1)), \\ &\quad \min((u_4,1),(u_4,0.4)), \min((u_5,0.3),(u_5,0.6)), \min((u_6,0),(u_6,0.8))\} \\ &= \{(u_1,0.5),(u_2,0.4),(u_3,0.4),(u_4,0.4),(u_5,0.3),(u_6,0)\} \end{aligned}$$

$$\begin{aligned} f_C(x_1x_3) &= f_A(x_1) \cap f_B(x_3) \\ &= \{min((u_1,0.5),(u_1,0.2)), \min((u_2,1),(u_2,0.3)), \min((u_3,0.4),(u_3,1)), \\ &\quad \min((u_4,1),(u_4,1)), \min((u_5,0.3),(u_5,1)), \min((u_6,0),(u_6,0))\} \\ &= \{(u_1,0.2),(u_2,0.3),(u_3,0.4),(u_4,1),(u_5,0.3),(u_6,0)\} \end{aligned}$$

$$\begin{aligned} f_C(x_4x_2) &= f_A(x_4) \cap f_B(x_2) \\ &= \{min((u_1,1),(u_1,1)), \min((u_2,0),(u_2,0.4)), \min((u_3,1),(u_3,1)), \\ &\quad \min((u_4,0.2),(u_4,0.4)), \min((u_5,1),(u_5,0.6)), \min((u_6,0.2),(u_6,0.8))\} \\ &= \{(u_1,1),(u_2,0),(u_3,1),(u_4,0.2),(u_5,0.6),(u_6,0.2)\} \end{aligned}$$

$$\begin{aligned} f_C(x_4x_3) &= f_A(x_4) \cap f_B(x_3) \\ &= \{min((u_1,1),(u_1,0.2)), \min((u_2,0),(u_2,0.3)), \min((u_3,1),(u_3,1)), \\ &\quad \min((u_4,0.2),(u_4,1)), \min((u_5,1),(u_5,1)), \min((u_6,0.2),(u_6,0))\} \\ &= \{(u_1,0.2),(u_2,0),(u_3,1),(u_4,0.2),(u_5,1),(u_6,0)\} \end{aligned}$$

$$\begin{aligned} F_C = \{&(x_1x_2, \{(u_1,0.5),(u_2,0.4),(u_3,0.4),(u_4,0.4),(u_5,0.3),(u_6,0)\}), \\ &(x_1x_3, \{(u_1,0.2),(u_2,0.3),(u_3,0.4),(u_4,1),(u_5,0.3),(u_6,0)\}), \\ &(x_4x_2, \{(u_1,1),(u_2,0),(u_3,1),(u_4,0.2),(u_5,0.6),(u_6,0.2)\}), \\ &(x_4x_3, \{(u_1,0.2),(u_2,0),(u_3,1),(u_4,0.2),(u_5,1),(u_6,0)\})\} \end{aligned}$$

The following section presents our proposed fuzzy soft association rules.

3 Fuzzy Soft Association Rules

At the first presented, the association rule mining issue was focused to "market basket analysis". This is the investigation at the store, investigation about things that was bought together by a consumer. thus, the term "market basket analysis" and the notation and semantic meaning. The corresponding association rules can be categorized as Boolean association rules. The semantic meaning of a Boolean association rule that denoted as X $\Rightarrow$ Y is not an implication. The meaning is the record containing X also contain Y. The support of the rule is defined as the percentage of the records that contain both X and Y. Consequently, the relationship between the antecedent and consequence is logical "AND" and not an implication. We can interpret the semantic meaning of quantitative association rules as, "The records with attribute X's value also have attribute Y's". However, if X and Y is fuzzy soft set F_A and F_B respectively then we can interpret the semantic mining of $F_A \Rightarrow F_B$ as "The record with attribute x's ($x \in$ A with fuzzy membership f_A) value also have attribute y's ($y \in$ B with fuzzy membership f_B)". Hence we can define fuzzy soft association rules as follows.

Definition 3.1. *Let F_A is fuzzy soft set over U that defined by a function f_A representing a mapping $f_A : A \to F(U)$ and F_B is fuzzy soft set over U that defined by a function f_B representing a mapping $f_B : B \to F(U)$ and A, B $\subseteq$ E, where A $\cap$ B = ϕ A fuzzy soft association rules between F_A and F_B is denote by the form $F_A \Rightarrow F_B$. The fuzzy soft set F_A is called antecedent and F_B is consequent.*

From Definition 3.1, we have the following notion of support in fuzzy soft association rules.

Definition 3.2. *The support of a fuzzy soft association rule $F_A \Rightarrow F_B$, denoted by sup($F_A \Rightarrow F_B$) is defined by*

$$sup(F_A \Rightarrow F_B) = \frac{|f_A(x) \cap f_B(y)|}{|U|} = \frac{|min(f_A(x), f_B(y))|}{|U|}$$

From Definition 3.2, we have the following notion of confidence in fuzzy soft association rules.

Definition 3.3 *The confidence of a fuzzy soft association rule $F_A \Rightarrow F_B$, denoted by conf($F_A \Rightarrow F_B$) is defined by*

$$conf(F_A \Rightarrow F_B) = \frac{|f_A(x) \cap f_B(y)|}{|f_A(x)|} = \frac{|\min(f_A(x), f_B(y))|}{|f_A(x)|}$$

Let us consider the following example to describes the idea of fuzzy soft association rules.

Example 3. Consider Example 1, let A = $\{x_2\}$, B = $\{x_1\}$, where A, B $\subseteq$ E. F_A and F_B *are* two *fuzzy soft set* over U that defined by a function f_A and f_B, respectively.

$$f_A(x_2) = \{(u_1, 1), (u_2, 0.4), (u_3, 1), (u_4, 0.4), (u_5, 0.6), (u_6, 0.8)\}$$
$$f_B(x_1) = \{(u_1, 0.5), (u_2, 1), (u_3, 0.4), (u_4, 1), (u_5, 0.3), (u_6, 0)\}$$

A fuzzy soft association rule $F_A \Rightarrow F_B$ {beautiful (x_2)} ⇒ {expensive (x_1)}

$$Sup(F_A \Rightarrow F_B) = \frac{|\min(0.5, 1), \min(1, 0.4), \min(0.4, 1), \min(1, 0.4), \min(0.3, 0.6), \min(0, 0.8)|}{|6|}$$
$$= \frac{|0.5, 0.4, 0.4, 0.4, 0.3, 0|}{|6|} = \frac{0.5 + 0.4 + 0.4 + 0.4 + 0.3 + 0}{6} = 0.33$$

$$Conf(F_A \Rightarrow F_B) = \frac{|\min(0.5, 1), \min(1, 0.4), \min(0.4, 1), \min(1, 0.4), \min(0.3, 0.6), \min(0, 0.8)|}{|10.410.40.60.8|}$$
$$= \frac{|0.5, 0.4, 0.4, 0.4, 0.3, 0|}{|1, 0.4, 1, 0.1, 0.6, 0.8|} = \frac{0.5 + 0.4 + 0.4 + 0.4 + 0.3 + 0}{|1 + 0.4 + 1 + 0.4 + 0.6 + 0.8|}$$
$$= \frac{2}{4.2} = 0.48$$

The rule means "*beautiful and expensive house*" has supported by 33% of data and 48% confidence level.

Example 4. Consider Example 1. Let A = $\{x_2, x_3\}$ Let B = $\{x_1\}$ where A, B ⊆ E. F_A and F_B *are* two *fuzzy soft set* over U that defined by a function f_A and f_B respectively.

$$f_A(x_2) = \{(u_1, 1), (u_2, 0.4), (u_3, 1), (u_4, 0.4), (u_5, 0.6), (u_6, 0.8)\}$$
$$f_A(x_3) = \{(u_1, 0.2), (u_2, 0.3), (u_3, 1), (u_4, 1), (u_5, 1), (u_6, 0)\}$$
$$f_B(x_1) = \{(u_1, 0.5), (u_2, 1), (u_3, 0.4), (u_4, 1), (u_5, 0.3), (u_6, 0)\}$$

a fuzzy soft association rule $F_A \Rightarrow F_B$: {beautiful (x_2), wooden(x_3)} ⇒ {expensive (x_1)}

$$Sup(F_A \Rightarrow F_B) = \frac{|\min(1, 0.2, 0.5), \min(0.4, 0.3, 1), \min(1, 1, 0.4), \min(0.4, 1, 1), \min(0.6, 1, 0.3), \min(0.8, 0, 0)|}{|6|}$$
$$= \frac{|0.2, 0.3, 0.4, 0.4, 0.3, 0|}{|6|} = \frac{0.2 + 0.3 + 0.4 + 0.4 + 0.3 + 0}{6}$$
$$= \frac{1.6}{6} = 0.27$$

$$Conf(F_A \Rightarrow F_B) = \frac{|\min(1, 0.2, 0.5), \min(0.4, 0.3, 1), \min(1, 1, 0.4), \min(0.4, 1, 1), \min(0.6, 1, 0.3), \min(0.8, 0, 0)|}{|\min(1, 0.2), \min(0.4, 0.3), \min(1, 1), \min(0.4, 1), \min(0.6, 1), \min(0.8, 0)|}$$
$$= \frac{|0.2, 0.3, 0.4, 0.4, 0.3, 0|}{|0.2, 0.3, 1, 0.4, 0.6, 0|} = \frac{0.2 + 0.3 + 0.4 + 0.4 + 0.3 + 0}{0.2 + 0.3 + 1 + 0.4 + 0.6 + 0} \frac{1.6}{2.5} = 0,64$$

The rule mean "*beautiful and wooden and expensive house*" has supported by 27% of data with confidence level = 64%.

4 Conclusion

This paper presented the idea of hybridization of fuzzy soft set in mining association rules with emphasize on real-valued Items. We have presented the notion of a fuzzy soft association rule concept. The idea is combination association rule and a hybrid fuzzy soft set. We also defined the notion of support and confidence association rules for fuzzy soft set. As a proof of concept we have also presented examples on mining fuzzy soft association rules. This concept has proven able to mine an interesting association between the real-valued items that was representation on fuzzy soft set. We will further examine the performance of the proposed method for mining interesting rules in text document.

References

1. Agrawal, R., Imielinski, T., Swami, A.: Mining association rules between sets of items in large databases. In: Proceeding of thet ACM SIGMOD International Conference on the Management of Data, pp. 207–216 (1993)
2. Rahman, C.M.: Text classification using the concept of association rule of data mining. In: Proceedings of International conference on Information Technology, pp. 234–241 (2003)
3. Agrawal, R., Srikant, R.: Fast algorithms for mining association rules. In: Proceedings of the 20th International Conference on Very Large Data Bases (VLDB), pp. 487–499 (1994)
4. Lopes, A.A.Ã., Pinho, R., Paulovich, F.V., Minghim, R.: Visual text mining using association rules. J. Comput. Graph. **31**, 316–326 (2007)
5. Haralambous, Y., Lenca, P.: Text classification using association rules, dependency pruning and hyperonymization. In: DMNLP2014: Workshop on Interactions Between Data Mining and Natural Language Processing. CEUR Workshop Proceedings, Nancy, France, pp. 65–80, September 2014
6. Liu, B.: Integrating classification and association rule mining. In: KDD-98 Proceeding (1998)
7. Doddi, S.: Discovery of Association Rules in Medical Data. US National Library of Medicine National Institut of health, pp. 1–17 (2001)
8. Kwasnicka, H., Switalski, K.: Discovery of association rules from medical data - classical and evolutionary approaches. In: Conference Proceeding: XXI Auntum Meeting of Polish Information Processing Society, pp. 163–177 (2005)
9. Simovici, D.A.: Data Mining of Medical Data : Opportunities and Challenges in Mining Association Rules, no. Dm, pp. 1–25 (1968)
10. Hu, R.: Medical Data Mining Based on Association Rules, www.ccsenet.org: computer and information science, vol. 3, no. 4, pp. 104–108 (2010)
11. Martin, A., Manjula, M., Venkatesan, P.: A business intelligence model to predict bankruptcy using financial domain ontology with association rule mining algorithm. IJCSI Int. J. Comput. Sci. Issues **8**(3), 211–218 (2011)

12. Xu, Z., Zhang, R.: Financial revenue analysis based on association rules mining. Comput. Intell. Ind. Appl. **1**, 220–223 (2009)
13. Kamruzzaman, S.M., Haider, F., Hasan, A.R.: Text classification using association rule with a hybrid concept of Naive Bayes classifier and genetic algorithm. In: Proceeding: 7th International Conference on Computer and Information Technology (ICCIT-2004), pp. 628–687 (2004)
14. Molodtsov, D.: Soft set theory-first result. Comput. Math. Appl. **37**, 19–31 (1999)
15. Herawan, T., Deris, M.M.: On multi-soft sets construction in information systems. In: Huang, D.-S., Jo, K.-H., Lee, H.-H., Kang, H.-J., Bevilacqua, V. (eds.) ICIC 2009. LNCS (LNAI), vol. 5755, pp. 101–110. Springer, Heidelberg (2009). doi:10.1007/978-3-642-04020-7_12
16. Qin, H., Ma, X., Herawan, T., Zain, J.M.: An adjustable approach to interval-valued intuitionistic fuzzy soft sets based decision making. In: Nguyen, N.T., Kim, C.-G., Janiak, A. (eds.) ACIIDS 2011. LNCS (LNAI), vol. 6592, pp. 80–89. Springer, Heidelberg (2011). doi:10.1007/978-3-642-20042-7_9
17. Herawan, T., Deris, M.M.: Soft decision making for patients suspected influenza. In: Taniar, D., Gervasi, O., Murgante, B., Pardede, E., Apduhan, B.O. (eds.) ICCSA 2010. LNCS, vol. 6018, pp. 405–418. Springer, Heidelberg (2010). doi:10.1007/978-3-642-12179-1_34
18. Maji, P.K., Biswas, R., Roy, A.R.: Soft set theory. Comput. Math. Appl. **1221**, 555–562 (2003)
19. Das, P.K., Borgohain, R., Pradesh, A.: An application of fuzzy soft set in medical diagnosis using fuzzy arithmetic operations on fuzzy number created by neevia personal converter trial version. SIBCOLTEJO 05, 107–116 (2010)
20. Roy, A.R., Maji, P.K.: A fuzzy soft set theoretic approach to decision making problems. Comput. Math. Appl. **203**, 412–418 (2007)
21. Kong, Z., Gao, L., Wang, L.: Comment on 'A fuzzy soft set theoretic approach to decision making problems'. Comput. Math. Appl. **223**(2), 540–542 (2009)
22. Alkhazaleh, S.: The multi-interval-valued fuzzy soft set with application in decision making. Appl. Math. **6**, 1250–1262 (2015)
23. Çağman, N., Enginoğlu, S.: Soft matrix theory and its decision making. Comput. Math Appl. **59**(10), 3308–3314 (2010)
24. Kharal, A.: Soft approximations and uni-int decision making. Hindawi: Sci. World J. 2014, no. 1999, 2014
25. Feng, F., Bae, Y., Liu, X., Li, L.: Journal of Computational and Applied An adjustable approach to fuzzy soft set based decision making. Comput. Math. Appl. **234**(1), 10–20 (2010)
26. Kong, Z., Wang, L., Wu, Z.: Journal of Computational and Applied Application of fuzzy soft set in decision making problems based on grey theory. Comput. Math. Appl. **236**(6), 1521–1530 (2011)
27. Handaga, B., Herawan, T., Deris, M.M.: FSSC: An Algorithm for Classifying Numerical, vol. 3 (2012)
28. Kalaiselvi, N., Hannah Inbarani, H.: Fuzzy soft set based classification for gene expression data. IJSER **3** (2012)
29. Hong, T., Lee, Y.: An overview of mining fuzzy association rules. In: Bustince, H., Herrera, F., Montero, J. (eds.) Fuzzy Sets and Their Extensions: Representation, Aggregation and Models, vol. 220, pp. 397–410. Springer, Heidelberg (2008)
30. Jiang, Y., Liu, H., Tang, Y., Chen, Q.: Semantic decision making using ontology-based soft sets. Math. Comput. Model **53**(5–6), 1140–1149 (2011)

31. Çağman, N.: Fuzzy parameterized fuzzy soft set theory and its applications. Iranian J. Fuzzy Syst. **1**(1), 21–35 (2010)
32. Glu, S.E.: Fuzzy soft set theory and its applications. Iranian J. Fuzzy Syst. **8**(3), 137–147 (2011)
33. Zadeh, L.A.: Fuzzy sets. Inform. Control **8,** 338–353 (1965)
34. Han, J., Pei, J., Yin, Y.: Mining frequent patterns without candidate generation. In: The 2000 ACM SIGMOD International Conference on Management of Data, 29(2), pp. 1–12 (2000)

Application of Wavelet De-noising Filters in Mammogram Images Classification Using Fuzzy Soft Set

Saima Anwar Lashari[1(✉)], Rosziati Ibrahim[1], Norhalina Senan[1], Iwan Tri Riyadi Yanto[2], and Tutut Herawan[2]

[1] Faculty of Computer Science and Information Technology, Universiti Tun Hussein Onn Malaysia, Batu Pahat, 86400 Parit Raja, Johor, Malaysia
hi120040@siswa.uthm.edu.my, {rosziati,halina}@uthm.edu.my

[2] Faculty of Computer Science and Information Technology, University of Malaya, Lembah Pantai, 50603 Kuala Lumpur, Malaysia
yanto.itr@is.uad.ac.id, tutut@um.edu.my

Abstract. Recent advances in the field of image processing have revealed that the level of noise in mammogram images highly affect the images quality and classification performance of the classifiers. Whilst, numerous data mining techniques have been developed to achieve high efficiency and effectiveness for computer aided diagnosis systems. However, fuzzy soft set theory has been merely experimented for medical images. Thus, this study proposed a classifier based on fuzzy soft set with embedding wavelet de-noising filters. Therefore, the proposed methodology involved five steps namely: MIAS dataset, wavelet de-noising filters hard and soft threshold, region of interest identification, feature extraction and classification. Therefore, the feasibility of fuzzy soft set for classification of mammograms images has been scrutinized. Experimental results show that proposed classifier FussCyier provides the classification performance with Daub3 (Level 1) with accuracy 75.64% (hard threshold), precision 46.11%, recall 84.67%, F-Micro 60%. Thus, the results provide an alternative technique to categorize mammogram images.

Keywords: Mammogram images · Feature extraction · Wavelet filters · Fuzzy soft set

1 Introduction

Digital mammograms have enhanced the aptitude to sense breast anomalies. Over the years, computer-aided systems have been in used to aid radiologists by improving the quality of images and identify the suspicious regions. However, yet, radiologist overlook breast cancer detection and identification between the range of 10%–30% during breast screening [1]. Consequently, quite a lot of researchers investigated the potentials of using data mining techniques to detect and predict the breast cancer [2]. However, the noise present in the mammogram images is subtle and varied in

T. Herawan et al. (eds.), *Recent Advances on Soft Computing and Data Mining*, Advances in Intelligent Systems and Computing 549, DOI 10.1007/978-3-319-51281-5_53

appearance which adversely affects classification accuracy of these images [3]. Moreover, when addressing the digital mammogram images, the emphasis has been to develop algorithms that attempt to improve the imaging quality [4, 5]. Thus, de-noising plays an imperative role in the field of image pre-processing, image analysis and classification. However, there has been relatively diminutive research on the noise removal using wavelet de-noising filters for mammogram images [6]. Although, much emphasis have been placed on standard images and other medical images such as (MRI, ultrasound, CT scan) [7, 8].

Meanwhile, medical diagnosis and prognosis problems are leading paradigm of decision making in the face of uncertainty [9]. Thus, fuzzy set theory plays a vital role in formalizing uncertainties for medical diagnosis and prognosis [10, 11]. To handle uncertainty in the decision making, the use of fuzzy set theory bring in a lot of new

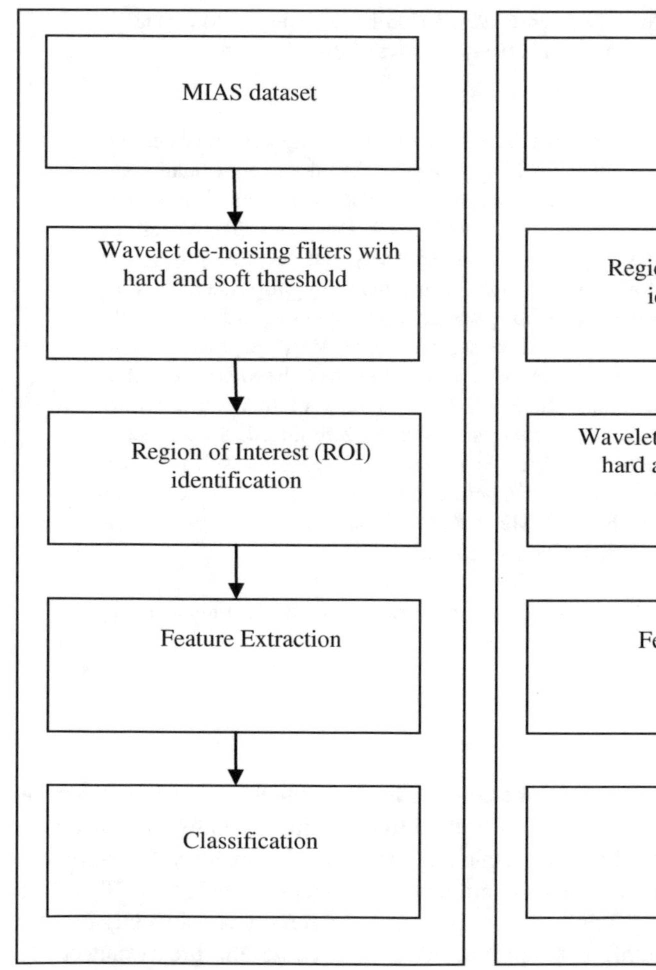

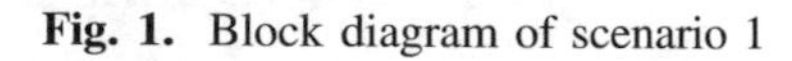
Fig. 1. Block diagram of scenario 1

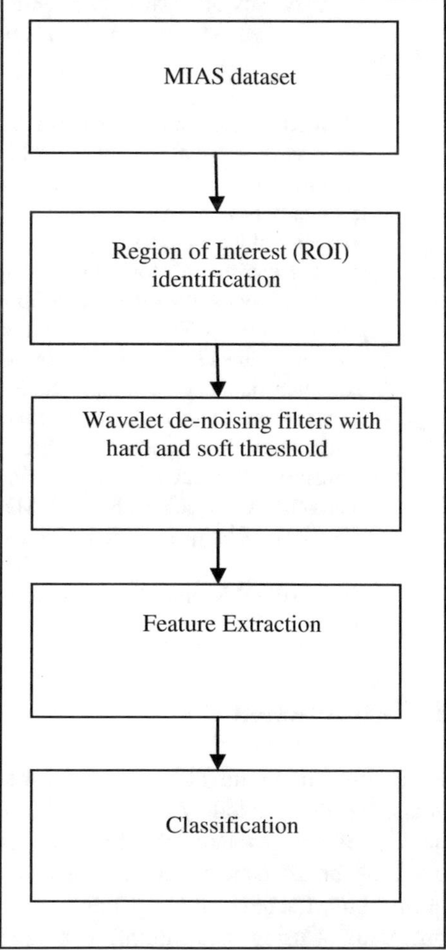

Fig. 2. Block diagram of scenario 2

methods of decision making such as Mushrif *et al.,* [12] offered a Soft Set Classifier (SSC) for natural textures using soft set theory. However, soft set theory is appropriate for binary numbers although still difficult to handle real numbers [13, 14]. For that reason, fuzzy soft set can handle fuzzy attributes (parameters in the form of real numbers) [15, 16]. Later, Handaga *et al.,* [17] demonstrated a new application of soft set for numerical data classification by offering a more general concept based on similarity measure between two fuzzy soft sets that is Fuzzy Soft Set Classifier (FSSC), which can handle parameters in the form of real numbers, yet, FSSC has high algorithm complexity.

Limitations of the earlier studies and lack of work on the mammogram images classification using similarity measure on fuzzy soft set motivated the present research. Thus, the present study is intended to increase the mammogram images quality by incorporating wavelet threshold de-noising functions (pre-processing phase) whilst introducing distance measure function for mammogram images classification and named the proposed classifier as FussCyier. Thus, to conduct this study, the proposed methodology involved two scenarios which are stated in Figs. 1 and 2. The reason for designing these two scenarios is to observe whether de-noising images is more effective or getting region of interest (ROI) first then de-noising images, which scenario provides better classification accuracy rate.

2 Wavelet Threshold De-noising

Wavelet threshold de-noising is a very efficient method in order to remove noise [18]. Wavelet threshold de-noising is mainly divided into two categories: hard thresholding and soft thresholding.

2.1 Hard Thresholding

Hard-thresholding is stated in Eq. 1 [18, 19].

$$\tilde{w}_{j,k} = \begin{cases} \tilde{w}_{j,k} & |w_{j,k}| \geq \lambda \\ 0 & |w_{j,k}| < \lambda \end{cases} \tag{1}$$

where

$\tilde{w}_{j,k}$ the signal
λ is the threshold

Universal threshold is stated in Eq. 2 [18]

$$\lambda = \sigma\sqrt{2InN} \tag{2}$$

where,

σ refers to standard deviation of the noise
N refers to number of data samples in signal

2.2 Soft Thresholding

The soft thresholding is stated in Eq. 3 [18–21]

$$\tilde{w}_{j,k} = \begin{cases} \mathrm{sgn}(w_{j,k})\left(|w_{j,k}| - \lambda\right) & |w_{j,k}| \geq \lambda \\ 0 & |w_{j,k}| \geq \lambda \end{cases} \quad (3)$$

where sgn(∗) is symbol function

$$\mathrm{sgn}(n) = \begin{cases} 1 & n > 0 \\ -1 & n < 0 \end{cases} \quad (4)$$

3 Proposed Methodology

The study was conducted considering two scenarios which are presented in Figs. 1 and 2. Figure 1 represents block diagram of scenario 1 which is comprises of five phases namely MIAS (Mammographic Image Analysis Society) dataset, Wavelet de-noising filters with hard and soft threshold, region of interest identification, feature extraction and classification. Figure 2 shows block diagram of scenario 2 which comprises of same five phases, only switching phase 2 and phase 3. The reason for designing these two scenarios is to observe whether de-noising images first is more effective and contributing factor towards better classification rate or getting region of interest (ROI) first then de-noising images, which scenario provides better classification accuracy rate.

Mammogram images were collected from the Mammographic Image Analysis Society (MIAS). MIAS dataset consists of 63 benign and 51 malign. Later, wavelet de-noising filters with hard and soft threshold has been done, pseudocode for de-noise mammogram images has been explained in Fig. 3 [22]. Afterwards, the Region of interest (ROI) has been calculated in order to focus on the important point solely on the appropriate breast region, which lessens the opportunity for erroneous classification. Soon after ROI identification, feature extraction step has done by extracting six features namely: mean, variance, skewness, kurtosis, contrast and smoothness respectively [21, 22]. For classification, a classifier has been proposed based on fuzzy soft as stated in Sect. 3.1.

3.1 Classification

Classification is based on the concept of distance measure between two fuzzy soft sets. A measure of similarity or dissimilarity defines the resemblance between twor objects. Thus, FussCyier comprises of three phases namely pre-processing phase, training phase and testing phase. Pre-processing phase has been incorporated that consists of two steps (a) de-noised images using wavelet hard and soft threshold functions as stated in Sects. 2.1 and 2.2(b) feature normalization as stated in Eq. 5 in both training and testing phases. For training phase, FussCyier is train by calculating the average value of each parameter from all objects with the same class label to construct fuzzy soft set

Algorithm: Pseudocode for De-Noise Mammogram Images
Input: Raw Mammogram Images
Output: De-noised Images

Begin

Step 1: Transform the images into Discrete Wavelet Transform (DWT)
Step 2: Estimate the threshold value using hard and soft threshold
Step 3: Calculate ROI
Step 4: Generate statistical features
Step 5: Compresses images and reconstruct images from the shrunken coefficients
Step 6: Carry out Inverse Discrete Wavelet Transform (IDWT)
Step 7: Calculate PSNR Values

End

Fig. 3. Pseudocode for de-noise mammogram images

model as shown in Eq. 6. For testing phase, FussCyier applied the distance between two fuzzy soft set as stated in the work of Baccour *et al.,* [23] as illustrated in Eq. 7. Since, FussCyier measures the distance between image features, intuitively, small

Pre-Processing phase

1. De-noised images with wavelet hard and soft threshold functions using Equations 1, 2 and 3 and obtain a feature vector E_{wi} for $i = 1,2,..,N$.
2. Feature normalization for all training and testing data using Equation 5.

Training phase

1. Given N samples obtained from the data class W.
2. Calculate the cluster center vector E_w $i = 1,2,..,N$ using Equation 6

$$E_w = \frac{1}{N}\sum_{i=1}^{N} E_{wi} \quad (6)$$

3. Obtain Fuzzy soft set model (F_w, E), is a cluster centre vector for every class w having D features
4. Repeat the steps 2 and 3 for all W classes

Testing phase

1. Take the unknown class data
2. Obtain a Fuzzy soft set model for unknown class data $(\widetilde{G}, E)$, compute similarity measure based on distance between $(\widetilde{G}, E)$ and (F_w, E) for each W using equation

$$S(F,G) = \frac{1}{n}\sum_{i=1}^{n}\left(1 - \left|\mu_F^i - \mu_G^i\right|\right) \quad (7)$$

3. Assign the unknown data to class w if distance measure is maximum $w = \arg\left[\max_{w=1}^{W} S(G, F_w)\right]$ (8)

Fig. 4. Mammogram images classification using FussCyier

distances correspond to higher similarity. Lastly, gives maximum score computed from distance measure to determine class label for the test data as shown in Eq. 8. Figure 4 shows the classifier FussCyier for mammogram images classification. Feature normalization is done by dividing each attributes value with the largest value at each attributes [17].

$$e_{fi} = \frac{e_i}{\max(e_i)} \tag{5}$$

where e_i, $i = 1, 2, \ldots..., n$ is the old attribute and e_{fi} is attribute with new value between $[0, 1]$

4 Results and Discussion

Pre-Processing for mammogram images based on the five wavelet de-noising filters namely: Sym8, Haar, Coif1, Daub3 and Daub4 whilst utilizing different levels of Gaussian noise with hard and soft threshold functions have been presented. Empirical results for these wavelet de-noising filters tested with MIAS (Mammographic Image Analysis Society) dataset are reported. Different Peak Signal-to-Noise Ratio (PSNR) values are calculated and compared by applying these wavelets filters techniques one after the other. Table 1 summarizes different wavelet de-noising filters namely: Sym8, Daub3, Daub4, Haar and Coif1 with different noise level $\sigma = 10$, $\sigma = 20$, and $\sigma = 40$. From these obtained results, it was found out that Daub3 wavelet de-noising filter is more efficient for the mammogram images.

Table 1. PSNR values for MIAS after processing through different wavelet filters

Mammogram images	Type of threshold	Filter Sym8	Filter Daub3	Filter Daub4	Filter Haar	Filter Coif1
	Hard	45.89395	46.36423	45.22268	45.74382	41.40651
	Soft	43.46429	43.66108	43.50071	43.2415	41.40651

Therefore, the applicability of the thresholding functions along with wavelet transforms is well established. When the overall mammogram images de-noising performance is measured, it is found that Daub3 offer better results while compared with the other wavelet filters. The best PSNR value was 46.36423 dB (hard thresholding) and 43.66108 dB (soft thresholding). Thus, the adoption of different wavelets in order to improve images quality and provides detail visibility without distorting their appearance and shapes were successfully achieved.

Table 2 illustrates the performance analysis of images de-noising with wavelet thresholding methods for different levels of wavelet decomposition for Scenario 1. Daub3 (Level 1) gives maximum classification rate with accuracy 75.87% (soft threshold), precision 40.56%, recall 84%, F-Micro 46.15% with CPU time 0.0029 s, whereas the highest classification rate occurs with filter Sym8 (Level 1) with accuracy

Table 2. Performance analysis of images de-noising with wavelet thresholding methods for different levels of decomposition for scenario 1

Wavelet de-noising filters with different decomposition levels		Accuracy (%)	Precision	Recall	F-Micro
Daub3 (Level 1)	Hard threshold	67.46	48.89	74.00	51.61
	Soft threshold	75.87	40.56	84.00	46.15
Daub3 (Level 4)	Hard threshold	67.57	45.00	73.33	58.06
	Soft threshold	71.15	38.89	82.67	64.29
Daub3 (Level 8)	Hard threshold	66.85	45.56	74.00	51.43
	Soft threshold	69.76	40.56	79.33	41.38
Sym8 (Level 1)	Hard threshold	61.92	53.33	64.00	56.25
	Soft threshold	75.86	33.89	86.67	46.15
Sym8 (Level 4)	Hard threshold	71.18	52.73	76.67	56.25
	Soft threshold	73.24	35.00	87.33	44.44
Sym8 (Level 8)	Hard threshold	68.20	47.22	73.33	64.71
	Soft threshold	71.67	35.56	84.00	39.20

75.86% (soft threshold), precision 33.89%, recall 86.67%, F-Micro 46.15% with CPU time 0.0028 s.

Table 3 demonstrate the performance analysis of images de-noising with wavelet thresholding methods for different levels of wavelet decomposition for Scenario 2. Daub3 (Level 1) offer the utmost classification rate with accuracy 75.64% (hard threshold), precision 46.11%, recall 84.67%, F-Micro 60% with CPU time 0.0032 s, whereas the highest classification rate occurs with filter Sym8 (Level 4) with accuracy 75.64% (hard threshold), precision 46.11%, recall 84.67%, F-Micro 51.43% with CPU time 0.0026 s.

Table 3. Performance Analysis of images de-noising with wavelet thresholding methods for different levels of decomposition for scenario 2

Wavelet de-noising filters with different decomposition levels		Accuracy (%)	Precision	Recall	F-Micro
Daub3 (Level 1)	Hard threshold	75.64	46.11	84.67	60.00
	Soft threshold	74.17	32.22	86.67	46.15
Daub3 (Level 4)	Hard threshold	65.61	42.22	77.33	58.06
	Soft threshold	73.70	33.89	82.67	55.17
Daub3 (Level 8)	Hard threshold	71.87	46.67	76.00	44.44
	Soft threshold	74.08	39.44	82.00	57.14
Sym8 (Level 1)	Hard threshold	75.64	46.11	84.67	44.44
	Soft threshold	74.04	37.33	85.33	58.06
Sym8 (Level 4)	Hard threshold	75.64	46.11	84.67	51.43
	Soft threshold	74.19	46.11	84.00	56.00
Sym8 (Level 8)	Hard threshold	68.20	47.22	73.33	53.33
	Soft threshold	70.49	35.56	80.00	46.67

Effectiveness of the proposed Scenario 2 have been thoroughly tested, it can observed from Table 3 that soft threshold provides better classification rate than hard threshold, even slightly better than Scenario 1. In general, de-noising filters perform well for both scenarios. As soft thresholding present more visually satisfying image and decrease the hasty sharp changes that took places in hard Thresholding [24]. Thus, this study can suggest that soft threshold function more appropriate when comes to classification of mammogram images.

5 Conclusion

This study applied a classification algorithm based on fuzzy soft set with wavelet de-noising filters. To observe the effect of de-noising before and after ROI, two scenarios were designed in order to observe their effect towards performance of classifier and from the obtained results, calculating ROI first and then filtering contribute toward high classification accuracy rate. The inclusion of pre-processing phase was done by incorporating hard and soft threshold functions with Daub3 and Sym8 filters with different orders of approximation levels on mammogram images where Daub 3 is more suitable filter for de-noising mammogram images. Moreover, this study contributes by extending the robustness of fuzzy soft theory into examining mammogram images within medical image classification domain.

Acknowledgment. The authors would like to thank office for Research, Innovation, Commercialization and Consultancy Management (ORICC) and Universiti Tun Hussein Onn Malaysia for supporting this research under Vote U110.

References

1. Mohanty, A.K., Champati, P.K., Swain, S.K., Lenka, S.K.: A review on computer aided mammography for breast cancer diagnosis and classification using image mining methodology. Int. J. Comput. Sci. Commun. **2**(2), 531–538 (2011)
2. Singh, N., Mohapatra, A.G., Kanungo, G.: Breast cancer mass detection in mammograms using K-means and fuzzy C-means clustering. Int. J. Comput. Appl. **22**(2), 0975–8887 (2011)
3. Naveed, N., Hussain, A., Jaffar, M.A., Choi, T.S.: Quantum and impulse noise filtering from breast mammogram images. Comput. Methods Programs Biomed. **108**(3), 1062–1069 (2012)
4. James, A.P., Dasarathy, B.V.: Medical image fusion: a survey of the state of the art. Inf. Fusion **19**, 4–19 (2014)
5. Saha, M., Naskar, M.K., Chatterji, B.N.: Soft, hard and block thresholding techniques for denoising of mammogram images. IETE J. Res. **61**(2), 186–191 (2015)
6. Sidh, K., Khaira, B., Virk, I.: Medical image denoising in the wavelet domain using Haar and DB3 filtering. Int. Refereed J. Eng. Sci. **1**, 1–8 (2012)

7. Taujuddin, N.S.A.M., Ibrahim, R.: Enhancement of medical image compression by using threshold predicting wavelet-based algorithm. In: Sulaiman, H.A., Othman, M.A., Othman, M.F.I., Rahim, Y.A., Pee, N.C. (eds.) Advanced Computer and Communication Engineering Technology. LNEE, vol. 315, pp. 755–765. Springer, Cham (2015). doi:10.1007/978-3-319-07674-4_71
8. Ramani, R., Vanitha, N.S., Valarmathy, S.: The pre-processing techniques for breast cancer detection in mammography images. Int. J. Image Graph. Sig. Process. **5**(4), 47 (2013)
9. Begum, S.A., Devi, O.M.: Fuzzy algorithms for pattern recognition in medical diagnosis. Assam Univ. J. Sci. Technol. **7**(2), 1–12 (2011)
10. Zadeh, L.A.: Fuzzy sets. Inf. Control **8**(3), 338–353 (1965)
11. Steimann, F.: On the use and usefulness of fuzzy sets in medical AI. Artif. Intell. Med. **21**(1), 131–137 (2001)
12. Mushrif, M.M., Sengupta, S., Ray, A.K.: Texture classification using a novel soft set theory based classification algorithm. In: Narayanan, P.J., Nayar, S.K., Shum, H.-Y. (eds.) ACCV 2006. LNCS, vol. 3851, pp. 246–254. Springer, Heidelberg (2006). doi:10.1007/11612032
13. Herawan, T., Deris, M.M., Abawajy, J.H.: Matrices representation of multi soft-sets and its application. In: Taniar, D., Gervasi, O., Murgante, B., Pardede, E., Apduhan, B.O. (eds.) ICCSA 2010. LNCS, vol. 6018, pp. 201–214. Springer, Heidelberg (2010). doi:10.1007/978-3-642-12179-1_19
14. Ma, X., Sulaiman, N., Qin, H., Herawan, T., Zain, J.M.: A new efficient normal parameter reduction algorithm of soft sets. Comput. Math Appl. **62**(2), 588–598 (2011)
15. Handaga, B., Deris, M.M.: Similarity approach on fuzzy soft set based numerical data classification. Commun. Comput. Inf. Sci. **180**(6), 575–589 (2011)
16. Roy, A.R., Maji, P.K.: A fuzzy soft set theoretic approach to decision making problems. J. Comput. Appl. Math. **203**(2), 412–418 (2007)
17. Handaga, B., Herawan, T., Deris, M.M.: FSSC: an algorithm for classifying numerical data using fuzzy soft set theory. Int. J. Fuzzy Syst. Appl. (IJFSA) **2**(4), 29–46 (2012)
18. Zang, H., Wang, Z., Zheng, Y.: Analysis of signal de-noising method based on an improved wavelet thresholding. In: 9th International Conference on Electronic Measurement & Instruments, ICEMI 2009, pp. 1–987. IEEE, August 2009
19. Donoho, D.L., Johnstone, I.M., Kerkyacharian, G., Picard, D.: Wavelet shrinkage: asymptopia? J. Roy. Stat. Soc. Ser. B (Methodological), 301–369 (1995)
20. Daubechies, I.: Ten Lectures on Wavelets, vol. 61, pp. 198–202. Society for Industrial and Applied Mathematics, Philadelphia (1992)
21. Aarthi, R., Divya, K., Kavitha, S.: Application of feature extraction and clustering in mammogram classification using support vector machine. In: Third International Conference on Advanced Computing (ICoAC), pp. 62–67 (2011)
22. Lashari, S.A., Senan, N., Ibrahim, R.: Effect of presence/absence of noise in mammogram images using fuzzy soft based classification. In: 2015 Second International Conference on Computing Technology and Information Management (ICCTIM), pp. 55–61. IEEE (2015)
23. Baccour, L., Alimi, A.M., John, R.I.: Some notes on fuzzy similarity measures and application to classification of shapes, recognition of arabic sentences and mosaic. IAENG Int. J. Comput. Sci. **41**(2), 81–90 (2014)
24. Xiao, F., Zhang, Y.: A comparative study on thresholding methods in wavelet-based image denoising. Procedia Eng. **15**, 3998–4003 (2011)

Design Selection of In-UVAT Using MATLAB Fuzzy Logic Toolbox

Haris Rachmat[1(✉)], Tatang Mulyana[1], Sulaiman bin H. Hasan[2], and Mohd. Rasidi bin Ibrahim[2]

[1] FRI Universitas Telkom, Jl. Telekominikasi no. 1, Bandung, Indonesia
harisrachmat@telkomuniversity.ac.id, tatang21april@gmail.com

[2] FKMP UTHM, 86400 Parit Raja, Batu Pahat, Johor Bahru, Malaysia
{sulaiman,rasidi}@uthm.edu.my

Abstract. The design of tool holder was a crucial step to make sure the tool holder is enough to handle all forces on turning process. Because of the direct experimental approach is expensive, a few design of innovative ultrasonic vibration assisted tuning (In-UVAT) has proposed. This design has analyzed using finite element simulation to predict feasibility of tool holder displacement and effective stress. SS201 and AISI 1045 materials were used with sharp and ramp corners flexure hinges on design. To decide which one the design is selected was used MATLAB Fuzzy Logic Toolbox. The result shows that AISI 1045 material and which has ramp corner flexure hinge was the best choice to be produced. It has the Eff. Stress Static equal 3, Displacement Static equal 17.5, Eff. Stress Dynamic equal 3, Displacement Dynamic equal 17.5 and Durability Value is 86.4.

Keywords: Finite element method · Piezoelectric actuators · Flexure hinge

1 Introduction

The present work is particularly focused on analysis of In-UVAT design using finite element method (FEM) and then will be select the best designed based on MATLAB Fuzzy Logic Toolbox. From the available literature it is obvious that FEM analysis is commonly used in modeling of tool holder design [1, 4, 5, 6]. The aim of the paper is to study and investigate specific features of the two geometry design of In-UVAT's head with two different materials. The results were choosing a design which has 11 microns displacement of In-UVAT's head i.e. same as maximum displacement of PPA 10 M piezo actuator and the factor of safety (FOS) is also high (1 < FOS < 8).

Theoretical models and instrumental implementation for vibration assisted micro-milling has been studied by [7]. The modeling of his studied is focused on establishing the scientific relationship between the process variables/tooling geometry and all the process outcomes including tool wear, surface roughness and material removal rate, and the instrumental implementation mainly includes the design of the vibration generation device and its control system. In the [8] has been presented an innovative design of fast tool servo (FTS) by combining compliant mechanism

T. Herawan et al. (eds.), *Recent Advances on Soft Computing and Data Mining*,
Advances in Intelligent Systems and Computing 549, DOI 10.1007/978-3-319-51281-5_54

(CM) and precision engineering (PE) so as to meet the stringent requirements of a one degree of freedom (1DOF) compact FTS. To refer this studied, the design requirements of the FTS include: (a) 1 nm resolution, (b) 10–20 μm of range, (c) first natural frequency of over 1400 Hz, (d) compatibility with holding different diamond cutting tools, and (e) without fatigue of the compliant mechanism. Then FEA is used to evaluate the static and dynamic performance of the FTS. The preliminary results show that both the static and dynamic performances are matched to the design objectives.

Many papers have used MATLAB Fuzzy Logic Toolbox on its studied, such as [9, 10, 11, 12, 13, 14, 15, 16, 17]. A way to enhance the performance of a model that combines genetic algorithms and fuzzy logic for feature selection and classification is proposed by [9]. The paper by [10] has presented the development and modeling of Water Tank System (WTS) for temperature control using system identification technique. The paper by [11] has explained the technique for tracking the progress of the software project being built and the technique for selecting an optimal PERT chart developed by using fuzzy logic and it also explains the important variables that effects the strategic decisions connected with the software project. In the paper by [12], an intelligent system of room temperature controller that based on fuzzy logic controller has been presented. The system is an intelligent autonomous control of the two control parameter that is room temperature and the humidity of a room. Motion control for intelligent ground vehicles based on the selection of paths using fuzzy inference has studied by [13]. The paper by [14] focused on selection of best materials for absorber tube and reflective surfaces of Solar Parabolic Collector (SPC) using fuzzy logic, after analyzing the material data. The paper by [15] addressed the selection of optimal shift numbers considering inventory information, customer requirements and machine reliability using fuzzy logic. The paper by [16] provided the design for the hypothetical system using fuzzy logic toolbox that will be modeled is a container of water where 2 chemicals A and B are mixed to obtain a product. This paper presents an optimum approach for designing of fuzzy controller for nonlinear system using Genetic Algorithms (GA). In the paper by [17], a magnetic levitation system is considered as a case study and the controller is designed to keep a magnetic object suspended in the air counteracting the weight of the object.

2 In-UVAT Design

Piezoelectric actuators are becoming increasingly popular in industrial applications. Because of their low displacement and high force outputs, they are commonly coupled through flexure-hinged mechanical displacement amplifiers. The flexure hinge has numerous advantages, such as smoothness of movement, no need of lubrication, zero backlash, and high precision. The use of a flexure-hinged mechanical displacement amplifier is the most appropriate approach to magnifying the output displacement of the piezo actuators [2].

To determine the performance design of In-UVAT i.e. maximum displacement and has a good safety factor required testing. Because of the direct experimental approach is expensive in terms of time, equipment, materials and manpower have, however, encouraged alternative methods of analysis to be explored, in particular computer based simulation. Within this area, finite element modeling (FEM) is pre-eminent [3]. Finite element analysis (FEA) is often referred when the FEM is applied to a specific field of analysis (like stress analysis, thermal analysis, or vibration analysis).

The basic concept behind the FEM is to replace any complex shape with the union (or summation) of a large number of very simple shapes (like triangles) that are combined to correctly model the original part. The smaller simpler shapes are called finite elements because each one occupies a small but finite sub-domain of the original part [18]. All aspects of a design have some degree of uncertainty, as does how the design will actually be utilized. For all the reasons cited above, design always employ a FOS and in practice we justified 1 < FOS < 8 [19]. The inverse piezoelectric effect, the key effect used to realize piezo actuator functions, consists in the deformation of a piezoelectric material due to the applied electric field. PZT that will be used in the simulation is Cedrat PPA 10 M supplied by Cedrat. Some parameters of this device are given in the supplier a data sheet is shown in Table 1 [23].

Table 1. PZT Specifications

Parameter Specifications	Unit	Ranges
Displacement	Micron	7–11
Blocked force	N	700–1000
Response time free – free	ms	0.01
Response time blocked	ms	0.02
Voltage range	VDC	−20–150
Height (in actuation direction)	mm	18
Base depth	mm	10
Base width (including wage and wires)	mm	9
Mass	gr	6

During simulations, fixed geometry was applied at the body of In-UVAT as show in Fig. 1 which has green area [23]. The PZT that embedded in the In-UVAT will create the force which will leads the head of the In-UVAT to the normal force direction (-z axis). The maximum force applied by the PZT was 1000 N. This force is applied to generate the displacement on In-UVAT's head as show in Fig. 1 which shows with purple arrow. The displacement has to in the range 7–11 microns. According to Figs. 2, 3, 4 and 5 are simulation result used FEM, and Table 2 shown summary of these result.

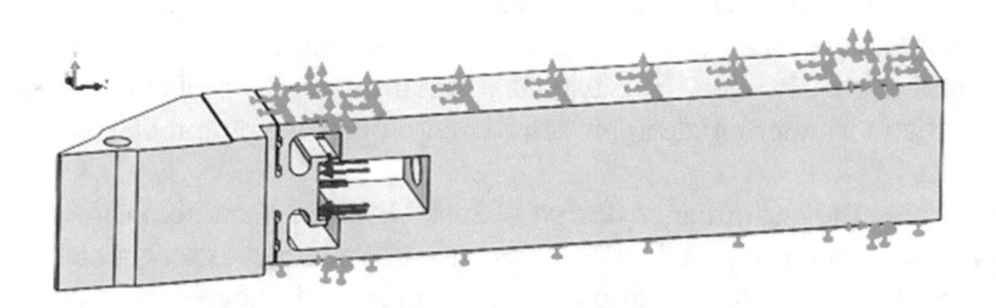

Fig. 1. Load and position of gripping In-UVAT

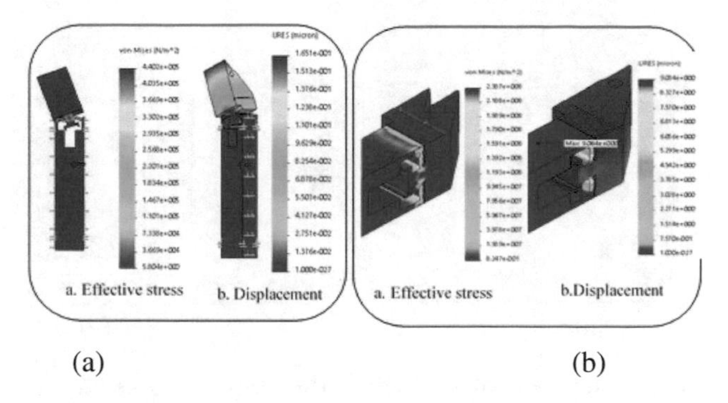

Fig. 2. (a) Ramp Static SS2011; (b) Ramp Dynamics SS201

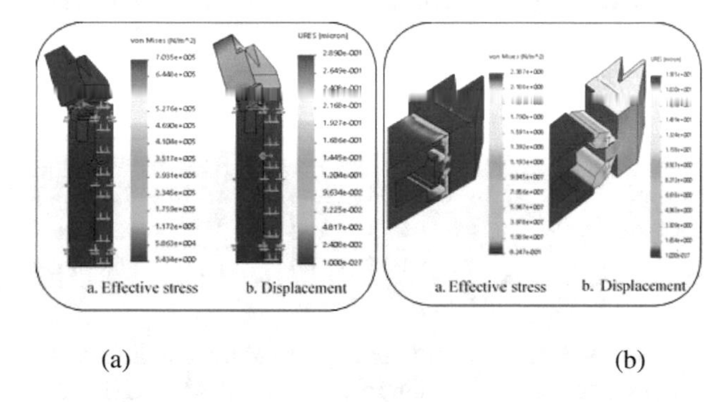

Fig. 3. (a) Sharp Static SS2011; (b) Sharp Dynamics SS201

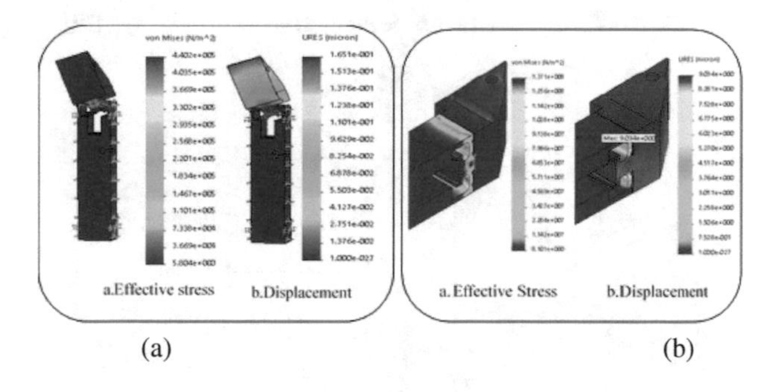

Fig. 4. (a) Ramp Static AISI 1045; (b) Ramp Dynamics AISI 1045

3 MATLAB Fuzzy Toolbox

The Fuzzy Logic Toolbox is a collection of functions built on the MATLAB® numeric computing environment. It provides tools to create and edit fuzzy inference systems within the framework of MATLAB, or integrate fuzzy systems into simulations with Simulink®. This toolbox relies heavily on graphical user interface (GUI) tools, and

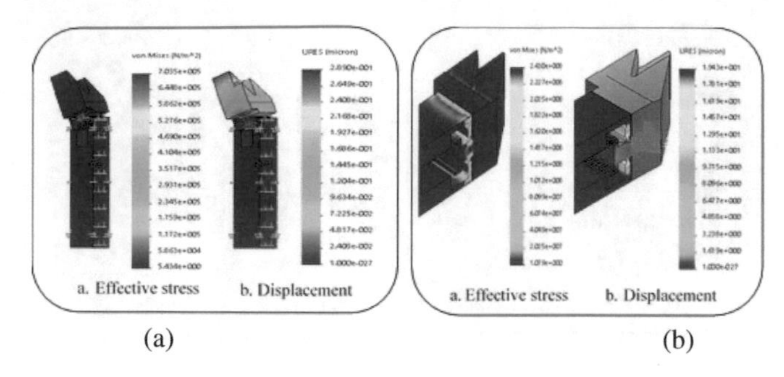

(a) (b)

Fig. 5. (a) Sharp Static SS2011; (b) Sharp Dynamics SS201

Table 2. Statics and dynamics analysis results of In-UVAT

Material & hinge shape	Static		Dynamic	
	Eff. Stress (N/m^2)	Displ'mnt (micron)	Eff. Stress (N/m^2)	Displ'mnt (micron)
SS201 sharp	3.06e + 008	24.11	3.06e + 008	24.38
SS201 ramp	1.77e + 008	11.62	1.74e + 008	11.36
AISI 1045 sharp	3.04e + 008	24.02	3.04e + 008	24.29
AISI 1045 ramp	1.74e + 008	11.60	1.71e + 008	11.3

provides three categories of tools: Command line functions, Graphical interactive tools, and Simulink blocks and examples. There are five primary GUI tools for building, editing, and observing fuzzy inference systems in the Fuzzy Logic Toolbox: the Fuzzy Inference System or FIS Editor, the Membership Function Editor, the Rule Editor, the Rule Viewer, and the Surface Viewer as demonstrated in Fig. 6 ([20, 21, 22]).

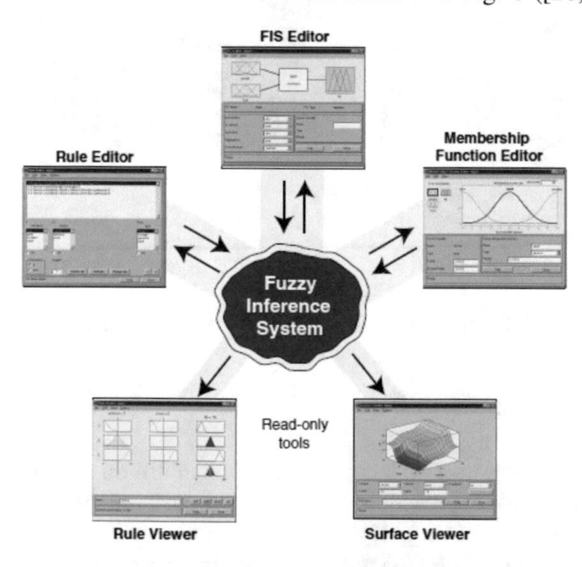

Fig. 6. The fuzzy logic toolbox

4 Result and Discussion

In-UVAT has been designed as shown in Table 3 will be selected using MATLAB Fuzzy Logic Toolbox. The result is demonstrated in Figs. 7, 8, 9, 10 and 11. According these result that the best design is AISI 1045 ramp since it has the Eff. Stress Static equal 3, Displacement Static equal 17.5, Eff. Stress Dynamic equal 3, Displacement Dynamic equal 17.5 and Durability Value is 86.4.

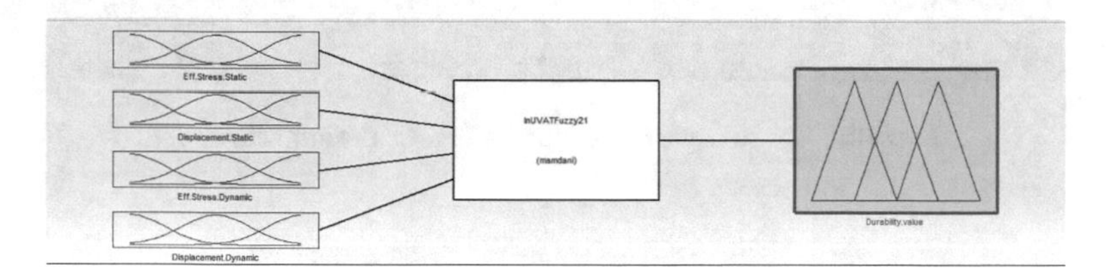

Fig. 7. The FIS editor of the In-UVAT design selection

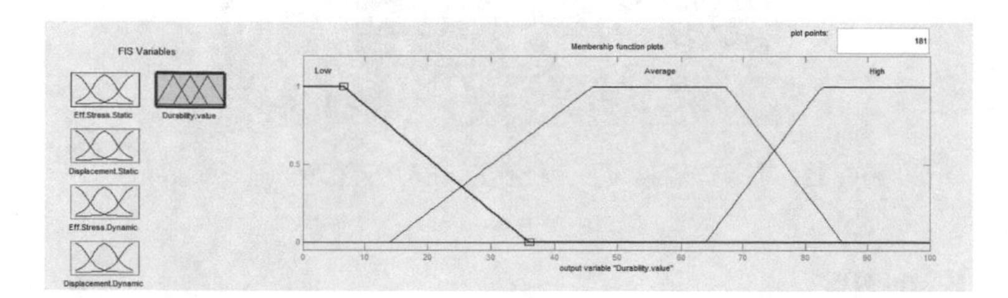

Fig. 8. The membership function editor of the In-UVAT design selection

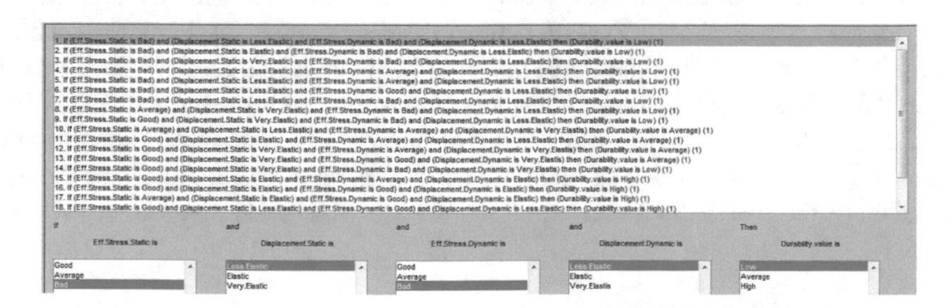

Fig. 9. The rule editor of the In-UVAT design selection

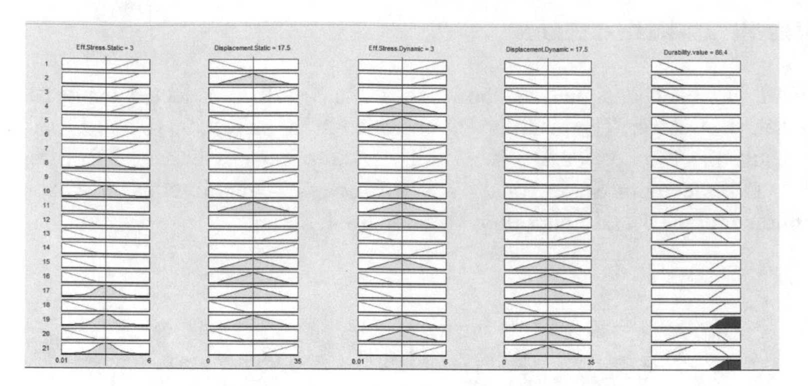

Fig. 10. The rule viewer of the In-UVAT design selection

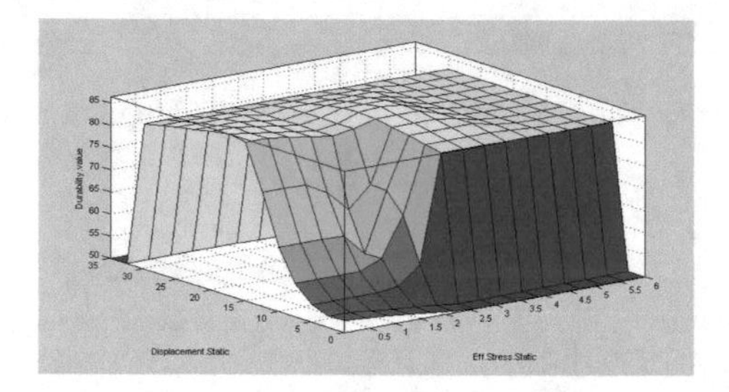

Fig. 11. The surface viewer of the In-UVAT design selection

5 Conclusions

The present study was designed to determine the effect of robustness In-UVAT design using computationally method. The results of this investigation show that the feasible design to be produced is design made from AISI 1045 material which has ramp corner of flexure hinges. It was shown that maximum displacement is 11.3 microns which is different 2.7% with PZT supplier data sheet and ratio between effective stress and yield strength 0.34. This shows that the robustness of design able restrain the 1000 N forces from PZT. This research will serve as a base for future studies on UVAT, and the current findings add to a growing body of literature on design of UVAT tool holder. An implication of this is the possibility that the design will be reliable when we produce to be a tool holder.

References

1. Ahmed, N., Mitrofanov, A.V., Babitsky, V.I., Silberschmidt, V.V.: 3D finite element analysis of ultrasonically assisted turning. Comput. Mater. Sci. **39**, 149–154 (2007)
2. Xu, W., King, T.: Flexure hinges for piezo actuator displacement amplifiers: flexibility, accuracy, and stress considerations. Precis. Eng., vol. 6359, no. 95

3. Soo, S.L., Aspinwall, D.K., Dewes, R.C.: 3D FE modelling of the cutting of Inconel 718. J. Mater. Process. Technol. **150**, 116–123 (2004)
4. Grossi, N., Montevecchi, F., Scippa, A., Campatelli, G.: 3D finite element modeling of holder-tool assembly for stability prediction in milling. Procedia CIRP **31**, 527–532 (2015)
5. Hayati, F., Gurgen, S., Alper, M., Nuri, O.: Finite element modeling of ultrasonic assisted turning of Ti6Al4 V alloy. Procedia - Soc. Behav. Sci. **195**(222), 2839–2848 (2015)
6. Kuo, K.: Design of rotary ultrasonic milling tool using FEM simulation, vol. 1, pp. 48–52 (2007)
7. Ibrahim, M.R., Cheng, K.: Vibration Assisted Micro-Milling: Theoretical Models and Instrumental Implementation. School of Engineering and Design, Brunel University Uxbridge, West London, UB8 3PH, UK
8. Li, H., Ibrahim, M.R., Cheng, K.: Design and principles of an innovative compliant fast tool servo for precision engineering. Mech. Sci. **2**, 139–146 (2011). http://www.mech-sci.net/2/139/2011/doi:10.5194/ms-2-139-2011
9. Ephzibah, E.P.: Cost effective approach on feature selection using genetic algorithms and fuzzy logic for diabetes diagnosis, Int. J. Soft Comput. (IJSC) **2**(1) (2011)
10. Aras, M.S.M., Basar, M.F., Hasim, N., Kamaruddin, M.N., Jaafar, H.I.: Development and modeling of water tank system using system identification method. Int. J. Eng. Adv. Technol. (IJEAT) **2**(6) (2013). ISSN 2249–8958
11. Kaur, A., Chopra, V.: Fuzzy model for optimizing strategic decisions using MATLAB, research cell. Int. J. Eng. Sci. **1** (2011). ISSN 2229-6913
12. Hasim, N., Aras, M.S.M.: Intelligent room temperature controller system using MATLAB fuzzy logic toolbox. Int. J. Sci. Res. (IJSR) (2012). ISSN 2319-7064, Impact Factor, 3.358
13. Wang, S.: Motion control for intelligent ground vehicles based on the selection of paths using fuzzy inference, Master Thesis, Worcester Polytechnic Institute (2014)
14. Reddy, S.P.M., Venkataramaiah, P., Reddy, D.V.V.: Selection of best materials and parametric optimization of solar parabolic collector using fuzzy logic. Energ. Power Eng. **6**, 527–536 (2014). In SciRes, http://www.scirp.org/journal/epe, http://dx.doi.org/10.4236/epe.2014.614046
15. Paul, S.K., Azeem, A.: Selection of the optimal number of shifts in fuzzy environment: manufacturing company's facility application. J. Ind. Eng. Manage. (JIEM) **3**(1), 54–67 (2010). Online ISSN 2013-0953, Print ISSN 2013-8423
16. Kaur, S.: Two inputs two output fuzzy controller system design using MATLAB. Int. J. Adv. Eng. Sci. Technol. (IJAEST) **2**(3) (2012). ISSN 2249-913X
17. Hamed, B., Elreesh, H.A.: Design of fuzzy logic controller for magnetic levitation using genetic algorithm. J. Inf. Commun. Technol. **2**(1) (2012)
18. Akin, J.E.: Finite Element Analysis Concepts via SolidWorks, Elements (2009)
19. Wu, Y., Zhou, Z.: Design calculations for flexure hinges. Rev. Sci. Instrum. **73**(9), 3101–3106 (2002)
20. MathWorks, Fuzzy Logic Toolbox User's Guide. MathWorks, Inc. (1995–2004)
21. MathWorks, Fuzzy Logic Toolbox User's Guide. MathWorks, Inc. (2013)
22. Chen, G., Pham, T.T.: Fuzzy Sets, Fuzzy Logic, and Fuzzy Control Systems. CRC Press LLC (2001)
23. Rachmat, H., Ibrahim, M.R., Hasan, S.: Design selection of an innovative tool holder for Ultrasonic Vibration Assisted Turning (In-UVAT) using finite element analysis simulation. In: ICME 2016, pp. 47–770 (2016)

Web Mining, Services and Security

A Web Based Peer-to-Peer RFID Architecture

Harinda Fernando[1(✉)] and Hairulnizam Mahdin[2]

[1] Asia Pacific Institute of Information Technology, Colombo, Sri Lanka
harinda@apiit.lk
[2] Faculty of Computer Science and Information Technology, Universiti Tun Hussein Onn Malaysia, Parit Raja, Malaysia
hairuln@uthm.edu.my

Abstract. To realize the maximum benefits of RFID technology in large scale distributed environments, the use of an architectural framework which fulfils the specific requirements of those systems is paramount. Unfortunately, the existing frameworks are designed at a high level to allow the development and deployment of a number of fundamentally different systems. Therefore, specialist systems based on this kind of framework will run into a number of issues due to the nature of those applications and their unique needs. In this paper, we present web based P2P architecture for distributed RFID systems specifically targeted at distributed RFID systems. We carry out a comparative analysis of the proposed which shows that our architecture has a number of significant advantages over other existing systems.

Keywords: RFID architecture · Distributed systems · Secure middleware

1 Introduction

In current ICT world there a major emphasis on Internet of Things (IOT) and in the future IOT technologies such as RFID will be used in fundamentally different systems. Due to the differences of these systems their specific functional and performance requirements will also change. The current globally accepted RFID architecture is a generic framework knows as the EPC Global Architecture Framework (EPCGAF) [1]. Therefore much of the specific requirements of specialist RFID systems cannot be met using it, creating a strong need for the creation of Specialist RFID architecture frameworks that can be used to build specialist systems such as global supply chain systems. The developed frameworks must also ensure compatibility with existing systems using the EPCGAF but must be able to provide additional functional and performance benefits as required by specific specialist systems.

By developing a Peer-to-Peer web based RFID architecture we fulfill the requirements of RFID enabled Global supply chain systems. Our approach increase both scalability and availability of the overall system while reducing the processing overhead required when using data provided by external partners. We have also leveraged the preexisting relationships between partners to simplify the issues of node churn and new partner identification for the Peer-to-peer system.

T. Herawan et al. (eds.), *Recent Advances on Soft Computing and Data Mining*,
Advances in Intelligent Systems and Computing 549, DOI 10.1007/978-3-319-51281-5_55

The remainder of this paper is structured as follows. A brief review of previous works is discussed in Sect. 2. In Sect. 3, we describe the proposed architecture. The results of are analyzed and presented in Sect. 4. Finally, some concluding remarks are given in Sect. 5.

2 Related Work

One of the first peer-to-peer RFID architectures is proposed in [2], and uses a hybrid method for peer resolution. Information discovery is done using the traditional EPC while service discovery is done using a DHT based system. The proposed architecture removes some of the bottlenecks and scalability and availability issues associated with the EPCGAF. But because the peers are networked by chain linking the address entries, if any participants are not available the chain would break and information access would be compromised.

In [1] the authors present a peer-to-peer, DHT based alternative to the EPCGAF. This architecture allows for the use of any type of tag identifier to allow greater interoperability with other architectures. The hash values of the tag identifier, which are also used to identify information sources about that tag, are mapped to a distinct location in the network where the participants can retrieve the entry directly. The actual data lookup is carried out using either a direct or indirect search. In direct search the object identifiers are used as keys and looked up in the DHT key space. For indirect searches indices have to be created and updated periodically. But all information associated with the tags remains in the participant's local system and other partners retrieve the required data from that one location. The proposed architecture has a number of improvements over the EPCGAF including greater scalability, because the data look-up is done in a distributed manner and interoperability. The main issue with this architecture is the partner data sources creating a single point of failure as well as scalability issues.

In [3] the authors present another peer-to-peer based RFID resolution framework which is based on the original proposal presented in [2] but without chain links. This system uses the EPC company prefix number (CPN), to map the keys to the nodes which contain data about it. The nodes in the system are then arranged in a logical circle based on their node ID. Because the first part of the node ID is based on the country this ensures that the nodes in the logical network are arranged with physically closer peers arranged logically closer to each other as well. The data resolution is done by going along the circle till a peer with the required information is found. While its scalability and resistance to failure is higher than the EPCGAF it still has scalability and availability issues at the actual data services. In addition, it also has issues with duplicate data look, retrieval and formatting as retrieved data is discarded once it's used and must be re-retrieved when needed again.

3 Proposed Architecture

The role of the proposed RFID architecture is to organize and manage RFID infrastructure throughout the enterprise in order to capture tag events, generate RFID data in real time, store it with minimal loss and share that data with partners using Peer-to-Peer web services (Fig. 1). The middleware in our architecture is developed to be modular and is required to carry out the tasks listed below:

- Filter and collect the data received from multiple readers.
- Carryout security tasks to ensure the integrity and confidentiality of data.
- Translate the tag identifier and data retrieved to information.
- Generate transaction data based on business events.
- Retrieve, aggregate, filter and format RFID tag data.
- Act as the communication hub for different components.

Data Cleaning and Filtering Module - The Data Cleaning and Filtering Module (DCFM) of the middleware is in charge of accepting RFID tags reads from multiple different readers, cleaning it by removing false reads and filtering duplicate reads and collating the reads from multiple readers [4]. Please note that there are a number of different data filtering and collection mechanisms proposed in recent literature that the developer could implement. For further details on the different challenges in RFID

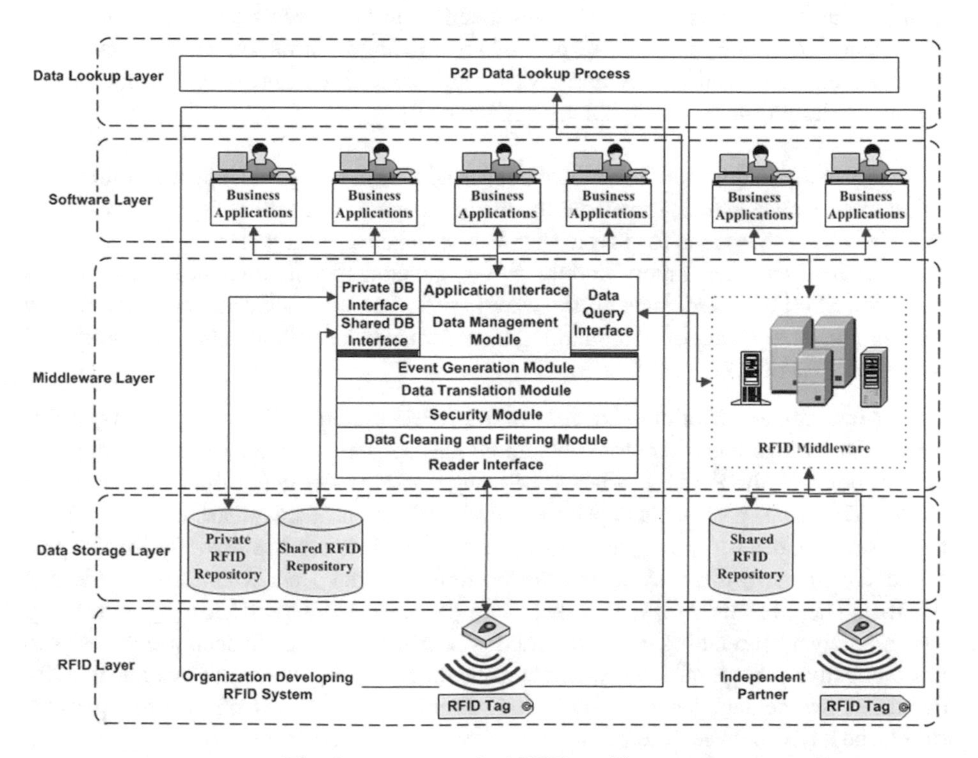

Fig. 1. P2P networked RFID architecture

data filtering and management and a comparison of the different possible approaches refer to [5].

Security Module - Overall, the minimum security functionality that the security module needs to provide includes mutual authentication of tags and readers, transmission confidentiality and integrity and tag anonymity. In addition, security requirements such as storage confidentiality and integrity, non-repudiation, tag malware protection and access control need to be implemented as well [6]. There are a number of security solutions and protocols, that can be used by this module, that offer varying levels of protection and features as discussed in [7].

Data Translation Module - The data translation module in our middleware needs to carry out two types of data translations. It needs to translate raw binary data from the tag into a format usable by the system and it also needs to translate the information received from the databases and business applications into raw binary data to be stored on the tag. For the first task the module splits tag data into different fields and for each field retrieves the data translation rules and applies them. To translate database data to tag storable data the module identifies which field each data should be stored to, retrieves the data translation rules and applies them to the received data. Once the translation is complete, the update is forwarded to readers.

Event Generation Module - The most important functions of an RFID system is the automated generation of transaction data concerning the tag's it identifies. Transaction data is created by associating EPCs with specific business events and transactions [8]. The event recognition module is responsible generating information based on tag reads and business rules and events and tasked with controlling and coordinating certain actions in the physical and digital environment [9].

Imagine the middleware receives the tag reads for new tags at a specific warehouse and it also receives information from the business applications that a logistics company X is delivering the goods for invoice for that warehouse from seller Y. By combining this data the Event Generation Module (EGM) generates the invoice received event for that transaction and associates all the newly picked up tags with that invoice. It may also initiate the opening of outbound logistics for sales for that specific good from that warehouse (Table 1).

Data Management Module - In networked RFID systems, different partners at different data storage locations store data about the tags used in the system. Additionally in our approach, the RFID data is shared using a P2P model rather than a client server model. Therefore, proper management and identification of this distributed data as its being saved and retrieved is required if the system is to work efficiently. In the proposed architecture, this task is the responsibility of the Data Management Module (DMM). Therefore the DMM is tasked with the responsibility of locating, retrieving, aggregating and formatting data from multiple different sources in such a manner as to most effectively respond to any single data request [10]. It is also responsible for deciding how the data generated by the EGM and data retrieved from external partners should be saved between the private repository and shared repository.

Table 1. P2P networked RFID architecture

Tag event detected	Actual business process	Business application data	Data stored in shared RFID repository	Business process triggered
Tags T_1 picked up by readers at warehouse	Receiving of new stock from logistics supplier S_1 delivered by truck TR_2	Goods for invoice I_1 received at warehouse W_1	Tag T_1 is at warehouse W_1 Tag T1 belongs to delivery invoice I_1	Start shipping out goods for orders which can be fulfilled Divert any more shipments if warehouse space is full
Tag T_1 leaves warehouse	Logistics supplier S_2 picks up stock for delivery using truck TR_1	Goods for sale SL_1 shipped from warehouse W_1	Tag T_1 is no longer at warehouse W_1 Tag T_1 belongs to sale invoice I_2	Request for more stock if extra space is available

3.1 P2P Technology in RFID Systems

The two big issues for P2P networks are node churn and security or privacy concerns [3]. In normal public P2P networks, partners/nodes are constantly joining and leaving the network. But in RFID systems, the network partners and therefore the P2P nodes are very stable. When one such node enters the system it is there permanently, except for occasional down time, till that partner leaves the supply chain. Therefore the issue of node churn does not apply to in the environment we are working in [3]. The other main drawback of P2P systems is security concerns because anyone can join the network. Therefore, access control, privacy and trust concerns come into play. However, partners in distributed RFID systems are business entities with existing business partnerships and connections. Therefore most of the security and privacy concerns plaguing public P2P networks do not apply to supply chain RFID P2P networks [2].

RFID data is currently categorized into two groups: Static data (data created at the birth of the object and which does not change) and Transaction data (data that is generated by different partners over the course of its lifetime).

However, sharing data that is constantly changing over a P2P network creates data synchronization problems while sharing only static data would defeat the purpose of using P2P technology, as it's only a very small percentage of the total data concerning any given RFID tag. Therefore, to remove data synchronization requirements and ensure that the highest amount of data can be shared via P2P we further split the transactional data into constant and updatable transaction data (as shown in Table 2).

In the proposed environment, the transaction data needs to be associated with the original partner who generated that data when it's stored at a different location. Therefore, in our system in addition to the tag identifier we also use the unique identifier of the partner who generated that data to identify transaction data. In addition,

Table 2. Types of data

	Generated at	Generated by	Updated	Example
Static data	Birth of object	Manufacturer	No	Batch number of item is 3476 Item expires on 14/08/2012
Constant transaction data	Over lifetime	Supply chain partners	No	Item was checked into warehouse ×23 on 21/10/2010 Item was sold to supply partner Y as part of invoice 211
Updatable transaction data	Over life time	Supply chain partners	Yes	The next destination for item is warehouse ×56 There are 1863 lots of model Z at warehouse 34

Table 3. Comparison on stored transaction data details

	Other system	Proposed system	Details
Tag identifier	Yes	Yes	Unique to the tag: must be given by a global authority
Original partner identifier	No	Yes	Unique to each partner for each supply chain
Date generated on	No	Yes	The date on which the information was generated or last updated
Data class	No	Yes	Number indicating the data class: 1- Static data, 2- Constant transaction data, 3- Updatable transaction data
Transaction information	Yes	Yes	The actual information that was generated concerning the tagged object

to allow for stronger and more granular identification and filtering were also store the date on which that information was actually generated (Table 3).

When a partner requests specific data all of the above information it will also be transmitted for each transaction event. When a business application requires data concerning an object it will relay that request to the data management module which will identify which data services of which partners might contain the information required and retrieve that data.

When requesting P2P data from external data services additional information provided by the application can be used to retrieve only a subset of the total data at the external service. The system can ask the P2P data service for data concerning tag X, which was generated by partner Y between two specific dates. Once data is retrieved from external partners it will be stored on the local servers and shared with other partners via the P2P network. The potions of the retrieved data that has been classed as static data or constant transactional data (indicated by either 1 or 2 for data class) will be saved in raw form in the shared data repository while any data classed as updatable transaction data will forwarded to the business applications and not stored and shared via the P2P network. All locally generated shared transaction data, regardless of data class, will be stored in the shared data repository and shared via the P2P network.

In our system the data services will be web based and will have a service profile, which contains Meta data about the service it offers. It also will generate its data profile, which contains information about the data it's sharing, and share these two profiles with other partners. The two extra data fields that are filled by the partners and allow the partner to track when he last downloaded the data profile for a particular external data service and when he should retrieve a newer data profile for that service. Each web data service will also create a list of all the tags the service has data about along with the original partner who generated that data and the last time data for that tag and partner combination was generated or updated.

The proposed service and data profile file sharing approach is based on the fact that partner chains pre-exist in distributed RFID systems. Once the initial discovery is done, the partners use direct communication to retrieve data from partner web data services. The partner profile distribution and lookup process is divided into two main parts: (1) partner data service discovery (2) partner data profile update.

When a completely new partner joins RFID system it is with the knowledge and approval of at least one existing partner and that existing partner will be able to directly get the service profiles of the new partner. That existing partner will then be tasked with distributing the new partner's service profiles to their up or down stream partners. When the other partners receive the new partners service profiles they will directly contact those services and retrieve their data profiles. The existing partner who initiated the new partner will also be in charge of forwarding all services profiles it has to the new partner. The process is shown in Fig. 2.

When a new data service is added that partner includes the service profile of the new service in all his existing data services. As all service-profiles have an expiry date partners need to regularly contact all the data services and refresh their service profiles. When this happens, any service profiles of new data services will be sent along with the current service profile for that particular data service. When the requesting partner receives the service-profiles they directly contact the new data services and request them for their data profiles as shown in Fig. 3.

Whenever a data service gets new data it will update its data profile to reflect the new data it has available. External partners will use the data profile time stamp and the data profile expiry fields in the service profile for each data service to update their data

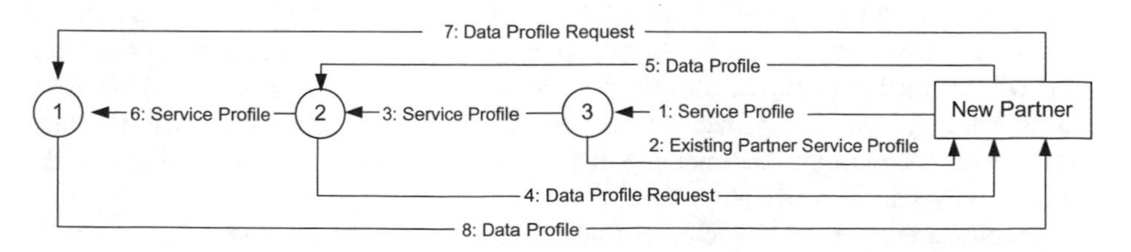

Fig. 2. New partner service profile sharing

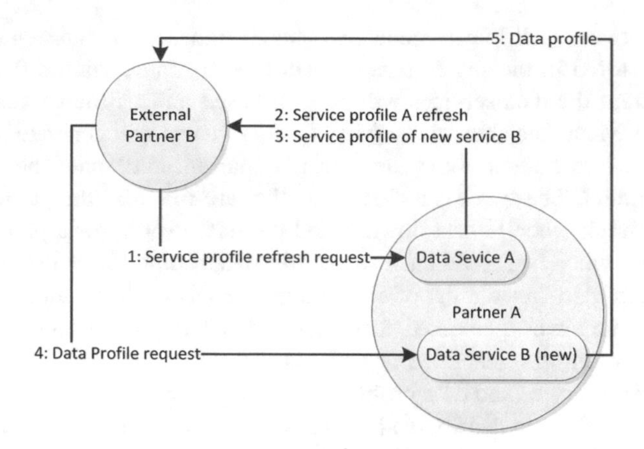

Fig. 3. New data service profile sharing

from a partner. Using these up-to-date data profiles partners can request the information they require from any number of data services rather than just the data service that originally generated it.

4 Comparative Analysis

In the EPCGAF, the lookup services offered by the EPCGlobal itself and the EPCIS are all centralized and based on client-server technology and scale badly. In the more recently proposed P2P based architectures [2, 3] the bottleneck that is created by the EPCGlobal lookup services is removed. However, the bottlenecks at the actual RFID data sources still exist because they are still a single server. In our architecture both the lookup process and the data sharing is done using modified P2P techniques. The comparison of architectures is presented in Table 4.

The structure of the EPCGAF introduces a number of Single Points of Failure (SPOF), at the lookup process and the EPCIS of partners, which affects the availability of the system [11]. In the more recent P2P architectures the SPOF at the data lookup is removed but the system still retains a SPOF at the data service components. In contrast, the architecture proposed by us removes both these SPOF and increase system availability and reliability in a number of ways. (1) By storing static data on the RFID tag in addition to the data service of the manufacturer, we reduce the dependency on external data sources and services, (2) By storing filtered, aggregated and formatted data in a local DB, we further minimize the dependence on all external services and (3) By using P2P technology, which is proven to have much better availability and reliability than client-server technology, we increase the overall reliability and availability of the networked system as a whole.

Unfortunately the security features that are provided by the EPCGAF are very basic [11]. In addition, the ONS service that it offers have a number of security issues such as vulnerabilities to DDoS attacks and cache poisoning [12]. The P2P architectures in [2, 3] remove the vulnerability to cache poisoning and partially removes the vulnerabilities to DDoS attacks. However, they do not have any security features built into them; neither do they discuss the potential threats to the proposed system. Our architecture offers a number of advantages when it comes to system security and the P2P technology that our architecture employs eliminates the vulnerability to cache poisoning and reduces vulnerability to DDoS attacks.

Table 4. Comparison of architectures

		Proposed	EPCGAF	P2P architecture	Peer resolution framework
Scalability	Data lookup	High (P2P)	Low (Client-Server)	High (P2P)	High (P2P)
	Data sharing	High (Client-Server)	Low (Client-Server)	Low (Client-Server)	Low (Client-Server)
Availability	Data lookup	High (P2P)	Low (Client-Server)	High (P2P)	High (P2P)
	Data sharing	High (Client-Server)	Low (Client-Server)	Low (Client-Server)	Low (Client-Server)
Performance	Partner/server discovery	Uses chain distribution, done one time for each data service	Hierarchy based ONS, is repeated each time data is required	Uses DHT tables	Peers are arranged in a circle based on location
	Data lookup	First time networked than local, based on simple list shared by partners	Networked, based on hierarchy based ONS	Networked, based on DHT tables	Networked, based on profiles published by service
	Data sharing	P2P	Client-Server	Client-Server	Client-Server
	Data retrieval	Only done once for each data set	Lots of duplicate work	Not indicated	Not indicated

The EPCGAF performs certain tasks quite inefficiently and most of the time, the system needs to at least access the manufacturer's EPCIS as well as the EPCIS of the partner containing the transaction data required when retrieving data. In addition the data recovered from another partner's EPCIS needs to be filtered, aggregated and formatted before it can be used by business applications [10]. As the EPCGAF requires that the data be retrieved from the partner's EPCIS each time it is needed, the system must filter, aggregate and format the same data whenever it is retrieved from partner's EPCIS. This creates unnecessary duplication of work, which affects system performance negatively. Because the EPCGlobal implements the ONS as a hierarchy, the system also has to complete a large number of processes to complete a data lookup. The P2P architectures reduce the number of steps required for the data lookup as its direct lookup. Additionally our architecture improves performance in a number of ways. Due to static data being stored on the tag, our architecture has a fewer number of situations requiring a data look-up. In our architecture, once data is retrieved and formatted its stored in the local private database. By doing this, our architecture significantly reduces the amount of duplicate filtering, aggregation and formatting done by the middleware component of the system compared to the EPCGAF. This significantly reduces the load on the middleware and therefore improves the overall system performance. We also use the chain distribution method for locating new data services and partners. This approach is a lot more efficient than typical decentralized P2P node discovery methods such flooding, because partners directly query each other. In addition the P2P data sharing balances the loads more efficiently and reduces bottlenecks therefore improving overall system performance.

5 Conclusion

In the future RFID will be used in large number of different systems. In this paper we design and present a peer-to-peer RFID architecture that can run as a web service for systems that are distributed and of very large scale. To do this an efficient manner we have used peer-to-peer technology as well as developed a novel way of classifying and identifying different types of RFID based data. We have also proposed a mechanism through which service and data profiles can be used to simplify the data discovery and retrieval process. The comparative analysis shows that it will provide greater scalability, availability and performance than currently existing RFID architectures for the type of system it's developed for.

References

1. Uckelmann, D.: Quantifying the Value of RFID and the EPCglobal Architecture Framework in Logistics. Springer Science & Business Media, Heidelberg (2012)
2. Wakayama, S., Doi, Y., Ozaki, S., Inoue, A.: Cost-effective product traceability system based on widely distributed databases. J. Commun. **2**(2), 45–52 (2007)
3. Shrestha, S., Kim, D.S., Lee, S., Park, J.S.: A peer-to-peer RFID resolution framework for supply chain network. In: IEEE Second International Conference on Future Networks, pp. 318–322 (2010)
4. Kamaludin, H., Mahdin, H., Abawajy, J.H.: Filtering redundant data from RFID data streams. J. Sens. (2016)
5. Mahdin, H.: A review on bloom filter based approaches for RFID data cleaning. In: Herawan, T., Deris, M.M., Abawajy, J. (eds.) DaEng-2013. LNEE, vol. 285, pp. 79–86. Springer, Singapore (2014). doi:10.1007/978-981-4585-18-7_9
6. Fernando, H.S., Abawajy, J.: A security framework for networked RFID. In: Abawajy, J.H., Pathan, M., Rahman, M., Pathan, A.-S.K., Deris, M.M. (eds.) Internet and Distributed Computing Advancements: Theoretical Frameworks and Practical Applications, p. 85. IGI Publishing, Hershey (2012)
7. Piramuthu, S.: RFID mutual authentication protocols. Decis. Support Syst. **50**(2), 387–393 (2011)
8. Jia, F., Jeon, S., Hong, B., Kwon, J., Kwak, Y.S.: Flexible capturing application for enhanced generation of EPCIS events. Int. J. Distrib. Sens. Netw. (2014)
9. Tan, J., Wang, H., Li, D., Wang, Q.: A RFID architecture built in production and manufacturing fields. In: Third International Conference on Convergence and Hybrid Information Technology, vol. 1, pp. 1118–1120 (2008)

10. Musa, A., Gunasekaran, A., Yusuf, Y.: Supply chain product visibility: methods, systems and impacts. Expert Syst. Appl. **41**(1), 176–194 (2014)
11. Armenio, F., Barthel, H., Burstein, L., Dietrich, P., Duker, J., Garrett, J., Suen, K.: The EPCglobal architecture framework. Rapport technique Version, 1 (2007)
12. Peris-Lopez, P., Hernandez-Castro, J.C., Estevez-Tapiador, J.M., Ribagorda, A.: 2 attacking RFID systems. In: Security in RFID and Sensor Networks, p. 29 (2016)

Performance-Aware Trust-Based Access Control for Protecting Sensitive Attributes

Mohd Rafiz Salji[1,2(✉)], Nur Izura Udzir[1], Mohd Izuan Hafez Ninggal[1], Nor Fazlida Mohd. Sani[1], and Hamidah Ibrahim[1]

[1] Faculty of Computer Science and Information Technology, Universiti Putra Malaysia, Seri Kembangan, Malaysia
mohdrafiz@sarawak.uitm.edu.my
[2] Faculty of Information Management, Universiti Teknologi MARA, Shah Alam, Malaysia

Abstract. The prevailing trend of the seamless digital collection has prompted privacy concern not only among academia but also among the majority. In enforcing the automation of privacy policies and law, access control has been one of the most devoted subjects. Despite the recent advances in access control frameworks and models, there are still issues that impede the development of effective access control. Among them are the lack of assessment's granularity in user authorization, and reliance on identity, role or purpose-based access control schemes. In this paper, we address the problem of protecting sensitive attributes from inappropriate access. We propose an access control mechanism that employs two trust metrics name experience and behavior. We also propose a scheme for quantifying those metrics in an enterprise computing environment. Finally, we show that these metrics are useful in improving the assessment granularity in permitting or prohibiting users to gain access to sensitive attributes.

Keywords: Behavior-aware · Trust-based access control · Sensitive attributes · Privacy protection

1 Introduction

Privacy is increasingly becoming one of the very important issues in data management. People are now more conscious about how their information are being secured and protected by service providers. This awareness has been getting more highlights when sharing and collecting of information become seamless and prevalent by the omnipresent of internet connection. In common situation, companies or data keepers are required to allow access to the information reside within the information systems to multitude of users. The administrator may allow the users to access to the information in supporting decision making or analysis activities.

Many efforts have been made in terms of enforcing the automation of privacy policies and law. In providing the solution, most of works have been focusing on access control in which the access authorization to a source is selectively permitted. It

T. Herawan et al. (eds.), *Recent Advances on Soft Computing and Data Mining*,
Advances in Intelligent Systems and Computing 549, DOI 10.1007/978-3-319-51281-5_56

is important that every information systems are equipped with an access control mechanism to ensure that access to personal information is in accordance with company policies [3, 5, 8, 9, 11, 16–18]. Despite the recent advances in access control frameworks and models, there are still issues that impede the development of effective access control models such as the lack of assess granularity in authorizing, and reliance on identity, role or purpose-based access control schemes.

One of the many access control mechanisms, Trust-based Access Control (TBAC) is an access control model that is inspired by an important role in human life, which is trust. By this concept, a user that is highly trusted will be granted more accessibility to a source as compared to lower thereof. However, trust is mutable in response to the changings of situations. Therefore, it is paramount important to design an efficient access control model that is able to capture the dynamic nature of user behavior with regards to trustworthiness.

This paper addresses the issue of protecting sensitive attributes from inappropriate access that can causes privacy disclosure. We propose an access control scheme that embraces two trust metrics named experience and behavior with respect to the user. In order to deal with the dynamic nature of trust, we design a scheme that engages with the continuous process of updating and measuring user behavior in an organization. This involves a comprehensive policy that is devised from the combination of existing access control policies and other resources for determining the level of trust. Three factors have taken into consideration to bridge the trust relationship between a user and the system; properties, experience and recommendations. By using the proposed mechanism, the system is able to identify whether an access request to sensitive attributes is permitted or denied. Authorized user with lower level of trust is still granted to access personal information, but user with preferred experience and behavior will be allowed to access to sensitive attributes. In summary, the main contributions of this paper are as follows:

(a) We propose a new access control model based on trust to protect sensitive attributes.
(b) We identify two trust metrics called behavior and experience to be used as decision factor in controlling access to sensitive attributes.
(c) We propose a quantification method to deal with the dynamic nature of trust.

The rest of this paper is organized as follows: Sect. 2 provides the related works. The proposed method is then presented in Sect. 3. We discuss the result in Sect. 4 and finally, Sect. 5 concludes the work.

2 Related Works

Trust-based access control models have been explored in many distributed computing environments.

In previous work, situational trust is defined as the security of a location by using a level of trust, which limits the documents that can be sent to or observed at that location [7]. The main focus of Performance-Aware Trust-Based Access Control for Protecting

Sensitive Attributes (PATBAC) is to secure sensitive attributes by using a level of seniority and behaviour as a trust.

To access high risk resource, the system needs to filter the user with a certain degree of trust. A multi delegation model with trust management has been proposed to permit or prohibit access to the access control system. Three levels of delegated tasks are organized; low (less trust), medium (intermediate trust) and high (highly trust) [12]. A higher level of delegation task is assigned to the delegate if they have a higher trust level. In PATBAC, the system have to check a user role performance *rp* which comprises with the levels of seniority and behaviour. Two levels of user seniority (junior (less trust) or senior (highly trust)) and three levels of user behaviour (mistrust (junior), trust (senior) or uncertainty (senior performing negative behaviours)) are organized. All authorized users are permitted to access personal information but the user with a higher level of *rp* (senior-with-trust) are able to access sensitive attributes.

In access control model with trust management, the user with a higher trust level have more privileges compared to other levels and the user who are unauthorized will be restricted access to the system. Trust into role based access control model (TRBAC) has been proposed where user with good behaviour will be rewarded with the higher level of trust and they are permitted to access more resources, while malicious users' authorizations may be revoked [22]. The same concept is proposed in PATBAC where the user who is assigned as a higher level of *rp* are able to access more resources.

To specify the user's trust value, the system needs to quantify their performance in substantive service. The user performance is calculated by using the history and recommendation [13, 14]. The history or experience of user is stored in the User Role History (URH) [20]. In PATBAC, URH is assigned to store and calculate automatically the user experience or activity in their substantive service. Moreover, Evaluation Form (EF) is assigned to evaluate the user behaviour and it is based on recommender evaluation. URH and EF may represent values in range [0, 1], which are taken directly from system measurements [2].

Generally, trust can be changed from time to time. This change may invoke user from ongoing access. It can be invoked manually or automatically, depending on the trust evaluation concept set by the administrator [19, 20]. In PATBAC, if the user performs negative behaviour, the administrator will change the user role trust attribute manually. It means that even the user role status is senior, if the role trust attribute is changed to uncertainty, the user is not permitted to access sensitive attribute. The user can apply for the role trust as trust after a certain period of time set by the administrator. If the user has attained a certain period of time, they are allowed to request for re-calculation of their behaviour.

3 Performance-Aware Trust-Based Access Control (PATBAC)

In this section, we propose our method. We first present the trust metrics and discuss about its function in building trust relationship between user and the system. We then present our method to quantify those metrics in enterprise computing system.

3.1 Trust Metrics

Each role in the organization requires certain properties of a user. The properties of a user in PATBAC are referring to the user experience and behaviour, and the explanation of those metrics is as follows:

a. **User Experience**

- Refers to the number of the user activities that is performed during their substantive service.
- It is assigned to specify the seniority of a user.
- It can be set at the role status attribute in the user personal details.
- Two levels of user seniority: junior (less trust) and senior (highly trust).

b. **User Behavior**

- Refers to the user attitude shown during their substantive service. The scope of the user behaviour in this model refers to the categories that is introduced by Bruhn [4] in Table 3.
- It is assigned to specify the behaviour of a user.
- Recommendations are assigned to quantify the user behaviour and the result is supplied in the role trust attribute at the user personal details.
- Three levels of user behaviour: mistrust (junior), trust (senior) and uncertainty (senior performing negative behaviours).

Role performance *rp* refers to the trust degree of a user based on the level of seniority and behaviour to access sensitive attributes. If the *rp* of a user is junior, the system will automatically assign as mistrust and s/he is not allowed to access sensitive attributes. Similar to the role *rp* of a user is senior-with-uncertainty, s/he is also restricted to access sensitive attributes. However, if the *rp* of a user is senior-with-trust, s/he is permitted to access sensitive attributes.

3.1.1 Quantification of User Experience

Experience refers to the number of activities calculated by a system regarding a user activity in their substantive service. The activities that is participated by a user for example, seminar, workshop, courses and others that is determined by the organization. Different department performs different activities. The calculation of a user's experience is perform by using weighing evidence [6].

Weighing Evidence

Weighing evidence is a decision process to specify the seniority of a user. The administrator needs to identify how many activities to be set to identify the activeness of a user. Each of these components has a value between [0, 1] and the sum of these components is 1. The minimum required weight should be set by the administrator to identify either a user is granted or denied to be a senior.

Let m denote the total amount of each activity and w is the total number of activities. The total sum of m is calculated $(m_i + \ldots + m_j)$. Then, sum of m is divided by w to obtain the result of a user activities ua. The result is in the range of [0, 1]. The ua is calculated as in Eq. 1.

$$\frac{\sum_{i=1}^{j} i}{w} \in [0, 1] \quad (1)$$

Hence, the administrator a have to decide the minimum required weight of ua. If the result of ua is more than the required weight set by a, user u able to be assigned as senior role.

Assume the minimum required weight set by the administrator is 0.4 and a user Alice's overall score is 0.5. This means that she is permitted to assign as senior role. Based on Table 1 [21], Alice's overall score are in Level 3, i.e. the activeness of Alice is average.

Table 1. Indicator of the user activeness

Value	Meaning	Activeness score
Level 0	Totally inactive	0
Level 1	Inactive	0.1–0.19
Level 2	Minimal	0.2–0.39
Level 3	Average	0.4–0.59
Level 4	Active	0.6–0.79
Level 5	Very active	0.8–1

In PATBAC, calculation of user's experience is not enough to assign a user as trustworthy. A user's behaviour will be evaluated by recommendations to permit access to sensitive attributes.

3.1.2 Quantification of User Behaviour

Recommendations are assigned by the administrator to evaluate a user behaviour. User behaviour is evaluated in the evaluation form (EF) (Table 3). A user behaviour categories is applied in this research to specify the user behaviour [4]. Table 2 has become an indicator to facilitate the recommender to evaluate a user trust behaviour based on categories [21]. The value of each category is between [0, 1] and the sum of these categories is 1. For example, if recommender A evaluates user B on the category of open, participative, accept responsibility, recommender A needs to place a mark in that category. Assume recommender A gives a score to a user B in that category is 0.5, it means that user B is in Level 3, which the score on that category is average. Scores will be placed in the user evaluation form as illustrated in Table 3.

Let b denote the total amount of each behaviour category and c is the total number of behaviour categories. The sum of b is $(b_i + \ldots + b_k)$. Then, total sum of b is divided by c to obtain the result of a user behaviour ub. The result is in the range of [0, 1]. The ub is calculated as in Eq. 2.

Table 2. Indicator of a user trusted behaviour based on categories

Level	Meaning	Trust range
Level 1	Very poor	0–0.19
Level 2	Poor	0.2–0.39
Level 3	Average	0.4–0.59
Level 4	Good	0.6–0.79
Level 5	Very good	0.8–1

Table 3. User behaviour evaluation form

No.	Categories	Mark
1.	Open, participative, accept responsibility	
2.	Highly productive	
3.	Loyalty to the organization	
4.	Not defensive	
5.	Cooperation, work teams	
6.	High job satisfaction	
7.	Problem-solving attitude	
8.	Involvement in decision-making	
9.	Sense of pride in work	
	Total mark	
	Total mark/9	

$$\frac{\sum_{i=1}^{k} i}{c} \in [0, 1] \tag{2}$$

Hence, the administrator *a* have to decide the minimum required weight of *ub*. If the result of *ub* is more than the required weight set by *a*, user *u* can be assigned as trust.

Scores for each category will be added first and divided by a number of categories to obtain an overall score. Combinations from the notions of Kim et al. and Vidyalakshmi et al. [10, 21], the level of a user trusted behaviour for the overall score is illustrated as in Table 4. For example, assume a user Carol obtains the overall score 0.7. Based on Table 4, Carol is in Level 4, which is good. If the minimum requirement set by the administrator is 0.6, she is qualified to be assigned as trust.

3.2 Access Control Mechanism

Figure 1 shows the process of access control model using *rp* as a trust to access sensitive attributes and the explanations are as follows:

1. User: User in this model refers to the staff. User is requested to access privacy in the system. First, user needs to sign in using user identification and password.

Table 4. Levels of a user trusted behaviour for overall score

Value	Meaning	Explanation	Trust range
Level 0	Distrust completely	Untrustworthy	0
Level 1	Ignorance	Cannot decide	0.1–0.19
Level 2	Minimal	Lowest trust	0.2–0.39
Level 3	Average	Mean trustworthiness	0.4–0.59
Level 4	Good	Trusted by major population	0.6–0.79
Level 5	Fully trust	Fully trustworthy	0.8–1

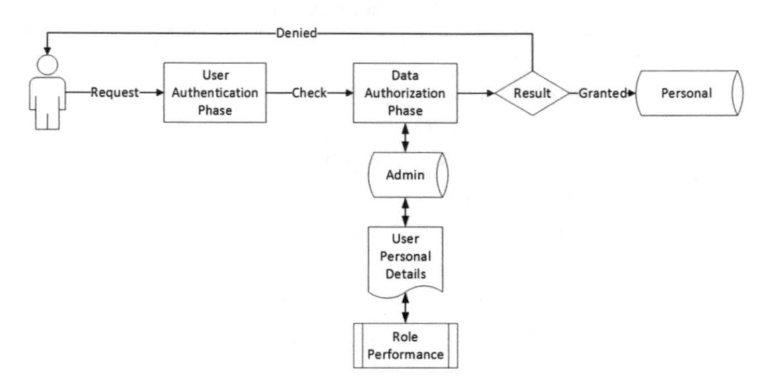

Fig. 1. Role performance trust-based access control

2. User authentication phase: This is the first stage in access control mechanism. In this stage, system authenticates the user identification, and password. If the user supply wrong user identification and password, they are denied further process.
3. Data authorization phase: This is the second stage in access control mechanism. This stage is assigned to identify the user's trust value either allowed or prohibited access to sensitive attributes. If the user role status is senior and role trust is a trust, s/he is prohibited to access sensitive attributes. Otherwise, they are allowed to access personal information without sensitive attributes.
4. Admin database: The authorization of user's *rp* in the user personal details is located in this database.
5. User personal details: User personal details (Table 5) includes the user information and the necessary attributes that are assigned for user authorization.

Table 5. The illustration of user personal details

User Personal Details

Name:	Caren	
Address:	4 July Ave. WA 11000	
Age:	40	
Email:	Caren@yahoo.com	
Department:	Human Resource	Updated by the administrator
Role Status:	Senior	Role performance
Role Trust:	Trust	

6. Role performance: Role performance is a role status and role trust attributes which are assigned to identify either user is permitted or prohibited to access sensitive attributes. It is used to identify the trustworthiness of the user.
7. Result: All authorized users are granted access to personal information. Moreover, user as a senior-with-trust can access sensitive attributes. User is denied access to personal information if the administrator may not state any values in their role status and/or role trust attributes.
8. Personal: Personal information is located on the personal database.

4 Results and Discussion

In this section, we discuss on how the user is either permitted or prohibited access to the sensitive attributes. In this model, if user request to access sensitive attributes, the parameter is assigned to identify the trust of the user. Four parameters are identified to permit access to sensitive attributes. The parameter is as follows; $< u, rp, a, o >$ where $u \in U$, $rp \in RP$, $a \in A$, $o \in O$. In PATBAC, action a refers to the user which allow to perform read privilege [15] or *select* operation (to retrieve data) [1]. The parameter stated a user u has a role performance rp with an action a to access object o. For example, if a user is granted to access personal information without sensitive attributes. The parameter is as follows:

$$< \text{Staff, Junior Mistrust, Select, Income} >$$

For example, based on the parameters above, the result of user Danny (Table 6) access to Bob Parker's personal information is shown in Table 7:

Table 6. The illustration of user personal details

User Personal Details		
Name:	Danny	
Address:	5 Aug Ave. WA 22000	
Age:	38	
Email:	Danny@yahoo.com	
Department:	Human Resource	
Role Status:	Junior	Role performance — Updated by the administrator
Role Trust:	Mistrust	Role performance — Updated by the administrator

In Table 7, Bob's income which is a sensitive attribute does not appear in the result due to Danny's *rp* does not allow access to sensitive attribute. In contrast, the parameter for a user to access personal information with sensitive attribute are as follows:

$$< \text{Staff, Senior Trust, Select, Income} >$$

Table 7. The result appear for junior-with-mistrust or senior-with-uncertainty

	Name	Age	Address
Result	Bob Parker	40	5 Aug Ave. WA 21000

This parameter is owned by the user with a higher level of *rp* to access sensitive attribute. The result has appeared as in Table 8:

In Table 8, Bob's income appears in the result as Caren's (Table 5) *rp* has attained a higher level of trust to access sensitive attribute.

Table 8. The result appear for senior-with-trust

	Name	Age	Address	Income
Result	Bob Parker	40	5 Aug Ave. WA 21000	10000

5 Conclusion and Future Work

In this paper, we propose a comprehensive policy to permit authorized user access sensitive attributes based on seniority and behaviour. To specify the user seniority and behaviour, the system will calculate seniority by using a user experience, and behaviour is evaluated by recommendations. Subsequently, our new trust-based access control model is designed to permit all authorized users access to personal information. However, authorized users with higher level of trust are permitted to access sensitive attributes. These two contributions show the issue of authorized user without trust to access sensitive attributes will be solved. Result shows PATBAC are able to permit or prohibit authorized users access to sensitive attributes.

Among the future work planned includes a prototype to implement the PATBAC. In addition, the model will be combined with purpose based access control (PBAC) to allow user access personal information with sensitive attributes based on trust and purpose.

Acknowledgments. The authors would like to thank the reviewers for their valuable comments to help improve this article. This work is partly sponsored by the Scholarship Department, Ministry of Education, Malaysia.

References

1. Abdul Ghani, N.: Credential purpose-based access control for personal data protection in web-based applications. Ph.D. thesis, Universiti Teknologi Malaysia, Faculty of Computing (2013)
2. Bernabe, J.B., Perez, G.M., Gomez, A.F.S.: Intercloud trust and security decision support system: an ontology-based approach. J. Grid Comput. 1–32 (2015)
3. Bertolissi, C., Fernandez, M.: A metamodel of access control for distributed environments: Applications and properties. Inf. Comput. **238**, 187–207 (2014)
4. Bruhn, J.G.: Trust and the Health of Organizations. Springer Science & Business Media, New York (2001)

5. Crampton, J., Sellwood, J.: Path conditions and principal matching: a new approach to access control. In: Proceedings of the 19th ACM Symposium on Access Control Models and Technologies, pp. 187–198. ACM (2014)
6. Gollmann, D.: From access control to trust management, and back – a petition. In: Wakeman, I., Gudes, E., Jensen, C.D., Crampton, J. (eds.) IFIPTM 2011. IAICT, vol. 358, pp. 1–8. Springer, Heidelberg (2011). doi:10.1007/978-3-642-22200-9_1
7. Heupel, M., Fischer, L., Kesdogan, D., Bourimi, M., Scerri, S., Hermann, F., Gimenez, R.: Context-aware, trust-based access control for the di.me userware. In: 2012 5th International Conference on New Technologies, Mobility and Security (NTMS), pp. 1–6. IEEE (2012)
8. Hung, P.C.: Towards a privacy access control model for e-healthcare services. In: PST (2005)
9. Kayes, A., Han, J., Colman, A.: A semantic policy framework for context-aware access control applications. In: 2013 12th IEEE International Conference on Trust, Security and Privacy in Computing and Communications (TrustCom), pp. 753–762 (2013). doi:10.1109/TrustCom.2013.91
10. Kim, M., Seo, J., Noh, S., Han, S.: Identity management-based social trust model for mediating information sharing and privacy enhancement. Secur. Commun. Netw. **5**(8), 887–897 (2012)
11. Lazouski, A., Martinelli, F., Mori, P.: Usage control in computer security: a survey. Comput. Sci. Rev. **4**(2), 81–99 (2010)
12. Li, M., Sun, X., Wang, H., Zhang, Y.: Multi-level delegations with trust management in access control systems. J. Intell. Inf. Syst. **39**(3), 611–626 (2012)
13. Li, M., Wang, H., Ross, D.: Trust-based access control for privacy protection in collaborative environment. In: IEEE International Conference on e-Business Engineering, ICEBE 2009, pp. 425–430. IEEE (2009)
14. Lin, G., Wang, D., Bie, Y., Lei, M.: Mtbac: a mutual trust based access control model in cloud computing. China Commun. **11**(4), 154–162 (2014)
15. Mirabi, M., Ibrahim, H., Mamat, A., Udzir, N.I.: Integrating access control mechanism with EXEL labeling scheme for XML document updating. In: Fong, S. (ed.) NDT 2011. CCIS, vol. 136, pp. 24–36. Springer, Heidelberg (2011). doi:10.1007/978-3-642-22185-9_3
16. Ruj, S., Stojmenovic, M., Nayak, A.: Privacy preserving access control with authentication for securing data in clouds. In: 2012 12th IEEE/ACM International Symposium on Cluster, Cloud and Grid Computing (CCGrid), pp. 556–563. IEEE (2012)
17. Samarati, P.: Protecting respondents identities in microdata release. IEEE Trans. Knowl. Data Eng. **13**(6), 1010–1027 (2001)
18. Sandhu, R., Ferraiolo, D., Kuhn, R.: The NIST model for role-based access control: towards a unified standard. In: ACM Workshop on Role-Based Access Control, vol. 2000 (2000)
19. Sarrouh, N.: Formal modeling of trust-based access control in dynamic coalitions. In: Computer Software and Applications Conference Workshops (COMPSACW), 2013 IEEE 37th Annual, pp. 224–229. IEEE (2013)
20. Toahchoodee, M., Abdunabi, R., Ray, I., Ray, I.: A trust-based access control model for pervasive computing applications. In: Gudes, E., Vaidya, J. (eds.) DBSec 2009. LNCS, vol. 5645, pp. 307–314. Springer, Heidelberg (2009). doi:10.1007/978-3-642-03007-9_22
21. Vidyalakshmi, B., Wong, R.K., Chi, C.H.: Decentralized trust driven access control for mobile content sharing. In: 2013 IEEE International Congress on Big Data (BigData Congress), pp. 239–246. IEEE (2013)
22. Yang, R., Lin, C., Jiang, Y., Chu, X.: Trust based access control in infrastructure-centric environment. In: 2011 IEEE International Conference on Communications (ICC), pp. 1–5. IEEE (2011)

E-Code Checker Application

Shahreen Kasim(✉), Ummi Aznazirah Azahar, Noor Azah Samsudin, Mohd Farhan Md Fudzee, Hairulnizam Mahdin, Azizul Azhar Ramli, and Suriawati Suparjoh

Faculty of Computer Science and Information Technology, Universiti Tun Hussein Onn Malaysia, Parit Raja, Malaysia
{shahreen, azah, farhan, hairuln, azizulr, suriati}@uthm.edu.my, ummiazahar@gmail.com

Abstract. Nowadays, many areas in computer sciences use ontology such as knowledge engineering, software reuse, digital libraries, web on the heterogeneous information processing, semantic web, and information retrieval. The area of halal industry is the fastest growing global business across the world. The halal food industry is thus crucial for Muslims all over the world as it serves to ensure them that the food items they consume daily are syariah compliant. However, ontology has still not been used widely in the halal industry. Today, Muslim community still have problem to verify halal status for halal products in the market especially in foods consisting of E number. In this paper, ontology will apply at E numbers as a method to solve problems of various halal sources. There are various chemical ontology and databases found to help this ontology construction. The E numbers in this chemical ontology are codes for chemicals that can be used as food additives. With this E numbers ontology, Muslim community could identify and verify the halal status effectively for halal products in the market.

Keywords: E number ingredients · Various database · Ontology · Chemical entities

1 Introduction

In general, ontology is a representation of knowledge. Ontology is an explicit formal specification of the terms in domain and relation among them [1, 2]. In computer sciences and information sciences areas, it defines a set of representational primitives with which to model a domain of knowledge or discourse. The representational primitives are typically classes (or sets), attributes (or properties), and relationships (or relations among class members). The definitions of the representational primitives include information about their meaning and constraints on their logically consistent application [3]. Nowadays, many areas in computer sciences use ontology such as knowledge engineering, software reuse, digital libraries, web on the heterogeneous information processing, information retrieval and semantic web [4].

However, the ontology in semantic web area has still not been used widely particularly in the halal industry. The halal industry has been the fastest growing global business in Malaysia. Today, Muslim community still have problem to verify halal

T. Herawan et al. (eds.), *Recent Advances on Soft Computing and Data Mining*,
Advances in Intelligent Systems and Computing 549, DOI 10.1007/978-3-319-51281-5_57

status for halal products in the market especially in foods consisting of E number [5]. Most of the information available on the internet simply displays a list of companies and list of products with the identification of their halal status. Nevertheless, there are also some that display status of materials used in the food on the internet but the status is always in conflict or not the same as other websites. Besides, halal certification logo is lack of security which makes it easier for this logo to be copied [6].

Due to this problem, varieties of food products that have halal status have been doubted. Therefore, to solve this problem, there is the need to establish a method by which the user can check the status of food and know the food source especially those oriented from E number. In this paper, ontology will apply at E numbers in which E numbers are codes for chemicals that can be used as food additives. As told, E number is chemical nature and to establish a database on the E number consists of many different types of resources. Due to the variety of sources used, ontology mapping technique was used to combine all types of databases. Section 2 methodology will describe the data collection phase and in the following Sect. 3 will describe early results in the developing of the ontology.

2 Related Works

Various applications have been developed and touted to provide a source of knowledge and information to the public in line with technological developments [12–14]. Information on the use of the code number E is also available through the Play Store for Android smartphone users and App Store for iPhone. Most applications are built using the search engine where it can allow users to search for E number easily. E-Code Checker is an application that is a search engine for halal food additive in the market. The data used is the code number E (E code), which is used for additional materials (additives) and it is often used on food labels in the European Union. There are many of searching applications are built regarding E number in Android platform such as E-Codes Free, E-Inspect and Halal Check.

E-Codes Free is an Android application that allow user to search for E number. This application provides the function of the search box to find information about a food additive which users only need to enter E number using the keyboard numbers that have been provided. E-Inspect is an application that is built on the Android platform. This application lists the E number in the form of a list. Halal Check app is an application based on the Android platform. This application provides three main menus, the E-Numbers, Salah, and the Quran.

3 Methodology

In this study, we used chemical data where E number represents chemical entities of the food. This section will describe on these data and its process in detail.

3.1 E Number as a Chemical Entities

E number is no longer a foreign ingredient in the food world. E number is used as food additives for flavor enhancers, stabilizers food, colours, preservatives, antioxidants and antibiotics. E number is made up of chemical substances which are permitted to be used in food. Nevertheless, the halal status is always at doubt, particularly for Muslim consumers. Various databases and chemical ontology have been created but its function does not describe the E number. Various chemical ontology and databases found to help this ontology construction, among which are Chemical Entities of Biological Interest (ChEBI), the National Center for Biomedical ontology (NCBO), Comparative Toxicogenomics Database (CTD), PubChem Bioassay Database (PubChem) and Human Metabolome Database (HMDB). Due to different sources of information to build E number, ontology was used to complete this study. Although various chemical databases have been found to help establish this ontology, not all the data will be used to build the E number ontology. Next, Sect. 3.2 will describe the process of collecting the E number.

3.2 Collection of E Number

Throughout this study, we have collected E numbers from four websites; http://www.guidedways.com [7], http://www.muslimtents.com [8], http://www.muslimconsumergroup.com/enumbers_list.html [9] and http://special.worldfislam.info/Food/numbers.html [10]. These E numbers are accompanied by their halal status.

Currently, on these websites, there are 516 E numbers which represents the E number from E100 to E1599. The E number was classified into 9 groups which are E100–E199 (colours), E200–E299 (preservatives), E300–E399 (antioxidants & acidity regulators), E400–E499 (thickeners, stabilizers & emulsifiers), E500–E599 (pH regulators & anti-caking agents), E600–E699 (flavour enhancers), E700–E799 (antibiotics), E900–E999 (miscellaneous) and E1100–E1599 (additional chemicals). The purpose of taking E number from different websites is due to the differences of halal status shown in each website. For instance, E101 representing Riboflavin (Vitamin B2) has different status between guidedways.com and muslimtents.com website. Next, Sect. 3.3 will describe the methods to filter these data.

3.3 Filtering Data that Related with E Number

There were 2 databases or ontology used in this study, but all the data requires data filtering in advance to avoid the new created database full of unwanted data.

3.3.1 ChEBI - Chemical Entities of Biological Interest

ChEBI is a freely available dictionary of 'small molecular entities' and ChEBI incorporate an ontological classification, whereby the relationships between compounds, groups or classes of compounds and their parents, children and or siblings are specified [11]. There are six types of formats which can be selected for downloading ChEBI data which include the SDF file, OWL file, OBO file, Flat file/tab delimited,

Oracle binary table dumps and Generic SQL (Structured Query Language) table dumps. In this study, the data was downloaded in the form of flat file which is easier and various spreadsheets tools available to import this into a relational database. The files were stored in the same structure as the relational database. As shown in Fig. 1 after downloading all the ChEBI data, the data was then separated according to necessary data and data that should not be used. The following was the method used to filter ChEBI data. Once the ChEBI data was downloaded using the flat file format, data inspection was conducted to identify the contents of each table and important keywords in the database. Here we notice that chebi_id and compound_id are the identifications which were used to ensure that they have relationships between tables. Each chemical has a chemical structure of its own id for example Riboflavin has CHEBI: 17015 and has compound_id 8843. Therefore, in order to identify any information in this database, we need to know chebi_id and compound_id for each material. The identification of chebi_id can be done using the search function in www.ebi.ac.uk by typing the name of ingredients. After the chebi_id is identified, it could be a reference to a table using the keyword chebi_id and separate the information into a new table with information about the E number only. Similarly, compound_id uses the same process as it is used in chebi_id but the difference is, it is used in a table that uses compound_id.

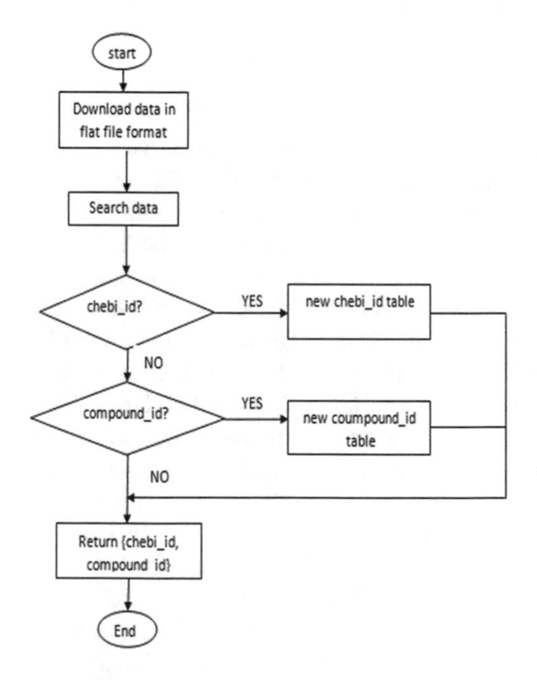

Fig. 1. Flowchart of filtering ChEBI

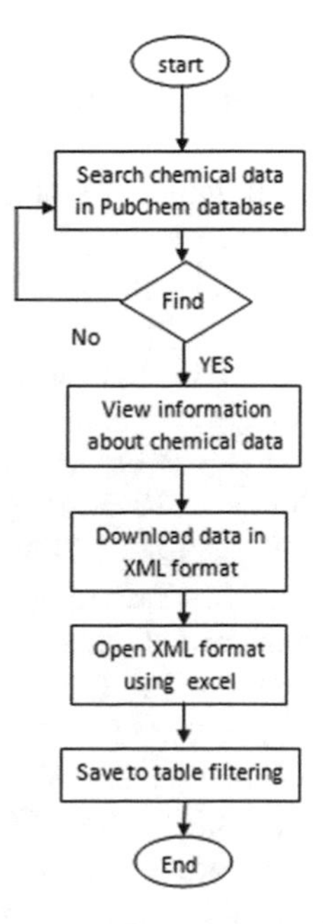

Fig. 2. Flowchart to download PubChem

3.3.2 PubChem - PubChem Compound Database

The PubChem Compound Database provides information on the biological activities of small molecules. PubChem is organized as three linked databases; PubChem Substance, PubChem Compound, and PubChem BioAssay. PubChem's chemical structure records links to other Entrez databases providing information on biological properties; PubMed scientific literature and NCBI's protein 3D structure resource. A part of the E number is from PubChem database. Data from these PubChem can be downloaded for adding information to build the E number ontology and these data can be downloaded in a variety of formats; Abstract Syntax Notation One (ASN.1), Extensible Markup Language (XML) and Standard Delay Format (SDF). The PubChem offers users to download data individually. This is easier to filter data by not having to download all the data in PubChem database but only download the required data. Besides, PubChem also provides a chemical structure search which user can use the names of the PubChem chemical or PubChem chemical id to find the desired information. Each chemical data in PubChem have CID code, for example; CID 6093240 representing Pigment Rubine (E180). After searching the required chemical data, information about that chemical will appear and it can be downloaded in a variety of formats. In this study, we downloaded these data in XML format. Figure 2 is a flow process to download the PubChem data.

4 Result and Discussion

To develop E number ontology, the selection of ontology editors is important. In this study, we chose to use Protégé as ontology editor. Ontology development can be started after the data is collected. Data are classified into several parts, according to the E number ingredients that have been set. As has been described earlier, the E number was classified into 9 groups. Figure 3(a) shows that each group represents a specific function such as E100–E199 represents the colours. Each representation still has other subclasses. Refer to Fig. 3(b), E100–E199 is classified into 7 groups which are E100–E109 (yellow), E110–E119 (orange), E120–E129 (red), E130–E139 (blue & violet), E140–E149 (green), E150–E159 (brown & black), E160–199 (gold & others). Each group of color has individual. Individual is the ground level components of ontology or specifying the actual value of specific instances of the class. Figure 3(c) shows colour group for E100–E109 (yellow) in which have 9 individuals; E100 (Curcumin), E101 (Riboflavin), E101a (Riboflavin-5'-Phosphate), E102 (Tartrazine), E103 (Alkannin), E104 (Quinoline Yellow WS), E105 (Fast Yellow AB), E106 (Riboflavin-5-Sodium Phosphate), E107 (Yellow 2G). In addition, ontology has annotation. Annotation for ontology is a vocabulary for performing several types of annotation such as comment, entities annotation, textual annotation, notes and example. Besides that, images and audio can also be annotation ontology. Annotation is the process of assigning E number terms and their synonyms. Synonyms are words with the same or similar meaning. For E number ontology, we use code like E101 but at the same time this E number is named as Curcumin. It also has synonyms such as "kacha haldi", "natural yellow 3", "turmeric" and "turmeric yellow". In the protégé editor, synonyms for every E number are placed in the (same individual as) partition.

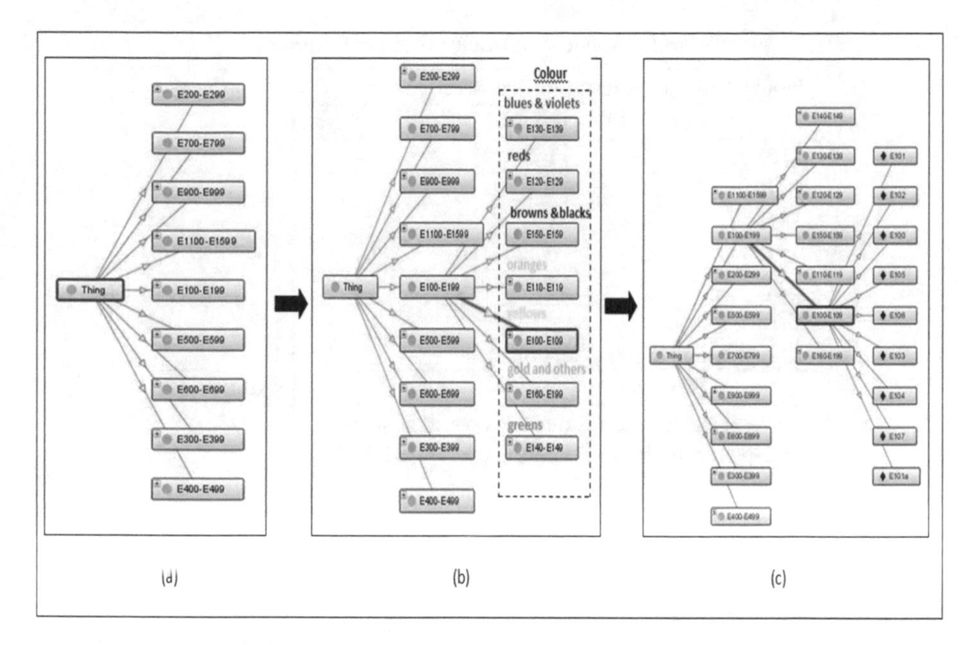

Fig. 3. The example visualization for E number ontology of (a) 9 classified E number (b) colour group E number (c) individual for colour group (Color figure online)

In this paper also, we presented the development of Android application called Halal Checker application that can proved the analysis and studies on ontology of E number which have various databases on food additives from four different websites. The purpose of development of this application is to help Muslim community verify the halal status information from data collected.

In Halal Checker apps the main interface presented with search box which the user can use it to search for E number and list of E number by key in the keyword such as "e" or "e104". User can directly choose E number from the list or insert the E number in search box to view the details of the E number. The user can view the details for E number by click on the E number. It will directed to details page which display the E number chosen, name of food additive, halal status information and also the function of the food additives. As for "Comment" button, user can drop a comment or their opinion about the E number provided. The button "Comment" function in this application is for user give their opinion about the information provided about E number by the application. When user hit the button, they will be directed to the "Sync Account" page which it syncing Google account from their smartphone to get the user information such as username, email, birthday, gender and more. For the "Comment" page, an email of the user will be directed filled in the place. For comment, user can drop any comment for the E number and hit the button "Submit".

Table 1 showed several attempts of E number from the user input. This application able to store a new E number into the database. User might give an unknown input. This situation might happen if the database is not updated. While in Table 2, authors showed the capability of this application to save the comment input. User can key-in

Table 1. Application result for insert E number.

Input	Application result for insert ecode					Consistency
	1st Run	2nd Run	3rd Run	4th Run	5th Run	
E103	1	1	1	1	1	1
E440	1	1	1	1	1	1
E338	1	1	1	1	1	1
E531	1	1	0	0	0	0
E923	1	1	1	1	1	1
Σ *Test*						5
Σ *Application Insert* (Total of "1")						4
Accuracy (%)						80.00

1 = Application insert E number operation success.
0 = Application insert E number operation failed.

Table 2. Application result for insert comment

Input	Application result insert data for comment					Consistency
	1st Run	2nd Run	3rd Run	4th Run	5th Run	
"nice"	1	1	1	1	1	1
"good"	1	1	1	1	1	1
"nicely done"	1	1	1	1	1	1
"very helpful"	0	0	1	1	1	0
"good knowledge.tq"	1	1	0	0	0	0
Σ *Test*						5
Σ *Application Insert* (Total of "1")						3
Accuracy (%)						60.00

1 = Application insert data operation for comment success.
0 = Application insert data operation for comment failed.

any comment and this comment also need to be stored in the database. Table 3 display the comparison results from four databases. There are ten E number that been compared to these database.

From the implementation and testing phase, we can conclude the advantages and disadvantages of this application that can be improvised in the future. The advantages of this application is user can use the search function to check on E number halal status information. This application also provide the information of E number used type in food additives. Besides, user also can comment or provide their opinion over the information of E number. There is also a limitation on information provide by E number, such as the details that did not display the daily maximum intake which the user can consume. E number or food additive can cause of some allergies to the consumer such as hyperactivity, asthma and skin disorder if they consume with the excessive amount. In the future, the application can be improved by provide the information for the allergies precaution for the user. Overall, the Halal Checker application helps users identify various food additives that have been used in the manufacture of food by using the E number.

Table 3. Table of E number check with four different website.

E number	Database from website specialworldofislam.info	Database from website muslimconsumergroup.com	Database from website muslimtents.com	Database from website guidedways.com
E103 (Chrysoine Resorcinol)	–	1	1	1
E150(b) (Caustic Sulphite Caramel)	–	–	1	1
E214 (Ethyl 4-hydroxybenzoate)	1	1	1	1
E240 (Formaldehyde)	–	–	1	1
E312 (Dodecyl Gallate)	1	1	1	1
E303 (Potassium Ascorbate)	–	–	1	1
E474 (Sucroglycerides)	1	1	1	1
E441 (Gelatine)	–	1	1	1
E938 (Argon)	–	1	1	1
E927(b) (Carbomide)	0	1	1	1

1 = Name check for the E number same.
0 = Name check for the E number different.
- = E number name does not exist in database.

5 Conclusion

Mobile application Halal Food Checker is developed to help people check the status of halal food supplements available in the market easily. The good news for those who are allergic to foods, this application also provides information about the code number E or ingredients that can cause allergic reactions such as asthma, rhinitis, dizziness, low blood pressure and so on. This application is developed based on Android. In addition to the use of smart phones, there are a few people who do not know about the code number E contained packaged product. Therefore, stick to the concept of search engines, this application allows users to check the status of a list of kosher food available on the market faster and more detailed information. The methodology used in the development of prototypes for process improvements can be made in the future to further enhance existing functionality. The existence of this system has contributed to the effective impact in our daily live.

Acknowledgments. The authors would like to thank Universiti Tun Hussein Onn Malaysia (UTHM) for providing the research facilities and research grant (Vot R045) to perform this research study. This research also supported by GATES IT Solution Sdn. Bhd under its publication scheme.

References

1. Noy, N.F., McGuinness, D.L.: Ontology Development 101: A Guide to Creating Your First Ontology, Stanford Knowledge Systems Laboratory Technical Report KSL-01-05 and Stanford Medical Informatics Technical Report (2001). http://protege.stanford.edu/publications/ontology_development/ontology101noy-mcguinness.html
2. Gruber, T.R.: A translation approach to portable ontology specification. Knowl. Acquisition **5**, 199–220 (1993)
3. Gruber, T.R., Liu, L., Ozsu, M.T.: Ontology. In: Liu, L., Tamer Özsu, M. (eds.) Encyclopedia of Database Systems. Springer, Heidelberg (2009)
4. Wei, Q.: Development and application of knowledge engineering based on ontology. In: Third International Conference on Knowledge Discovery and Data Mining, Phuket, Thailand, pp. 518–521 (2010)
5. Kassim, M., Che Ku Yahaya, C.K.H., Zaharuddin, M.H.M., Bakar, Z.A.: A prototype of halal product recognition system. In: International Conference on Computer & Information Science (ICCIS), pp. 990–994 (2012)
6. Zailani, S., Arrifin, Z., Wahid, N.A., Othman, R., Fernando, Y.: Halal traceability and halal tracking systems in strengthening halal food supply chain for food industry in Malaysia (a review). J. Food Technol. **8**, 74–81 (2010)
7. Guidedways Technology. http://www.guidedways.com
8. Muslim Hosting Community. http://www.muslimtents.com
9. Muslim Consumer. http://www.muslimconsumergroup.com/enumbers_list.html
10. Food Ingredients Number. http://special.worldofislam.info/Food/numbers.html
11. Chemical entities of Biological Interest (ChEBI). http://www.ebi.ac.uk/chebi
12. Kamaludin, H., Kasim, S., Selamat, N., Hui, B.C.: M-learning application for basic computer architecture. In: 2012 International Conference on Innovation Management and Technology Research (ICIMTR), pp. 546–549. IEEE, May 2012
13. Kasim, S., Zakaria, F.A.: Daily calorie manager for basic daily use. In: 2013 Third International Conference on Innovative Computing Technology (INTECH), pp. 437–442. IEEE, August 2013
14. Kasim, S., Wai, B.S.: Multilingual phrasebook for Android (MPA). In: 2013 Third International Conference on Innovative Computing Technology (INTECH), pp. 443–448. IEEE, August 2013

Factors Influencing the Use of Social Media in Adult Learning Experience

Masitah Ahmad[1(✉)], Norhayati Hussin[1], Syafiq Zulkarnain[1], Hairulnizam Mahdin[2], and Mohd Farhan Md. Fudzee[2]

[1] Faculty of Information Management, Universiti Teknologi MARA, 40150 Selangor, Malaysia
{masitah, norhayati}@salam.uitm.edu.my, syzmy90@yahoo.com.my

[2] Faculty of Computer Science and Information Technology, Universiti Tun Hussein Onn Malaysia, 83000 Batu Pahat, Johor, Malaysia
{hairuln, farhan}@uthm.edu.my

Abstract. The uses of social media are very popular nowadays as a platform of informations exchanges among the users. It has been used to support e-learning, where lecturers uploading lecture notes and tutorial videos on a group page that joined by their students. For young generations, this is something that they expect from their lecturer. However it is different with adult learner who are not all used to today's technologies. This paper study the factors of that influenced the use of social media in adult learning. By understanding the factors, practitioners can used it as a basis to build a more specific social media tools that is intended for adult learner. To understand the factors, a survey has been distributed to a group of postgraduate students, and the data is analyzed by using IBM SPSS 2.0. The research suggested that technology acceptance factor as most dominant factor among the others that influenced the use of social media in adult learning. The results of study also suggested there is significant relationship of all factors specified.

Keywords: Adult learning · Social media · Technology acceptance

1 Introduction

The emergence of the Internet applications has impacted the ways people learn and interact with others [1]. The use of the Internet has made a learning process becoming easier to be implemented in transferring the knowledge, teaching and learning and coaching [2]. Younger generations use social media as their primary source of information and they also support it as a tool of learning. In survey presented in [3], the study found younger generations had spent their time about 3.3 h just to access a social network [3]. The result indirectly has connection with the evolution of Web 2.0 that coming with social media component that directly introduced the elements of social networks that made information and knowledge available in just at one click [4].

However, capabilities of social media as the popular platform for information and knowledge dissemination creates other problem such as degrading social interaction,

T. Herawan et al. (eds.), *Recent Advances on Soft Computing and Data Mining*,
Advances in Intelligent Systems and Computing 549, DOI 10.1007/978-3-319-51281-5_58

increasing of digital divide, privacy and safety issues, psychological and learning [5–7]. Social media are not excluded from abuse by the users, which some users intend to use advantages of social media application for unethical purpose, for example use fake identity to post irresponsible comments.

This study attempted to investigate factor influences adult learner use of social media and its outcomes will demonstrate the impact of social media use. This study helps to reveal a solution to overcome information explosion issue, addiction to social media and unethical use with extended learns to control and diminish the issue through understanding on factor of social media use in learning knowledge.

2 Related Work

2.1 Adult Learning

Basically, social media is web-based platforms, applications and technologies which enable user to socially interact with other user while adult learning involve the learning through formal, non-formal and informal learning activities, which the activities are used by an after a break since leave formal education and training process, which results in the acquisition of new knowledge and skills [8].

The adult learner can be describes as undergraduate or graduate students that are lifelong learners who generally are 25 years or older with additional responsibilities such as family, career or community [9]. Adult learning also could be then seen as, for example, the process of managing the external conditions that facilitate the internal change in adults called learning [9]. In other words, it is a relationship that involves a conscious effort to learn something.

Hartshorn [11] defines social networks as an act of engagement, where inside of social networks, groups of people with common interests, or like-minds engage together on social networking sites such as Facebook and develop relationships through community.

2.2 Social Media

History of social media was started in Compuserve era and at the beginning of ARPANET project in 1969 [12]. Until now the advancement of technology and the introduction of Web 2.0 has brought new perspectives on how people engage to study and learns [4]. Social media is one of Web 2.0 technologies, which is very popular has been used nowadays. Facebook, Twitter, Wikipedia, and YouTube are several social networks that hold thousands of billion users use the service for many purposes.

The popularity of social media application has changed in how people getting information and knowledge. In another aspect, the popularity of social media application changes several values of relationship towards individual level, for example, social interaction, academic developments, issues of privacy, content ownership opportunities and e-learning [5, 13, 14]. The social network is also viewed as a version of a new pool of knowledge and information that changes people done on knowledge sharing and knowledge transfer and become factors that ignite the knowledge seeking.

2.3 Social Media Use

A driven factor of intention to use social media application can be viewed in many aspects. [15] for example, highlight the intention of the use of Facebook by academic and professional circles because the ability of social media application to analyze the sociology, finance, marketing and public relation matters. For businesses, they can be used to understand stakeholders' need, directly able to increase the productivity, sales, and reputation of the organization. Most of Small Medium Enterprises (SME) in Malaysia uses Facebook, because it helps to cut cost in advertising and promotional activities and improves customer relationship [16]. In the other study that focuses the uses of social media among students reveal that students use social media because they want to interact with others to gain something, share and getting a feedback [17].

In one's learning process, the attitude of learner plays a main role in achieving the goal of learning. The variable of Technology Acceptance Model (TAM) as introduced by [18] suggests the acceptance of technology such as perceived usefulness, perceived ease of use and attitude can be measured. Social media allow user specifically teens to share their feelings, engage with communities, enhances collective learning and expand critical thinking [6] that can enhance learning experiences. In teaching and learning process, innovation is important to ensure learning objective reaches its goal. A study conducted by [12] found social media application open the opportunities for students to enhance skills in communication and interaction.

Many research reveals the use of social media as a learning platform among adult nowadays are driven by four main factors which are technology acceptance, cost effectiveness, interactivity and compatibility [14, 15, 19–21].

3 Research Methodology

The study is conducted at one of the Public Malaysian University in Selangor. The research target is postgraduate students and Flexible Learning Mode (FLP) students as they are literally familiar with the uses of social media platform use for learning and they also meet with definition of adult learning. The nature of this study applied descriptive study and use quantitative approach in getting the results. This study investigation is based on conceptual framework which several variables are derived from previous studies with little modification to suit with current studies.

The conceptual framework consists of four independent variables (IV) and one dependent variable (DV) which independent variable consists technology acceptance, cost effectiveness, interactivity and compatibility and for dependent variable consists of objectives in this study which focused on the use of social media among adult learners. Each dimension of variable is included in technology acceptance are perceived usefulness, perceived ease to use and technical support. For cost effectiveness, dimensions involve are cost, user relationship and information accessibility. Interactivity's dimensions are communication features, sharing support and current features. The next dimension is compatibility which comprised of accessible, availability of infrastructure and infrastructure itself. For dependent variable which, dimensions involved are frequency of use, purpose of use and benefit of use as shown in Fig. 1.

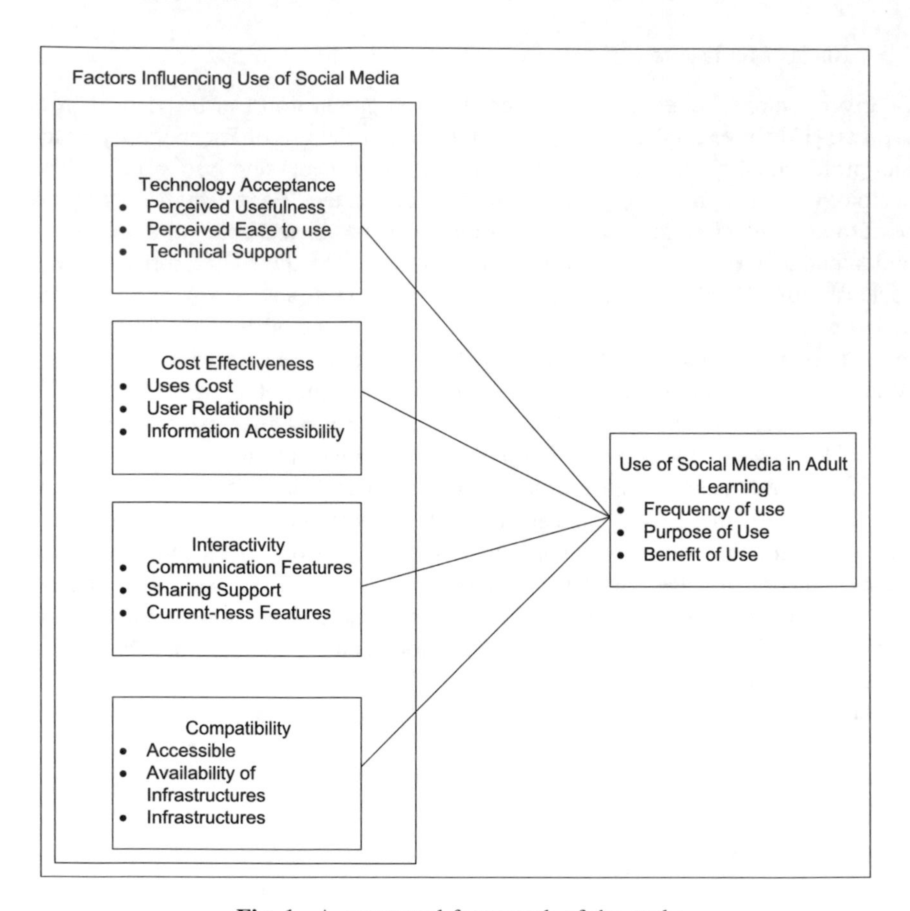

Fig. 1. A conceptual framework of the study

Data collection was carried out using a questionnaire constructed based on literature reviews and distributed among research targets randomly. The research strategy was adopting survey research and the questions in the survey instruments used self-administered and web-based questionnaires. The setting of this study is based on non-contrived setting as it was conducted in Selangor area with a time horizon for this study adopted cross-sectional study or one-shot studies that involved one day data collection.

The target population in this study was postgraduate student and Flexible Learning Program (FLP) student, who studied in academic year September 2015 with total number of 190 students. The sample size used in this study is 127 students, from the group that the met the study requirements.

Probability sampling, specifically simple random sampling is used as sampling technique and unit of analysis in this study is individual as the current research is focused on user's use of social media. A research instrument used for data collection is a questionnaire that developed based on research objective and research question as well to examine the relationship between variables. The questionnaire constructed in single language which is English and all topics in the questionnaire are arranged page-by-page.

Data analysis is performed by using Statistical Package of Social Science (SPSS) software, version 22.0 and the result represented using descriptive statistics. The analyses used to investigate the findings are demographic profile respondent analysis, reliability test, descriptive statistics, Pearson correlation analysis and multiple regression analysis. Pre-test was conducted to measure reliability and trustworthiness of research instruments by generating the result by using SPSS and measured by using Cronbach's alpha that measured each item in the questionnaire that reflected the relationship as positively correlated to each other [22].

4 Analysis and Findings

Based on data collection conducted, out of 190 questionnaires distributed, only 120 questionnaires were returned and usable for analysis with response rate is 63.5%. Thirty-four respondents are male and the rest is female with 51.7% of them are married. Most of them have a bachelor degree with 64 frequencies and 56 frequencies for master degree. The majority of respondent has working experiences for 1–5 years and another 40 respondents have 6–10 years working and only 24 respondents have more than 10 years working experience. From 8 types of social media application asked, most of respondent have a Facebook account which record over 90% (n = 108), and followed by Instagram - 66.7% (n = 80), Twitter - 50% (n = 60), Google+ - 32.5% (n = 39), and the rest Tumblr - 15.8% (n = 19) and MySpace - 7.5% (n = 9).

In terms of experience of using social media, most of respondents answered, they already used social media for more than 4 years and above, which recorded 88.3% (n = 106) of them and the rest respondents answered only using social media since last 2–3 years ago and last 1–6 months with 8.3% (n = 10) and 3.3% (n = 4) in total. Only twenty-two respondents spent less than one hour per day on social media with other 30 respondents spent about three hours, 26 respondents spent in 2–3 h and others spent 4 h and above on social media daily. Reliability of research using Cronbach's Alpha is accepted when the variables of research exceed 0.7%. Table 1 shown the Reliability statistics of variables to fulfill the requirement.

Table 1. Reliability statistics of variables to fulfill the requirement

Variables	Cronbach's alpha (%)	No. of items
Technology acceptance	0.915	9
Cost effectiveness	0.908	8
Interactivity	0.925	9
Compatibility	0.935	7
Use of social media in learning	0.921	8

Relationship of factors influence (IV) and social media use in adult learning (DV) can be determined using two inferential analyses which are Pearson Correlation analysis and multiple regression analysis that are applied to analyze data answered from questionnaires.

Table 2 shown that a values of Pearson R range 0.396 to 0.760, this showing that there is no overlapping variables and positive relationship is exist. The lowest correlation value is r = 0.396 between Interactivity factor and social media use, while the highest correlation value, r = 0.760 is between technology acceptance and social media use. The relationship between technology acceptance and user use of social media in learning, knowledge can be accepted as it has 0.760** correlation score, positive sign of relation and strong level of relation. Moderate level of relation, positive sign of relation and 0.604** correlation score able hypotheses: cost effectiveness of social media positively influences social media use in learning knowledge to be accepted. Third hypotheses which is a interactivity of social media positively influence social media use in learning, knowledge is accepted even though it has a weak level of relation and has 0.396** correlation score with positive signs of relation. Last hypotheses stated that compatibility has positive relationship influences user use social media in learning, knowledge can be accepted with correlation score 0.548**, positive sign of relation and moderate level of relation.

Table 2. Correlations analysis factor influencing use adult learning and use of social media.

Factor influencing use in adult learning	Use of social media
Technology acceptance	0.760
Cost effectiveness	0.604
Interactivity	0.396
Compability	0.548

**Correlation is significant at the 0.01 level (2-tailed)

Based on Table 3, values of Pearson R range 3.73 to 3.93 showing that there is no overlapping variables and positive relationship is exist. The lowest correlation value is r = 3.73 between Interactivity factor and social media use, while the highest correlation value, r = 3.93 is between technology acceptance and social media use.

Table 3. Overall mean scores and standard deviation for research variables.

Variables	Mean	Standard deviation
Technology acceptance	3.93	.869
Cost effectiveness	3.69	.816
Interactivity	3.73	.763
Compatibility	3.77	.787
Social media use	3.87	.88

On Table 4, values of Pearson R ranged 0.396 to 0.760 showing that there is no overlapping variables and positive relationship is exist. The lowest correlation value is r = 0.396 between Interactivity factor and social media use, while the highest correlation value, r = 0.760 is between technology acceptance and social media use.

Table 4. Correlations analysis factor influencing use adult learning and use of social media.

Factor influencing use in adult learning	Use of social media
Technology acceptance	0.760
Cost effectiveness	0.604
Interactivity	0.396
Compability	0.548

**Correlation is significant at the 0.01 level (2-tailed)

The relationship between technology acceptance and user use of social media in learning, knowledge can be accepted as it has 0.760** correlation score, positive sign of relation and strong level of relation. Moderate level of relation, positive sign of relation and 0.604** correlation score able hypotheses: cost effectiveness of social media positively influences social media use in learning knowledge to be accepted. Third hypotheses which is a interactivity of social media positively influence social media use in learning, knowledge is accepted even though it has a weak level of relation and has 0.396** correlation score with positive signs of relation. Last hypotheses stated that compatibility has positive relationship influences user use social media in learning, knowledge can be accepted with correlation score 0.548**, positive sign of relation and moderate level of relation.

Based on Table 5, the first dominant predictor of factor influencing use of social media in adult learning is the technology acceptance factor that proves it's influencing in adult learning. Cost effectiveness becomes a second dominant predictor influencing use of social media in adult learning and followed by compatibility factor.

Table 5. Coefficients

Model		Unstandardized Coefficients		Standardized coefficients	t	Sig.
		B	Std. error	Beta		
1	(Constant)	1.212	.287		4.226	.000
	All technology	.783	.103	.757	7.634	.000
	All Cost effectiveness	.135	.107	123	1.256	212
	All interactivity	−.294	.111	−.249	−2.654	.009
	All_compatibility	.103	.111	098	928	.355

(Dependent Variable: ALl_UseSociaMedia)

Last dominant predictor of factor influencing use of social media in adult learning is interactivity. For multiple regression analysis, p value equals to 0.000 ($p = 0.000$, $p < 0.05$) gives significant regression model and t value more than 16.45 ($t > 16.45$) concludes that technology acceptance factor is the most dominant predictor influencing use of social media in adult learning compare to other independent variables. The results indicate that the technology acceptance factor has strong positive relation towards the use of social media in adult learning. Technology acceptance becomes the dominant influence because many users feel social media is flexible and easy to use, and feel that social media useful for learning process. Four factors investigated can be ranked in order of technology acceptance, cost effectiveness, compatibility and interactivity.

5 Discussion

This study was carried out with a purpose to determine whether four factors that have been highlighted in the literature are the most influential in the use of social media among adult learning. For the first objective, this study has significant found that technology acceptance factor ($t = 7.634$, $t > 1.645$) is the most dominant influence among the four factors varied and these findings match with previous studies done by [14, 19, 20]. The results indicate that the technology acceptance factor has strong positive relation towards the use of social media in adult learning. Technology acceptance becomes the most dominant influence because many users feel that social media is flexible and easy to use, and feel that social media is useful for learning process. Four factors that have been investigated can be ranked in order of technology acceptance, cost effectiveness, compatibility and interactivity.

The findings found in this study also to fulfill the second objective as all variables or factors correlate each other and have positive relation. All the independent variables (factor) reflected very well as items used are measured using reliability analysis using Cronbach's alpha technique as shown before and indicate that the items are consistent and reliable in measuring all the factors. The relationship between technology acceptance and use of social media in adult learning are relevant and significant based on evidence shown in correlating analysis before with significant at 0.01 level ($r = 0.760$, $p < 0.01$). From descriptive analysis with overall mean 3.73 represents a good sign of technology acceptance that became a factor chosen by most respondents based on their intention to use social media in learning. Multiple regression analysis gives the relationship between technology acceptance and use of social media in adult learning as it has significant predictors that could significantly influence adult students in using social media ($t = 7.634$, $t > 1.645$).

The relationship between cost effectiveness and use of social media in adult learning are relevant and significant as shown in correlating analysis before with significant at 0.01 level ($r = 0.604$, $p < 0.01$). Descriptive analysis represents evidence that cost effectiveness factor specifically in dimension of user relationship, social media minimize a cost create and share information to others (highest mean) became a factor why most respondents chose cost effectiveness factor that influences social media use for adult learning.

For the last objective, only 15 respondents answered related questions as it is not a compulsory question. Most of them agree that social media is a good platform to support adult learning and only two respondents gave different answers. One respondent gave answer regarding user concern on information content as the respondent where the respondent believe that some information might not be trusted while another respondent gave opinion on how learning using social media ease knowledge sharing like sharing of recipes between her/his friends.

6 Conclusion

In this paper, we presented data which obtained from a survey questionnaire that collected from postgraduate and FLP (Flexible Learning Program) students. The survey asked about four factors which have been identified from the literature which are technology acceptance, cost effectiveness, interactivity and compatibility. We also proposed conceptual framework with several variables that derived from previous studies and modification had been made to suit with current studies. The results show that social media is a good platform to support adult learning performances. The most significant factor among the four factors is the technology acceptance. In future work, we want to use the conceptual framework that has been suggested in this study on actual system.

Acknowledgments. This research was partially supported by MARA through Eduloan scheme 2015 and self-funded.

References

1. Brown, J.S.: Growing up digital. Change **32**(2), 10–20 (2000)
2. Baumgartner, P., Payr, S.: Learning with the internet a typology of applications. In: World Conference on Educational Multimedia and Hypermedia, pp. 124–129. AACE, Charlottesville (1998)
3. Kemp, S.: Social, digital & mobile worldwide in 2014. Wearesocial.net (2015). http://wearesocial.net/blog/2014/01/social-digital-mobile-worldwide. Accessed 1 June 2015
4. WhatIs.com.Social networking. WhatIs.com (2015). http://whatis.techtarget.com/definition/social-networking. Accessed 6 June 2015
5. Ahn, J.: The effect of social network sites on adolescents' social and academic development: current theories and controversies. J. Am. Soc. Inf. Technol. **62**(8), 1434–1445 (2011)
6. O'Keefee, S.G., Clarke-Pearson, K.: The impact of social media on children. Adolesc. Fam. Pediatr. **12**(4), 800–804 (2011)
7. Shelke, P., Badiye, A.: Social networking: its uses and abuses. Res. J. Forensics Sci. **1**(1), 2–7 (2013)
8. Devoe, N.: 17 things your parents had to do before the internet. Seventeen (2015). http://www.seventeen.com/life/friends-family/a24295/crazy-things-parents-did-before-the-internet/. Accessed 22 Dec 2015
9. European Commission. European Terminology in Adult Learning for a common language and common understanding and monitoring of the sector. London: National Research and Development Centre for adult literacy and numeracy (2008)
10. Brookfield, S.D.: Understanding and Facilitating Adult Learning: A Comprehensive Analysis of Principles and Effective Practices. Open University Press, Milton Keynes (1986)
11. Hartshorn, S.: 5 differences between social media and social networking. SocialMedia Today.com (2010). http://www.socialmediatoday.com/content/5-differences-between-social-media-and-social-networking. Accessed 2 June 2015
12. Bennett, S.: A brief history of social media. SocialTimes (2013). http://www.adweek.com/socialtimes/social-media-1969-2012/487353. Accessed 1 June 2015

13. Jonnavithula, L., Tretiakov, A.: A model for the effects of online social networks on learning, pp. 1–3 (2012)
14. Siemens, G., Weller, M.: Higher education and the promises and perils of social network. RUSC, Univ. Knowl. Soc. J. **8**(1), 164–170. doi:10.7238/rusc.v8i1.1076 (2011)
15. Bonson, E., Escobar, T., Ratkai, M.: Testing the inter-relations of factors that may support continued use intention: the case of Facebook. Soc. Sci. Inf. **53**(3), 293–310 (2014). doi:10.1177/0539018414525874
16. Sulaiman, A., Farzana, P., Moghavvemi, S., Noor, I.J., Nor, L.M.S.: Factors influencing the use of social media by SMEs and its performance outcomes. Ind. Manage. Data Syst. **115**(3), 570–588 (2015)
17. Voss, K.A., Kumar, A.: The value of social media: are universities successfully engaging their audience? J. Appl. Res. High. Educ. **5**(2), 156–172 (2013). doi:10.1108/JARHE-11-2012-0060
18. Davis, F.D.: Perceived usefulness, perceived ease of use, and user acceptance of information technology. MIS Q. **13**(3), 319–340 (1989). doi:10.2307/249008
19. Lee, W., Tyrrell, T., Erdem, M.: Exploring the behavioral aspects of adopting technology. J. Hospit. Tourism Technol. **4**(1), 6–22 (2013). doi:10.1108/17579881311302329
20. Shittu, A.T., Basha, K.M., AbdulRahman, N.S.N., Ahmad, T.B.T.: Investigating students' attitude and intention to use social software in higher institution of learning in Malaysia. Multicult. Educ. Technol. J. **5**(3), 194–208 (2011). doi:10.1108/17504971111166929
21. Senge, M.P.: The Fifth Discipline: The Art and Practice of the Learning Organization. Currency/Doubleday, New York (1990)
22. Sekaran, U., Bugie, R.: Research Methods for Business. Wiley, West Sussex (2013)

A Framework to Analyze Quality of Service (QoS) for Text-To-Speech (TTS) Services

Mohd Farhan Md Fudzee[1], Mohamud Hassan[1], Hairulnizam Mahdin[1(✉)], Shahreen Kasim[1], and Jemal Abawajy[2]

[1] Faculty of Computer Science and Information Technology, Universiti Tun Hussein Onn Malaysia, Parit Raja, Malaysia
{farhan, hairuln, shahreen}@uthm.edu.my

[2] School of Information Technology, Deakin University, Burwood, Australia
jemal@deakin.edu.au

Abstract. Quality of service (QoS) evaluation is vital for text-to-speech (TTS) web service applications. Most of the current solutions focus on either evaluating functional or nonfunctional attributes of the TTS. In this paper, we propose a QoS framework to evaluate and analyze the perceived QoS that combines general and specific mechanisms for measuring both functional and nonfunctional requirements of speech quality. General mechanism measures the response time of TTS services while specific mechanism measures intelligibility and naturalness through subjective quality measurements, which are mapped onto mean opinion score (MOS). The result shows the workability of the framework, tested by predetermined users to three services: service1 (From-texttospeech) resulting 47.84%; service2 and service3 (NaturalReader and Yakitome) are 31.62 and 21.53% respectively. The TTS services evaluation can be to enhance the user experience.

Keywords: Quality of Services (QoS) · Text To Speech (TTS) · Mean Opinion Score (MOS) · Intelligibility · Naturalness · Response time · Quality attributes

1 Introduction

Text to speech systems are vital in our everyday activities. The use of speech synthesis and building voices became common due to the rapid advancement in information technology and communications. The Voice User Interface (VUI) plays huge role in technology such as computer systems, mobile multimedia and voice-enabled equipment. Speech Understanding and Synthesis Technology are among the frequently used technology to support users. Ultimately, high quality synthesized outputs are preferred. Thus, evaluating the quality of web text to speech (TTS) services are important. TTS is useful in the areas like disabled, education, consumer, computer interface and telecommunications. The Quality of Service (QoS) in multimedia conversion of TTS examines the performance in terms of accessibility, media conversion availability, conversion accuracy, and user satisfactions. QoS determines how well a service performs while functionality determines what a service does [1, 2].

T. Herawan et al. (eds.), *Recent Advances on Soft Computing and Data Mining*,
Advances in Intelligent Systems and Computing 549, DOI 10.1007/978-3-319-51281-5_59

Quality evaluations of TTS web services can be classified into functional and nonfunctional requirements. Functional requirement focuses on what TTS service does, while nonfunctional requirements also known as quality attributes is used to determine the quality of services requirements. There are general and specific QoS mechanisms to evaluate functional and nonfunctional requirements of TTS services. Currently, there is less effort to integrate both mechanisms into single solution to analyze QoS from the end-user perspective to provide users with capacity to enhance their experience.

In this paper, a QoS analysis framework for text to speech services is proposed. This framework is aimed to analyze and examine the quality of services of the multimedia conversion text to speech on the web and measure TTS performance in term of content accessibility, response time, and voice intelligibility and naturalness by comparing three web TTS services to enhance the quality of experience of the online users. The remainder of this paper is structured as follows. A brief review of previous works is discussed in Sect. 2. In Sect. 3, we describe the proposed framework. The results of the web QoS for TTS are analyzed in Sect. 4. Finally, some concluding remarks are given in Sect. 5.

2 Related Work

Recently, a number of works have focused on developing subjective QoS evaluation frameworks. For instance, a probabilistic model was introduced by Wang et al. [3]. Based on the received speeches, the system will calculate the confidence score of multiple different levels using a constrained generalized posterior probability (CGPP) algorithm. This method calculates the received speech input in multiple different levels (e.g. phoneme, syllable, and word and/or utterance level) using CGPP algorithm. This QoS service evaluation needs complex data processing, which consumes a lot of time. Also, it is very expressive for end users.

On the other hand, Remes et al. proposed frequency-weighted segmental Signal-to-noise ratio (SNR) quality measurement that exhibited a performance using standardized perceptual evaluation of speech quality (PESQ) objective evaluation measure [4]. It will allow capturing the automated quality evaluation. This method has a drawback for its relaying automation but it performs when it comes to intelligibility and naturalness, which reflect the user experience. Other author developed TTS quality evaluation using E-model. E-model is a computation model which takes into account all links between transmission parameters [5]. This model requires the individual transmission path parameters not being assessed separately but rather all their possible combinations and corresponding interaction are considered. This can be achieved using quality estimation based on system approach of computation model. The computation model approach has downward because it depends on transmission path instead of the experience of the quality services.

As mentioned earlier, the significant issue of TTS services is QoS. Even though there are some researches on how to determine the qualities of TTS media conversions however, previous researches on quality have concentrated only on server side and media conversion stage which involves implementation part of text to speech applications. Alternatively, we highlight the crucial needs for a model to be used as guidelines for service users to get the best TTS application services and enhance the user experience. It is not easy for Internet users to captivate the TTS online services. Thus, it's very significant to write a generic tool that provides client-side measurements for the performance of the TTS services available online [6].

3 TTS Evaluation Framework

In this section, the framework for QoS analysis for TTS media conversion has been proposed. The two elements in this framework are user's requirement and perception to deliver quality of services of text to speech services. To evaluate non-functional quality attributes for TTS web services, we analyze and measure general and specific QoSs. The proposed framework gives great flexibility, which provides analytical solutions for the end users as illustrated by the Fig. 1.

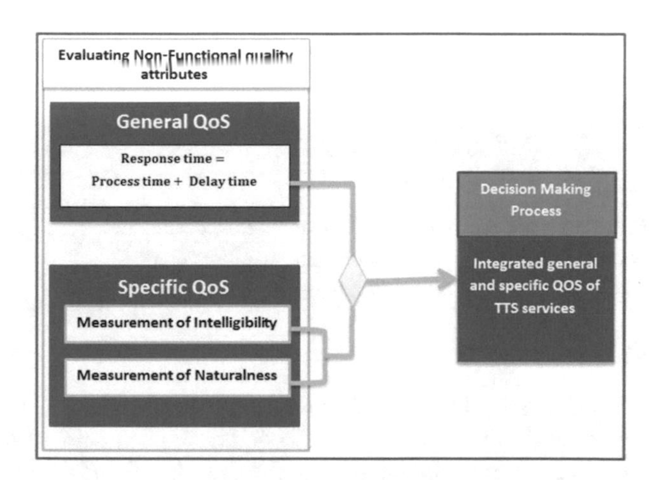

Fig. 1. Proposed evaluation framework for QoS of TTS.

3.1 General QoS Attributes

General QoS attributes estimates the response time of the services. Measuring response time performance is one of the most significant factors in the servers since service providers need to manage and enhance the delay.

$$\mathrm{R(t)} = P(t) + D(t) \tag{1}$$

where R(t) = response time, P(t) is process time and D(t) delay time.

3.2 Specific QoS Attributes

The second part of the framework is the specific QoS attributes, that consists of evaluations for intelligibility and naturalness of synthetic speech using subjective measures. In this study we use Mean Opinion Scores (MOS) in calculating and measuring the performance of the intelligibility and naturalness of TTS service.

The framework combines the attributes of general and specific QoS. Specifically, the response time is captured for general QoS while a five point mean opinion score is used to estimate the scale for user experience of the intelligibility and naturalness of text to speech (TTS) services.

3.3 Experimental Setup

To implement the QoS analysis for TTS online services, we consider three TTS online services to examine the quality of the speech that online system generates. Table 1 shows the list of TTS services that we use to examine the QoS analysis. We compare these services in term of file size, duration, delay time and then the response time is calculated.

Table 1. List of used TTS services.

Service name	Link
FromTextToSpeech	http://www.fromtexttospeech.com/
Natural Reader	http://www.naturalreaders.com/
Yakitome	https://www.yakitome.com/upload/from_text

3.4 Calculating Response Time

The length of the TTS speech is the duration of the audio where it is achieved by capturing the length or the duration of the audio that have been converted. An array of arbitrary parameter values from starting time of the speech and end of the speech audio time for the duration is used. The duration parameter is the amount of time speech in seconds values that will be calculated according to the parameter values. To get the response time of the speech of the TTS service, we calculate the processed time and deduct the delay time of the audio speech read.

$$\mathrm{R(t)} = P(t) + D(t) \tag{2}$$

R(t) represents the response time of the converted speech in seconds, P(t) is the process time or the duration time of the TTS in seconds and the D(t) stands for the amount of seconds delayed from the started period.

3.5 MOS Design and Structure

Mean opinion scores (MOS) of quality of services (QoS) analysis measure Listening Quality of mean opinion score (MOS-LQ) of the speech audio that is being listened by the users [7]. This value takes into consideration the speech intelligibility and naturalness and from this data calculates how users would rate the audio quality they hear. MOS-LQ scale is divided into two major areas, intelligibility and naturalness of the text to speech application. To minimize the complex nature of the MOS, we use various question logic and rating scales ranging from "strongly agree" to "disagree" and "not at all" to "very often" as part of ITU recommendation.

3.5.1 MOS Intelligibility

Mean opinion scores (MOS) consist of five parts that focus on intelligibility of the web speech for TTS services. Intelligibility test will measure how the speech is comprehensible to the users and in what degree that can the TTS speech be understood. The users will scale questions on quality of services on spoken clarity, explicitness, comprehensibility, perspicuity, and precision. Intelligibility consists of five parts which are:

- Overall impression: rates the quality of the TTS audio
- Listening effort: rates the effort that are required to make in order to understand the message
- Pronunciation; irregularities in pronunciation of the TTS speech
- Speaking rate: rates the average delivery of the speed of TTS speech
- Pleasantness: measures the pleasantness of the voice.

3.5.2 MOS Naturalness

Questions on naturalness will examine the TTS web service's ability on how it is perceived as natural degrade human perception of quality. Naturalness essentially considers if the text is read in a natural human sounding manner. Naturalness questions consist of:

- Naturalness: rates the overall quality of the naturalness of the audio.
- Audio flow: rates the continuity or flow of the audio.
- Ease of listening: how easy or difficult to listen to the voice of TTS for long period of time.
- Comprehension problems: considers if it finds certain word hard to understand.
- Articulation: if the sounds in the audio distinguishable
- Acceptance: do you think that this voice can be used for TTS service users.

Client rates the services using MOS scale and the client scaled the preferred services by comparing the response time of the services. Table 2 shows some examples of categorization of quality of services (QoS).

Table 2. Quality of Services (QoS) categorization.

QoS attributes	Service's QoS
Response time	Negative relations
MOS	Rating

MOS data is computed as normalized score between 1 to 5 to calculate the aggregate score and represented as following [8]:

$$n_l = \sum_{m=1}^{j} q_m \times w_m \tag{3}$$

n_l is the node score for response time of the services, where q_m is the quality of services (QoS) relation associated from the rated data and w_m denoted as weight of the QoS relation.

$$q_m = \frac{v_i min}{v_i} \quad if\ v_i \min > 0 \tag{4}$$

where v_i represents the quality of services (QoS) value for specific services *(i)*. In the case for *if* v_i min > 0 response time must be greater than zero, because it is impossible to have zero or less than zero for the response time. q_m is quality measurement of specific services.

Twenty users were initially asked to evaluate three online TTS services, from which all of the participants completed the quality evaluations of TTS services. The participants consisted undergraduate as well as post graduate students and independent parties, and were asked to take MOS questions to scale the quality of services for three different TTS web services. The experiment was carried out in the university in non-soundproof space with basic laptop, headphone and Wi-Fi internet to do the quality analysis testing for TTS services [9].

4 Result

To implement the QoS analysis, we consider 3 TTS online services to examine the quality of the speech that online system generates. Table 1 shows the list of TTS services that we use to examine the QoS analysis. We compare these services in term of file size, duration, delay time and then the response time is calculated.

4.1 TTS Speech Test

Application Response records TTS speech response times for actual end-user activity to provide insight to the user's experience. It collects the speech data file name, size and processing time. After receiving collected information about converted speech file it calculates the delay time and response times to compare the performance of the text to speech (TTS) web services. Table 3 shows the example of the mean for response time. In this experiment, the file size used are (408, 591 and 151) KB.

4.2 MOS Analysis Result

The outcome of the mean opinion scores (MOS) is analyzed by comparing the three different services. The main aim is to give the users an overall outlook of the quality of

Table 3. Quality of Services (QoS) categorization.

Service	Process time	Delay time	Response time
fromtexttospeech.com	00:1:12	00:00:6	00:1:18
Natural Reader	00:1:15	00:00:17	00:1:32
Yakitome	00:1:35	00:00:26	00:2:01

the services (QoS) of the TTS web applications. The collected user observation on TTS services aimed to measure the intelligibility, naturalness and reading comprehension of the web text to speech TTS service.

4.2.1 MOS Intelligibility Result Analysis

Mean opinion score for the text to speech performance summarized in the Table 4 below. The overall average score is shown from 1 to 5 MOS scale.

Table 4. MOS intelligibility performance of compared online TTS services.

MOS intelligibility	fromtexttospeech.com	Natural Reader	Yakitome
Overall impression	4.2	2.2	1.4
Listening effort	3.4	2.3	2
Pronunciation	3.5	1.2	3
Speaking rate	3	1.5	2
Pleasantness	3.5	2	2.3

Table 4 illustrates that the overall impression of the quality of the TTS speech, fromtexttospeech.com is high compared to other services by obtaining 4.2 while Natural Reader and Yakitome obtain 2.2 and 1.4 rate respectively. Intelligibility performance for fromtexttospeech.com service surpasses both Natural Reader and Yakitome on listening effort, pronunciation, speaking rate, as well as pleasantness. MOS Naturalness result analysis.

Table 5 shows the metric performance mean opinion score (MOS) for naturalness. It compares all three text to speech web services (fromtexttospeech.com, Natural Reader and Yakitome).

Table 5 shows that text to speech (TTS) services that we used for this testing. Although the two services have the same audio flow Fromtexttospeech: 3.5, NaturalReader: 3.5 while Yakitome: 3.5. The TTS web service Fromtexttospeech has more human understanding and it's more preferred by the participants compared to Natural Reader and Yakitome. NaturalReader has the least comprehension rate, where it contains certain words which are very difficult to understand. Although, Yakitome, is less articulate than the rest of web TTS services (Articulation: Fromtexttospeech: 4, NaturalReader: 2.4 and Yakitome: 2.1). It is more efficient that Fromtexttospeech in overall quality of services in MOS naturalness. Participants can easily read and understand the outcome of the quality of service analysis for TTS web services and suggested that Fromtexttospeech voice can be useful for TTS service users.

Table 5. MOS naturalness performance of compared online TTS services.

MOS intelligibility	fromtexttospeech.com	Natural Reader	Yakitome
Naturalness	3.6	1.5	2.2
Audio flow	3.5	3.5	3.2
Easy listening	4	3	2.3
Comprehension	2.5	2	1
Articulation	3	1.2	1
Acceptance	4	2.4	2.1

4.3 Overall Quality Analysis

The result of the TTS services comparison shows that the overall quality of services (QoS) performance where Q = {intelligibility, naturalness, responseTime} such that intelligibility = (service1 = 1, service2 = 0.75, and service3 = 0.25), naturalness = (service1 = 1, service2 = 0.7, and service3 = 1), responseTime = (service1 = 1, service2 = 0.85, and service3 = 0.64), response time has negative relationship quality attributes while intelligibility and naturalness both QoS attributes are based on rating [10]. All QoS attributes are shown in Table 6 below.

$$AgS(q) = \sum\nolimits_{m=1}^{k} n_m \quad (5)$$

where $AgS(q)$ is the aggregate score for given quality attribute, k is maximum number of q.

The overall quality observations and the conclusion indicates that service1 (Fromtexttospeech) has 47.84% is acceptable TTS service provider where service2 and service3 (NaturalReader and Yakitome) are close with 31.62 and 21.53% respectively and less preferred because of the focus and attention needed from general comprehension for the speech.

Table 6. QoS performance with the computed aggregate score.

Service	Intelligibility	Naturalness	Response time	Aggregate score
fromtexttospeech	1	1	1	47.48%
Natural Reader	0.57	0.5	0.85	31.62%
Yakitome	0	0.4	0.64	21.53%

5 Conclusion

This research work covers text to speech media conversion users on the internet speech quality of services measurement. This work proposed client side QoS analysis framework for TTS services on web environment. This framework is designed to give a comparative estimation for specific quality attributes such as intelligibility, reading comprehension, and naturalness, as well as general quality attributes for performance requirements including accessibility and response time of the TTS services. It gives the capability of measuring the quality speech of TTS and estimate user perception of the services to provide feedback to the end users.

The analysis for speech quality could be extended with other adjustment variables such as QoS estimation for speech to text (STT), video quality analysis and other media conversion to be integrated with the user experience to provide better online media services.

Acknowledgments. The authors would like to acknowledge the Malaysian Ministry of Higher Education for the Fundamental Research Grant Scheme vot 1238. This research also supported by GATES IT Solution Sdn. Bhd under its publication scheme.

References

1. Patil, M., Kawitkar, R.S.: "Syllable" concatenation for text to speech synthesis for Devnagari script. Int. J. Adv. Res. Eng. Comput. Sci. Softw. **2**(9), 180–184 (2012)
2. Md Fudzee, M.F., Abawajy, J.: A protocol for discovering content adaptation services. In: Xiang, Y., Cuzzocrea, A., Hobbs, M., Zhou, W. (eds.) ICA3PP 2011. LNCS, vol. 7017, pp. 235–244. Springer, Heidelberg (2011). doi:10.1007/978-3-642-24669-2
3. Wang, L., et al.: Evaluating text-to-speech intelligibility using template constrained generalized posterior probability. U.S. Patent Application (2012)
4. Remes, U., Reima, K., Mikko, K.: Objective evaluation measures for speaker adaptive HMM-TTS systems. In: Proceedings of 8th ISCA Speech Synthesis Workshop (2013)
5. Möller, S., Wai, Y.C., Cote, N., Falk, T., Raake, A., Waltermann, A.: Speech quality estimation: models and trends. IEEE Sign. Process. Mag. **28**, 18–28 (2011)
6. Egger, S., et al.: Waiting times in quality of experience for web based services. In: 2012 Fourth International Workshop on Quality of Multimedia Experience (QoMEX). IEEE (2012)
7. Streijl, C.R., Winkler, S., Hands, D.S.: Mean Opinion Score (MOS) revisited: methods and applications, limitations and alternatives. Multimedia Syst. **22**, 213–227 (2014)
8. Md Fudzee, M.F., Abawajy, J.: Request-driven cross-media content adaptation technique. In: Ragab, K., Helmy, T., Hassanien, A.E. (eds.) Developing Advanced Web Services Through P2P Computing and Autonomous Agents: Trends and Innovations, chap. 6, pp. 91–113. IGI Global (2010)
9. Eyben, F., et al.: Unsupervised clustering of emotion and voice styles for expressive TTS. In: International Conference on IEEE Acoustics, Speech and Signal Processing (ICASSP) (2012)
10. Md Fudzee, M.F., Abawajy, J.: Management of Service level agreement for service-oriented content adaptation platform. In: Network and Traffic Engineering in Emerging Distributed Computing Applications, pp. 21–42 (2012)

Indoor Navigation Using A* Algorithm

Shahreen Kasim(✉), Loh Yin Xia, Norfaradilla Wahid, Mohd Farhan Md Fudzee, Hairulnizam Mahdin, Azizul Azhar Ramli, Suriawati Suparjoh, and Mohamad Aizi Salamat

Faculty of Computer Science and Information Technology, Universiti Tun Hussein Onn Malaysia, Parit Raja, Malaysia
{shahreen, faradilla, farhan, hairuln, azizulr, suriati, aizi}@uthm.edu.my, yxloh@outlook.com

Abstract. This paper introduced an indoor navigation application that helps junior students in Faculty of Computer Science and Information Technology (FSKTM) to find their classroom location. This project implements the A* (pronounced A Star) path finding algorithm to calculate the shortest path for users. Users can choose to view the floor plan of the building or start navigation. Users can choose their starting point from the list and set their destination to start navigation. The application is then calculate the shortest path for users by implement the A* algorithm. The route path will show on the floor plan after calculation done. Thus, users will find this project is easy to use and time saving.

Keywords: Android · Shortest path · A* algorithm · 2D map

1 Introduction

A navigation application is used to provide instructions on how to arrive at a given destination. The application requires connection to Internet data and normally uses a Global Positioning System (GPS) satellite connection to determine its location. A user can enter a destination into the application, which will plot a path to it. The application displays the user's progress along the route and issues instructions for each turn. Every September of the year, there are thousands of new students in University Tun Hussein Onn Malaysia (UTHM). New students from different state in the country are unfamiliar of the environment of the campus. Students getting lost while finding their classroom location. Some of them even using GPS navigation applications such as Google Maps for searching the location. As we know that GPS navigation application is specified for searching global area location and may not function well when searching tiny site such as classroom location. So, an in-campus navigation application is necessary for students' especially junior students of UTHM. The Android based FSKTM 2D map using A* Algorithm is an Android application for students and visitors of UTHM. It is a navigation application that is used to find the classroom location by the students. It allows students to choose their destination (the classroom) and provide the route map to students. It calculates the shortest path for students and navigates students to their destination.

T. Herawan et al. (eds.), *Recent Advances on Soft Computing and Data Mining*,
Advances in Intelligent Systems and Computing 549, DOI 10.1007/978-3-319-51281-5_60

The main problem addressed in this paper is how to find the exact location of the classroom. While students' classroom located in the building which has complex structure, they getting lost. Senior students are familiar with the environment of the campus; they will take a shortcut to classroom for saving time. But for the junior students, they are still new in the campus so finding classroom is always troublesome them. The major contributions are summarized as follows:

1. Design an Android based FSKTM 2D map that is able to find classroom location.
2. Develop an Android based FSKTM 2D map that is able to provide route map from starting point to the destination.
3. Develop an Android based FSKTM 2D map that is able to find the shortest path to the classroom using A* Algorithm.

The rest of the paper is organized as follows: Sect. 2 describes the related work on path finding algorithm and navigation application. Section 3 presents the proposed classification model for Android based FSKTM 2D map. Section 4 shows the evaluation methodologies and experimental results. Finally, Sect. 5 concludes the work and highlights a direction for future research.

2 Related Work

Navigation application is being widely used by many smart phone users. Nowadays, over billion of smart phone users around the world are utilize the uses of navigation application. The number of smartphone users worldwide will surpass 2 billion in 2016, according to new figures from eMarketer-after nearly getting there in 2015 [1].

2.1 Path Finding Algorithm

Path finding or pathing is the plotting, by a computer application, of the shortest route between two points [2]. Path finding is a fundamental component of many important applications in the fields of GPS, video games, robotics, logistics and crowd simulation and can be implemented in static, dynamic, and real-time environment. Path finding was first used extensively in real-time strategy games [3]. The path finding algorithms from computer science textbooks work on graphs in the mathematical sense – a set of vertices with edges connecting them [4]. Path finding algorithms can be divided in two broad categories: local and global in which local approaches try to find a path to a destination by analysing the surroundings of the current position only [5]. A path finding algorithm searches a graph by starting at one vertex and exploring adjacent nodes until the destination node is reached. Some common example of graph-based path finding algorithm:

(a) Dijkstra's Algorithm [6], Best-First-Search Algorithm [7] and A* Algorithm

A* (pronounced as "A Star") is a path finding algorithm that is widely used in games. It is an extension of Edsger Dijkstra's 1959 algorithm.

A* algorithm assigns a weight to each open node equal to the weight of the edge to that node plus the approximate distance between that node and the goal. The approximate distance is found by using heuristic function, and represents a minimum possible distance between that node and the goal. This ensures the algorithm to eliminate longer paths once an initial path is found.

In the standard terminology used when talking about A*, *g(n)* represents the exact cost of the path from the starting point to any vertex *n*, and *h(n)* represents the heuristic estimated cost from vertex n to the goal. Each time through the main loop, it examines the vertex n that has the lowest heuristics cost as shown in Eq. (1).

$$f(n) = g(n) + h(n) \tag{1}$$

2.2 Existing Navigation Applications

Currently, there are numbers of android application [13–15] and navigation applications in the android mobile market such as Google Maps, Waze Social GPS, HERE Maps and others. Google Maps is the most used smartphone application in the world [8]. Meanwhile, Waze is a GPS based geographical navigation application program for smart phones with GPS [10]. The Global Positioning System (GPS) is a satellite-based navigation system that was developed by the U.S. Department of Defense (DoD) in the early 1970s [11]. Another application is HERE. HERE is a mapping mobile application originally developed by Nokia for Android, IOS and Windows Phone platforms [12]. HERE Maps seems to be a rich features navigation application compare to Google Maps and Waze. However, HERE currently lack of ability in provides real-time traffic information to the users.

3 Methodology

In this section, the methodology used for develop the proposed project will be discussed in detail. Android based FSKTM 2D map is an application developed for the junior students in FSKTM. It is an indoor navigation application in which users can search the location inside the building. The proposed application is differ from the traditional navigation applications in which it is more likely an indoor navigation application. In order to fulfil all the requirements of Android based FSKTM 2D map, the development of the application requires a systematic approach methodology to produce a satisfying result.

The proposed application will cover this lack of functionality in the existing navigation applications by providing indoor navigation to users. In advanced, it provides the shortest path for users to reach their destination in shorter time. In order to provide indoor navigation, the proposed application is in pedestrian navigation mode. The hardware required to develop the proposed application is personal computer and android smartphone while the software required are Windows operating system, Android integrated development environment (Android IDE), Android mobile operating system (Android OS) and floor plan drawer.

The path finding algorithm chosen to implement in the proposed application is A* algorithm. A* algorithm is a combination of heuristic approaches like Best-First-Search (BFS) algorithm and formal approaches like Dijkstra's algorithm. This combination makes A* algorithm perform better than BFS algorithm and Dijkstra's algorithm. In this project, Unified Modelling Language (UML) was used to present the use case diagram to illustrate each function in the system.

At the beginning of system design, floor planes were sketched on drawing papers before mapping it into computer by using Microsoft Paint. This is to ensure all the dimensions, distances, and their ratios to the actual building measurement were correct. Then, the floor planes were mapped into computer by using the software – Microsoft Paint. Figure 1 shows the process of drawing floor plan.

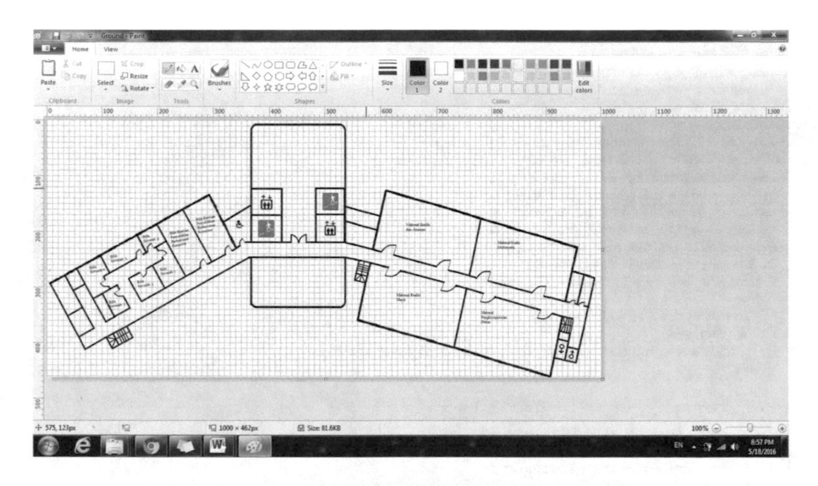

Fig. 1. Drawing Floor Plan using Microsoft paint

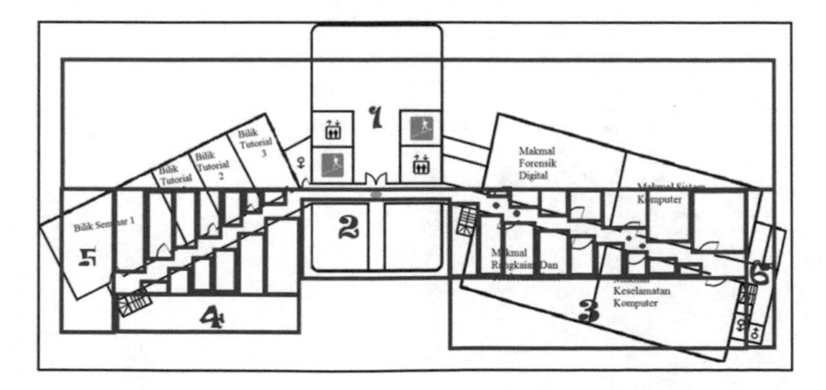

Fig. 2. Set blocked cell corresponding to wall in the Floor Plan

Next, the design of the user interface of the proposed application was created. Android Studio was used to develop the proposed application in the interface design and programmed the requirements into enrich functions (Fig. 2).

In order to calculate the shortest path, A* algorithm is chosen. The pseudo code of the A* algorithm implementation in the application is shown as below:

```
Input: two non-negative integers
Output: floor plan with shortest path
Start
// Declaration for A* Class
Declare class of Cell as two dimension array with heuristic cost, final cost
(according to Eq. (1)) and coordinates
Declare open list as priority queue
Declare closed list as Boolean type
add the start node to open
loop
   current = node in open with the lowest final cost
   remove current from open list
   add current to closed list
   if current is the target node //path has been found
      return
   for each neighbour of the current node
      if neighbour is not traversable or neighbour is in closed
         skip to the next neighbour
   if new path to neighbour is shorter or  neighbour is not in open
         set final cost of neighbour
         set parent of neighbour to current
         if neighbour is not in open
            add neighbour to open
// Check User Input
   Clear floor number, Clear classroom number
   Set coordinates according to classroom number
      Test coordinates against floor number inputted
         Set ei, ej to be the ending point's coordinates
         Initialize x, y to be the floor plan dimension
         Initialize si, sj to be the starting point's coordinates
         Declare grid as new Cell with dimension x, y
         Declare new closed list with dimension x, y
         Declare new open list with comparator to compare final cost
            loop
               set grid as new Cell with dimension x, y
               for each grid
               calculate heuristic cost using Manhattan Distance Equation shown in
               Eq. (2)
                                                                          (2)
                     where          (x1, y1) = coordinates of node n
                                    (x2, y2) = coordinates of ending point
            loop
               initialize final cost of starting point to zero
               set wall of floor plan to blocked Cell which is equal to null value in grid
               (the coordinates is corresponding to the pixel numbers and the blocked
               cell set as shown in Fig.2)
         Call A* class to calculate shortest path
//Display Output
   Test floor number
   Retrieve floor plan image according to floor number
         Declare bitmap of floor plan image
         Create a copy of mutable bitmap
         Declare new canvas for new bitmap
         Declare new paint to set color for new bitmap
           if coordinates of ending point in closed list trace back the path from ending
           point
              loop
                 set color of the path to red on the bitmap
           Set color of starting point to green
           Set color of ending point to blue
           Draw new bitmap into image view to show output
    Else show error message. End
```

Table 1. Functional test on android based FSKTM 2D map

Module tested	Expected result	Final result
Navigation Button	Bring users to Navigation page to choose the starting point and destination	Success
Dropdown list for Starting Point	Shows the list of Starting Point for users to choose	Success
Dropdown list for Destination	Shows the list of Destinations for users to choose	Success
Back Button in *Navigation* page	Bring users back to the Home Page	Success
View Floor Plan Button	Bring users to View Floor Plan page to choose the floor plan to view	Success
Dropdown list for Floor number	Shows the list of Floor number for users to choose	Success
Back Button in *View Floor Plan* page	Bring users back to the Home Page	Success

4 Result

In implementation phase, there are two vital system design parts which are user interface and code implementation. The user interface design will be written using Extensible Markup Language (XML). The language used in code implementation is Java Programming Language. The code implementation is critically important in order to ensure the application is function well. Table 1 shows the testing on each module of Android Based FSKTM 2D Map and its result.

The accuracy test is carried out in order to test the accuracy of the application. The shortest path calculated by the application is compared to the actual shortest distance and the accuracy is calculated using Eq. (3). Table 2 shows the accuracy test on the Android Based FSKTM 2D Map Using A* Algorithm and the result.

$$\text{Accuracy} = \frac{\sum Application\ ShortestPath}{\sum Test} \tag{3}$$

Table 2. Accuracy test on android based FSKTM 2D map using A* algorithm

Starting point	End point	Application result compared to actual shortest distance					
		1st Run	2nd Run	3rd Run	4th Run	5th Run	Consistency
Lift	*Makmal Grafik dan Animasi*	1	1	1	1	1	1
Lift	*Makmal Realiti Maya*	1	1	1	1	1	1
Lift	*Makmal Studio Multimedia*	1	1	1	1	1	1
Lift	*Makmal Pengkomputeran Pintar*	1	1	1	1	1	1
Lift	*Makmal Kejuruteraan Perisian*	1	1	1	1	1	1
Lift	*Makmal Pembangunan Perisian*	1	1	1	1	1	1
Lift	*Pusat Data*	1	1	1	1	1	1
Lift	*Makmal Pengatucaraan*	1	1	1	1	1	1
Lift	*Makmal Infosis*	1	1	1	1	1	1
Lift	*Ruang Membaca*	1	1	1	1	1	1
Lift	*Makmal Forensik Digital*	1	1	1	1	1	1
Lift	*Makmal Rangkaian dan Telekomunikasi*	1	1	1	1	1	1
Lift	*Makmal Sistem Komputer*	1	1	1	1	1	1
Lift	*Makmal Keselamatan Komputer*	1	1	1	1	1	1
Lift	*Bilik Tutorial 3*	1	1	1	1	1	1
Lift	*Bilik Tutorial 2*	1	1	1	1	1	1
Lift	*Bilik Tutorial 1*	1	1	1	1	1	1
Lift	*Bilik Seminar 1*	1	1	1	1	1	1
Lift	*Auditorium*	1	1	1	1	1	1
Lift	*Makmal Penyelidikan*	1	1	1	1	1	1
Lift	*Makmal Pengatucaraan Internet*	1	1	1	1	1	1
Lift	*Makmal Teknologi Web*	1	1	1	1	1	1
Lift	*Makmal Perisian dan Multimedia*			1	1	1	1
Lift	*Bilik Tutorial 5*	1	1	1	1	1	1
Lift	*Bilik Tutorial 4*	1	1	1	1	1	1
Makmal Grafik dan Animasi	*Ruang Membaca*	1	1	1	1	1	1
Makmal Realiti Maya	*Bilik Tutorial 3*	1	1	1	1	1	1
Makmal Studio Multimedia	*Bilik Tutorial 5*	1	1	1	1	1	1

(*continued*)

Table 2. (*continued*)

Starting point	End point	Application result compared to actual shortest distance					
		1st Run	2nd Run	3rd Run	4th Run	5th Run	Consistency
Makmal Kejuruteraan Perisian	*Makmal Infosis*	1	1	1	1	1	1
Makmal Pembangunan Perisian	*Bilik Tutorial 2*	1	1	1	1	1	1
Pusat Data	*Makmal Perisian dan Multimedia*	1	1	1	1	1	1
Ruang Membaca	*Makmal Forensik Digital*	1	1	1	1	1	1
Bilik Tutorial 1	*Makmal Sistem Komputer*	1	1	1	1	1	1
Bilik Tutorial 4	*Makmal Keselamatan Komputer*	1	1	1	1	1	1
Makmal Infosis	*Auditorium*	1	1	1	1	1	1
Bilik Seminar 1	*Makmal Penyelidikan*	1	1	1	1	1	1
Makmal Perisian danMultimedia	*Makmal Teknologi Web*	1	1	1	1	1	1
Ruang Membaca	*Bilik Seminar 1*	0	0	0	0	0	0
Ruang Membaca	*Bilik Tutorial 4*	0	0	0	0	0	0
Pusat Data	*Makmal Sistem Komputer*	0	0	0	0	0	0
Pusat Data	*Makmal Teknologi Web*	0	0	0	0	0	0
Makmal Realiti Maya	*Makmal Kejuruteraan Perisian*	0	0	0	0	0	0
Makmal KejuruteraanPerisian	*Makmal Forensik Digital*	0	0	0	0	0	0
Makmal Infosis	*Bilik Tutorial 3*	1	1	1	1	1	1
Makmal Infosis	*Makmal Perisian dan Multimedia*	1	1	1	1	1	1
Makmal Pembangunan Perisian	*Makmal Teknologi Web*	1	1	1	1	1	1
Makmal Grafik dan Animasi	*Makmal Keselamatan Komputer*	1	1	1	1	1	1
Makmal Grafik dan Animasi	*Makmal Penyelidikan*	1	1	1	1	1	1
Makmal Grafik dan Animasi	*Makmal Pengatucaraan Internet*	1	1	1	1	1	1
$\sum$*Test*							49
$\sum$*Application Shortest* (Total of "1")							43
Accuracy (%)							87.75

0 = Application Result is longer than actual shortest distance
1 = Application Result is equal to actual shortest distance

5 Conclusion

Android based FSKTM 2D Map Using A* Algorithm is a mobile application that can help user to search their classroom location in FSKTM. This application provides navigation by shows the path on the floor plan. In addition, the application will calculate the shortest path for user.

Although the objectives of this project have been achieved, there are limitations in the actual system. Firstly, the application is unable to locate the user; it can only know which floor the user currently locates through the input by the user. The application only provides navigation in two-dimension (2D). Besides, the floor plan shown in the application cannot be zoom in or zoom out.

During the application developments process, there are variety of constraints had faced. First of all, the constraints had faced is drawing the floor plan. Initially, it is planned to have a three-dimension (3D) floor plan in the application for better users' experience. However, it is difficult to draw the floor plan in 3D. Besides, the implementation of A* algorithm is hard to apply on 3D floor plan. In order to solve the problem, the two dimension (2D) floor plan was used in the application.

In future, Android based FSKTM 2D map using A* Algorithm can be improved by automatic locate user position. Therefore, users no need to input the starting point and can directly navigated by the application after chosen the destination. The application can provide 3D navigation for a better users' experience. Besides, the application can also be improved by make the floor plan zoom able.

Acknowledgement. The authors would like to thank Universiti Tun Hussein Onn Malaysia (UTHM) for providing the research facilities and research grant (Vot U540) to perform this research study. This research also supported by GATES IT Solution Sdn. Bhd under its publication scheme.

References

1. 2 Billion Consumers Worldwide to Get Smart(phones), Emarketer.com (2016). http://www.emarketer.com/Article/2-Billion-Consumers-Worldwide-Smartphones-by-2016/1011694. Accessed 27 Oct 2015
2. Delling, D., Sanders, P., Schultes, D., Wagner, D.: Engineering route planning algorithms. In: Lerner, J., Wagner, D., Zweig, Katharina, A. (eds.) Algorithmics of Large and Complex Networks. LNCS, vol. 5515, pp. 117–139. Springer, Heidelberg (2009). doi:10.1007/978-3-642-02094-0_7
3. Millington, I., Funge, J.: Artificial Intelligence for Games, p. 275. Morgan Kaufmann/Elsevier, Burlington (2009)
4. Patel, A.: Introduction to A*, Theory.stanford.edu (2016). http://theory.stanford.edu/~amitp/GameProgramming/AStarComparison.html. Accessed 27 Oct 2015
5. Sánchez-Crespo Dalmau, D.: Core Techniques and Algorithms in Game Programming, p. 228. New Riders Education, Indianapolis (2004)
6. Russell, J., Cohn, R.: Dijkstra's algorithm, [S.l.]: Book On Demand (2012)
7. Pearl, J.: Heuristics. Addison-Wesley Pub. Co., Reading (1984)

8. Richter, F.: Infographic: Google Maps is the Most-Used Smartphone App in the World, Statista Infographics (2013). http://www.statista.com/chart/1345/top-10-smartphone-apps-in-q2-2013/. Accessed 27 Oct 2015
9. McNamara, J.: GPS for Dummies, p. 47. Wiley Pub., Hoboken (2004)
10. Waze: Wikipedia (2016). https://en.wikipedia.org/wiki/. Accessed 27 Oct 2015
11. El-Rabbany, A.: Introduction to GPS. Artech House, Boston (2002)
12. Here Maps (app), Wikipedia (2015). https://en.wikipedia.org/wiki/HERE(app). Accessed 27 Oct 2015
13. Kamaludin, H., Kasim, S., Selamat, N., Hui, B.C.: M-learning application for basic computer architecture. In: 2012 International Conference on Innovation Management and Technology Research (ICIMTR), pp. 546–549. IEEE, May 2012
14. Kasim, S., Zakaria, F.A.: Daily calorie manager for basic daily use. In: 2013 Third International Conference on Innovative Computing Technology, pp. 437–442. IEEE, August 2013
15. Kasim, S., Wai, B.S.: Multilingual phrasebook for android (MPA). In: 2013 Third International Conference on Innovative Computing Technology (INTECH), pp. 443–448. IEEE, August 2013

Mining Significant Association Rules from on Information and System Quality of Indonesian E-Government Dataset

Deden Witarsyah Jacob[1(✉)], Mohd Farhan Md Fudzee[2], Mohamad Aizi Salamat[2], Rohmat Saedudin[1], Zailani Abdullah[3], and Tutut Herawan[4,5,6]

[1] Department of Industrial Engineering, Telkom University, Bandung, West Java, Indonesia
{dedenw, rdrohmat}@telkomuniversity.ac.id
[2] Faculty of Computer Science and Information Technology, Universiti Tun Hussein Onn Malaysia, Parit Raja, Malaysia
{farhan, aizi}@uthm.edu.my
[3] Faculty of Entrepreneurship and Business, Universiti Malaysia Kelantan, Pengkalan Chepa, 16100 Kota Bharu, Kelantan, Malaysia
zailania@umk.edu.my
[4] Faculty of Computer Science and Information Technology, University of Malaya, Pantai Valley, 50603 Kuala Lumpur, Malaysia
tutut@um.edu.my
[5] Faculty of Business and Information Technology, Universitas Teknologi Yogyakarta, Yogyakarta, Indonesia
[6] AMCS Research Center, Yogyakarta, Indonesia

Abstract. Electronic government (e-government) refers to how to apply the information and communication technologies (ICT) to improve the efficiency, effectiveness, transparency and responsibility of public governments. They are usually adopted in complex setting influenced not only the infrastructure factor but also the others factor such as end user satisfaction. Information and system quality are often seen as a key antecedent of user satisfaction. This paper presents an application of data mining technique based on association rules mining to capture interesting rules on information and system quality of Indonesian e-Government dataset. It is based on Least Frequent Items method by embedding FP-Growth algorithm. The rules are formed by implementing the relationship of an item or many items to an item (cardinality: many-to-one). The rule is categorized as interesting if it has a highest critical relative support, positive correlation and confidence. The results show that the total number of significant rules is 256 which is 14% from the overall rules captured i.e. 1811 on information quality data, meanwhile for system quality the total number of significant rules is 1790 which is 21% from the overall rules captured i.e. 18414.

Keywords: Data mining · Association rules · Information and system quality · E-Government

T. Herawan et al. (eds.), *Recent Advances on Soft Computing and Data Mining*,
Advances in Intelligent Systems and Computing 549, DOI 10.1007/978-3-319-51281-5_61

1 Introduction

Not only the businesses sector, but also governments have also been enhancing their services through Internet technology (Lee *et al.* 2011; Schaupp *et al.* 2010; Shareef *et al.* 2011) increasing effectiveness, efficiency (Cegarra-Navarro *et al.* 2012; Karkin and Janssen 2013; Rowley 2011), and convenience (Shareef *et al.* 2011), especially at the transaction level (Horst *et al.* 2007). Typically, going online enables governments to publish information that increases their citizens' access to information, to interact with its citizens, and to better transact with them (Cegarra-Navarro *et al.* 2012; Schaupp *et al.* 2010). Awareness and readiness of a country and its citizens are very important two aspects in implementing and developing e-government (Srivastava and Teo 2009; Teo *et al.* 2008; Venkatesh *et al.* 2012). Some scholar states the major issue such as design inadequacies (Aladwani 2013), insufficient system capabilities, low network security, shortage of information technology literacy (Venkatesh *et al.* 2012), lack of security policies and regulations, absence of implementation guidelines, associated social, management, and financial issues (Aladwani 2013; Lee *et al.* 2011), and ultimately citizen adoption issues (Schaupp *et al.* 2010) hound e-government development and use.

Furthermore, the main considerations in evaluating a success of e-government is website design and service delivery are usually (Aladwani 2013; Lee and Koubek 2010). In this context, quality factors based on the IS Success Model (DeLone and McLean 1992; 2003) and the nature of the e-government website (Aladwani 2013; Floropoulos *et al.* 2010; Teo *et al.* 2008) will affect its success. It means that e-government success depends on the purpose of users.

The IS Success Model outlines how to evaluate e-success in this manner (DeLone and McLean 1992; 2003). Critical to this model is to distinguish the frame of reference and the level of analysis (Petter and McLean 2009), since net benefits are the most important success measure (DeLone and McLean 2003; Petter *et al.* 2008; Petter and McLean 2009). Information and system quality relates to measures of the system's output. It is a function with the value at the output produced by a system as perceived by the user (DeLone and McLean, 2004; DeLone and McLean, 2003; Negash *et al.* 2003). Information and system quality is the "characteristics of the output offered by the IS, such as accuracy, timeliness, and completeness" (Petter *et al.* 2008).

Information and system quality is often seen as a key antecedent of user satisfaction (Urbach and Müller 2010). It subsumes measures focusing on the quality of the information that constitutes the desirable characteristics of IS output that the system produces and its usefulness. Issues, such as relatedness, clearness and goodness of the information delivered, are important features of websites (McKinney *et al.* 2002). DeLone and McLean (2003). Many researchers as mentioned above find themselves overwhelmed with data concerning end users adoption of e-government system, however they lack the information they need to make informed decisions based on information and system quality. Currently, there is an increasing interest in data mining and adoption of e-government systems.

This paper presents an application of data mining technique based on association rules mining to capture interesting rules on information and system quality of

Indonesian e-Government dataset. In summary, the contributions of this work are described as follow:

a. We present a method for mining association rules for e-government data which based on Least Frequent Items method by embedding FP-Growth algorithm.
b. We show that rules are formed by implementing the relationship of an item or many items to an item (cardinality: many-to-one).
c. We categorize captured rules as interesting if it has a highest critical relative support, positive correlation and confidence.
d. We show that all these rules have the value "1.00" for critical relative support and all of them have the positive correlation.

The remainder of this paper is organized as follows. Section 2 describes the related work and the basic concepts and terminology of association rules mining. Section 3 describes scenario on capturing rules. Section 4 describes the results and following by discussion. Finally, the conclusions of this work are reported in Sect. 5.

2 Association Rules Mining

Until this recent, association rules (ARs) mining has attracted much research. It is also considered as one of the widely discussed topics in data mining. Its objectives are tried to find the correlations, associations or casual structures among sets of items in the data repository. Since introduced by Agrawal *et al.* (1994), it has been employed in various applications such as in e-government, manufacturing, and etc. The process of association rule is defined as follows. Let a set $I = \{i_1, i_2, \cdots, i_{|A|}\}$, for $|A| > 0$ refers to the set of literals called set of items and the set $D = \{t_1, t_2, \cdots, t_{|U|}\}$, for $|U| > 0$ refers to the data set of transactions, where each transaction $t \in D$ is a list of distinct items $t = \{i_1, i_2, \cdots, i_{|M|}\}$, $1 \leq |M| \leq |A|$ and each transaction can be identified by a distinct identifier TID. A set $X \subseteq I$ is called an itemset. An itemset with k items is called a k-itemset. The support of an itemset $X \subseteq I$, denoted $\text{supp}(X)\text{supp}(X)$ is defined as a number of transactions contain X. Let $X, Y \subseteq I$ be itemset. An association rule between sets X and Y is an implication of the form $X \Rightarrow Y$, where $X \cap Y = \phi$. The sets X and Y are called antecedent and consequent, respectively. The support for an association rule $X \Rightarrow Y$, denoted $\text{supp}(X \Rightarrow Y)$, is defined as a number of transactions in D contain $X \cup Y$. The confidence for an association rule $X \Rightarrow Y$, denoted $\text{conf}(X \Rightarrow Y)$ is defined as a ratio of the numbers of transactions in D contain $X \cup Y$ to the number of transactions in D contain X. Thus $\text{conf}(X \Rightarrow Y) = \text{supp}(X \Rightarrow Y)/\text{supp}(X)$.

3 Scenario on Capturing Rules

3.1 Dataset Information and System Quality

The dataset was taken from a survey exploring six types of IS success model among citizens in Bandung, West Java, Indonesia. A total 200 participants have participated in this survey. The majority of respondents are female i.e. 105 persons and the rest are

Table 1. Reliability Statistics

Cronbach's alpha	Cronbach's alpha based on standardized items	Range of items
.802	.821	12

Table 2. Scale Statistics

Mean	Variance	Std. deviation	Range of items
43.890	29.515	5.4328	12

male i.e. 95 persons. In this survey, the Indonesian e-government dataset has been test for reliability with alpha score yielded 0.802 and accessing content validity. Tables 1 and 2 describe each attribute of information and system quality study includes the mean, standard deviation, variance, and range of item.

3.2 Information Quality

The survey's findings indicate that information quality among citizens is manifested through seven dimensions: (a) No errors, (b) Up to date, (c) Relative to the need, (d) Information for the needs, (e) Related to subject matter, (f) Contains necessary issue. For this, we have a dataset, which comprises the number of transactions (citizen), which is 200, and the number of items (attributes), which is 6 (refer to Table 3).

Table 3. Information quality dataset

Dataset	Size (kb)	#Transaction	#Item
Information Quality	15	200	6

3.3 System Quality

System quality among citizen in Bandung city focuses on 6 dimensions: (a) Downloadable, (b) Responsiveness, (c) Ease of use, (d) Accessibility, (e) Reliability, and (f) Easy go back forth between pages. For this, we have a dataset, which comprises the number of transactions (citizen), which is 200, and the number of items (attributes), which is 6 (refer to Table 4).

3.4 Design

Throughout this section, some basic definitions and related algorithms in mining association rules are discussed. In addition, Critical Relative Support (CRS) (Abdullah

Table 4. System quality dataset

Dataset	Size (kb)	#Transaction	#Item
System Quality	15	200	6

et al. 2010; Abdullah *et al.* 2011a; Abdullah *et al.* 2012a; Herawan and Abdullah 2012; Abdullah *et al.* 2012b; Herawan *et al.* 2012; Abdullah *et al.* 2011b; Abdullah *et al.* 2011c) is also elaborated.

3.4.1 Definition

Definition 1 (Least Items). *An itemset X is called least item if* $\alpha \leq \text{supp}(X) \leq \beta$, *where* α *and* β *is the lowest and highest support, respectively.*

The set of least item will be denoted as Least Items and

$$\text{Least Items} = \{X \subset I | \alpha \leq \text{supp}(X) \leq \beta\}$$

Definition 2 (Frequent Items). *An itemset X is called frequent item if* $\text{supp}(X) > \beta$, *where* β *is the highest support.*

The set of frequent item will be denoted as Frequent Items and

$$\text{Frequent Items} = \{X \subset I | \text{supp}(X) > \beta\}$$

Definition 3 (Merge Least and Frequent Items). *An itemset X is called least frequent items if* $\text{supp}(X) \geq \alpha$, *where* α *is the lowest support.*

The set of merging least and frequent item will be denoted as LeastFrequent Items and

$$\text{LeastFrequent Items} = \{X \subset I | \text{supp}(X) \geq \alpha\}$$

LeastFrequent Items will be sorted in descending order and it is denoted as

$$\text{LeastFrequent Items}^{\text{desc}} = \left\{ \begin{array}{l} X_i | \text{supp}(X_i) \geq \text{supp}(X_j), 1 \leq i, j \leq k, i \neq j, \\ k = |\text{LeastFrequent Items}|, x_i, x_j \subset \text{LeastFrequent Items} \end{array} \right\}$$

Definition 4 (Ordered Items Transaction). *An ordered items transaction is a transaction which the items are sorted in descending order of its support and denoted as* t_i^{desc}, *where*

$$t_i^{\text{desc}} = \text{LeastFrequentItems}^{\text{desc}} \cap t_i, \; 1 \leq i \leq n, \left|t_i^{least}\right| > 0, \left|t_i^{frequent}\right| > 0.$$

An ordered items transaction will be used in constructing the proposed model, so-called LP-Tree.

Definition 5 (Significant Least Data). *Significant least data is one which its occurrence less than the standard minimum support but appears together in high proportion with the certain data.*

Definition 6 (Critical Relative Support). *A Critical Relative Support (CRS) is a formulation of maximizing relative frequency between itemset and their Jaccard similarity coefficient.*

The value of Critical Relative Support denoted as CRS and

$$\begin{aligned}\mathrm{CRS}(I) &= \max\left(\left(\frac{\mathrm{supp(A)}}{\mathrm{supp(B)}}\right),\left(\frac{\mathrm{supp(B)}}{\mathrm{supp(A)}}\right)\right)\\ &\times\left(\frac{\mathrm{supp}(A \Rightarrow B)}{\mathrm{supp(A)}+\mathrm{supp(B)}-\mathrm{supp}(A \Rightarrow B)}\right)\end{aligned}$$

CRS value is fallen in the range of 0 and 1, and is determined by multiplying the highest value either supports of antecedent divide by consequence or in another way around with their Jaccard similarity coefficient. It indicates the level of CRS between combination of the both Least Items and Frequent Items either as antecedent or consequence, respectively.

3.4.2 Algorithm Development

Determine Interval Least Support. Let I is a non-empty set such that $I = \{i_1, i_2, \cdots, i_n\}$, and D is a database of transactions where each T is a set of items such that $T \subset I$. From Definition 3, an itemset is said to be least if its occurrence within a range of α and β, respectively. In other words, it fulfils the predefined Interval Least Support (*ILSupp*).

Construct LP-Tree. A Least Pattern Tree (LP-Tree) is a compact representation of the least itemset. It is constructed by scanning every line of transaction at a time and then mapping onto a new or existing path in the LP-Tree. However, only items that satisfy the ILSupp are employed in constructing the LP-Tree.

Mining LP-Tree. Once the LP-Tree is fully constructed, the mining process will begin by employing hybrid 'Divide and conquer' method decompose the tasks of mining desired pattern. LP-Tree utilizes the strength of hash-based method during constructing itemset in descending order. Intersection technique from Definition 4 is used to minimize computational cost and the same time to reduce the complexity.

4 Experiment Results

4.1 Information Quality

Information quality is defined as the extent to which the information provided best fits customer needs (Chang *et al.* 2005), usually based on measures on how accurate, relevant, timely, and complete the information is to address such needs (DeLone and McLean 2003). The information quality dataset comprises of 200 transactions with 6

Table 5. The mapping of the attributes

Description	Value	Attribute Id
Information on the government website is free from errors (has no errors),covers all information needed	1–5	1
Information on the government website is up-to-date (New)	1–5	2
Information presented on the government website is relative to my needs	1–5	3
Government online services provide me with the information according to my needs	1–5	4
Information presented on this website is related to the subject matter I am look for	1–5	5
Information on this website contains all necessary issues to complete tasks I need	1–5	6

items (attribute). Table 5 displays the mapped of description, value and new attribute id for dataset.

The experiment have been conducted on Intel® Core™ 2 Quad CPU at 2.33 GHz speed with 4 GB main memory, running on Microsoft Windows Vista. All algorithms have been developed using C# as a programming language.

Item in dataset is constructed based on the combination of description and its value (existence). For simplicity, let consider a description "Information on the government website is free from errors (has no errors), covers all information needed" with value "1" for strongly disagree. Here, an item "11" will be constructed by means of a combination of an attribute id (first characters) and its description (second character).

By embedding FP-Growth algorithm, 1811 ARs are produced. The ARs are formed by applying the relationship of an item or many items to an item (cardinality: many-to-one). Figure 1 depicts the correlation's classification of interesting ARs. For this dataset, the rule is categorized as interesting if it has a highest CRS, positive

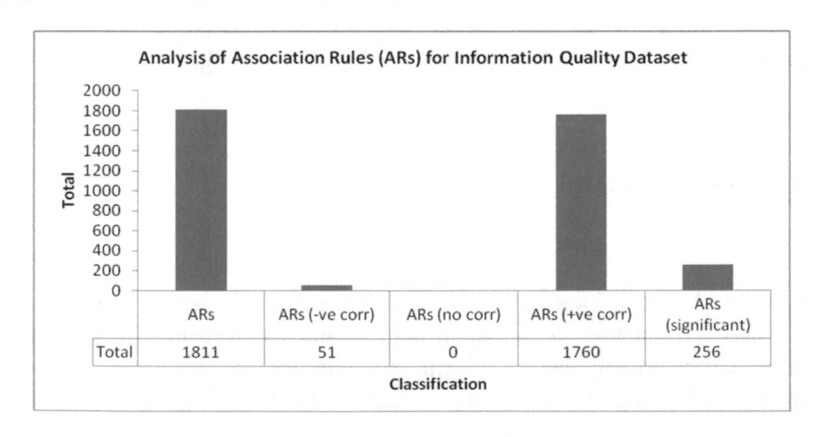

Fig. 1. Classification of ARs for Information Quality Dataset

correlation and confidence. The total number of significant ARs is 256 which is 14% from the overall captured of ARs (1811). All these ARs have the value "1.00" for CRS and 51 of them have a value less than "1.00" in term of the correlation. In addition, 186 of the significant ARs also have the positive correlation.

4.2 System Quality

System quality is defined as the degree in which the functionalities of the system can best address customer needs, with as much ease and as minimal problems encountered as possible (Chang *et al.* 2005; DeLone and McLean 2003). Examples of such functionalities include user interface consistency, ease of use, response rates, and program management; can best address customer needs (Chang *et al.* 2005; DeLone and McLean 2003; Wang 2008; Wang and Liao 2008).

We have a dataset of information quality and information system comprises of 200 transactions, respectively. Each of dataset contains 6 items (attribute). Tables 3 and 4 displays the mapped of description, values with a new attribute id for both dataset, respectively (Table 6).

Table 6. The mapping between system quality, value and a new attribute Id

Description	Value	Attribute Id
Government online services provide necessary information and forms to be downloaded	1–5	1
Government online services loads all texts and graphics quickly (Responsiveness)	1–5	2
It is easy to navigate within this website (Navigation) (Ease of Use)	1–5	3
Government website provides fast information access (Accessibility)	1–5	4
This website is available all the time (Reliability)	1–5	5
It is easy to go back and forth between pages (Easiness)	1–5	6

As similar to Information Quality Dataset, all items are constructed based on the combination of description and its value (existence). By embedding FP-Growth algorithm, 18414 ARs are produced. The ARs are formed by implementing the relationship of an item or many items to an item (cardinality: many-to-one). Figure 2 depicts the correlation's classification of interesting ARs. The rule is categorized as interesting if it has a highest CRS, positive correlation and confidence. The total number of significant ARs is 1790 which is 21% from the overall captured of ARs (18414). All these ARs have the value "1.00" for CRS and all of them have the positive correlation.

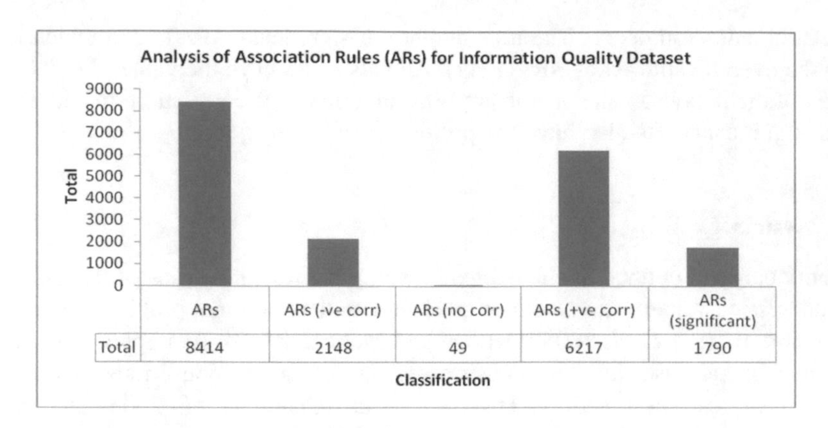

Fig. 2. Classification of ARs using correlation analysis

5 Conclusion

This paper has presented an application of data mining technique based on association rules mining to capture interesting rules on information and system quality of Indonesian e-Government dataset. We have shown that by embedding FP-Growth algorithm on Least Frequent Items method, we can capture interesting rules from the dataset. We have formed rules by implementing the relationship of an item or many items to an item (cardinality: many-to-one). We have categorized all rules as interesting if they have a highest critical relative support, positive correlation and confidence. We have shown that the results of the total number of significant rules is 256 which is 14% from the overall rules captured i.e. 1811 on information quality data. Meanwhile for system quality the total number of significant rules is 1790 which is 21% from the overall rules captured i.e. 18414. Finally, we have also shown that all these rules have the value “1.00” for critical relative support and all of them have the positive correlation.

Acknowledgments. The authors would like to thank the Telkom University Indonesia for supporting this work. The work of Tutut Herawan is supported by Excellent Research Grant Scheme no vote O7/UTY-R/SK/0/X/2013 from Universitas Teknologi Yogyakarta, Indonesia.

References

Abdullah, Z., Herawan, T., Deris, M.M.: Scalable model for mining critical least association rules. In: Zhu, R., Zhang, Y., Liu, B., Liu, C. (eds.) ICICA 2010. LNCS, vol. 6377, pp. 509–516. Springer, Berlin (2010). doi:10.1007/978-3-642-16167-4_65

Abdullah, Z., Herawan, T., Noraziah, A., Deris, M.M.: Mining significant association rules from educational data using critical relative support approach. Procedia Soc. Behav. Sci. **28**, 97–191 (2011a)

Abdullah, Z., Herawan, T., Noraziah, A., Deris, M.M.: Detecting critical least association rules in medical databasess. Int. J. Modern Phys. Conference Ser. **9**, v464–479 (2012a). World Scientific

Herawan, T., Abdullah, Z.: CNAR-M: a model for mining critical negative association rules. In: Li, Z., Li, X., Liu, Y., Cai, Z. (eds.) ISICA 2012. CCIS, vol. 316, pp. 170–179. Springer, Berlin (2012). doi:10.1007/978-3-642-34289-9_20

Herawan, T., Noraziah, A., Abdullah, Z., Deris, M.M., Deris, M.M.: IPMA: Indirect Patterns Mining Algorithm. In: Nguyen, N.T., Trawinski, B., Liu, Y., Cai, Z. (eds.) ISICA 2012. CCIS, vol. 457, pp. 170–179. Springer, Berlin (2012). doi:10.1007/978-3-642-34289-9_20

Herawan, T., Vitasari, P.. Abdullah, Z.: Mining interesting association rules of student suffering mathematics anxiety. In J.M. Zain et al. (Eds.): ICSECS 2011, Communication of Computer and Information Sciences, vol. 188, II, pp. 495–508. Springer-Verlag (2011)

Abdullah, Z., Herawan, T., Deris, M.M.: Efficient and scalable model for mining critical least association rules. J. Chin. Inst. Eng. **35**(4), 547–554 (2012b). A special issue from AST/UCMA/ISA/ACN 2010, Taylor and Francis

Abdullah, Z., Herawan, T., Noraziah, A., Deris, M.M.: Extracting highly positive association rules from students' enrollment data. Procedia Soc. Behav. Sci. **28**, 107–111 (2011b)

Abdullah, Z., Herawan, T., Noraziah, A., Deris, M.M.: Mining significant association rules from educational data using critical relative support approach. Procedia Soc. Behav. Sci. **28**, 97–101 (2011c)

Agrawal, R., Srikant, R.: Fast algorithms for mining association rules. In: Proceedings of the 20th International Conference on Very Large Data Bases, pp. 487–499 (1994)

Aladwani, A.M.: A cross-cultural comparison of Kuwaiti and British citizens' views of e-government interface quality. Gov. Inf. Q. **30**(1), 74–86 (2013)

Cegarra-Navarro, J.G., Pachón, J.R.C., Cegarra, J.L.M.: E-government and citizen's engagement with local affairs through e-websites: the case of Spanish municipalities. Int. J. Inf. Manage. **32**(5), 469–478 (2012)

Chang, I.-C., Li, Y.-C., Hung, W.-F., Hwang, H.-G.: An empirical study on the impact of quality antecedents on tax payers' acceptance of internet tax-filing systems. Gov. Inf. Q. **22**(3), 389–410 (2005)

DeLone, W.H., McLean, E.R.: Information system success: the quest for the dependent variable. Inf. Syst. Res. **3**(1), 60–94 (1992)

DeLone, W.H., McLean, E.R.: The DeLone and McLean model of information systems success: a ten-year update. J. Manage. Inf. Syst. **19**(4), 9–30 (2003)

Floropoulos, J., Spathis, C., Halvatzis, D., Tsipouridou, M.: Measuring the success of the Greek taxation information system. Int. J. Inf. Manage. **30**(1), 47–56 (2010)

Horst, M., Kuttschreuter, M., Gutteling, J.M.: Perceived usefulness, personal experiences, risk perception and trust as determinants of adoption of e-government services in The Netherlands. Comput. Hum. Behav. **23**(4), 1838–1852 (2007)

Karkin, N., Janssen, M.: Evaluating websites from a public value perspective: a review of Turkish local government websites. Int. J. Inf. Manage. **34**(3), 351–363 (2013)

Lee, J., Kim, H.J., Ahn, M.J.: The willingness of e-government service adoption by business users: The role of offline service quality and trust in technology. Gov. Inf. Q. **28**(2), 222–230 (2011)

Petter, S., DeLone, W., McLean, E.: Measuring information systems success: models, dimensions, measures and interrelationship. Eur. J. Inf. Syst. **17**(3), 236–263 (2008)

Petter, S., McLean, E.R.: A meta-analytic assessment of the DeLone and McLean IS success model: an examination of IS success at the individual level. Inf. Manage. **46**(3), 159–166 (2009)

Rowley, J.: E-government stakeholders—Who are they and what do they want? Int. J. Inf. Manage. **31**(1), 53–62 (2011)

Schaupp, L.C., Carter, L., McBride, M.E.: E-file adoption: a study of U.S. taxpayers' intentions. Comput. Hum. Behav. **26**(4), 636–644 (2010)

Shareef, M.A., Kumar, V., Kumar, U., Dwivedi, Y.K.: E-government adoption model (GAM): Differing service maturity levels. Gov. Inf. Q. **28**(1), 17–35 (2011)

Srivastava, S.C., Teo, T.S.H.: Citizen trust development for e-government adoption and usage: insights from young adults in Singapore. Commun. Assoc. Inf. Syst. **25**(31), 359–378 (2009)

Teo, T.S.H., Srivastava, S.C., Jiang, L.: Trust and electronic government success: an empirical study. J. Manage. Inf. Syst. **25**(3), 99–131 (2008)

Venkatesh, V., Chan, F.K.Y., Thong, J.Y.L.: Designing e-government services: key service attributes and citizens' preference structures. J. Oper. Manage. **30**(1–2), 116–133 (2012)

Wang, Y.-S.: Assessing e-commerce systems success: a respecification and validation of the DeLone and McLean model IS success. Inf. Syst. J. **18**(5), 529–557 (2008)

Wang, W., Benbasat, I.: Attributions of trust in decision support technologies: a study of recommendation agents for e-commerce. J. Manage. Inf. Syst. **24**(4), 249–273 (2008)

Wang, Y.-S., Liao, Y.-W.: Assessing e-government systems success: a validation of the DeLone and McLean model of information system success. Govern. Inf. Q. **25**(4), 717–733 (2008)

A Feature Selection Algorithm for Anomaly Detection in Grid Environment Using *k-fold* Cross Validation Technique

Dahliyusmanto[1(✉)], Tutut Herawan[2], Syefrida Yulina[3], and Abdul Hanan Abdullah[4]

[1] Department of Computer Science, Faculty of Engineering, Universitas Riau, 28293 Pekanbaru, Indonesia
dahliyusmanto@lecturer.unri.ac.id

[2] Department of Information Systems, University of Malaya, 50603 Kuala Lumpur, Malaysia
tutut@um.edu.my

[3] Department of Computer Engineering, Polytechnic Caltex of Riau, 28265 Pekanbaru, Indonesia
syefrida@pcr.ac.id

[4] Department of Computer Science, Faculty of Computing, Universiti Teknologi Malaysia, 81310 Skudai, Johor, Malaysia
hanan@utm.my

Abstract. An Intrusion Detection System (IDS) seeks to identify unauthorized access to computer systems' resources and data. The spreading of a data set size, in number of records as well as of attributes, as trigger the development of a number of big data platforms as well as parallel data analysis algorithms. This paper proposed a state-of-the-art technique to reduce the number of input features in dataset by using the Sequential Forward Selection (SFS) with *k*-Fold Cross Validation Model. Before reaching the feature reduction stage, the pre-processing analysis for detecting unusual observations that do not seem to belong to the pattern of variability produced by the other observations. The pre-processing analysis consists of outlier's detection and Transformation. Outliers are best detected visually whenever this is possible. This paper explains the steps for detecting outliers' data and describes the transformation method that transforms them to normality. The transformation obtained by maximizing Lamda functions usually improves the approximation to normality.

Keywords: IDS · Outliers · Transformation · *k*-fold · Cross validation

1 Introduction

The methodology of intrusion detection can be divided into two-category: anomaly intrusion detection and misuse intrusion detection. Anomaly intrusion detection refers to detecting intrusion based on the anomalous behavior of the attackers. Therefore, the distinction by categorizing the good or acceptable behavior is very important. In the anomaly detection method, a statistical approach and neural net approach are usually

T. Herawan et al. (eds.), *Recent Advances on Soft Computing and Data Mining*,
Advances in Intelligent Systems and Computing 549, DOI 10.1007/978-3-319-51281-5_62

taken to detect intrusion attempts. There are many ways in which dataset could be used to characterize the normal behaviour of programs, each of which involves building or training a model using traces of normal processes. The enumerating sequence method [1, 2] depends only on enumerating sequences that occur empirically in traces of normal behaviour and subsequently monitoring for unknown patterns. Two different methods of enumeration were tried, each of which defines a different model, or generalization, of the data. There was no statistical analysis of these patterns in the earlier work.

Data mining approaches are designed to determine what features are most important out of a large collection of data. In the current problem, the idea is to discover a more compact definition of normal than that obtained by simply listing all patterns occurring in normal. Also, by identifying just the main features of such patterns, the method should be able to generalize to include normal patterns that were missed in the training data. Lee and others used this approach to study a sample of system call data [3, 4]. They used a program called "RIPPER" to characterize sequences occurring in normal data by a smaller set of rules that capture the common elements in those sequences. During monitoring, sequences violating those rules are treated as anomalies. Because the results published in [3] on synthetic data were promising, we chose this method for further testing.

Feature selection is the process of selecting a subset of input variables or attributes, using only the subset as features fed into classification methods. Feature selection serves two main purposes: First, it makes training and applying a classifier more efficient by reducing the high dimensionality of feature sets. Second, feature selection improves classification accuracy by eliminating irrelevant or noisy features. Moreover, as high dimensional and large data sets become increasingly common in computational fields, feature selection plays a key role for machine learning or data mining algorithms to make them become efficient and scalable.

The paper is organized as follows: the dataset extraction; the steps for detecting outlier data and Box-Cox power transformation standardizes values, generalized square distances and transformation to near normality. Finally, features selection using SFS with *k-fold* cross validation are performed.

2 Related Work

Feature selection is the process of selecting a subset of input variables or attributes, using only the subset as features fed into classification methods. Feature selection serves two main purposes: First, it makes training and applying a classifier more efficient by reducing the high dimensionality of feature sets. Second, feature selection improves classification accuracy by eliminating irrelevant or noisy features. Moreover, as high dimensional and large data sets become increasingly common in computational fields, feature selection plays a key role for machine learning or data mining algorithms to make them become efficient and scalable.

The feature selection has been studied by machine learning communities for many years. Selecting significant features are important, especially when dealing with large feature space [5]. Feature selection leads to a simplification of the problem, faster and more accurate in detection). The definitions for irrelevance and for two degrees relevant (weak and strong) [6]. They also stated that the features selected should depend not only on the features and the target concept, but also on the induction algorithm. A method for hybrid feature selection based on information theories had been examined by [7]. They presented a theoretically hybrid model for optimal feature selection, based on using cross-entropy to minimize the amount of predictive information lost during feature elimination. The study of feature selection metrics proposed by Forman in 2003 while variables for feature selection introduced by [8].

3 Methodology

It is necessary to do the data preprocessing to extract the feature attributes from the dataset, and then, the outliers data and data normalization will be processed to analyse whole features to a unit range. In this study, data preprocessing is focused on the dataset.

The extraction of feature attributes of a data set is the foundation of machine learning algorithms in anomaly detection. Moreover, detection models or algorithms must be combined with the rational feature vector extraction to improve the attack recognition capability. The dataset features should prefer to differentiate normal and intrusive activity profiles. The aim of feature extraction is to achieve the maximum difference degree between normal and intrusive activities in Grid.

This section discusses the data preprocessing that will be divided into three steps; the dataset extraction process, outliers detection, and transformation data.

3.1 Dataset Extraction

The experiment's data for Grid computing intrusion detection model were dataset obtained from Computer Immune Systems (CIS) Lab and Grid Lab test. These datasets were appropriate for host-based intrusion detection aspects in the Grid. CIS has collected several data sets of system calls executed through active process, which include different kinds of programs (e.g. A program that runs as daemons and those that do not), programs that vary widely in their size and complexity, different kinds of intrusions (buffer overflow, symbolic link attacks, and Trojan programs). Some of the normal data are "*Synthetic*" and some are "*Live*". *Synthetic* traces are collected in production environments by running a prepared script; the program options are chosen solely for the purpose of exercising the program, and not to meet any real user requests. *Live* normal data are traces of programs collected during normal usage of a production computer system, while, the data from Grid Lab test are collected from the programs running on Grid services (e.g. *globus-gsiftp*, *globus-url-copy*, *globus-ws-submit*, etc.). Nonetheless, both of the datasets will be integrated. Afterward, the datasets have to be extracted to obtain the characteristics of systems calls. The extraction procedures consist of the number of system calls, the number of processes, and the characteristic system calls itself.

Traces of each program's data sets are recorded in *.int and gzipped files, because of that extraction process are needed. This research must consider how to extract a file from data sets easily. This research considers the extraction method using an internal viewer program on windows operating system to obtain the number of system call, number of processes, and to identify the characteristic of system call as shown in Fig. 1.

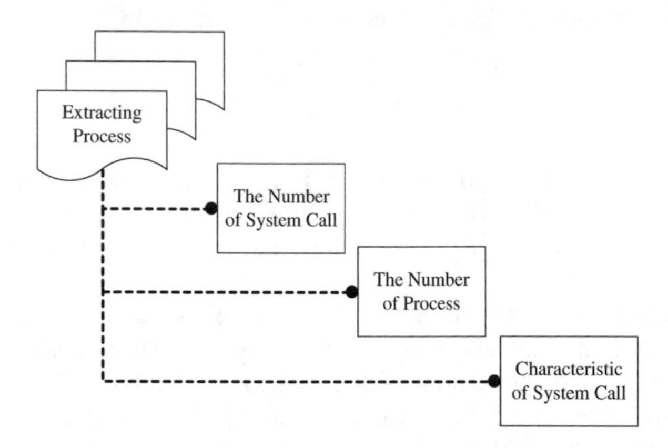

Fig. 1. Extracting process data set of system calls

The Number of System Call. Certain of number system call are extracted in a certain number of processes that are in a parent-child relation in the system log. The algorithm used to count system call is shown below.

```
Procedure Counting the Number of System Call;
   var l, i, a, n, c, j;
          l  =  170;  %maximum  number  of  system  call  in  each
          activity
   read ( data );
   for i = 1 to l
          a ( i ) = find ( data ( : , 1 ) = 1 );
          [ n , c ] = size a ( i );
          j ( i ) = n;
   end;
```

The Number of Processes. All system calls are extracted within a number of processes that are in a parent-and-child relationship. The numbers of process can be counted manually because they are a little resulted in extracting process.

Characteristic of System Call. System calls are extracted in one or more processes that are in a parent-and-child relationship. The extraction range is based on certain characteristic system calls (e.g., fork, exit, open, etc.) to an activity.

3.2 Outliers Detection

An outlier is a data point that is located far from the rest of the data. Given a mean and standard deviation, a statistical distribution expects data points to fall within a specific range. Many researchers have used this approach in statistical data analysis. Generally, the contemporary algorithms are hybrid algorithms based on mathematical considerations and random restarts. They exploit the concept of discriminant analysis, which for an outlier is expected to appear big and significant. There are four steps and algorithms for detecting outliers:

Step 1. To make a dot plot for each variable.

```
Procedure Making a Dot Plot;
   read ( data ), ( syscall );
   Boxplot ( data, 0, '+', 0 );
   Set ( gca, 'YTicklabel', syscall );
```

Step 2. The algorithm to make a scatter plot.

```
Procedure Making a Scatter Plot;
    var i, d, xchi, xbar, cov, invcov;
    read ( data )
       xbar    = mean ( data );
       cov     = cov ( data );
       invcov = inv ( cov );
       d = ( x - xbar ) * invcov * ( x - xbar )';
       Sort ( d );
       xchi = [ ];
       For i = 1 to 49 % 49 are sum of activities
         xchi = [ xchi  chi2inv ( ( i - 0.5 ) / 49, 11 );
       end;
       scatter ( xchi, d )
```

Step 3. To calculate the standardized values z on each column and examine these standardized values for large or small values.

```
Procedure Calculating Standardized Value;
   var  n, c, xbar, std, datcen, datstd;
   read ( data );
    xbar = mean ( data );
    std   =  std ( data );
    [ n , c ] = size ( data );
    datcen = data - repmat ( xbar, n, 1);
    datstd  = datcen ./ repmat ( std, n, 1 );
```

Step 4. To calculate the generalized square distances $(x - \bar{x}).S^{-1}(x - \bar{x})$ and examine these distances for unusually large values. In a Scatter plot, these would be the points farthest from the origin.

3.3 Transforming Data

The transformations are nothing more than *re-expressions* of the data in different units. Appropriate transformations are suggested by (1) theoretical considerations or (2) the data themselves (or both). It has been shown theoretically that the data which are counted can often be made more normal by taking their square roots. Similarly, the *largest transformation* applied to proportions and Fisher's z-transformation applied to correlation coefficients yield quantities that are approximately normally distributed.

To select a power transformer, an investigator looks at the marginal dot diagram or histogram, and decides whether large values have to be "pulled" or "pushed out" to improve the symmetry about the mean. A Q-Q plot or other will check to see whether the tentative normal assumption is satisfactory, and should always examine the final choice. A convenient analytical method is available for choosing a power transformation. Box and Cox (1964) had considered the slightly modified family of power transformations.

$$x^{(\lambda)} = \begin{cases} x^{\lambda} - 1/\lambda & \lambda \neq 0 \\ \ln x & \lambda = 0 \end{cases} \tag{1}$$

which is continuous in λ for $x > 0$. Given the observations x_1, x_2, x_n, the Box-Cox solution for the choice of an appropriate power, λ is the solution that maximizes the expression. With multivariate observations, a power transformation must be selected for each of the variables. Let $\lambda_1, \lambda_{2,\ldots,}\lambda_p$ be the power transformations for the p measured characteristics. Each λ_k can be selected to maximize:

$$\ell_k(\lambda) = -n/2 \ln\left[1/n \sum_{j=1}^{n} (x_{jk}^{(\lambda_k)} - x_k^{(\bar{\lambda}_k)})^2\right] + (\lambda_k - 1)\sum_{j=1}^{n} \ln x_{jk} \tag{2}$$

is the arithmetic average of the transformed observations. The jth transformed multivariate observation is

$$x_j^{(\hat{\lambda})} = \begin{bmatrix} x_{j1}^{(\hat{\lambda}_1)} - 1/\hat{\lambda}_1 \\ x_{j2}^{(\hat{\lambda}_2)} - 1/\hat{\lambda}_2 \\ \vdots \\ x_{jp}^{(\hat{\lambda}_p)} - 1/\hat{\lambda}_p \end{bmatrix} \tag{3}$$

where $S(\lambda)$ is the sample covariance matrix computed from

$$x_j^{(\hat{\lambda})} = \begin{bmatrix} x_{j1}^{(\lambda_1)} - 1/\lambda_1 \\ x_{j2}^{(\lambda_2)} - 1/\lambda_2 \\ \vdots \\ x_{jp}^{(\lambda_p)} - 1/\lambda_p \end{bmatrix} \quad j = 1, 2, \tag{4}$$

Maximizing Eq. (3) not only is substantially more difficult than maximizing the individual expressions, but is also unlikely to yield remarkably better results. The selection method based on Eq. (3) is equivalent to maximizing a multivariate likelihood over μ, Σ, and λ, whereas the method based on Equation A.4 corresponds to maximizing the kth univariate likelihood over μ_k, σ_{kk}, and λ_k. The latter likelihood is generated by pretending there are some λ_k for observations $\left(x_{jk}^{\lambda_1} - 1\right)/\lambda_k, j = 1, 2, \ldots, n$ have a normal distribution.

4 Results and Discussions

4.1 Extracting Process of Dataset

There were several different programs used for analysis (normal and intrusion programs). There are three steps to perform an extracting process to obtain characteristics of system calls: the number of system calls triggered by programs running on active processes, the number of processes identified, and the characteristic of system calls to an activity. The research performed extraction process on normal activities and intrusive activities are divided into 38 normal activities and 18 intrusive activities. The statistical summaries of extracting results are shown in Table 1.

Table 1. Numerical summaries data sets

Variable	N	Sum	Mean 1.0e+003	Std deviation 1.0e+004
chdir	49	2668	0.054	0.023
geteuid	49	1604	0.033	0.021
open	49	20033	0.409	0.116
read	49	135335	2.762	1.031
setgid	49	135	0.003	0.000
setuid	49	299	0.006	0.001
exit	49	4919	0.100	0.069
getpgrp	49	2664	0.054	0.037
unlink	49	107	0.002	0.000

From the Table 1, the research selected a few system calls as variables for analysis during extraction process. These variables were selected based on some of the system calls that correlate to intrusions as introduced by IPA (International of Promotion Agent) of Japan and SNARE of Australia's group research. This research combined these system calls and obtained nine of them that will be in the analysis (i.e., *chdir*, *geteuid*, *open*, *read*, *setgid*, *setuid*, *exit*, *getpgrp*, and *unlink*).

4.2 Detecting Outliers Data

This research, performed the steps to detect outliers which were based from the designed algorithms, they are:

To Make a Dot Plot for Each Variable. One of the difficulties inherent in multivariate statistics is the problem of visualizing multidimensionality. To rectify this problem, this research used Matlab plot command to display a graph of the relationship between system call variables as shown in Fig. 2.

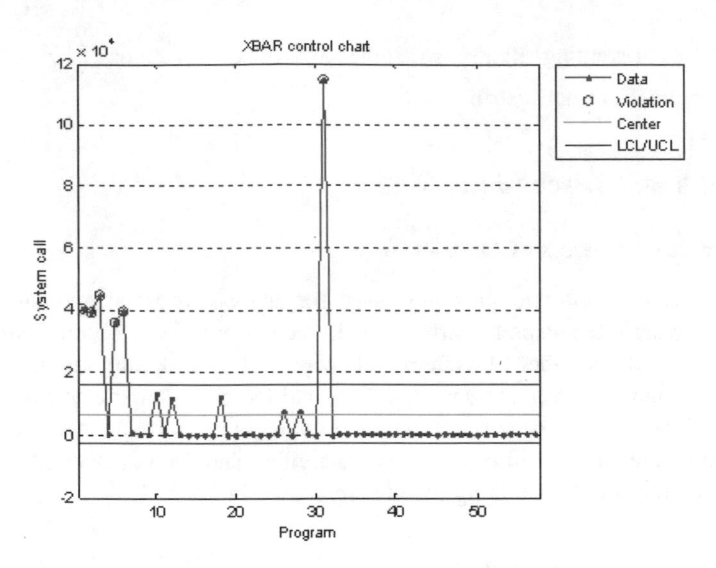

Fig. 2. System calls relationship

As can be seen in Fig. 2, there were substantially more varying variables in *named-unm1*, *named-unm2*, *named-unm3*, *named-unm4*, *ftp-nonself1*, and *sendmail-daemon* programs than in the others. Some of features were far from others. There is substantially more variability in the system calls of the *read* (135335) and *open* (20033) than in the other system call. This was caused by the presence of some redundant features of system calls in the original datasets. Therefore, to rectify this problem, the evaluations had to exclude redundant features and find out the most independent and the best.

To Make a Scatter Plot for Each Pair of Variables. To construct the scatter plot, the distance of each activity must be arranged from smallest to largest. Figure 3 shows the scatter plot for each pair of variables.

The points as in Fig. 3 were found not lying long the line with slope1. The smallest distance was *xloc15* (*d2* = 3.4398).

To Calculate the Standardized Values. The standardized values are based on the sample mean and variance, calculated from 49 activities. They are calculated for each column of system call variables and examine these standard values for large or small value. In this research, "large" must be interpreted relative to the sample size and the number of variables. There is *no* X *p* standardized values. In this paper $n = 49$ and $p = 9$, there are 441 values. The value 6.8571 on standardized data might be considered large for moderate sample size.

To Calculate the Generalized Square Distance. The generalized distances are calculated using Eq. (4) and they are examined for unusually large values. In a scatter

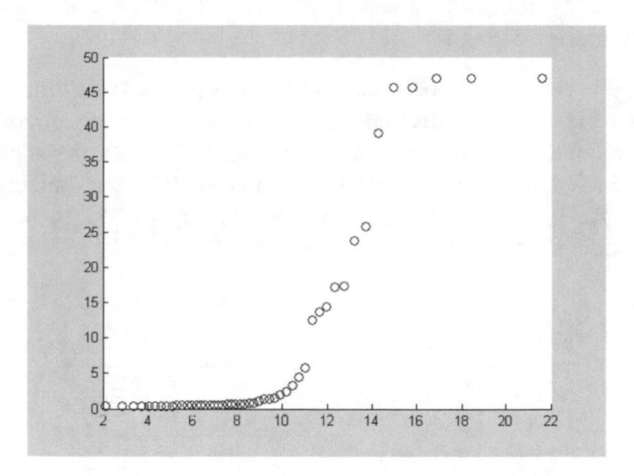

Fig. 3. Scatter plots for the system call data

plot, these would be the points farthest from the origin. The last column reveals that the activities *login1*, *ftp*, *xlock17*, *inetd-i*, *login1-i*, *ps2-i*, *xlock1-i*, and *ftp-i*, are a multivariate outlier, since $X_9^2(.005) = 23.59$; yet all of individual measurements are well within their respective univariate scatter. Activities *login2*, *ps1*, *xlock13*, *xlock16*, and *login2-i*, also have large square distance values as shown in Table 2.

The thirteen activities (*login1*, *ftp*, *xlock17*, *inetd-i*, *login1-i*, *ps2-i*, *xlock1-i*, and *ftp-i*, *login2*, *ps1*, *xlock13*, *xlock16*, and *login2-i*) with large squared distances (Table 2) stand out from the rest of the pattern in Fig. 3. Once these thirteen points are removed, the remaining pattern conforms to the expected straight-line relation.

Table 2. The activities with large squared distance

No.	Activities	Squared distance
1	login1	39.086
2	ftp	46.944
3	xlock17	23.763
4	inetd-i	46.940
5	login1-i	45.639
6	ps2	25.748
7	xlock1-i	47.020
8	ftp-i	45.655
9	login2	14.468
10	ps1	13.675
11	xlock13	17.257
12	xlock16	17.178
13	login2-i	12.560

4.3 Transformation Process

Since all the observations are positive, it performs a power transformation of the data which, it hope will produce results that are more nearly normal. Restricting the research attention to the family of Box-Cox transformations in Eq. (3), the analysis must find that value of $\lambda_1, \lambda_2, \ldots, \lambda_p$ (p = measured characteristics) maximizing the function $l_1(\lambda), l_2(\lambda), .., \ell_k(\lambda)$ (k = system call variables), in Eq. (4). The pairs of $(\lambda, l(\lambda))$ are listed in the following Table 3 for several values of λ.

Table 3. The value of λ maximizing the function $\ell_k(\lambda)$

λ	$\ell_1(\lambda)$	$\ell_2(\lambda)$	$\ell_3(\lambda)$	$\ell_4(\lambda)$
−1	−72.465	**−21.070**	−308.623	−467.927
−0.6	**−69.150**	−24.197	−251.904	−385.222
−0.2	−76.115	−35.112	**−225.136**	−329.613
−0.1	−80.807	−40.667	−225.885	**−323.068**
3.003	−47.884	**11.659**	**38.402**	**30.542**
−1.976	**−44.842**	4.632	27.333	21.881

The curve of $\ell_k(\lambda)$ versus λ that allows the more exact determination $\lambda_1 = -0.6, \lambda_2 = -1, \lambda_3 = -0.2, \lambda_4 = -0.1, \lambda_5 = -1, \lambda_6 = -0.6, \lambda_{text7} = -1, \lambda_8 = -1$, and $\lambda_9 = -1$. It is evident from both the Table 3 and the plot values of λ maximize $\ell_k(\lambda)$. A scatter plot was constructed from the transformed quantities. This plot is shown in Fig. 4.

As shown in Fig. 4(b), the quantile pairs fall very close to a straight line, this research would conclude from this evidence that $x_j^{-0.6}$, x_j^{-1}, $x_j^{-0.2}$, $x_j^{0.1}$, x_j^{-1}, $x_j^{-0.6}$, x_j^{-1}, x_j^{-1}, and x_j^{-1} are approximately normal.

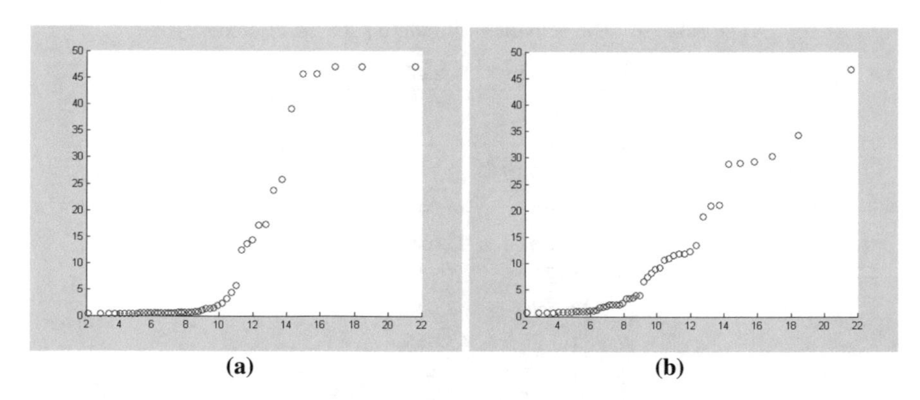

Fig. 4. Scatter plots of (a) the original and (b) the transformed system call

4.4 Feature Selection and Dimension Reduction

In this experiment, we performed a cross validation and focused on Sequential Forward Selection method, and employed 27 attributes and integrated datasets (dataset from CIS and from Grid lab test).

We applied 10-fold cross validation in all the experiments. The Matlab syntax was used to construct a feature selection method, as shown in procedure below.

```
        Procedure Feature Selection:
          load dataset
              data = dataset(:,2:end);
         group_data = dataset(:,1);
         T = num2 str(group_data);
          cell = cellstr(T);
          y = cell;
          X = data;
     c = cvpartition(y,'k',10);
     opts = statset('display','iter');
     fun = @(XT,yT,Xt,yt)(sum(~strcmp(yt,classify
     (Xt,XT,yT,'quadratic'))));
      [fs,history]=sequentialfs(fun,X,y,'cv',c,'options',opts)
End;
```

The sample of features selection and the experimental results by applying the 10 *k*-fold cross validation are listed in Table 4.

```
Star forward sequential feature selection:
         Initial columns included: none
         Column that cannot be included: none
         Step 1, added column    7, criterion value
0.3684
         Step 1,  added  column 14,  criterion  value
0.3333
         Step 1,  added  column 17,  criterion  value
0.2456
         Final columns included: 7 14 17
```

The iterative display showed a decrease in the criterion value as each new feature was added to the model. The final result was a reduced model with only ten (as listed in Table 4) of the original twenty-seven features. These features were indicated in the logical vector returned by sequential feature selection.

Table 4. Dataset feature testing by applying 10 *k*-fold cross validation

Test 1	Test 2	Test 3	Test 4	Test 5	Test 6	Test 7	Test 8	Test 9	Test 10
2	2	2	2	2	2	2	8	2	2
7	7	7	7	10	3	4	7	3	10
14	10	10	10	12	5	3	10	10	16
17	14	12	12	14	15	17	12	14	17
–	18	15	18	18	17	18	15	15	18
–	–	16	–	20	–	20	–	16	20

5 Conclusion

This paper has explored a commonly used technique for dimensionality reduction: removing data columns, with too many missing values, removing low variance columns, and reducing highly correlated columns. The first stage preceded by outlier's detection and transformation processes. This paper has showed how this technique can be implemented within SFS with *k*-fold cross validation model. This paper also has demonstrated how to select dataset. With datasets becoming bigger and bigger in terms of dimensionality and a high number of features meaning longer computation times, yet no guarantee of better performances, the dimensionality reduction is currently increasing in popularity as a means to speed up execution times, simplify problems, and optimize algorithm performances.

Acknowledgments. The authors would like to thank the Riau University Indonesia for supporting this work. The work of Tutut Herawan is supported by Excellent Research Grant Scheme no vote O7/UTY-R/SK/0/X/2013 from Universitas Teknologi Yogyakarta, Indonesia.

References

1. Forrest, S., Hafmeyr, S.A., Somayaji, A., LongStaff, T.A.: A sense of self for UNIX processes. IEEE Trans. Softw. Eng. **13**(2), 222–232 (1996). Proceedings of the IEEE Symposium on Computer Security and Privacy
2. Hofmeyr, S.A., Forrest, S., Somayaji, A.: Intrusion detection using sequences of system calls. J. Comput. Secur. **6**, 151–180 (1998)
3. Lee, W., Stolfo, S.J., Chan, P.K.: Learning patterns from UNIX process execution traces for intrusion detection. In: AAAI Workshop on AI Approaches to Fraud Detection and Risk Management, pp 50–56. AAAI Press (1997)
4. Lee, W., Stolfo, S.J.: Data mining approaches for intrusion detection. In: Proceedings of the 7th USENIX Security Symposium (1998)
5. Chebrolu, S., Abraham, A., Thomas, J.P.: Feature deduction and ensemble design of intrusion detection system. J. Comput. Secur. **24**, 295–307 (2005). Elsevier Publisher
6. Kayacik, H.G., Heywood, A.N.Z., Heywood, M.I.: Selecting features for intrusion detection: a feature relevance analysis on KDD 99. In: Proceedings of the 3rd Annual Conference on Privacy, Security and Trust (PST), New Brunswick, Canada (2005)
7. Das, S.: Filter, wrappers and boosting-based hybrid for feature selection. In: Proceedings of the 18th International Conference on Machine Learning (ICML 2001), San Fransisco, CA, USA, pp. pp. 74–81. Morgan Kaufmann Publisher Inc. (2001)
8. Guyon, I., Elisseeff, A.: An introduction of variable feature selection. J. Mach. Learn. Res. **3**, 1157–1182 (2003)

Soft Set Approach for Clustering Graduated Dataset

Rd Rohmat Saedudin[1(✉)], Shahreen Binti Kasim[2], Hairulnizam Mahdin[2], and Muhammad Azani Hasibuan[1]

[1] Department of Information System, Telkom University, Bandung, West Java, Indonesia
{rdrohmat,muhammadazani}@telkomuniversity.ac.id
[2] Faculty of Computer Science and Information Technology, UTHM, Parit Raja, Malaysia
{shahreen,hairuln}@uthm.edu.my

Abstract. Every university has objectives to make sure their students graduate on time. This objective can be achieved by using early warning system (EWS). Through EWS, students who will graduate late can be recognized in advance. Thus, appropriate interventions can be given to the student so that they can graduate on time. The predictive model is the core of an EWS, that built based on the graduated student data. The problem that often arises in a predictive model is the degree of accuracy. In order to increase the accuracy of the prediction, the clustering of attribute selection need to be conducted first. One of approach that can be used to cluster attribute selection is by using Maximum Degree of Domination in Soft Set Theory (MDDS) algorithm. This article implements the MDDS algorithm to cluster the attributes from student datasets. The results obtained from this research is the dominant attributes that can be used as a foundation to develop a predictive model of student graduation time.

Keywords: Soft set · Student graduated data · Attributes/features selection · MDDS

1 Introduction

As an education institution, a university has objectives to make sure students graduate on time and also keep the minimum number of student drop out. Many approach was implemented by several universities to reach this objectives, one of popular approach is using the early warning system (EWS) that embody the predictive analytic model [1, 2]. This model usually using student data as dataset to be analyzed. The dataset may have several attribute that are consist of the GPA score for each semester, demography data, GRE and GMAT score [1].

Prior to develop the early warning system, a predictive model need to be developed first. The accuracy of this model will determine the effectivity of an early warning system. One way that can be done to increase the accuracy of a predictive model is to identify the most discriminant feature a dataset [3]. This kind of process also known as feature selection.

T. Herawan et al. (eds.), *Recent Advances on Soft Computing and Data Mining*,
Advances in Intelligent Systems and Computing 549, DOI 10.1007/978-3-319-51281-5_63

Data mining techniques is one of the popular approach that can be used to select the most determinant features in a dataset. In the data mining, there are two types of categories in term of feature selection there are filter and wrapper [3]. In the filter-based feature selection, the evaluation of the features are based on the general characteristics of the data without considering type of data mining algorithm that will be used [4]. Whereas in the wrapper-based feature selection, the evaluation of features consider one mining algorithm and uses the result of mining algorithm to determine the quality of feature sets [4].

In the context of student dataset problem encountered was the uniformity of features, for example, student GPA score in every semester. Therefore, requires an approach that deal with this kind of uncertainty. One of the technique that can accommodate this kind of situation is by using soft set theory [5]. This paper, using the Maximum Degree of Domination in soft set theory (MDDS)—as an alternative to implement feature selection on a dataset based on soft set theory [6]. The MDDS used to identify the most discriminant attributes/features in order build a model to determine whether a student will graduate on time or not.

This paper is organized as follows. In Sect. 2, the basic knowledge about soft set theory, multi-valued soft sets and MDDS will be described in brief. In Sect. 3, will be described the characteristic of the dataset. Section 4 will discuss the experiment results and an evaluation of the result. Finally, in Sect. 5 will be explained the conclusion of this paper.

2 Preliminary

This section will review some basic notion of information system, soft set theory, multi-valued soft set and MDDS. Through this section, let U be an initial universe set, E be a set of parameters, and $P(U)$ is the power set of U [7].

2.1 Information System

An information system can be seen as the representation of objects in terms of their attribute values. An information system is a 4-tuple (quadruple) notated as $S = (U, A, V, f)$, where $U = \{u_1, u_2, \ldots, u_i\}$ is a non-empty limited set of objects, $A = \{a_1, a_2, \ldots, a_{|A|}\}$ is a non-empty limited set of attributes, $V = \cup_{a \in A} V_a$, V_a is the domain (value set) of attribute a, $f : U \times A \rightarrow V$ is an information function such that $f(u, a) \in V_a$, for every $(u, a) \in U \times A$, called information (knowledge) function [7].

From the notion of an information system above, in the next section we will review the basic idea of soft set theory form the point of view of data set.

2.2 Soft Set Theory

This sub section, will review the fundamental concept of soft set theory as a preliminary to easily understand the basic concept of MDDS which will discussed in later sub section.

Definition 1 (see [5]). A pair (F,E) is called soft set (over U) if and only if F is a mapping of E into the set of all subsets of the set U, notated as $F{:}E \rightarrow P(U)$. In other words, a soft set (F,E) over U is parameterized family of the universe U. For $\alpha \in E$. F (α) may be considered as the set of α-elements of the set (F,E) or the α approximate elements of set (F,E).

2.3 Multi-valued Soft Set

This sub-section will explain briefly the idea of decomposing a multi-valued information system $S = (U, A, V, f)$ into A numbers of binary-valued information system $S_i = (U, a_i, V_{\{0,1\}}, f)$, where A is the cardinality of A. The decomposition of information system $S = (U, A, V, f)$ is based on decomposition of set of attributes $A = \{a_1, a_2, \ldots, a_{|A|}\}$ into the disjoint singleton attribute $\{a_1\}, \{a_2\}, \ldots, \{a_{|A|}\}$. This decomposition process can only be performed on an information system that all object has no missing attributes, or known as complete information system.

Let $S = (U, A, V, f)$ be an complete information system such that for every $a \in A$, $V_a = f(U, A)$ is a limited non-empty set and for every $u \in U$, $|f(u,a)| = 1$. For every a_i under ith-attribute consideration, $a_i \in A$ and $v \in V_a$, we define the map $a^i_v : U \rightarrow \{0, 1\}$ such that $a^i_v(u) = 1$ if $f(u, a) = v$, otherwise $a^i_v(u) = 0$. The next result, we define a binary-valued information system as a quadruple $S^i = (U, a_i, V_{\{0,1\}}, f$ which is the result of decomposition of a multi-valued information system $S = (U, A, V, f)$ into A binary-valued information system. The result of decomposition of an information system into a multi-valued information system can be seen as follow [6].

$$
\begin{aligned}
S &= (U, A, V, f) \\
&= \left\{ \begin{array}{l}
S^1 = (U, a_1, V_{\{0,1\}}, f) \leftrightarrow F, a_1) \\
S^2 = (U, a_2, V_{\{0,1\}}, f) \leftrightarrow (F, a_2) \\
\quad\quad\quad\quad . \\
\quad\quad\quad\quad . \\
\quad\quad\quad\quad . \\
S^{|A|} = (U, a_{|A|}, V_{\{0,1\}}, f) \leftrightarrow (F, a_{|A|})
\end{array} \right. \\
&= ((F, a_1), (F, a_2), \ldots, (F, a_{|A|}))
\end{aligned}
$$

2.4 Maximum Degree of Domination in Soft Set Theory (MDDS)

Basic concept of Maximum Degree of Domination in Soft Set Theory (MDDS) is built on two principle theories, multi soft set and Maximum Domination in Soft Set. Maximum Domination can be explained using following definition.

Definition 2 (see [6]). Let (F, A) be multi-soft sets over U representing information system $S = (U, A, V, f)$ and $(F, a_i), (F, a_j) \in (F, A)$. (F, a_j) is referred to as dominated in degree k by (F, a_i), denoted $(F, a_i), \leq k\ (F, a_j)$, where $k = |\cup X : X \subseteq Y|/|U|$ and, $X \in C_{(F,a_i)}$ and $Y \in C_{(F,a_j)}$.

Value of k is range between 0 and 1. If k equals 1 then it means (F, a_i) is dominated totally by (F, a_j). Otherwise if k equals 0 then it means (F, a_j) is dominated totally by (F, a_i).

Based on above definition, then a MDDS can be defined as follow.

Definition 3 (see [6]). Let (F,A) is a multi soft sets over U, and represent as S = (U, A, V, f), the soft set (F, a_i) with maximum degree of domination will be considered as attributes that has more predictive power, i.e., max $(k_1, k_2, \ldots, k_n)$.

3 Datasets

This section will describe the characteristics of the datasets that used in this paper. The datasets are taken from undergraduate student data that have already graduated from school of industrial engineering in Telkom University Indonesia. The datasets consist of 200 unique student's data. Where each data/tuple has 8 attribute, they are studentId, 1st GPA, 2nd GPA, 3rd GPA, 4th GPA, 5th GPA, 6th GPA, and student Gradute Status.

For the purpose of building the predictive model, data have been used is limited to the first 6 semester of GPA. By doing so, the model can predict the graduation of students in the third year of their college.

The metadata of each attribute can be seen in the following Table 1.

Table 1. Attribute metadata

Attribute name	Description	Attribute set value
StudentId	Id of student	{1, 2, 3, …, 200}
1st GPA	Letter representation of student GPA in their first semester	{A, AB, B, BC, C, D, E}
2nd GPA	Letter representation of student GPA in their second semester	{A, AB, B, BC, C, D, E}
3rd GPA	Letter representation of student GPA in their third semester	{A, AB, B, BC, C, D, E}
4th GPA	Letter representation of student GPA in their forth semester	{A, AB, B, BC, C, D, E}
5th GPA	Letter representation of student GPA in their five semester	{A, AB, B, BC, C, D, E}
6th GPA	Letter representation of student GPA in their six semester	{A, AB, B, BC, C, D, E}
Student Gradute Status	Graduate status of student	{ON TIME, LATE}

The values of GPA column are a letter representation of their actual numeric score (4.0 scale). The conversion of the actual score of GPA to letter representation based on standard that shown in the following Table 2.

The field student Graduate Status represent the status of student graduation period. On time value mean that student graduated from university at least at their 8th semester,

Table 2. Standard of GPA mapping

Range GPA point	GPA letter	Category
3.51–4	A	Excellent
3.01–3.5	AB	Very good
2.51–3.0	B	Good
2.01–2.5	BC	Fair
1.51–2.0	C	Satisfactory
1.1–1.5	D	Passing
0–1	E	Poor

on the other hand the late value means that the student completed their study more than 8 semesters.

Data that used in this paper mixed between student that graduated on time and those who graduated late. The following is sample of 10 out 200 of student data that used as a dataset in this paper (Table 3).

Table 3. Sample student data

StudentId	1st GPA	2nd GPA	3rd GPA	4th GPA	5th GPA	6th GPA	Student Gradute Status
1	BC	BC	AB	AB	AB	A	ON TIME
2	AB	B	AB	AB	AB	A	ON TIME
3	A	A	A	AB	AB	A	ON TIME
4	AB	B	AB	B	AB	AB	ON TIME
5	BC	C	C	BC	C	B	LATE
6	AB	AB	B	A	B	A	LATE
7	AB	AB	AB	AB	AB	A	LATE
8	BC	C	B	B	B	B	LATE
9	B	BC	B	C	B	C	LATE
10	B	B	AB	A	AB	A	LATE

4 Results and Discussion

The MDDS technique is used to cluster the most determinant attribute in the datasets. The technique implemented using Mat lab programming language. The result of this program is a matrix that consist of degree of dominant value for each attribute, as shown in Table 4.

The visualization of the MDDS result can be shown in following Fig. 1.

Table 4. MDDS result

1st GPA	0	0.005	0.005	0.005	0	0.015	0.015
2nd GPA	0.005	0	0.005	0.005	0.015	0.01	0.015
3rd GPA	0.005	0.005	0	0.005	0	0	0.005
4th GPA	0.005	0.005	0.005	0	0	0	0.005
5th GPA	0.005	0.005	0.005	0.005	0	0.01	0.01
6th GPA	0.005	0.005	0.005	0.005	0	0	0.005

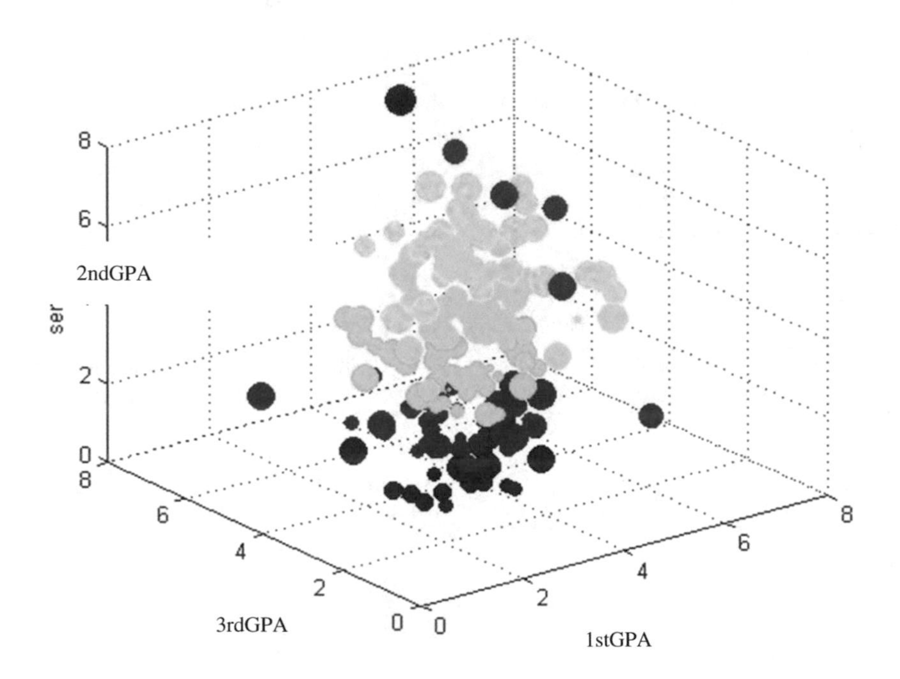

Fig. 1. Visualization of MDDS result

5 Conclusion

With regard to establish the accurate model for student lateness early warning system, the most determinant attributes need to be selected first. Data mining technique can be used to cluster the most determinant feature. The data mining technique that used in this paper is MDDS (Maximum Degree of Domination in Soft Set Theory).

The dataset that used to modeling the student graduation lateness is taken from GPA of student of School of Industrial Engineering, Telkom University Indonesia. The dataset mixed the student that graduate on time and the student that graduate late. From the results, can be concluded that the most determinant attributes that has high correlation to student graduate status is first semester GPA and second semester GPA attributes. These results can be used as a basis to develop a predictive model to develop in order to build the early warning system on student graduation lateness.

References

1. Beck, H.P., Davidson, W.D.: Establishing an early warning system: predicting low grades in college students from survey of academic orientations scores. Res. High. Educ. **42**(6), 709–723 (2001)
2. Hu, Y.-H., Lo, C.-L., Shih, S.-P.: Developing early warning systems to predict students' online learning performance. Comput. Human Behav. **36**, 469–478 (2014)
3. Jothi, G., Hannah Inbarani, H.: Soft set based feature selection approach for lung cancer images. IJSER **3**(10), 7 (2012)
4. Fu, X., Tan, F., Wang, H., Zhang, Y., Harrison, R.W.: Feature similarity based redundancy reduction for gene selection. Dmin **9912**(404), 357–360 (2006)
5. Molodtsov, D.: Soft set theory—first results. Comput. Math Appl. **37**(4–5), 19–31 (1999)
6. Herawan, T., Yanto, I.T.R., Zain, J.M., Hongwu, Q., Abdullah, Z.: A soft set approach for clustering student assessment datasets. J. Comput. Theor. Nanosci. **12**(12), 5928–5939 (2015)
7. Maji, P.K., Biswas, R., Roy, A.R.: Soft set theory. Comput. Math. Appl. **45**(4–5), 555–562 (2003)

An Application of Rough Set Theory for Clustering Performance Expectancy of Indonesian e-Government Dataset

Deden Witarsyah Jacob[1(✉)], Mohd Farhan Md. Fudzee[2], Mohamad Aizi Salamat[2], Rd Rohmat Saedudin[1], Iwan Tri Riyadi Yanto[3], and Tutut Herawan[4,5,6]

[1] Department of Industrial Engineering, Telkom University, Bandung, West Java, Indonesia
{dedenw, rdrohmat}@telkomuniversity.ac.id

[2] Faculty of Computer Science and Information Technology, Universiti Tun Hussein Onn Malaysia, Parit Raja, Malaysia
{farhan, aizi}@uthm.edu.my

[3] Department of Information Systems, University of Malaya, Kampus III UAD, Jalan Prof. Dr. Soepomo, Yogyakarta, Indonesia
zailania@umk.edu.my

[4] Faculty of Computer Science and Information Technology, University of Malaya, Pantai Valley, 50603 Kuala Lumpur, Malaysia
tutut@um.edu.my

[5] Faculty of Business and Information Technology, Universitas Teknologi Yogyakarta, Yogyakarta, Indonesia

[6] AMCS Research Center, Yogyakarta, Indonesia

Abstract. Performance expectancy has been studied as an important factor which influences e-government. Therefore, grouping of e-government users involving performance expectancy factor is still challenging. Computational model can be explored as an efficient clustering technique for grouping e-government users. This paper presents an application of rough set theory for clustering performance expectancy of e-government user. The propose technique base on the selection of the best clustering attribute where the maximum dependency of attribute in e-government data is used. The datasets are taken from a survey aimed to understand of the adoption issue in e-government service usage at Bandung city in Indonesia. At this stage of the research, we point how a soft set approach for data clustering can be used to select the best clustering attribute. The result of this study will present useful information for decision maker in order to make policy concerning theirs people and may potentially give a recommendation how to design and develop e-government system in improving public service.

Keywords: Clustering · Rough set theory · Performance expectancy · e-Government

T. Herawan et al. (eds.), *Recent Advances on Soft Computing and Data Mining*,
Advances in Intelligent Systems and Computing 549, DOI 10.1007/978-3-319-51281-5_64

1 Introduction

Growth in public familiarity with information and communication technologies (ICTs) in the world, the internet in particular, has opened up opportunities for the public sector to embrace the technologies and use them to better serve citizens. The implementation of e-government systems has been attracting increased research interest, and is believed to constitute one of the most important IT implementation and organizational change challenges of the future (Warkentin et al. 2002). Electronic government is designed as a process of interaction between government and society. Carter and Belanger (2003, 2005), Pavlou (2003), and Gefen et al. (2003) states that one important factor for the success of e-government services is the acceptance and willingness of people to use e-government services.

Venkatesh et al. (2003) defines "Performance Expectancy" as the degree to which one believes in using the system will help the person to gain performance on the job. In this concept there is a combination of variables obtained from the model of previous studies of the model acceptance and use of technology. The variables are: 1. Perceived usefulness, 2. Extrinsic Motivation, 3. Job Fit, 4. Relative advantage, and the last, Outcome Expectations. In this concept there is a combination of variables obtained from the model of previous studies of the model acceptance and use of technology. The clear explanation about performance expectance could be seen in Table 1:

Meanwhile, Davis (1989); Adams et al. (1992) defined performance expectancy as a level where a person believes that the use of a particular subject will be able to improve the work performance of the person. Chin and Todd (1995) adds the dimension of

Table 1. Variables in performance expectancy

No.	Variable	Definition	Studies
1	Perceived usefulness	The extent to which a person believes that using a particular system would enhance his performance	Venkatesh et al. (2003) and Davis et al. (1989)
2	Extrinsic motivation	The perception that the user wants to perform an activity because it is considered as a tool in achieving valuable results that differ from the activity itself	Venkatesh et al. (2003) and Davis et al. (1992)
3	Job fit	How the capabilities of a system increases the performance of individual work	Venkatesh et al. (2003) and Davis et al. (1992)
4	Relative advantage	The extent to which use innovation something perceived to be better than using its predecessor	Venkatesh et al. (2003) and Benbasat and Moore (2001)
5	Outcome expectations	Expectations are the result (outcome expectations) is associated with the consequences of his behavior	Venkatesh et al. (2003), Compeau and Higgins (1995) and Compeau et al. (1999)

expediency TI, which makes the work easier, rewarding, increase productivity, enhance the effectiveness of, and improve job performance. It can be concluded that a person's trust and feel by using an information technology will be very useful and can enhance performance and job performance.

Huang (1998) states that the clustering operation is required in a number of data analysis tasks, such as unsupervised classification and data summation, as well as segmentation of large homogeneous data sets into smaller homogeneous subsets that can be easily managed, separately modeled and analyzed. Meanwhile, a well-known approach for data clustering is using rough set theory (Pawlak 1982, 1991; Pawlak and Skowron 2007). For example, Mazlack et al. (2000) had developed a rough set approach in choosing partitioning attributes. One of the successful pioneering rough clustering for categorical data techniques is Minimum–Minimum Roughness (MMR) proposed by Parmar et al. (2007).

However, pure rough set theory is not well suited for analyzing noisy information systems. A knowledge discovery system must be tolerant to the occurrence of noise. For example, in the previous work on constructing student models through mining students classification-test answer sheets by Wang and Hung (2001), much noise was found in the classification tables, either the feature values or the class values, created by students. Their empirical results showed that attention should be paid to handle the noisy information in order to reach a satisfactory prediction accuracy (Wang 2005).

In this paper we present a real dataset of the users of e-government services. This data were taken from a survey aimed to identify of citizen behavior in using e-government. Descriptive statistics is used to find out the Mean (M) and Standard Deviation (SD) to identify the potential sources of study behavior. It is ran in SPSS version 22.0 and the results show that there are 5 potential sources of study performance expectancy.

The traditional main objectives of grouping awareness citizen in using e-government service are to deal with the uncertainty due to design intervention, to conduct a treatment to reduce a lack of awareness and further to improve citizen's public service. To achieve this objective, certain clustering techniques are also being applied. Clustering a set of objects into homogeneous classes is an important Data mining operation.

The remainder of this paper is organized as follows. Section 2 describes the related work and prose method. Section 3 describes the study's performance expectancy of e government data set. Section 4 describes experiment result. Finally, the conclusions of this work are reported in Sect. 5.

2 Proposed Method

Using MDA.

3 The Study's Performance Expectancy of e-Government Dataset

The data set was taken from a survey in Bandung. A total population were 200 people take part in this survey. The profile of the respondents is used to provide a description of the characteristics of the sample, so it is very useful in the discussion of the results of the study investigators. The majority of respondents were women, i.e. 105 people, and the respondents were male is as much as 95 peoples. To analysis the data, for distribution of study performance expectancy scores, it follows likert-scale, i.e., 1 very not agree; 2 not agree; and, 3 neutral; 4 Agree and 5 very agree. In this survey, the study performance expectancy questionnaire has been test for reliability with alpha score yielded 0.699 and accessing content validity. Table 2 describes each attribute of performance expectancy study include the mean, standard deviation, variance and range.

Table 2. Summary of the study's performance expectancy of e-government dataset

		Perceived usefulness	Extrinsic motivation	Job-fit	Relative advantage	Outcome expectations
N	Valid	200	200	200	200	200
	Missing	0	0	0	0	0
Mean		4.010	3.680	3.675	3.685	3.170
Std. deviation		.4701	.6078	.6720	.7673	.7708
Variance		.221	.369	.452	.589	.594
Range		3.0	3.0	3.0	4.0	3.0

3.1 Perceived Usefulness

Perceived usefulness is a leading source with M = 4.010 and SD = 0.4701. Perceived usefulness refers to is the extent to which the person believes that using a particular system would enhance his job performance (Davis 1989). Table 3 describe data distribution include frequency and percent.

3.2 Extrinsic Motivation

The second source is extrinsic motivation, it refers to the perception that users would and want to do an activity because it is considered a valuable role in achieving a

Table 3. Summary of perceived usefulness data distribution

		Frequency	Percent	Valid percent	Cumulative percent
Valid	2.0	1	.5	.5	.5
	3.0	18	9.0	9.0	9.5
	4.0	159	79.5	79.5	89.0
	5.0	22	11.0	11.0	100.0
	Total	200	100.0	100.0	

Table 4. Summary of extrinsic motivation data distribution

		Frequency	Percent	Valid percent	Cumulative percent
Valid	2.0	4	2.0	2.0	2.0
	3.0	67	33.5	33.5	35.5
	4.0	118	59.0	59.0	94.5
	5.0	11	5.5	5.5	100.0
	Total	200	100.0	100.0	

different result from the activity itself, such as improved job performance, earnings, or promotions (Davis et al. 1992). This variable has M = 3.680 and SD = 0.680 (Table 4).

3.3 Job Fit

The third source is job fit with M = 3.675 and SD = 0,6720. This variable describes how to improve individual performance base on the system capabilities. (Thompson et al. 1991). Table 5 is the result of data distribution of job fit.

Table 5. Summary of job-fit data distribution

		Frequency	Percent	Valid percent	Cumulative percent
Valid	2.0	5	2.5	2.5	2.5
	3.0	73	36.5	36.5	39.0
	4.0	104	52.0	52.0	91.0
	5.0	18	9.0	9.0	100.0
	Total	200	100.0	100.0	

3.4 Relative Advantage

The fourth source is relative advantage with M = 3.685 and SD = 0.7673. it refers to the extent to which use of an innovation is considered to be better than using its predecessor (Compeau and Higgins 1995; Compeau et al. 1999). Table 6 portray the result of distribution data.

Table 6. Summary of relative advantage data distribution

		Frequency	Percent	Valid percent	Cumulative percent
Valid	1.0	3	1.5	1.5	1.5
	2.0	14	7.0	7.0	8.5
	3.0	40	20.0	20.0	28.5
	4.0	129	64.5	64.5	93.0
	5.0	14	7.0	7.0	100.0
	Total	200	100.0	100.0	

3.5 Outcome Expectations

The last source is outcome expectations, this variable refers to dealing with the consequences of behavior, based on empirical evidence, is separated into performance expectations (job-related) and personal expectations (individual goals) (Compeau and Higgins Compeau and Higgins 1995; Compeau et al. 1999). Table 7 representative of the data distribution include frequency and percent.

Table 7. Summary of outcome expectations data distribution

		Frequency	Percent	Valid percent	Cumulative percent
Valid	2.0	40	20.0	20.0	20.0
	3.0	91	45.5	45.5	65.5
	4.0	64	32.0	32.0	97.5
	5.0	5	2.5	2.5	100.0
	Total	200	100.0	100.0	

4 Experiment Results

In order to apply the proposed technique, a prototype implementation system is developed using MATLAB version 7.6.0.324 (R2008a). The algorithm is executed sequentially on a processor Intel Core 2 Duo CPUs. The total main memory is 1G and the operating system is Windows XP SP3.

Table 8. MDA results of e-government performance expectancy dataset

Cluster number	Number of objects
1	5
2	73
3	104
4	18

Table 9. MDA results of e government performance expectancy dataset

Attribute with respect to	Mean dependency				Max
	EM	JF	RA	OE	
PU	0.02	0	0.015	0.025	0.025
	PU	JF	RA	OE	
EM	0.005	0	0.015	0	0.015
	PU	EM	RA	OE	
JF	0.005	0.075	0.015	0	**0.075**
	PU	EM	JF	OE	
RA	0.005	0.055	0.025	0.025	0.055
	PU	EM	JF	RA	
OE	0.005	0.02	0	0.015	0.02

There are five attributes of e government performance expectancy; Perceived Usefulness (PU), Extrinsic Motivation (EM), Job-Fit (JF), Relative Advantage (RA), Outcome Expectations (OE). The MDA result is shown in Tables 8 and 9. The selected attribute is Job-fit with the value 0.075. For attribute Job-fit, we have four clusters as follows.

The visualization of the clusters is captured in below figures.

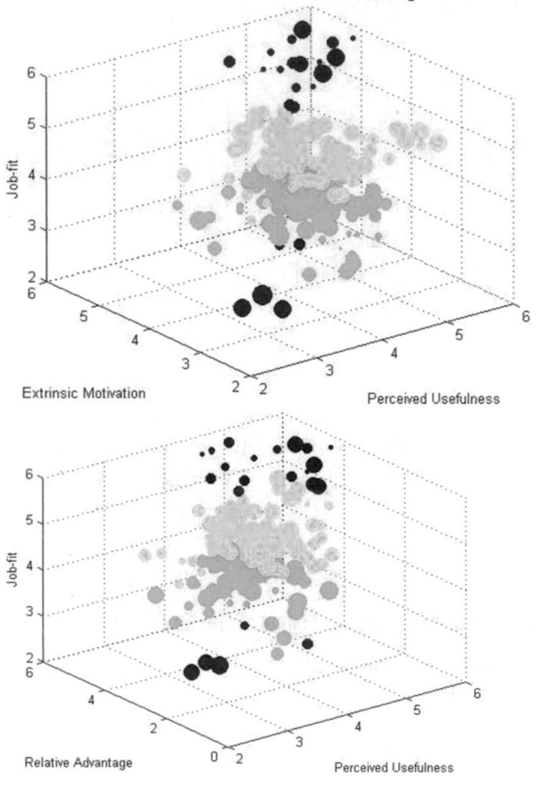

5 Conclusion

In this paper, the variable precision rough set has been used as attribute selection to study performance expectancy. The technique is based on the mean of accuracy of approximation using variable precision of attributes. We elaborate the technique approach through five of variable sources among people in Bandung, i.e., perceived usefulness, extrinsic motivation, job fit, relative advantage, and the last outcome expectations. The results show that variable precision rough set can be used to groups people in each study's performance expectancy.

Acknowledgments. The authors would like to thank the Telkom University for supporting this work.

References

Adams, D.A., Nelson, R.R., Todd, P.A.: Perceived usefulness, ease of use, and usage of information technology: a replication. MIS Q. **16**, 227–247 (1992). doi:10.2307/249577

Barney, J., Hansen, M.: Trustworthiness as a source of competitive advantage. Strat. Manag. J. **15**, 175–190 (1994)

Carter, L., Bélanger, F.: The utilization of e-government services: citizen trust, innovation and acceptance factors. Inf. Syst. J. **15**(1), 5–25 (2005)

Carter, L., Belanger, F.: Citizen adoption of electronic government initiatives. In: 37th Hawaii International Conference on System Sciences, Hawaii (2004)

Carter, L., Belanger, F.: Diffusion of innovation & citizen adoption of e-government. In: The Fifth International Conference on Electronic Commerce (ICECR-5), Pittsburg, PA, pp. 57–63 (2003)

Carter, L., Belanger, F.: Trust, and risk in e-government adoption. J. Strat. Inf. Syst. **17**(2), 165–176 (2008)

Chen, J.V., Jubilado, R.J.M., Capistrano, E.P.S., Yen, D.C.: Factors affecting online tax filing – an application of the IS success model and trust theory. Comput. Hum. Behav. **43**, 251–262 (2015)

Chin, W.W., Todd, P.A.: On the use, usefulness, and ease of structural equation modeling in MIS research: a note of caution. MIS Q. **19**(2), 237–246 (1995)

Compeau, D., Higgins, C.: Computer self-efficacy: development of a measure and initial test. MIS Q. **19**(2), 189–211 (1995)

Davis, F.D.: Perceived usefulness, perceived ease of use, and user acceptance of information technology. MIS Q. **13**(3), 318–340 (1989)

Diaz, M.C., Loraas, T.: Learning new uses of technology while on an audit engagement: contextualizing general models to advance pragmatic understanding. Int. J. Acc. Inf. Syst. **11**, 61–77 (2010). doi:10.1016/j.accinf.2009.05.001

DeLone, W.H., McLean, E.R.: The DeLone and McLean model of information systems success: a 10-year update. J. Manag. Inf. Syst. **19**(4), 9–30 (2003)

DeLone, W.H., McLean, E.R.: Information systems success: the quest for the dependent variable. Inf. Syst. Res. **3**(1), 60–95 (1992)

Ghobakhloo, M., Zulkifli, N.B., Abdul Aziz, F.: The interactive model of user information technology acceptance and satisfaction in small and medium-sized enterprises. Eur. J. Econ. Finan. Adm. Sci. **19**, 7–27 (2010)

Gupta, B., Dasgupta, S., Gupta, A.: Adoption of ICT in a government organization in a Developing Country: an empirical study. J. Strat. Inf. Syst. **17**, 140–154 (2008). doi:10.1016/j.jsis.2007.12.004

Hofstede, G.: Cultures and Organizations: Software of the Mind. McGraw-Hill, New York (1997)

Keong, M.L., Ramayah, T., Kurnia, S., Chiun, L.M.: Explaining intention to use an enterprise resource planning (ERP) system: an extension of the UTAUT model. Bus. Strat. Ser. **13**(4), 173–180 (2012). doi:10.1108/17515631211246249

Langton, N., McKnight, H.: Using expectation disconfirmation theory to predict technology trust and usage continuance intentions (2006). http://misrc.umn.edu/workshops/2006/spring/harrison.pdf

McKinney, V., Yoon, K., Zahedi, F.M.: The measurement of web-customer satisfaction: an expectation and disconfirmation approach. Inf. Syst. Res. **13**(3), 296–315 (2002)

Moore, G.C., Benbasat, I.: Development of an instrument to measure the perceptions of adopting an information technology innovation. Inf. Syst. Res. **2**(3), 192–222 (2001)

Pavlou, P.: Consumer acceptance of electronic commerce: integrating trust and risk with the technology acceptance model. Int. J. Electron. Commerce **7**(3), 101–134 (2003)

Räckers, M., Hofmann, S., Becker, J.: The influence of social context and targeted communication on e-government service adoption. In: Wimmer, M.A., Janssen, M., Scholl, H.J. (eds.) EGOV 2013. LNCS, vol. 8074, pp. 298–309. Springer, Heidelberg (2013). doi:10.1007/978-3-642-40358-3_25

Sa, F., Rocha, A., Cota, M.P.: From the quality of traditional services to the quality of local e-government online service: a literature review. Gov. Inf. Q. **33**, 149–160 (2015). doi:10.1016/j.giq.2015.07.004

Shin, D.H.: Towards an understanding of the consumer acceptance of mobile wallet Original Research Article. Comput. Hum. Behav. **25**, 1343–1354 (2009). doi:10.1016/j.chb.2009.06.001

Schaupp, L.C., Carter, L., McBride, M.E.: E-file adoption: a study of US taxpayers' intentions. Comput. Hum. Behav. **26**(4), 636–644 (2010)

Susanto, T.D., Goodwin, R.: User acceptance of SMS-based e-government services: differences between adopters and non-adopters. Gov. Inf. Q. **30**(4), 486–497 (2013)

Schepers, J., Wetzels, M.: A meta-analysis of the technology acceptance model: investigating subjective norm and moderation effects. Inf. Manag. **44**(1), 90–103 (2007). doi:10.1016/j.im.2006.10.007

Titah, R., Barki, H.: e-Government adoption and acceptance: a literature review. Int. J. Electron. Gov. Res. **2**(3), 23–57 (2006)

Urbach, N., Müller, B.: The updated DeLone and McLean model of information systems success. In: Dwivedi, Y.K., Wade, M.R., Schneberger, S.L. (eds.) Information Systems Theory: Explaining and Predicting Our Digital Society, pp. 1–18. Springer Science+Business Media, LLC, Hamburg (2010)

Venkatesh, V., Morris, M., Davis, G., Davis, F.: User acceptance of information technology: toward a unified view. MIS Q. **27**(3), 425–478 (2003)

Warkentin, M., Gefen, D., Pavlou, P., Rose, G.M.: Encouraging citizen adoption of e-government by building trust. Electron. Markets **12**(3), 157–162 (2002)

Erratum to: Text Detection in Low Resolution Scene Images Using Convolutional Neural Network

Anhar Risnumawan[1(✉)], Indra Adji Sulistijono[2], and Jemal Abawajy[3]

[1] Mechatronics Engineering Division, Politeknik Elektronika Negeri Surabaya (PENS), Kampus PENS, Surabaya, Indonesia
[2] Graduate School of Engineering Technology, Politeknik Elektronika Negeri Surabaya (PENS), Kampus PENS, Surabaya, Indonesia
{anhar, indra}@pens.ac.id
[3] School of Information Technology, Deakin University, Geelong, Australia
jemal.abawajy@deakin.edu.au

Erratum to: Chapter "Text Detection in Low Resolution Scene Images Using Convolutional Neural Network" in: T. Herawan et al. (eds.), *Recent Advances on Soft Computing and Data Mining*, Advances in Intelligent Systems and Computing, DOI 10.1007/978-3-319-51281-5_37

The original version of the chapter "Text Detection in Low Resolution Scene Images Using Convolutional Neural Network" was published with the co-author name "Younes Saadi" which has to be deleted. The erratum chapter and the book have been updated with the changes.

The updated original online version for this chapter can be found at
DOI: 10.1007/978-3-319-51281-5_37

T. Herawan et al. (eds.), *Recent Advances on Soft Computing and Data Mining*,
Advances in Intelligent Systems and Computing 549, DOI 10.1007/978-3-319-51281-5_65

Author Index

T. Herawan et al. (eds.), *Recent Advances on Soft Computing and Data Mining*,
Advances in Intelligent Systems and Computing 549,
DOI 10.1007/978-3-319-51281-5

Printed in the United States
by Baker & Taylor Publisher Services